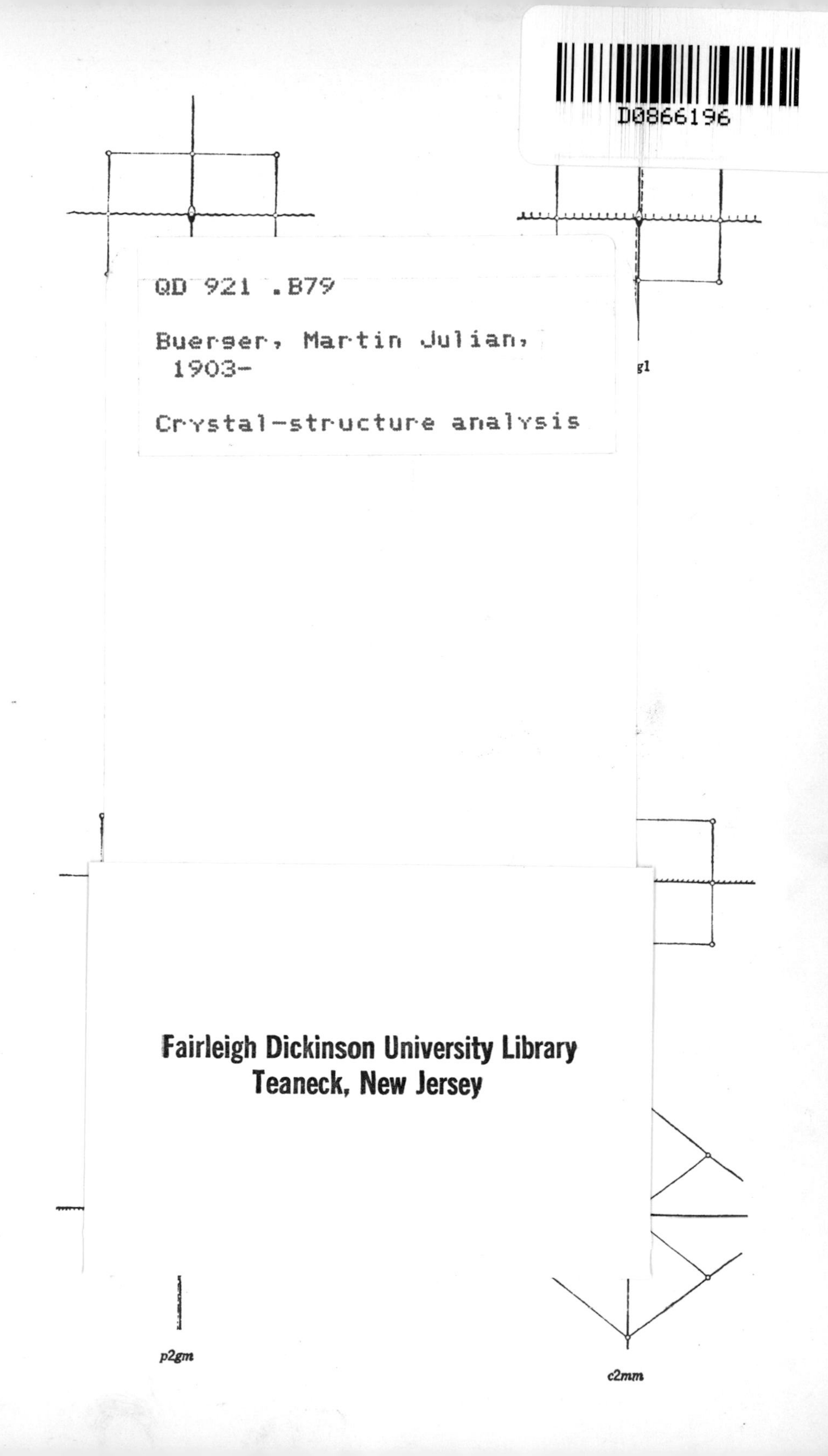

gl
p2gm
c2mm

Crystal-structure analysis

Crystal-structure analysis

Martin J. Buerger

Professor of Mineralogy and Crystallography
Massachusetts Institute of Technology

John Wiley & Sons, Inc. New York • London • Sydney

LIBRARY OF CONGRESS CATALOG CARD NUMBER: 60-10310

PRINTED IN THE UNITED STATES OF AMERICA

SECOND PRINTING, MARCH, 1967

To

Richard and Lucille Webster

Preface

Several of those who wrote reviews of my book *X-ray crystallography* expressed the hope that it would be followed by another volume, in which the methods of finding the locations of the atoms in the cell would be discussed. Many other readers also expressed the same hope privately. There were, of course, several books already available which treated the subject, but at that time these treatments were contained in more general works on crystals and their investigation. There were no books designed to furnish the reader with specific equipment for handling a problem in crystal-structure analysis.

Nevertheless, I did not seriously entertain the idea of writing such a book until I found myself, immediately after the war, with a comparatively large number of students and postdoctoral fellows whose main interest was the study of crystal-structure analysis. It then became important for me to formulate for them, in some way, the methods used and the theories back of them. From time to time, therefore, I wrote notes on various limited parts of the field, and eventually it seemed desirable to complete these and to put them into a more convenient form. This book is the result. I hope it will serve as an introduction to the subject of crystal-structure analysis.

This book does not contain a development of the Patterson synthesis and vector space, because a companion volume† (which is really volume two of this book) has already been published. The crystal-structure analyst will find several other topics not included. Most important of these is reflection statistics. This is a somewhat advanced topic, and, in any case, is a large and self-contained subject and deserves to be treated in a separate monograph by an expert in statistics. The subject of optical Fourier transformers, such as the "fly's eye," and their application, was omitted (although written in an early draft) because, with the increasing importance of numerical values and the increasing use of high-speed digital computers to produce them, the use of optical analogue methods is waning. But most of the other subjects which are normally encountered in a crystal-structure analysis are included. Each chapter

† M. J. Buerger. *Vector space and its application in crystal-structure investigation.* (John Wiley and Sons, New York, 1959)

ends with a literature list, usually subdivided according to the topics considered. These are reasonably complete, especially as regards recent literature, so that those who wish to browse further are supplied with a guide.

I am indebted to several persons for help in preparing the manuscript. Mrs. Jean Breaghy and Mrs. Delphine Radcliffe prepared various drafts of the difficult typescript, and Mrs. Radcliffe checked most of the tables in proof. Mr. Tibor Zoltai patiently made many versions of the final drawings from my rough pencil sketches. The typescript was read by Mr. Bernhardt Wuensch, to whom I am also indebted for suggesting improvements in the mathematical form in many places. Parts or all of the typescript were also read by Prof. Leonid V. Azaroff, Prof. I. Fankuchen, and Dr. Howard Evans, all of whom contributed to its improvement. Professors Azaroff and Fankuchen again read the book in page proof. I am indebted to all of these friends for their interest and help.

MARTIN J. BUERGER

Massachusetts Institute of Technology
May 1960

Contents

1

Introduction

Historical background

Early crystallography was a rather empirical science. It was largely supported by laws which were generalizations of observations, such as the law of constancy of interfacial angles, the law of rationality of intercepts, Bravais's "law" of prominence of crystal forms, Mallard's laws of twinning, etc. Fortunately a few crystallographers concerned themselves with the reasons for the laws. These thinkers succeeded in erecting a framework of theory which associated the unique properties of a crystal with an orderly pattern of molecules which they supposed to constitute the ultimate structure of the crystal.[1] This phase of crystallography was complete before the turn of the twentieth century. By about 1890 it had already been thoroughly established that, if crystals were indeed composed of molecules in ordered array, such three-dimensional patterns were limited to 230 possibilities. This conclusion had been reached independently and about the same time by Fedorov,[2,3,5,7,9] Schoenflies,[4] and Barlow.[6,8,10,11]

This result was an enormous extrapolation based upon pyramided theories. At its foundation was the conjecture that the properties of crystals were to be accounted for by the symmetrical ordered arrangements of their molecules, yet nobody had as yet proven that there were such things as molecules. This aspect of crystallography was therefore highly theoretical.

The big turning point in crystallography came in 1912. At that time an exceptional group of scientists was in residence in Munich. This included Professor Paul von Groth (the dean of crystallographers), Professor Wilhelm Conrad von Röntgen (the discoverer of x-rays), and Professor A. Sommerfeld. In addition there were a number of instructors, assistants, and graduate students. These included Privatdozent

Max Laue, P. Debye (who was Sommerfeld's assistant), P. P. Koch and E. Wagner (who were Röntgen's assistants), W. Friedrich and P. Knipping (graduate students who were doing experimental work towards their doctor's degrees in Röntgen's laboratory), P. P. Ewald (who was writing his doctor's thesis under Sommerfeld), as well as P. S. Epstein and O. R. Glocker. The assistants and graduate students usually had coffee at Café Lutz in the Hofgarten near Odeonsplatz. Undoubtedly each of these scientists had his background of knowledge colored by the knowledge and ideas of the others. For example, Groth was a protagonist of the crystallographic theory of the periodic arrangement of molecules in crystals. Crystals were supposed to be composed of three-dimensional patterns, and by this time there was some evidence that the chemical molecule ought to have dimensions of the order of 10^{-8} cm.

In his thesis, Ewald was attempting to account for the optical properties of crystals. Specifically he was studying the behavior of a set of dipoles in the anisotropic environment of an orthorhombic crystal. He had derived rigorous equations for the interaction of such an array with light. When he consulted Laue about his manuscript, Laue raised the question as to what might be expected if the wavelength of the light were of the same order as the separation between dipoles.

Meanwhile, Sommerfeld and Koch believed that x-rays were wave-like radiations, and their discussion of Walter and Pohl's experiments on passing these rays through various slits led them to the view that x-rays probably had wavelengths of the order of 10^{-9} cm. With this environment of ideas, Laue thought it would be interesting to see how x-rays would interact with a crystal.

Friederich and Knipping performed the experiment suggested by Laue and, after a first disappointment, obtained the remarkable result that x-rays were indeed diffracted by crystals of copper sulfate.[12–14] In one stroke, then, crystals were shown to be triperiodic arrangements of matter with the periods on a molecular scale, and x-rays were proved to be wave-like radiation similar to visible light.

This was not only a turning point in crystallography, it was a landmark in modern science as well. For from this time forth matter of molecular dimensions could be probed by studying the way it scattered x-rays. Bragg[15–20] quickly improved on the Laue experiment, chiefly by substituting monochromatic for polychromatic x-radiation, but also by simplifying the theory of the scattering experiment. With these tools a number of simple crystal structures were completely analyzed.

In his early work, Bragg was fortunate in being able to build upon a foundation which had been unused for about a generation. Barlow, one of the three who had independently developed the theory of space groups, had applied his own ideas to suggesting specific structures for some simple compounds.[11] In particular, he derived likely models of the crystal structures for compounds like NaCl, CsCl, and ZnS. Barlow's structures contained no discernible molecules in spite of the view, then current, that such compounds were composed of molecules. In attempting to find a reason for the differences in the intensities of the spectra caused by the diffraction of x-rays by ZnS, and later the alkali halides, Bragg was urged by Pope[23] to consider the Barlow models of these simple structures. Bragg then investigated the diffraction intensities to be expected from Barlow's models and found them to check the observed intensities. In this way the correctness of Barlow's generation-old conjectures about simple crystal structures was confirmed, and the science of crystal-structure analysis was established.

The early crystal-structure investigations made a minimum of use of the beautiful space-group theory which had been developed many years earlier. The reason for this was partly that the results had not been tabulated in a very useful form, and partly that no method had been devised for distinguishing space groups from one another experimentally. Crystallographers are indebted to Niggli[21] for showing how the space group of a crystal could be determined with the aid of extinctions (*Auslöschungen*) of certain classes of spectra. He also first compiled the results of space-group theory in a form useful for x-ray diffraction purposes. Niggli effected a fortunate redirection of crystal-structure analysis, and now it is a part of normal routine to first determine the unit cell and space group of a crystal preliminary to any attempt to locate its atoms.

With the tools of diffraction theory and space-group theory in its possession, modern science as we know it unfolded rapidly in the hands of many who played the part of crystallographers. They were recruited from the ranks of physics, chemistry, metallurgy, and mineralogy. Between the two world wars, thousands of simple crystal structures were completely established by x-ray diffraction. The results had impacts upon every science, but they especially changed the course of chemistry. In fact, chemistry, metallurgy, and mineralogy, as we know them today, have a solid core of classified knowledge which is based specifically on arrangements of atoms as revealed in crystal-structure studies. The chemistry of molecules gave way to the chemistry of interatomic relations, in which measured interatomic distances and bond angles became important parameters. Theories of chemical (including metallic) bonding have been completely revised in the light of the revelations effected by crystal-structure analysis.

The spread of knowledge stemming from x-ray crystallography was rapid during this interwar period. But during the Second World War x-ray crystallography really came into its own. The urgency of getting results pressed the crystallographer into the service of industry as well as science. The ability of the crystallographer to produce results of importance during this period is evidenced by the great increase in the membership of societies at this time. The American Society for X-Ray and Electron Diffraction (succeeded in 1950 by the American Crystallographic Association) was initiated in 1942 with 133 charter members. At the end of the war it had some 600 members. A similar increase in interest in crystallography had developed in other countries, and, as a consequence, an International Union of Crystallography arose from an international gathering of delegates in London in 1946. The 1951 congress of this Union in Stockholm was attended by some 3000 members from all parts of the world.

Crystallography was once an obscure science; now it is a very prominent one. What caused this change in position? The reason is apparently that once crystallography was a pedantic discipline; now it is an experimental science dealing with the matter of the universe. As such it has made important contributions to the status of science as a whole, and has also been applied to many problems of industry. This change was brought about by one tool, x-ray diffraction. It is with this tool that this book is concerned.

It should not be supposed that crystal-structure analysis is a dead science because it is essentially a completed framework. This is far from true. It will become evident in subsequent chapters that crystal-structure analysis struggles with a problem in which it has only half of the data. It attempts to find a crystal structure from diffraction spectra. The structure can be determined almost as a routine from a knowledge of the intensities and phases of the diffraction spectra. But only the intensities can be found experimentally. The problem appears, offhand, to be indeterminate by reason of the missing phases. This constitutes the so-called "phase problem" of present-day crystallography, discussed in more detail in one of the chapters of this book. This is a fascinating field in which many crystallographers have made contributions. Because of their efforts, not only have a number of methods been discovered for the direct solution of fairly simple crystal structures, but our knowledge of crystallography in general has been enriched and deepened.

Plan of the book

In order to introduce the connection between diffraction and the structure which produces it, this book starts with an elementary treatment of diffraction in Chapters 2 and 3. With this as background, the general steps in a crystal-structure analysis are outlined in Chapter 4. In Chapters 5 through 8 the student is taken through the necessary stages of gathering data and transforming it into a useful form. He is then in a position to carry out a crystal-structure analysis for his particular crystal. If his crystal has a simple structure, he may be able to solve it with the aid of the additional matter discussed in Chapters 10 and 11. Chapter 12 is devoted to some examples of structure determination utilizing only qualitative considerations, which are adequate for certain simple structures.

Most simple structures have already been solved, however, so the crystal-structure analyst is usually faced with problems requiring the use of Fourier syntheses and, more generally, an understanding of reciprocal space. A discussion of various aspects of reciprocal space and how it is used in crystal-structure analysis occupies most of the remainder of the book.

In a book of these small dimensions, the treatment cannot be complete. An attempt has been made, therefore, to provide the student with an introduction to the topics considered most useful. On the other hand, each chapter ends with a list of relevant literature, and it is believed these lists are reasonably complete and up-to-date. Thus, if the student wishes to go further into the subject matter of the chapter, the direction of further study is indicated. (Incidental references, not properly part of the literature of the chapter, are given in footnotes.)

The determination of the unit cell and space group is not considered part of the subject of this book. It is assumed, therefore, that the reader already has the background given in the author's earlier volume.[†] References are occasionally given to subject matter already developed there.

Literature

[1] William Barlow. *Probable nature of the internal symmetry of crystals.* Nature **29** (1883) 186–188, 205–207.

[2] E. S. Fedorov. *An attempt to express by means of an abbreviated sign the symbols of all equal directions of a given sub-division of a symmetry system* (in Russian). Trans. Imperial St. Petersburg Mineralogical Soc. **23** (1887) 99–116.

[3] E. S. Fedorov. *Notice of the success of theoretical crystallography in the past decade* (in Russian). Trans. Imperial St. Petersburg Mineralogical Soc. **26** (1890) 345–377.

[4] Arthur Schoenflies. *Krystallsysteme und Krystallstructur.* (B. G. Teubner, Leipzig, 1891)

[5] E. S. Fedorov. *The beginning of the study of figures* (in Russian). Trans. Imperial St. Petersburg Mineralogical Soc. **21** (1885) 1–279. [Also appearing as a long abstract in German, as follows: E. von Fedorow. *Elemente der Gestaltenlehre.* Z. Krist. **21** (1893) 679–694.]

[6] William Barlow. *Ueber die geometrischen Eigenschaften homogener starrer Structuren und ihre Anwendung auf Krystalle.* Z. Krist. **23** (1894) 1–63.

[†] M. J. Buerger. *X-ray crystallography.* (John Wiley and Sons, New York, 1942)

[7] E. von Fedorow. *Theorie der Krystallstructur.* Z. Krist. **24** (1895) 209–252.

[8] W. Barlow. *Nachtrag zu den Tabellen homogener Structuren und Bemerkungen zu E. von Fedorow's Abhandlung über regelmässige Punktsysteme.* Z. Krist. **25** (1896) 86–91.

[9] E. von Fedorow. *Theorie der Krystallstructur.* Z. Krist. **25** (1896) 113–224.

[10] William Barlow. *On homogeneous structures and the symmetrical partitioning of them, with application to crystals.* Mineral. Mag. **11** (1896) 119–136. [German translation: *Ueber homogene Structuren und ihre symmetrische Theilung, mit Anwendung auf die Krystalle.* Z. Krist. **27** (1897) 449–467.]

[11] William Barlow. *A mechanical cause of homogeneity of structure and symmetry geometrically investigated; with special application to crystals and to chemical composition.* Sci. Proc. Roy. Dublin Soc. **8** (1898) 527–689. [Also published as a German translation: *Geometrische Untersuchung über eine mechanische Ursache der Homogenität der Structur und der Symmetrie; mit besonderer Anwendung auf Krystallisation und chemische Verbindung.* Z. Krist. **29** (1898) 433–588.]

[12] W. Friedrich, P. Knipping, and M. Laue. *Interferenz-Erscheinungen bei Röntgenstrahlen.* Sitzungsberichte der mathematisch-physikalischen Klasse der Königlich Bayerischen Akademie der Wissenschaften zu Munchen, 1912, 303–322; reprinted in Naturwiss. (1952) 361–367.

[13] M. Laue. *Eine quantitative Prüfung der Theorie fur die Interferenz-Erscheinungen bei Röntgenstrahlen.* Sitzungsberichte der mathematisch-physikalischen Klasse der Königlich Bayerischen Akademie der Wissenschaften zu Munchen, 1912, 363–373; reprinted in Naturwiss. (1952) 368–372.

[14] M. v. Laue. *Röntgenstrahlinterferenzen.* Physik. Z. **14** (1913) 1075–1079.

[15] W. L. Bragg. *The diffraction of short electromagnetic waves by a crystal.* Proc. Cambridge Phil. Soc. **17** (1913) 43–57.

[16] W. H. Bragg and W. L. Bragg. *The reflection of x-rays by crystals.* Proc. Roy. Soc. London (A), **88** (1913) 428–438.

[17] W. L. Bragg. *The structure of some crystals as indicated by their diffraction of x-rays.* Proc. Roy Soc. (London) (A) **89** (1913) 248–277.

[18] W. H. Bragg and W. L. Bragg. *The structure of diamond.* Proc. Roy. Soc. (London) (A) **89** (1913) 277–291.

[19] W. Lawrence Bragg. *The analysis of crystals by the x-ray spectrometer.* Proc. Roy. Soc. (London) (A) **89** (1914) 468–489.

[20] W. H. Bragg and W. L. Bragg. *X rays and crystal structure* (G. Bell and Sons, London, 1915) especially 8–21.

[21] Paul Niggli. *Geometrische Kristallographie des Diskontinuums.* (Gebrüder Borntraeger, Leipzig, 1919)

[22] W. H. Bragg and W. L. Bragg. *The discovery of x-ray diffraction.* Current Sci. (India), special number on "Laue Diagrams," (1937) 9–10.

[23] Lawrence Bragg. *The history of x-ray analysis.* (Address given before First Conference on X-Ray Analysis in Industry, Institute of Physics, Cambridge, England, 1942. Issued by the British Council as part of a pamphlet series entitled "Science in Britain") (Longmans, Green and Co., London, 1943) especially 8.

2

Some fundamental diffraction relations

This chapter is designed as an introduction to the theory of diffraction, as used in crystal-structure analysis, but, as far as possible, stripped of a number of technical matters which are reserved for subsequent chapters. Diffraction can be briefly described as the scattering of waves by various parts of an object, and the recombination of the scattered waves in various directions. A discussion of diffraction, therefore, requires a technique for studying the combination of wave motions.

Description and combination of waves

Wave motion. As a wave advances past a point, its effect at the point can be studied in terms of a vector of length f rotating at a constant

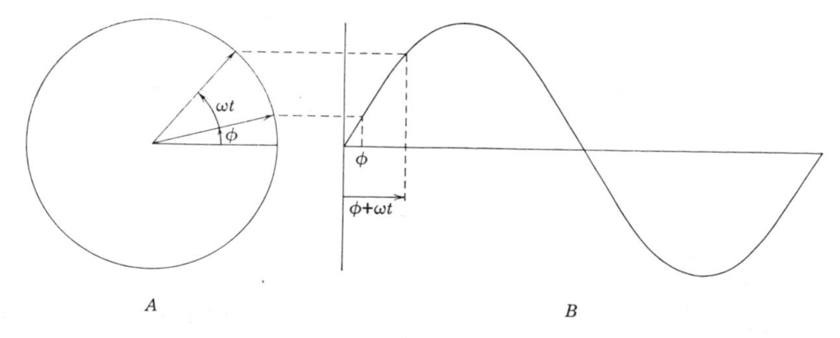

Fig. 1.

angular rate ω, Fig. 1A. For example, in an elastic medium, the displacement at the point varies sinusoidally, that is, as the projection of the rotating vector, Fig. 1B. The angle which the rotating vector makes with the origin line is said to be its *phase*. If the phase of the wave is initially ϕ, then at the end of time t it is $\phi+\omega t$.

If several waves are advancing past a point, the amplitude and phase of the resulting wave can be derived from vectors on such a diagram. For example, suppose there are two waves whose amplitudes, initial phases, and angular phase changes are f_1, ϕ_1, and $\omega_1 t$, and f_2, ϕ_2, and $\omega_2 t$. Then the resultant wave can be represented by the resultant R of the vectors which represent the individual waves, Fig. 2.

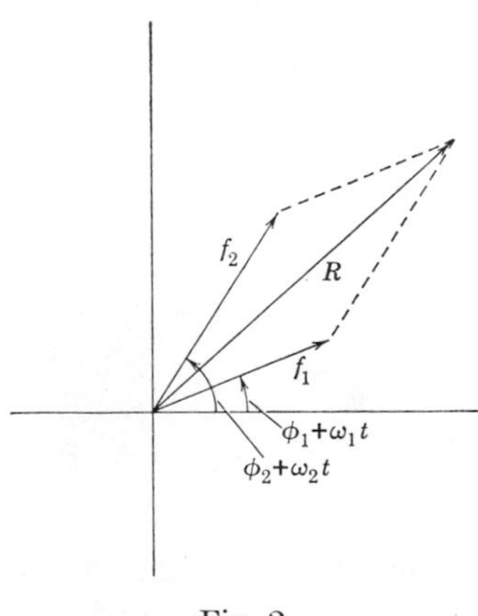

Fig. 2.

In diffraction problems the several wavelets to be combined arise from the scattering of the same original wave by different points of the object, so that ω is constant for the problem. If one is interested merely in the general way that the wavelets combine, he is interested in the resultant of the vectors representing the individual amplitudes, and not in the variation of the projection with time. In this case the term ωt can be neglected. For this purpose a wave of amplitude f and initial phase ϕ can be represented by the static vector shown in Fig. 3, and two such waves can be combined as shown in Fig. 4.

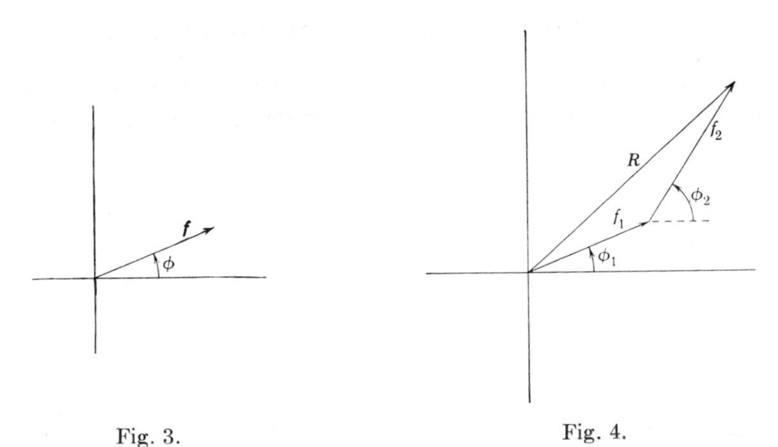

Fig. 3. Fig. 4.

Notation of the complex plane. It is customary to regard the vectors representing waves as occurring in the complex plane. Not only does this divorce the phase from any connection with the coordinates of ordinary space, but it offers a compact notation. Since this is the usual way of treating combinations of waves, the reader is advised to review the algebra of the complex plane.

A set of wave vectors in the complex plane can be resolved into their real and imaginary components. In Fig. 5, the real component of each wavelet has the form $f \cos \phi$, and the imaginary component has the form

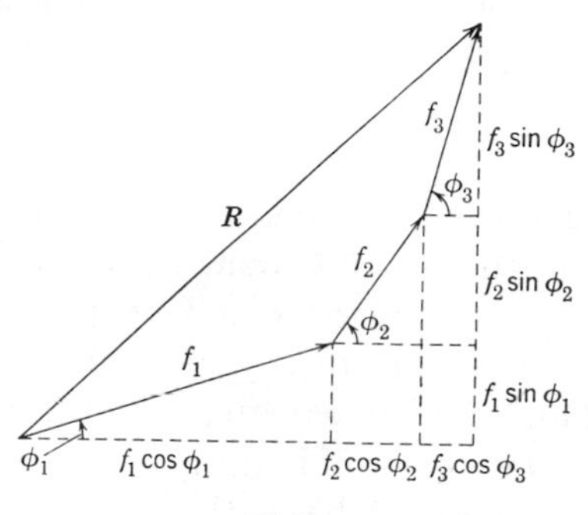

Fig. 5.

$if \sin \phi$. The complex resultant R can also be expressed in terms of its real and imaginary parts:

$$R = |R| \cos \phi + i|R| \sin \phi. \tag{1}$$

Since

$$|R| \cos \phi = f_1 \cos \phi_1 + f_2 \cos \phi_2 + f_3 \cos \phi_3, \tag{2}$$

and

$$i|R| \sin \phi = if_1 \sin \phi_1 + if_2 \sin \phi_2 + if_3 \sin \phi_3, \tag{3}$$

it follows from (1) that

$$\begin{aligned} R = \; & f_1 \cos \phi_1 + f_2 \cos \phi_2 + f_3 \cos \phi_3 \\ & + if_1 \sin \phi_1 + if_2 \sin \phi_2 + if_3 \cos \phi_3, \\ = \; & f_1(\cos \phi_1 + i \sin \phi_1) \\ & + f_2(\cos \phi_2 + i \sin \phi_2) \\ & + f_3(\cos \phi_3 + i \sin \phi_3). \end{aligned} \tag{4}$$

Now, Euler's relation provides that

$$e^{i\phi} = \cos \phi + i \sin \phi, \tag{5}$$

so that (4) can be more compactly represented by

$$R = f_1 \, e^{i\phi_1} + f_2 \, e^{i\phi_2} + f_3 \, e^{i\phi_3}, \tag{6}$$

or, more generally, as

$$R = \sum_j f_j \, e^{i\phi_j}. \tag{7}$$

The factor $e^{i\phi}$. The nature of relations like (6) are readily appreciated if it is recognized that the factor $e^{i\phi}$ behaves as an operator which rotates the other factor of the term through the angle ϕ. To see this, first consider a real quantity f, represented in the complex plane by a vector of length f, parallel to the axis of reals. If this term is multiplied by $e^{i\phi}$, the meaning of the new quantity $fe^{i\phi}$ can be appreciated by multiplying the left and right of (5) by f:

$$fe^{i\phi} = f \cos \phi + if \sin \phi. \tag{8}$$

This expresses the fact that the quantity $fe^{i\phi}$ has a component $f \cos \phi$ along the axis of reals, and a component $f \sin \phi$ along the axis of imaginaries. That is, the quantity $fe^{i\phi}$ can be represented in the complex plane by a vector whose magnitude is f and which makes an angle ϕ with the axis of reals. Thus, if the real quantity f is multiplied by $e^{i\phi}$, the effect is the same as if the vector of length f were rotated through angle ϕ.

Not only does the factor $e^{i\phi}$ have this effect upon a real quantity; it affects a complex quantity in the same way: Consider the complex quantity $fe^{i\phi_1}$. It has just been shown that this can be represented in the complex plane by a vector of length f making an angle ϕ_1 with the axis of reals. Let this complex quantity be multiplied by $e^{i\phi_2}$. The result is

$$(fe^{i\phi_1})e^{i\phi_2} = fe^{i\phi_1+i\phi_2} \tag{9}$$

$$= fe^{i(\phi_1+\phi_2)}. \tag{10}$$

According to (8) and the discussion following it, the right of (10) can be represented by a vector in the complex plane whose magnitude is f and which makes an angle $\phi_1+\phi_2$ with the axis of reals. Since $fe^{i\phi_1}$ is a vector making an angle ϕ_1 with the axis of reals, the term $e^{i\phi_2}$ on the left of (9) has the property of rotating the vector $fe^{i\phi_1}$ through angle ϕ_2. This property of the factor $e^{i\phi}$ can accordingly be summarized as:

Theorem 1: The factor $e^{i\phi}$ *behaves as an operator which rotates the other factor of the term through angle* ϕ *in the complex plane.*

Diffraction by one-dimensional patterns

Some of the fundamental features of diffraction which permit its use in the study of crystal structures can be appreciated with a minimum of complication by considering diffraction in a plane by a one-dimensional pattern. The basic ideas can be readily generalized to diffraction by a three-dimensional pattern of atoms.

Diffraction by a row of translation-equivalent points. Consider, first, a row of equally spaced, identical points each capable of scattering a wave, Fig. 6. First, suppose that the wave front is parallel to the row, and let the amplitude of the wavelet scattered by a point be f. How do the wavelets scattered by the several points interact with one another?

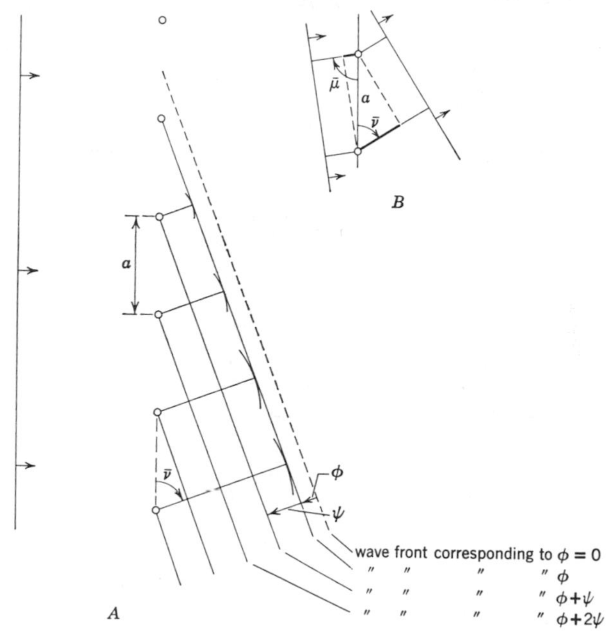

Fig. 6.

Consider the phases of these wavelets along a wave front in some arbitrary direction defined by an angle $\bar{\nu}$, Fig. 6*A*. These phases are proportional to the distance of the wave front from the scattering points, and therefore increase linearly with the distance of the point considered, from some location whose phase is taken as a reference phase. Suppose that the phase difference between waves scattered by neighboring points is ψ. Then if the phase of the wave scattered by one of the points is ϕ, the phase of the wave scattered by its first neighbor is $\phi+\psi$, and that scattered by its nth neighbor is $\phi+n\psi$.

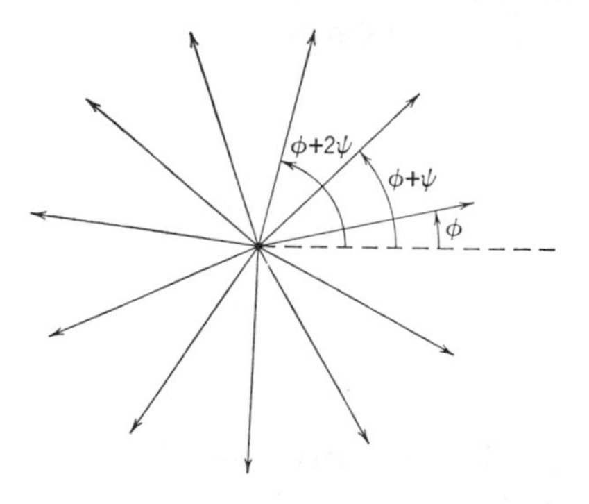

Fig. 7.

The resultant of the wavelets scattered by the series of points can be studied by plotting in the complex plane the vectors representing the amplitudes and phases of the wavelets, that is, by plotting an *Argand diagram.* This is shown in Fig. 7. The resultant can also be represented analytically by using the notation of the complex plane. In this case the resultant wave, according to (7), is represented by

$$R = fe^{i\phi} + fe^{i(\phi+\psi)} + fe^{i(\phi+2\psi)} \cdots + fe^{i(\phi+[N-1]\psi)}$$
$$= f \sum_{n=0}^{N-1} e^{i(\phi+n\psi)}. \tag{11}$$

By manipulating the exponents according to the rules of ordinary algebra, this can be written

$$R = f \sum_{n=0}^{N-1} e^{i\phi}\, e^{in\psi}$$
$$= fe^{i\phi} \sum_{n=0}^{N-1} e^{in\psi} \tag{12}$$
$$= fe^{i\phi}\, P, \tag{13}$$

where
$$P = \sum_{n=0}^{N-1} e^{in\psi}. \tag{14}$$

The term $\sum_{n=0}^{N-1} e^{in\psi}$ in (12) is a function well known in diffraction theory. It can be easily evaluated as follows: In the complex plane, Fig. 8, this function represents the resultant of N unit vectors, each making an angle ψ with its neighbor. The resultant P, and the individual unit vectors, are related to the radius r of the circle enclosing the partial polygon of vectors as follows:

$$\sin\frac{\psi}{2} = \frac{\tfrac{1}{2}}{r}, \qquad r = \frac{\tfrac{1}{2}}{\sin\frac{\psi}{2}}; \tag{15}$$

$$\sin\frac{N\psi}{2} = \frac{\tfrac{1}{2}P}{r}, \qquad r = \frac{\tfrac{1}{2}P}{\sin\frac{N\psi}{2}}. \tag{16}$$

If r is eliminated from (15) and (16), the value of the resultant is seen to be

$$P = \frac{\sin\frac{N\psi}{2}}{\sin\frac{\psi}{2}}. \tag{17}$$

It is apparent from Fig. 8 that the phase of the resultant, P, is the same as that for the middle vector in the sequence when N is odd, or the average of the two middle vectors when N is even, that is $(N-1)\psi/2$. This phase information can be combined with (17) by writing

$$\mathbf{P} = \frac{\sin \frac{1}{2} N\psi}{\sin \frac{1}{2}\psi} e^{i(N-1)\psi/2}. \tag{18}$$

In (17), P is a magnitude; in (18) it is a complex quantity.

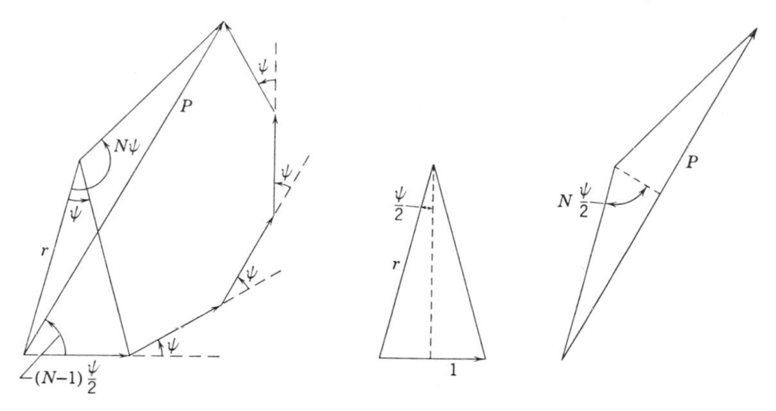

Fig. 8.

For the purpose of studying the properties of this function it is convenient to scale it so that its maximum value is unity. This is done by dividing both sides of (17) by N, which gives

$$\frac{P}{N} = \frac{\sin \frac{1}{2} N\psi}{N \sin \frac{1}{2}\psi} \tag{19}$$

The function $\dfrac{\sin \frac{1}{2} N\psi}{N \sin \frac{1}{2}\psi}$**.** Function (19) has the form $(\sin nx)/(n \sin x)$. Some properties of this function and its square are shown in Fig. 9. Figure 9*A* shows the numerator part of the function, namely $\sin nx$, and Fig. 9*B* shows the denominator part of the function, namely $n \sin x$. Their quotient, $(\sin nx)/(n \sin x)$, is shown in Fig. 9*C*. This has a relatively large maximum at $x = 0$ and then drops rapidly to zero, becomes negative, and subsequently oscillates between positive and negative values with decreasing amplitude. The oscillation behavior follows that of the numerator, $\sin nx$, except that the amplitude at all points is scaled down by a factor $1/(n \sin x)$ due to the denominator. Thus the general location of peaks, troughs, and zeros follows that of the numerator. The zeros occur whenever $\sin nx = 0$. These occur for $nx = m\pi$ (where m is an integer), so that zeros occur at locations $x = m\pi/n$.

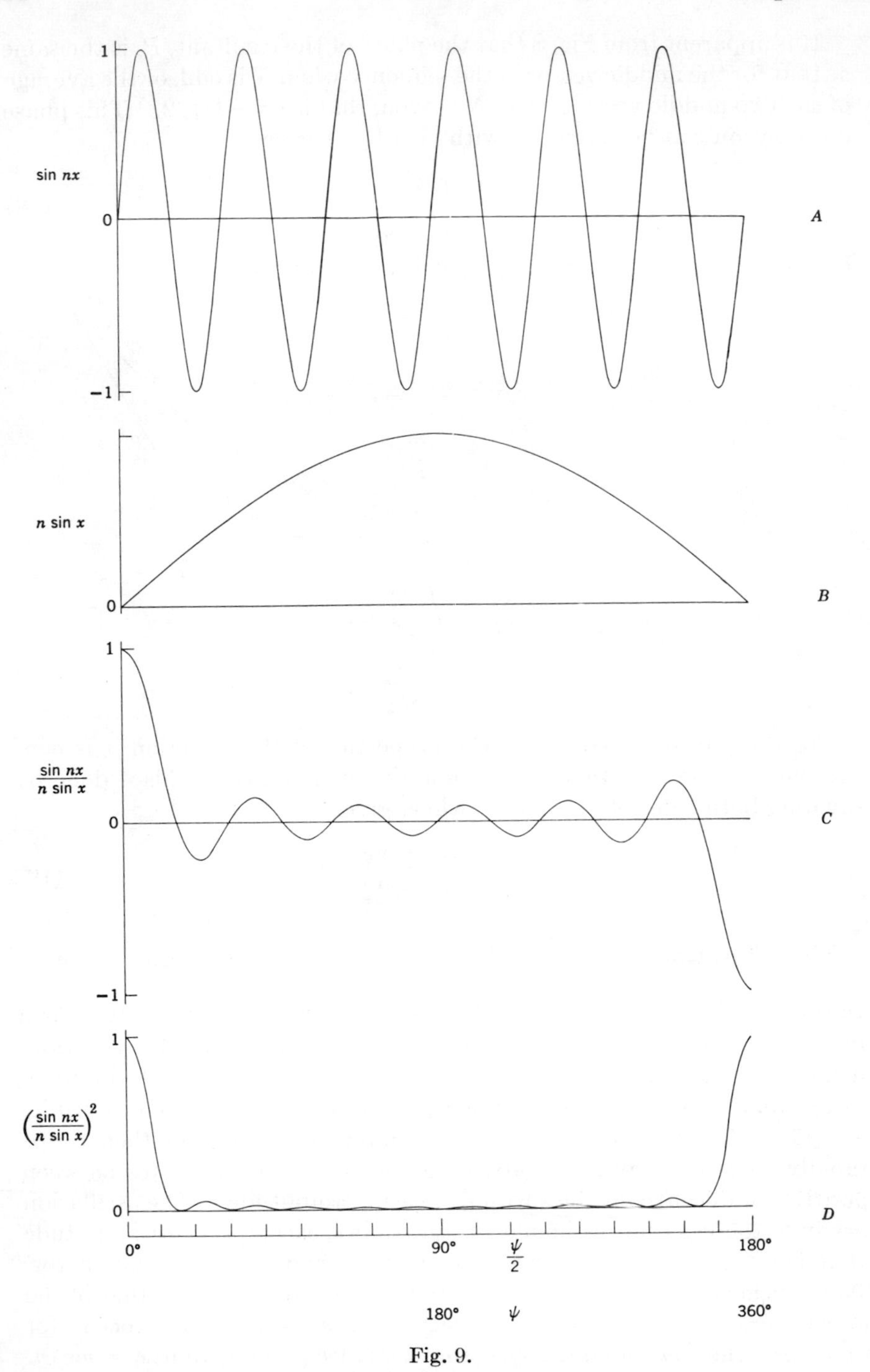

sin nx
1
0
−1
A
n sin x
0
B
sin nx / n sin x
1
0
−1
C
(sin nx / n sin x)²
1
0
D
0°
90°
ψ/2
180°
180°
ψ
360°

Fig. 9.

The peaks and troughs also occur in the neighborhood of peaks and troughs in $\sin nx$, except that the peak of $\sin nx$ next to the origin is not separately represented in the function $(\sin nx)/(\sin x)$, but is part of the main peak.

It is interesting to follow the geometrical behavior of P as the function passes through zero. It was noted in the last section that the phase of P is the same as the phase of the middle vector of the sequence, that is, $(N-1)\psi/2$. Therefore the phase of P increases linearly with ψ. When the sequence of little vectors just closes in an approximation to a

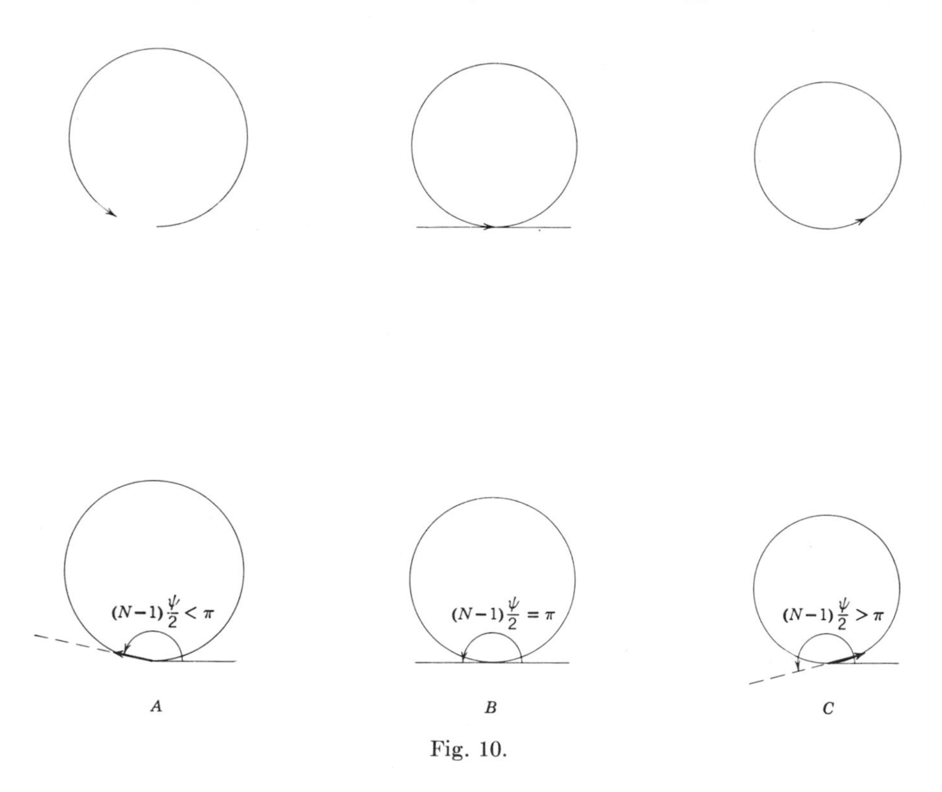

Fig. 10.

circle, Fig. 10B, the resultant is zero. Just before closing, Fig. 10A, the resultant is small and positive, and its phase just less than π. Just after closing, Fig. 10C, if the phase is regarded as increasing linearly, it is just greater than π, and the resultant must be regarded as negative. But if, as the arc closes, the phase should be regarded as changing discontinuously from π to 0, the resultant is positive. This situation is annoying at $\psi = 2\pi$: When N is odd, a large maximum occurs which is a duplicate of that at $\psi = 0$, but when N is even, there occurs a corresponding minimum. This situation is avoided if one wishes merely the absolute value of the resultant. This can be obtained by using $\left(\frac{\sin nx}{n \sin x}\right)^2$ instead of $\frac{\sin nx}{n \sin x}$. The squared function is shown plotted in Fig. 9*D*. It is characterized by a large central peak at $x = 0$. This is flanked by a series of rapidly declining small satellitic maxima, the greatest of which has a magnitude of only a few per cent of that of the main peak.

The number of scattering points in the grating is the integer N of (17) and (18). For $N = \infty$, function (18) has a value of zero everywhere except at $\psi = m2\pi$ when the absolute value is unity, as shown in Fig. 11. For moderately large values of N, the function behaves very nearly the same as when $N = \infty$. The behavior of the function in ordinary problems concerned with the diffraction of x-rays by crystals, where N is 10^6 or more, is substantially the same as if N were ∞. When N is so large, each

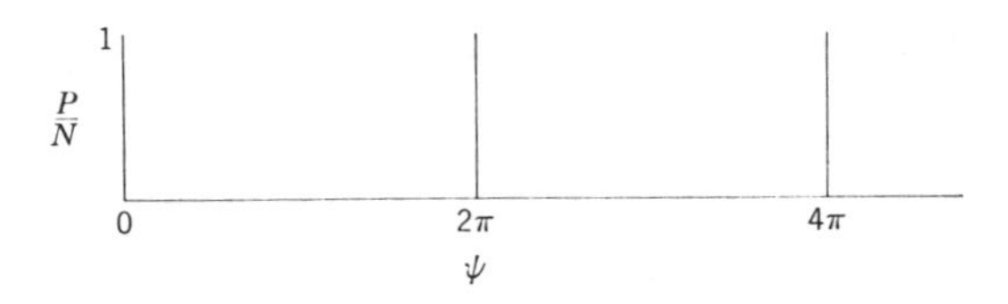

Fig. 11.

of the contributing vectors in Fig. 8 is so small that together they constitute a virtually smooth curve. The angle ψ opposite each vector is so small that $\sin \psi \approx \psi$. Under these circumstances $\sin(\psi/2)$ in (17) can be replaced by $\psi/2$, so that a very approximate form of (19) is

$$\frac{P}{N} \approx \frac{\sin \frac{1}{2} N\psi}{\frac{1}{2} N\psi} \tag{19A}$$

Tables of the numerical value of this function are available.[1]

When (19) has the value unity, the geometrical reason is that all the vectors in Fig. 7 or 8 are aligned parallel to one another. This occurs, for example, when $\psi = 0$, but it also occurs when $\psi = m(2\pi)$. For all values of ψ other than $\psi = m(2\pi)$, the vectors of Fig. 7 have a relatively uniform distribution in all directions, and their sum is substantially zero. Under these circumstances, the points of the grating of Fig. 6 scatter in such a way that their wavelets give rise to *destructive interference*. But for those directions $\bar{\nu}$ for which $\psi = m(2\pi)$, the wavelets reinforce one another to cause *constructive interference*, and a *maximum* is produced. The maximum is given a designation corresponding with the value of m. When $m = 1$, the maximum is said to be a *first-order maximum;* when $m = 2$, it is said to be a *second-order maximum;* etc.

Directions of maxima. It is easy to relate the direction $\bar{\nu}$ to the phase angle expressed in terms of the wavelength λ of the wave. When the mth-order maximum is produced so that $\psi = m(2\pi)$, the path difference from neighboring scattering points to the wave front is $m\lambda$. If the translation period in Fig. 6 is a, then

$$\cos \bar{\nu} = \frac{m\lambda}{a}.$$

More generally, Fig. 6B, if the normal to the incoming wave front makes an angle $\bar{\mu}$ with the row of scattering points, and the normal to the diffracted wave front makes an angle $\bar{\nu}$ with the row, then the condition that wavelets be in phase along the scattered wave front is

$$a \cos \bar{\nu} - a \cos \bar{\mu} = m\lambda \tag{21}$$

or

$$\cos \bar{\nu} - \cos \bar{\mu} = \frac{m\lambda}{a}. \tag{22}$$

The nature of more general one-dimensional patterns. The pattern of points whose scattering has just been considered was specialized in that all points were identical and equally spaced along a line. In other words, the points were translation equivalent with respect to a one-dimensional lattice. Such a set of points may be called a *lattice array*.

In the more general one-dimensional pattern there are several nonidentical points per translation period, Fig. 12. Let any desired location along the line of points be chosen as the origin. Then the absolute coordinate of the jth point with respect to the origin is X_j. If the translation period is a, then there are translation-equivalent points at X_j+a, X_j+2a $\cdots$ X_j+ma.

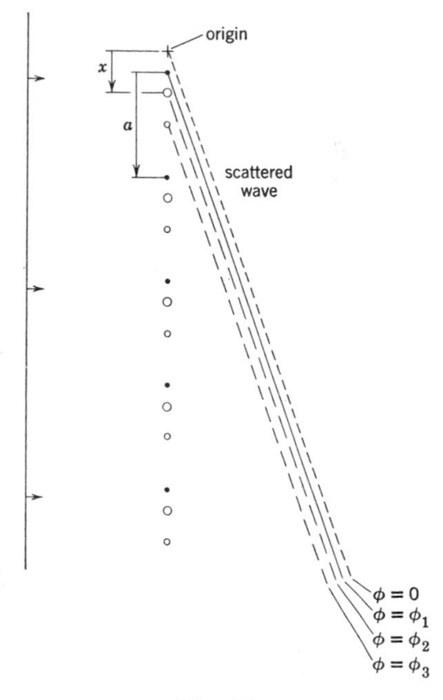

Fig. 12.

Scattering by general one-dimensional patterns. It is evident that each set of translation-equivalent points scatters independently as in Fig. 6, and that its independent behavior is described by a relation like (12). Each such grating, however, has a ϕ characteristic of its location with respect to the chosen origin, and, in general, a scattering power f_j characteristic of the jth scattering point. The combined wave is accordingly described by

$$R = f_1 e^{i\phi_1} \sum_{n=0}^{N-1} e^{in\psi} + f_2 e^{i\phi_2} \sum_{n=0}^{N-1} e^{in\psi} \cdots f_J e^{i\phi_J} \sum_{n=0}^{N-1} e^{in\psi}$$

$$= (f_1 e^{i\phi_1} + f_2 e^{i\phi_2} \cdots + f_J e^{i\phi_J}) \sum_{n=0}^{N-1} e^{in\psi} \tag{23}$$

The expression in parentheses represents the combination of wavelets from the points within a translation period of length a. Let this combined wave of the mth maximum be represented by F_m, so that the wave scattered by a single period of the pattern is

$$F_m = f_1 e^{i\phi_1} + f_2 e^{i\phi_2} \cdots + f_J e^{i\phi_J}. \tag{24}$$

The combination of contributions from different periods is given by the summation term in (23). This is the same as P, which was defined in (14) and evaluated in (17). With these abbreviations, (23) can be expressed

$$R = F_m \cdot P. \tag{25}$$

In order to evaluate the term in parentheses in (23), it is necessary to relate the scattering phases ϕ, of the several sets of points, to the coordinates X of the points. This can be done as follows.

For the mth-order maximum, neighboring translation-equivalent points scatter with a phase difference of $m(2\pi)$. These points have a linear separation a. The question is, what phase ϕ corresponds to a linear separation X? Since phases are obviously proportional to linear separations, it is evident that

$$\frac{\phi}{X} = \frac{m(2\pi)}{a}, \tag{26}$$

so that

$$\phi = m \frac{X}{a} 2\pi. \tag{27}$$

If this value is substituted into (24), the result for the mth maximum is

$$F_m = f_1 e^{im(X_1/a)2\pi} + f_2 e^{im(X_2/a)2\pi} \cdots + f_J e^{im(X_J/a)2\pi}. \tag{28}$$

This expression is simplified if the scattering points are referred to coordinates expressed as a fraction of the period a. These *fractional coordinates* x are defined by

$$x = \frac{X}{a}, \tag{29}$$

and the diffracted wave is represented, in terms of it, by

$$F_m = f_1\, e^{imx_1 \cdot 2\pi} + f_2\, e^{imx_2 \cdot 2\pi} \cdots + f_3\, e^{imx_J \cdot 2\pi}. \tag{30}$$

Interpretation of the scattering expressions. The expressions just derived show that the scattering by a general one-dimensional pattern has something in common with the scattering by a simple lattice array of identical scattering points. The last factor in (12) and (13), and (23) and (25) is

(14):
$$P = \sum_{n=0}^{N-1} e^{in\psi}.$$

For gratings with a reasonably large number N of translation periods, this expression is zero except for $\psi = m(2\pi)$, when a maximum occurs. This means that both the simple one-dimensional lattice array and the general one-dimensional pattern have maxima in the same directions, namely those defined by (22), specifically

$$\cos \bar{\nu} = \cos \bar{\mu} + \frac{m\lambda}{a} \qquad (m = 1, 2, 3, \cdots). \tag{31}$$

This interpretation, generalized to three dimensions, would correspond to saying that the directions of maxima depend only on the dimensions of the unit cell, and not on the arrangement of the atoms in the cell.

Now, comparing the remaining term in (12) and (23), but substituting the atomic coordinates x for ϕ, according to (27) and (29), the terms for the two cases turn out to be:

Lattice array:

$$F_m = fe^{imx \cdot 2\pi}, \tag{32}$$

General pattern:

$$F_m = f_1\, e^{imx_1 \cdot 2\pi} + f_2\, e^{imx_2 \cdot 2\pi} \cdots + f_J\, e^{imx_J \cdot 2\pi}. \tag{33}$$

These expressions give the amplitude of the resultant wavelet scattered by the set of points within one translation period, Fig. 13. The resultant, F_m, is a complex quantity with both magnitude, $|F_m|$, and phase. In the case of the simple lattice array, (32), all maxima have the same magnitude because

$$\begin{aligned} |F_m| &= |fe^{imx \cdot 2\pi}| \\ &= f. \end{aligned} \tag{34}$$

But for the general pattern, (33), the various maxima have different magnitudes because the angles between the vectors of Fig. 13 depend on the ϕ's of the vectors, and these, in turn, depend upon the coordinates of the corresponding points, according to (27). It follows that the amplitudes

F_m of the various maxima diffracted by a general pattern are, in general, different and are characteristic both of the order m and of the coordinates of the set of scattering points in the translation period. This interpretation, generalized to three dimensions, would correspond to saying that the amplitudes of the maxima depend on the relative arrangement of the scattering points in the unit cell, and not on the dimensions of the cell.

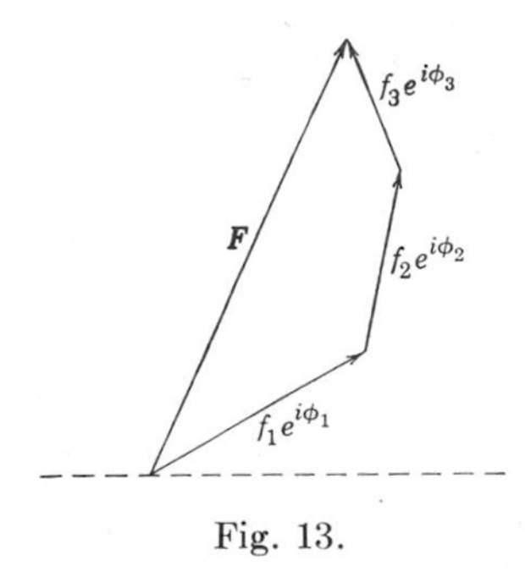

Fig. 13.

It is this correspondence between coordinates of the scattering elements and the amplitudes of the maxima which gives rise to the possibility of finding the first if the second is known. If the *complex* amplitudes F_m are indeed known, it is a routine matter to find the distribution of scattering matter in a translation period, as will be seen from Chapter 13. But only the magnitudes, $|F_m|$, of the complex amplitudes, F_m, can be experimentally observed, so that complete advantage cannot, in general, be taken of this relationship. This is the epitome of the *phase problem*, discussed in Chapter 21.

Diffraction by three-dimensional patterns

The scattering by a three-dimensional pattern has a number of fundamental features in common with that of the simple one-dimensional pattern just discussed. Certain complications occur, however, which are characteristic of the additional number of dimensions involved.

Diffraction by a lattice array of points. The simplest kind of three-dimensional pattern is a set of points, each point located at a lattice point. Such a simplified pattern is called a *lattice array*. Let each point be capable of scattering identically. In order that all the points of a lattice array scatter in phase, it is necessary that a condition like (22) be satisfied along each of three non-equivalent rows of the array. These rows may be taken along the translations of the lattice which are used to define the edges of the cell. In this case the rows have periods a, b, and c, and the three conditions to be satisfied are

$$\begin{aligned} \cos \bar{\nu}_a - \cos \bar{\mu}_a &= \frac{h\lambda}{a}, \\ \cos \bar{\nu}_b - \cos \bar{\mu}_b &= \frac{k\lambda}{b}, \\ \cos \bar{\nu}_c - \cos \bar{\mu}_c &= \frac{l\lambda}{c}. \end{aligned} \tag{35}$$

The integers h, k, and l are, in general, different, and are the number of wavelengths path difference from the original wave front to the diffracted wave front for the rows a, b, and c respectively. It is shown elsewhere† that this condition is equivalent to the first-order Bragg reflection of the incoming wave by a plane having indices (hkl).

The nature of more general three-dimensional patterns. In more general three-dimensional patterns, there are several sets of points which are not translation equivalent. But all points of each set are translation equivalent. That is, each set lies on a lattice. This means that the general pattern may be thought of as several lattice arrays which are parallel but mutually displaced.

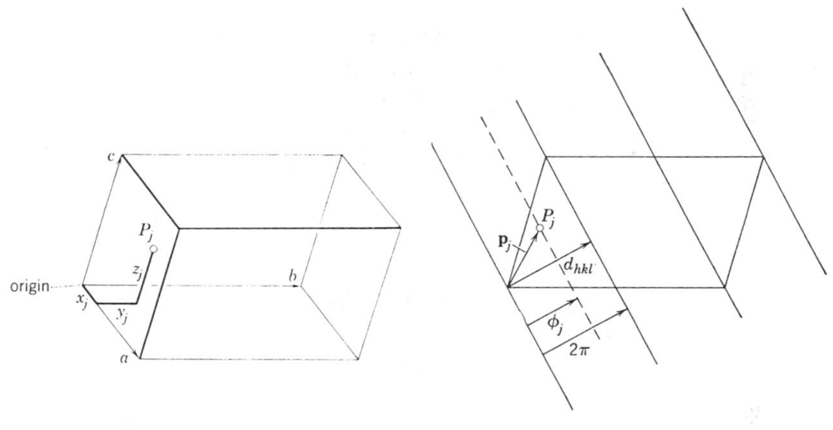

Fig. 14. Fig. 15.

Let any desired point in the space occupied by the pattern be chosen as the origin. Then a unit cell can be chosen with the three cell edges a, b, and c as coordinate axes, radiating from this origin, Fig. 14. The absolute coordinates of the jth point with respect to these axes are $X_j\ Y_j\ Z_j$, while the fractional coordinates, each referred to the magnitude of the translation in its direction as unity, are

$$x_j = \frac{X_j}{a},$$
$$y_j = \frac{Y_j}{b}, \tag{36}$$
$$z_j = \frac{Z_j}{c}.$$

The phase of a wavelet scattered by a point in a pattern. Since a maximum hkl is equivalent to a first-order reflection from a plane (hkl), the spacing d_{hkl} corresponds to a phase difference of 2π, Fig. 15. The

† M. J. Buerger, *X-ray crystallography.* (John Wiley and Sons, New York, 1942) 42–43.

phase scattered by any irrational plane between the origin and the first rational plane, is directly proportional to the distance of the irrational plane from the origin. Since the scattering point P_j lies in a plane whose distance from the origin is the projection of its vector, $\mathbf{p}_j$, on d, the proportion can be written

$$\frac{\phi_j}{2\pi} = \frac{\text{proj. } \mathbf{p}_j}{d_{hkl}} \tag{37}$$

$$= \frac{1}{d_{hkl}} (\text{proj. } \mathbf{p}_j). \tag{38}$$

It is convenient to express the right of (38) in vector notation. The vector of length $1/d_{hkl}$, and having the direction of d_{hkl}, is the vector in reciprocal space from the origin to the reciprocal-lattice point hkl; this vector is customarily written $\boldsymbol{\sigma}_{hkl}$. The right of (38) is the scalar product of this vector with $\mathbf{p}_j$, so that (38) can be written

$$\frac{\phi_j}{2\pi} = \boldsymbol{\sigma}_{hkl} \cdot \mathbf{p}_j. \tag{39}$$

To find the numerical value of this, one can substitute the values of $\boldsymbol{\sigma}_{hkl}$ and $\mathbf{p}_j$ in terms of their components:

$$\boldsymbol{\sigma}_{hkl} = h\mathbf{a}^* + k\mathbf{b}^* + l\mathbf{c}^*, \tag{40}$$

$$\mathbf{p}_j = x_j\, \mathbf{a} + y_j\, \mathbf{b} + z_j\, \mathbf{c}. \tag{41}$$

The scalar product of these is

$$\boldsymbol{\sigma}_{hkl} \cdot \mathbf{p}_j = (h\mathbf{a}^* + k\mathbf{b}^* + l\mathbf{c}^*) \cdot (x_j\, \mathbf{a} + y_j\, \mathbf{b} + z_j\, \mathbf{c})$$

$$= hx_j + ky_j + lz_j. \tag{42}$$

If this is substituted in the right of (39), then the value of the phase scattered by the jth point is seen to be

$$\phi_j = 2\pi(hx_j + ky_j + lz_j). \tag{43}$$

Scattering by general three-dimensional patterns. If the amplitude of the wavelet scattered by the jth point of the pattern is f_j, then the scattered wavelet is described in amplitude and phase by the term $f_j\, e^{i2\pi(hx_j+ky_j+lz_j)}$. At the maximum of the spectrum hkl the wave scattered by all the atoms in one unit cell is accordingly described by

$$F_{hkl} = f_1\, e^{i2\pi(hx_1+ky_1+lz_1)} + f_2\, e^{i2\pi(hx_2+ky_2+lz_2)} \cdots + f_J\, e^{i2\pi(hx_J+ky_J+lz_J)} \tag{44}$$

$$= \sum_j f_j\, e^{i2\pi(hx_j+ky_j+lz_j)}. \tag{45}$$

The expression in (44) corresponds to the one for the one-dimensional case in (33). The two expressions are analogous, and the comparative complexity of (44) is due to the two additional dimensions.

Application to crystals

The amplitude scattered by a "point." The discussion in this chapter has been carried out without regard to the specific nature of the "scattering points," and without supplying a mechanism for scattering. The details of scattering are considered in the next chapter, where it is shown that the individual electrons of the atoms of the crystal are the scattering units.

It is usually more convenient, however, to think of the scattering units of a crystal as its chemical atoms. Of course, the electrons of an atom are responsible for its scattering power, so, to first approximation, the scattering power of an atom is proportional to the number of electrons it contains, that is, to its atomic number Z. This approximation was commonly used in the early days of crystal-structure analysis. Actually, the separation of the various electrons within the volume of the atom causes phase differences among the wavelets they scatter, so the scattering power of an atom is, in general, less than that of Z electrons. The amplitude scattered by an atom falls off with $(\sin \theta)/\lambda$. A further discussion of the scattering power of an atom is given in Chapter 10.

The possibility of crystal-structure analysis by diffraction. The discussion following (34) can be readily generalized to three dimensions and applied to (44): The wave scattered in each maximum hkl is characterized by the complex quantity F_{hkl}. This can be written as an explicit function of the coordinates of each of the J atoms in the unit cell. For a particular set of such coordinates, a specific set of the several F_{hkl}'s is determined. This implies that for a particular arrangement of atoms in a cell, there is a specific set of diffraction spectra F_{hkl}.

Suppose that, in studying the arrangement of atoms in a given crystal, a particular model is proposed. The set of the F_{hkl}'s of the model can be computed by (44). A necessary condition for the model to be the correct one is that the calculated maxima match those which are experimentally observed. It will appear that this is also a sufficient condition, so that one can state:

Theorem 2: The necessary and sufficient condition for a crystal-structure model to be correct is that the set of $\mathrm{F_{hkl}}$*'s computed for it match the observed* $\mathrm{F_{hkl}}$*'s.*

This statement contains a misleading feature. Each F_{hkl} is complex; that is, it is characterized both by a magnitude, $|F_{hkl}|$, and a phase, ϕ_{hkl}.

The magnitude can be experimentally observed, but so far no means has been found to observe the phase. If means could be found, every crystal structure could be solved as a routine by using the methods of Chapter 13. On the other hand, the situation is not quite so serious as this might appear to imply; indeed in Chapter 21 it will be pointed out that, except for certain "homometric mates," the magnitudes alone of the set of F_{hkl}'s determine a unique arrangement of atoms. This permits statement of a further theorem:

Theorem 3: Except for homometric mates, a necessary and sufficient condition for a crystal-structure model to be correct is that the magnitude of the calculated maxima match those which are experimentally observed.

By one means or another, then, crystal-structure analysis requires finding a model for which calculated $|F_{hkl}|$'s match the set experimentally observed. How the model is arrived at is immaterial. It can be found by clever guess work, by trial-and-error, or by a systematic study. Most of this book is concerned with the theory and practice of systematic study.

Literature

[1] J. Sherman. *A four place table of* $\frac{\sin x}{x}$. Z. Krist. **85** (1933) 404–419

3

Some quantitative aspects of the diffraction of x-rays by crystals

In the last chapter a picture was sketched of the general nature of diffraction as divorced from most of its technical details. In this chapter, the diffraction of x-radiation by crystals is examined somewhat more closely. This introduces the technical details in their most general form, and provides a background for understanding how allowance can be made for some of the factors which affect the intensities of spectra.[9]

Unfortunately, one cannot conveniently observe the absolute value of the diffraction amplitude, $|F_{hkl}|$. Instead, a related quantity commonly known as the *integrated intensity* is customarily observed. In this chapter the relation between amplitude and integrated intensity is developed, and in Chapters 7 and 8 some routine devices for deducing one from the other are described. A further purpose of this chapter is to put the matter of x-ray diffraction on a somewhat more quantitative basis. This requires a knowledge of how a single electron behaves in an x-ray beam. With this background, the combined behavior of all the electrons in the crystal can be considered.

Scattering by a single electron[3]

If an electron is located in the path of an x-ray beam, it is forced into oscillation by the electromagnetic field of the x-rays impinging upon it. Due to this acceleration, the electron in turn becomes a source of radiation, and in this way, the electron is said to scatter the impinging radiation.

Suppose that the intensity of the electric vector of the primary x-ray beam is $\mathcal{E}_0$. The force experienced by the electron of charge e is $\mathcal{E}_0 e$, and the acceleration as the result of this force is

$$a = \frac{\mathcal{E}_0 e}{m}, \tag{1}$$

where m is the mass of the electron. The electron may now be supposed to undergo forced oscillation in which the acceleration vector follows the vector of the electromagnetic field.

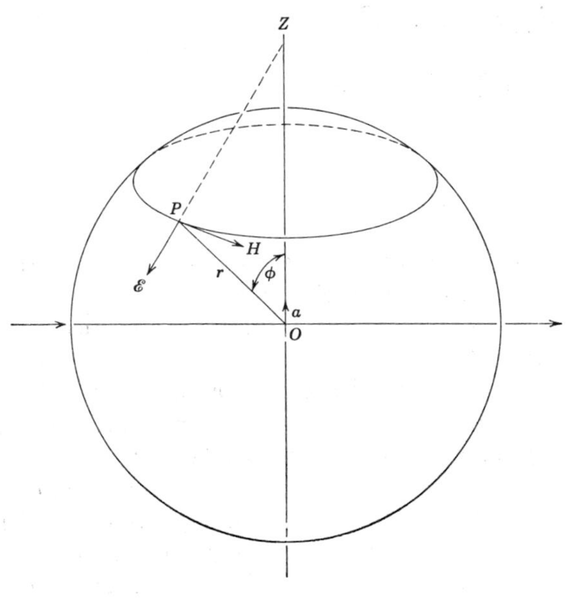

Fig. 1.

Now an accelerating charged particle is itself a source of electromagnetic radiation, which spreads as a set of spherical waves from the source. In Fig. 1, let the electron be located at O, and let it have an acceleration a along the Z axis. At any point P, both an electric and a magnetic field are experienced due to the accelerating electron. Let P be at a distance r from the electron, and let the vector from the electron to P make an angle ϕ with the Z axis. The directions of the electric- and magnetic-field vectors at P are as shown in Fig. 1, and their magnitudes are given, according to classical theory, by

$$\mathcal{E} = H = \frac{ae}{rc^2} \sin \phi, \tag{2}$$

where c is the velocity of light. The value of a is given by (1), so (2) may be written

$$\mathcal{E} = \frac{\mathcal{E}_0 e^2}{rmc^2} \sin \phi. \tag{3}$$

The energy per unit cross-sectional area of a beam, known as its *intensity*, is proportional to the square of the electric vector, and specifically equal to

$$I = \frac{c}{4\pi} \mathcal{E}^2. \tag{4}$$

Thus, the ratio of the intensities of the original x-ray beam and the scattered beam is

$$\frac{I}{I_0} = \frac{\frac{c}{4\pi} \mathcal{E}^2}{\frac{c}{4\pi} \mathcal{E}_0{}^2} = \left(\frac{\frac{\mathcal{E}_0 e^2}{rmc^2} \sin \phi}{\mathcal{E}_0} \right)^2 = \left(\frac{e^2}{rmc^2} \sin \phi \right)^2. \tag{5}$$

The polarization factor

In the discussion just given, it was assumed, for simplicity, that the electric vectors of all the x-rays impinging on the electron are confined to a single direction, i.e., that the original x-ray beam is plane polarized. In all the usual experimental arrangements, however, the x-ray beam is unpolarized, which means that the azimuth of the electric vector assumes all directions with time. The effective amplitude of the radiation after it is reflected by the crystal at the angle 2θ consists only of the components of these azimuths after reflection. This feature has the effect of reducing the intensity of the x-ray beam by a factor, p, known as the *polarization factor*.

Let the primary beam striking the crystal be assumed to be unpolarized; that is, the electric vector of the direct x-ray beam is in a random direction. Let this vector be represented by $\mathcal{E}_0$, Fig. 2. Resolve $\mathcal{E}_0$ into its components perpendicular and parallel to the plane of the direct beam and the reflected beam, namely $\mathcal{E}_{0,\perp}$ and $\mathcal{E}_{0,\parallel}$. Since $\mathcal{E}_0$ is random, these two components occur with equal frequency, and the intensities associated

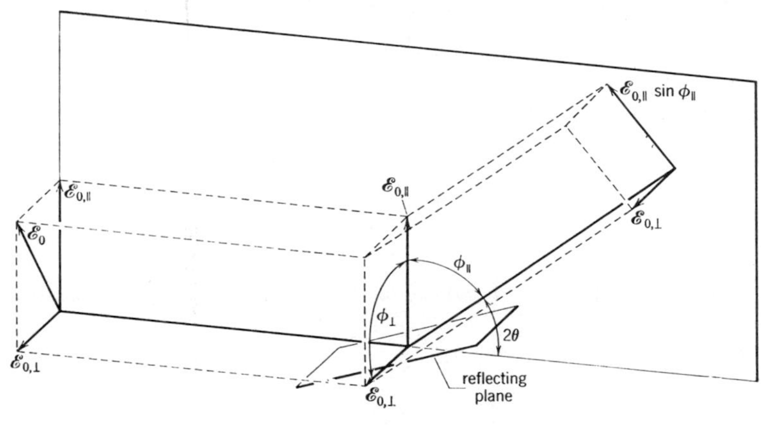

Fig. 2.

with these component $\mathcal{E}$'s are equal. Since the sum of the two intensity components is equal to the total intensity,

$$I_{0,\perp} = I_{0,\parallel} = \tfrac{1}{2}I_0, \tag{6}$$

and

$$I_{0,\perp} + I_{0,\parallel} = I_0. \tag{7}$$

According to (5), the intensity scattered by an electron depends on the angle ϕ. For scattered intensities $I_\perp$ and $I_\parallel$ these are

$$I_\perp = I_{0,\perp}\left(\frac{e^2}{rmc^2}\right)^2 \sin^2\phi_\perp. \tag{8}$$

Substituting for $I_{0,\perp}$ from (6), and also substituting 90° for $\phi_\perp$ gives

$$I_\perp = \tfrac{1}{2}I_0\left(\frac{e^2}{rmc^2}\right)^2. \tag{9}$$

Also,

$$I_\parallel = I_{0,\parallel}\left(\frac{e^2}{rmc^2}\right)^2 \sin^2\phi_\parallel. \tag{10}$$

Substituting for $I_{0,\parallel}$ from (6), and also substituting $90° - 2\theta$ for $\phi_\parallel$ gives

$$I_\parallel = \tfrac{1}{2}I_0\left(\frac{e^2}{rmc^2}\right)^2 \cos^2 2\theta. \tag{11}$$

The total intensity of the reflected beam is

$$
\begin{aligned}
I_{\text{reflected}} &= I_\perp + I_\parallel \\
&= \tfrac{1}{2} I_0 \left(\frac{e^2}{rmc^2}\right)^2 + \tfrac{1}{2} I_0 \left(\frac{e^2}{rmc^2}\right)^2 \cos^2 2\theta \\
&= \left[I_0 \left(\frac{e^2}{rmc^2}\right)^2 \right] (\tfrac{1}{2} + \tfrac{1}{2}\cos^2 2\theta) \qquad (12) \\
&= \left[I_0 \left(\frac{e^2}{rmc^2}\right)^2 \right] p, \qquad (13)
\end{aligned}
$$

where $p = \tfrac{1}{2} + \tfrac{1}{2}\cos^2 2\theta.$ (14)

Now the term in square brackets in (12) is the magnitude the intensity would have if there were no reduction of intensity due to reduction of the component of $\mathcal{E}_\parallel$ with the angle $\phi = 90° - 2\theta$. The term in parentheses gives the factor by which the intensity of the reflected beam is reduced due to this effect. It is called the polarization factor and varies only in the limited numerical range of 1 to $\tfrac{1}{2}$. Note that the polarization factor depends only on the angle θ (or 2θ) and is unaffected by the x-ray method employed.

The reflected x-ray beam does not have its electric vector in a random direction. Rather there is an excess of intensity having its electric vector perpendicular to the plane containing the direct beam and the reflected beam. In other words, the reflected beam is "polarized." The efficiency of polarization can be measured by the excess of one component over the other, compared with the total intensity, namely

$$\frac{I_\perp - I_\parallel}{I} = \frac{I_\perp - I_\parallel}{I_\perp + I_\parallel}. \qquad (15)$$

This ratio varies from 0 when 2θ is 0° or 180°, to 1 when 2θ is 90°. In other words, at $2\theta = 90°$, the reflected x-ray beam is 100% polarized.

The scattering by a crystal[4–8]

Resolution of a crystal into planes. The theoretical amount of radiation scattered by the entire crystal can be found by several methods. A convenient treatment is to regard the three-dimensional pattern of the crystal as resolved into a sequence of parallel planes of translation-equivalent atoms. Then each plane behaves as a plane diffracting source, and can be referred to Fresnel-zone theory.[1, 2, 10] This approach has the advantage of being rather readily pictured, and of being related to ordinary optical diffraction theory. This is also the classical treatment of Bragg, James, and Bosanquet.[7, 8]

Fresnel-zone theory. In Fig. 3, light emanating from point O impinges on a plane. Each point of this plane acts as a new source giving rise to a new wavelet. Fresnel diffraction theory considers the

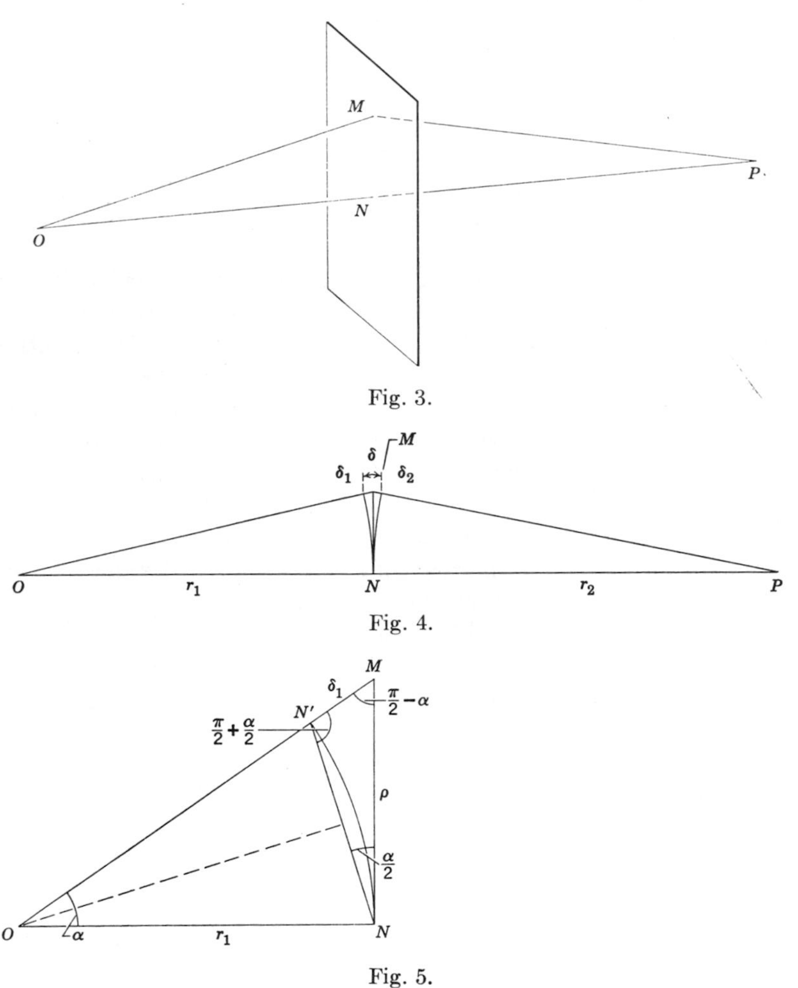

Fig. 5.

resultant of all these wavelets as received at a point, P, beyond the plane, and on a line from O at right angles to the plane. Since the various paths OMP have various lengths, the phases of the wavelets reaching P along these paths represent a variable which must be evaluated before the resultant effect at P can be found.

Figure 4 is a section in plane OMP of Fig. 3. The path length OMP is greater than that of the straight line ONP by the amount δ, which is composed of two parts, δ_1 and δ_2. The magnitude of δ_1 can be determined with the aid of Fig. 5, which shows only the left side of Fig. 4, with the angle $NOM = \alpha$ exaggerated. In Fig. 5, draw the chord NN' perpendicular to the bisector of $\angle NOM$. Then

$$\angle N'NM = \frac{\alpha}{2},$$

$$\angle N'MN = \frac{\pi}{2} - \alpha,$$

and consequently

$$\angle MN'N = \pi - \left(\frac{\pi}{2} - \alpha\right) - \frac{\alpha}{2} = \frac{\pi}{2} + \frac{\alpha}{2}.$$

Applying the law of sines to the scalene triangle $NN'M$, there results

$$\frac{\delta_1}{\sin(\alpha/2)} = \frac{\rho}{\sin\left(\frac{\pi}{2} + \frac{\alpha}{2}\right)} = \frac{\rho}{\cos\frac{\alpha}{2}}; \tag{16}$$

$$\delta_1 = \rho \frac{\sin\frac{\alpha}{2}}{\cos\frac{\alpha}{2}} = \rho \tan\frac{\alpha}{2}. \tag{17}$$

Provided that angle α is small, its tangent is approximately equal to the angle, so that a good approximation is

$$\tan\frac{\alpha}{2} \approx \tfrac{1}{2}\tan\alpha. \tag{18}$$

Figure 5 shows that

$$\tfrac{1}{2}\tan\alpha = \tfrac{1}{2}\frac{\rho}{r_1} \tag{19}$$

Substituting (18) and (19) in (17), there results

$$\delta_1 = \rho\tan\frac{\alpha}{2} \approx \rho\,\tfrac{1}{2}\frac{\rho}{r_1}, \tag{20}$$

so that

$$\rho^2 \approx 2r_1\,\delta_1. \tag{21}$$

A similar relation applies to the right side of Fig. 4, namely

$$\delta_2 \approx \frac{\rho^2}{2r_2}. \tag{22}$$

And, since

$$\delta = \delta_1 + \delta_2, \tag{23}$$

it follows that

$$\delta \approx \frac{\rho^2}{2r_1} + \frac{\rho^2}{2r_2}$$

$$\approx \frac{\rho^2}{2}\left(\frac{1}{r_1} + \frac{1}{r_2}\right) \tag{24}$$

$$\approx \frac{\rho^2}{2}\left(\frac{r_1+r_2}{r_1 r_2}\right). \tag{25}$$

This relation provides the phase difference, δ, as a function of the distances, r_1 and r_2, from the plane to the source and to the point, and as a

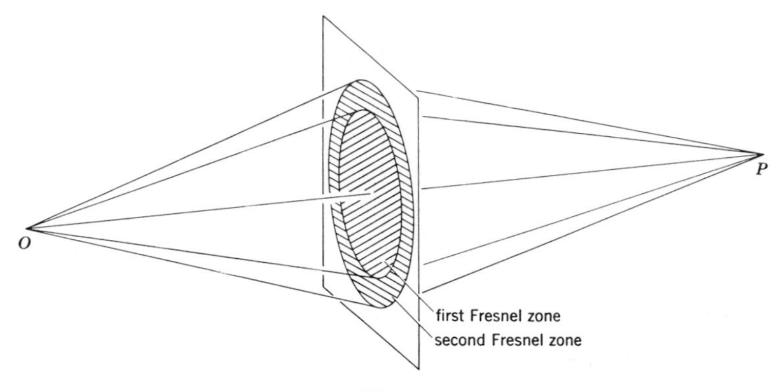

Fig. 6.

function of the distance, ρ, of the scattering point from the axis. When the plane is normal to OP, evidently points of equal retardation, δ, occur on a circle of radius

$$\rho \approx \sqrt{2\delta \Big/ \left(\frac{1}{r_1} + \frac{1}{r_2}\right)}$$

$$\approx \sqrt{2\delta\left(\frac{r_1 r_2}{r_1+r_2}\right)}. \tag{26}$$

The first *Fresnel zone*, Fig. 6, is defined as the area (in this case, a circle) within which the retardation δ ranges between 0 and $\lambda/2$; the second Fresnel zone as the area (in this case, a ring) within which δ ranges between $\lambda/2$ and $2\lambda/2$; and, in general, the nth Fresnel zone as the area (in this case, a ring) within which δ ranges between $(n-1)\lambda/2$ and $n\lambda/2$. Thus, for the outer edge of the nth Fresnel zone,

$$\delta_n = n\frac{\lambda}{2}. \tag{27}$$

The radius of this edge in terms of λ is given by substituting (27) in (26):

$$\rho_n \approx \sqrt{\frac{r_1 r_2 n\lambda}{r_1+r_2}}. \tag{28}$$

The area of the nth zone is

$$\begin{aligned}\sigma_n &= \pi\rho_n^2 - \pi\rho_{n-1}^2 \\ &\approx \pi\frac{r_1 r_2 n\lambda}{r_1+r_2} - \pi\frac{r_1 r_2(n-1)\lambda}{r_1+r_2} \\ &\approx \frac{\pi r_1 r_2 \lambda}{r_1+r_2} \\ &\approx \pi\lambda\left(\frac{1}{\dfrac{1}{r_1}+\dfrac{1}{r_2}}\right).\end{aligned} \tag{29}$$

This indicates that all zones have approximately the same area. In the same way, it can be shown that if a zone is subdivided into zonelets whose edges correspond with a chosen retardation difference other than the fraction $\lambda/2$ (used to divide the plane into Fresnel zones), the areas of these zonelets are equal. This equality depends on the approximation used in (18). Actually, if the characteristics of this approximation are taken into account, it becomes evident that the zones shrink in area at an ever-increasing pace with increasing δ.

Now, suppose that the first Fresnel zone is subdivided into zonelets whose edges correspond to the same retardation, δ. Each of these zonelets is at substantially the same distance, r_1, from the source, O, Fig. 4, so that the radiation from the source has substantially the same amplitude as it reaches each zonelet. Furthermore, each zonelet has the same area (to the first approximation), so that all zonelets scatter radiation with the same resultant amplitude. But the phase of the radiation received at P from any zonelet differs from that of its neighboring zonelet by some small angle

$$\Delta\psi = \frac{\Delta\lambda}{\lambda}2\pi. \tag{30}$$

The resultant scattering reaching P from all the zonelets can be found by plotting on the Argand diagram the equal amplitudes received from each zonelet at phase angles which increase uniformly by the amount given in (30). This summation is shown in Fig. 7. The phase of the wavelet

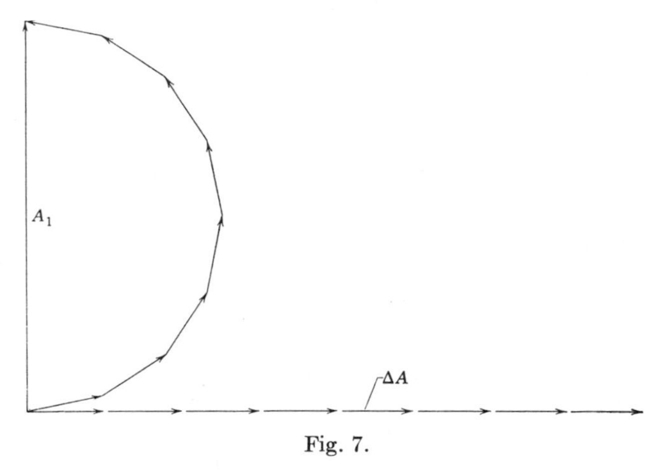

Fig. 7.

contributed by the last zonelet of the first zone differs from that of the first zonelet of the first zone, according to (30), by an amount

$$\Delta\psi = \frac{\frac{1}{2}\lambda}{\lambda} 2\pi = \pi. \tag{31}$$

In this discussion, the first Fresnel zone was assumed to be divided into zonelets of finite width differing in phase according to (30). If the zonelets are taken to be delimited by an infinitely small phase difference, $d\psi$, the separate arrows of Fig. 7 become infinitely small and their concatenation becomes a semicircle. Now the ratio of the resultant amplitude actually scattered by the zonelets of the first zone, to the amplitude they would scatter if they did so with no phase difference, Fig. 7, is the ratio of the diameter of a circle to half its circumference, namely

$$\frac{A_{\text{actual}}}{A_{\text{in phase}}} = \frac{2}{\pi}. \tag{32}$$

If A_1 is taken to represent the resultant amplitude scattered to P by the first Fresnel zone, (32) can be rewritten

$$A_1 = \frac{2}{\pi} \Sigma \, \Delta A. \tag{33}$$

As mentioned above, it has been assumed that the amplitudes contributed by the various zonelets are equal. This followed partly because the zonelets have approximately the same area, as indicated by (29).

A more exact treatment shows that the area decreases with increasing zonelet number, at first slowly, but at an ever-increasing pace. As a consequence, the decreasing amplitudes of the zonelets cause the summation curve of Fig. 7 to spiral toward the center of the circle, as shown in Fig. 8. This causes the resultant to have a simple form, which can be expressed as follows:

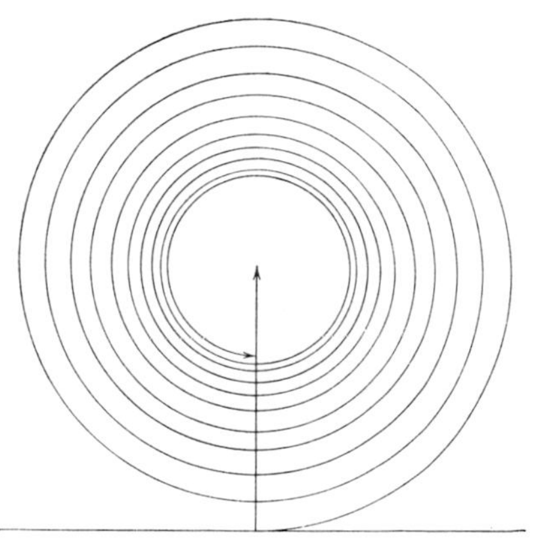

Fig. 8.

Theorem: The resultant amplitude due to all the Fresnel zones is half the resultant due to the first Fresnel zone.

Suppose that the diffracting plane is tilted out of normal to the axial line of radiation, OP, Fig. 9, so that the plane makes an angle θ with that line. Then a point on a circle on the normal plane, and its projection on the ellipse of the tilted plane, have substantially equal retardations, δ, because $OMP \approx O'M'P'$, Fig. 10. Thus, the Fresnel zone on the tilted plane has an elliptical border, and its area is

$$\sigma_{\text{tilted}} = \frac{\sigma_{\text{normal}}}{\cos\left(\frac{\pi}{2} - \theta\right)} = \frac{\sigma_{\text{normal}}}{\sin\theta}. \tag{34}$$

According to (29), this area is specifically

$$\begin{aligned}\sigma &= \frac{\pi\lambda r_1 r_2}{r_1 + r_2}\frac{1}{\sin\theta}\\ &= \pi\lambda\frac{1}{\left(\frac{1}{r_1} + \frac{1}{r_2}\right)\sin\theta}.\end{aligned} \tag{35}$$

Because the direct beam and the reflected beam are symmetrical with respect to the reflecting plane, as shown in Fig. 11, this relation is directly applicable to reflection.

Diffraction by a plane of atoms. Since atoms contain electrons, and electrons scatter x-radiation as discussed in the first section of this chapter, it follows that a plane of atoms scatters x-radiation. Consider a plane of atoms in lattice array. Although the scattering matter is distributed non-uniformly in the plane, in the application of Fresnel-zone theory the distribution of matter may be regarded as essentially uniform.

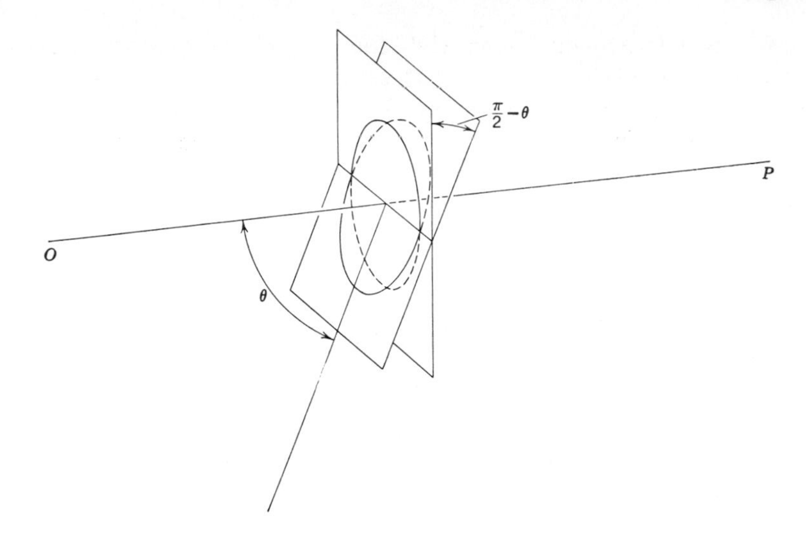

Fig. 9.

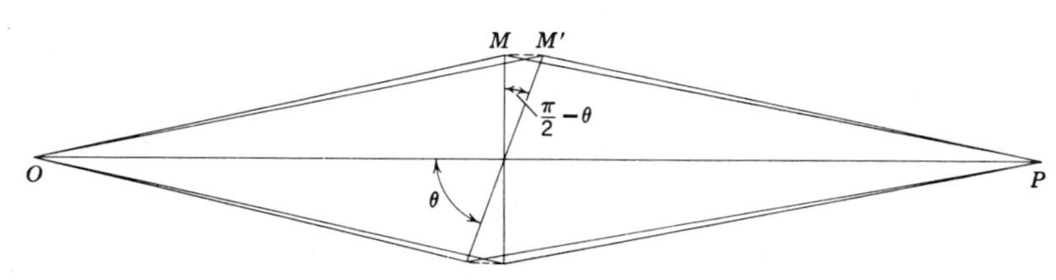

Fig. 10.

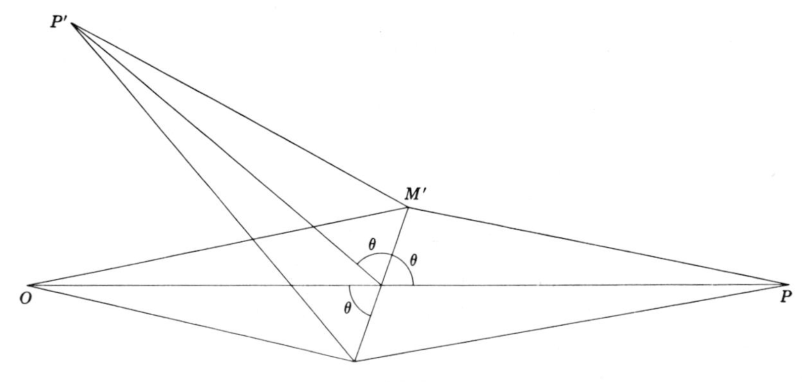

Fig. 11.

This is because the distances between neighboring atoms on the points of the two-dimensional lattice in the plane are very small compared with the distances, r_1 and r_2 (of the last section), which actually arise in practical experiments. Let S equal the area of a cell of the two-dimensional lattice in the plane. Then the density of lattice points (and of atoms in a lattice array) is one per cell area, or $1/S$. If the scattering amplitude of the atom is f, the scattering amplitude of a unit of area of the plane is f/S.

The nature of f will be discussed in more detail in Chapter 10. It is essentially the ratio of the amplitude scattered by the atom to the amplitude scattered by a single electron. If all electrons in the atom scattered in phase with one another, f would equal Z, the number of electrons in the atom. Actually, at all angles except $\theta = 0$ there is a phase difference in the scattering by electrons in the several regions of the atom, and this feature reduces the effective scattering of the collection to some fraction of the arithmetic sum Z. Thus, for $\theta = 0$, $f = Z$, but the value of f declines continuously for increasing angles θ. As a consequence of this situation, if A_e is the amplitude scattered by a single electron under a given set of conditions, and A_a is the amplitude scattered by an atom of Z electrons,

$$A_a = fA_e. \tag{36}$$

In the first section of this chapter, the magnitude of the electric vector of the radiation scattered by a single electron was derived. For present purposes, it is convenient to speak of amplitude instead of electric vector, and with this change of language, the amplitude A_e at a distance r_2 scattered by a single electron is provided by

$$\frac{A_e}{A_0} = \frac{\mathcal{E}}{\mathcal{E}_0} = \frac{I^{1/2}}{I_0^{1/2}}. \tag{37}$$

If the ratio of I's is obtained from (5), and recalling that the averaging of $\sin \phi$ over all azimuths results simply in the polarization factor p of (14), the value of A_e is given by

$$A_e = A_0 \frac{e^2}{r_2\, mc^2} p^{1/2}. \tag{38}$$

This gives the amplitude due to a single electron. According to (36) the amplitude scattered by the entire atom is therefore

$$A_a = fA_0 \frac{e^2}{r_2\, mc^2} p^{1/2}. \tag{39}$$

There is one atom to every S units of area of the plane, where S is the area of the cell of the plane lattice on which the atoms are located. Thus the density of atoms per unit area is

$$a = \frac{1}{S}. \tag{40}$$

The scattering amplitude of a unit area of the plane is therefore

$$A_u = afA_0 \frac{e^2}{r_2\, mc^2}\, p^{1/2}. \tag{41}$$

The scattering amplitude of the entire plane A_p is equal to half that due to the first Fresnel zone:

$$A_p = \tfrac{1}{2}A_{\text{zone 1}}. \tag{42}$$

According to (33), the resultant of this first Fresnel zone is $2/\pi$ times the arithmetic sum of the scattering from all the units in the area, i.e.,

$$A_{\text{zone 1}} = \frac{2}{\pi} A_u \cdot \sigma_{\text{zone 1}}, \tag{43}$$

where, from (35),

$$\sigma_{\text{zone 1}} = \pi\lambda \frac{1}{\left(\dfrac{1}{r_1} + \dfrac{1}{r_2}\right) \sin\theta}. \tag{44}$$

Starting with (42) and successively substituting necessary values from (43), (41), and (44), the amplitude, A_p, scattered by the entire plane at a distance r_2 is

$$\begin{aligned}
(42):\qquad A_p &= \tfrac{1}{2}A_{\text{zone 1}} \\
&= \tfrac{1}{2}\cdot\frac{2}{\pi} A_u\, \sigma_{\text{zone 1}} \\
&= \tfrac{1}{2}\cdot\frac{2}{\pi}\left(afA_0 \frac{e^2}{r_2\, mc^2}\, p^{1/2}\right) \pi\lambda \frac{1}{\left(\dfrac{1}{r_1} + \dfrac{1}{r_2}\right) \sin\theta} \\
&= A_0 f\lambda \frac{e^2}{mc^2}\, p^{1/2} \frac{a}{r_2\left(\dfrac{1}{r_1} + \dfrac{1}{r_2}\right) \sin\theta}.
\end{aligned} \tag{45}$$

In practical experimental work, the distance of the crystal from the source of x-rays is so large that $1/r_1$ can be taken as zero. With this simplification, (45) reduces to

$$A_p = A_0 f \frac{\lambda e^2}{mc^2}\, p^{1/2} \frac{a}{\sin\theta}. \tag{46}$$

Diffraction by a sequence of identical planes. A sequence of parallel identical planes of interval d diffracts x-radiation so that all planes scatter in phase at the Bragg glancing angle θ_0, conforming to Bragg's law,

$$n\lambda = 2\,d \sin \theta_0. \tag{47}$$

The contribution from all the s planes of the sequence is s times that from a single plane, given by (46), namely

$$A_L = sA_p = sA_0 f \frac{\lambda e^2}{mc^2} p^{1/2} \frac{a}{\sin \theta}. \tag{48}$$

Diffraction by a crystal. Actually, the stack of lattice planes just considered, Fig. 12*A*, represents an oversimplified crystal structure. In general, crystal structures are much more complex than this, but can always be resolved into a set of J such stacks, where J is the total number of atoms in the unit cell. This is illustrated for a comparatively simple structure in Fig. 12*B*. Each lattice stack of the entire collection scatters

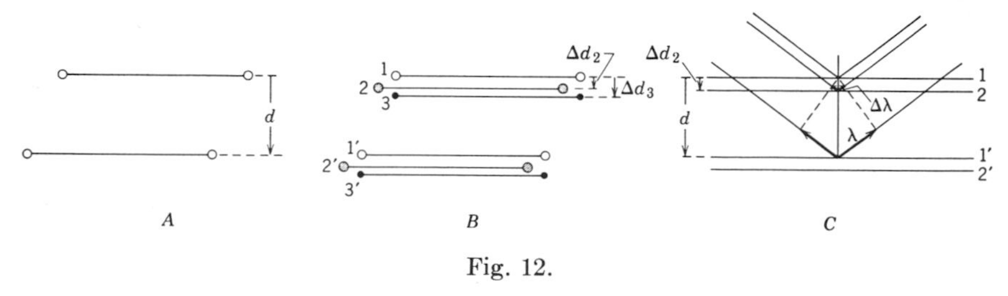

Fig. 12.

in phase according to (47), but the phases scattered by the separate lattice stacks differ. Fig. 12*C* shows a pair of planes of each of two stacks. It will be noted that, according to the Bragg-law derivation, a pair of neighboring planes of a stack, namely 1 and 1′ of Fig. 12*C*, separated by spacing d, reflect x-rays with a path difference of λ and a phase difference of 2π. A pair of planes 1 and 2 of separation Δd reflects x-rays with a path difference $\Delta\lambda$ and a phase difference supplied by the proportion

$$\frac{\Delta\phi}{2\pi} = \frac{\Delta\lambda}{\lambda} = \frac{\Delta d}{d}, \tag{49}$$

from which

$$\Delta\phi = \frac{\Delta d}{d} \cdot 2\pi. \tag{50}$$

To find the composite scattering from the atoms in the separate stacks, first consider the scattering due to a single cell. This cell contains J atoms of scattering power $f_1, f_2, f_3 \cdots f_J$. Since, in general, these atoms occur on planes of different levels, they scatter with different phases. The net scattering from the cell can be found by compounding the amplitudes, f, at phases, ϕ, on the Argand diagram, as shown in

Fig. 13 of Chapter 2. The resultant scattering is F, a complex number involving both magnitude and phase. Analytically, F can be compounded as follows:

$$F = f_1 e^{2\pi i\phi_1} + f_2 e^{2\pi i\phi_2} \cdots + f_J e^{2\pi i\phi_J} \tag{51}$$

$$= \sum_j f_j e^{2\pi i\phi_j}. \tag{52}$$

Returning now to the scattering by the collection of stacks, evidently this consists of different components each like (48), but with the separate components phased in a manner identical with the f's of (51). Therefore, the composite scattering of the several stacks is

$$A = \left(A_0 \frac{s\lambda e^2}{mc^2} p^{1/2} \frac{a}{\sin\theta}\right)(f_1 e^{2\pi i\phi_1} + f_2 e^{2\pi i\phi_2} \cdots + f_J e^{2\pi i\phi_J})$$

$$= A_0 \left(\frac{s\lambda e^2}{mc^2} p^{1/2} \frac{a}{\sin\theta}\right) F. \tag{53}$$

It is convenient to replace a, the number of pattern units per unit area, by a function of the pattern units per unit volume. This transformation is

$$a = \frac{\text{units}}{\text{area}} = \frac{\text{units}}{\text{volume/spacing}} = \left(\frac{\text{units}}{\text{volume}}\right) \cdot \text{spacing} \tag{54}$$

$$= N \cdot d,$$

where N = the number of pattern units in a unit volume

It is convenient to define $q = A/A_0$. When this transformation is made, (53) takes the form

$$q = \frac{A}{A_0} = \left(\frac{sNd}{\sin\theta} \lambda \frac{e^2}{mc^2} p^{1/2}\right) F. \tag{55}$$

It is desirable for subsequent development to define the relative amplitude reflected from a unit slab, for which $s = 1$. This is, in effect, the relative amplitude A, reflected from a beam of amplitude A_0 by a unit slab which, if repeated s times, would comprise a crystal composed of s sheets, each of spacing d. For such a unit slab, $s = 1$, and

$$q_0 = \frac{A_{s=1}}{A_0} = \frac{Nd}{\sin\theta} \lambda \frac{e^2}{mc^2} p^{1/2} F. \tag{56}$$

The integrated reflection. The square of (55) is the ratio of the intensity of the reflected beam to the intensity of the incident beam. It represents the reflecting efficiency of a tiny piece of crystal so small that absorption and other complications are not involved. Unfortunately this measure of reflecting efficiency is not a good one to use experimentally. A major reason for this is that it applies to a crystal composed of absolutely parallel volume units. But real crystals differ in respect to this kind of perfection. If imperfection exists, adjacent volume units depart slightly from parallelism, and the crystal must be turned slightly to bring each such volume unit into perfect Bragg-reflection condition. This causes the reflection peak to be drawn out over an angular range due to angular imperfection of the crystal. Furthermore, the degree of this imperfection may vary with direction in the crystal and, therefore, differ for reflections from different crystal planes. Consequently the "peak" intensity is not a very reliable measure of $|F_{hkl}|^2$.

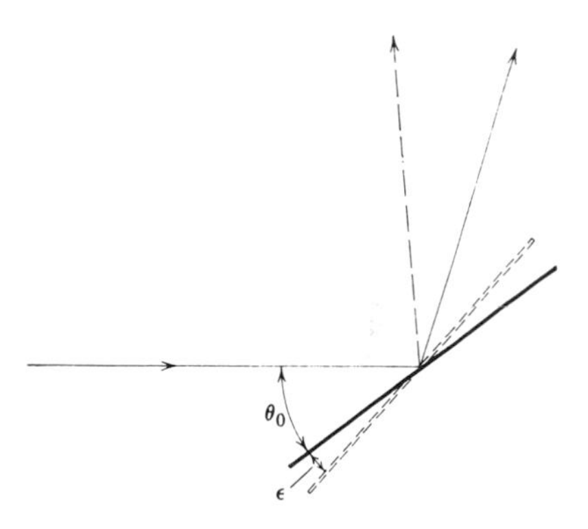

Fig. 13.

A better measure of the reflecting power can be obtained by adding the reflections as the crystal is set at various positions near the Bragg angle, or, more elegantly, by integrating the reflection as the crystal is uniformly rotated through the region of the Bragg reflection. In this way, the contribution of each volume unit in even an imperfect crystal can be counted.

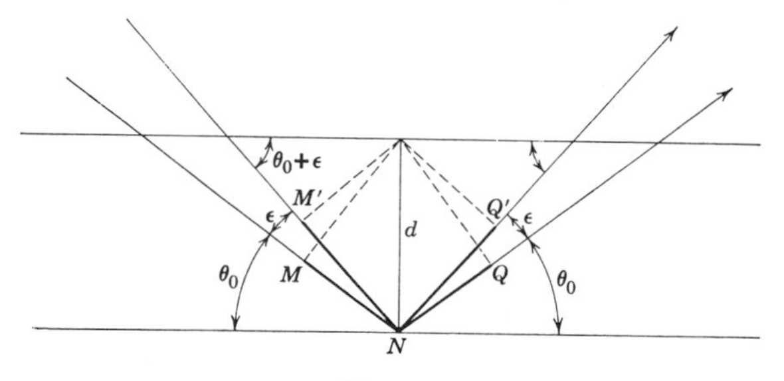

Fig. 14.

To develop the relations involved in this method, return to the tiny perfect piece of crystal whose amplitude efficiency q was given by (55). The maximum intensity of reflection occurs at the ideal Bragg angle, θ_0, given by the Bragg law

$$n\lambda = 2\,d \sin \theta_0. \tag{57}$$

Suppose that the orientation of the crystal is displaced from this orientation by a very small angle, ϵ, Fig. 13. The planes now reflect at a glancing angle $\theta_0+\epsilon$. Under these new conditions the path difference between planes, which is ideally $n\lambda$, is changed. The original path difference for glancing angle θ_0 is MNQ, Fig. 14, while the new path difference for glancing angle $\theta_0+\epsilon$, is $M'NQ'$. If p is the length of the path, then

$$p = 2\,d \sin\theta \tag{58}$$

The change in p due to a change in θ is

$$\frac{dp}{d\theta} = 2\,d \cos\theta, \tag{59}$$

so that

$$\begin{aligned}\Delta p &= 2\,d\,\Delta\theta \cos\theta \\ &= 2\,d\,\epsilon \cos\theta. \end{aligned} \tag{60}$$

In the notation of (30) this corresponds to $\Delta\lambda$. Thus, the change in phase occasioned by the change in path, according to (30), is

$$\delta = \frac{2\,d\,\epsilon\cos\theta_0}{\lambda}\cdot 2\pi \tag{61}$$

$$= 2\frac{2\pi\,d\cos\theta_0}{\lambda}\,\epsilon$$

$$= 2B\epsilon, \tag{62}$$

$$\text{where}\quad B = \frac{2\pi\,d\cos\theta_0}{\lambda}. \tag{63}$$

Each of the s different planes of the crystal now scatters slightly out of phase with its nearest neighbor by a phase difference δ, and the first and nth plane scatter out of phase by the phase difference $(n-1)\delta$. The resultant scattering due to these s different planes can be readily represented on the Argand diagram, Fig. 15. For convenience, let the amplitude scattered by each plane of the crystal be represented by a vector of unit length. The amplitudes of the s different planes combine to form an arc of length A and radius r, whose resultant is the chord C. This chord and arc subtend an angle $s\delta$ at the center of the circle of which the arc is a partial circumference. The magnitude of the resultant C is given by (17) of Chapter 2, which can be rewritten in terms of the symbols used here, as

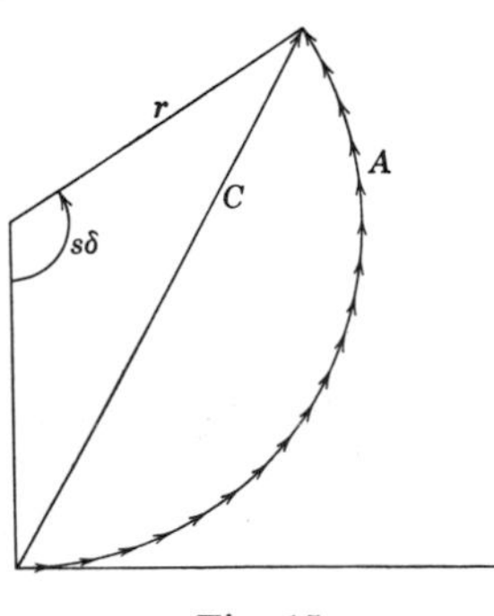

Fig. 15.

$$C = \frac{\sin \frac{s\delta}{2}}{\sin \frac{\delta}{2}}. \tag{64}$$

Since δ is a very small angle, a convenient approximation, following that of (19*A*) of Chapter 2, for the integration to be given beyond is

$$C = \frac{\sin \frac{s\delta}{2}}{\frac{\delta}{2}}. \tag{65}$$

The derivation of (64) was based upon amplitude vectors of unit length. The actual length of a vector of interest is the relative amplitude q_0, reflected by a single slab of the crystal, (56). Applied to the result in (65), this would scale up the resultant, C, by the factor q_0, to a quantity which will be designated D. Making this scale change in (65) and also substituting the value of δ given by (62) provides the particular resultant amplitude reflected from the crystal as a function of the angle of deviation, ϵ, from the ideal Bragg angle, θ_0:

$$\frac{A}{A_0} = D = \frac{A_{s=1}}{A_0} \frac{\sin sB\epsilon}{B\epsilon} \tag{66}$$

$$= q_0 \frac{\sin sB\epsilon}{B\epsilon}. \tag{67}$$

This is the amplitude of the reflected radiation. Its intensity is

$$\frac{I}{I_0} = D^2 = \frac{A_{s=1}^2}{A_0{}^2} \frac{\sin^2 sB\epsilon}{(B\epsilon)^2} = \frac{I_{s=1}}{I_0} \frac{\sin^2 sB\epsilon}{(B\epsilon)^2}$$

$$= q_0{}^2 \frac{\sin^2 sB\epsilon}{(B\epsilon)^2}. \tag{68}$$

A standard way of measuring the relative reflecting powers of various crystal planes is to rotate the crystal at a constant angular velocity of ω radians per second about an axis in the plane and perpendicular to the plane of the incident and reflected beam, Fig. 16. The reflected radiation is allowed to fall on photographic film or quantum counter as the plane passes through the neighborhood of the position corresponding to the Bragg angle θ_0. Figure 16 shows a broad parallel beam striking an extended crystal surface and being reflected toward a device arranged to receive the radiation and record it. Consider a small area, G, on the receiving device and at right angles to the direction of the reflected beam. The total energy received in area G is measured while the reflection waxes and wanes as the angle made by the reflecting plane approaches and recedes from the ideal Bragg-reflection condition. That part of the crystal having G as a base and extending backward along the reflection direction is the only part concerned in contributing reflected radiation to the area G.

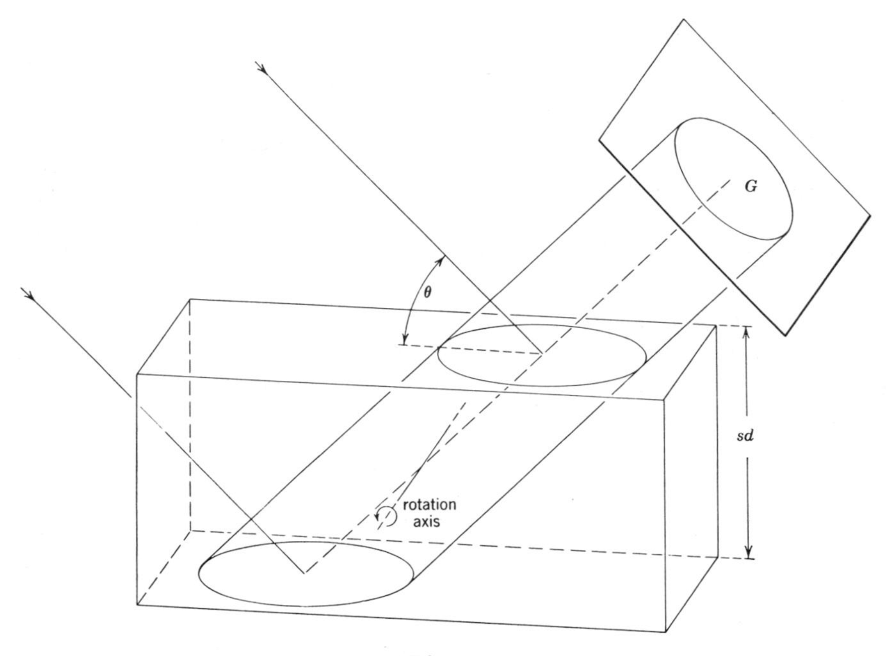

Fig. 16.

If the rate of rotation is ω, then the plane turns through the angular range $d\epsilon$ in time $d\epsilon/\omega$. The total energy reaching the area G in this time is then the intensity times the area, times the time, namely,

$$dE = IG\frac{d\epsilon}{\omega}. \tag{69}$$

According to (68),

$$I = I_0\,q_0{}^2\left(\frac{\sin^2 sB\epsilon}{(B\epsilon)^2}\right). \tag{70}$$

Substituting this in (69),

$$dE = \left(I_0\,q_0{}^2\,\frac{\sin^2 sB\epsilon}{(B\epsilon)^2}\right)G\,\frac{d\epsilon}{\omega} \tag{71}$$

The total energy reflected to area G as the crystal passes through the Bragg-reflection region is therefore

$$E = \int dE = \int I_0 q_0{}^2 \frac{\sin^2 sB\epsilon}{(B\epsilon)^2} G \frac{d\epsilon}{\omega}. \tag{72}$$

Since $I_0 G$ and ω are constant, this can be written

$$E = \frac{I_0 G}{\omega} \int q_0{}^2 \frac{\sin^2 sB\epsilon}{(B\epsilon)^2} d\epsilon \tag{73}$$

$$= \frac{I_0 G}{\omega B} \int q_0{}^2 s \frac{\sin^2 sB\epsilon}{(sB\epsilon)^2} d(sB\epsilon). \tag{74}$$

Reference to (56) shows that q_0 is actually a function of θ, and therefore of ϵ, since $\theta - \theta_0 = \epsilon$. For practical purposes, the reflection is complete in a very small range of ϵ, during which time θ, and hence q, are substantially constant. Also, s is a constant. Thus (74) can be reduced to

$$E = \frac{I_0 G q_0{}^2 s}{\omega B} \int \frac{\sin^2 (sB\epsilon)}{(sB\epsilon)^2} d(sB\epsilon). \tag{75}$$

For the same reason, the limits of integration of (75) are unimportant, since the function only has appreciable values when ϵ is very small. Since

$$\int_{-\infty}^{\infty} \frac{\sin^2 x}{x^2} dx = \pi, \tag{76}$$

in integrating (75) where $sB\epsilon = x$, it is convenient to choose the limits $-\infty$ to $+\infty$. The integration of (75) thus gives

$$E = \frac{I_0 G q_0{}^2 s}{\omega B} \pi. \tag{77}$$

The terms q_0 and B are merely shorthand for the lengthy expressions in (56) and (63). When these values are introduced, (77) becomes

$$E = \frac{I_0}{\omega} G q_0{}^2 \frac{1}{B} s\pi$$

$$= \frac{I_0}{\omega} G \left(\frac{Nd}{\sin\theta} \lambda \frac{e^2}{mc^2} p^{1/2} F \right)^2 \left(\frac{\lambda}{2\pi d \cos\theta} \right) s\pi$$

$$= \left(\frac{I_0}{\omega} \frac{e^4}{m^2 c^4} N^2 \lambda^3 \right) \frac{1}{2 \sin\theta \cos\theta} p|F|^2 \left(\frac{Gsd}{\sin\theta} \right). \tag{78}$$

Figure 16 shows that the volume of the crystal which contributes reflected radiation to the area G is

$$dV = \frac{Gsd}{\sin \theta}. \tag{79}$$

For this term may be substituted, therefore, dV, the volume of the crystal concerned with the experiment. When this substitution is made and also the substitution $\sin 2\theta$ for $2 \sin \theta \cos \theta$, (79) takes the form

$$E = \frac{I_0}{\omega} \frac{N^2 \lambda^3 e^4}{m^2 c^4} \frac{1}{\sin 2\theta} p|F|^2 \, dV \tag{80}$$

$$= \frac{I_0}{\omega} Q \, dV, \tag{81}$$

where $$Q = \frac{N^2 \lambda^3 e^4}{m^2 c^4} \frac{1}{\sin 2\theta} p|F|^2. \tag{82}$$

A quantity which has a form invariant with some of the conditions of the experiment is defined by rewriting (81) as

$$\frac{E\omega}{I_0} = Q \, dV. \tag{83}$$

The quantity $E\omega/I_0$ is known as the "integrated reflection." In (80), if the rotation rate is doubled, the energy, E, received from the volume dV, is halved, but, in (83), the product $E\omega$ remains unchanged. The quantity $E\omega$ has the dimensions energy·radians/time = power·radians. Thus $E\omega$ represents power integrated over an angular range, and could be called appropriately the *integrated power of the reflection.*

Practical forms of the relation between E and F. Another arrangement of (80) is important for many present-day experimental purposes. The E of (80) is the total energy reflected by the volume dV in the course of a single pass of the crystal through the region of the Bragg reflection. Let E_1 specifically represent this total energy reflected by dV in a single pass. Then, by rearranging (80) so that experimental constants are placed together in a group (the terms in the second parentheses), and also by letting $1/\sin 2\theta = L$ (since this geometrical factor will be shown in Chapter 7 to be the "Lorentz factor"), this important relation can be written

$$E_1 = \left(\frac{I_0 \lambda^3 N^2 \, dV}{\omega}\right)\left(\frac{e^4}{m^2 c^4}\right) Lp|F|^2 \tag{84}$$

$$= K_1 \, Lp|F|^2, \tag{85}$$

where $$K_1 = \left(\frac{I_0 \lambda^3 N^2 \, dV}{\omega}\right)\left(\frac{e^4}{m^2 c^4}\right). \tag{86}$$

This relation between E_1 and $|F|^2$ is convenient for certain experimental arrangements (such as measurement of the energies reflected from the several planes of a single crystal by means of a quantum counter) where a single pass of the plane through the reflecting condition produces sufficient energy to be detected.

The energies in the several x-ray reflections *hkl* are commonly recorded by photographic means. In order to cut down absorption errors and other disturbing effects, a crystal is ordinarily selected which is as small as is convenient to handle. Since the energy of the reflection is proportional to dV, which is the volume of this crystal according to (84), and since the volume is deliberately chosen very small, the energy of the resulting reflection is usually not sufficient to give a reasonable blackening on the film in one pass through the reflecting condition of each plane. But the energy can be built up, and the blackening on the film therefore enhanced, by causing the crystal to pass through the condition for reflection many times, i.e., by many rotations. Then the total energy reflected in these n rotations of the crystal is obviously n times that given in (84), namely

$$E_n = \left(\frac{I_0\lambda^3 N^2 dV}{\omega}\right)\left(\frac{e^4}{m^2c^4}\right) Lp|F|^2 n. \tag{87}$$

This can be put in simpler form by expressing the rate of rotation as

$$\omega = \frac{\text{radians}}{\text{sec}} = \frac{2\pi n}{t} \tag{88}$$

where t is the duration of the experiment. If this is substituted in (84) there results

$$E_n = \left(\frac{I_0\lambda^3 N^2 dV}{2\pi n/t}\right)\left(\frac{e^4}{m^2c^4}\right) Lp|F|^2 n$$

$$= \left(\frac{I_0\lambda^3 N^2 t\, dV}{2\pi}\right)\left(\frac{e^4}{m^2c^4}\right) Lp|F|^2 \tag{89}$$

$$= K_n Lp|F|^2 \tag{90}$$

where $$K_n = \left(\frac{I_0\lambda^3 N^2 t\, dV}{2\pi}\right)\left(\frac{e^4}{m^2c^4}\right). \tag{91}$$

The practical utilization of (89) will be discussed in other chapters. The meanings of the symbols involved in the results given in (84), (86), (89), and (91) are as follows:

e = the charge on the electron,
m = the mass of the electron,
c = the velocity of light,
I_0 = the intensity of the radiation striking the crystal,
λ = the wavelength of the x-radiation,
N = the number of the unit cells per unit volume,
dV = the volume of the crystal (assumed to be very small so that absorption, extinction, etc., which are discussed in Chapter 8, can be neglected),
t = the time of the experiment,
ω = the rate of rotation of the crystal,
L = the "Lorentz factor" (discussed in Chapter 7),
p = the polarization factor, $\frac{1+\cos^2 2\theta}{2}$,
F = the factor representing the resultant scattering by a single unit cell, (52).

Literature

[1] Henri de Senarmont, Emile Verdet, and Leonor Fresnel. *Oeuvres completes d'Augustin Fresnel.* Vol. 1. (Imprimerie Imperiale, Paris, 1866) especially 247–382.

[2] A. Fresnel. *Memoir on the diffraction of light.* In Vol. 10 of Scientific Memoirs: *The wave theory of light,* edited by Henry Crew (American Book Co., New York, 1900) 79–144.

[3] J. J. Thomson. *Conduction of electricity through gases.* (Cambridge University Press, Cambridge, England, 1st Ed., 1903) 268–273, (or 2nd Ed., 1906) 321–327, (or 3rd Ed., Vol. 2, 1933) 256–259.

[4] C. G. Darwin. *The theory of x-ray reflection.* Phil. Mag. (6) **27** (1914) 315–333, 675–690.

[5] Arthur H. Compton. *The intensity of x-ray reflection, and the distribution of the electrons in atoms.* Phys. Rev. **9** (1917) 29–57.

[6] P. P. Ewald. *Zur Begrundüng des Kristalloptik, Teil III: Die Kristalloptik des Röntgenstrahlen.* Ann. Physik. **54** (1917) 519–597.

[7] W. Lawrence Bragg, R. W. James, and C. H. Bosanquet. *The intensity of reflexion of x-rays by rock-salt.* Phil. Mag. (6) **41** (1921) 309–337.

[8] W. L. Bragg, R. W. James, and C. H. Bosanquet. *The intensity of reflexion of x-rays by rock-salt.* Phil. Mag. (6) **42** (1921) 1–17.

[9] F. C. Blake. *On the factors affecting the reflection intensities by the several methods of x-ray analysis of crystal structures.* Rev. Mod. Phys. **5** (1933) 169–202.

[10] Charles F. Meyer. *The diffraction of light, x-rays, and material particles.* (University of Chicago Press, Chicago, 1934) 28–45.

4

Outline of a crystal-structure analysis

The last two chapters have dealt with the basic theory of x-ray diffraction by crystals. In Chapter 2 it was seen that each crystal structure is characterized by a set of diffraction spectra *hkl*. Crystal-structure analysis consists in finding a structure whose diffraction spectra match the observed set. More specifically, the absolute magnitudes of the F_{hkl}'s can be derived from x-ray diffraction experiments, and the analysis comprises finding a structure model which would duplicate them.

In Chapter 3 it was seen that the kind of diffraction experiment to be performed provides a set of "integrated reflections" for the various *hkl*'s. These cannot be used as they stand, but must be corrected in various ways to provide an appropriate starting point for analyzing the structure of the crystal. To the uninitiated, the procedure may seem a complicated one, and many technical points must be considered in some detail if the matter is to be handled intelligently. Before taking the plunge into these technical questions and procedures, it seems appropriate to pause and consider the course ahead in outline. This should provide the crystal-structure analyst with perspective.

While the analysis may be varied a great deal to fit the case, most present-day analyses of crystals of moderate complexity follow a standard pattern which may be divided into several stages.

1. Before any experimental work is done, an appropriate crystal, upon which all the subsequent work will be based, must be selected. The importance of this stage is not frequently emphasized, and it is all too frequently slighted by the uninitiated. If one does not start with appropriate material, then what follows may well be labor in vain. The selection and preparation of the crystal are discussed in Chapter 5.

2. Before the investigation of the structure proper is undertaken, the

unit cell and space group of the crystal should, if possible, be known. Before any part of the actual work of crystal-structure analysis begins, it is a good plan to consult the available references to see if unit-cell and space-group information is available, since these data have been recorded for many crystals whose structures have not yet been determined. A general reference for such information is *Crystal data.*[†] Data for minerals are recorded in more detail in Dana's *System of mineralogy.*[§]

Since so much depends on the correctness of this basic information, data found in compilations should be regarded as preliminary, and should be checked by experimental work before the crystal-structure analysis proceeds. The theory and practice of unit-cell and space-group determination is thoroughly discussed elsewhere.[¶‡] It should be observed that unit-cell and space-group information can be obtained without quantitatively accurate intensity control. In order to record any weak spectra, therefore, it is wise to overexpose photographs made purely for unit-cell and space-group information. The crystal which is used for this purpose, however, may be the same one which is to be used later for the accurate intensity record.

3. The experimental part of a crystal-structure analysis consists in the gathering of accurate values of the intensities of the many spectra. This stage is discussed in detail in Chapter 6.

4. It is most convenient to work with a set of amplitudes, $|F_{hkl}|$, of the spectra rather than the intensities themselves. This requires the conversion of intensities to amplitudes for all spectra. Unfortunately the "integrated intensities" derived from the experiment are not merely the squares of the corresponding amplitudes. They contain, in addition, several factors which must be allowed for. These can be classified into certain factors having a geometrical basis and certain others having a physical basis. The origins of the geometrical factors are discussed in Chapter 7, and the origins of the physical factors in Chapter 8. These chapters also discuss practical ways of allowing for the factors. After appropriate allowances have been made, there is available a set of $|F_{hkl}|$'s.

5. With all experimental data from stage 4 on hand, one seeks an arrangement of atoms consistent with them. This amounts to seeking, within the symmetry limitations of the space group, sets of coordinates

[†] J. D. H. Donnay and Werner Nowacki. *Crystal data.* Geol. Soc. Am. Mem. **60** (1954).

[§] *The system of mineralogy of James Dwight Dana and Edward Salisbury Dana* 7th Ed. revised by Charles Palache, Harry Berman, and Clifford Frondel. (John Wiley and Sons, New York, Vol. 1, 1944, Vol. 2, 1951; Vol. 3 in preparation)

[¶] M. J. Buerger. *X-ray crystallography.* (John Wiley and Sons, New York, 1942)

[‡] M. J. Buerger. *The photography of the reciprocal lattice.* Monograph No. 1 (1944), Am. Soc. for X-ray and Electron Diffraction.

xyz, for each of the atoms, which are consistent with the $|F_{hkl}|$'s. This stage may take various forms depending on the complexity of the structure being analyzed. The complexity of the structure, from the point of view of crystal-structure analysis, depends upon the number of parameters which must be fixed, as well as on their relation to one another. This information is often available after an investigation of how the atoms may be distributed among the equipoints of the space group. This matter, which should always be investigated, is discussed in Chapter 9. While it may lead to rather indefinite conclusions, it may, on the other hand, lead to information which is helpful in the structure analysis.

For example, the equipoint analysis may show that the atoms of the structure are symmetry fixed. If so, this stage of the analysis degenerates into a selection among several alternatives, all consistent with the symmetry, but only one of which is consistent with the $|F_{hkl}|$'s. Such analyses are simple indeed, and can even be completed by using mere qualitative estimates of the intensities. Qualitative intensities can also be used to solve structures in which there is only one x, one y, and one z coordinate to be fixed, as well as in some slightly more complicated cases. Simple analyses of this sort are discussed in Chapter 12.

But in the more usual case the structure is too complicated for such a simple analysis. In the more complicated cases it is necessary to have a good set of $|F_{hkl}|$'s, and, at some stage, to employ the methods of Fourier syntheses. More generally it is necessary to have a deeper understanding of reciprocal space than is required for unit-cell and space-group determination. The required background and how it is used in Fourier synthesis is discussed in Chapters 13 through 18.

It turns out, as shown in Chapter 13, that the electron density of a crystal can be computed throughout the cell with the aid of Fourier series. The coefficients of the series are the F_{hkl}'s. Unfortunately these are quantities involving both magnitude and phase. If both could be measured in the diffraction experiment, all crystal structures could be solved as a routine. But only the magnitudes are accessible from the diffraction experiment, so that the solution of a crystal structure is not a matter of routine. There are certain kinds of crystals for which the additional phase information can be readily determined. The nature of these cases and how they are treated are discussed in Chapters 19 and 20. Most of the moderately complicated crystal structures which have been solved in the past have been solved by the methods treated in Chapter 19.

Just after the Second World War it was discovered that the magnitude and phases of the F's are not independent, but are related. This brings about the possibility of direct solutions of crystal structures. The several kinds of direct solution are discussed in Chapter 21.

6. Having found a structure which gives a reasonable explanation of

the intensities, it is ordinarily desirable to refine it, i.e., to vary the locations of the atoms in the structure slightly, so that it gives the best fit with the experimental $|F_{hkl}|$'s. This refinement may take a number of different forms. These are discussed in Chapter 22, along with tests for the correctness of the structure, and the accuracy of the coordinates to be expected from the analysis.

7. When the structure has been thoroughly established, the distances between neighboring atoms, and the angles subtended by such neighbors, can be computed. This information is of great interest in understanding the bonding in the crystal, and is ordinarily reported as an integral part of the results of crystal-structure analysis. The nature of these computations is outlined in Chapter 23.

5

Selection of material suitable for crystal-structure analysis

When one plans to measure the intensities of diffraction from a grating, it is important that he use a reasonably perfect grating. This remark is intended as a word of warning. A good deal of time is usually spent in measurement of intensities, and an even greater amount of time is spent on attempting to deduce a structure consistent with these intensities. It therefore behooves one to make sure that the crystal upon which all this effort is expended is an appropriate one.

There are two ways in which a crystal commonly differs from an ideal three-dimensional grating in such a way that intensity measurements derived from it are useless: The crystal may be twinned, or it may be excessively imperfect. Of these two, the twinned material is the most insidious, and results obtained from twinned crystals may affect such fundamental information as the dimensions of the unit cell, the point-group symmetry, and the space-group symmetry.

These two types of defective gratings are discussed in the following sections.

Twinning[7]

It is vital that one should not obtain experimental data for a crystal-structure investigation from a twin under the supposition that he is using a single crystal. If one is so unfortunate as to make this error, sooner or later the investigation is bound to come against a barrier beyond which there is no proceeding until it is recognized that twinned material has been used.

The basic difficulty with twins is that, in effect, one examines the interpenetrating reciprocal lattices of the individuals of the twin under the impression that he is dealing with the reciprocal lattice of a single crystal. One therefore interprets a composite set of reciprocal lattices as a single reciprocal lattice. This implies that he misjudges the point-group symmetry, space-group symmetry, the $|F_{hkl}|$'s of the crystal, and that he may misjudge the unit cell also.

That this is so can be readily appreciated by referring to Mallard's empirical theory of twinning,[1, 4–6] which has a rational explanation.[6]

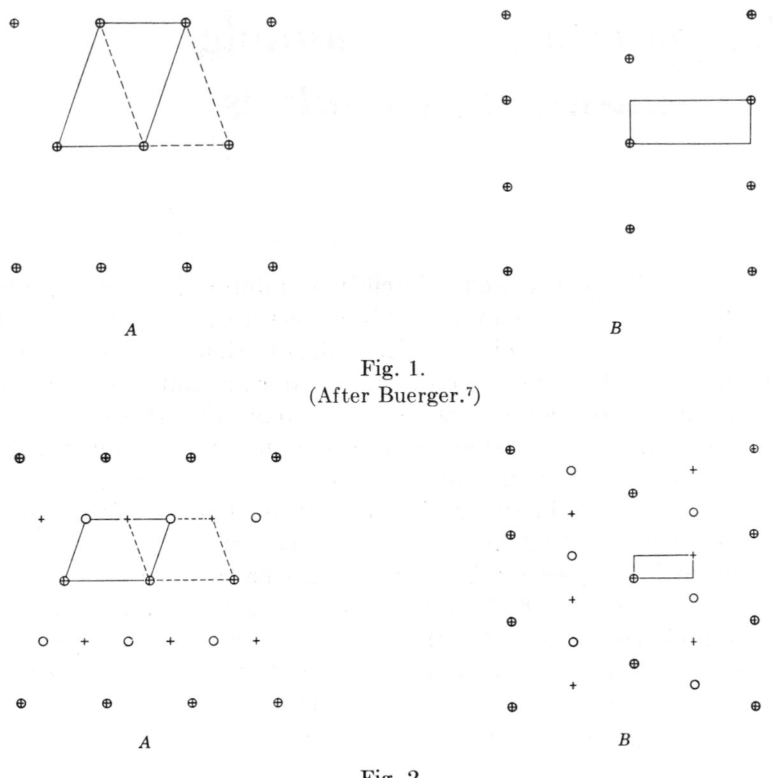

Fig. 1.
(After Buerger.[7])

Fig. 2.
(After Buerger.[7])

Mallard noticed that twinning is probable in a crystal species if its lattice has a certain dimensional specialization. He described this condition by saying that if either the cell of the lattice, or a supercell, had greater dimensional symmetry than the true symmetry of the point group of the crystal, twinning was likely. These symmetry elements of the cell over and above those of the point group of the crystal are possible twin elements. Mallard's description amounts to a dimensional specialization such that there is a registry of certain of the points of the original lattice with the lattice in twinned position.[6] This is illustrated in Figs. 1*A* and

$2A$. In both illustrations one is viewing a monoclinic crystal along the unique (2-fold) axis. The lattice points of the original crystal are shown as circles; those of the twinned crystal as crosses. In Fig. $1A$, the dimensions of the lattice are such that all lattice points of both crystals coincide. In Fig. $2A$, the points of every other horizontal row coincide. The greater the fraction of the points which register, the greater the probability of twinning, according to the Mallard theory. In the illustrations the registry of points is perfect. The registry may, in a given instance, be more or less exact. The more exact it is, the more probable is the occurrence of twins, according to Mallard's theory.

Now, whereas the direct lattices are merely in contact with one another, the extent of a reciprocal lattice is independent of the size of the crystal, and pervades all reciprocal space. Therefore the reciprocal lattices of the several individuals of the twin occupy the same space. If the registry of points is rather exact for the direct lattices, it is just as exact in the reciprocal lattices. Figures $1B$ and $2B$ show the reciprocal lattices of Figs. $1A$ and $2A$ respectively. If the registry of the two reciprocal lattices is so exact that one cannot observe that he is dealing with two lattices, they may be said to be unresolved; then the difficulties discussed beyond are encountered. If, on the other hand, the registry of the two lattices is sufficiently imperfect that they can be seen to be resolved, no insurmountable difficulties due to twinning are present. In this event, one simply draws a net through each lattice of the twin and fixes attention on one member of the twin only. The chief difficulty in this case is a loss of convenience of interpretation. It is also difficult to allow for a correction of the absorption by one twin of the radiation diffracted by the other.

In case the twins are unresolved, the difficulties discussed in the succeeding sections, under *unit cell*, *point-group symmetry*, and *space-group symmetry*, apply.

Kinds of crystals likely to offer trouble due to twinning. Twins occur with great frequency in derivative structures.[†][§] A *derivative structure* is one which can be derived by generalization of another structure known as a *basic structure*. This relation can be expressed in terms of symmetry. Specifically, a derivative structure can be derived from a basic structure by suppressing one or more of its symmetry operations, Fig. 3. It follows that the symmetry content per unit volume of the derivative structure is a submultiple of that of the basic structure. A particular and well-recognized variety of a derivative structure is a superstructure, but it should be understood that the category of derivative

[†] M. J. Buerger. *Derivative crystal structures.* J. Chem. Phys. **15** (1947) 1–16.

[§] M. J. Buerger and Newton W. Buerger. *Low-chalcocite and high-chalcocite.* Am. Mineralogist **29** (1944) 55–65.

structures is much more general than this and includes superstructures as a special case.

Any crystal which has high- and low-temperature forms related by a disordering or displacive transformation invariably has a basic structure

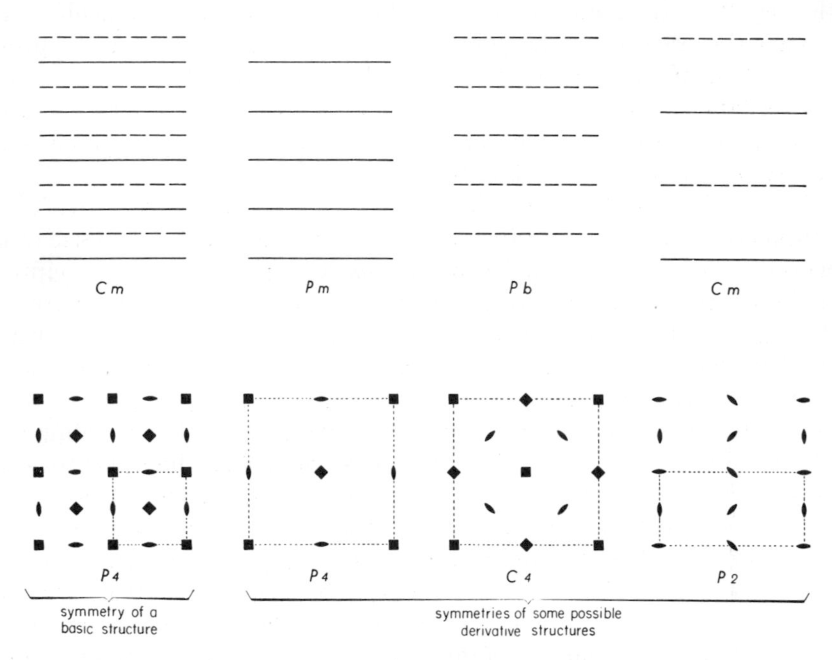

Fig. 3.
(After Buerger.[7])

for the high-temperature form and a derivative structure for the low-temperature form.[†§¶] Since the low-temperature form is a derivative structure, it almost invariably occurs as a twinned intergrowth whether the particular specimen has descended from the high-temperature form by inversion or not. Crystals having high-low inversions are not the only examples of derivative structures; the category is much more general. It also includes crystals of double salts.

† M. J. Buerger *Disorder in crystals of non-metals.* Anais acad. brasil. cienc. **21** (1949) 245–266, especially 252–253.

§ M. J. Buerger. *Crystallographic aspects of phase transformations.* Chapter 6 of *Phase transformations in solids,* edited by R. Smoluchowski, J. E. Mayer, and W. A. Weyl. (John Wiley and Sons, New York, 1951) 183–209, especially 195–196.

¶ M. J. Buerger. *Precipitation of segregate phases from solid solution.* Proc. 2nd International Symposium on the Reactivity of Solids, Gothenburg, 1952. Part I, 225–235. (Published by Elanders Boktryckeri Aktiebolag, Gothenburg, Sweden, 1953.)

Another version of this same subject which does not involve the notion of derivative structure but which covers partly similar ground may be had by adopting Mallard's view on twinning. According to Mallard's theory,[1,4,5] the greater the fraction of points on the twinned lattice which register with the untwinned lattice, the greater is the probability of occurrence of twins. In the merohedral crystal classes,† all the points

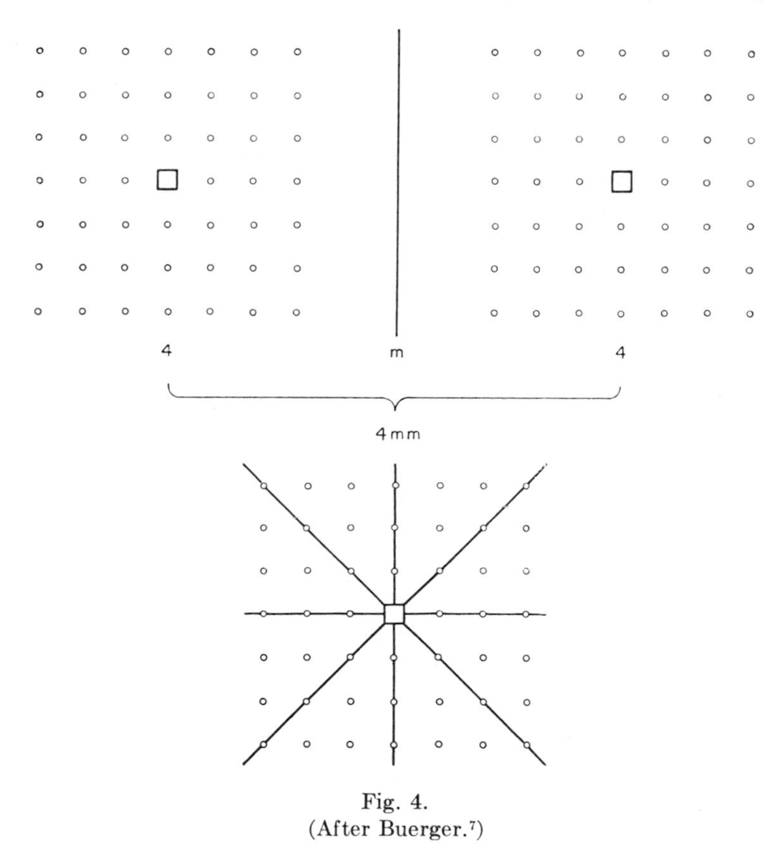

Fig. 4.
(After Buerger.[7])

of the lattice of a structure register with the lattice points of the reversed (and therefore twinned) structure, so this fraction is unity. Therefore twinning should have the highest probability of occurrence in crystals of the merohedral crystal classes. This is borne out by experience.

Unit cell. It will be observed from Fig. 1 that if there is an exact registry of all points of the reciprocal lattice, the unit cell inferred for the composite lattice is also the true cell (although its symmetry is altered as discussed in the next paragraph). On the other hand, Fig. 2*B* shows

† A merohedral crystal class is one having a symmetry which is a proper subgroup of the symmetry of its lattice.

that if not all points of the two lattices register, the reciprocal cell as inferred from the composite reciprocal lattice is smaller than the true reciprocal cell. This means that one infers too large a direct cell by some integral multiple. *If the reciprocal lattice contains strange extinctions in the spectra which affect the lattice type, this is a key to the detection of its composite character.*† For example, in Fig. 2*B*, every alternate vertical

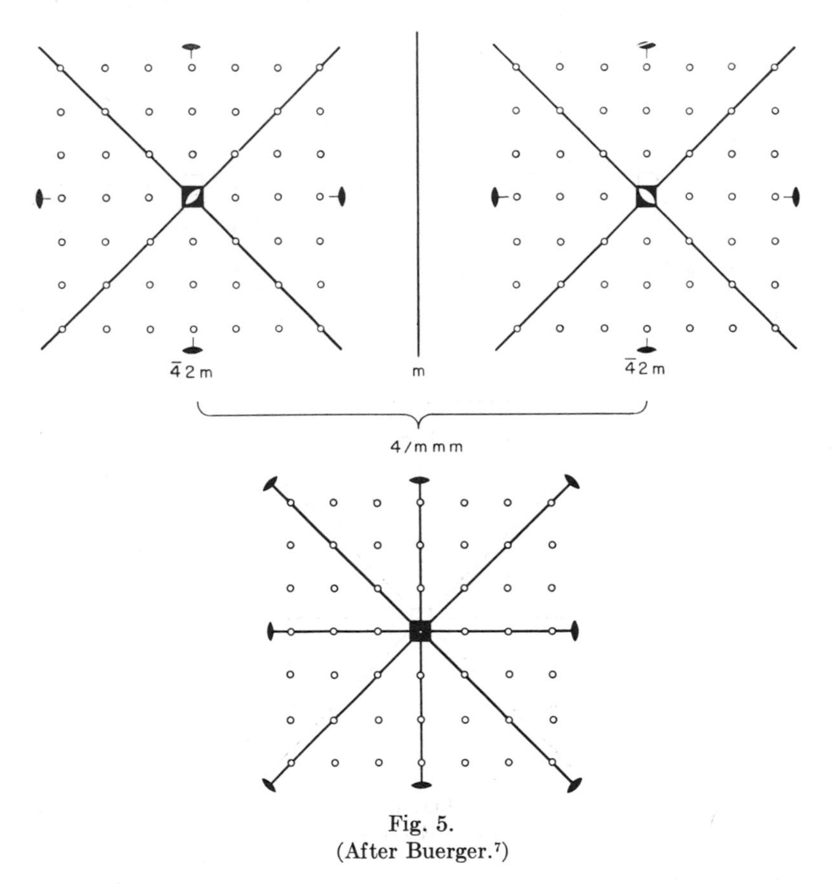

Fig. 5.
(After Buerger.[7])

row has missing points. This curious condition could be caused by no possible lattice requirement.

Point-group symmetry. If the several individuals of the twin are equally developed, the symmetry information obtained from the twin is, in general, different from that obtained from the single crystal. It is a simple matter to predict the symmetry of the twin composite. Consider

† An interesting example has been recently discovered by Gabrielle Donnay, J. D. H. Donnay, and G. Kullerud. *Crystal and twin structure of digenite,* Cu_9S_5. Am. Mineralologist **43** (1958) 230–242.

Fig. 4, which shows separately the reciprocal lattices of the two parts of a twin in twin orientation. Each has a 4-fold axis. When the reciprocal lattices are superposed so that they have the same origin, the 4-fold axes of the separate individuals coincide, so that the twin also has a 4-fold axis. But the superposition causes further symmetry in the composite. Any

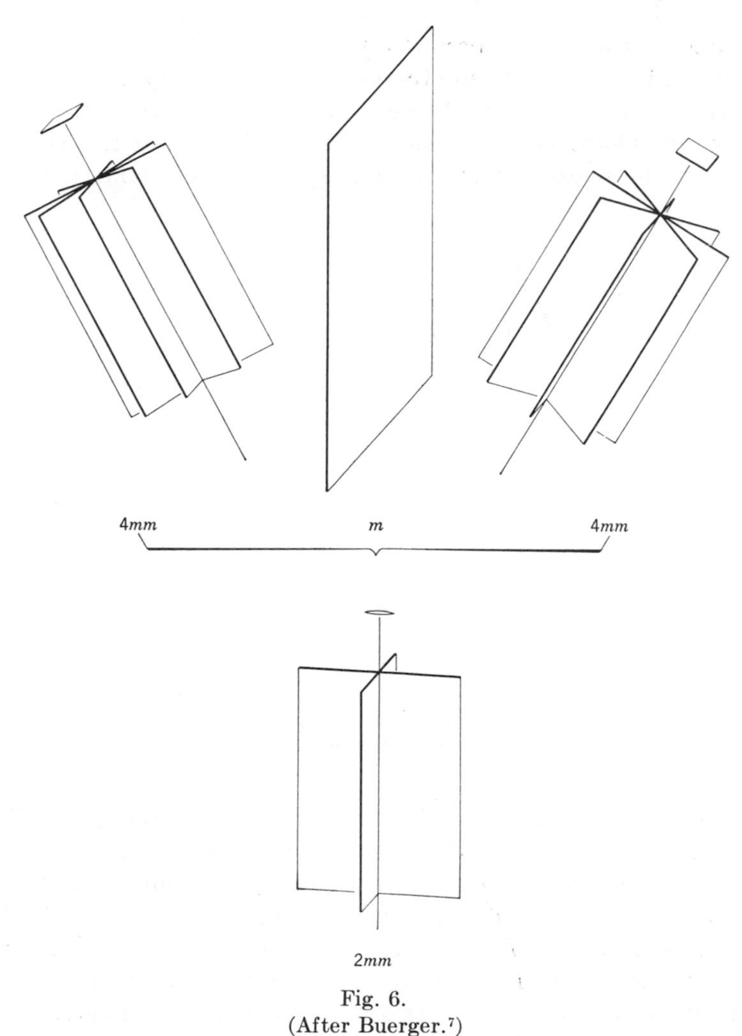

Fig. 6.
(After Buerger.[7])

point on the left (say) of one lattice is related to a similar point on the right of the other lattice by the symmetry operation which relates the two individuals of the twin. Therefore the *twin composite has the common symmetry of the individuals of the twin in twin orientation, augmented by the operation of the twin law.*

A somewhat more complicated example is shown in Fig. 5. Here the single crystal has symmetry $\bar{4}2m$ and the individuals are twinned by reflection across (010). Not only the $\bar{4}$ axis, but also the 2-fold axes and the two diagonal planes are parallel in the two individuals of the twin. The twin composite therefore includes the common symmetry elements, namely the $\bar{4}$ axis, the 2-fold axes, and two diagonal planes, and this is augmented by the mirror operation which relates the two individuals of the twin. The twin composite therefore has symmetry $4/m\ 2/m\ 2/m$.

It might appear from these two examples that the symmetry of a twin is always an enhancement of that of the individual, but this is not generally so. The reason for this is that symmetry elements not parallel

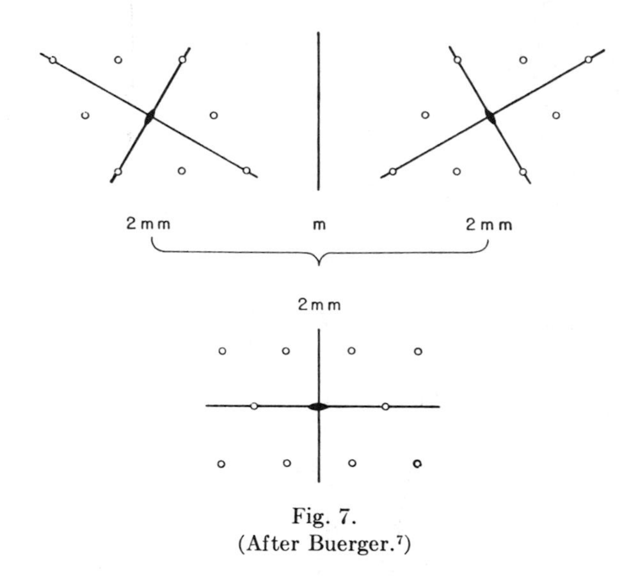

Fig. 7.
(After Buerger.[7])

to each other in the individuals of the twin are suppressed. This is demonstrated in Fig. 6. In this illustration the individual has the symmetry $4mm$. But the only parallel symmetry element in the two individuals is the symmetry plane parallel to the plane of the paper. Therefore all others are suppressed. The twin composite has this common symmetry element augmented by the operation of the mirror which relates the individuals. These two perpendicular mirrors give the twin composite the net symmetry $2mm$.

It is possible for the twin to have the same symmetry as the individual. This is shown with the aid of Fig. 7. Here the individual has the symmetry $2mm$. Both symmetry planes are suppressed in the twin, although the 2-fold axis is retained. When this is augmented by the mirror operation of the twin law, the twin again has the symmetry $2mm$ which is the same as the symmetry of the individual. But the orientation of the twin symmetry is quite different from that of either individual.

It is thus evident that the twin can have either a lower or a higher symmetry than that of the single crystal. The requirement of common symmetry elements tends to suppress symmetry elements so that they

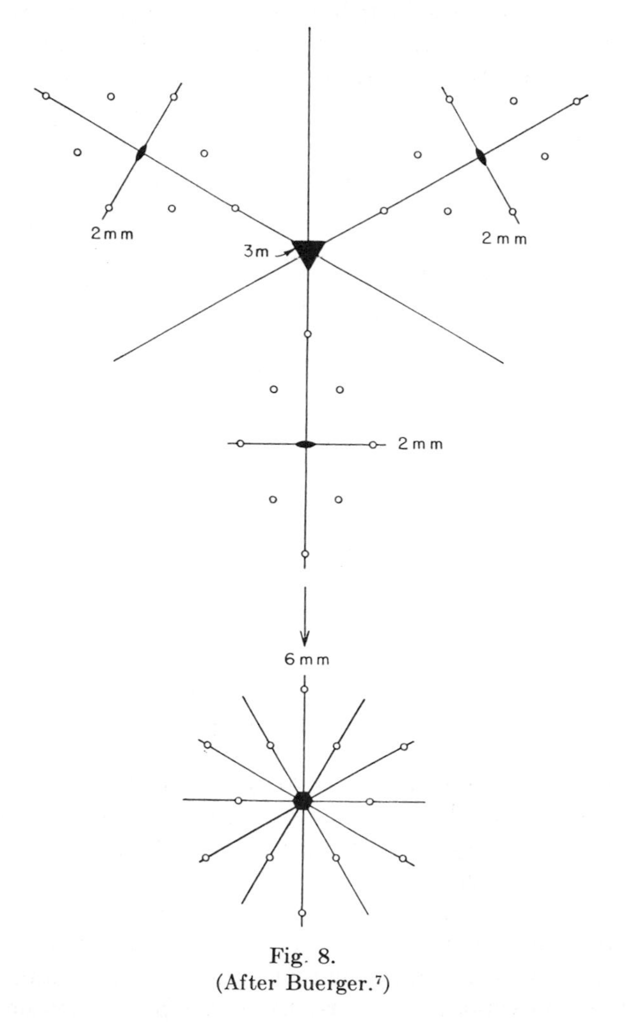

Fig. 8.
(After Buerger.[7])

are missing in the twin, yet the addition of the symmetry operation of the twin law tends to increase the symmetry elements present in the individual.

The analysis of the symmetry of repeated twins is a little more complicated but follows the same general principles. Two examples may be cited. Figure 8 shows a common type of repeated twinning known as a *trilling*. Three orthorhombic individuals of symmetry $2mm$ are intergrown so that each pair of neighbors is related by a reflection. This is equivalent to saying that the three individuals are related by the complex symmetry law $3m$, indicated in the illustration. The only common element among the three individuals is the 2-fold axis. When this is augmented by symmetry $3m$, the trilling is seen to have the symmetry $6mm$. It is in this way that orthorhombic crystals of pseudohexagonal cell dimensions commonly form trilled intergrowths to mimic hexagonal crystals.

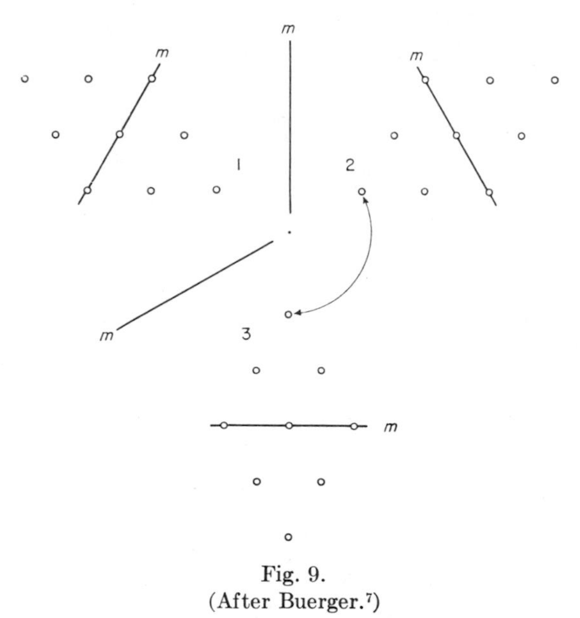

Fig. 9.
(After Buerger.[7])

But such trillings need not necessarily have hexagonal symmetry. A hypothetical example where this does not occur is shown in Fig. 9. Here monoclinic crystals of symmetry m are trilled. Individual 1 is related by reflection to both individual 2 and 3, but 2 and 3 cannot also be related by a reflection, although they are related by a 120° rotation. The mirrors of the individuals are suppressed because they are not parallel. The unlike twin operations also are suppressed because each only relates part of the whole set of individuals. The resulting composite has no symmetry. Another way of deriving this result is to first combine individual 1 with 2 according to the standard procedure for finding the symmetry of a twin. The symmetry of this pair is m, a vertical mirror. If now this twin pair with a vertical mirror is combined with individual 3, which has a horizontal mirror, these mirrors are suppressed. No other relation exists between the 1+2 combination and 3. They are not twins with respect to one another and they have no parallel symmetry elements, so that if combined, they have together no symmetry.

Note that in determining symmetry by x-ray means, the diffraction introduces an inversion center into the symmetry collection.[†] Thus the symmetry of a twin always appears to have an inversion center unless the wavelength employed is near the absorption edge of an element present in the crystal.

As a matter of practical procedure one must determine the symmetry by one of the standard x-ray techniques. If the investigator has the equipment, he will certainly make this determination by one of the moving-film methods, presumably by investigating the reciprocal lattice, a level at a time. The relation between the level symmetries and the symmetry as a whole is discussed elsewhere.[§] By far the most convenient way of investigating symmetry is by means of the precession method. The Weissenberg method provides the same information somewhat less conveniently. The oscillating-crystal method provides very little symmetry information indeed,[¶] while the rotating-crystal method and powder method are capable of giving no direct symmetry information whatsoever.

Space-group symmetry. Space-group determination, certainly in the preliminary stages of a crystal-structure analysis, is carried out with the aid of characteristic extinctions (absent spectra). The reciprocal lattice of a crystal whose symmetry includes translation-bearing symmetry elements contains central lines or planes, or both, which are characterized by systematically missing reciprocal-lattice points. The twin operation transforms the reciprocal lattice so that these central lines or planes of one reciprocal lattice superpose with some other lines or planes of the twinned reciprocal lattice. In general, the tendency in such superposition is for an "absent" point of one lattice to fall upon a present point of the other lattice, and this annuls the absence. This causes the extinction rule to be violated and the presence of a characteristic screw axis or glide plane is hidden. This alteration of the apparent symmetry is always in the direction of removing an extinction. It is therefore in the direction of replacing a screw by an apparent rotation axis, and replacing a glide plane by a reflection plane (also of replacing a centered unit cell by a primitive one). For this reason any crystal which appears to belong to a space group with no extinctions, or even very few extinctions, is subject to the suspicion that the symmetry data may have been derived from twinned material. Space groups with all reflection

[†] M. J. Buerger. *X-ray crystallography.* (John Wiley and Sons, New York, 1942) 55–58.

[§] *Ibid.* Chapter 22, especially Tables 28 and 29.

[¶] *Ibid.* pages 204–206.

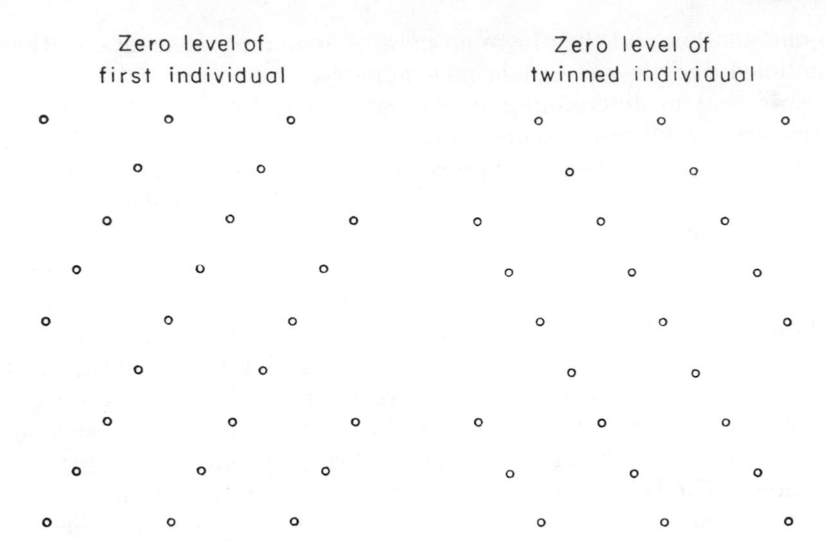

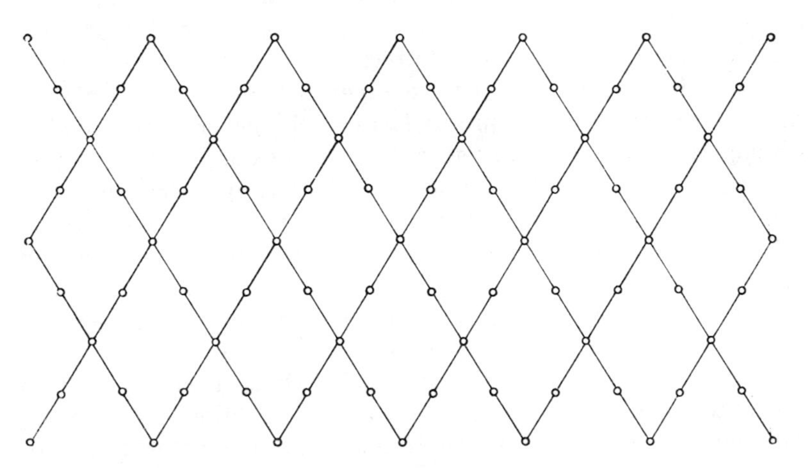

Fig. 10.
(After Buerger.[2])

planes or all rotation axes, such as *Pmmm*, should be regarded with suspicion.

In case the masking of extinctions is incomplete, there may remain a set of systematically absent spectra which correspond to no extinction rule. This is a warning that the extinction cannot be relied upon. This has been illustrated by arsenopyrite,[2] FeAsS, which is monoclinic, $P\,2_1/c$. This crystal is a derivative of marcasite, FeS_2, and twins according to the rules discussed in a foregoing section. From the Mallard point of view, this crystal should twin readily because its dimensional symmetry is such that an orthogonal supercell can be found. But, more fundamentally, it should twin readily because it is a derivative structure based upon the orthorhombic marcasite, FeS_2. In fact it can be regarded as the double salt $FeS_2 \cdot FeAs_2$. In any event, it can be referred to space group $B\,2_1/d$ to bring out this relationship. The reciprocal lattices of the two individuals of a twin are shown separately and superposed in Fig. 10. The extinction rules are shown analytically for both $P\,2_1/c$ and $B\,2_1/d$ in Table 1. Figure 10 shows that, in accordance with the

Table 1
Apparent extinctions from a twin of arsenopyrite, FeAsS
(After Buerger[2])

Primitive lattice, $P\,2_1/c$

	First individual	Twinned individual indexed on reference frame of first individual
Extinction rule for individual	$h0l$ absent when l is odd	$h0l$ absent when h is odd
Extinction rule for twinned composite indexed on reference frame of first individual	$h0l$ absent when both h and l are odd	

B-centered lattice, $B\,2_1/d$

	First individual	Twinned individual indexed on reference frame of first individual
Extinction rule for individual	$h0l$ absent when $h+l = 4-2n$	$h0l$ absent when $-h+l = 4-2n$ or $h-l = 4-2n$
Extinction rule for twinned composite indexed on reference frame of first individual	$h0l$ absent when $\pm(h \pm l) = 4-2n$	

discussion given in a foregoing section, the point-group symmetry of the reciprocal lattice is $2/m\,2/m\,2/m$. Although the composite reciprocal lattice has orthorhombic symmetry, the extinction rule given in Table 1 cannot be recognized as any orthorhombic rule, and the crystal would appear to have no extinctions, and therefore to belong to space group $P\,2/m\,2/m\,2/m$. The fact that the apparent space group is one without any screw axes or glide planes, and yet there are systematic absences of a meaningless type, is an inherent warning that the specimen is a twin.

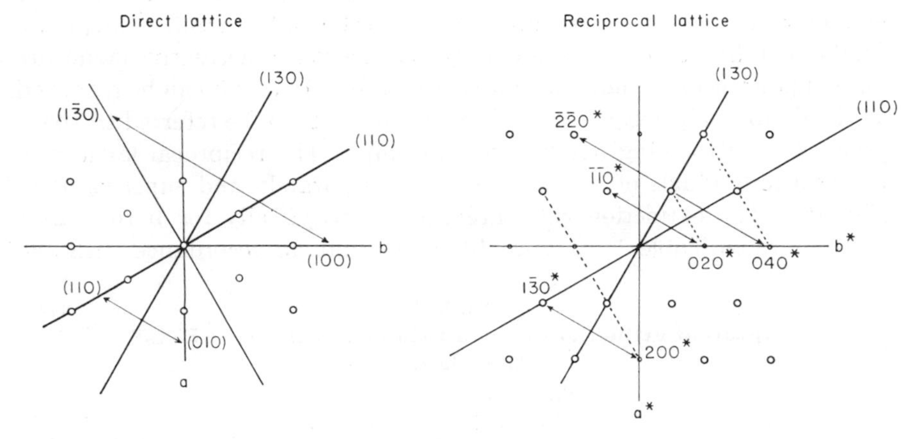

Fig. 11.
(After Buerger.[7])

In other instances there may be no warning. This is unfortunately true in trillings of orthorhombic crystals which mimic a hexagonal individual, an illustration of which was given earlier in Fig. 8. In that example the twin plane was (130). If Mallard's rule is accepted, the twin plane could be either (130) or (110), as shown in Fig. 11, since reflection across either plane throws the net into coincidence with itself. Figure 11 shows both direct lattice and reciprocal lattice. Let the planes of the reciprocal lattice which contain extinctions be (100) and (010). Consider where these are located after twinning. If (130) is the twin plane, (100) is reflected across (130) so that it comes to coincide with $(1\bar{3}0)$ of the untwinned individual, and (010) comes to coincide with (110) of the untwinned individual. In other words, the planes of the twin which contain extinctions come to coincide with planes of the untwinned individual in which all points are, in general, present. Therefore the extinction disappears except along the line of intersection [001]*. That is, the only extinction of the twin is $00l$ when l is odd, which is a perfectly normal hexagonal extinction corresponding to a 6_3 screw. If the orthorhombic crystal is repeatedly twinned to simulate a hexagonal crystal, then one would mistake this twinned aggregate for a normal hexagonal crystal of diffraction symbol[†] $6/mP6_3$ -, which contains space groups $P6_3$ and $P6_3/m$, or diffraction symbol $6/mmmP6_3$ - - -, which contains space group $P6_3 2$. These conclusions are wrong. An identical conclusion would have been obtained had (110) been the twin plane. Crystals which appear hexagonal but which have only extinction of $00l$ when l is odd should be regarded with suspicion.

Fig. 12.
Device for grinding and polishing surfaces on single crystals. (After Buerger and Lukesh.[3])

Detection of twinning. There are several means of detecting twinning. If the crystal is transparent it should certainly be examined with a polarizing petrographic microscope for areas with different optical orientations. This technique is too well known to require description here.[§ ¶]

It is less well known that opaque crystals can be examined for optical properties with the aid of the polarizing metallographic microscope.[‡] By means of this instrument any smooth face of a crystal can be examined for optically distinct areas. Since twin boundaries often occur at the

[†] See M. J. Buerger, *X-ray crystallography* (John Wiley and Sons, New York, 1942), page 514, and change C to P for modernization of symbols.

[§] Alexander N. Winchell. *Elements of optical mineralogy.* Part I. (John Wiley and Sons, New York, 5th Ed., 1937)

[¶] Ernest E. Wahlstrom. *Optical crystallography.* (John Wiley and Sons, New York, 2nd Ed., 1951)

[‡] J. Orcel. *Notes sur les caractères microscopiques des minéraux opaques, principalement en lumière polarisée.* Bull. soc. franc., minéral. **51** (1928) 197–210.

edge between two faces of the twin, it is even better to grind and polish an artificial plane on the presumably single crystal for optical examination. Figure 12 shows an apparatus especially designed to grind and polish such surfaces on small crystals.[3] It consists chiefly of a motor whose shaft bears a small disk on which can be placed various grinding and polishing surfaces, and a crystal mount which carries a ruggedly constructed goniometer head to which the crystal is cemented. When the motor is running, the crystal can be gently forced against the rotating lap and caused to range over the area of the lap. Figure 13 shows an example of a crystal surface prepared by this instrument. The crystal is one of arsenopyrite, FeAsS, which had been thought to be a single individual. That the crystal is really a twin is obvious when examined

Fig. 13.
Polished surface of Spindelmühle arsenopyrite crystal viewed between crossed nicols in reflected, polarized light, showing twinning.
(After Buerger.[2])

by reflected polarized light, as shown in Fig. 13. One can observe that the corner of the supposed single crystal is actually determined by the twin boundary. Another example of a crystal prepared by this apparatus is shown in Fig. 14. This shows polysynthetic twinning in a danaite crystal, (Fe, Co)AsS, as revealed by reflected polarized light.

Unfortunately, optical methods are powerless to reveal certain kinds of twins. For example, twins of merohedry (according to the Mallard nomenclature) in the tetragonal and hexagonal systems cannot be distinguished because the obverse and reverse forms both have their uniaxial indicatrices parallel. Twins of this sort can, however, be revealed by suitable etching.

Crystal imperfection

Imperfection resulting from growth. Crystals vary in perfection. The nature of the imperfection which spoils some of them for use in intensity studies can be appreciated if it is present in sufficiently exaggerated form. It then becomes apparent that what one first takes to be a single crystal really behaves as an aggregate of smaller crystals in subparallel orientation, Fig. 15. If the crystal is examined more closely, it is usually evident that what should be a single termination actually is replaced by several separate terminations having slightly different orientations. As

Fig. 14.
Polished surface of Sulitjelma danaite crystal viewed between crossed nicols in reflected, polarized light, showing polysynthetic twinning.

these subindividuals are followed back toward the center of the crystal, the deviations in orientation become less and less marked, and, if they can be followed back to the original nucleus of the crystal, these separate branches of the structure merge with one another to form a true single crystal. This is an aspect of the *lineage structure* [10, 11] of crystals. This name is a shorthand way of saying that a crystal tends to be continuous but branched. An idealized two-dimensional diagram of the nature of lineage structure for a symmetrical crystal is shown in Fig. 16. Evidently as the crystal grows, portions of its structure advance independently. These separate *lineages*, in growing, may twist slightly with respect to one another, as shown in Figs. 17 and 18.

Fig. 15.
Pyrite crystal from Elba which appears to be composed of a subparallel intergrowth of smaller crystals.
(After Buerger.[10])

Fig. 16.
Lineage structure. *A*. Diagram of the cells of a small, perfect crystal. *B*. Diagram of the cells of a larger real crystal; branching into *lineages* is apparent. *C*. Discontinuities in the structure of *B*.
(After Buerger.[11])

The single-crystal x-ray photographs of such "subparallel" aggregates have a characteristic appearance. Usually each reflection consists of a major reflection accompanied by various satellites trailing from it, Fig. 19. Unfortunately the degree and direction of dispersion of the satellites varies with the rotation coordinate, ω, of the Weissenberg photograph. This is because the angular departure from exact parallelism of the sepa-

Fig. 17.
Galena from Joplin, showing lineage structure.
(After Buerger.[10])

rate lineages occurs in three dimensions and is not uniform in the several dimensions.

X-ray photographs showing such satellitic reflections are most inappropriate for intensity measurement. If an integrated-intensity method is used in intensity determination, it is necessary, with such a photograph, to be certain that the entire cluster of satellites is included in the integrated intensity measured. Unfortunately this includes a great deal of background of variable amount with each reflection. On the other hand, if a peak-intensity method is used, the height of the peak for a given integrated intensity varies inversely but in an unpredictable way with the amount of dispersion of the satellites, which is itself variable with the rotation angle, ω, of the Weissenberg photograph. In other words, if this

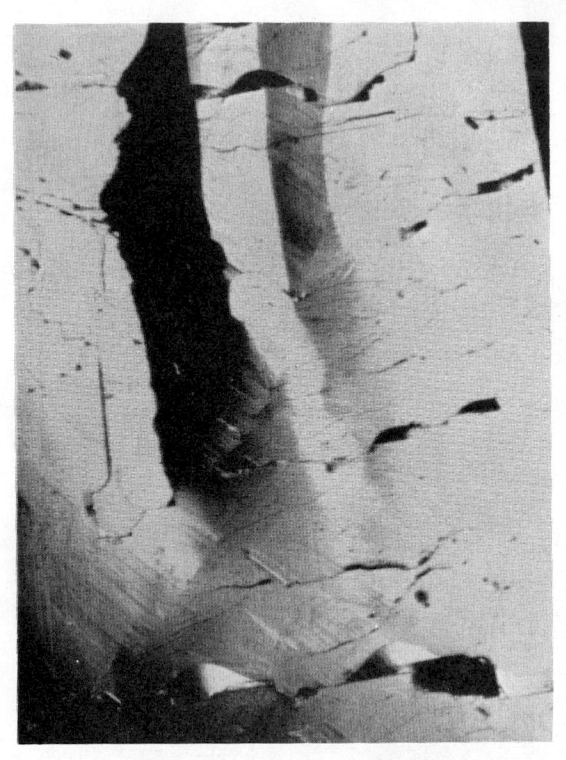

Fig. 18.
Cleavage surface of galena from Joplin, showing the coalescence of individual lineages toward the nucleus, below.
(After Buerger.[11])

effect is present at all, the peak height is by no means a measure of the integrated intensity even for the same value of sin θ.

Imperfection due to plasticity. Plastic crystals[9] offer great difficulties in intensity determination and even in space-group determination.† To investigate such a crystal requires finding an individual crystal not previously deformed and then making sure that in subsequent handling no plastic deformation is incurred. Since some plastic crystals are as plastic as butter, it is no simple matter to arrange for

† Newton W. Buerger. *The unit cell and space group of sternbergite,* $AgFe_2S_3$. Am. Mineralogist **22** (1937) 847–854.

appropriate material for x-ray analysis. Crystals having layer structures are particularly prone to become plastically deformed. Such crystals have one plane along which translation gliding is extremely easy. Ordinarily plastic deformation is accompanied by bending about an axis in this plane so that the crystal grating loses its true three-dimensional translational periodicity due to curvature. Crystals bent in this fashion

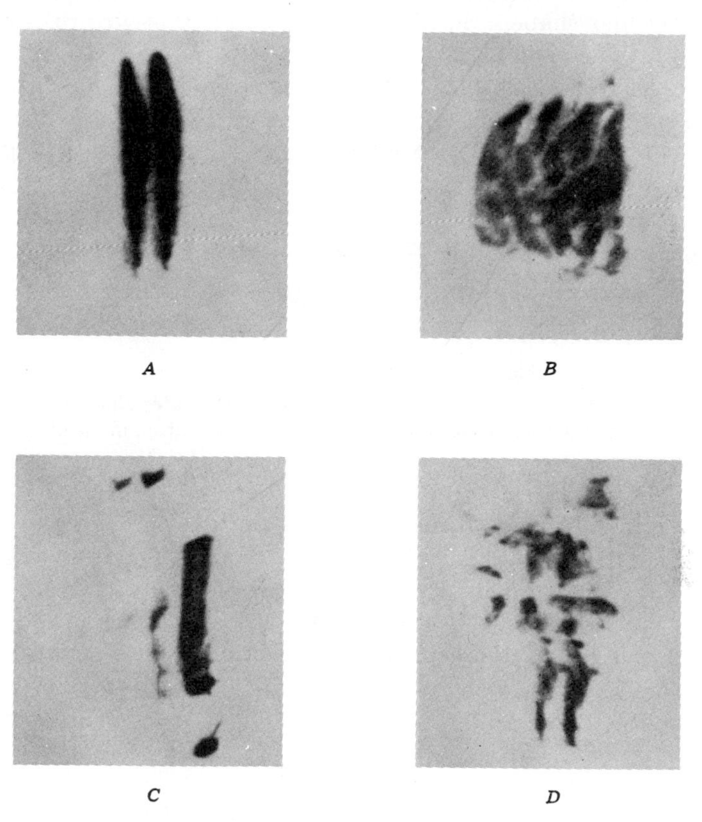

Fig. 19.
X-ray reflections split in various ways due to lineage structure.
(After Leonhardt and Tiemeyer, Z. Physik **102** (1936) 784.)

give reflections most of which are defocused, or drawn out, by unpredictable amounts. Photographs of such crystals are nearly useless for intensity measurement, and even space-group determination may be hampered because one cannot always be sure whether an apparently absent reflection is really absent or merely appears to be absent due to weakness resulting from defocusing. Quantum-counter intensity work based upon such a crystal can give spurious results because an unpredictable amount of the defocused reflection does not enter the counter.

Preferred orientation

If intensities are determined by reflection from a powdered specimen, it is vital that there is no preferred orientation in the powder. This undesired condition arises readily if the crystal fragments composing the powder have a non-equidimensional shape, particularly if they have a platy or acicular shape. Such shapes respectively occur in the crushed fragments if the crystal has a single perfect cleavage, or two or more perfect cleavages in a single zone. Micaceous minerals and amphiboles are examples of crystals having such cleavages. Preferred orientation may also occur in a powder which is composed of tiny uncrushed crystals of platy or acicular habit.

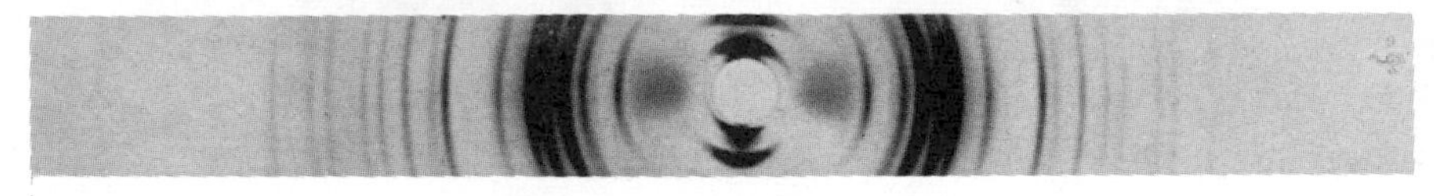

Fig. 20.
Powder photograph of α 1:1 stearic acid : sodium stearate, showing shaded arcs indicative of preferred orientation caused by plastic deformation induced in preparing the powder sample.

Preferred orientation also commonly develops in powders composed of crystals of extreme plasticity regardless of cleavage or habit. This is particularly common in powders of layer-structure crystals such as graphite, molybdenite, and soap, Fig. 20. The preparation of the powder sample causes the crystal to deform and this is attended by a reorientation.[9]

Size and shape of crystal

It will become evident from the discussion in Chapter 8 that it is difficult to correct diffraction intensities for the absorption of the crystal. If the absorption coefficient of the crystal is not too high, the absorption is often neglected. But when this is done, it is important to use a very small crystal so that whatever absorption does exist is minimized. Usually crystals about 0.1 or 0.2 mm. in cross-section diameter are appropriate for intensity determination. That this is small enough should be checked by taking a preliminary zero-layer Weissenberg photograph. If the reflections near the center line of the photograph show less intense centers, this is an indication that the absorption is still extreme and a smaller crystal should be used.

A crystal which is needle shaped parallel to the rotation axis is a desirable form since it provides large intensities without large absorptions. Since the correction of absorptions for pure cylindrical specimens is comparatively simple (Chapter 8), the crystal can be ground into the form of a cylinder for intensity measurements.[12, 14, 15]

A simple and practical technique has recently been developed for grinding a spherical surface onto a crystal.[13, 16] A spherical crystal has the advantage that the absorption correction is the same for any level for the same value of sin θ. The same spherical crystal can be used for all rotation axes so that all reflections are based upon the same volume of crystal.

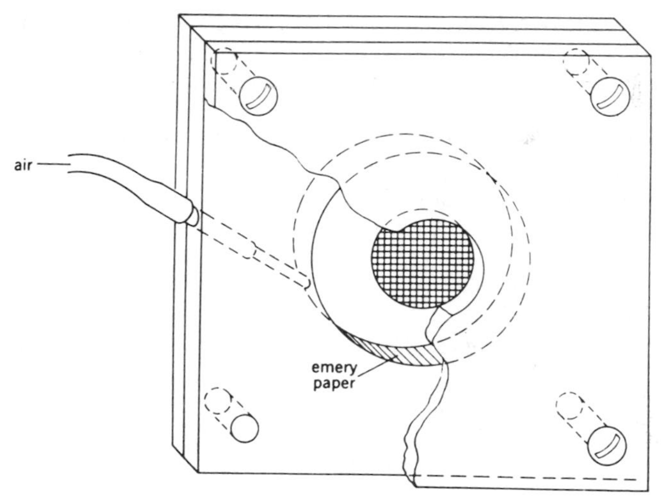

Fig. 21.
Device for grinding spherical surfaces on crystals.
(After Bond,[13] modified.)

Figure 21 shows the apparatus for grinding a spherical surface onto a crystal. The crystal is placed in a disk-shaped chamber whose circumference is lined with fine emery paper. The crystal is tumbled about the periphery of the chamber by blowing a stream of compressed air tangentially into the chamber. The crystal acquires a spherical surface in a matter of seconds.

The resulting sphere must be oriented. Ordinarily the sheen of the spherical surface differs with the orientation of the plane tangent to the surface. The sheen therefore reveals the symmetry of the crystal in a rough way. If the crystal is transparent and non-isometric, the actual orientation can be carried out optically with the aid of the polarizing petrographic microscope. This preliminary orientation can be refined on the precession apparatus.

Literature

Twinning

[1] Georges Friedel. *Leçons de cristallographie.* (Berger-Levrault, Paris, 1926), especially 421–483.

[2] M. J. Buerger. *The symmetry and crystal structure of the minerals of the arsenopyrite group.* Z. Krist. (A) **95** (1936) 83–113, especially 98–102.

[3] M. J. Buerger and J. S. Lukesh. *The preparation of oriented polished sections of small single crystals.* Am. Mineralogist **21** (1936) 667–669.

[4] J. D. H. Donnay. *Width of albite-twinning lamellae.* Am. Mineralogist **25** (1940) 578–586.

[5] J. D. H. Donnay. *Plagioclase twinning.* Bull. Geol. Soc. Am. **54** (1943) 1645–1652.

[6] M. J. Buerger. *The genesis of twin crystals.* Am. Mineralogist **30** (1945) 469–482.

[7] M. J. Buerger. *The diffraction symmetry of twins.* Anais acad. brasil. cienc. **26** (1954) 111–121.

[8] S. C. Nyburg and J. Hilton. *Molecular structure of cyclooctatetraene dimers. I. The 1:1 silver nitrate adduct of the dimer which melts at 38.5° C.* Acta Cryst. **12** (1959) 116–121.

Crystal imperfection

[9] M. J. Buerger. *Translation-gliding in crystals.* Am. Mineralogist **15** (1930) 45–64.

[10] M. J. Buerger. *The significance of "block structure" in crystals.* Am. Mineralogist **17** (1932) 177–191.

[11] M. J. Buerger. *The lineage structure of crystals.* Z. Krist. (A) **89** (1934) 195–220.

Size and shape

[12] H. Kersten and W. Lange. *Method for preparing crystals for rotation photographs.* Rev. Sci. Instr. **3** (1932) 790–791.

[13] W. L. Bond. *Making small spheres.* Rev. Sci. Instr. **22** (1951) 344–345.

[14] Ray Pepinsky. *Method for cutting and shaping fragile crystals.* Rev. Sci. Instr. **24** (1953) 403.

[15] F. Barbieri and J. Durand. *Method of cutting cylindrical crystals.* Rev. Sci. Instr. **27** (1956) 871–872.

[16] K. S. Revell and R. W. H. Small. *The preparation of spherical single crystals for x-ray diffraction work.* J. Sci. Instr. **35** (1958) 73–74.

6

Measurement of intensities

Intensity measurement may be carried out by either of two different kinds of instrumentation, the photographic film or, more directly, the quantum counter. With either of these devices it is possible to measure either the peak intensity or the integrated intensity. The measurement of the integrated intensity is the more difficult undertaking.

The advantage of the photographic instrumentation is that the primary record is made automatically by the Weissenberg apparatus (for example) without requiring any supervision or control by the investigator. Counter methods, on the other hand, ordinarily require constant supervision and control, and a permanent primary record may or may not be automatically made. Up to the present time the photographic instrumentation has been by far the most common, partly because the apparatus required is no different from that used in the determination of the unit cell and space group, and therefore is available in every laboratory.

Powder versus single-crystal methods

It is possible to make intensity determinations using either the powder method or a single-crystal method. The powder method offers two advantages. Corrections for absorption are very simple with the powder method, and if a fine enough powder (composed of grains of the order of 10^{-4} cm. diameter) is used, no correction is necessary for primary "extinction", as pointed out in Chapter 8.

The powder method, however, has many disadvantages. The greatest disadvantage is lack of resolving power, for reflections with nearly the same value of the reciprocal-lattice distance, σ, fall together on the powder record, as pointed out in Chapter 7. For this reason the intensities of individual reflections from crystals with large cells cannot be investigated by the powder method.

A special case of lack of resolving power occurs with crystals belonging to the classes of the tetragonal and hexagonal crystal systems lacking "vertical" symmetry planes. For such crystals the reflections $\bar{h}kl$ and hkl have different intensities, yet they superpose identically in the powder method. Fankuchen† has suggested that the difficulty in such cases can be mitigated by the following procedure: The composite intensity from $\bar{h}kl$ and hkl is furnished by the powder method. The ratio of these intensities can be rather closely determined from a single-crystal photograph, for example by comparing the blackening of $\bar{h}kl$ and hkl as they occur on an l-level, c-axis Weissenberg photograph. This determination is readily made, particularly from a symmetrically developed crystal. Since all corrections to be applied to $\bar{h}kl$ and hkl are known, the composite intensity of $\bar{h}kl + hkl$ can be readily apportioned to the two reflections. While this method was suggested specifically for application of the powder method to the determination of the intensities of tetragonal and hexagonal crystals lacking "vertical" symmetry planes, obviously the scheme can be extended to any crystal which gives unresolved powder reflections.

Photographic methods

Fundamentals of the photographic process.[3–5] The sensitive material of a photographic film is silver bromide containing a few per cent of silver iodide. Grains of silver bromide suspended in gelatin constitute the photographic emulsion. For most x-ray purposes both sides of the photographic film are coated with this emulsion; film to be used for focusing powder cameras should be coated on one side only, because the rays strike the film at very oblique angles in such cameras.

Exposure to light or x-rays causes a change in certain grains of the emulsion. In each of these changed grains there arises a small nucleus of metallic silver which constitutes a latent-image nucleus. When the emulsion is treated with developer, these particular grains gradually change from silver halide to metallic silver. It should be emphasized that the development proceeds by transforming sensitized grains only, leaving unchanged those other grains which have not been sensitized by the exposure. The total amount of silver produced by developing depends, first, upon the number of grains in the emulsion affected by light and, secondly, upon the extent that the developer is allowed to transform these specific grains from silver halide to metallic silver.

It is believed that the absorption of 1 quantum of light of the visible region produces a free silver atom, but that about 300 quanta need to be absorbed to produce a silver nucleus which is developable. In the x-ray region, however, only 1 quantum is believed necessary to produce a

† Personal communication.

developable silver nucleus. To the first approximation the total number of nuclei formed is thus proportional to the number of quanta received by the film. For emulsions exposed to x-rays it has been well established that the reciprocity law is obeyed, i.e., that exposure is equal to the product of intensity and time, so that intensity and time enter as equal factors. Thus the amount of developable silver nuclei produced by x-rays (in a fixed time, t) in a film is proportional to the intensity of the x-rays falling upon it.

The amount of developed silver is commonly measured by the blackness produced on the film.[1] To discuss this it is convenient to consider how much light the developed film transmits, or how much light it blocks. Let a beam of light of initial intensity L_0 fall on the film. In general the intensity of the beam is diminished in passing through the film so that it has a lesser intensity L on emerging. The transmission of the film is defined as

$$T = \frac{L}{L_0}. \tag{1}$$

T may be regarded as a measure of the efficiency of transmission of the film. The opacity is defined as

$$O = \frac{L_0}{L}. \tag{2}$$

If the transmitted intensity, L, is unity, (2) yields $O = L_0$; thus O may be thought of as the intensity of light which must fall on the film in order that a unit intensity emerges through it.

Consider a small thickness dt of developed emulsion containing N silver atoms per unit area. Let light of initial intensity L_0 fall upon this small layer. The absorption by the silver reduces the intensity by an amount dL. This reduction is proportional to the amount of silver, to the initial intensity, and to the thickness of the film; i.e.,

$$-dL = kNL\,dt, \tag{3}$$

where k is a proportionality constant (the linear absorption coefficient, expressed in cm.$^{-1}$ units). Therefore,

$$\frac{-dL}{L} = kN\,dt. \tag{4}$$

On integration this gives

$$\ln L_0 - \ln L = kNt,$$

or,

$$\ln \frac{L_0}{L} = kNt. \tag{5}$$

The quantity Nt is the number of silver atoms in a volume bounded by a unit area and the total thickness of the film, so that kNt is proportional to the density of silver in the film.

In photographic practice it is customary to express (5) in terms of common logarithms. Since common logarithms and natural logarithms are related by

$$\log_{10} x = 0.4343 \log_e x, \tag{6}$$

(5) is equivalent to

$$\log_{10}\left(\frac{L_0}{L}\right) = 0.4343kNt \tag{7}$$

$$= D. \tag{8}$$

*The quantity 0.4343*kNt is proportional to the density of silver in the film and *is defined as the density*, D, for photographic purposes.

From (1), (2), and (8) the following relations between transmission, opacity, and density can be deduced:

From (1), (2): $$O = \frac{1}{T}, \tag{9}$$

From (2), (8): $$\log_{10} O = D, \tag{10}$$

From (9), (10): $$\log_{10} T = \log_{10} \frac{1}{O} = -\log_{10} O \tag{11}$$

From (10): $$O = 10^D, \tag{12}$$

From (9), (12): $$T = 10^{-D}. \tag{13}$$

This discussion shows that the density of the silver precipitated can be measured by $\log_{10}(L_0/L)$. Under the simple conditions considered, this measurable quantity is proportional to the x-ray exposure. Consequently if one plots $\log_{10}(L_0/L)$ against the x-ray intensity, or exposure E, a straight line should result. This is actually true up to a limited exposure corresponding to a density value which may be as high as 1.4 with normal development, but which depends upon conditions of development. At density values beyond a certain point the density does not increase as rapidly as exposure. The smaller increments of blackening with larger exposures may be attributed largely to absorption of quanta by grains which have already been sensitized by quanta received earlier. Repeated absorption by the same grain does not increase the number of developable grains.

Typical density-exposure curves[3] for x-rays and visible light are shown for comparison in Fig. 1. The curve for visible light differs chiefly by having a region of "inertia" where exposure does not give rise to any

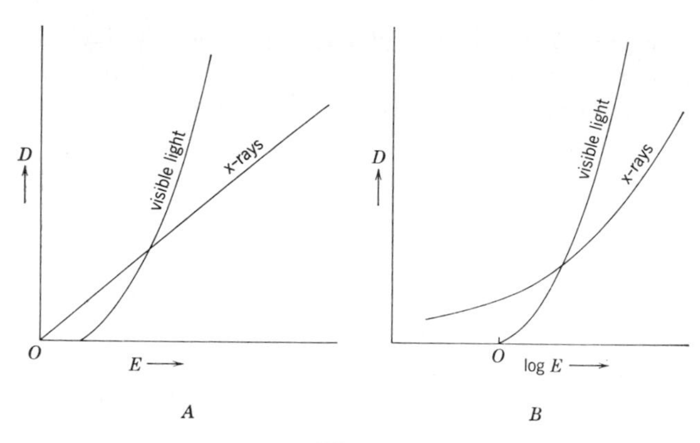

Fig. 1.
(After Charlesby,[3] modified.)

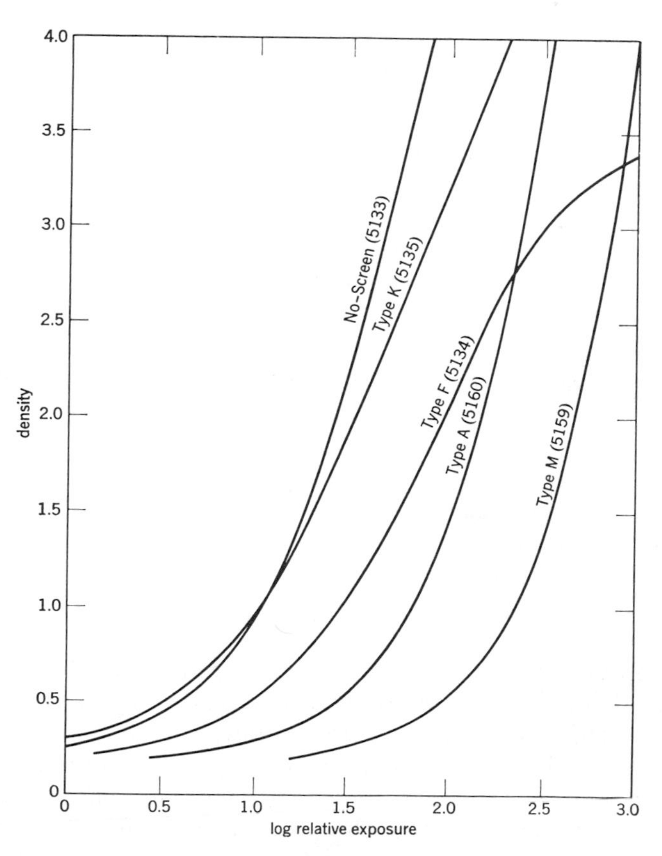

Fig. 2.
(Courtesy of Eastman Kodak.)

blackening. Figure 2 shows the relation between density and exposure for some photographic materials used in the United States.

As pointed out above, the density is a linear function of exposure for x-rays, up to some limiting value of the density. Van Horn[6] has pointed out how this linear relation can be extended to much higher densities. Fundamentally the linearity is limited by a certain exposure level,

Table 1
Increase in slope at low densities with extent of development
(After Van Horn[6])

Development time in minutes	Slope (No-Screen film at 5 min. = 100)	
	Kodak Industrial X-ray Film, Type K	Kodak No-Screen Medical X-ray Film
3	89	87
5	104	100
8	120	122
10	127	130
13	132	146
18	139	158

Table 2
Linear ranges of density-exposure relation
(After Van Horn[6])

Film	Range of relative exposure†	Range of density	Development time in minutes at 68° F
Kodak Industrial X-ray, Type K	1–38	0.61–2.5	18
Kodak No-Screen Medical X-ray	1–64	0.41–4.0	18
Kodak Industrial X-ray, Type A	1–36	0.15–3.0	5
Kodak Industrial X-ray, Type M	1–56	0.17–3.0	3

† Unit exposure is that which produces a density of 0.05 above fog.

beyond which second quanta begin to reach grains already sensitized, and therefore fail to leave a record of their arrival on the emulsion. But density is a function of both exposure and development time. Density can also be enhanced by increased development so that it increases strictly proportionally to the exposure. Great densities can therefore be attained by moderate exposures followed by long developments. Now, provided that the exposures do not overstep the linear region, the linear density:exposure relation can be extended to large densities simply by long development. For example, normal development for Eastman No-Screen film is 5 minutes at 68° F in Eastman rapid x-ray developer. But so far as the weak exposures are concerned, the development can be extended to 18 minutes with profit. This gives the linear part of the curve a steeper slope, and the density:exposure relation remains linear

Table 3
Speeds of films at different wavelengths
(After Van Horn[6])

Film	Speed for		
	Cr$K\alpha$	Cu$K\alpha$	Mo$K\alpha$
Kodak Industrial X-ray, Type K	150	130	110
Kodak No-Screen Medical X-ray	100	100	100
Kodak Industrial X-ray, Type F	85	60	40
Kodak Single-Coated X-ray	80	45	25
Kodak Industrial X-ray, Type A	22	17	14

up to densities of about 4. Unfortunately these long exposure times increase the background of fog density of the film.

Some data showing the properties of some commercial x-ray film as given by Van Horn[6] are reproduced in Tables 1, 2, and 3.

Use of intensifying screens. It is well known that a specified density can be achieved with a shorter exposure if an intensifying screen is placed behind the x-ray film. A careful quantitative study of the

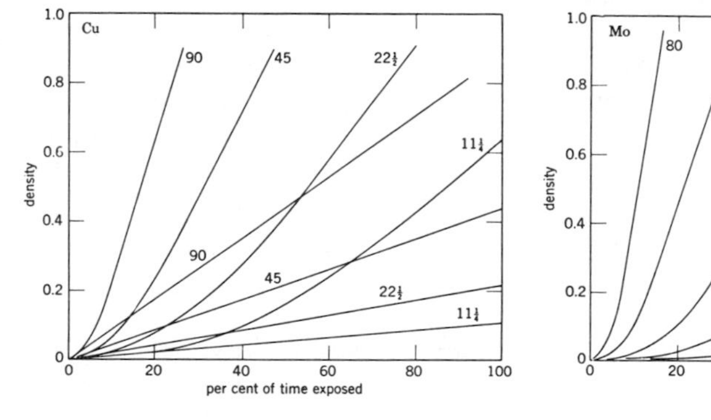

Fig. 3.
(Filtered Cu radiation.)

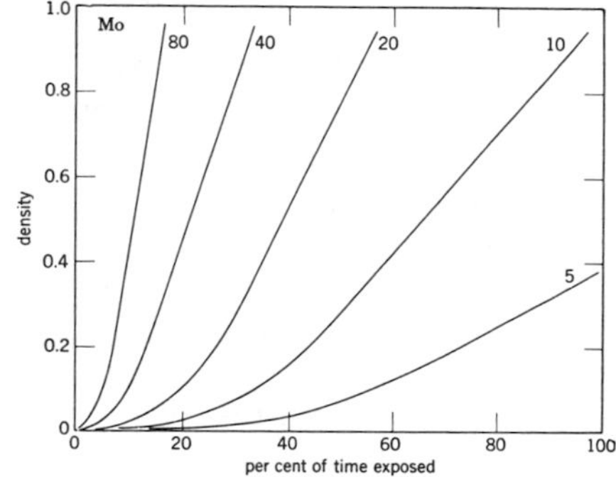

Fig. 4.
(Filtered Mo radiation.)

Density versus per cent of time exposed, for various exposure times (in seconds). Straight lines correspond to relations without intensifying screen, curved lines to relations with intensifying screen.

(After Gamertsfelder and Gingrich.[2])

intensifying effect has been made by Gamertsfelder and Gingrich.[2] The effect of the Fluorazure intensifying screen on enhancing density, as found by them, is illustrated in Figs. 3, 4 and 5. It will be observed that intensifying screens contribute an "inertia" region to the density:exposure relation. The reason for this is probably that the intensifying action consists of converting x-ray energy to light, which then exposes the film. The film thus behaves partly as if it were being used with visible light, the action of which is characterized by an "inertia" region. It is plain from these results that the intensifying screen provides enhanced density at the cost of destroying the beautiful linearity between the density and exposure which is characteristic of exposure to pure x-rays. For this reason it is undesirable to use intensifying screens in intensity determinations.

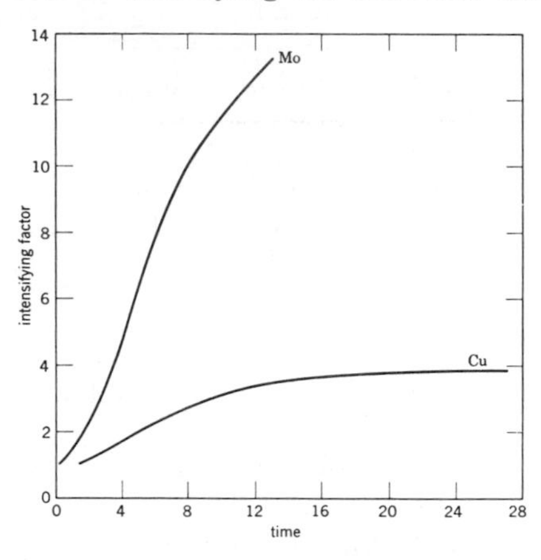

Fig. 5.
Intensification factor versus time of exposure (in seconds).
(After Gamertsfelder and Gingrich.[2])

Incidentally, Gamertsfelder and Gingrich[2] point out that greater densities for a given exposure can be produced by using two films in contact in place of a single film. If the combined density of the pair of films is measured, exposures can be reduced by a factor of nearly $\frac{1}{2}$.

In the pages which immediately follow, some methods in common use for applying photographic methods to the determination of x-ray intensities are described.

Peak-intensity methods. It would be an easy matter to determine integrated intensities if one could find an automatic device for adding log L's. Since this is not accomplished in any simple fashion, many investigators content themselves by determining the intensity at the maximum blackening of the x-ray reflection. To accept this for the integrated intensity is to imply that the integrated intensity is proportional to the peak intensity. Actually this is not so† for three reasons:

1. Unless the x-ray beam is composed of parallel rays (i.e., if the beam has any appreciable divergence), then there is a tendency to focus

† There is, however, a mitigating circumstance. When a collimator with large pinholes ($\frac{1}{2}$ to 1 mm. diameter) is used, a convergence of $\frac{3}{4}°$ to $1\frac{1}{2}°$ is permitted to the beam incident upon the crystal. This is enough to produce a "plateau" in the center of the spot, as discussed in page 104. For this reason, what appear to be measurements of peak intensities are often measurements of the integrated intensity, provided a collimator of large aperture is used.

reflections† more with increasing value of θ. Thus the same integrated intensity for small-θ and large-θ regions has different peak heights, as illustrated in Fig. 6. This is further complicated by resolution of the $K\alpha$ reflection into an α_1 α_2 doublet at large values of θ.

2. If the crystal is imperfect, the same area of a peak shown in Fig. 6 may well have different peak heights for different reflections of the same θ value, as explained in Chapter 5.

3. Reflections reaching the film from different angles (i.e., zero and n levels of Weissenberg or rotation photographs) must be corrected for the increase in spread of the reflection with increasing angle.[40, 42, 43, 52] Reflections on the upper and lower halves of n-level Weissenberg photographs must also be corrected for compaction and attenuation of the spots.[§ 48, 55]

In spite of this non-proportionality, peak intensities are very commonly accepted in lieu of integrated intensities. Some methods commonly used for peak-intensity measurements are described below.¶

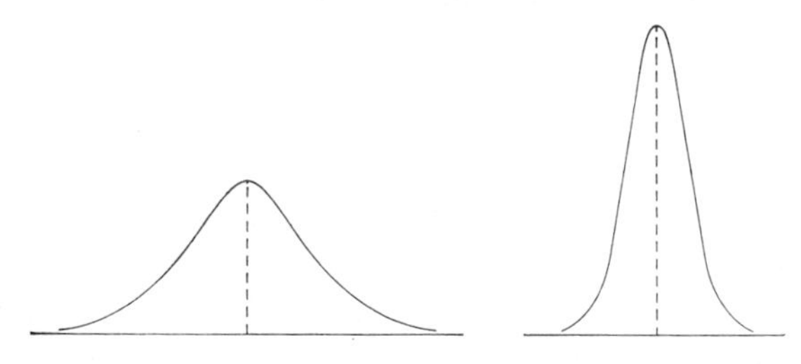

Fig. 6.
Peak shapes for small values of θ (left), and large values of θ (right).

Density methods. It has been pointed out in an earlier section that for density values which are not too great the density is a linear function of exposure, and therefore of intensity. This gives rise to the possibility of determining the x-ray intensity at the peaks of the reflections by merely photometering them. It should be observed that this procedure depends on the linearity of the density-exposure curve, and this depends on appropriate development. In any case the intensities depart from linearity with density beyond some limiting density value. If the method is to be used it is important that the particular combination of x-ray film,

† M. J. Buerger. *X-ray crystallography.* (John Wiley and Sons, New York, 1942) Fig. 203, page 398.

§ M. J. Buerger. *X-ray crystallography.* (John Wiley and Sons, New York, 1942) Fig. 125, page 229.

¶ When any of the peak-intensity methods are used, allowance must be made for the angle at which the x-ray beam strikes the film (see reference 4, Chapter 7). When using the Weissenberg method, allowance must also be made for contraction or extension of the spot on upper levels.[48, 55]

developer, developing time, and developing temperature be standardized for the laboratory, and that the density:exposure curve be determined experimentally for these conditions. After such a calibration it is, of course, possible to determine peak intensities regardless of linearity by consulting the density-exposure curve.

Standard scale. Another method of determining peak intensities is to prepare a standard exposure scale and compare the blackness of the various x-ray reflections visually with the blacknesses on the standard scale. This method makes no specific use of measured density. The standard scale can be readily prepared by allowing a weak, uniform beam of x-rays, limited in cross-section by an appropriate aperture, to fall on part of an x-ray film. A graded exposure series can be arranged by using the usual rotating sector.

Alternatively a graded series can be prepared by allowing the x-ray beam to fall through a small aperture on a film for a seconds, then, after displacing the film by a little more than the width of the aperture, exposing the film for ar seconds, then, after displacing the film again, exposing for ar^2 seconds, etc.† In this way a set of exposures in geometrical progression is made. The value of r is chosen so that the blackness of the spot for exposure ar seconds is just distinguishably greater than the blackness of the spot made in a seconds.

A useful variation of this procedure is to use a strong reflection from the crystal being investigated as the weak source of x-rays. For example, a strong reflection $h_1\,k_1\,0$ may be selected from a zero-level Weissenberg or precession photograph. The instrument is then set so that this $(h_1\,k_1\,0)$ plane is in reflecting position. The reflection $h_1\,k_1\,0$ then comes through the layer line screen to the film in the camera of the Weissenberg or precession instrument. By interrupting the x-ray beam with a lead screen, exposures of a, ar, $ar^2 \cdots$ seconds can be recorded in successive positions as the film is moved by a definite interval between exposures. Such a graded series, made with the crystal being investigated, has the advantage that the spots on the graded series and the spots on the x-ray photograph being calibrated have approximately the same shape and so are more readily compared

It is usually recommended that the graded series be recorded on the same sheet of film as the x-ray record to be interpreted. This is an unnecessary precaution provided that the photographic film is obtained from the same box, that the developing procedure is standardized as to time and temperature, and provided that fresh developer is always used.

Multiple-film technique. Suppose that one wishes to record $|F|$'s to about 5% accuracy. This can be accomplished by using a scale in which

† A similar method is to find an exposure time a, which provides a perceptible blackening of the film. Then a set of n exposures are made which conform to $a2^{n/2}$.

the maximum value of the density represents the maximum value of $|F|$, and dividing the range from zero to this value into 20 parts. But then the scale of $|F|^2$'s varies from 1^2 to 20^2, i.e., from 1 to 400. The most intense of the reflections is further likely to be scaled up by the Lorentz-polarization factor so that a range of 1 to 1000 is easily needed for the blackening scale. The ordinary photographic film cannot render this range, and furthermore it is almost impossible to distinguish between various degrees of blackness for very black spots.

Table 4
Transmissions of some specific x-ray films

Radiation	λ	X-ray film		
		Ilford†	Eastman No-Screen§	DuPont Type 508§
Ag$K\alpha$	0.56 Å		82.9%	93.2%
Mo$K\beta$	0.63		77.0, 80%¶	90.7
Mo$K\alpha$	0.71		70.3	88.5
Cu$K\beta$	1.39		33.8	63.8
Cu$K\alpha$	1.54	50%	23.8, 27%‡	54.9
Ni$K\alpha$	1.66		17.5	48.4
Co$K\alpha$	1.79		11.8	41.3
Fe$K\alpha$	1.94		7.2	32.8
Cr$K\alpha$	2.29		1.7	17.6

† From Robertson.[25]
§ From Taylor and Parrish.[56]
¶ B. E. Warren, 1946 (private communication).
‡ W. Nowacki, 1937 (private communication).

The multiple-film technique[20, 25, 42, 43, 45, 47, 52] was devised to overcome the lack of range of a single film. In this method, a pack of 3 to 6 photographic films is used for each x-ray photograph instead of the customary single film. Since modern photographic films are made with a very uniform coating, each film of the pack acts as a uniform absorber and reduces the x-ray reflection intensities which arrive at the film underneath it by a factor which is said to be constant to within $\frac{1}{2}$% over the area of the film and also from film to film.[25] By this arrangement the limited range of one film is supplemented by the ranges of others of the pack. Specifically, the intensities of the strong reflections can be estimated by examining the films of the pack farthest from the crystal, and the weak reflections can be estimated from the films nearest the crystal.

Some values of the transmission of x-ray films in common use are given in Table 4. In the event that the emulsion of the x-ray film contains any fluorescent material placed there to enhance the speed of the film, the several films of the pack must be separated by black paper.

Robertson[25] recommends developing all the films, representing a single exposure, in a holder capable of suspending them all at the same time in the developing tank. A deep tank must be used, and the films should be continuously agitated. A blank film may be used at each end to prevent unusual treatment of the films suspended on the outside of the rack, although this precaution is unnecessary if the films are continuously agitated.

While the multiple-film technique is ordinarily used for measuring peak intensities, it may also be used for measuring integrated intensities by various methods. It has been used in connection with the Dawton scan photometer described later.

The multiple-film technique cannot be readily adapted to the de Jong-Bouman or precession methods because the exactness of the film location required in these methods does not tolerate the varying distances from the crystal of the several films of the pack. But for these methods, and other methods as well, a variation may be used. Two or more photographs may be taken of the same level, but with different timings chosen at will. This variation has the advantage that the intensity ratio on the several films may be given any desired value.

A variation of the multiple-film technique is to compare $K\alpha$ and $K\beta$ reflections on the same film. The relative intensities of these reflections for copper radiation is in the neighborhood of 6:1.

Lukesh's method. A novel method of estimating peak intensities was developed by Lukesh.[23] In Lukesh's method the original x-ray "negative" is printed on high-contrast printing paper. The same negative is printed on, say, 10 sheets of paper using exposures of $\log n$, $\log 2n$, $\log 3n \cdot \cdot \cdot \log 10n$ seconds, respectively. The prints are all given identical development. The finished prints have a black background upon which appear the white x-ray reflections. The particular print exposed $\log n$ seconds shows very many white spots, the print exposed $\log 2n$ seconds shows fewer white spots, and the print exposed $\log 10n$ seconds shows no white spots. The exposure required to first start to darken a spot is a measure of the density of the spot.

To consider this process quantitatively, let ρ_1 be the value of the density of the positive print which can just be discerned as incipient darkening. This density is caused by a particular printing exposure E_1, which is the product of the intensity of light L, through the "negative," times the printing time t, and this exposure is constant for each print; i.e.,

$$Lt = \text{constant}. \tag{14}$$

The light L transmitted by the negative to the print is related to the density D of the negative by relations (2) and (12), which can be combined to give

$$\frac{L_0}{L} = 10^D. \tag{15}$$

If this is combined with (14) there results

$$\frac{L_0}{10^D} t = \text{constant}. \tag{16}$$

Since the intensity of the printing light, L_0, is constant, this can be written

$$t = \text{constant} \times 10^D, \tag{17}$$

and taking logs of both sides,

$$D = \log\,(\text{constant} \times t). \tag{18}$$

Lukesh recommends using the extremely contrasty Eastman Kodak Kodalith and developing 10 minutes.

Integrated-intensity methods. Integrated intensities are more difficult to determine than peak intensities because the x-ray spot is blacker in the central region and lighter toward the edges. Each region of the spot has thus received different intensities. Each of these intensities must be added to make up the total "integrated intensity." The chief difficulty in doing this is that intensity cannot be measured directly but is ordinarily arrived at by measuring the light transmission of the blackened photographic film. Unfortunately this is not a linear function of the intensity received by the region. Thus it is not possible to simply add together the light transmitted through the several regions of the spot.

A number of methods have been devised to circumvent this difficulty. These are described in the following sections. A comparison of some of these methods has been made by Robertson and Dawton.[22]

Photometer-trace method. A relatively simple method is frequently used to obtain integrated intensities from powder photographs. The photograph is placed in a recording photometer, which provides a graphical record of the transmission along the center line of the powder photograph. The result is usually termed a *photometer trace* of the photograph. The ordinates of the trace are proportional to the transmission, T, while the abscissas are distances along the film. To obtain integrated intensities from this, it is first necessary to transform the ordinates from light transmitted to x-ray exposure. Relation (8) gives

$$D = \log_{10} \frac{L_0}{L}. \tag{19}$$

Insofar as the density is proportional to exposure, and since L_0 is constant, one can deduce that

$$E = kD = k(\log L_0 - \log L)$$
$$= k(\text{constant} - \log L). \tag{20}$$

This provides the relation between L, the light transmitted by the film,

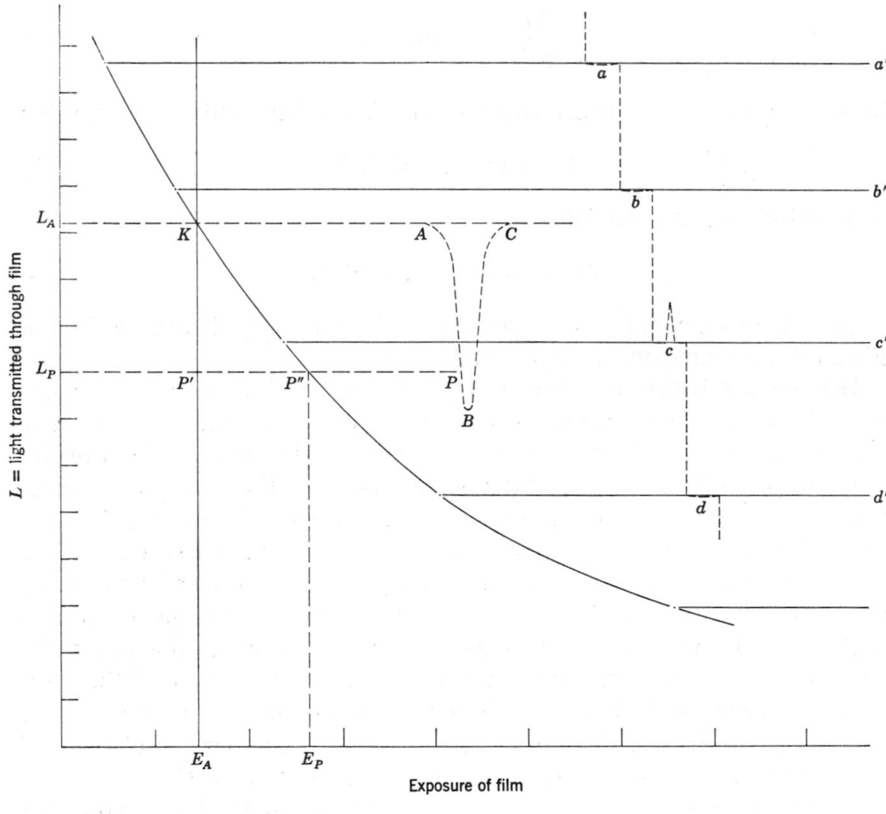

Fig. 7.
(After Brentano.[28])

and E, the exposure responsible for the blackening. A curve representing this relation for a particular L_0 (which is characteristic of the photometer) and k (characteristic of the development process) enables one to calibrate the whole process.

Brentano[28] has discussed the precautions which should be observed when using this method. Figure 7, reproduced from Brentano with slight modification, illustrates the method to be used. The long sloping line is the calibration curve corresponding to (20) for the specific process. Brentano recommends that the transmission of a stepped exposure series (the steps outlined at the right of Fig. 7) be superposed in correct relation to the trace of the x-ray line (the inverted peak of Fig. 7). The level L_A represents the light transmitted through the background, while L_P represents the light transmitted at some point P of the diffraction line being examined. Then the exposure above background exposure is simply the difference $E_P - E_A$. To determine the integrated intensity corresponding to the x-ray line, one divides the distance across the line

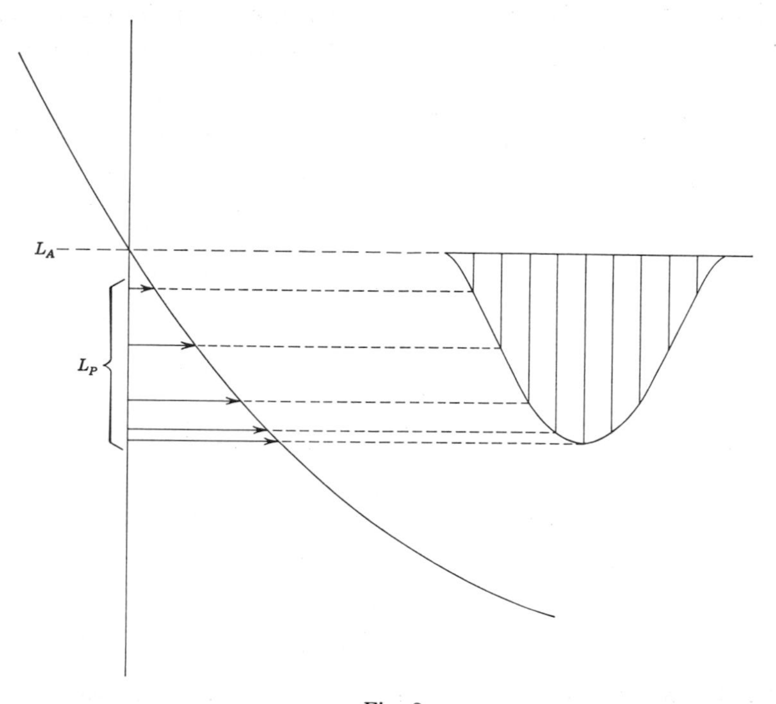

Fig. 8.

into equal parts (Brentano recommends 20 parts), and determines the sum of the distances $E_P - E_A$, shown as arrows in Fig. 8, for these positions on the line.

This method can be applied to diffraction spots on single-crystal photographs. Because of the many readings required for each spot, however, this method is seldom used for single-crystal intensity determination.

Density-wedge methods. A neat way of determining densities is to balance the density on the photograph with a known density, for example a location on a density "wedge." To determine the point of balance it is customary to take two equal beams of light (for example, from the same source) and pass one through the photograph and the other through the density "wedge," adjusting the position of the latter until the two densities are equal. This equality can be judged by eye, for example, by moving the wedge until the region observed is just as black as the photograph. The balance can be elegantly effected by means of a photocell arrangement. Figure 9 shows a diagram of an arrangement used by Dobson.[8] The light from the lamp is passed over two equivalent paths. The photograph is placed in one path, a density "wedge" in the other. The two light paths enter a photocell. By means of a shutter the light from the two paths is caused to enter the photocell alternately. When the light from these two beams is unequal, the output of the photocell, as indicated by a galvanometer, is fluctuating. The wedge is then

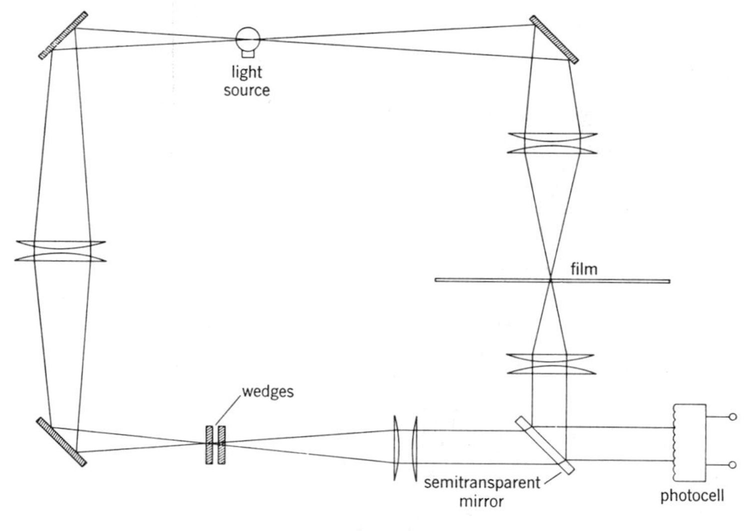

Fig. 9.
(After Dobson.[8])

adjusted to a point of no fluctuation. The density of the wedge is then equal to the density of the photograph. This principle was used by Taylor[36] in the design of an elaborate microdensitometer.

Robinson[15] also used the general scheme for an integrating photometer intended for single crystals. Before Robinson's photometer for single crystals is described a simpler hypothetical device for powder photographs using Robinson's principle, will first be discussed. In this simpler device the powder photograph is mounted on a movable carriage. The principal feature of the photometer is two identical beams of light which are caught on identical photocells. One beam is focused on a small spot ($\frac{1}{10}$ mm. in diameter) of the x-ray film; the other beam is sent through a calibrated exposure strip made from the same x-ray film and developed by the same developing technique. The outputs of the two photocells are wired together so as to oppose one another; no signal is given unless unequal amounts of light are received by each photocell. The x-ray film is set in a position so that the light beam goes through the film at a point having linear coordinate x. In general, the signal indicates that the photocells are not balanced. The calibrated wedge is then moved along axis Y until a balance is achieved at coordinate y. The coordinate y thus gives the density, and therefore the exposure, at film coordinate x. The curve resulting from plotting y versus x is the relative x-ray intensity received at points along the center line of the powder photograph.

Robinson's instrument[15] uses this principle for measuring the integrated intensities of single-crystal reflections or "spots." A calibration wedge is exposed on the x-ray film to be measured. The film can be moved along two coordinate directions so that the focused beam of light can be transmitted through any part of the two-dimensional "spot." Robinson recommends dividing the spot into 100 to 300 periodically spaced samples. At each point the wedge is moved to effect a balance of the photocells. The position of the wedge is proportional to its density, and therefore to the x-ray exposure at that point. The wedge positions for the various points of the x-ray spot are added on a mechanical counter. The sum recorded for a spot, with a correction for the surrounding background level, is proportional to the "integrated intensity" of the particular spot.

According to Robinson, the reflection intensities of one Weissenberg film can be measured with this photometer in a "good day's work."

Density-transforming methods. An entirely different principle for determining integrated intensities was devised by Robertson and Dawton.[22] An image of the spot to be photometered is projected onto an aperture in a screen. By means of a scan drum behind the aperture, the light from only a small area in the reflection is allowed to reach a photocell. The scan drum thus allows a small beam of light to scan the area of the x-ray spot. The output of the photocell is proportional to the transmission of light through the various parts of the spot. The unusual feature of the device is that this output is fed into a non-linear electrical circuit which, in effect, transforms the light received by the photocell to the density, and therefore to the x-ray intensity which fell on that portion of the spot. In this way the converter solves (20). The entire spot is scanned in about $\frac{1}{10}$ sec., and the output of the photocell is introduced into a milliammeter having a period of about $\frac{1}{2}$ sec. The steady reading of this instrument therefore gives the "integrated intensity" of the x-ray reflection. The behavior of this instrument is said to be reliable.

A densitometer has been devised by Brown and Birtley[35] which makes use of a "function transformer." As in the Dawton scan photometer, the light transmitted by the photograph is transformed into the corresponding density.

The Astbury α-ray method. Astbury devised a photographic method[9–12] for obtaining "integrated intensities" based upon novel principles. The original x-ray photograph is a "negative." Astbury obtained a positive print of it by the bichromated-gelatin process.† This positive print constitutes a record of the x-ray exposure, and can be stripped free from its backing. The transmission of the positive film for α particles from polonium, as measured by an electrometer, is proportional to the original x-ray exposure.

For faithful reproduction of the integrated intensity, the development and exposure of the gelatin film must be properly matched to the characteristics of the x-ray film and its development. Astbury's method does not seem to have come into very general use.

Photometry based on scattering. Brentano[14] devised a method of photometry based on the amount of light scattered by a photographic

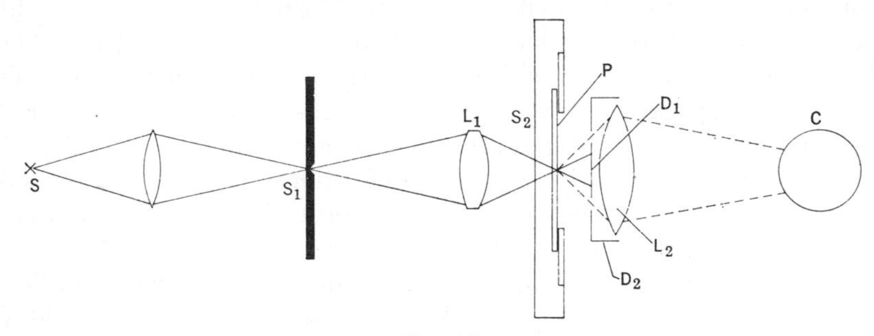

Fig. 10.
(After Brentano, Baxter, and Cotton.[16])

film. Brentano and co-workers have investigated this effect in some detail and designed a photometer to make use of it.[16, 24] The amount of light scattered by the silver particles is proportional to the number of particles. According to photographic theory, this number is also proportional to the number of quanta received by the film, and therefore proportional to the x-ray intensity, up to a limited value of the exposure where more than one quantum is received by the same grain. In this favorable region then, the light scattered by a given area of the film is proportional to the "integrated intensity" received on that area. Unfortunately there are complicating factors. The intensity of scattered light varies with the angle of scattering and with the wavelength used. Brentano's microphotometer[16] is shown diagrammatically in Fig. 10. The light from a source S is focused on an adjustable slit S_1, which is then focused by a lens L_1 onto the photographic film P. The direct light

† The carbon printing paper or "tissue" of this process contains a colloidal pigment dispersed in gelatin. The tissue is sensitized in a dilute solution of potassium dichromate and then dried. Exposure to light renders part of the gelatin insoluble. The rest of the gelatin is dissolved off in water, leaving a layer of gelatin having a thickness which increases with the exposure it has received.

beam is caught on a diaphragm D_1 and the light scattered by the emulsion is focused by a lens L_2 and received on a photocell C. Some typical results for various photographic materials are shown in Fig. 11. It will be observed that at low values of the exposure there is a linear relation of scattering to exposure, which deteriorates at high exposures. There is always a comparatively large amount of scatter for zero exposures, which represents both fog and scatter by imperfections, scratches,

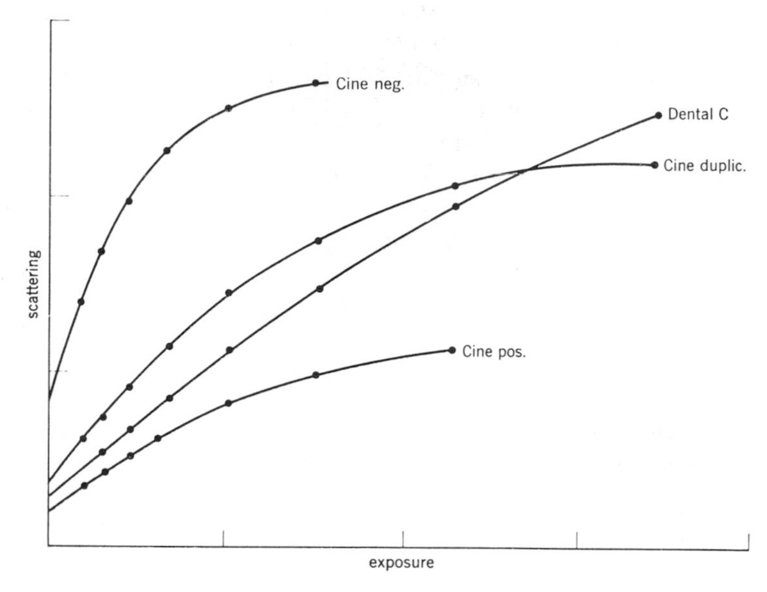

Fig. 11.
(After Brentano, Baxter, and Cotton.[16])

and other non-photographic features of the film. This is a disadvantage of the method. The shape of the curve depends markedly on photograpic film and development technique. Brentano's method has not come into general use.

Positive-film photometry. Dawton invented an inherently simple method of determining "integrated intensities."[19] This method has the same basis as that of making a perfect picture of an object by photography. If the entire photographic procedure is properly controlled, including making the original exposure, developing the negative, exposing the printing paper through the negative, and developing the positive print, the result should be a print each point of which reflects the same as the respective point of the original object. The corresponding process for x-ray photography involves taking the original x-ray "negative" on specific material, developing it in a specific way, printing this negative on a second appropriate film (which becomes a positive), and developing the positive in a controlled way. If the negative and positive processes are appropriately matched, the resulting positive print of the original single-crystal x-ray negative consists of a collection of more-or-less transparent spots on a dense black background, and the total transmission of each spot is directly proportional to the total energy ("integrated intensity") reaching the spot in the original x-ray exposure. Once the development procedure has been properly established, it is a very simple matter to determine the light transmission of each spot for an x-ray film. The method was published by Dawton[19] in 1938. It has been used in several laboratories in the United States and England,[32] each with its own particular developing procedure.

The general theory for making perfect prints is well known.[4] It can be outlined in the following way: When one makes a print, the exposure of the positive depends on the density of the negative and the amount of light, L_0, used as the printing source. Specifically,

$$\begin{aligned} E_P = L_N &= L_0 \frac{L_N}{L_{0,N}} \\ &= L_0 \, T_N, \end{aligned} \tag{21}$$

where subscripts N and P refer to negative and positive, respectively. If logarithms of both sides of (21) are taken, there results

$$\begin{aligned} \log E_P &= \log L_0 + \log T_N \\ &= K_1 \quad + \log \frac{1}{O_N} \\ &= K_1 \quad - D_N. \end{aligned} \tag{22}$$

Otherwise expressed, log exposure of the positive is minus the density of the negative, plus a constant depending on the amount of light used in making the print. The printing process can therefore be graphically demonstrated by combining the characteristic curves of negative and positive (in the form of D versus log E, not D versus E) as shown in Fig. 12, so that the log E axis of the positive is parallel to, but increasing in the opposite direction to, that of the density of the positive. A vertical shift of the curve of the positive material of Fig. 12 corresponds to a change of K_1 in (22) and therefore to the amount of printing light used; according to (22), more printing light shifts the curve downward, i.e., in the plus direction along the log E_P axis. For perfect reproduction it is required that the transmission of the positive be proportional to the exposure of the original negative, i.e.,

$$T_P = k_2 \, E_N. \tag{23}$$

Taking logarithms of both sides,

$$\begin{aligned}\log T_P &= \log\ k_2 + \log E_N \\ &= K_2 + \log E_N.\end{aligned} \tag{24}$$

Since, according to (11) and (10),

$$\log T = -\log O = -D, \tag{25}$$

(24) is equivalent to

$$-D_P = K_2 + \log E_N,$$

or

$$D_P = -K_2 - \log E_N. \tag{26}$$

This result simply requires the final density of the positive to be a linear function of the logarithm of the original exposure. This is illustrated in

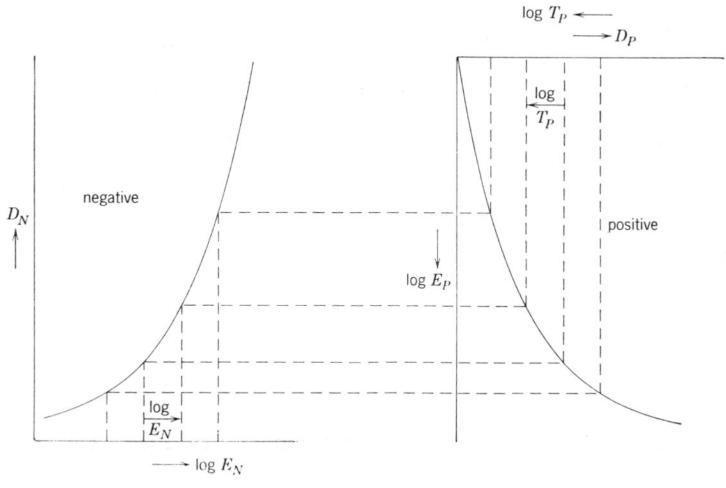

Fig. 12.

Fig. 12. An interpretation of this is that if equal intervals of increasing $\log E_N$ are taken, equal intervals of decreasing D_P result. This, of course, is equivalent to equal increments of increasing $\log T_P$, according to (25). The condition that the transmission of the positive should linearly reproduce the exposure in the original x-ray negative can be given a graphical significance in terms of Fig. 12: Perfect reproduction requires that the left and right curves of Fig. 12 be reflections of each other in a vertical line. Note that this can remain true if either curve is shifted horizontally. This shift is dependent on k_2 and K_2 of (24). The constant k_2 is simply the scale by which the transmission of the positive represents the exposure of the original x-ray negative. Increasing k_2 or K_2 represents increasing scale, and corresponds to a shift of the positive curve toward the left in Fig. 12.

If γ is the slope of the straight central part of the curve D versus $\log E$, a condition for perfect reproduction can be shown to require that $\gamma_P \gamma_N = 1$. That this is so is evident from the following considerations: Let NA and NB be two points on the straight section of the negative curve, and PA and PB be the two points on the straight section of the positive curve. Then the definition of γ gives

$$\gamma_N = \frac{D_{NA} - D_{NB}}{\log E_{NA} - \log E_{NB}}, \tag{27}$$

and

$$\gamma_P = \frac{D_{PA} - D_{PB}}{\log E_{PA} - \log E_{PB}}. \tag{28}$$

Equations (22) and (26) yield:

(22): $\log E_P = K_1 - D_N$ (general condition for printing),

(26): $\log E_N = -K_2 - D_P$ (condition for perfect reproduction).

Substituting for the $\log E$'s in (27) and (28) from (26) and (22) gives:

$$(27): \quad \gamma_N = \frac{D_{NA} - D_{NB}}{(-K_2 - D_{PA}) - (-K_2 - D_{PB})} = \frac{D_{NA} - D_{NB}}{D_{PB} - D_{PA}}, \tag{29}$$

$$(28): \quad \gamma_P = \frac{D_{PA} - D_{PB}}{(K_1 - D_{NA}) - (K_1 - D_{NB})} = \frac{D_{PA} - D_{PB}}{D_{NB} - D_{NA}}$$

$$= \frac{D_{PB} - D_{PA}}{D_{NA} - D_{NB}}. \tag{30}$$

It will be observed that the right members of (29) and (30) are reciprocals, so that

$$\gamma_P \gamma_N = 1. \tag{31}$$

The Dawton positive-print method has been in use in the author's laboratory since 1941. The matching of characteristic curves for negative and positive films was carefully investigated by a former assistant, John E. Tyler. The process was adjusted so that Kodak No-Screen x-ray film could be used for the original negative material, since this film was normally used for other purposes in the laboratory. (When this brand of film was taken off the market, it was found that with adjustment of the printing time the detailed directions gave good results with Ansco Non-Screen film.) The characteristic curves of many candidates for the positive-print film were investigated. After considerable study, Kodak Commercial Ortho, developed to a γ of 1, was chosen as the positive material. To match the curves of the two materials, the x-ray film was given a low-contrast development, and its characteristic curve was shifted by adding small amounts of KBr and KI to the developer. A specific procedure was found for Mo$K\alpha$ radiation by John E. Tyler, and subsequent adaptations of his technique to Cu$K\alpha$ and Co$K\alpha$ radiation were made by Joseph S. Lukesh and Gilbert E. Klein in the author's laboratory. The following schedule gives specific directions for making suitable positive prints.

I. Preparation of x-ray negative:
  Film: Eastman No-Screen x-ray film.
  Developer: Eastman D 76 to which have been added 2 cc. of 1% KI solution and 20 cc. of 2.5% KBr solution per gallon of developer.
  Developing temperature: 65° F.
  Developing time:
    For photographs taken with Mo$K\alpha$ radiation: 6 minutes, with constant agitation.
    For photographs taken with Cu$K\alpha$ or Co$K\alpha$ radiation, 8 minutes, with constant agitation.
  Fixing solution: Hypo for x-ray film.

II. Preparation of positive print:
  Film: Eastman Commercial Ortho.
  Printing exposure: 18 seconds with printing box described below.
  Developer: Eastman DK50
  Developing temperature: 65° F.
  Developing time: $3\frac{3}{4}$ minutes with constant swabbing.
  Fixing solution: Regular hypo (not x-ray hypo), to assure removal of red dye on back of film.

Since correct results depend on exact processing, these directions should be carefully followed, especially with respect to temperature and time of development, and with respect to printing exposure. Development is most conveniently carried out in small stainless-steel tanks (just large enough to hold the standard 5 x 7 in. film) immersed in a water bath maintained at 65° F. The positive requires constant swabbing to continuously replace the developer over the entire surface of the film, almost all of which area is developed to a dense black. The development of the negative does not require swabbing since only the inconsequential areas of the x-ray spots become black. Agitation is sufficient to freshen the developer in contact with these spots. In the case of the negative, the swabbing is conveniently done with the aid of a wad of cotton carried on a hoe-shaped holder. The cotton is trimmed straight along the cutting edge of the hoe, and this edge is used for swabbing the film while it is in the tank.

To assure accurate exposures in printing, this is best carried out in a specially constructed box. Such a printing box is shown in Fig. 13. The box is about 7 ft. long and arranged so that a standard 5 x 7 in. printing frame can be inserted at one end. The frame is fitted with a metal slide to protect the film until it is placed in the box and ready for the exposure. Direct shadows of dust particles and imperfections in the glass of the printing frame are prevented by diffusing the printing light by means of a

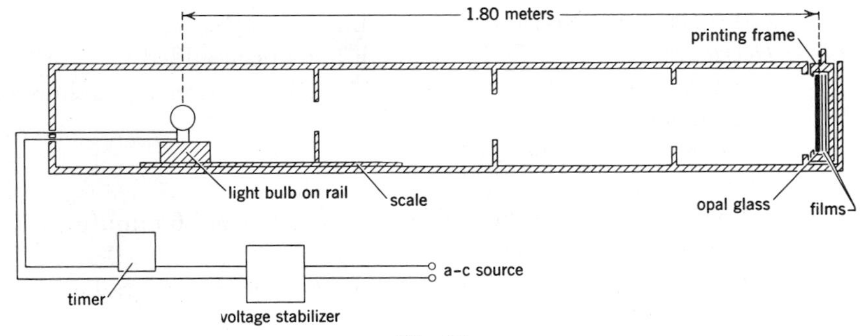

Fig. 13.

piece of opal glass. (The particular piece of opal glass in use transmits 21% of the light which falls on it.) The light source is a lamp which can be adjusted to appropriate distance from the film (to give a predetermined illumination there) by sliding its base on a track with graduated scale. This lamp is a clear, 6-watt, 115-volt Mazda bulb previously aged to uniform light output. It is fed with constant voltage from a commercial voltage stabilizer to assure invariable illumination on all films. With the light located 1.8 meters from the film and the opal glass (transmission 21%) interposed before the printing frame, the light falling on the x-ray negative is 1.13 lumens/meter^2. With this illumination an 18-second exposure is correct for the positive-print procedure described. If the lamp must be changed to another of different rating, its distance from the film can be adjusted by applying the inverse-square law.

Results of transmission of positive films versus original x-ray exposure are shown in Fig. 14 for various printing times. It will be observed that the curve labeled 120 is essentially a straight line tending almost exactly toward the origin.

The photometer for determining integrated intensities by the positive-film method is shown diagrammatically in Fig. 15. An overall view of the equipment is also shown, in Fig. 16. In Fig. 15, the light source is fed from a constant-voltage transformer to assure a uniform line voltage during a long series of readings. The intensity of the light is regulated

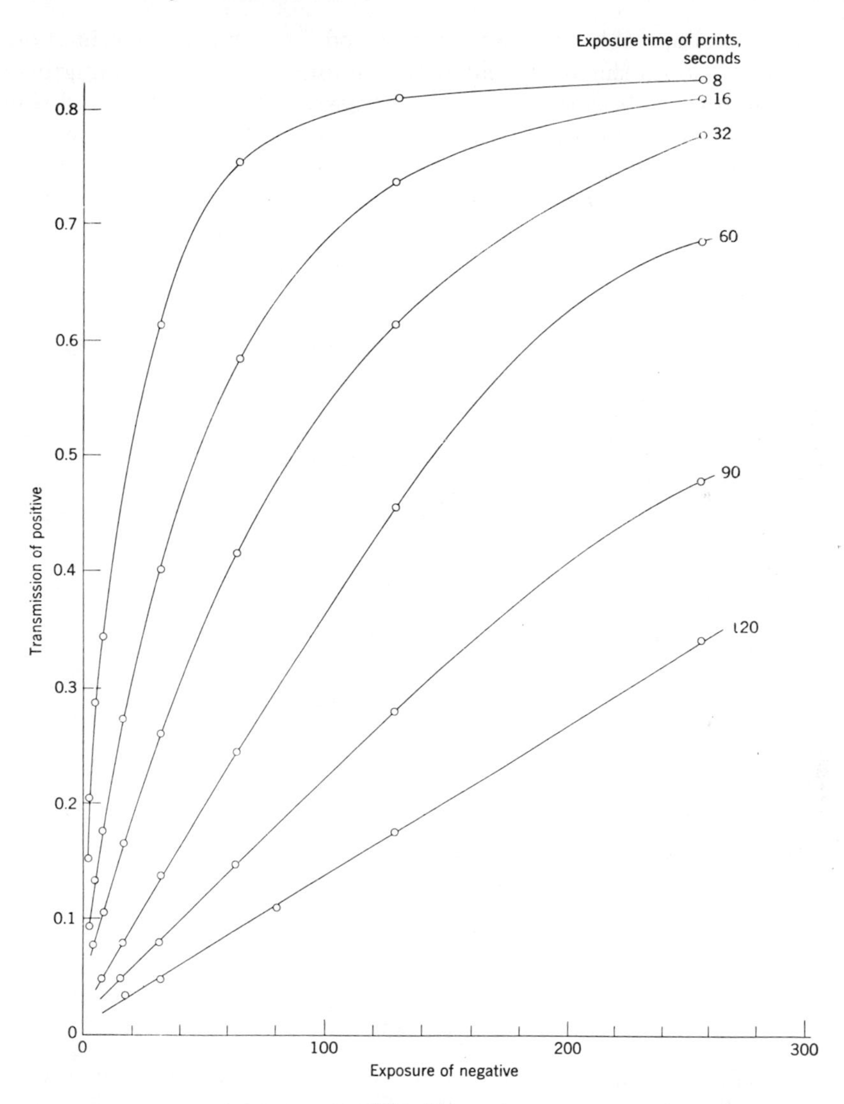

Fig. 14.
Variation of transmission of positive Dawton film with exposure time of negative, for various exposure times of positives. The 120-second exposure gives the required linear relation between transmission and x-ray exposure.

for a particular film by means of a small autotransformer such as a Variac. Light from the lamp bulb is first passed through a special glass filter to remove heat, and then concentrated on the film with a condenser. It is important that the concentrated beam which is passed by the aperture just ahead of the film is of uniform intensity in the cross-section. The aperture is removable and can be replaced by others. One is chosen which just frames the x-ray spot with a minimum of surrounding background in order to assure that the transmission of the background times the area illuminated is small compared with the total transmission of the spot being measured. Beyond the film is placed a photocell. A complete unit containing photocell with its amplification circuit is available

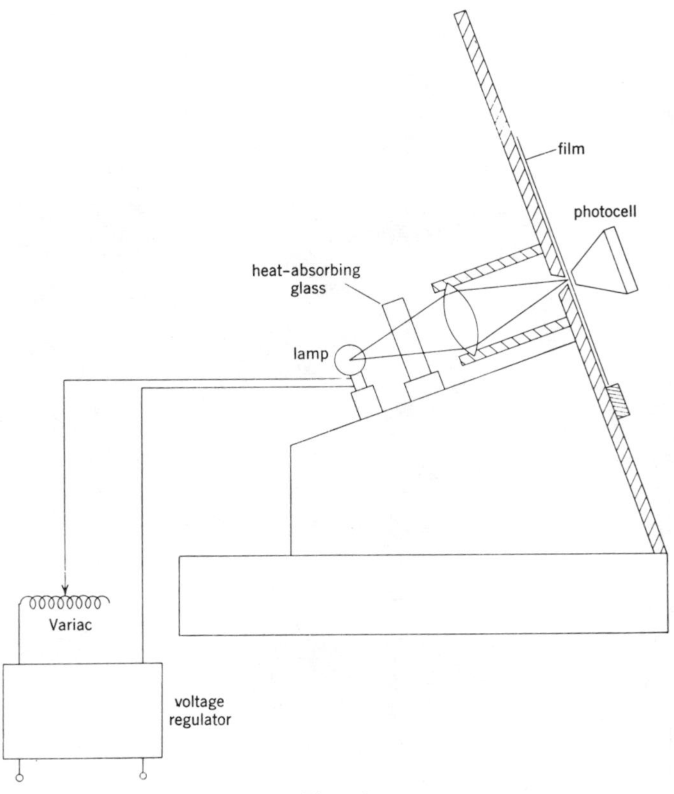

Fig. 15.

commercially.† An appropriate unit contains a galvanometer giving a linear reading of light received by the photocell (and hence of comparative x-ray energy received by the spot on the original x-ray negative), and provision for different scales of amplification.

When a reading is made, the light received consists of the light representing integrated intensity, plus light transmitted by background caused by fog, etc. This background reading must be determined from a transmission reading of nearby clear background, and subtracted from the reading given by the spot itself.

† Electronic exposure photometer, model 501-M, made by Photovolt Corporation, 95 Madison Avenue, New York 16, New York.

The plateau method. The plateau method seems to have been discovered about four times. Apparently its first discovery occurred during a general discussion which arose in the Conference on X-ray Diffraction held at the New York Academy of Sciences, January 10, 1941. In discussing the problem of determining integrated intensities, Debye suggested that if the powder camera were rotated slightly but uniformly about its

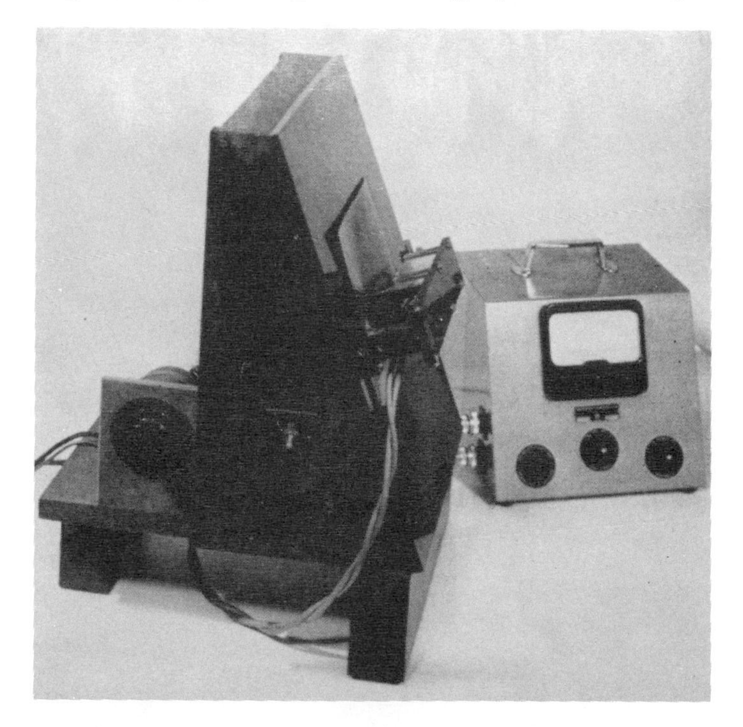

Fig. 16.

axis during exposure, the powder line would be smeared out. The center of the smeared-out line would have a plateau-like profile and the height of the plateau would be proportional to the integrated intensity. The method appears to have been developed somewhat by Brentano and Froula.[27] The principle was rediscovered by Wiebenga[31] and again by Cochran.[74]

Figures 17 and 18 illustrate the occurrence of the plateau. Figure 17 shows a profile of a powder line, i.e., intensity versus position. The "integrated intensity" is the area under the curve. This area under the curve can be approximated by measuring the sum of the heights of a number of equally spaced vertical samples, Fig. 17, and multiplying by the spacing of the samples. An alternative method of determining the sum of the samples is shown in Fig. 18*A*. If the intensity profile is laid down at equally spaced intervals to cover a sufficient width, a vertical line near the center samples the ordinates of the profiles. The sum of the ordinates of the several curves along this vertical line is the same as the sum of the ordinates in Fig. 17. This sum is constant in the neighborhood of the center, and produces a "plateau" in the curve which is the sum of the

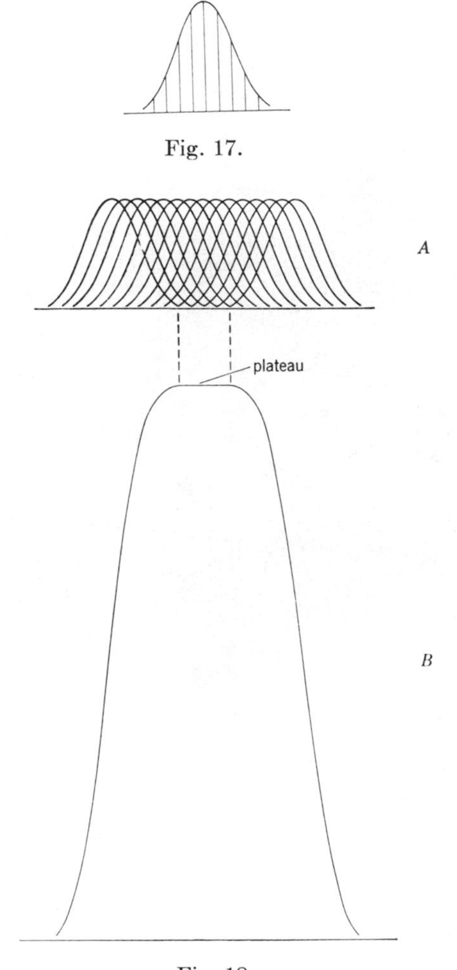

Fig. 17.

Fig. 18.

ordinates Fig. 18*B*. (Actually it is the integral $\int y\, dx$, which is the area under the curve of Fig. 17.) Thus the height of the "plateau" is a measure of the "integrated intensity."

There are two methods of smearing out a reflection so that it has a profile produced as diagrammatically illustrated in Fig. 18. Either the film can be moved uniformly during the exposure, or a convergent beam may be used so that the crystal continuously passes through the Bragg

A

B

Fig. 19.
Views of the Wiebenga integrating Weissenberg.
(Courtesy of Nonius, Delft, Holland.)

reflecting condition for various ray directions, after the method of Kratky.† This experimental condition obtains whenever the crystal can be placed quite near the source of x-radiation. For example, the plateau condition is usually observed on precession photographs taken with a Machlett x-ray tube. In Wiebenga's first device,[29] he exaggerated this normal convergence by using the broad-focus port of a line-focus x-ray tube. The convergent beam from this source was introduced by special collimator into a Weissenberg camera.

The use of a convergent beam requires a uniform emanation of x-rays from a substantial length of the focal spot of the x-ray tube. Since this condition may not occur experimentally, Wiebenga[31, 34, 41] used the alternative technique of rotating the film. He designed a Weissenberg apparatus in which the cylindrical camera undergoes a slight rotation about the axis of the cylinder and also a small auxiliary translation parallel to the axis of the cylinder. Each time the ordinary camera translation reaches one end of its motion, a pin actuates a ratchet and causes the ratchet to rotate $\frac{360}{14}$ degrees. This rotation is converted into the rotation and auxiliary translation of the camera, the magnitude of which is adjustable.

Figure 19 shows a view of Wiebenga's integrating mechanism. Note that the camera is mounted on wheels so that it can rotate on the carriage. The rotation is caused by the somewhat slanting lever, which pushes on the horizontal lug attached to the camera. The ratchet is about to be rotated as the lower-left pin engages the stationary pin. The spring and the ball-bearing disk at the left of the ratchet fix the position of the ratchet in one of 14 positions. Figure 20 shows how the position of the x-ray

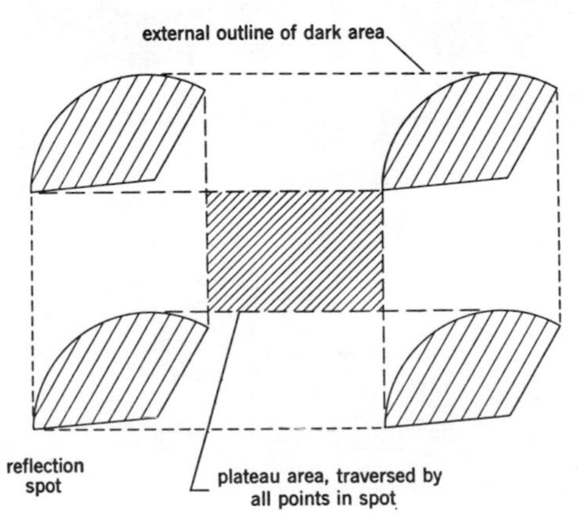

Fig. 20.

† M. J. Buerger. *X-ray crystallography.* (John Wiley and Sons, New York, 1942) 193, and literature pages 212–213.

spot migrates over a small area of the film with this arrangement. The plateau region is the shaded portion in the center of the migration area. Within this region the film has received contributions from all parts of the migrating spot.

Fig. 21.
(After Nordman, Patterson, Weldon, and Supper.[51])

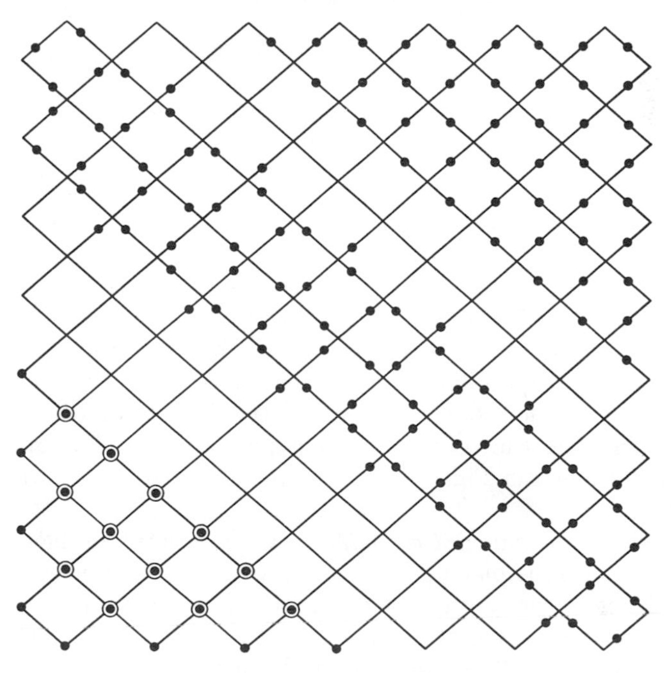

Fig. 22.
(After Nordman, Patterson, Weldon, and Supper.[51])

The Wiebenga camera requires an increased exposure, said to be about twice the normal exposure, under favorable conditions. The background receives an increased background exposure by the same factor.

Whittaker[38] and Stanley[49] have suggested using only the Wiebenga translation without the rotation motion. This draws each Weissenberg spot out into a streak. At the center of the streak the cross-section of the spot provides a true profile, comparable with the profile of a powder line.

Wiebenga's general scheme for integrating intensities by moving the film has been applied by Nordman et al.[51] to the precession camera. A photograph of the device is shown in Fig. 21, and the scanning pattern for a spot is shown in Fig. 22.

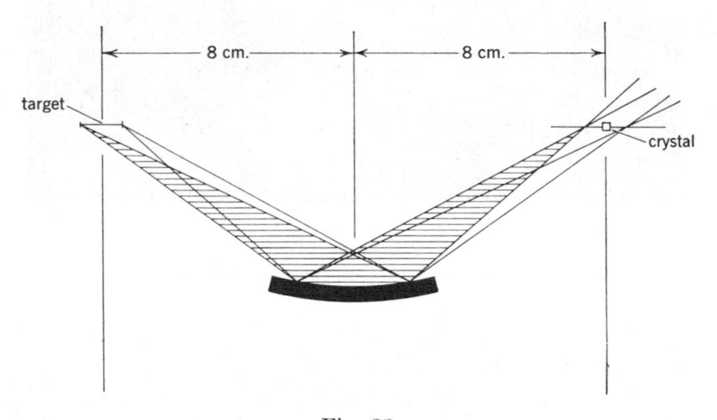

Fig. 23.
(After Azaroff,[58] modified.)

As pointed out earlier, a plateau is also achieved by using convergent radiation. Azaroff[58] combined this with the advantages of monochromated radiation by an interesting arrangement, shown diagrammatically in Fig. 23. A bent crystal focuses the target of the x-ray tube on the crystal. Every point of the crystal is therefore radiated from different directions from different points on the target. Provided that the radiation from these different points of the focal spot of the target is uniformly intense, the reflection has a plateau. A photograph of the apparatus is shown in Fig. 24, and the results it produces are shown in Fig. 25*A*.

Comparison with standard reflection. An important problem in intensity determination is placing the arbitrary scale of the intensities on an absolute basis. One method of doing this is to compare the reflection intensities of the sample of crystal under consideration with those of a standard material. This can be done both for powders and for single crystals. Since the powder method offers the possibility of avoiding correction for primary "extinction," a considerable literature exists for reflection comparison using the powder method. There are two methods of making the comparison for powders. In one method, called the mixed-powder method, a single powder sample is made by mixing the powder of the calibrating material with the powder of the material whose reflection intensities are desired. In a second method, called the substitution method, a special camera is usually used which is so designed that the two

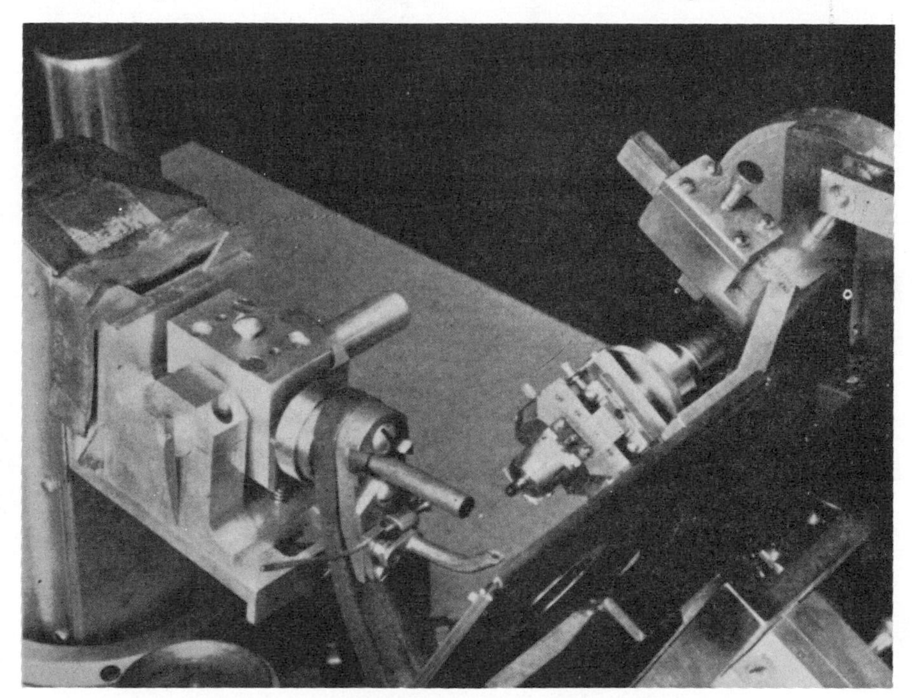

Fig. 24.
(After Azaroff.[58])

separate powder samples are alternately placed in the x-ray beam, the spectra of both being recorded on the same film so that the spectra to be compared receive identical development.

Both of these methods are subject to complications. The mixed-powder method is useful when the absorptions of the two powders are negligible, or when they are nearly the same. When either of these conditions obtains, the considerable difficulties due to differential absorption are avoided. The substitution method avoids the differential-absorption difficulty, although it does require a detailed knowledge of the absorptions of the two powders. Both methods suffer from the inherent disadvantages of the powder method and are useless if either of the powders is characterized by preferred orientation.

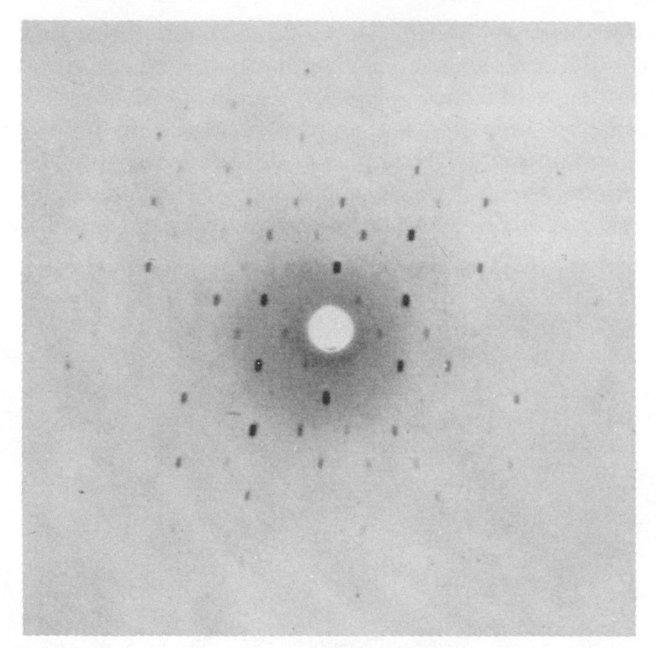

A

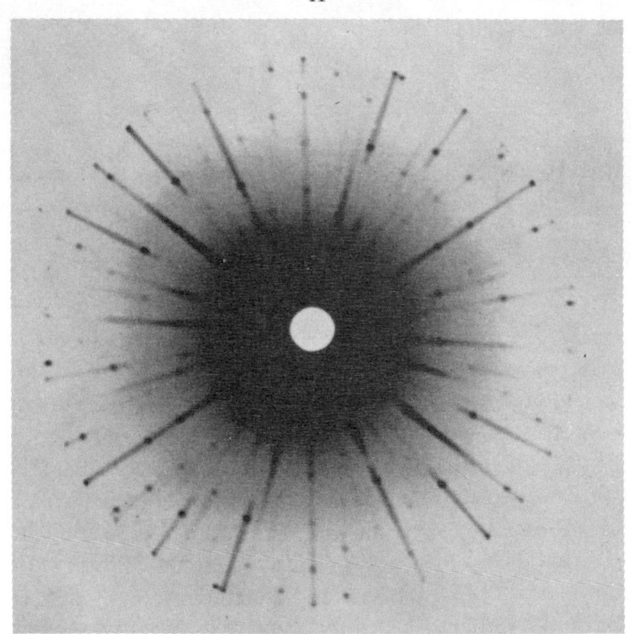

B

Fig. 25.
A. Precession photograph made with apparatus of Figs. 23 and 24.
B. Comparable photograph made in the usual way.
(After Azaroff.[58])

Single-crystal comparison methods. The substitution method was first adapted to single crystals by Robertson,[17] who designed a special photographic spectrometer for the purpose. In this apparatus the crystal whose reflections are to be calibrated, and the calibrating crystal, are alternately placed in the x-ray beam and rotated. The reflections are resolved in Weissenberg fashion, but on a flat film, and the reflections from both crystals are received on the same film.

Wooster and Martin[21] designed a conventional Weissenberg-type apparatus in which two crystals could be substituted in the x-ray beam. One goniometer head with its crystal enters each open end of the cylindrical camera, and a mechanism introduces one after the other in succession into the x-ray beam. The records of both crystals appear on the upper half of the film, but the lower half receives only the record of the crystal to be calibrated. It should be observed that the upper and lower halves of the film are different, in general, for the upper levels of crystals for which the rotation axis is not an evenfold axis, unless a full 360° rotation of the crystal is used.†

Stadler[33, 37] has suggested a calibration method related to the two-crystal substitution method. When a standard Weissenberg apparatus is set at equi-inclination for an upper level, the zero-level cone is at anti-equi-inclination. If a slit is provided for each of these cones in the layer-line screen, both can be recorded on the same film.

All such apparatus, which is designed to effect a substitution of crystals so that the records of two crystals are placed on the same film during the same run, is justified only if the output of the x-ray tube is erratic. A much simpler way of attaining the same result is to assure constant x-ray output, and record the intensities from the two crystals at different times. Ordinary commercial sealed-off tubes fed with a voltage-regulated power supply obviate the necessity of complicated two-crystal devices, and are necessary in any case when intensity data of all levels and different rotation axes are required on the same scale. The same scale is essential for ordinary three-dimensional analysis, for example.

Robertson has suggested specific crystals for standards. In his investigations he found a "mosaic"-type diamond suitable. By spectrometer methods a particular 111 reflection from this diamond had been found to have an integrated intensity of 18.6×10^{-7} for unpolarized Cu$K\alpha$ radiation. He also recommends as good standards anthracene, the orthorhombic modification of 1,2,5,6-dibenzanthracene, and pthalocyanines, whereas he found the sugars and similar compounds to be unsatisfactory. Wooster[30] adds rock salt and urea as suitable standards.

† M. J. Buerger. *X-ray crystallography.* (John Wiley and Sons, New York, 1942) Chapter 22.

Spectrometer methods

Early apparatus. In the first crystal-structure investigations undertaken by the Braggs[60–62] the intensities were measured with the aid of the ionization spectrometer. These crystals had relatively simple structures which were determined by using only a few reflections. As more complicated structures were investigated, it was necessary to measure the intensities of many reflections. Because of the supervision required, the spectrometer method became tedious. Wooster[64, 71] attempted to remove this disadvantage by designing instrumentation with the aid of which the crystal was automatically set at the several reflecting positions, the spectrometer chamber was set so as to receive the reflections, and the ionization current in the chamber was automatically recorded on 35-mm. photographic paper.

But the measurement of intensities by the ionization spectrometer had already declined and given way to photographic methods. A key reason for this is that the leakage of charge from the electrometer sets a lower limit on the diffraction intensity which can be measured. This required the use of crystals having faces of considerable areal extent. Most crystal species cannot be obtained readily in this form.

Not only is the photographic method suited to small crystals, but it also has other important advantages. It provides a permanent record automatically. Accordingly it requires no supervision from the investigator while the record is made and therefore removes him from the danger of accidental x-ray exposure. It also requires no specialized diffraction apparatus beyond that required for the ordinary unit-cell and space-group investigation. It is natural, therefore, that crystal structures have been determined almost exclusively from intensities derived by photographic methods until very recently.

Quantum counters. Geiger-Müller counters have been used for some time in the measurement of x-ray intensities[66, 67] but their use did not become popular until a reliable commercial unit became available.[68, 69] Such apparatus is a direct descendent of equipment devised in the United States during the last war to control the cutting of quartz-crystal oscillator plates. Counter spectrometers are characterized by high sensitivity, and are therefore inherently suited to measuring the intensities of reflections from small crystals. In the determination of intensities by photographic methods, about 10^4 to 10^5 quanta are required to produce a minimum detectable blackening of the photographic film. Geiger counters can detect this level of x-ray energy with an accuracy of better than $\frac{1}{10}\%$ by taking sufficiently long counts. Early forms of Geiger-Müller tubes were able to detect about 4 to 15% of the quanta entering the counter tube, but modern tubes can detect better than 50%.

The Geiger-Müller counter tube is a comparatively simple affair, Fig. 26. It consists essentially of a cylindrical metal cathode within which is located a coaxial metal wire anode, both immersed in a controlled gas atmosphere. The space within the cathode cylinder is accessible to radiation which arrives through a window at the base of the cylinder. This window is composed of material that is transparent to the particular radiation to be detected. The immediate electrical circuit is diagrammatically shown in Fig. 27.

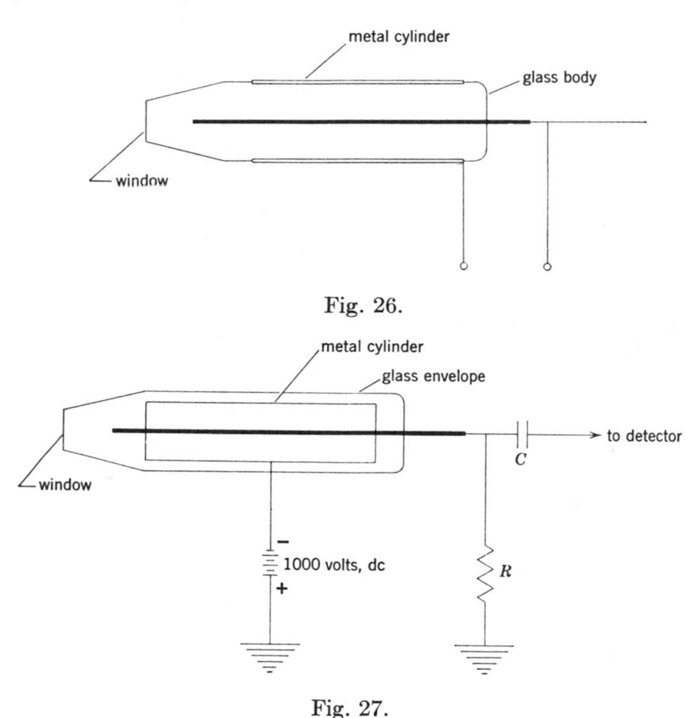

Fig. 26.

Fig. 27.

A beam of radiation is a flow of discrete quanta. When a quantum of radiation having appropriate energy enters the window of the tube, it ionizes the gas within. The formation of a single ion pair triggers a discharge which develops a voltage pulse at C, Fig. 27. This constitutes a single "count." This impulse is normally smoothed and amplified in two stages before being passed on to recording apparatus. The relation between counting rate and applied voltage is shown in Fig. 28. It will be observed that for a given x-ray intensity there is a "plateau" region over which the counting rate is substantially independent of changes of voltage.

Geiger-counter tubes have been much improved recently. Originally the ionizing radiation was permitted to fall upon the metal surfaces which released photoelectrons. An improvement in efficiency resulted from allowing the photoelectric effect to occur in the gas of the counter. Recently the form of the tube has been further improved by making the windows of mica and sealing this directly to the cathode cylinder. This construction reduces the dead space directly behind the window. Geiger-counter tubes using mica windows can count about 55% of the quanta reaching the tube.

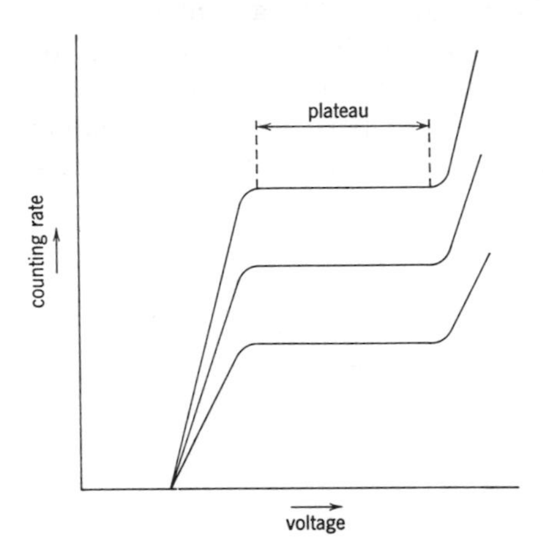

Fig. 28.
Responses of a Geiger counter. Each curve represents a different intensity of radiation.
(After Friedman.[68])

The gas within the tube is commonly one of the noble gases. Their responses to x-radiation of various wavelengths is shown in Fig. 29. In order to quench the discharge evoked by receipt of a quantum, an organic vapor is customarily mixed with the gas of the tube. Formerly alcohol was used for this purpose, but tubes with longer lives are now made by incorporating methylene bromide as quenching agent.

In order for individual quanta to be distinguished by separate counts, their arrivals must be separated by a certain time interval. This is known as the *dead time* of the counter and depends on the time required to quench the discharge; dead time is of the order of 10^{-4} sec. A cor-

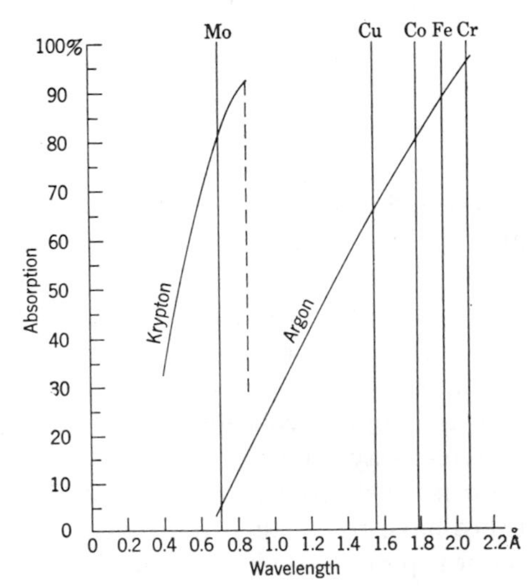

Fig. 29.

responding measure applied to the entire circuit is called *resolving time;* this is usually set by some unit of the circuit and sometimes can be varied. In American commercial equipment all units of the circuit are faster than the counts so that the resolving time of the circuit is actually the dead time of the counter.

The intensity of a beam of monochromatic radiation is proportional to its flow of quanta. The arrival of each quantum is independent of the arrival of others; i.e., the arrival of quanta is a random process. When the intensity of the beam is small, the quantum current is small, and quanta generally arrive at rather widely spaced intervals. The arrival of each quantum then ideally evokes a count, so the counting rate is proportional to the intensity of the beam. As the intensity increases,

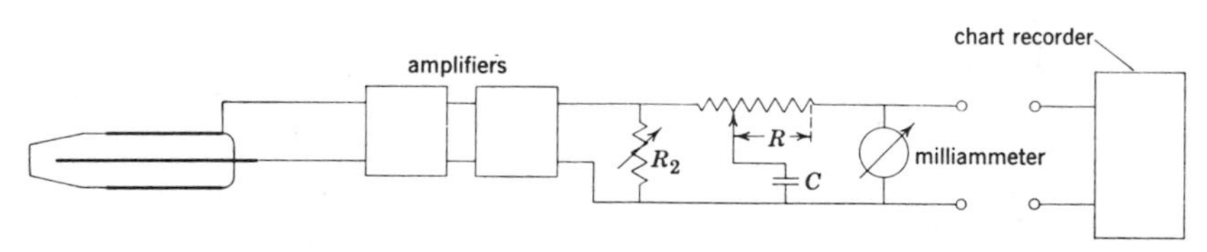

Fig. 30.
(After Bleeksma et al,[69] modified.)

the quantum current increases, and the chance increases that pairs of quanta arrive at intervals shorter than the dead time of the counter. Whenever this occurs the second quantum of the pair is not counted. On increasing intensity a state is eventually reached when the proportion of uncounted quanta begins to appreciably affect the linearity between intensity and counting rate. This occurs in the neighborhood of 100 to 150 counts/sec. (Bleeksma et al.[69] give 600 counts/sec.) If the nonlinearity of the relation is taken into account, it can be said that present-day Geiger-Müller tubes are capable of responding to 3000 counts/sec.

The output of the Geiger tube with its amplifying circuit may be introduced into a circuit which provides the mean value of the rate of arrival of counter pulses. Such a circuit is shown diagrammatically in Fig. 30. The resistance R_2 is simply an adjustable leak which permits the amplitude of the pulse reaching condenser C to be controlled. The voltage rise in C, within limits, is proportional to the pulses per second. This voltage is measured by the discharge current it causes through the milliammeter. Alternatively, the discharge can be made to actuate a graphical recorder. The resistance R is adjustable to the scanning speed of the recorder. The recording apparatus appears in the photograph shown in Fig. 32.

Unfortunately the linearity of the relation between the recorder amplitude and counting rate is limited to low counting rates.[70] For accurate intensity work the number of pulses arriving in a given time interval must be actually counted. Mechanical counters are not fast enough to count directly the impulses arriving from x-ray beams of useful intensities, so an electric device is customarily employed for reducing the number of counts. The unit electrical device commonly used for this purpose is a *scale-of-two*, or *flip-flop*, circuit.[63, 65] This circuit has the characteristic that when it is given an input of $2n$ pulses, it presents an output of n pulses. It therefore functions as a device for scaling down the number of pulses by a factor of 2. Scale-of-two units can be connected in sequence. Each unit passes half of the pulses it receives on to the next

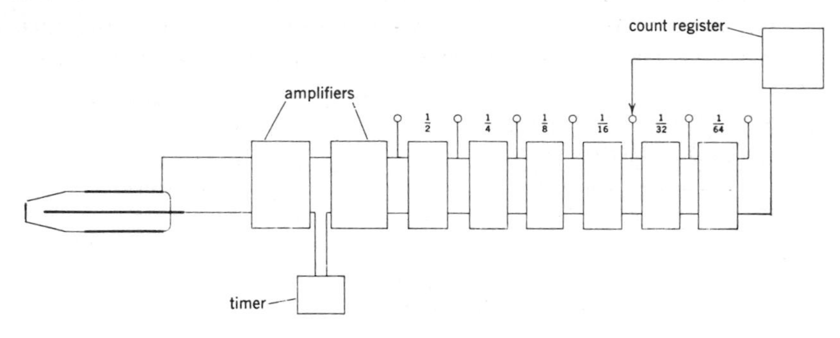

Fig. 31.

unit. The last unit in the sequence reduces the number of pulses it receives by $\frac{1}{2}$; the last two units together reduce the number of pulses they receive by $\frac{1}{2} \cdot \frac{1}{2} = \frac{1}{4}$; and, in general, p units in sequence reduce the number of pulses they receive by $1/2^p$. Commonly six units in sequence are used, the combination being known as a *scale-of-64*. When 64 pulses are received by the first unit of the sequence, the successive units reduce the pulses in six stages as follows: $64 \rightarrow 32 \rightarrow 16 \rightarrow 8 \rightarrow 4 \rightarrow 2 \rightarrow 1$. In this general way the high rate of arrival of quanta can be scaled down so that a mechanical counter can be actuated by the pulses. A counter circuit is shown in Fig. 31.

Other quantum counters. In recent years two other types of counters have been used for x-ray detection:[87–89] the *proportional counter* and the *scintillation counter*. These have certain advantages over the Geiger counter. In the first place the linear response of the Geiger counter is limited for large intensities by the fact that its resolving time (or dead time) is rather high, namely 270 μsec. Under the same conditions the proportional counter and scintillation counter have a dead time of about 0.25 μsec. This enables these counters to handle much larger intensities within their linearity range. They have the further advantage that they can be made to distinguish energies by pulse-height discrimination, which makes it possible to count x-rays of certain wavelengths and disregard other wavelengths. This presents a kind of monochromatation during detection. The discrimination is not sufficiently good, however, to separate $K\alpha$ and $K\beta$ radiations, so that a filter should be used even with these counters. Its main advantages are the elimination of harmonics and fluorescent background.

Both proportional counter and scintillation counter have the disadvantage that they require additional accessories and circuitry, which

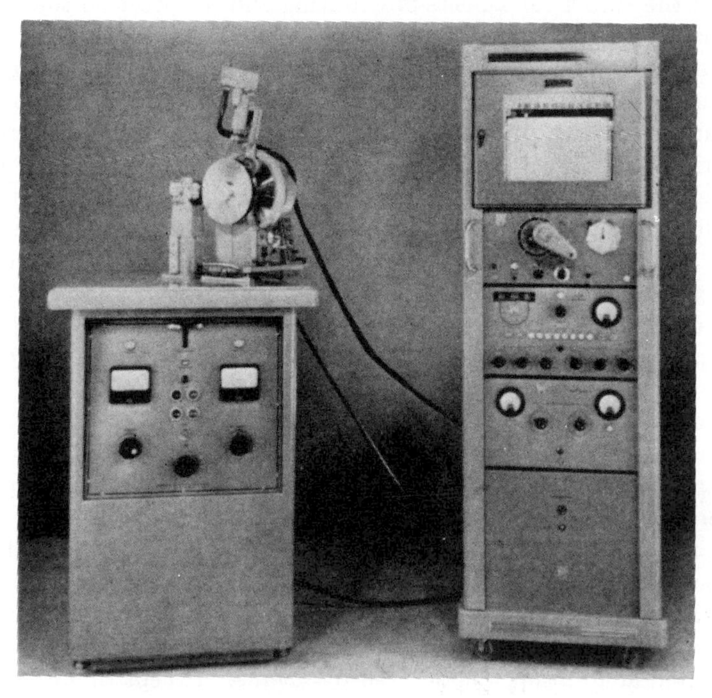

Fig. 32.
(Courtesy of Philips Electronics.)

not only use valuable space in the neighborhood of the counter, but also constitute additional complexity which must be understood and, when a failure occurs, serviced.

Counter apparatus for powders. Apparatus is commercially available for applying counter methods in the determination of the intensities of diffraction from powders. An example of this apparatus is shown in Fig. 32. Commercial apparatus is not designed to use the cylindrical powder specimen ordinarily used in Debye-Scherrer-type powder cameras. Instead a large flat powder specimen is used so that advantage may be taken of the increase of intensity of the reflected beam due to "focusing." A Soller slit system permits the x-ray beam from a broad source to have sufficient divergence to cover the specimen, yet limits the divergence in the direction at right angles to this.

Application of counters to measurement of intensities from single crystals. Preliminary investigations of the use of counters in the determination of the intensities of reflections from single crystals were made by Lonsdale[70] and by Wooster, Ramachandran, and Lang.[72] A study of the statistical aspects of counting and its effect on the precision of intensity determination has been given by Cochran.[74]

In the early applications[70–72, 74, 77] (and some later ones) of counters in intensity measurement from single crystals, only the reflections from the planes in a zone were considered. The experimental equipment for measuring reflections in a zone is inherently simple, and is usually arranged by attaching a single crystal to a spectrometer or diffractometer. Such simple arrangements, however, do not permit measuring the intensities of the full set of spectra which are required for three-dimensional analyses. For such measurement, more complicated apparatus, such as described in the next two sections, is required.

Adaptation of counters to Weissenberg apparatus. In order to include reflections not in the zero level of the reciprocal lattice, Clifton, Filler, and McLachlan[75] attached an arm bearing a counter tube to a Weissenberg apparatus. The arm pivots about the rotation axis of the crystal. The upper-level reflections are brought into the plane of the counter by using the flat-cone technique.† This technique has the disadvantage of having a central blind area on each upper level of the reciprocal lattice. At best, several mountings of the crystal for rotations about different axes are required for a full exploration of the reciprocal lattice.

A related adaptation was devised by Evans,[76] except that the counter tube was further adjustable so as to receive reflections making an angle $\bar{\nu}$ with the axis of crystal rotation. This permitted using the normal-beam, flat-cone, or equi-inclination techniques. Evans gives an excellent account of the precautions which must be observed in aligning the instrument in the x-ray beam.

Single-crystal apparatus specifically designed for counters.[91] When the intensities from single crystals are to be measured by counters it is desirable to use apparatus especially designed for the purpose rather than to attempt to adapt a counter to apparatus designed for other purposes. The key to appropriate design of apparatus suitable to measuring the intensities of all *hkl*'s is that the Laue cones are coaxial with the

† M. J. Buerger. *X-ray crystallography.* (John Wiley and Sons, New York, 1942) Chapter 15, especially pages 301–304.

crystal-rotation axis.[†] Diffraction directions are generators of these cones, Fig. 33. To collect diffracted radiation the counter should therefore be pointed along the generator of the appropriate cone. The counter should therefore always be pointed at the crystal, but able to rotate about the axis of rotation of the crystal. The crystal-rotation axis should be capable of being varied with respect to the direct x-ray beam by a measured amount $\bar{\mu}$, and the counter should be capable of being varied in orientation with respect to the crystal-rotation axis by

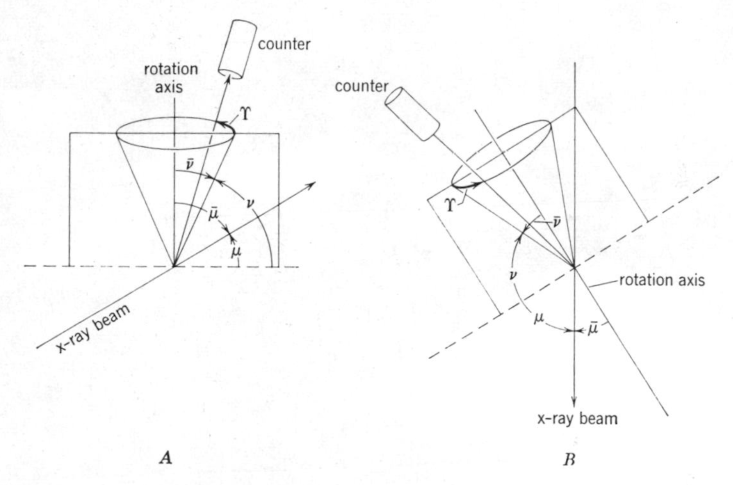

Fig. 33.
A. Scheme for setting counter, rotation axis vertical.
B. Scheme for counter apparatus shown in Fig. 34, viewed from above.

measured angle $\bar{\nu}$. Apparatus based upon such a general design can be used to collect intensity data by any of the inclined-beam techniques, including the equi-inclination technique.[§] The equi-inclination technique has many advantages, including absence of blind spot, great range of reflections which can be collected with one crystal mounting, and availability of pertinent data such as Lp corrections.

An instrument based upon this theory of collecting intensity data is shown in Fig. 34. The upper part of the instrument can be set so that the crystal-rotation axis makes an angle $\bar{\mu}$ with the x-ray beam. The crystal, mounted on a standard goniometer head and previously adjusted on an optical (or other) goniometer, can be rotated on its axis through any angle ω, the coordinate being measured on a dial. Any reflection belonging to a particular cone, for which $\bar{\mu}$ and $\bar{\nu}$ are set, can be brought to

[†] M. J. Buerger. *X-ray crystallography.* (John Wiley and Sons, New York, 1942) page 33, Fig. 15.

[§] M. J. Buerger. *X-ray crystallography.* (John Wiley and Sons, New York, 1942) Chapter 14, pages 252–295.

reflecting position by turning the crystal-rotation dial through an angle ϕ, which is a coordinate characteristic of a particular reflecting plane. When the reflection is picked up, the crystal is then oscillated about its rotation axis, through a small angle $\Delta\omega$. This is accomplished by a small reversible synchronous motor which drives the crystal through a gear

A

Fig. 34.
A, *B*. Two views of single-crystal counter diffractometer.
C. Detail in neighborhood of crystal, showing adjustable direct-beam stop.

train in which the gears can be interchanged for any desired crystal-rotation speed. The motor is reversed by a hand switch.

This instrument is designed to be set and operated by hand, or to be set and operated automatically by a suitable guide mechanism. To accomplish this dual purpose, the spindles which vary the ν, Υ, and ϕ settings are arranged so that there may be plugged in each of them either a hand crank or a motor, operated and controlled by the guide mechanism.

Crystal and counter settings.[101] The same reflection parameters occur when the reflection is recorded by Weissenberg photography† or by

† M. J. Buerger, *X-ray crystallography.* (John Wiley and Sons, New York, 1942) especially Chapters 13, 14, and 15.

a quantum counter. The angles μ and ν are related to each other in a manner depending on the technique employed (for example, for the equi-inclination technique $\mu = -\nu$). For a rational rotation axis, many reflections (all those on the same level of the reciprocal lattice) require

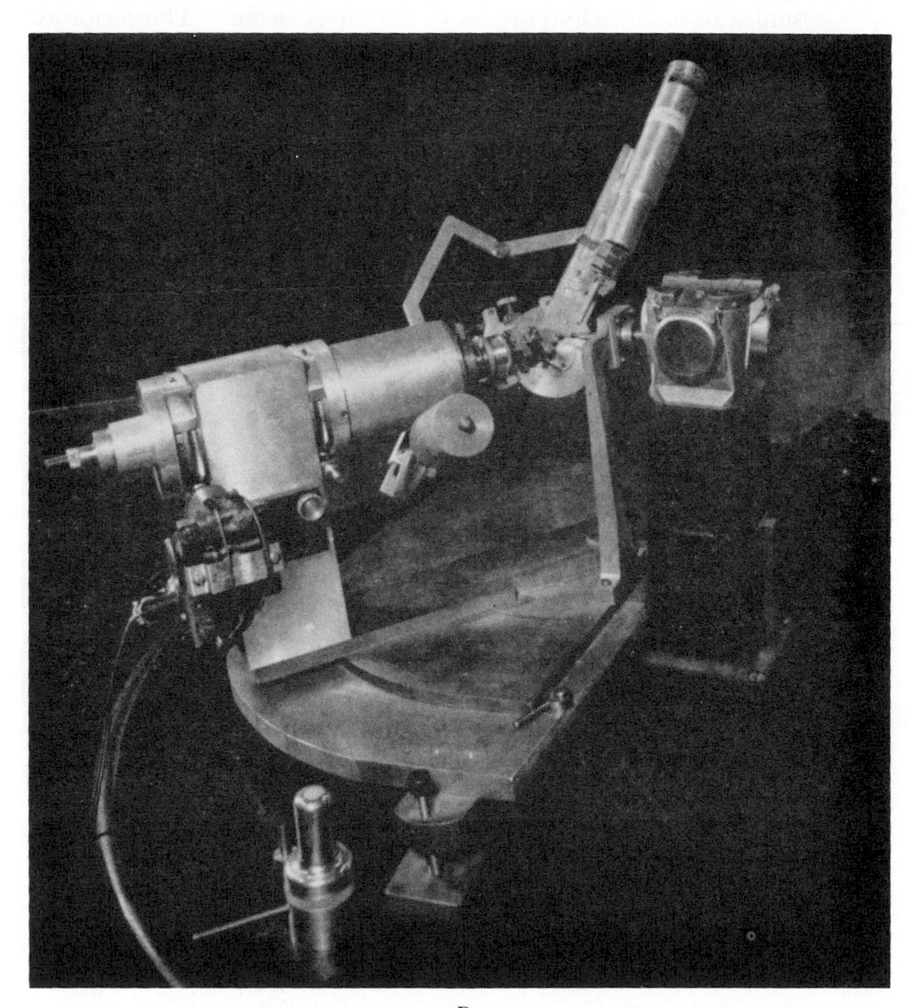

B

Fig. 34. (*Continued*)

the same angle ν. For all points in the same level of the reciprocal lattice, therefore, ν is fixed, and a reflection is caused to occur by turning the crystal through angle ϕ; the reflection occurs at an azimuth angle Υ (measured about the rotation axis). On a Weissenberg photograph, the resulting reflection occurs at coordinates ϕ, Υ on the Weissenberg film. When using counter recording, the crystal must be set at angle ϕ and the counter must be set at angle Υ.

The geometry is illustrated in Fig. 35. The reciprocal lattice, whose origin is at O, is referred to reciprocal-cell axes a^*, b^*, and c^*. The axis a^* is assumed to be initially parallel to the direct beam. The reciprocal lattice is assumed to rotate about the normal to the $a^*\ b^*$ plane, that is,

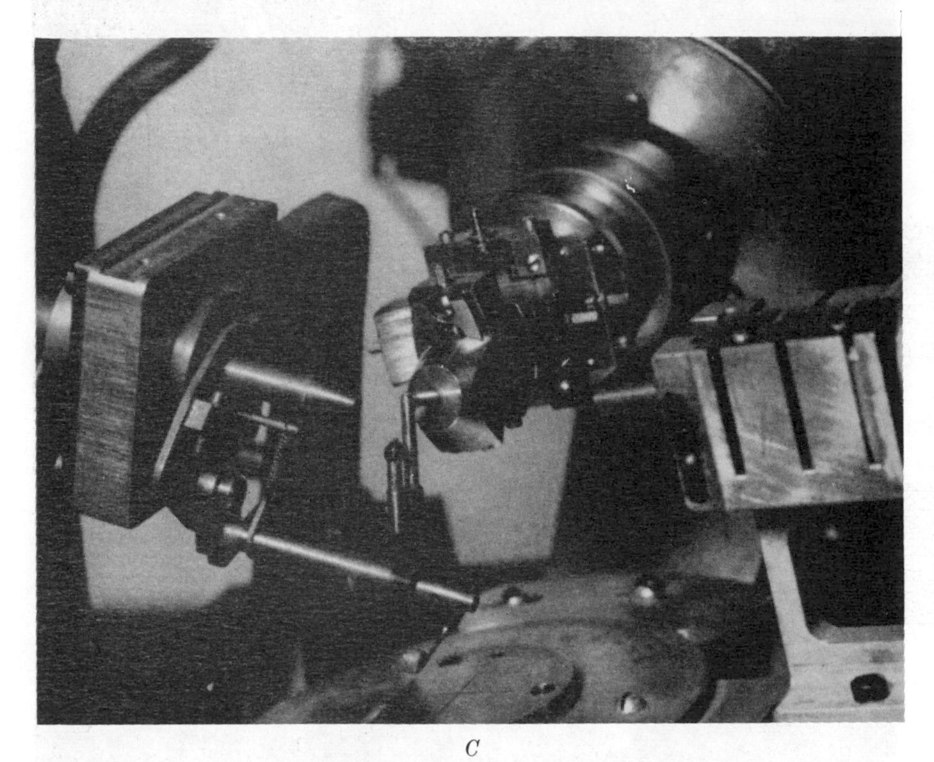

C

Fig. 34. (*Continued*)

about the c axis of the crystal. The indices of an arbitrary point P are hkl. The point is therefore at the end of a vector ξ_{hkl} on the lth level of the reciprocal lattice. A rotation through ϕ brings P to reflecting condition, and the reflection is directed at an azimuth angle Υ. In these general terms,

$$\sin\frac{\Upsilon}{2} = \frac{\frac{1}{2}\xi}{R}, \tag{32}$$

so

$$\Upsilon = 2\sin^{-1}\frac{\xi}{2R} \tag{33}$$

For equi-inclination,

$$R = \cos \nu \tag{34}$$

$$= \cos \sin^{-1} \frac{\zeta}{2}. \tag{35}$$

Thus, (33) and (35) provide the setting angle for the Geiger counter. The value of ξ is discussed subsequently.

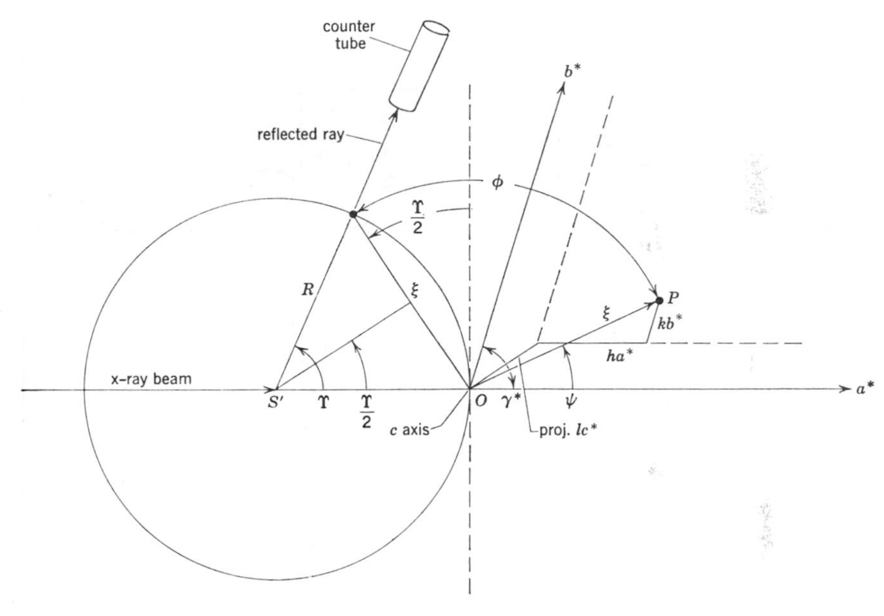

Fig. 35.

To produce this reflection, the crystal must be rotated about (say) its c axis through an angle ϕ. It can be seen from Fig. 35 that for equi-inclination,

$$\phi = \frac{\Upsilon}{2} + 90° - \psi \tag{36}$$

The values of ξ and ψ depend upon both hkl and the nature of the coordinate system $a^*\ b^*\ c^*\ \alpha^*\ \beta^*\ \gamma^*$. The vector ξ can be resolved into proj. $(l\mathbf{c}^*)+h\mathbf{a}^*+k\mathbf{b}^*$. The values of ξ and ψ for rotation about various axes of the several crystal systems are given in Table 5.

While the more general formulae in Table 5 appear cumbersome, the computations can be readily programmed for a high-speed digital computer. These computations involve the cell dimensions, which must be accurately known. To the same program which computes Υ and ϕ for each reflection there may be added a computation for the Lorentz-factor

Table 5
Values of ξ and ψ for various crystal systems

Crystal system	$\lvert\xi\rvert$	ψ (a^* initially $\parallel$ beam if rotation axis is c, c^* " " " " " " b.)
Triclinic rotation axis: c	$\left\{(h^2 a^{*2} + k^2 b^{*2}) + 2(hka^* b^* \cos\gamma^* + hla^* c^* \cos\beta^* + klb^* c^* \cos\alpha^*) + \frac{l^2 c^{*2}}{\sin^2\gamma^*}(\cos^2\alpha^* - 2\cos\alpha^*\cos\beta^*\cos\gamma^* + \cos^2\beta^*)\right\}^{1/2}$	$\tan^{-1}\left\{\frac{kb^*\sin\gamma^* + lc^*\left(\frac{\cos\alpha^* - \cos\beta^*\cos\gamma^*}{\sin\gamma^*}\right)}{ha^* + kb^*\cos\gamma^* + lc^*\cos\beta^*}\right\}$
Monoclinic (*1st setting*) rotation axis: c	$(h^2 a^{*2} + k^2 b^{*2} + 2hka^* b^* \cos\gamma^*)^{1/2}$	$\tan^{-1}\left(\frac{kb^*\sin\gamma^*}{ha^* + kb^*\cos\gamma^*}\right)$
rotation axis: b	$(h^2 a^{*2} + k^2 b^{*2}\cos^2\gamma^* + l^2 c^{*2} + 2hka^* b^* \cos\gamma^*)^{1/2}$	$\tan^{-1}\left(\frac{ha^* + kb^*\cos\gamma^*}{lc^*}\right)$
Hexagonal rotation axis: c	$(h^2 + k^2 + hk)^{1/2} a^*$	$\tan^{-1}\left(\frac{\sqrt{3}\, kb^*}{2ha^* + kb^*}\right)$
rotation axis: b	$(h^2 a^{*2} + \frac{1}{4}k^2 b^{*2} + l^2 c^{*2} + hka^* b^*)^{1/2}$	$\tan^{-1}\left(\frac{ha^* + \frac{1}{2}kb^*}{lc^*}\right)$
Orthorhombic rotation axis: c	$(h^2 a^{*2} + k^2 b^{*2})^{1/2}$	$\tan^{-1}\left(\frac{kb^*}{ha^*}\right)$
Tetragonal rotation axis: c	$(h^2 + k^2)^{1/2} a^*$	$\tan^{-1}\left(\frac{k}{h}\right)$
rotation axis: b	$(h^2 a^{*2} + l^2 c^{*2})^{1/2}$	$\tan^{-1}\left(\frac{ha^*}{lc^*}\right)$
Isometric rotation axis: a	$(h^2 + k^2)^{1/2} a^*$	$\tan^{-1}\left(\frac{k}{h}\right)$

correction, which is a simple function of Υ, and for the polarization-factor correction, which is a function of σ_{hkl}.

Evans[76] has described a simple graphical method of finding Υ and ϕ from a plot of the reciprocal lattice. While this is a relatively convenient way of finding the required settings when high-speed digital computing is not available, experience in the author's laboratory indicates that the results are not sufficiently precise for those reciprocal lattice points whose motion vectors make small angles with the tangent of the circle, and the accuracy of the resulting angles is further reduced if either the reciprocal-cell dimensions or the radius of the circle are inaccurately reproduced.

Mathieson[98] has described a mechanism which amounts to an analogue computer for Υ and ϕ. It is based upon a mechanical linkage which essentially links Υ and ϕ of Fig. 35 for each non-central line of reciprocal-lattice points. With the aid of this mechanism the coupling of Υ and ϕ causes the crystal and counter to continuously follow the reflections occurring along a Weissenberg reciprocal-lattice line chart.†

The "Eulerian cradle" apparatus. Furnas and Harker[84] devised a different type of apparatus for measuring the diffraction intensities

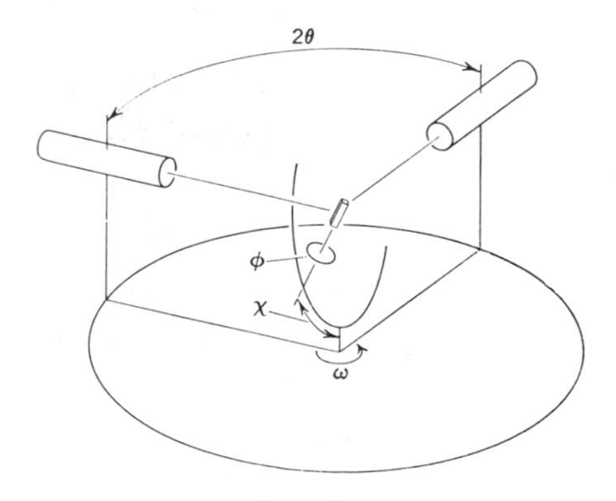

Fig. 36.

from a single crystal. The key to the design is that the counter rotates only about a vertical axis, so that the incident and diffracted rays always lie in a horizontal plane. The crystal occupies the point of intersection of these two rays, and is oriented with the aid of rotation about three axes. This crystal support is called an *Eulerian cradle* because it permits rotation about each of the Eulerian axes. A diagram of the cradle is shown in Fig. 36, and a photograph is given in Fig. 37. This apparatus is available commercially.[95] Because of the close dimensional tolerances of

† M. J. Buerger. *X-ray crystallography.* (John Wiley and Sons, New York, 1942) Fig. 148, page 275; and Fig. 149, page 280.

the apparatus, the crystal must be mounted on a special goniometer head in which it has already been adjusted to exactly correct height.

To measure the intensity of any reflection with this apparatus, the crystal must be so adjusted by manipulating the angle ω, χ, and ϕ that the reciprocal-lattice point corresponding to this reflection is brought into the zero level with respect to the axis of ω (the vertical axis). The energy of the reflection can then be gathered by the counter tube with a

Fig. 37.
(Courtesy of General Electric Co.)

short oscillation of the crystal about the vertical axis.† Alternatively, by using a beam which converges in the horizontal plane, the crystal can be set at the correct angle ω, and the convergence permits an automatic integration by the plateau method. (The general conditions under which such a stationary-counter method may be used are outlined by Lang.[81])

With this apparatus, it is necessary to precompute the angular settings ω, χ, and ϕ for all reflections, since these are, in general, different for each reflection. The determination of these values is discussed by Arndt and Phillips.[94] The setting problem is a more complicated one than with the instrument described in the previous section. With the equi-inclination

† Arrangement for oscillation is not now supplied in the apparatus as sold, but can be added.

technique reflections occur in levels of the reciprocal lattice for which there is no change in ν, so that only one setting angle, Υ, is required for the counter tube, and one more, ω, for the crystal.

Use of counter methods in intensity measurement. When no x-rays are entering the counter tube the counter records a background count due to quanta received by the tube from various extraneous sources. This is ordinarily in the neighborhood of 8 to 20 counts per minute. It depends upon the amount of extraneous radiation of high energy in the laboratory, and should be checked periodically during a run. The count is likely to be increased when other nearby x-ray apparatus is being operated. When picking up reflections from a crystal there is additional background due to x-rays scattered by the air in the x-ray path, and due to x-rays scattered incoherently by the crystal. There is also a contribution to the background by unwanted radiation in the spectrum of the x-ray source. This is discussed in a subsequent section.

It was pointed out above that, because of the dead time of the counter tube, the linearity of counting rate with intensity is limited to regions below a certain counting rate. Lonsdale[70] found the limit to occur for Geiger tubes in the neighborhood of 100 to 150 counts/sec. The counter itself may fail to resolve about 2000 counts/sec. Lonsdale recommends determining the limit of linearity of each specific instrument by a simple experiment: Metal foils (preferably nickel when using Cu radiation) of equal thickness are cumulatively interposed somewhere in the path of the beam from the x-ray tube to the counter. Each foil absorbs and reduces the intensity of the beam by e^{-x}. When n foils are used the beam is reduced by e^{-nx}. A plot of log e^{-nx} against n is therefore linear. With ideal apparatus the counts per second should be proportional to the intensity. By plotting the logarithm of the counts recorded, versus n, the practical point of departure from linearity can be observed. The result obtained by Lonsdale for a particular set of equipment is shown in Fig. 38. For intense reflections the counting rate is beyond the linearity limit, so the intensity of the direct beam should be reduced by interposing a suitable number of calibrated foils.

Lonsdale[70] found that a range of 10^5:1 is possible using a Geiger counter. On the other hand, only a range of about 300:1 is possible with automatic (graphical) recording because of the low limit of linearity of deflection versus intensity. Because of this limited range, automatic recording is unsuitable for very precise intensity measurement.

Because of the statistical nature of the arrival of quanta, the uncertainty of counting rises rapidly as the intensity falls. This means that if one determines intensities by counts in a given time (say 60 seconds) the precision of determination of the large intensities is good, but that of the smaller intensities is very poor. Lonsdale[70] therefore suggested that a better method is to determine the time required for a given number of counts. This is much slower for reflections of low intensity but it is much more precise.

A detailed discussion of the factors entering into intensity determination by counter methods has been given by Cochran.[74] In particular Cochran considers the allowances which must be made for lost counts. If N_0 is the number of counts recorded in one second, and τ the resolving

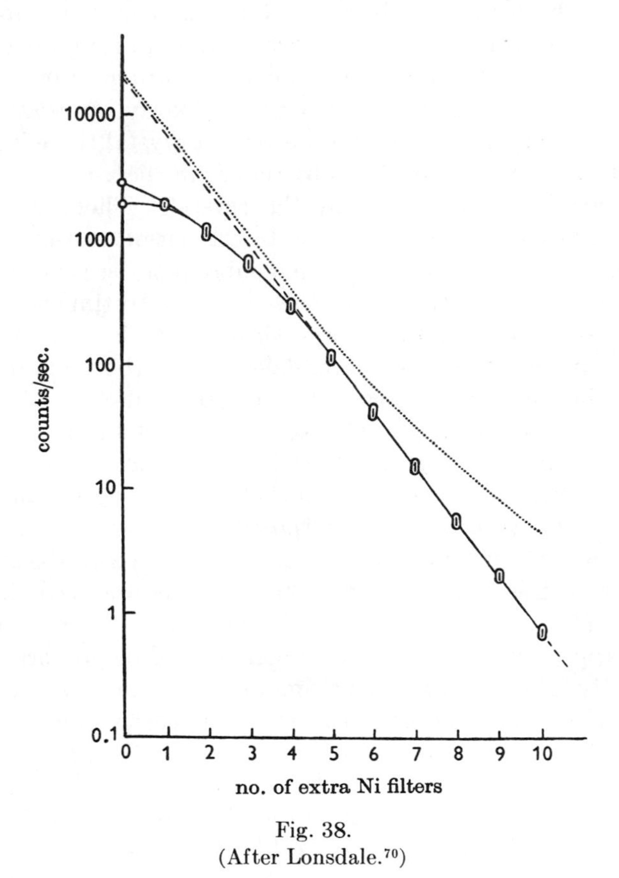

Fig. 38.
(After Lonsdale.[70])

time, then the corrected counting rate is given by

$$N = \frac{N_0}{1 - N_0 \tau}, \tag{37}$$

provided that the intensity of the source is constant. If it is not constant, a "form factor," defined by

$$K = \left(\frac{RMS \text{ intensity of source}}{\text{mean} \quad " \quad " \quad "} \right)^2, \tag{38}$$

must be introduced. In this case,

$$N = \frac{N_0}{1 - N_0 K\tau}. \tag{39}$$

If the circuit contains provision for varying the resolving time, the constant K can be found by performing counts N_1 and N_2 for two selected resolving times τ_1 and τ_2. Then, according to (39),

$$\begin{aligned} N &= \frac{N_1}{1 - N_1 K\tau_1} = \frac{N_2}{1 - N_2 K\tau_2} \\ &= \frac{1}{\frac{1}{N_1} - K\tau_1} = \frac{1}{\frac{1}{N_2} - K\tau_2} \end{aligned} \tag{40}$$

Therefore,

$$\frac{1}{N_1} - K\tau_1 = \frac{1}{N_2} - K\tau_2,$$

so that

$$K(\tau_2 - \tau_1) = \frac{1}{N_2} - \frac{1}{N_1}. \tag{41}$$

A non-uniform source also gives rise to complications in determining the integrated intensity; for during the time, T, that the crystal moves through the angular range α over which the reflection takes place, during which the intensity is variable, the measure of the integrated intensity is

$$M = \int_0^T N\, dT. \tag{42}$$

But the record is

$$M_0 = \int_0^T N_0\, dT. \tag{43}$$

Under these circumstances,

$$M = \frac{M_0}{1 - M_0 K_1 \frac{\tau}{T}}, \tag{44}$$

where

$$K_1 = K \left(\frac{RMS \text{ reflecting power of crystal over range } \alpha}{\text{mean}\ \text{"}\ \ \text{"}\ \ \text{"}\ \ \text{"}\ \ \text{"}\ \ \text{"}\ \ \text{"}} \right)^2. \tag{45}$$

K_1 can be determined in a manner analogous to K in (41).

Cochran[74] rediscovered the plateau method, discussed in an earlier section, and recommends its use. He also recommends the use of balanced filters (discussed in a subsequent section) to eliminate the effects of unwanted components of the radiation of the x-ray tube.

Unwanted x-radiation

Nature of characteristic radiation. A source of complication in intensity determination is that the x-ray tube does not emit pure monochromatic x-radiation. Unfortunately the desired $K\alpha$ radiation is accompanied by a $K\beta$ component of somewhat shorter wavelength, and both of these "characteristic" components are superposed on a varying background of general radiation, Fig. 39.

Filtered radiation. It is customary to differentially suppress the $K\beta$ component and shorter-wavelength general radiation in favor of the characteristic $K\alpha$ component by "filtering" the output of the x-ray tube through a thin sheet, or foil, of material containing the element of one (or two) atomic numbers less than the radiating element.[†] Suitable filtering elements are shown in the middle column of Table 6. This filtering action takes advantage of the specific discontinuity in the variation of absorption with wavelength known as the K absorption edge, Fig. 40. This absorption edge for the element $Z-1$ (or $Z-2$) is located so as to pass the $K\alpha$ radiation from element Z, but differentially suppress shorter radiation. Unfortunately, when the $K\beta$ component is sufficiently suppressed, a good deal of shorter-wavelength general radiation still gets through the filter. This radiation causes a "general-radiation streak" on zero-level moving-film photographs[§] such that each strong reflection is accompanied by a streak nearer the center of the film. The difficulty does not arise on upper-level photographs.[¶] The streaks on the zero-level photographs frequently superpose on Bragg reflections nearer the center. Conversely, in determining the intensity of a spot on the zero-level photograph made with filtered radiation, the reflection always contains a contribution, sometimes very strong, to its immediate background from general-radiation streaks corresponding to Bragg reflections nearer or farther from the center of the film. This background varies as it passes the spot, so that it is impossible to properly allow for it by subtracting the background above, below, or to one side of the spot.

Balanced filters. Ross devised an ingenious filtering arrangement, known as *balanced filters*, with which the unwanted components of an x-ray beam may be discarded.[102, 103] Unfortunately Ross's trick can only be used in connection with a diffractometer. Neighboring elements of the periodic system have absorptions which vary with wavelength in a

† See also: M. J. Buerger. *X-ray crystallography.* (John Wiley and Sons, New York, 1942) pages 176–178.

§ See: M. J. Buerger. *X-ray crystallography.* (John Wiley and Sons, New York, 1942) Fig. 140*A*, page 256.

¶ See: M. J. Buerger. *X-ray crystallography.* (John Wiley and Sons, New York, 1942) Fig. 140*B*, page 256.

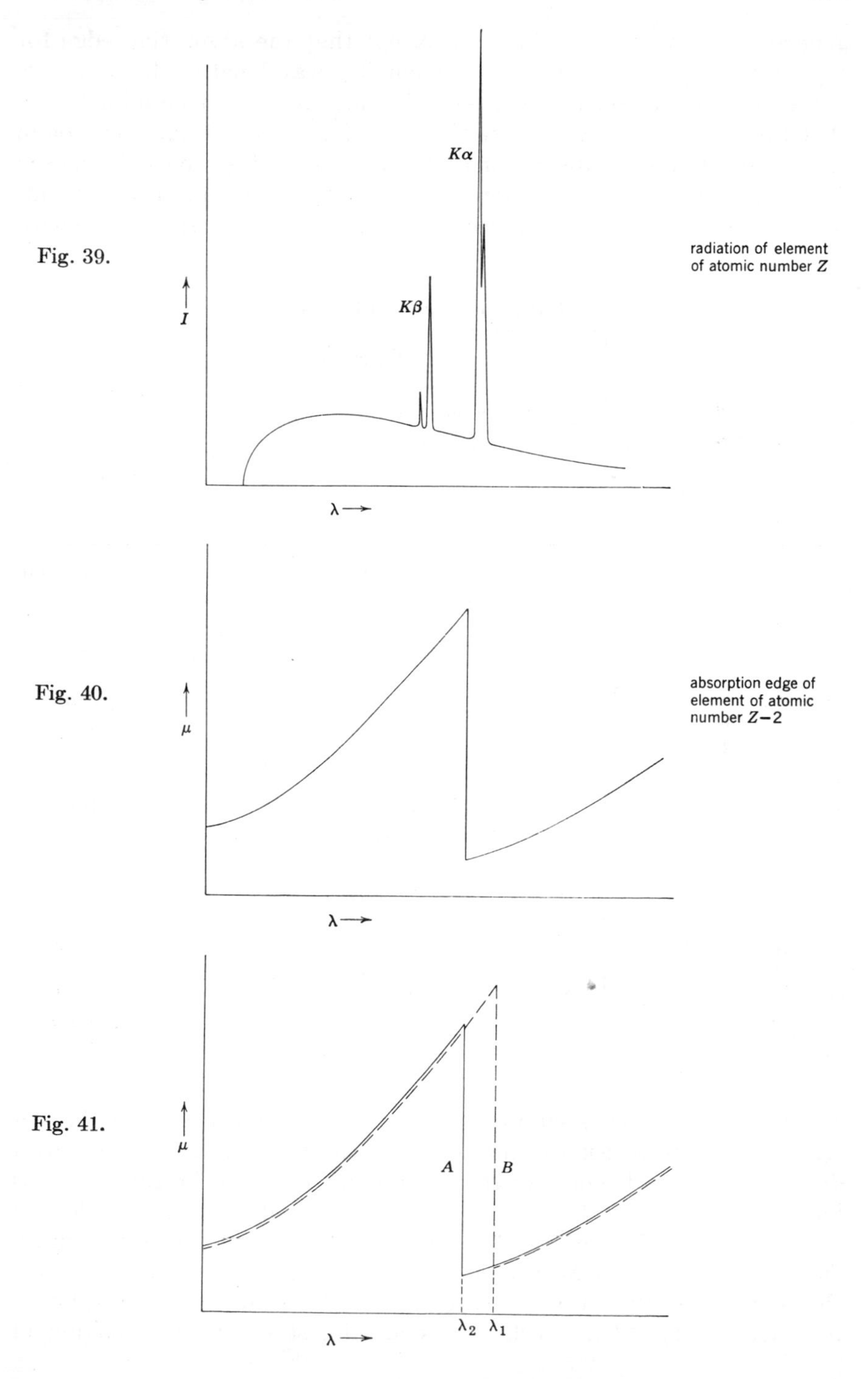
Fig. 39.
Kα
Kβ
I
λ→
radiation of element of atomic number Z
Fig. 40.
μ
λ→
absorption edge of element of atomic number Z−2
Fig. 41.
μ
A
B
λ2 λ1
λ→

generally similar manner, Fig. 40, except that the absorption edge for the heavier element is shifted to a smaller wavelength. If filters are made of a pair of such elements, their thicknesses can be adjusted so that they absorb almost exactly the same for all wavelengths except in the gap between their absorption edges, Fig. 41. This gap can be chosen so as to bracket the characteristic emission line of another element. Thus, in Fig. 41, filter *A* passes the desired characteristic wavelength

Table 6
Data for balanced filters

Radiating element			Filter elements					
			(*a*) Component for simple filter or (*b*) First component of balanced filter			Second component of balanced filter		
Z	Element	$K\alpha$ wavelength	$Z-2$	Element	Absorption edge	$Z-3$	Element	Absorption edge
47	Ag	0.5608 Å	45	Rh	0.5338 Å	44	Ru	0.5605 Å
46	Pd	0.5869	44	Ru	0.5605	43	Tc	?
45	Rh	0.6147	43	Tc	?	42	Mo	0.6198
44	Ru	0.6445	42	Mo	0.6198	41	Nb	0.6529
43	Tc	?	41	Nb	0.6529	40	Zr	0.6888
42	Mo	0.7107	40	Zr	0.6888	39	Y	0.7276
41	Nb	0.7476	39	Y	0.7276	38	Sr	0.7697
40	Zr	0.7873	38	Sr	0.7697	37	Rb	0.8155
			$Z-1$			$Z-2$		
30	Zn	1.4364	29	Cu	1.3804	28	Ni	1.4880
29	Cu	1.5418	28	Ni	1.4880	27	Co	1.6081
28	Ni	1.6591	27	Co	1.6081	26	Fe	1.7433
27	Co	1.7902	26	Fe	1.7433	25	Mn	1.8964
26	Fe	1.9373	25	Mn	1.8964	24	Cr	2.0701
25	Mn	2.1031	24	Cr	2.0701	23	V	2.2690
24	Cr	2.2909	23	V	2.2690	22	Ti	2.4973

of Fig. 39 while filter *B* suppresses it. If filter *A* and *B* are used in succession, the difference in the intensities which they pass, as measured by a diffractometer, is due only to the wavelength range in the gap, which is known as the passband.[111] This wavelength is substantially the desired wavelength only, usually the $K\alpha$ component. An example of the differential transmission of Mo$K\alpha$ radiation by balanced Zr and Y filters is shown in Figs. 42*C* and *B* respectively. The two results are plotted together in Fig. 42*A*, which shows an almost exact superposition of intensities except in the pass band, where the Mo$K\alpha$ radiation is the chief contributor. Table 6 shows pairs of elements appropriate for filters for various characteristic $K\alpha$ radiations.

Balanced filters have been discussed in some detail by Kirkpatrick.[104–106] The appropriate balance is arranged by choosing thicknesses so that the

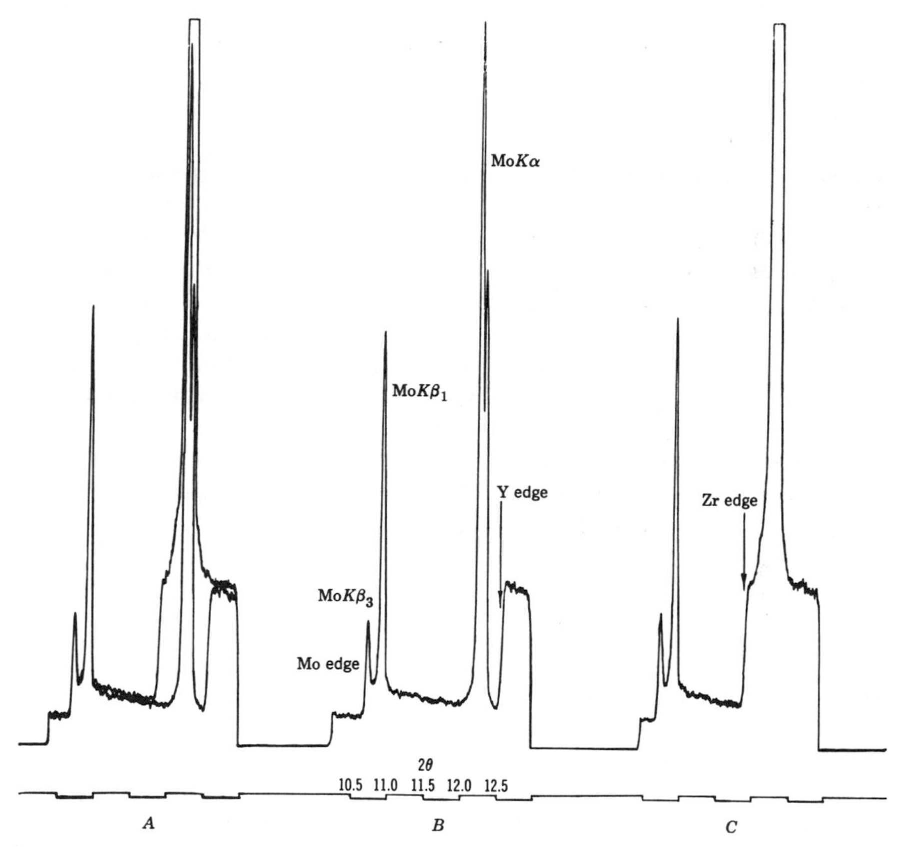

Fig. 42.
(After Dr. Howard Evans, courtesy of Philips Electronics.)

intensities transmitted by both filters are equal at some wavelength not in the pass band. This occurs when

$$e^{-\mu_A t_A} = e^{-\mu_B t_B}. \tag{46}$$

This calls for

$$\frac{t_B}{t_A} = \frac{\mu_A}{\mu_B}. \tag{47}$$

In the measurement of intensities required in crystal-structure analysis, it is appropriate to make this match for wavelengths corresponding to the

$K\beta$ component of the characteristic radiation, and also for $\lambda/2$, where λ is the wavelength of the desired $K\alpha$ component. The reason for selecting this last wavelength is discussed in the next section.

Difficulties due to harmonics. Unless the x-radiation is strictly monochromatic, the presence of harmonics of λ in the general radiation given out by the x-ray tube complicates the determination of intensities. For, while one is measuring the intensity of the reflection *hkl* using characteristic wavelength λ, the reflection *nh nk nl* is simultaneously occurring in the same direction due to the harmonic wavelength λ/n. This unfortunate situation is not necessarily corrected by using radiation monochromated by reflecting the beam from a crystal (as described in the subsequent section). Fortunately, when using a Geiger-Müller counter, the characteristics of the counting tube are such (Fig. 29) that the particular gas filling is sensitive only to radiation whose wavelength exceeds a critical value. Thus, the characteristic $K\alpha$ radiation from Cu, Ni, Co, Fe, Mn, and Cr ionizes argon. On the other hand the harmonics in the general radiation excited from copper targets have wavelengths too small to ionize argon, so they cause no trouble when the radiation is detected by an argon-filled counter tube.

All harmonics can be eliminated if the x-ray tube is operated at a potential too small to excite the $\lambda/2$ harmonic. The potential required to excite a specific wavelength of the continuous spectrum is given by the fundamental quantum relation,

$$\lambda V = \frac{hc}{e} = \text{constant.} \tag{48}$$

When λ is expressed in Ångströms, and V in kilovolts, this is equivalent to

$$\lambda V = 12.39. \tag{49}$$

If (49) is applied to the $\lambda/2$ harmonic, one finds that only about 35 kv. are required to excite it for Mo$K\alpha$ and 16 kv. for Cu$K\alpha$. Unfortunately these are very conservative voltages.

Monochromated radiation. A way of achieving clean radiation is to reflect the x-rays from a crystal, using one of its strong Bragg reflections. The reflected beam still contains harmonics unless these are eliminated as discussed in the last section. Alternatively, the $\lambda/2$ harmonic can be eliminated in monochromated radiation if the monochromating crystal is chosen so that it has a weak or absent second-order reflection.

The monochromating is carried out by a small device called a crystal monochromator. A number of types have been described.[112, 113] The design of one used in the author's laboratory for many years is shown in

Fig. 43. Some important characteristics are as follows: 1. It is attached directly to the x-ray tube so that the source-to-crystal distance is very short. 2. It is entirely enclosed, so that only the comparatively weak monochromated radiation enters the room. 3. The crystal is readily adjusted to reflecting position by a tangent screw. 4. The crystal is quickly replaced with any other. 5. When the crystal is replaced by

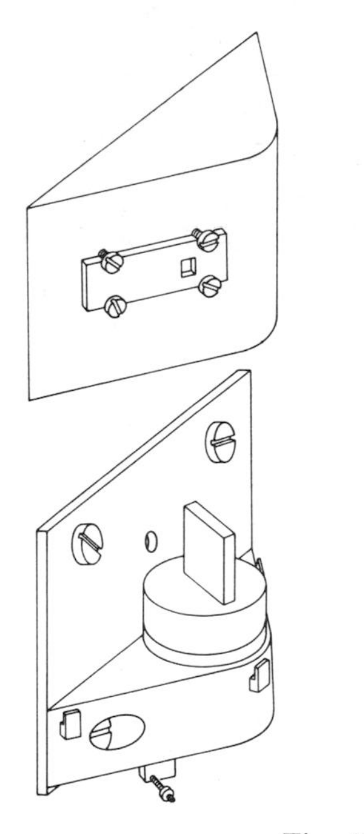

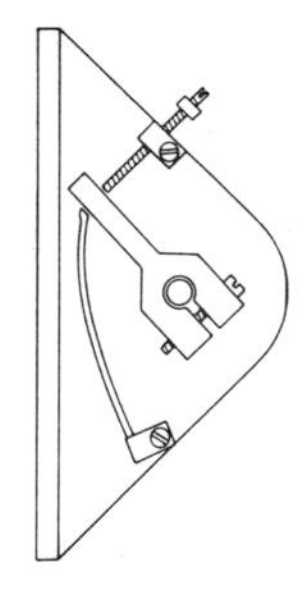

Fig. 43.

another of different Bragg glancing angle, the new reflected beam is quickly isolated by adjusting the position of an exit aperture.

The merits of a number of crystals for use in monochromators have been listed by Lipson, Nelson, and Riley.[113] Some of this material is shown in Table 7.

One of the disadvantages of using monochromated radiation is that very long exposures are required. Fankuchen[109] has designed a condensing monochromator, in which the intensity is increased over that of the ordinary crystal monochromator by a factor said to be of the order of 2. This is based on the principle of intensity increase by projection, suggested by Stephen and Barnes.[108] Fankuchen's arrangement is shown diagrammatically in Fig. 44. Some doubt has been expressed that appreciable intensity gain can be achieved by the condensing monochromator.[110, 111, 114]

Table 7
Characteristics and properties of various crystals for monochromator purposes
(After Lipson, Nelson, and Riley[113])

Crystal	Reflexion	Spacing d in kX	Properties of reflexion			Properties of crystal		Special uses
			Peak intensity	Breadth	Multiplicity	Stability	Mechanical properties	
β Alumina	0002	11.22	Weak			Perfect	Hard, brittle	For long wavelengths
β Alumina	0004	5.61	Weak-medium	Moderate	Great			
Gypsum	020	7.58	Medium-strong	Very small	Negligible	Poor	Soft, flexible	For small-angle scattering, focusing, long wavelengths
Pentaerythritol	002	4.39	Very strong	Moderate	Great	Poor	Soft, easily deformed	General purposes
Quartz	$10\bar{1}1$	3.34	Weak-medium	Very small	Negligible	Perfect	Can be elastically bent	For small-angle scattering, focusing
Fluorite	111	3.15	Medium-strong	Moderate	Small	Perfect	Moderately hard	For eliminating harmonics, general purposes, short wavelengths
Urea nitrate	002	3.13	Strong	Very large	Very great	Very poor	Very easily deformed	For large specimens
Calcite	200	3.03	Medium	Small	Negligible	Perfect	Moderately soft	For small-angle scattering, isolation of α_1
Rock salt	200	2.81	Medium-strong	Large	Great	Slightly deliquescent	Can be plastically bent in warm water	For focusing
Diamond	111	2.05	Weak	Very small	Negligible	Perfect	Very hard	For eliminating harmonics

A second disadvantage of using monochromatic radiation is that the radiation is partially polarized. The unpolarized fraction behaves normally; the polarized fraction† behaves in a manner which is difficult to compute due to analyzer action on the part of the crystal being examined. The polarization factor caused by one crystal is given by the usual relation§

$$p = \tfrac{1}{2}(1 + \cos^2 2\theta) \tag{50}$$

$$= \tfrac{1}{2}\{1 + (\cos\mu\cos\nu\cos\Upsilon + \sin\mu\sin\nu)^2\}. \tag{51}$$

If a monochromator is used whose Bragg angle is α, then if the incident

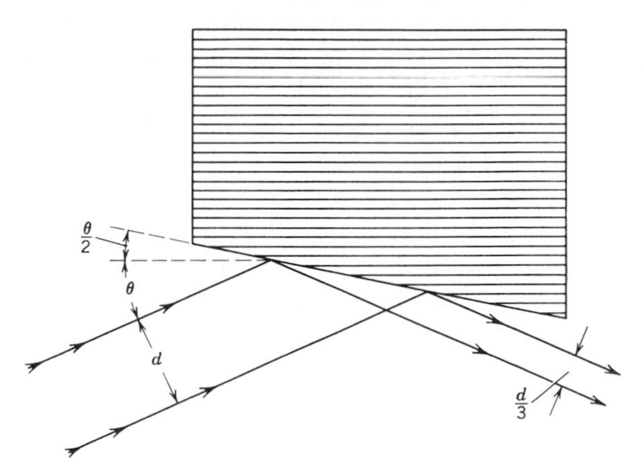

Fig. 44.
(After Fankuchen.[109])

and reflected rays of both crystal and monochromator lie in the same plane, the polarization factor is[115]

$$p = \frac{1 + \cos^2 2\alpha \cos^2 2\theta}{1 + \cos^2 2\alpha} \tag{52}$$

The case for inclined beams is more complicated: If the Bragg angle, α, of the monochromator lies parallel to the angle Υ, the polarization factor is[115]

$$\begin{aligned} p = \{&\cos^2 2\alpha + \cos^2\mu\cos^2\nu + \sin^2\mu\sin^2\nu \\ &+ (\sin^2 2\alpha - \cos^2\mu)\cos^2\nu\sin^2\Upsilon \\ &+ \tfrac{1}{2}\sin^2 2\mu\sin^2 2\nu\cos\Upsilon\}/(1 + \cos^2 2\alpha). \end{aligned} \tag{53}$$

The complexity of this expression suggests avoiding the use of monochromators in gathering intensity data unless α is so small that (50) is substantially true. It follows that crystals used for monochromators

† Chapter 3, equation (15).
§ Chapter 3, equation (14).

should have an appropriate strong reflection with a small 2θ. Azaroff[117] has discussed the polarization correction, especially for the precession method.

Not all crystals with strong reflections at small 2θ's are suitable monochromators for intensity determination. Some crystals, especially organic crystals like pentaerythritol, quickly deteriorate and therefore give radiation of variable intensity, and so are useless.

Notes on collecting the intensity record

In this section a discussion is given of some of the practical points which arise in collecting the set of intensities necessary for a crystal-structure analysis. The discussion applies specifically to intensities derived from a photographic record, but in many respects the general theories may be extended to diffractometer methods.

It is assumed that ordinarily the investigator will choose to gather data from single-crystal reflections. It is further assumed that reflections will be gathered using one of the moving-film methods. It is possible (and was once usual) to obtain intensity records from rotating-crystal photographs, or sets of oscillation photographs. The time spent in indexing, and the necessity for correlating intensities from one film to another, makes the use of the oscillation method most undesirable in these days of modern moving-film methods.

While the unit-cell and space-group determination require that a reasonable portion of the three-dimensional reciprocal lattice be recorded, not all reflections need necessarily be recorded in a sufficiently careful manner that acceptable intensities can be determined from them. Thus, it may be convenient to treat photographs intended only for space-group determination with a contrasty development, whereas the specific reflections needed for precise intensity determination may be recorded a second time using a special technique suited especially to intensity determination, for example, the Dawton positive-print technique.

The structure of a crystal with a small cell is usually sufficiently well fixed by the projections of the structure in the directions of the crystallographic axes. The projections in the directions a, b, and c are fixed by the intensities of $0kl$, $h0l$, and $hk0$, respectively. Thus, if the cell is small, the structure can often be determined by using intensities from these special sets of reflections only. Each of the sets $hk0$, $h0l$, and $0kl$ can be recorded on a single zero-level photograph taken with one of the moving-film techniques. In this case, the most convenient method to use is the precession method or the de Jong-Bouman method. Each such film contains the entire $hk0$ record (for example), arranged in a convenient reciprocal-lattice pattern, so that indexing is unnecessary.

When a structure with a reasonably large cell is to be investigated, such projection methods usually fail, and it is then necessary to measure the intensities of all reflections, i.e., to determine the relative weights of all points of the three-dimensional reciprocal structure. This complete set of intensities may be used, for example, to provide the coefficients for a three-dimensional Patterson synthesis (Chapter 21) or they may be used in inequality relations (Chapter 21) in an attempt to find the phases of the reflections.

Strategy with the Weissenberg apparatus. When the weights of all reciprocal-lattice points are to be determined, it is necessary to take

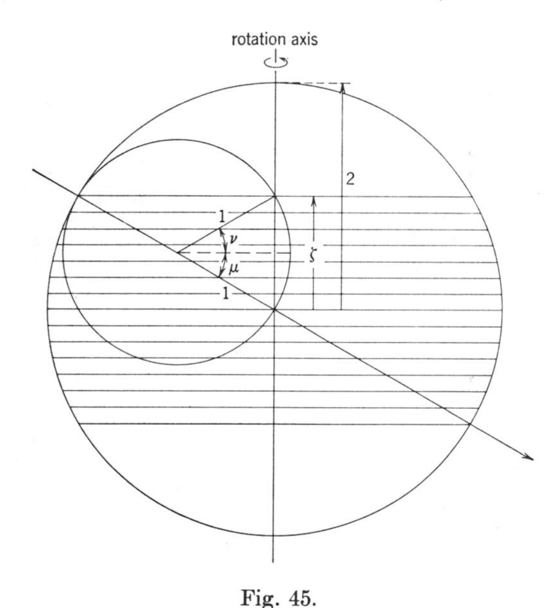

Fig. 45.

account of some of the physical limitations of the instruments used for recording. The equi-inclination technique is commonly used for gathering the record, for only this technique lacks the blind spot in the centers of the upper levels. Furthermore, the application of the various corrections (discussed in the next two chapters) is simplified for the equi-inclination technique.

Usually the Weissenberg apparatus is used for this purpose. If μ_{max} is the maximum inclination angle permitted by the design of the apparatus, then the reciprocal lattice can be explored to a level (Fig. 45)

$$\zeta_{max} = 2 \sin \mu_{max}. \tag{54}$$

For instruments commercially available in the United States at the present time $\mu_{max} = 30°$, so that $\zeta_{max} = 1$, which means that only half the

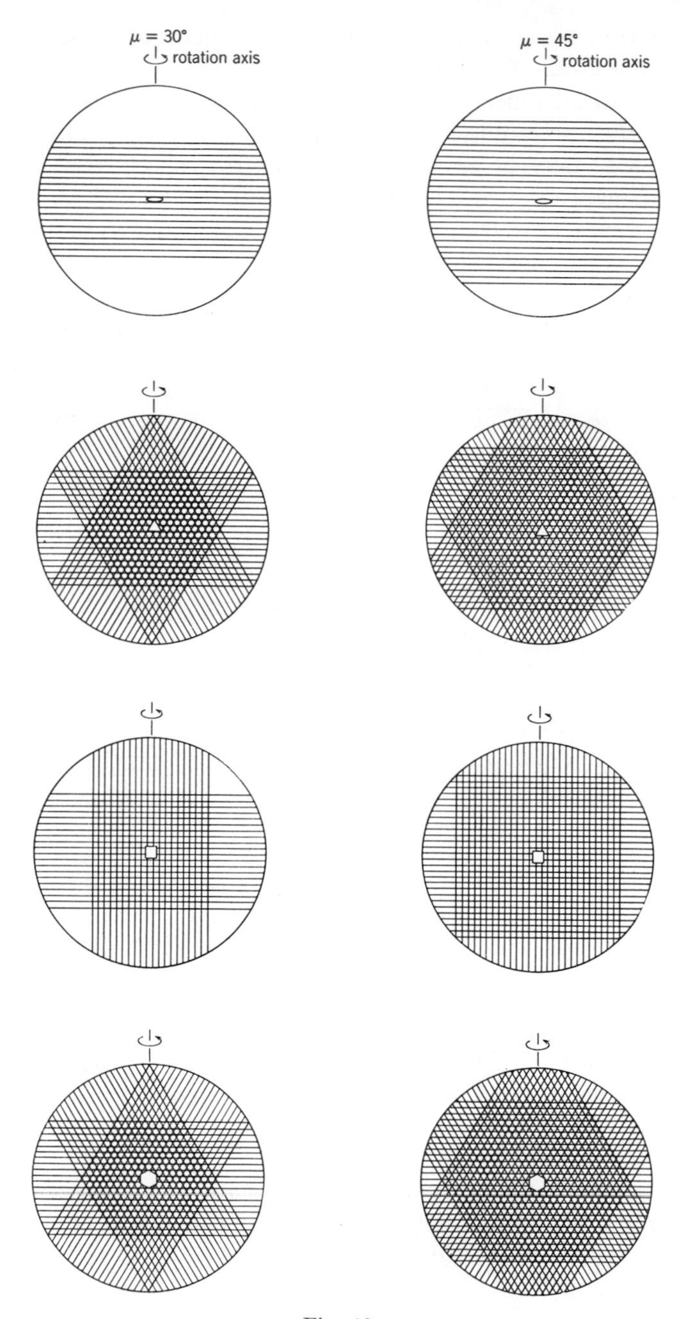

Fig. 46.

total recordable height of the reciprocal lattice can be explored in one setting of the crystal.

In Fig. 45, the unshaded region of the sphere represents reciprocal-lattice volume unrecorded in one crystal setting. This volume must be explored by remounting the crystal for rotation about other axes, unless the unshaded region is supplied by symmetry. Figure 46 shows how symmetry supplies information in the unrecorded regions. The left column of illustrations are drawn for $\mu_{\max} = 30°$. The pertinent symmetry is axial symmetry with the axis normal to the rotation axis. It will be observed that if a 3-fold or a 6-fold axis occurs at right angles to the rotation axis, the symmetry of the reciprocal lattice is such that the

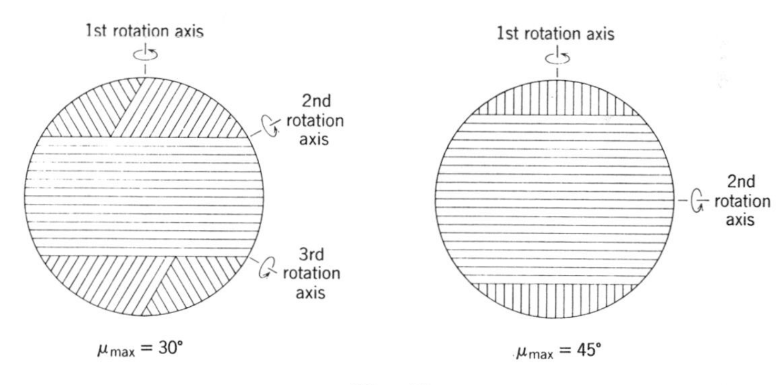

Fig. 47.

missing region is completely supplied by symmetry. On the other hand, if a 2-fold or a 4-fold axis is at right angles to the rotation axis, there still remains an unexplored region. The unexplored region vanishes for the 4-fold case if $\mu_{\max}$ is made 45°, as shown in the right column.

Figure 47 shows that to fill in the remaining unexplored region for a 2-fold axis requires mounting the crystal for at least three rotation axes separated by 60° intervals. If $\mu_{\max} = 45°$, two orthogonal mountings suffice.

In making use of symmetry, one should bear in mind that the total translation of most Weissenberg cameras is 180° (plus a small additional margin to bring the range up to, perhaps, 220°), not 360°. The upper and lower halves of the Weissenberg record are thus generally different. Suppose, now, that one is recording reflections from a monoclinic crystal by rotation about an axis in the symmetry plane, say the *a* or *c* axis. The entire record can be recorded on the film only if the record is properly arranged. Figure 48 shows two ways of recording an upper level for rotation about the *c* axis (second setting). Thus, in Fig. 48*A*, the symmetry line of the level appears in the center of the film. During the counterclockwise rotation of 180°, the portion of the reciprocal lattice shaded vertically passes through the upper semicircumference of the reflecting circle, producing reflections in the upper half of the 180° Weissenberg record. Meanwhile the portion of the reciprocal lattice shaded horizontally passes through the lower semicircumference of the

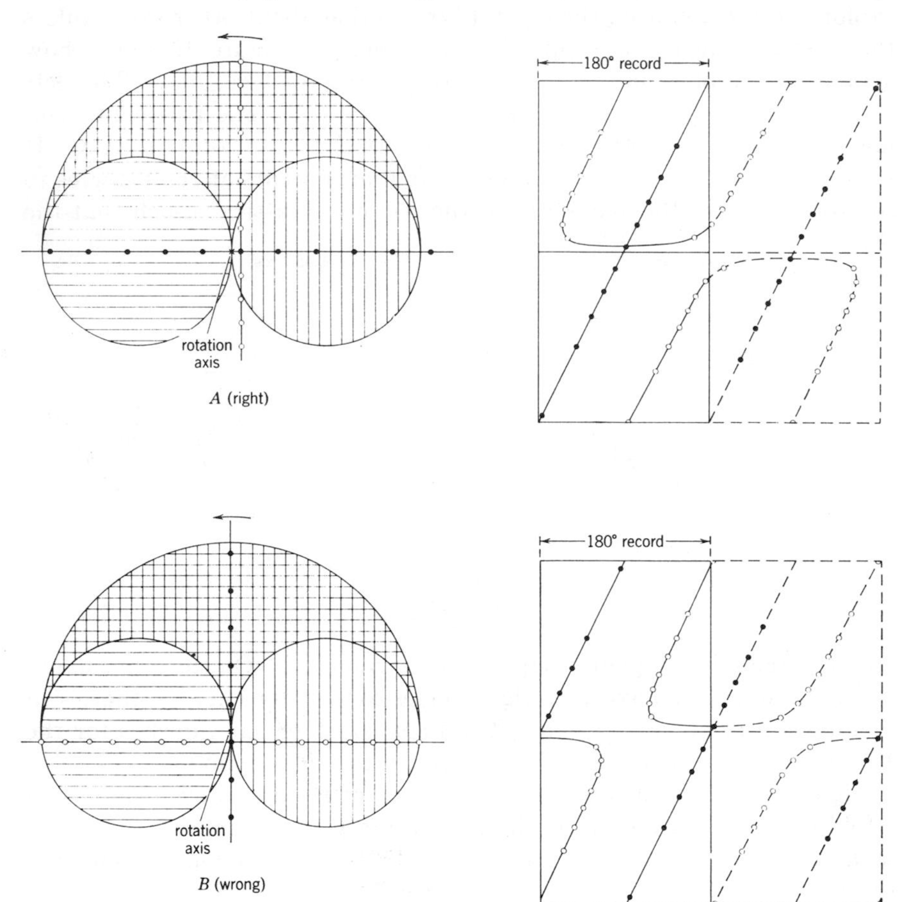

Fig. 48.

reflecting circle, producing reflections in the lower half of the 180° Weissenberg record. Thus a motif half of the reciprocal lattice is recorded. But in Fig. 48*B*, the symmetry line of the level starts at the edge of the film. In this case the upper and lower halves of the film can record equivalent portions of the reciprocal lattice related by the symmetry line, but neither record contains a complete motif half of the reciprocal lattice. (The records on the upper and lower halves of the film are related by an inversion center in the middle of the center line of the film.)

Suppose a preliminary photograph shows that this orientation of the crystal on the Weissenberg spindle has not been achieved. It is then necessary to shift a reflection occuring on the symmetry line so that its central lattice line moves to the center of the film. To perform such a shift with correct magnitude and direction it is necessary to understand the coupling of the particular Weissenberg camera being used. Figure 49 illustrates the transformation which should be studied. The left

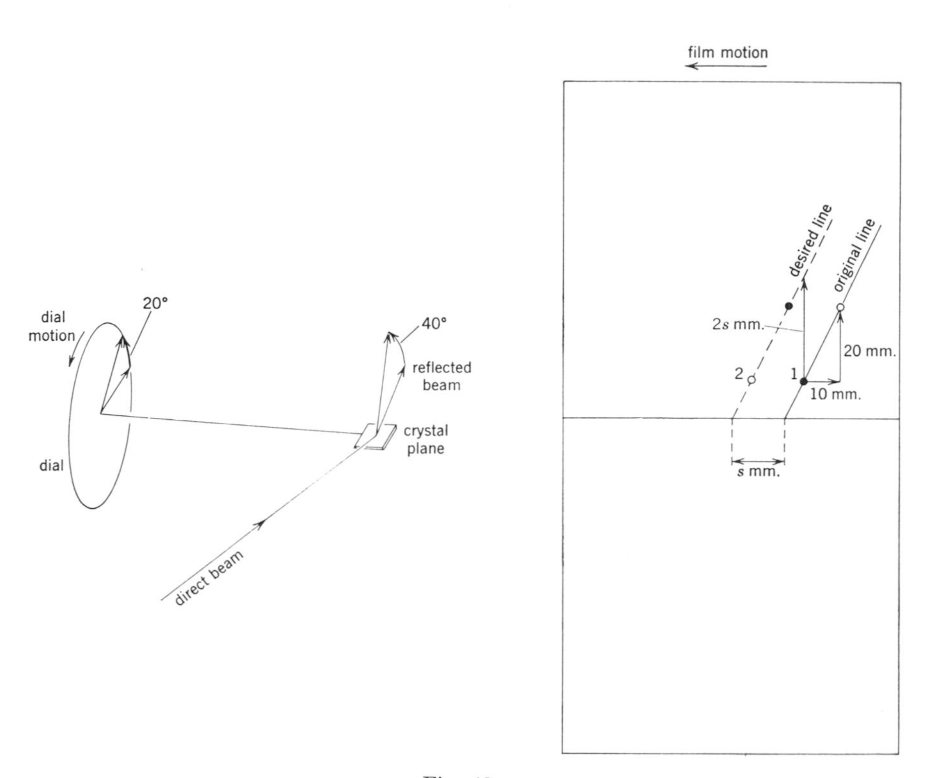

Fig. 49.

side of Fig. 49 represents the Weissenberg as seen from the x-ray tube. A study of the coupling shows (in this instance) that when the top of the dial is moved, say 20°, toward the x-ray tube, the film is moved 10 mm. toward the left. Meanwhile the trace of the line of reflections from the crystallographic plane shown on the rotation axis travels up and to the right on the film. Now, returning to the original setting, suppose that, in placing the crystal on the spindle, a reflection is found to occur on the film at 1, and it is desired on a second try to have it record at 2 by an adjustment of the spindle. If the film is left fixed and the spindle is rotated so that the top of the dial moves forward until the reflection is on the reflection path of the desired point 2, and the camera is then coupled to the rotation motion, the desired shift is achieved. In other words, to move a record from point 1 leftward to point 2, separated by s mm. on the film, uncouple the camera and rotate the spindle so that the dial moves $C_2\, s$ degrees toward the x-ray tube. Here C_2 is the coupling constant of the instrument.†

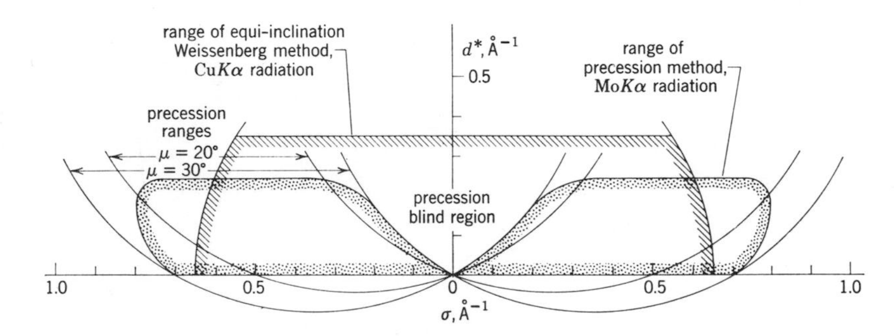

Fig. 50.
(Courtesy of Dr. Howard Evans.)

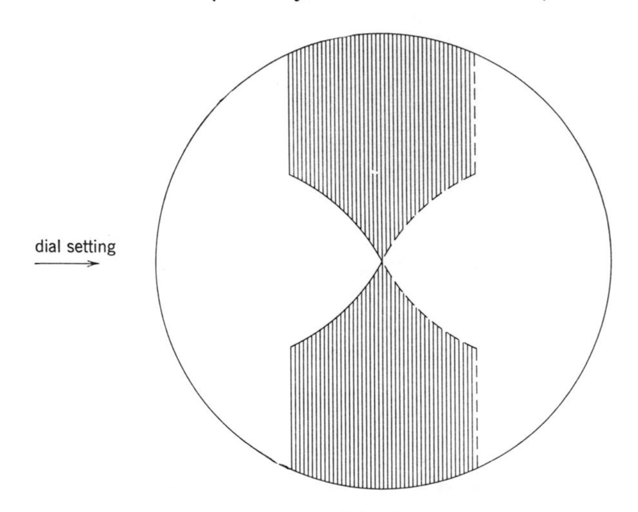

Fig. 51.

Strategy with the precession apparatus. When gathering three-dimensional photographic intensity data, it is convenient to use the precession method. When using Mo$K\alpha$ radiation, this method records about the same number of reciprocal-lattice points as does the Weissenberg method when using Cu$K\alpha$ radiation, Fig. 50. Unfortunately,

† M. J. Buerger, *X-ray crystallography.* (John Wiley and Sons, New York, 1942) 224.

the precession method suffers from a blind spot in the center of the record for the upper levels.[118, 120] The volume of the reciprocal lattice which can be recorded by a single dial setting is shown diagrammatically in Fig. 51. Azaroff[119] has pointed out that with a very limited number of

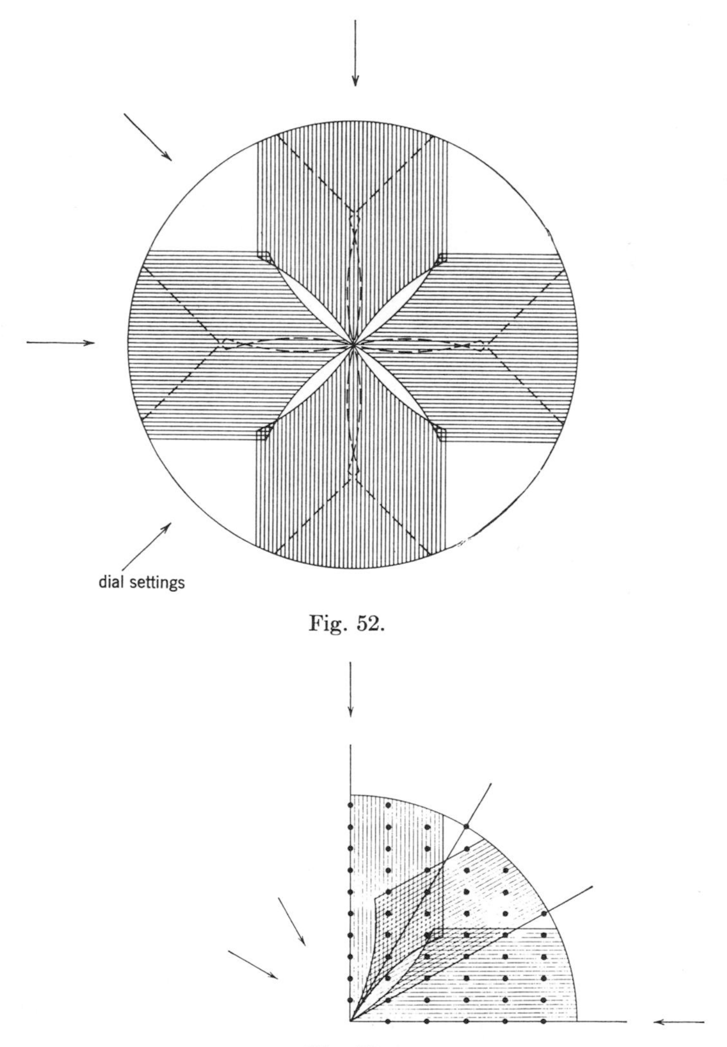

Fig. 52.

Fig. 53.

dial settings (but with no change of the crystal mounting or its adjustment) all points of the reciprocal lattice can be recorded. Figures 52 and 53 show that even for an unsymmetrical crystal, only three or four dial settings are required. To take full advantage of this strategy, the crystal is best mounted so that its axis of greatest symmetry is parallel to the dial axis. For hexagonal or trigonal crystals, one dial setting suffices for all reflections.

Literature

Basic photographic theory

[1] Ferdinand Hurter and V. C. Driffield. *Photo-chemical investigations and a new method of determination of the sensitiveness of photographic plates.* J. Soc. Chem. Ind. (London) **9** (1890) 455–469.

[2] Carl Gamertsfelder and Newell S. Gingrich. *The use of intensifying screens in x-ray diffraction work.* Rev. Sci. Instr. **9** (1938) 154–159.

[3] A. Charlesby. *The action of electrons and x-rays on photographic emulsions.* Proc. Phys. Soc. **52** (1940) 657–700.

[4] C. E. Kenneth Mees. *The theory of the photographic process.* (The Macmillan Co., New York, 1942, revised, 1952)

[5] Thomas H. James and George C. Higgins. *Fundamentals of photographic theory.* (John Wiley and Sons, New York, 1948)

[6] M. H. VanHorn. *The use of film in x-ray diffraction studies.* Rev. Sci. Instr. **22** (1951) 809–811.

[7] Commission on Crystallographic Apparatus, International Union of Crystallography. *A comparison of various commercially available x-ray films.* Acta Cryst. **9** (1956) 520–525.

Photographic methods

[8] G. M. B. Dobson. *A flicker type of photo-electric photometer giving high precision.* Proc. Roy. Soc. (London) (A) **104** (1923) 248–251.

[9] W. T. Astbury. *A simple radioactive method for the photographic measurement of the integrated intensity of x-ray spectra.* Proc. Roy. Soc. (London) (A) **115** (1927) 640–657.

[10] W. T. Astbury. *A new integrating photometer for x-ray crystal reflections, etc.* Proc. Roy. Soc. (London) (A) **123** (1929) 575–602.

[11] W. T. Astbury. *A new integrating microphotometer for x-ray crystal reflections.* J. Sci. Instr. **6** (1929) 2–11.

[12] W. T. Astbury. *An integrating microphotometer for x-ray crystal analysis.* Trans. Faraday Soc. **25** (1929) 1–4.

[13] B. W. Robinson. *An integrating photometer for x-ray crystal analysis.* Proc. Roy. Soc. (London) (A) **130** (1930) 120–133.

[14] J. Brentano. *Über ein Verfahren zur direkten Photometrierung geringer Schwärzungen und dessen Anwendung bei der Auswertung von Röntgenstrahl-F-Werten.* Z. Physik. **70** (1931) 74–83.

[15] B. Wheeler Robinson. *An integrating photometer for x-ray crystal analysis.* J. Sci. Instr. **10** (1933) 233–242.

[16] J. Brentano, A. Baxter, and F. W. Cotton. *Photographic photometry based on scattering.* Phil. Mag. **17** (1934) 370–397.

[17] J. Monteath Robertson. *A two-crystal moving film spectrometer for comparative intensity measurements in x-ray crystal analysis.* Phil. Mag. **18** (1934) 729–745.

[18] E. O. Wollan. *A technique for obtaining the integrated intensity from x-ray powder photographs.* Rev. Sci. Instr. **9** (1938) 79–81.

[19] Ralph H. V. M. Dawton. *The integration of large numbers of x-ray crystal reflections.* Proc. Phys. Soc. (London) **50** (1938) 919–925.

[20] J. J. deLunge, J. Monteath Robertson, and I. Woodward. *X-ray crystal analysis of trans-azobenzene.* Proc. Roy. Soc. (London) (A) **171** (1939) 398–410, especially 404–405.

[21] W. A. Wooster and A. J. P. Martin. *A two-crystal Weissenberg x-ray goniometer.* J. Sci. Instr. **17** (1940) 83–89.

[22] J. Monteath Robertson and R. H. V. M. Dawton. *Photometry of x-ray crystal diffraction diagrams.* J. Sci. Instr. **18** (1941) 126–128.

[23] Joseph S. Lukesh. *The estimation of intensities from x-ray films.* J. Chem. Phys. **9** (1941) 659–660.

[24] J. C. M. Brentano. *The measurement of photographic densities by the scatter of x-rays* (Abstract). Phys. Rev. **63** (1943) 64–65.

[25] J. Monteath Robertson. *Technique of intensity measurements in x-ray crystal analysis by photographic methods.* J. Sci. Instr. **20** (1943) 175–179.

[26] J. A. Wasastjerna. *An improved photographic method for the quantitative study of the reflexion of x-rays by crystals.* Kl. Svenska Vetenskapsakad. Handl. **20** (1944), No. 11, 26 pages.

[27] J. C. M. Brentano and H. Froula. *The evaluation of the intensities of line patterns recorded on a moving film in its application to the measurement of x-ray intensities* (Abstract). Phys. Rev. **65** (1944) 254.

[28] J. C. M. Brentano. *The quantitative evaluation of photographic line patterns.* J. Opt. Soc. Am. **35** (1945) 382–389.

[29] E. H. Wiebenga and C. J. Krom. *X-ray investigation of* d-α-*Br, Cl, and CN-camphor. A direct determination of a molecular structure by comparison of isomorphous crystal structures.* Rec. trav. chim. **65** (1946) 663–681, especially 666–668.

[30] W. A. Wooster. *The measurement of the intensities of x-ray reflections from crystals.* Acta Fac. Rerum Nat. Univ. Carolinae **175a** (1947) 13 pages.

[31] E. H. Wiebenga. *An integrating Weissenberg-apparatus for x-ray analysis.* Rec. trav. chim. **66** (1947) 746–748.

[32] R. G. Wood. *The positive print method of measuring x-ray reflections from a single crystal.* J. Sci. Instr. **25** (1948) 202–204.

[33] H. P. Stadler. *A new Weissenberg technique using a double slit.* Acta Cryst. **3** (1950) 262–264.

[34] E. H. Wiebenga and D. W. Smits. *An integrating Weissenberg apparatus for x-ray analysis.* Acta Cryst. **3** (1950) 265–267.

[35] Walter N. Brown, Jr. and Willard B. Birtley. *A densitometer which records directly in units of emulsion exposure.* Rev. Sci. Instr. **22** (1951) 67–72.

[36] A. Taylor. *An improved direct-reading microdensitometer.* J. Sci. Instr. **28** (1951) 200–205.

[37] W. R. Ruston. *Note on Stadler's double-slit Weissenberg technique.* Acta Cryst. **4** (1951) 473.

[38] E. J. W. Whittaker. *Two unconventional uses of a Weissenberg goniometer.* Acta Cryst. **6** ((1953) 93.

[39] Karl E. Beu. *On the use of x-ray film for the quantitative measurement of diffraction line intensity.* Rev. Sci. Instr. **24** (1953) 103–108.

[40] E. J. W. Whittaker. *The Cox & Shaw factor.* Acta Cryst. **6** (1953) 218.

[41] D. W. Smits and E. H. Weibenga. *Measurement of x-ray intensities on integrated Weissenberg photographs.* J. Sci. Instr. **30** (1953) 280–281.

[42] M. M. Qurashi. *A discussion of the Cox & Shaw factor for oblique incidence and the film-to-film factor in multiple-film exposures.* Acta Cryst. **6** (1953) 668–669.

[43] G. J. Bullen. *The multiple-film technique: The effect of angle of incidence on the correlating factor.* Acta Cryst. **6** (1953) 825–826.

[44] E. Alexander, B. S. Fraenkel, A. Many, and I. T. Steinberger. *An integrating photometer for x-ray intensity measurements.* Rev. Sci. Instr. **24** (1953) 955–960.

[45] J. Iball. *The use of multiple films for measuring intensities of x-ray diffraction spots.* J. Sci. Instr. **31** (1954) 71.

[46] S. C. Wallwork and K. J. Standley. *Photometry of single-crystal x-ray photographs.* Acta Cryst. **7** (1954) 272–275.

[47] R. J. Davis and W. E. Armstrong. *Multiple film methods in x-ray powder photography.* J. Sci. Instr. **31** (1954) 305–306.

[48] D. C. Phillips. *On the visual estimation of x-ray reflection intensities from upper-level Weissenberg photographs.* Acta Cryst. **7** (1954) 746–751.

[49] E. Stanley. *A one-dimensional integrating method for estimating the intensities on upper-level equi-inclination Weissenberg photographs.* Acta Cryst. **8** (1955) 58–59.

[50] J. R. Brown, H. K. Moneypenny, and R. J. Wakelin. *A servo-controlled microdensitometer for x-ray diffraction photographs.* J. Sci. Instr. **32** (1955) 55–59.

[51] Christer E. Nordman, A. L. Patterson, Alice S. Weldon, and Charles E. Supper. *Integrating mechanism for the Buerger precession camera.* Rev. Sci. Instr. **26** (1955) 690–692.

[52] H. J. Grenville-Wells. *Photographic intensity scales for use with three-dimensional data.* Acta Cryst. **8** (1955) 512–513.

[53] Kathleen Lonsdale and H. Judith Grenville-Wells. *Large increase of light sensitivity at low temperatures for certain types of x-ray films.* Brit. J. Appl. Phys. **7** (1956) 380.

[54] Michael G. Rossmann. *The absorption of x-rays by photographic films.* Acta Cryst. **9** (1956) 819.

[55] D. C. Phillips. *On the visual estimation of x-ray reflection intensities from upper-level Weissenberg photographs. II. Charts for the correction of reflection spot extension.* Acta Cryst. **9** (1956) 819–821.

[56] Jeanne Taylor and William Parrish. *Transmission of Kodak No Screen and DuPont Type* 508 *film.* Acta Cryst. **9** (1956) 971.

[57] C. E. Nordman and A. L. Patterson. *Integrating attachment for the Weissenberg camera.* Rev. Sci. Instr. **28** (1957) 384–385.

[58] Leonid V. Azaroff. *A new method for measuring integrated intensities photographically.* Acta Cryst. **10** (1957) 413–416.

[59] A. W. Hanson. *A double layer-line screen for Weissenberg photography.* J. Sci. Instr. **35** (1958) 180, 288.

Spectrometer methods

[60] W. H. Bragg and W. L. Bragg. *The reflection of x-rays by crystals.* Proc. Roy. Soc. (London) (A) **88** (1913) 428–438.
[61] W. Lawrence Bragg. *The analysis of crystals by the x-ray spectrometer.* Proc. Roy. Soc. (London) (A) **89** (1914) 468–489.
[62] W. H. Bragg and W. L. Bragg. *X rays and crystal structure.* (G. Bell and Sons, London, 1915) 22–37.
[63] C. E. Wynn-Williams. *A thyratron "scale-of-two" automatic counter.* Proc. Roy. Soc. (London) (A) **136** (1932) 312–324.
[64] W. A. Wooster and A. J. P. Martin. *An automatic ionization spectrometer.* Proc. Roy. Soc. (London) (A) **155** (1936) 150–172.
[65] Harold Lifschutz. *A complete Geiger-Müller counting system.* Rev. Sci. Instr. **10** (1939) 21–26.
[66] Harris M. Sullivan. *Quantum efficiency of Geiger-Müller counters for x-ray intensity measurements.* Rev. Sci. Instr. **11** (1940) 356–362.
[67] R. Lindemann and A. Trost. *Das Interferenz-Zählrohr als Hilfsmittel der Feinstrukturforschung mit Röntgenstrahlen.* Z. Phys. **115** (1940) 456–468.
[68] H. Friedman. *Geiger counter spectrometer for industrial research.* Electronics **18** (1945) 132–137.
[69] J. Bleeksma, G. Kloos, and H. J. di Giovanni. *X-ray spectrometer with Geiger counter for measuring powder diffraction patterns.* Philips Tech. Rev. **10** (1948) 1–12.
[70] Kathleen Lonsdale. *Geiger counter measurements of Bragg and diffuse scattering of x-rays by single crystals.* Acta Cryst. **1** (1948) 12–20.
[71] W. A. Wooster and G. L. MacDonald. *Crystalline texture and the determination of structure amplitudes.* Acta Cryst. **1** (1948) 49–54.
[72] W. A. Wooster, G. N. Ramachandran, and A. Lang. *A note on the use of x-ray counter-spectrometers for single-crystal measurements.* J. Sci. Instr. **25** (1948) 405–407.
[73] Leon F. Curtiss. *The Geiger-Müller counter.* Nat. Bur. Standards. (U. S.) Cir. 490 (1950).
[74] W. Cochran. *A Geiger-counter technique for the measurement of integrated reflexion intensity.* Acta Cryst. **3** (1950) 268–278.
[75] D. F. Clifton, Aaron Filler, and D. McLachlan. *The adaptation of a Geiger counter to the Weissenberg camera.* Rev. Sci. Instr. **22** (1951) 1024–1025.
[76] Howard T. Evans. *Use of a Geiger counter for the measurement of x-ray intensities from small single crystals.* Rev. Sci. Instr. **24** (1953) 156–161.
[77] S. C. Abrahams and H. J. Grenville-Wells. *A single-crystal adaptor for the Norelco high-angle diffractometer.* Rev. Sci. Instr. **25** (1954) 519–520.
[78] R. J. Weiss, J. J. deMarco, and G. Weremchuk. *Conversion of Norelco fluorescent spectrograph to an x-ray diffractometer.* Acta Cryst. **7** (1954) 599–600.
[79] W. Parrish, E. A. Hamacher, and K. Lowitzsch. *The "Norelco" x-ray diffractometer.* Philips Tech. Rev. **16** (1954) 123–133.

[80] P. J. A. McKeown and A. R. Ubbelohde. *A Geiger counter x-ray crystal spectrometer.* J. Sci. Instr. **31** (1954) 321–326.

[81] A. R. Lang. *Suggested necessary conditions for successful use of the stationary crystal integrated reflection measuring method with the counter spectrometer.* Rev. Sci. Instr. **25** (1954) 1039–1040.

[82] B. E. Warren. *Peak areas with a recording diffractometer.* Norelco Reptr. **2** (1955) 63.

[83] J. Taylor and W. Parrish. *Absorption and counting-efficiency data for x-ray detectors.* Rev. Sci. Instr. **26** (1955) 367–373.

[84] Thomas C. Furnas and David Harker. *Apparatus for measuring complete single-crystal x-ray diffraction data by means of a Geiger counter diffractometer.* Rev. Sci. Instr. **26** (1955) 449–453.

[85] W. L. Bond. *A single-crystal automatic diffractometer. I.* Acta Cryst. **8** (1955) 741–746.

[86] T. S. Benedict. *A single-crystal automatic diffractometer. II.* Acta Cryst. **8** (1955) 747–752.

[87] W. Parrish. *X-ray intensity measurements with counter tubes.* Philips Tech. Rev. **17** (1956) 206–221.

[88] A. R. Lang. *A versatile x-ray diffractometer for single-crystal and powder studies.* J. Sci. Instr. **33** (1956) 138–141.

[89] J. W. Hughes and E. R. Pike. *A monitored Geiger-counter x-ray powder diffractometer with automatic recording.* J. Sci. Instr. **33** (1956) 204.

[90] R. Bones. *A two-circle Geiger counter mounting.* J. Sci. Instr. **33** (1956) 241–243.

[91] M. J. Buerger. *New single-crystal counter-tube technique.* Acta Cryst. **9** (1956) 834.

[92] W. Parrish and T. R. Kohler. *Use of counter tubes in x-ray analysis.* Rev. Sci. Instr. **27** (1956) 795–808.

[93] P. H. Dowling, C. F. Hendee, T. R. Kohler, and W. Parrish. *Counters for x-ray analysis.* Philips Tech. Rev. **18** (1956/57) 262–275.

[94] U. W. Arndt and D. C. Phillips. *On the determination of crystal and counter settings for a single-crystal x-ray diffractometer.* Acta Cryst. **10** (1957) 508–510.

[95] Thomas C. Furnas, Jr. *Single crystal orienter instruction manual.* (General Electric Company, Milwaukee, 1957).

[96] G. E. B. Barstad and A. F. Andresen. *Single crystal goniometer for x-ray and neutron diffraction.* Rev. Sci. Instr. **28** (1957) 916–918.

[97] A. J. van Bommel and J. M. Bijvoet. *The crystal structure of ammonium hydrogen* D*-tartrate.* Acta Cryst. **11** (1958) 61–70.

[98] A. McL. Mathieson. *A guide mechanism for a single-crystal x-ray counter goniometer.* Acta Cryst. **11** (1958) 433–436.

[99] I. D. Brown. *Determination of triclinic crystal setting for a single-crystal diffractometer.* Acta Cryst. **11** (1958) 510–511.

[100] Yōichi Iitaka. *A single-crystal diffractometer.* Mineral Soc. Japan (1959).

[101] Charles T. Prewitt. *The parameters* Υ *and* ϕ *for equi-inclination, with application to the single-crystal counter diffractometer.* Z. Krist. **13** (1960).

Balanced filters

[102] P. A. Ross. *Polarization of x-rays* (Abstract). Phys. Rev. **28** (1926) 425.

[103] P. A. Ross. *A new method of spectroscopy for faint x-radiations.* J. Opt. Soc. Am. **16** (1928) 433–437.

[104] Paul Kirkpatrick. *On the theory and use of Ross filters.* Rev. Sci. Instr. **10** (1939) 186–191.

[105] Paul Kirkpatrick. *Theory and use of Ross filters. II.* Rev. Sci. Instr. **15** (1944) 223–229.

[106] Paul Kirkpatrick and C. K. Chang. *X-ray monochromatization by four balanced filters* (Abstract). Phys. Rev. **66** (1944) 159.

[107] Jack A. Soules, William L. Gordon, and C. H. Shaw. *Design of differential x-ray filters for low-intensity scattering experiments.* Rev. Sci. Instr. **27** (1956) 12–14.

Monochromators

[108] R. A. Stephen and R. J. Barnes. *New technique for obtaining x-ray powder patterns.* Nature **136** (1935) 793–794; **137** (1936) 532–533.

[109] I. Fankuchen. *A condensing monochromator for x-rays.* Nature **139** (1937) 193–194.

[110] R. M. Bozorth and F. E. Haworth. *Focusing of an x-ray beam by a rocksalt crystal.* Phys. Rev. **53** (1938) 538–544, especially 544.

[111] I. Fankuchen. *Intense monochromatic beams of x-rays.* Phys. Rev. **53** (1938) 910.

[112] J. D. H. Donnay and I. Fankuchen. *A simple crystal monochromator for x-rays.* Rev. Sci. Instr. **15** (1944) 128–129.

[113] H. Lipson, J. B. Nelson, and D. P. Riley. *Monochromatic x-radiation.* J. Sci. Instr. **22** (1945) 184–187.

[114] R. C. Evans, P. B. Hirsch, and J. N. Kellar. A *'parallel-beam' concentrating monochromator for x-rays.* Acta Cryst. **1** (1948) 124–129.

[115] E. J. W. Whittaker. *The polarization factor for inclined-beam photographs using crystal-reflected radiation.* Acta Cryst. **6** (1953) 222–223.

[116] B. E. Warren. *Monochromatic x-rays for single crystal diffuse scattering.* J. Appl. Phys. **25** (1954) 814–815.

[117] Leonid V. Azaroff. *Polarization correction for crystal-monochromatized x-radiation.* Acta Cryst. **8** (1955) 701–704.

Recording range

[118] M. J. Buerger. *The photography of the reciprocal lattice.* Am. Soc. for X-ray and Electron Diffraction, Monograph No. 1 (1944) 36.

[119] Leonid V. Azaroff. *Crystal settings for upper level photography, precession method.* Rev. Sci. Instr. **25** (1954) 928–929.

[120] Gabrielle Donnay and J. D. H. Donnay. *Domain of reciprocal space accessible to precession photography.* Rev. Sci. Instr. **26** (1955) 610–612.

7

Some geometrical factors affecting intensities

In Chapter 3 an outline was given of some of the quantitative aspects of x-ray diffraction. The amount of energy E_{hkl} diffracted in the spectrum hkl as a crystal rotates uniformly about an axis normal to the x-ray beam can be written, from (85), Chapter 3, as

$$E_{hkl} = KL_{hkl}\, p_{hkl}\, |F_{hkl}|^2. \tag{1}$$

Here K is a constant for the experiment, but L_{hkl}, the Lorentz factor, and p_{hkl}, the polarization factor, differ from reflection to reflection.

The polarization factor is a simple function of 2θ and, according to (14) of Chapter 3, is given by

$$p = \frac{1 + \cos^2 2\theta}{2} \tag{2}$$

This factor does not depend on the details of the experiment.

The Lorentz factor, on the other hand, is concerned with the specific motion of the crystal, which was assumed in Chapter 3 to be a uniform rotation about an axis at right angles to the x-ray beam. This experimental arrangement is not sufficiently general to cover present-day experimental procedures, so that a more general formulation of the Lorentz factor is required.

In many x-ray diffraction experiments, another factor, the multiplicity factor, must be recognized. The Lorentz factor and the multiplicity factor are discussed in this chapter.

The rotation factor

Relation (80) of Chapter 3 can be described as applying directly only to reflections appearing on the zero level of a rotating-crystal photograph

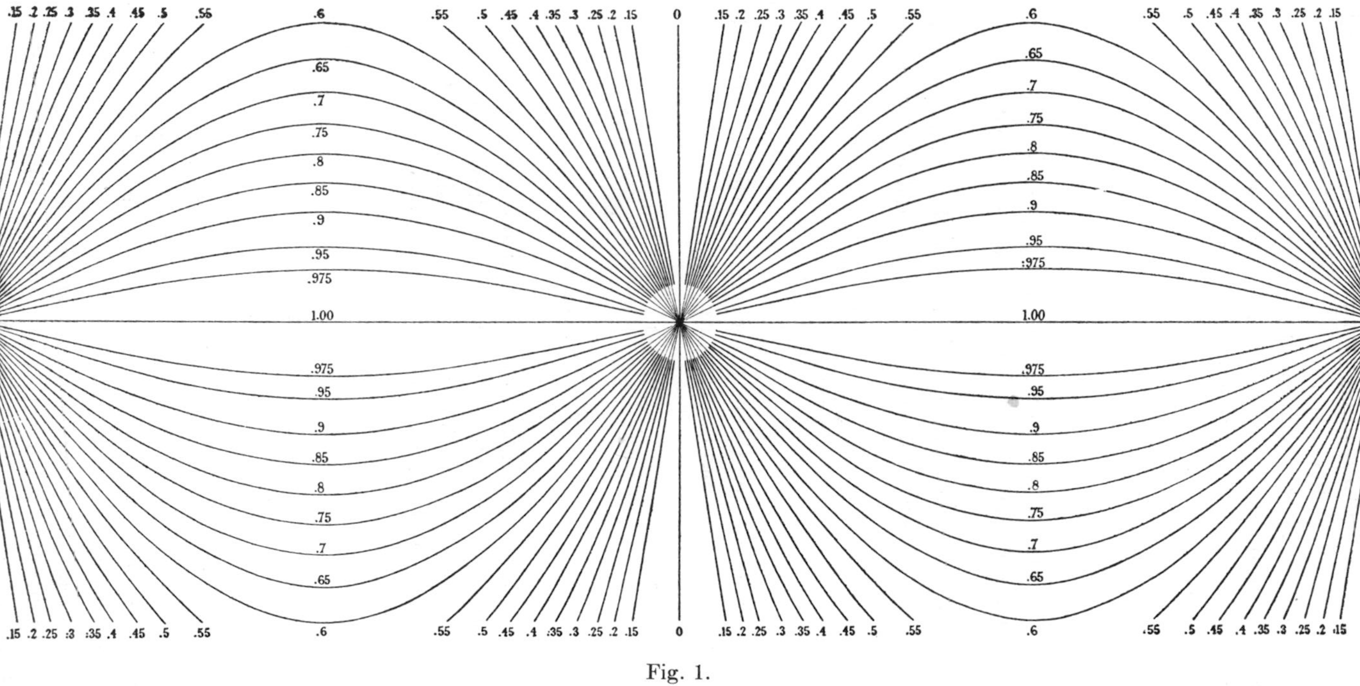

Fig. 1.
The rotation factor for the normal-beam rotating-crystal method for reflections recorded on a cylindrical film of diameter 57.3 mm. (After Cox and Shaw.[4])

taken by the normal-beam method. If it is to be applied to reflections on upper levels an allowance is required for the fact that, if the speed of rotation of the crystal is ω, planes inclined to the rotation axis pass through the condition of Bragg reflection at a smaller velocity ω'. In the early days of crystal-structure analysis, the correction for this velocity was called the *rotation factor*, and was defined as

$$D = \frac{\omega'}{\omega}. \tag{3}$$

Since this viewpoint is a less general one than given later, the theory of the Lorentz factor in terms of a rotation factor is given here only in outline.

Ott,[2] and later Cox and Shaw,[4] showed that the value of the rotation factor for the normal-beam rotating-crystal method is

$$D_{\perp} = \frac{\sqrt{\sin^2 \alpha - \sin^2 \theta}}{\cos \theta}, \tag{4}$$

where α is the angle between the rotation axis and the normal to the crystal plane. This can be converted to the form

$$D_{\perp} = \frac{1}{\sqrt{1 - \left(\dfrac{\sin \nu}{\sin 2\theta}\right)^2}}, \tag{5}$$

where ν is the angle between the equatorial plane and a generator of the cone of the nth level on which the reflection lies. Cox and Shaw,[4] published charts for the variation of $D_{\perp}$ over a cylindrical film and over a flat film. The chart for the cylindrical film, reduced for a camera of diameter 57.3 mm., is shown in Fig. 1.

Tunell[6] evaluated (3) for the equi-inclination arrangement and obtained

$$D_e = \frac{\sqrt{\cos^2 \mu - \cos^2 \theta}}{\sin \theta}. \tag{6}$$

The graph of this function is shown in Fig. 2. To use this chart, it is necessary to know $\sin \theta$ and μ for each reflection. The value of $\sin \theta$ may be found graphically from a plot of the reciprocal lattice with the aid of a simple device due to Booth.[12,17]

It will be evident in subsequent sections that the rotation factor is a part of the Lorentz factor correction. Unfortunately it was defined by Cox and Shaw[4] in inverse form. If the Lorentz factor is designated by

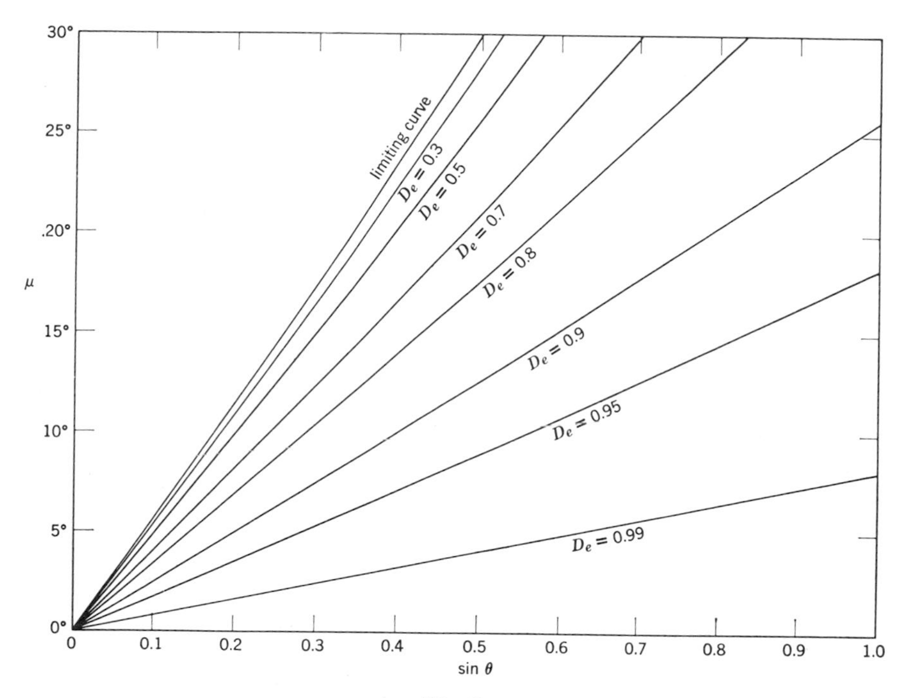

Fig. 2.
The rotation factor as a function of μ and $\sin\theta$ for the equi-inclination technique. (After Tunell.[6])

L_0 for the zero level, and by L_n for the nth level, then

$$L_n = \frac{1}{D} L_0. \tag{7}$$

Therefore:

For normal beam:

$$L_n = \frac{\cos\theta}{\sqrt{\sin^2\alpha - \sin^2\theta}} \quad \frac{1}{\sin 2\theta}$$

$$= \frac{1}{2\sin\theta\sqrt{\sin^2\alpha - \sin^2\theta}}, \tag{8}$$

and,

For equi-inclination:

$$L_n = \frac{\sin\theta}{\sqrt{\cos^2\mu - \cos^2\theta}} \quad \frac{1}{\sin 2\theta}$$

$$= \frac{1}{2\cos\theta\sqrt{\cos^2\mu - \cos^2\theta}}. \tag{9}$$

These are inconvenient functions. The more general view of the Lorentz factor given next not only lends itself to simpler expressions, but permits application in even more general cases.

Theory of the Lorentz factor

The Lorentz factor in terms of the reciprocal lattice. The factor which appeared as $1/\sin 2\theta$ in (80) of Chapter 3 is known as the *Lorentz factor*, because a particular form of it applicable to the Laue method was first derived by H. A. Lorentz in his classes. The form derived in Chapter 3 is a specialized form applicable only when two conditions are met: 1, that the reflection is from a plane parallel to the rotation axis and, 2, that the x-ray beam is directed perpendicular to the axis. There is evidently still needed the appropriate form of the factor for planes of any slope with respect to the rotation axis, as well as for any angle between the rotation axis and x-ray beam. In order to establish these forms it is important to inquire into the significance of the Lorentz factor.

If the term $1/\sin 2\theta$ in (80) of Chapter 3 is traced back, it is found that it was not in any way concerned with the integration in (75). This means that each variable element undergoing integration is also affected by this factor. Thus, the factor not only applies to integrated reflections, but it applies to peak reflections as well.

In making substitutions from (77) through (79) of Chapter 3 it is evident that the Lorentz factor is composed of two parts:

$$L = \frac{1}{\sin 2\theta}$$

$$= \left(\frac{1}{2}\frac{1}{\sin \theta}\right)\left(\frac{1}{\cos \theta}\right). \tag{10}$$

The first factor comes via q^2, and is a measure of the relative intensity reflected by a unit volume at glancing angle θ. The second factor comes via $1/B$, and is an inverse measure of the relative rate of change, with glancing angle, of the path length.

These two parts of the Lorentz factor can be identified as features of the reciprocal-lattice explanation of reflection. Figure 3 shows the situation for a point P on the zero level of the reciprocal lattice, the levels being taken normal to the rotation axis. These conditions correspond with the special case which was developed, namely, reflection by a plane parallel to the rotation axis. In Fig. 3,

$$\frac{\xi}{2} = \sin \theta. \tag{11}$$

Therefore, the trigonometric identity

$$\sin 2\theta = 2 \sin \theta \cos \theta \tag{12}$$

can be expressed as

$$\sin 2\theta = \xi \cos \theta, \tag{13}$$

so that

$$L = \frac{1}{\sin 2\theta} = \frac{1}{\xi \cos \theta}. \tag{14}$$

Now, let the crystal, together with its reciprocal lattice, rotate at a constant angular velocity ω. The velocity of a point P at a distance ξ

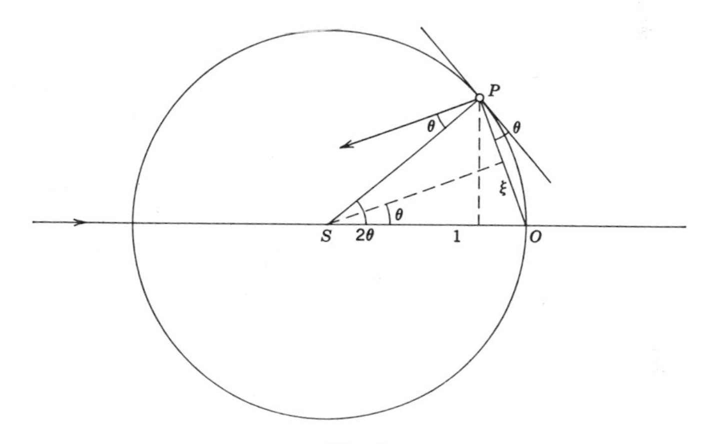

Fig. 3.

from the rotation axis is $\omega\xi$. Figure 3 shows that this velocity vector makes an angle θ with the normal to the sphere. Therefore the component of the velocity of P normal to the sphere is

$$V = \omega\xi \cos \theta. \tag{15}$$

The time taken for a reciprocal lattice point, P, to pass through the condition of reflection is inversely proportional to its velocity, so that the time during which the reflection occurs is

$$t = C\frac{1}{V} \tag{16}$$

$$= \frac{C}{\omega}\frac{1}{\xi \cos \theta}. \tag{17}$$

In (16) and (17), the constant C depends on the "size" of the reciprocal-lattice point[†] and on the width of the band making up the circle of reflection of the beam. The first term of (17) is a constant for the experiment, and the second term represents the variation which occurs in time-of-reflection opportunity for the various planes of the crystal. This second term is just the Lorentz factor, $1/\sin 2\theta$, for the special case, Fig. 4, according to (14). According to (15),

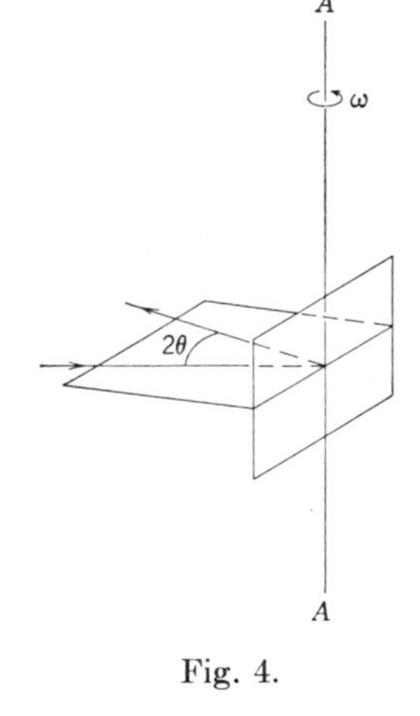

Fig. 4.

$$\frac{\omega}{V} = \frac{1}{\xi \cos \theta}, \tag{18}$$

and, according to (14) this is L, so that

$$L = \frac{\omega}{V}. \tag{19}$$

From this analysis, it can be concluded that the factor $1/\sin 2\theta$ can also be interpreted as the relative time opportunity for the various planes of the crystal to reflect. This factor can therefore be investigated for any desired set of conditions by making use of the reciprocal lattice and simply solving for the components of the factor. For this purpose the Lorentz factor for the special case just discussed can be represented by

$$L = \frac{1}{\sin 2\theta} \tag{20}$$

$$= \frac{1}{\xi \cos \theta}. \tag{21}$$

In this case the first component of the factor is the reciprocal of ξ and descends from q^2. The second is the reciprocal of $\cos \theta$, which descends from B and represents the ratio of the velocity of the point P to its normal component through the sphere of reflection.

Lorentz factor for methods employing a rotating crystal.[§] A more general treatment of the Lorentz factor than the one involving a "rotation factor," and at the same time one affording a simpler derivation, and achieving more general results, can be had by making use of the reciprocal-lattice concepts[5,8] of the last section. In this case it is chiefly

[†] In Chapter 2 it was seen that if the diffracting pattern consists of an infinite number of units, the diffraction maxima are infinitely sharp. Under such circumstances the reciprocal-lattice points are (ideally) infinitely small. When the number of pattern units is finite, the diffraction maxima have a finite width, and the reciprocal-lattice points have a finite size.

[§] Specifically, the rotation, oscillation, Weissenberg, Schiebold, Schiebold-Sauter, and de Jong-Bouman methods.

necessary to generalize the ratio of the velocity of the point P to its component normal to the sphere of reflection. Since the cases discussed in the next several sections are of great practical importance in the routine of crystal-structure analysis, they are treated rather fully.

In Fig. 5, the important reciprocal-lattice aspects of the problem are shown. Figure 5A shows the plan, Fig. 5C an elevation, and Fig. 5B an intermediate view of the sphere of reflection. The reciprocal lattice is rotating with an angular velocity ω. Reciprocal lattice point, P, on the nth level, has just reached the n-layer reflecting circle and is in the process of reflecting. P is moving in the direction PU and has a linear velocity $\omega\xi$. The velocity with which it passes through the sphere, however, is the component of this velocity on the normal to the sphere at the point P, namely, its component on PS. Designating as η the angle between PS and PU, the speed with which P passes through the condition of reflection is given by

$$V = \omega\xi \cos \eta. \tag{22}$$

To evaluate $\cos \eta$, pass a plane through the center of the sphere S, and normal to the velocity direction PU. It is evident that

$$\cos \eta = q. \tag{23}$$

Substituting this in (22) gives

$$V = \omega\xi q. \tag{24}$$

Referring to Fig. 5A, it will be observed that the last two terms of (24) represent twice the area, A, of triangle $O_n\, PS_n$; i.e.,

$$V = \omega \cdot 2A. \tag{25}$$

Now, according to (19), the Lorentz factor is simply the ratio of ω to V. Therefore

$$\frac{1}{L} = \frac{V}{\omega}. \tag{26}$$

Substituting from (25), it is seen that a general expression of the Lorentz factor correction is evidently

$$\frac{1}{L} = 2A, \tag{27}$$

A being the area of triangle whose sides are ξ, R_0 and R_n.

If the sides of a scalene triangle are a, b, and c, the area of the triangle is given by

$$A = \sqrt{s(s-a)(s-b)(s-c)}, \tag{28}$$

where

$$s = \tfrac{1}{2}(a+b+c). \tag{29}$$

By substituting from (29) into (28), the area can also be explicitly expressed

$$A = \tfrac{1}{4}\sqrt{(a+b+c)(-a+b+c)(a-b+c)(a+b-c)}. \tag{30}$$

In the present connection, the sides a, b, and c, have the following specific values (Fig. 5A and C):

$$\begin{aligned} a &= \xi, \\ b &= R_0 = \cos\mu, \\ c &= R_n = \cos\nu. \end{aligned} \tag{31}$$

When these are substituted into (30), and then this value of the area is substituted into (27), the general value of the Lorentz correction is seen to be

$$\frac{1}{L} = \tfrac{1}{2}\sqrt{(\xi+\cos\mu+\cos\nu)(-\xi+\cos\mu+\cos\nu)(\xi-\cos\mu+\cos\nu)(\xi+\cos\mu-\cos\nu)}. \tag{32}$$

This rather cumbersome form can be simplified as follows for certain special cases:

$$\left.\begin{array}{ll} \textit{Equi-inclination:} & \mu = -\nu \\ \textit{Anti-equi-inclination:} & \mu = \nu \\ \textit{Any zero level:} & \mu = \nu \end{array}\right\} \therefore \cos\mu = \cos\nu.$$

$$\frac{1}{L} = \tfrac{1}{2}\sqrt{(\xi+2\cos\nu)(-\xi+2\cos\nu)\xi^2} \tag{33}$$

$$= \frac{\xi}{2}\sqrt{4\cos^2\nu - \xi^2}. \tag{34}$$

The foregoing development provides a measure of the Lorentz factor or its correction in terms of the reciprocal-lattice coordinate ξ. The Lorentz factor can also be expressed in terms of the film coordinate Υ, as follows.

Returning to (27), it will be observed that the area, A, of the triangle $O_n PS_n$, Fig. 5A can be expressed as

$$A = \tfrac{1}{2}wR_0. \tag{35}$$

Substituting for R_0 from (31) and then substituting (35) in (27) gives

$$\frac{1}{L} = w\cos\mu. \tag{36}$$

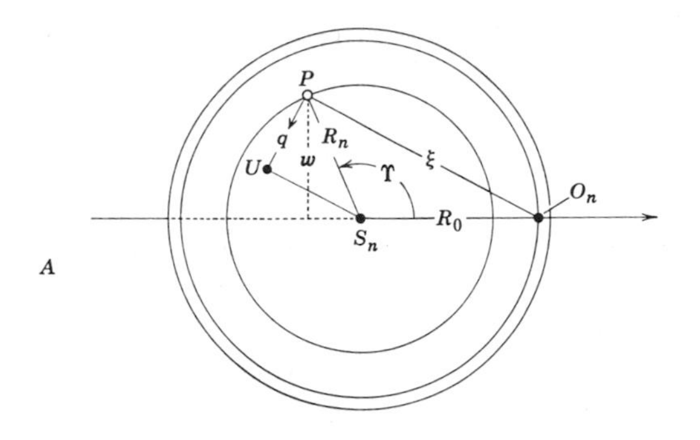

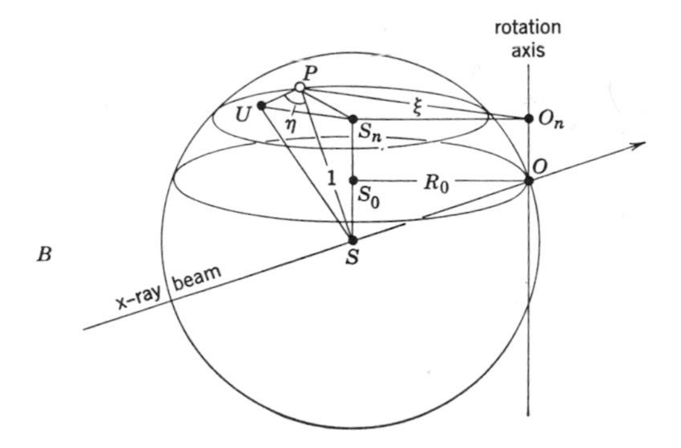

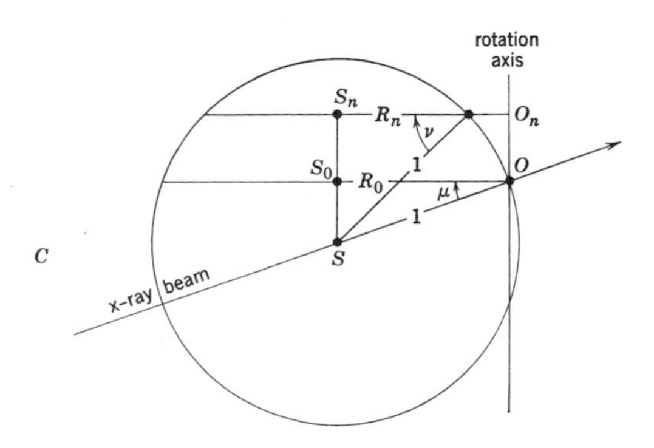

Fig. 5.
(After Buerger.[8])

For photographs made with a cylindrical film, the following relation provides a useful substitute for w (Fig. 5A):

$$\sin \Upsilon = \frac{w}{R_n}, \tag{37}$$

where R_n is given by (31). Substituting for w in (36) and for R_n in (37) gives

$$\frac{1}{L} = \cos \mu \cos \nu \sin \Upsilon. \tag{38}$$

In this relation, the angle Υ is directly proportional to the distance of a diffraction spot from the center line of the cylindrical film, so that Υ is a film coordinate.† With an appropriate camera diameter, the angle Υ in degrees is equal to the distance of the spot from the center line of the film, measured in half-millimeters. This camera diameter, namely 57.3 mm., is a standard one for Weissenberg and rotating-crystal cameras in the United States. The angle Υ is also a setting coordinate for the single-crystal quantum-counter apparatus, Chapter 6.

Relation (38) reduces to a simpler form for special arrangements of recording the reflections, as follows:

Equi-inclination, $\mu = -\nu$:
Anti-equi-inclination $\mu = \nu$:

$$\frac{1}{L} = \cos^2 \nu \sin \Upsilon, \tag{39}$$

Normal beam, $\mu = 0$:

$$\frac{1}{L} = \cos \nu \sin \Upsilon, \tag{40}$$

Flat cone, $\nu = 0$:

$$\frac{1}{L} = \cos \mu \sin \Upsilon. \tag{41}$$

For the practical application of a correction for the Lorentz factor, it should be noted that the forms of all the above relations—(38) through (41)—are the same except for the cosine term. This represents a scale factor

$$S = \cos \mu \cos \nu, \tag{42}$$

which is a function of the method of recording and of the particular level only. This means that all reflections recorded on a photograph for a particular level have a Lorentz factor

$$L = \frac{1}{S \sin \Upsilon}. \tag{43}$$

This scale factor can be removed as each level is photographed by timing the exposure of the photograph of that level by a time proportional to

† M. J. Buerger. *X-ray crystallography.* (John Wiley and Sons, New York, 1942) 223.

(42). If this is done, the exposure time cancels the scale factor (42) characteristic of the level, and the practical correction for all photographs is simply sin Υ.

The Lorentz factor for pure precession motion. In order to derive the Lorentz factor appropriate to the precession method,[11] it is necessary to inquire first into the character of the motion of precession. At the time this book is being written, precession instruments are still made in essentially the original design, in which the film holder and crystal are mounted on a universal joint. This mounting gives rise to a certain minor lack of symmetry of motion, which will be discussed later, and

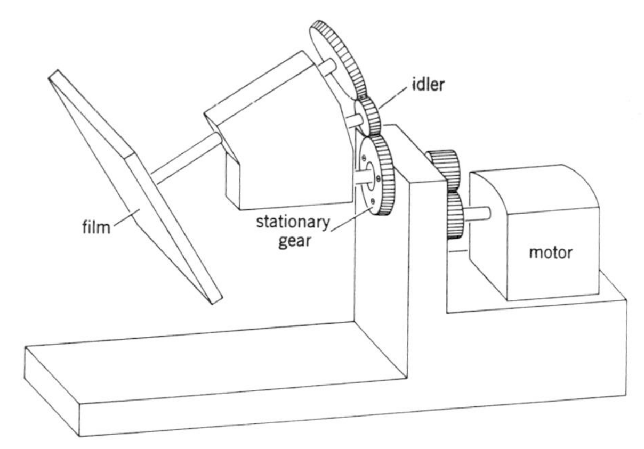

Fig. 6.

which complicates consideration of the motion in detail. In order to understand the derivation of the Lorentz factor readily, it is desirable to neglect these complications initially by considering a "pure" precession motion. This motion can be actually generated as indicated in Fig. 6, as well as in several other simple ways.

The motion of pure precession can be described in the following way: For precession angle $\bar{\mu}$, the zero-level plane touches the surface of a cone, Fig. 7, of half opening angle $\pi/2 - \bar{\mu}$. During the precession, the plane rolls on the surface of the cone, the contact migrating clockwise at angular rate ω. But at the same time there is a slippage at the contact that can be described as a counter-clockwise rotation of the plane about its normal. The reason for the slippage is that, at a distance R from the center, the cone has a radius of only $R \sin \bar{\mu}$. Since a point on the plane must touch the same point on the cone during each cycle, the precession motion must be such that the circumference $2\pi R$ of the plane must slip uniformly to match the circumference $2\pi R \sin \bar{\mu}$ of the cone. This requires a linear slippage of $2\pi R - 2\pi R \sin \bar{\mu}$ per cycle, which is equivalent to an angular slippage of $2\pi(1 - \sin \bar{\mu})$ per cycle, or an angular rate of $1 - \sin \bar{\mu}$. If the angular velocity of precession is ω, the angular slippage velocity is $(1 - \sin \bar{\mu})\omega$.

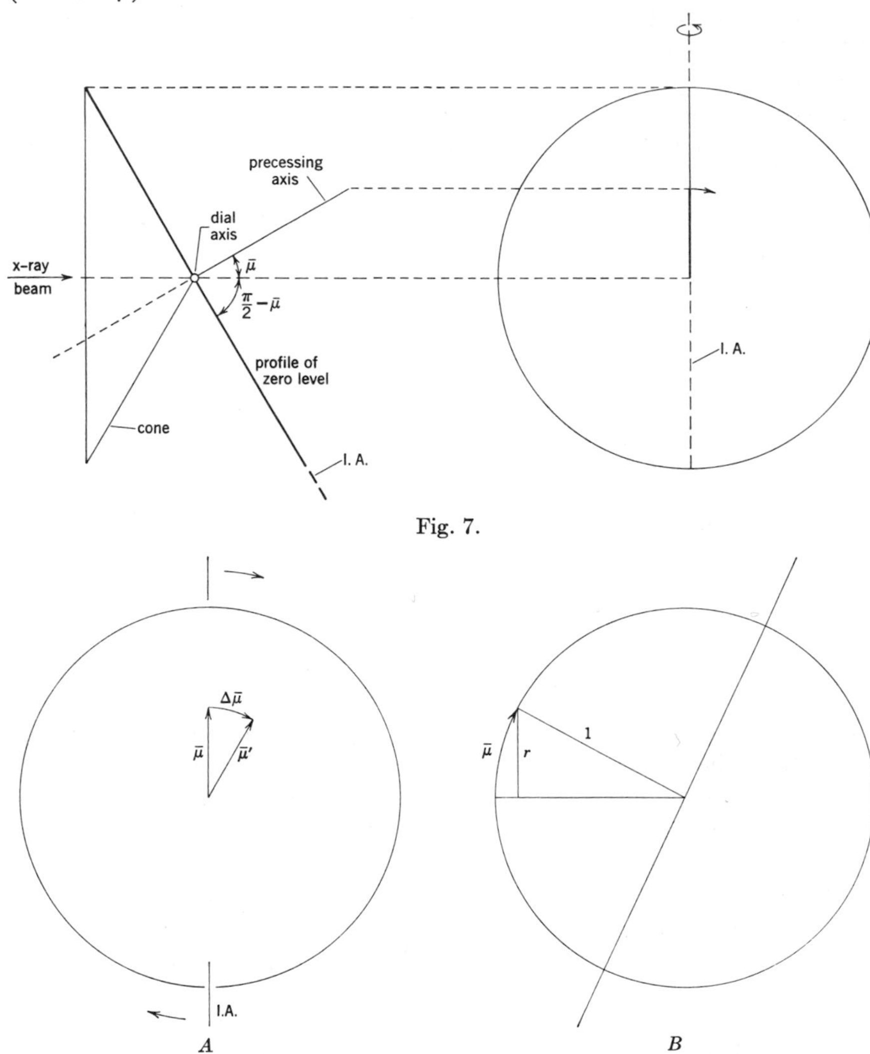

Fig. 7.

Fig. 8.

The total instantaneous motion of pure precession is thus a rolling motion about an axis in the plane of the zero level, plus a rotation at an angular rate $(1 - \bar{\mu})\omega$ about an axis normal to the zero level. Both of these axes precess. Each motion contributes a velocity to a reciprocal lattice point, and each therefore contributes to the Lorentz factor of the point.

Rolling component. This motion can be appreciated by representing angular displacements on the surface of a unit sphere. Suppose one looks along the x-ray beam toward the source. Then the precession angle, $\bar{\mu}$, can be represented by the displacement, $\bar{\mu}$, along a great circle on the sphere in Fig. 8A. (This is a reference sphere and not the sphere of reflection.) After the precession motion has proceeded a very short interval of time, the angular displacement is then represented by $\bar{\mu}'$. The small arrow connecting the ends of these displacement vectors represents the change, $\Delta\bar{\mu}$. Evidently when $\Delta\bar{\mu}$ is infinitesimally small, it is a vector equivalent to a rotation about the axis I.A., 90° away from the displacement $\Delta\bar{\mu}$.

In Fig. 8A, the arc $\bar{\mu}$ is rotating about the center of the diagram at the precession velocity ω. In Fig. 8B, both the arc $\bar{\mu}$ and the vertical line segment of length r are rotating about the horizontal line at velocity ω. The common end of both arc and line segment are therefore moving toward the observer at linear rate ωr. This motion is equivalent to an instantaneous rotation about the sloping line (I.A.) with angular velocity

$$\Omega = (\omega r) \cdot 1 \tag{44}$$

From Fig. 8B,

$$r = \sin \bar{\mu}. \tag{45}$$

Consequently

$$\Omega = \omega \sin \bar{\mu}. \tag{46}$$

The rolling component for reciprocal lattice point, P, which occurs, in general, on an n level, can be derived with the aid of Fig. 9. Figure 9A shows a view of the sphere of reflection, with the x-ray beam coming from left to right. The point of exit of the x-ray beam from the sphere is the origin O_0. The zero layer, tipped to the left by an angle $\bar{\mu}$, intersects the sphere of reflection in the zero-level circle. The diameter of the zero-level circle containing the origin is also the I.A. The relations just discussed are also shown looking normal to the levels in Fig. 9C. Most of the rest of the relations to be discussed are best appreciated from Fig. 9B, which is a view along the I.A. of Fig. 9A in the direction O_0S_0Q.

To derive the rolling component for reciprocal lattice point, P, it should be observed that the entire reciprocal lattice has an angular velocity Ω about the instantaneous axis, I.A. The velocity of the point P (Fig. 9B) at a distance j from the I.A. is Ωj. If the velocity vector makes an angle ψ with the radius of the sphere at P, the component of P's velocity normal to the surface of the sphere is

$$V = \Omega j \cos \psi \tag{47}$$

$$= (\omega \sin \bar{\mu}) j \cos \psi. \tag{48}$$

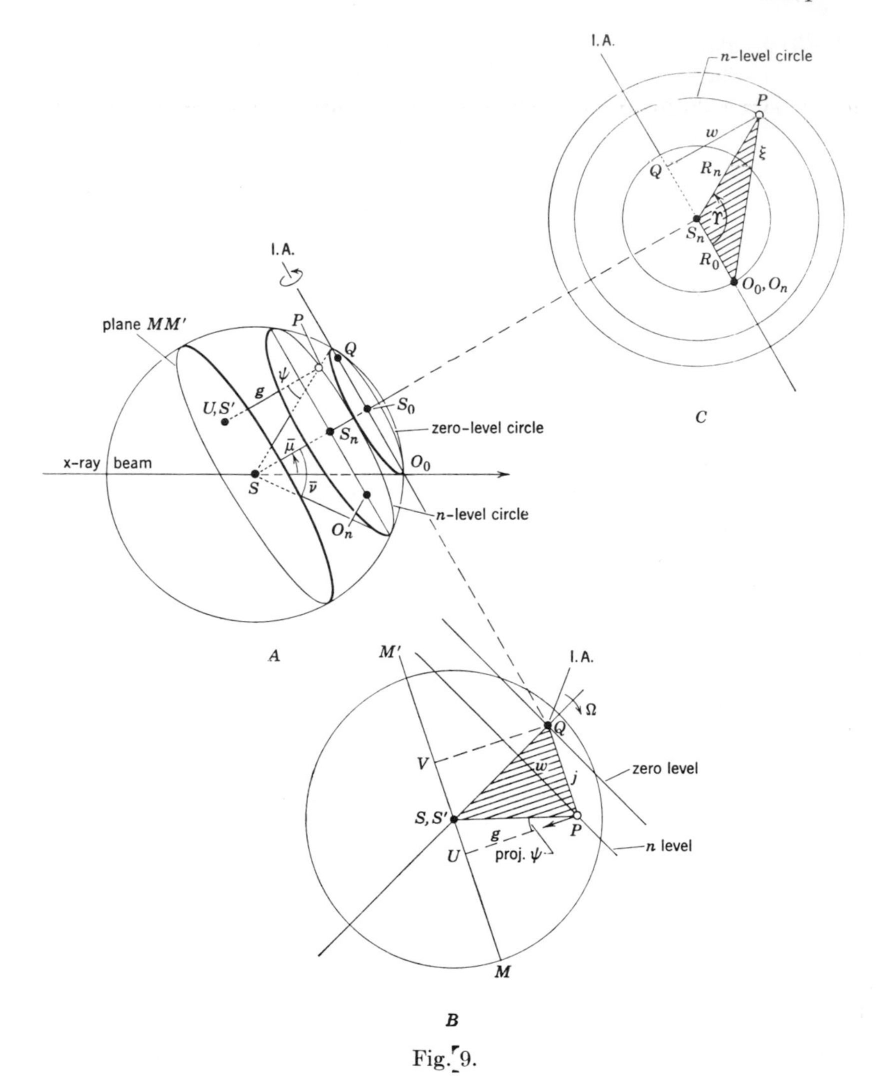

Fig. 9.

Since, according to (19), the Lorentz factor is the ratio of ω to V, it follows that

$$\frac{1}{L_{\|}} = j \sin \bar{\mu} \cos \psi. \tag{49}$$

To evaluate $\cos \psi$, pass a plane MM', Fig. 9B, through S normal to the velocity direction PU. Then PUS is a right angle, and

$$\cos \psi = \frac{PU}{PS} = \frac{PU}{1} = g. \tag{50}$$

Relation (49) is therefore equivalent to

$$\frac{1}{L_{\|}} = (jg) \sin \bar{\mu}. \tag{51}$$

The product in parentheses is the area of the rectangle $QPUV$, Fig. 9B. It is therefore twice the area of the triangle QPS'. A measure of the area of this triangle is $\frac{1}{2}\overline{QS'} \cdot w$. Inserting twice this value for jg in (51) one obtains

$$\frac{1}{L_{\|}} = \overline{QS'} \cdot w \sin \bar{\mu}. \tag{52}$$

From Fig. 9A, the distance

$$QS' = S_0S = \cos \bar{\mu}. \tag{53}$$

Therefore (52) can be rewritten

$$\frac{1}{L_{\|}} = w \sin \bar{\mu} \cos \bar{\mu} \tag{54}$$

$$= \tfrac{1}{2} w \sin 2\bar{\mu}. \tag{55}$$

Here w is the distance of the reciprocal-lattice point, P, from the projection of the I.A. on that level.

There are several ways of handling the development of the Lorentz factor at this point:

1. Relation (55) can be transformed as follows: Angle Υ is the projection of angle 2θ on the levels of the reciprocal lattice, Fig. 9C. The normal, w, can be expressed in terms of it:

$$\sin \Upsilon = \frac{w}{R_n}. \tag{56}$$

But

$$R_n = \sin \bar{\nu}, \tag{57}$$

so (54) and (55) can be rewritten

$$\frac{1}{L_{\|}} = \cos \bar{\mu} \sin \bar{\mu} \sin \bar{\nu} \sin \Upsilon \tag{58}$$

$$= \tfrac{1}{2} \sin 2\bar{\mu} \sin \bar{\nu} \sin \Upsilon. \tag{59}$$

It is interesting to compare (58), which applies to the rolling component of pure precession motion, with (38), which applies to rotation motion. The relation between the angles is

$$\bar{\mu} = \frac{\pi}{2} - \mu,$$
$$\bar{\nu} = \frac{\pi}{2} - \nu, \tag{60}$$

so that

$$\sin \bar{\mu} = \cos \mu,$$
$$\sin \bar{\nu} = \cos \nu. \tag{61}$$

Therefore the Lorentz factor for the two cases are the same except for the additional factor $\cos \bar{\mu}$ which occurs in the expression for the rolling component of pure precession motion.

2. But (54) can also be developed in a different way so as to display the relation between the Lorentz factors for the rolling component of precession and the rotation methods. In Fig. 9*C*,

$$wR_0 = 2A, \tag{62}$$

where A is the area of the triangle $O_n\ PS_n$, so

$$w = \frac{2A}{R_0} = \frac{2A}{\sin \bar{\mu}}. \tag{63}$$

This can be substituted into (54), giving

$$\frac{1}{L_{\|}} = \frac{2A}{\sin \bar{\mu}} \sin \bar{\mu} \cos \bar{\mu}$$
$$= 2A \cos \bar{\mu}. \tag{64}$$

Since the triangle $O_n\ PS_n$, whose area is given by A in (64), corresponds exactly to the triangle of the same labeling of Fig. 5*A*, and since (64) and (27) are the same except for the factor $\cos \bar{\mu}$, it is again evident that the Lorentz factors for rotation and the rolling component of pure precession are the same except for the factor $\cos \bar{\mu}$ in the precession case. The results of the last section on Lorentz factors for the rotation method can therefore be transformed readily into results applicable to pure precession motion. To do this, the customary designation for inclination angle and cone angle of the rotation methods should be transformed according to (61) into the angles customarily used in the precession method. If one makes this transformation of (32), the Lorentz correction for the rolling component of pure precession motion can be written

$$\frac{1}{L_{\|}} = \frac{\cos \bar{\mu}}{2} \sqrt{(\xi + \sin\bar{\mu} + \sin\bar{\nu})(-\xi + \sin\bar{\mu} + \sin\bar{\nu})(\xi - \sin\bar{\mu} + \sin\bar{\nu})(\xi + \sin\bar{\mu} - \sin\bar{\nu})}. \tag{65}$$

The complete Lorentz factor. The Lorentz factor for the motion of pure precession can be obtained by properly combining the separate factors due to the rolling and rotational components. The proper method of combination can be seen as follows: Each motion gives rise to a velocity V of a point P normal to the sphere. The algebraic sum of these velocities is the total velocity. Since $1/L = V/\omega$, each velocity component gives rise to a component of $1/L$. The rolling component of $1/L$ was derived in (58). The rotational component was shown to have the same form as (58) except that it lacks the term $\cos \bar{\mu}$. The general expression, which is a transformation of (38) by (61), is referred to a velocity V based upon the angular velocity of rotation. When the velocity is based upon the angular velocity of precession, then $V \to V(1 - \sin \bar{\mu})$ and the rotation component can be written

$$\frac{1}{L_\perp} = (1 - \sin \bar{\mu}) \sin \bar{\mu} \sin \bar{\nu} \sin \Upsilon. \tag{66}$$

If this is combined with (58), the result is

$$\frac{1}{L} = \frac{1}{L_\parallel} + \frac{1}{L_\perp} = (\cos \bar{\mu} + 1 - \sin \bar{\mu}) \sin \bar{\mu} \sin \bar{\nu} \sin \Upsilon. \tag{67}$$

This is the complete Lorentz factor for the motion of pure precession. The term in parentheses serves to distribute the correction between rolling and rotational components; when $\bar{\mu} = 0$ it is unity, and when $\bar{\mu} = 90°$ it is zero (so that $L = \infty$).

The Lorentz factor for precession motion with universal-joint suspension. At the time this book is being written, commercially made precession instruments conform to the original design in having crystal and film holder free to move in universal joints. This gives rise to interesting minor departures of the Lorentz factor from the values it has for pure precession motion. The reason for this is as follows.

Figures 10 and 11 next page show diagrams of the film holder in two positions. The film holder is free to rotate about a horizontal axis H, and the horizontal axis, in turn, is free to rotate about a vertical axis, V. In Fig. 10A, the normal to the film is horizontal. The arrow at the end of the normal shows the instantaneous direction of motion of that point of the normal. The instantaneous motion can be described as a rotation about an I.A. in the plane of the film, and this corresponds to the rolling component of pure precession motion. But when the precession has proceeded 90° further, as shown in Fig. 11, the situation is different. In this case the I.A. is the vertical axis V of the universal joint, and it does *not* lie in the plane of the zero level but makes an angle $\bar{\mu}$ with it. Such motion is *not* pure precession motion.

Figures 10 and 11 show two extreme cases. At any intermediate point

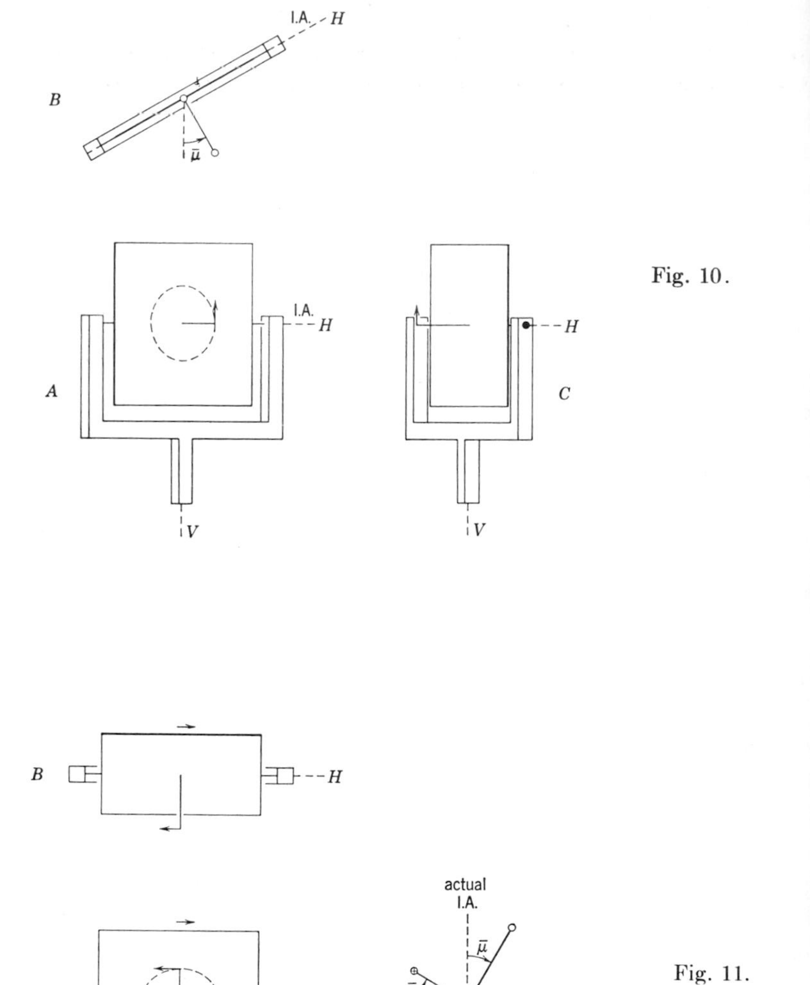

Fig. 10.

Fig. 11.

in the precession motion, the resultant motion has the effect of producing a Lorentz factor which varies with the aximuth τ, Fig. 12, (facing page), of the reciprocal-lattice point considered. Waser[24] has evaluated the function. This result, normalized to the definition of Lorentz factor used here, can be written

$$\frac{1}{L} = [\cos \bar{\mu} \sin \bar{\mu} \sin \bar{\nu} \sin \Upsilon]$$

$$\div \tfrac{1}{2}\left[\frac{1}{1 + \tan^2 \bar{\mu} \sin^2 (\tau + \eta)} + \frac{1}{1 + \tan^2 \bar{\mu} \sin^2 (\tau - \eta)}\right], \quad (68)$$

$$\text{where} \qquad \eta = \sin^{-1}\left(\frac{\sin \bar{\nu} \sin \Upsilon}{\xi}\right). \quad (69)$$

When $\bar{\mu} = 0$ this degenerates to

$$\frac{1}{L} = \cos \bar{\mu} \sin \bar{\mu} \sin \bar{\nu} \sin \Upsilon, \quad (70)$$

which is the value for the rolling component of pure precession motion.

Expression (68) gives numerical values of the Lorentz factor which increasingly deviate from those for the rolling component of pure precession motion with increasing precession angle $\bar{\mu}$. When $\bar{\mu} = 10°$, the maximum deviation is only about $\bar{\mu} = 30°$, 2%; when $\bar{\mu} = 20°$, it is 7%; but when the maximum deviation reaches about 16%.

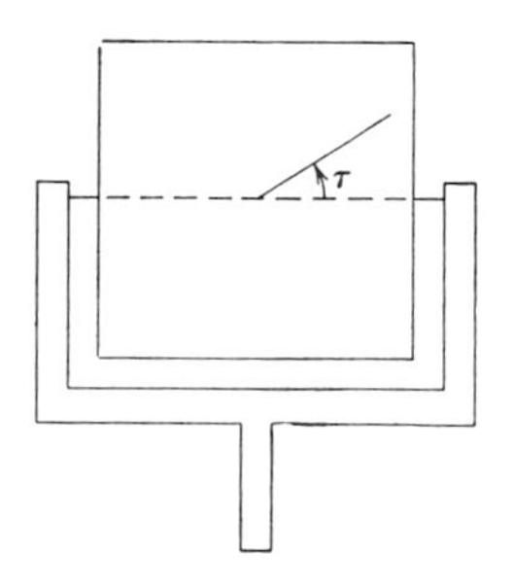

Fig. 12.

The Lorentz factor for the powder method. The Lorentz factor applicable to the powder method can be derived by a treatment similar to that used for the rotation and precession methods. In those methods, the element of volume of the crystal, dV, was represented by the reciprocal lattice point P_{hkl}, which had a specific location for a given orientation of the crystal. In the powder method, this volume is composed of a large number of crystallites in random orientation. The same volume, dV, can therefore be represented by a reciprocal lattice point P_{hkl} distributed over these orientations. In reciprocal space, P_{hkl} lies at a distance σ_{hkl} from the origin, so that the distribution of P_{hkl} is a uniform distribution over a sphere of radius σ_{hkl}.

As noted in the footnote on page 158, it is important to recognize that a reciprocal-lattice "point" is not a dimensionless point, but, rather, is a small volume whose linear dimensions in any direction are inversely proportional to the dimensions of the physical crystal in that direction. All reciprocal-lattice points of a specific crystal have the same dimensions. With this proviso, it is evident that the reciprocal-lattice point P_{hkl} for the powder method actually occupies a shell of thickness $d\sigma$ covering the surface of a sphere of radius σ as suggested by Fig. 13*A*

The shell thickness depends on the size of the reciprocal-lattice point of the average crystal of the powder. The entire shell is thus the statistical location of the reciprocal-lattice point P_{hkl}, and consequently represents the volume dV_{hkl}.

Only a part of this volume dV is in a position to reflect, namely that represented by the fraction of the shell of the sphere which is in contact with the sphere of reflection. Figure 13A shows that the fraction of the spherical shell P_{hkl} in condition to reflect can be represented by the ratio

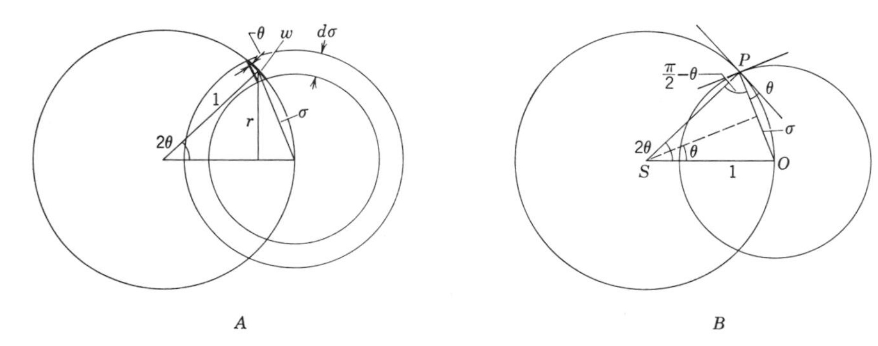

Fig. 13.

of the area of the ring of radius r and width w to the volume of the entire shell. Specifically, the fraction of the volume dV in reflecting condition is measured by

$$f = \frac{\text{Area of ring}}{\text{Volume of shell}} = \frac{w \cdot 2\pi r}{4\pi\sigma^2\, d\sigma}. \tag{71}$$

It is necessary to evaluate w, r, and σ as functions of something measureable in powder photographs, specifically the glancing angle, θ.

Figure 13B shows the way the two spheres intersect. Triangle PSO is isosoles, and $\angle PSO = 2\theta$. By bisecting this angle, it becomes evident that $\angle SPO = (\pi/2) - \theta$. The tangents at P are normal to the legs of this angle; hence the tangents intersect at $(\pi/2) - \theta$ and θ. Returning to Fig. 13A, it is now obvious that

$$w = \frac{d\sigma}{\cos\theta}. \tag{72}$$

Furthermore,

$$r = \sin 2\theta. \tag{73}$$

In Fig. 13B, it is seen that

$$\sigma = 2\sin\theta. \tag{74}$$

The values supplied by the last three relations can be substituted into (71), yielding

$$f = \frac{d\sigma}{\cos\theta}\,\frac{2\pi \sin 2\theta}{4\pi\cdot 4 \sin^2\theta\, d\sigma}$$

$$= \frac{\sin 2\theta}{8 \cos\theta \sin^2\theta}$$

$$= \frac{2 \sin\theta \cos\theta}{8 \cos\theta \sin^2\theta}$$

$$= \frac{1}{4 \sin\theta}. \tag{75}$$

This fraction of the volume element dV is in reflecting condition. Since this is a measure of the relative opportunity of the plane hkl to reflect, it plays a role similar to that of the Lorentz factor in the rotation of a single crystal. For, it will be observed that if one attempts to apply (89) of Chapter 3 to the powder method, the Lorentz factor, L, which was shown in (19) to be the ratio ω/V_{hkl}, is indeterminate, since both factors are zero. Since L merely represents the relative time opportunity of the various planes to reflect, it can be conveniently set equal to unity, since all opportunities are equal. But the various planes do not have equal volumes in reflecting position. Thus, $L\,dV$ for rotating crystals is replaced effectively in powder photographs by $f\,dV$, so (75) can therefore be regarded as the Lorentz factor for the entire Debye ring. The energy of the entire Debye ring is then found by substituting $f\,dV$ for $L\,dV$ in (89) and (90) of Chapter 3, giving

$$E_{\text{ring}} = K \frac{1}{4 \sin\theta}\, p|F|^2\, dV. \tag{76}$$

It is usually more convenient to measure the energy in some definite length. This can be derived from (76) from the following considerations. The total length of the Debye ring is $2\pi R \sin 2\theta$, where R is the radius of the camera. Therefore the energy caught per unit length of the ring is

$$E_u = \frac{K\left(\dfrac{1}{4 \sin\theta}\right) p|F|^2\, dV}{2\pi R \sin 2\theta}$$

$$= K\left(\frac{1}{8\pi R \sin 2\theta \sin\theta}\right) p|F|^2\, dV$$

$$= K\left(\frac{1}{8\pi R \cdot 2 \sin\theta \cos\theta \sin\theta}\right) p|F|^2\, dV$$

$$= K\left(\frac{1}{16\pi R \sin^2\theta \cos\theta}\right) p|F|^2\, dV. \tag{77}$$

Omitting the numerical factors, which are constant for the experiment, the factor in parentheses, namely

$$L_{\text{powder}} = \frac{1}{\sin^2 \theta \cos \theta} \tag{78}$$

$$= \frac{2}{\sin \theta \sin 2\theta}, \tag{79}$$

can be regarded as the practical Lorentz factor for the powder method.

It should be observed that (77) and (78) were derived on the assumption that only one reflection *hkl* is contributed to a Debye ring. It is necessary to recognize that a Debye ring always contains more than one reflection. In any case, it contains the contribution from *hkl* and $\overline{hkl}$, and it may contain many more. Further aspects of this are discussed beyond under "The multiplicity factor" and under "Reflection coincidences."

The polarization factor

In the foregoing discussion the form of the Lorentz factor was found in terms of instrumental coordinates and also as a function of the cylindrical coordinate ξ. For practical application, the polarization factor is required in these forms also.

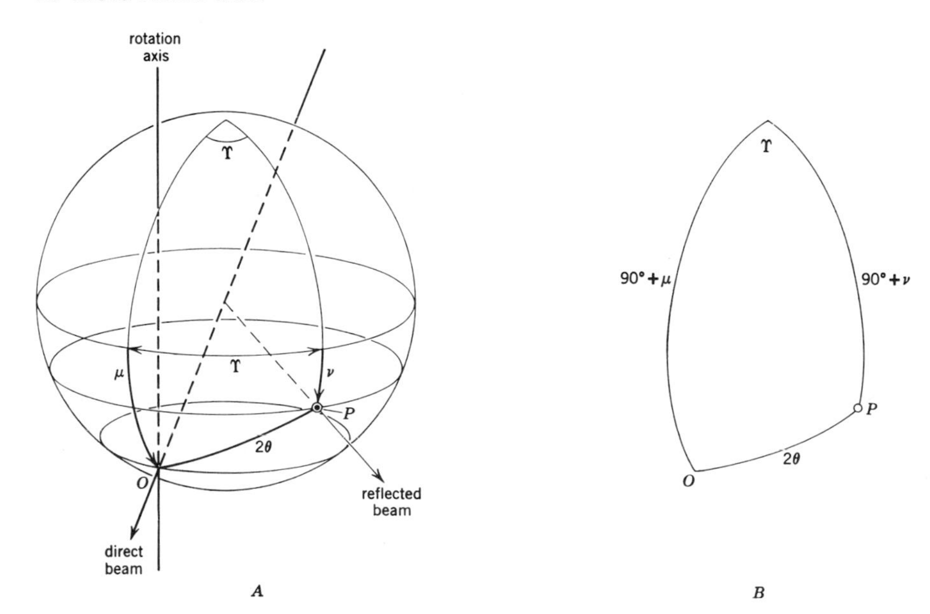

Fig. 14.

The polarization factor as a function of instrumental coordinates. The polarization factor is a simple function of θ:

$$p = \tfrac{1}{2}+\tfrac{1}{2}\cos^2 2\theta. \tag{80}$$

It is often desirable to express this in terms of μ, ν, and Υ. The relation between 2θ, μ, ν, and Υ for the general case is shown in Fig. 14*A*. If the law of cosines is applied to the spherical triangle shown in Fig. 14*B*, there results

$$\begin{aligned}\cos 2\theta &= \cos(90°+\mu)\cos(90°+\nu) + \sin(90°+\mu)\sin(90°+\nu)\cos\Upsilon \\ &= \sin\mu\sin\nu + \cos\mu\cos\nu\cos\Upsilon. \end{aligned} \tag{81}$$

If one substitutes this in (80), the general polarization factor is found to be

$$p = \tfrac{1}{2}+\tfrac{1}{2}\{\sin^2\mu\sin^2\nu + 2(\sin\mu\cos\mu)(\sin\nu\cos\nu)\cos\Upsilon + \cos^2\mu\cos^2\nu\cos^2\Upsilon\} \tag{82}$$

$$= \tfrac{1}{2}+\tfrac{1}{2}\sin^2\mu\sin^2\nu + \tfrac{1}{4}\sin 2\mu\sin 2\nu\cos\Upsilon + \tfrac{1}{2}\cos^2\mu\cos^2\nu\cos^2\Upsilon. \tag{83}$$

This reduces to the following special cases:

Normal beam, $\mu = 0$:

$$p = \tfrac{1}{2}+\tfrac{1}{2}\cos^2\nu\cos^2\Upsilon, \tag{84}$$

Flat cone, $\nu = 0$:

$$p = \tfrac{1}{2}+\tfrac{1}{2}\cos^2\mu\cos^2\Upsilon, \tag{85}$$

Equi-inclination, $\mu = -\nu$:

$$p = \tfrac{1}{2}+\tfrac{1}{2}\sin^4\nu - \tfrac{1}{4}\sin^2 2\nu\cos\Upsilon + \tfrac{1}{2}\cos^4\nu\cos^2\Upsilon, \tag{86}$$

Anti-equi-inclination, $\mu = \nu$:

$$p = \tfrac{1}{2}+\tfrac{1}{2}\sin^4\mu + \tfrac{1}{4}\sin^2 2\nu\cos\Upsilon + \tfrac{1}{2}\cos^4\nu\cos^2\Upsilon. \tag{87}$$

The polarization factor as a function of cylindrical coordinates. Figure 15 shows the direct and diffracted beams in relation to the sphere of reflection for general inclination. Since $SO = SP = 1$, triangle OSP is isoseles, and

$$\sin\theta = \frac{\sigma}{2}, \tag{88}$$

where σ is the vector in the reciprocal lattice to point P. If this is substituted in the trigonometric identity

$$\cos 2\theta = 1 - 2\sin^2\theta,$$

there results

$$\cos 2\theta = 1 - 2\left(\frac{\sigma}{2}\right)^2$$

$$= 1 - \frac{\sigma^2}{2}. \tag{89}$$

The polarization factor (80) can now be evaluated in terms of σ:

$$p = \tfrac{1}{2}+\tfrac{1}{2}\left(1 - \frac{\sigma^2}{2}\right)^2$$

$$= \tfrac{1}{2}+\tfrac{1}{2}\left(1-\sigma^2 + \frac{\sigma^4}{4}\right)$$

$$= \tfrac{1}{2}+\tfrac{1}{2} - \frac{\sigma^2}{2} + \frac{\sigma^4}{8}$$

$$= 1 - \frac{\sigma^2}{2} + \frac{\sigma^4}{8}$$

$$= \frac{8-4\sigma^2+\sigma^4}{8}. \tag{90}$$

The polarization factor and its reciprocal have been tabulated against σ as argument by Buerger and Klein.[13, 14]

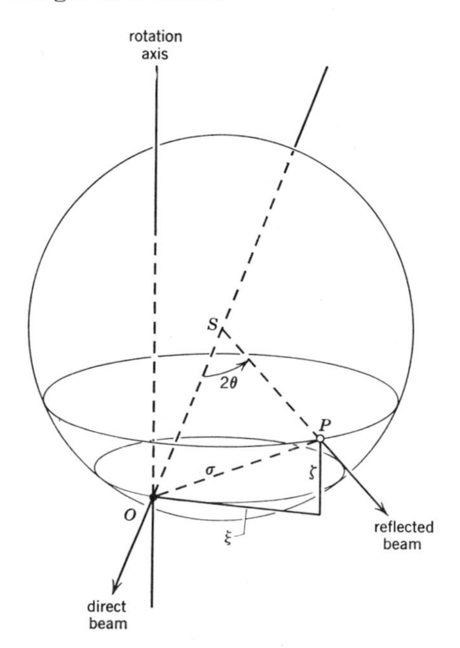

Fig. 15.

It is often desirable to express p as a function of ξ. Figure 15 shows that

$$\sigma^2 = \xi^2+\zeta^2. \tag{91}$$

Therefore

$$p = \frac{8-4\xi^2-4\zeta^2+\xi^4+2\xi^2\zeta^2+\zeta^4}{8}. \tag{92}$$

Practical methods for correcting for Lorentz and polarization factors

In a foregoing part of this chapter, the general theory of the Lorentz factor was discussed, and the forms of the Lorentz factors were derived for the general diffraction methods useful in crystal-structure analysis. The forms of $1/L$ for the several methods are listed in Table 1. In this section the practical methods of allowing for the Lorentz and polarization factors are considered.

The crystal-structure analyst starts with a set of "intensities." One of his first problems is to transform this to a set of $|F_{hkl}|^2$'s. This is done by solving (1) for $|F|^2$:

$$|F_{hkl}|^2 = \frac{1}{K}\frac{1}{L_{hkl}\,p_{hkl}}\,E_{hkl}. \tag{93}$$

The scale factor may be neglected since E_{hkl} is ordinarily not measured in absolute units anyway, and the $|F|^2$'s can be placed on an absolute basis by the method discussed in the next chapter. Thus each E must be multiplied by $1/Lp$ to obtain $|F|^2$. There are two essentially different methods of correcting E for the Lp factors:

1. The factors L and p can be removed in the process of recording, so that $|F|^2$'s are actually measured instead of E's.
2. (*a*) The measured E's can be multiplied by $1/Lp$ as obtained from a calculation tabulation, or chart based upon instrumental coordinates as arguments.

   (*b*) The measured E's can be multiplied by $1/Lp$ as obtained from a calculation, tabulation, or chart based upon reciprocal-lattice coordinates as arguments.

These methods are discussed briefly below.

Correction for *Lp* during recording. The Lorentz and polarization factors may be removed by suppressing the diffracted beam differentially with a function of the appropriate variable. The Lorentz factor is a simple function of Υ and the polarization is a simple function of θ. In changing from one level of the reciprocal lattice to another, these change in different ways, so it is best to set up a device which suppresses a reflection

Table 1
Forms of the Lorentz-factor correction

Methods involving a rotating crystal

General inclination, $\mu = \nu$:

$$\frac{1}{L} = \cos\mu \cos\nu \sin\Upsilon \tag{38}$$

$$\frac{1}{L} = \tfrac{1}{2}\sqrt{(\xi+\cos\mu+\cos\nu)(-\xi+\cos\mu+\cos\nu)(\xi-\cos\mu+\cos\nu)(\xi+\cos\mu-\cos\nu} \tag{32}$$

Equi-inclination, $\mu = -\nu$, and
Anti-equi-inclination, $\mu = \nu$:

$$\frac{1}{L} = \cos^2\nu \sin\Upsilon \tag{39}$$

$$\frac{1}{L} = \frac{\xi}{2}\sqrt{4\cos^2\nu - \xi^2} \tag{34}$$

Normal beam, $\mu = 0$:

$$\frac{1}{L} = \cos\nu \sin\Upsilon \tag{40}$$

$$\frac{1}{L} = \tfrac{1}{2}\sqrt{(\xi+1+\cos\nu)(-\xi+1+\cos\nu)(\xi-1+\cos\nu)(\xi+1-\cos\nu)}$$

Flat cone, $\nu = 0$:

$$\frac{1}{L} = \cos\mu \sin\Upsilon \tag{41}$$

$$\frac{1}{L} = \tfrac{1}{2}\sqrt{(\xi+1+\cos\mu)(-\xi+1+\cos\mu)(\xi-1+\cos\mu)(\xi+1-\cos\mu)}$$

Pure precession motion

$$\frac{1}{L} = (\cos\bar{\mu} + 1 - \sin\bar{\mu}) \sin\bar{\mu} \sin\bar{\nu} \sin\Upsilon \tag{58}$$

Precession motion with universal-joint suspension

$$\frac{1}{L} = \cos\bar{\mu} \sin\bar{\mu} \sin\bar{\nu} \sin\Upsilon \div$$

$$\tfrac{1}{2}\left[\frac{1}{1+\tan^2\bar{\mu}\sin^2(\tau+\eta)} + \frac{1}{1+\tan^2\bar{\mu}\sin^2(\tau-\eta)}\right] \tag{68}$$

where $\eta = \sin^{-1}\left(\dfrac{\sin\bar{\nu}\sin\Upsilon}{\xi}\right)$

Powder

$$\frac{1}{L} = \sin^2\theta \cos\theta \tag{78}$$

$$= \tfrac{1}{2}\sin\theta \sin 2\theta \tag{79}$$

Table 2
Charts for the correction of Lorentz-polarization factors

Methods involving a rotating crystal

Equi-inclination, $\mu = \nu$:
  As a function of ζ, ξ
    Lu,[10] Fig. 1
    Cochran[16] (X axis should be labeled ξ)
  As a function of μ, ξ:
    Boström[31] (gnomograph)
  As a function of ζ, Υ
    Lu,[10] Fig. 2 (X axis should be labeled Υ)
  As a function of μ, Υ
    Kaan and Cole[18] (axis labeled $1/Lp$ should be labeled Υ [use of chart presupposes exposures of levels have been timed by $\cos^2 \nu$])
    Bond[15] (gnomograph)
Anti-equi-inclination, $\mu = \nu$:
  As a function of ζ, ξ
    Kartha[25] (X axis should be labeled ξ)
  As a function of ζ, Υ
    Kartha[28] (X axis should be labeled Υ)
Normal beam, $\mu = 0$:
  As a function of ζ, ξ
    Cochran[16] (X axis should be labeled ξ)
  As a function of ζ, Υ
    (none)
  As a function of film coordinates x, y
    Kaan and Cole[18] (X axis is equal to ξ)

Precession motion with universal-joint suspension

Sections normal to levels
  $\bar{\mu} = 30°$: Burbank[26]
  $\bar{\mu} = 10°$, $15°$, and $21°$: Atoji and Lipscomb[29,30]
Sections parallel to levels
  $\bar{\mu} = 30°$: Grenville-Wells and Abrahams[27]

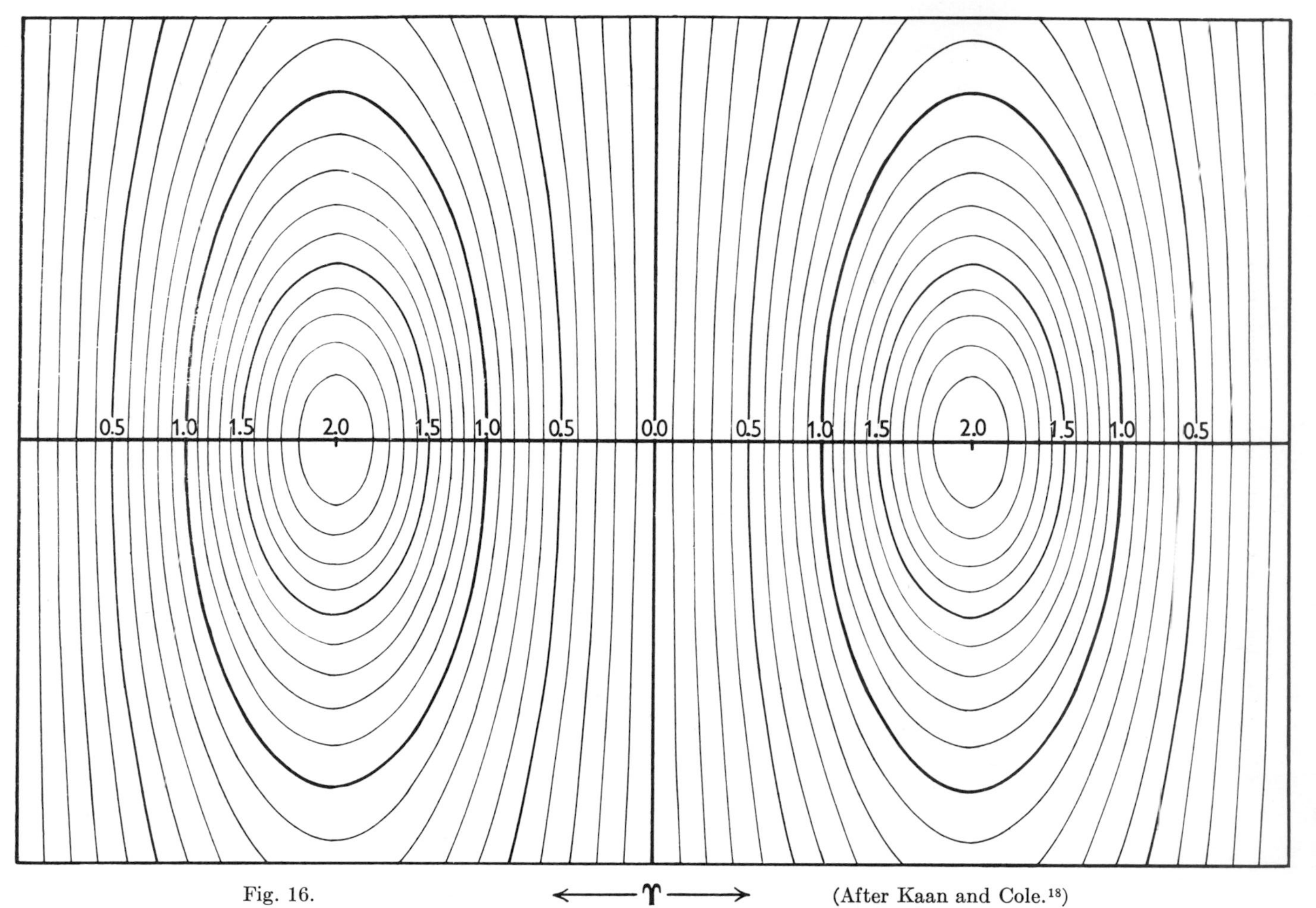

Fig. 16.

(After Kaan and Cole.[18])

The correction factor $1/Lp$ for the normal-beam rotating-crystal method for reflections recorded on a cylindrical film of diameter 57.3 mm.

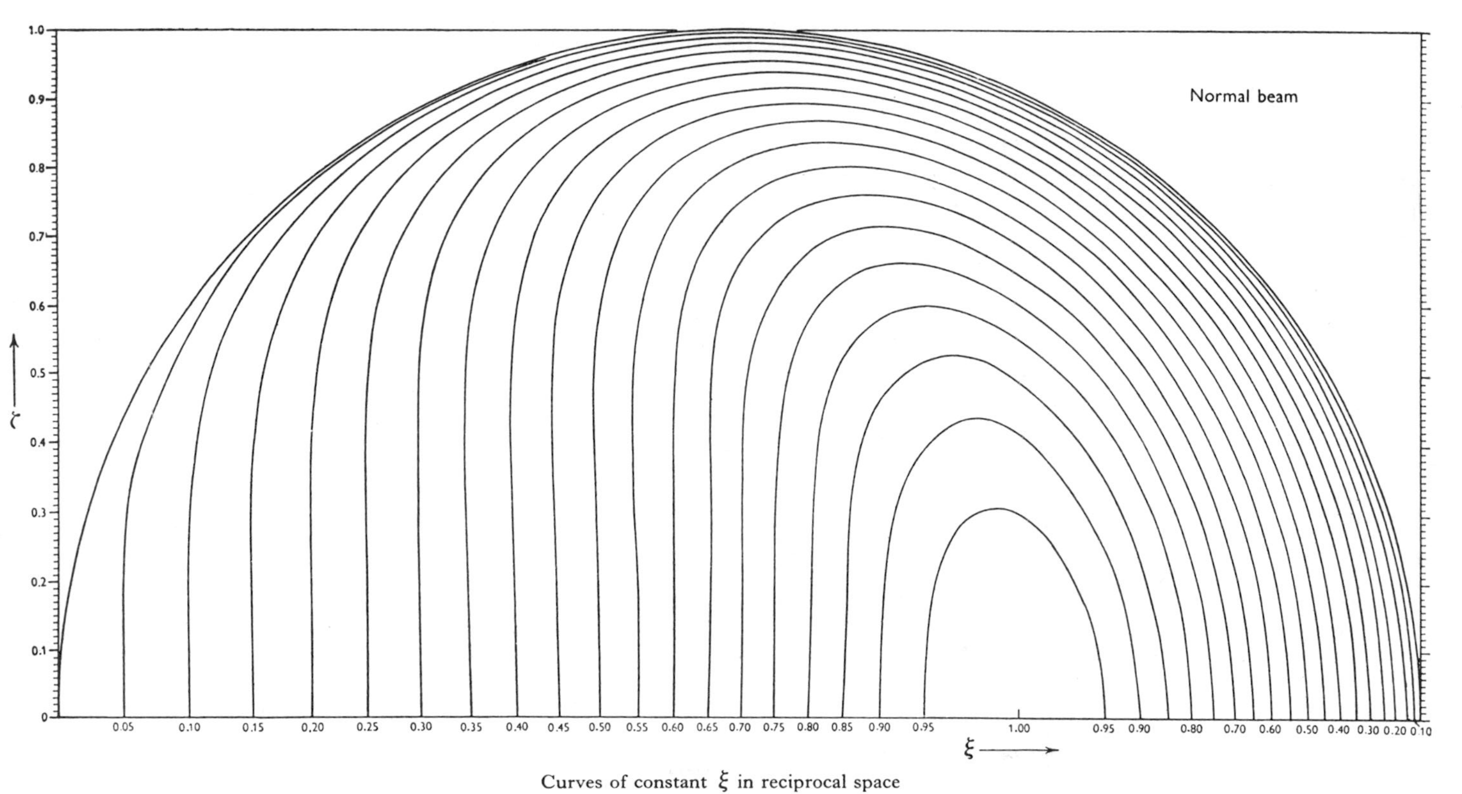

Fig. 17.
The correction factor $1/(2Lp)$ as a function of ζ and ξ, for the normal-beam technique.
(After Cochran.[16])

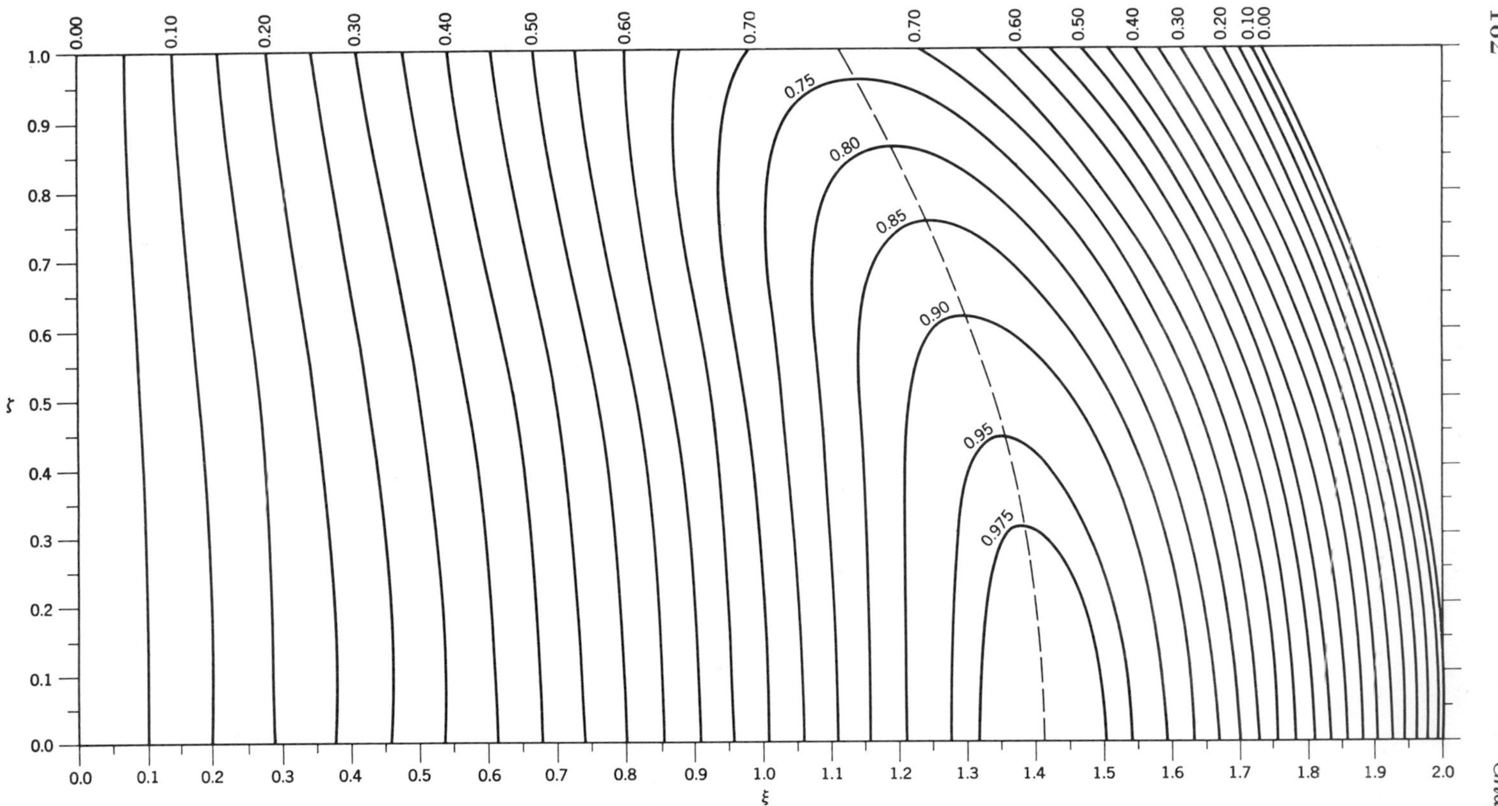

Fig. 18.
The correction factor $1/(2Lp)$ as a function of ζ and ξ for the equi-inclination technique.
(After Lu[10].)

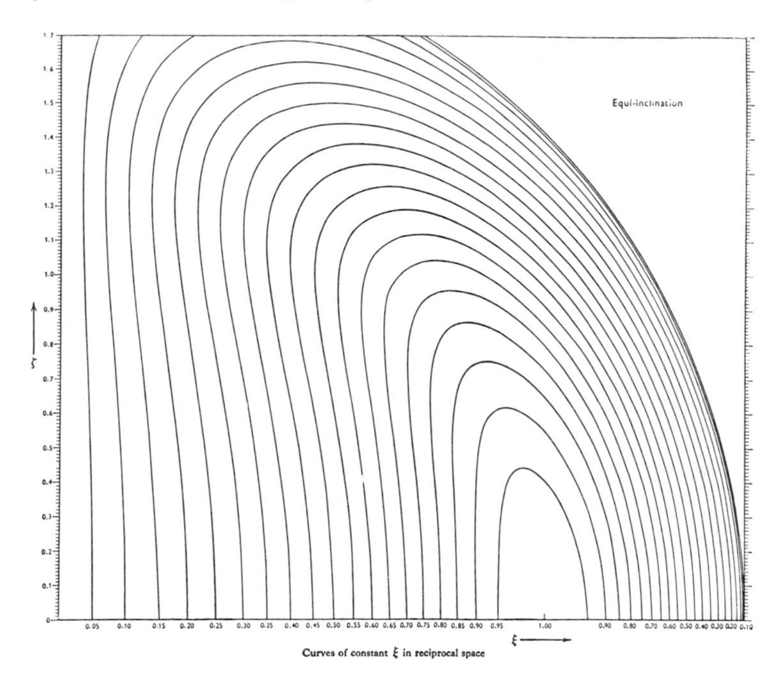

Fig. 19.
The correction factor $1/(2Lp)$ as a function of ζ and ξ for the equi-inclination technique. (After Cochran.[16])

as a function of Υ to correct for the Lorentz factor, and another device which suppresses a reflection as a function of θ to correct for the polarization factor.

Table 1 shows that regardless of the method, the Lorentz factor correction has the form $C \sin \Upsilon$, where C is $\cos \mu \cos \nu$, $\cos^2 \mu$, $\cos \mu$, or $\cos \nu$, depending on the inclination of direct and diffracted beams. A rotating sector[7] can be easily designed which permits transmission of the diffracted ray for a time proportional to $\sin \Upsilon$. This can be set, for example, in front of the cone opening of the de Jong-Bouman camera. The rotating sector is made of thin sheet metal and spun at high speed by a jet of compressed air directed against its edge. The spinning rate is so high that each reflection is interrupted by the sector blades many times while passing through the reflecting position. To apply the correction factor C, the exposure of each level is timed proportional to C.

The polarization correction, $2/(1 + \cos^2 2\theta)$, is radially symmetrical about the direction of the direct beam. It varies in the narrow range 1 to 2. It can be applied by any device which works well in this range. This can be arranged for each characteristic x-radiation by a symmetrical absorber. Such a device consists of a shell of absorbing material (Lucite has appropriate absorption for Mo$K\alpha$) radially symmetrical about the

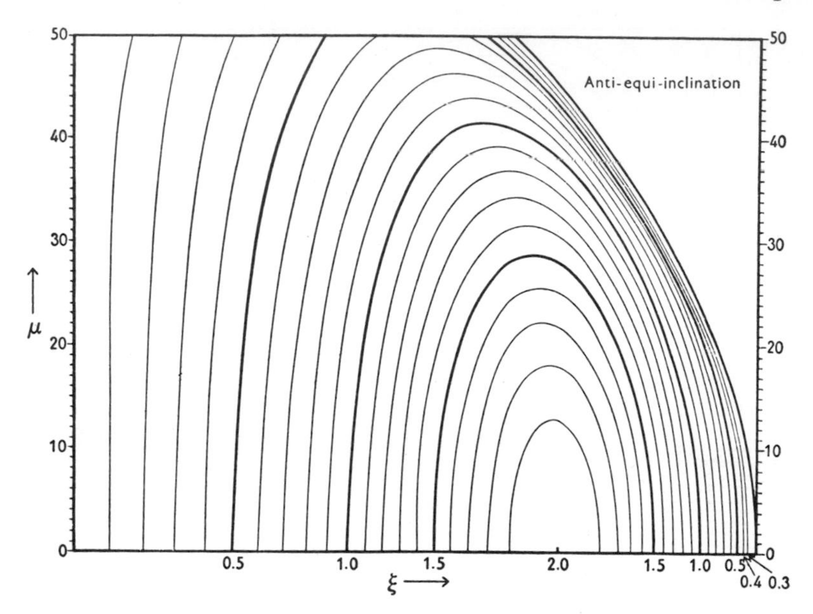

Fig. 20.
The correction factor $1/Lp$ as a function of μ and ξ for the anti-equi-inclination technique.

(After Kartha.[25])

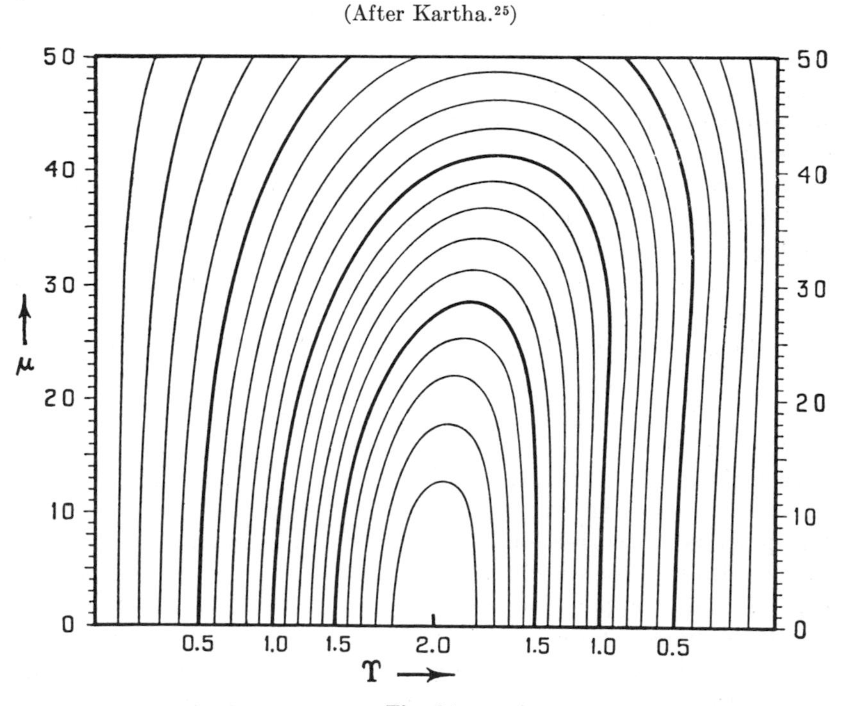

Fig. 21.
The correction factor $1/Lp$ as a function of μ and Υ for the anti-equi-inclination technique.

(After Kartha.[28])

x-ray beam and centered on the crystal. The thickness of the shell is such that its transmission is proportional to $1/p = 2/(1 + \cos^2 2\theta)$.

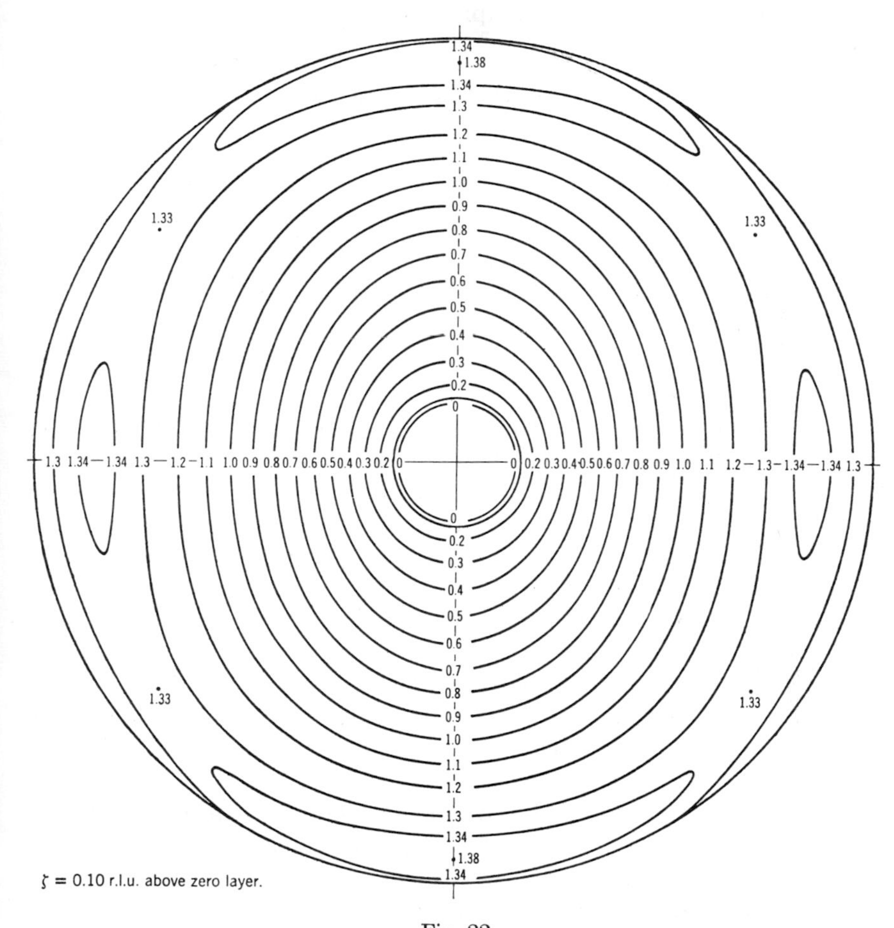

Fig. 22.
The correction factor $1/Lp$ for a precession photograph taken with $\mu = 30°$, $\zeta = 0.10$ above zero level.
(After Grenville-Wells and Abrahams.[27])

Correction for *Lp* by computing. A number of writers have computed the *Lp* correction for various methods, and presented the variation of the function in either scale form or chart form. The chart form permits one to prepare a scale for each layer line for any specific crystal and apply the correction (*a*) according to the position of the spot on the film, or the ϒ setting of a counter diffractometer, or (*b*) according to the location of the point on the reciprocal lattice. Table 2 outlines the charts available for use. The most important charts for use with normal beam are reproduced in Figs. 16 and 17, for equi-inclination in Figs. 18 and 19, and for anti-equi-inclination in Figs. 20 and 21. An example of the charts for precession photographs is illustrated in Fig. 22.

The multiplicity factor

By some of the methods for recording x-ray reflections it is impossible to separate certain reflections $h_1 k_1 l_1$, $h_2 k_2 l_2 \cdots h_m k_m l_m$, and as a consequence, the total energy in that portion of the record is

$$\begin{aligned} E_{\text{total}} &= E_{h_1k_1l_1} + E_{h_2k_2l_2} + E_{h_3k_3l_3} \cdots + E_{h_mk_ml_m} \\ &= \sum_j E_{h_jk_jl_j}. \end{aligned} \tag{94}$$

Allowance must be made for this in considering reflection intensities on either an absolute or a relative basis. The character of the allowance depends on whether the several *hkl*'s involved in (94) are symmetrically equivalent or distinct. In the event that the reflections are symmetrically equivalent, their energies are equal, and (94) becomes simply

$$E_{\text{total}} = mE_{hkl}. \tag{95}$$

The factor m is known as the *multiplicity factor*. It is an integer whose magnitude varies with the method of recording and with the index of the reflection involved. It is discussed for particular methods of recording below.

Weissenberg, Sauter, and Schiebold methods. In these methods, only one plane of the crystal reflects to a particular position on the photographic film. Therefore, for these methods, $m = 1$.

de Jong-Bouman and precession methods. In these two methods, only one plane of the crystal reflects to a particular spot on the film, but there is a complication which can be taken care of formally by an increased multiplicity factor. In any method involving rotation, the reciprocal-lattice point comes into reflecting position twice during a rotation, namely, once on entering the sphere of reflection and again on leaving it. In the Weissenberg method, the reflection arising from the first condition is recorded on the upper half of the film, while that arising from the second condition is recorded on the lower half, so that the reflections developed during these two times of attaining reflecting condition are recorded as two separate spots on the film. In the Schiebold-Sauter and in the Schiebold method, the reflection arising from the second condition is suppressed by closing half the opening of the layer-line screen so that each spot represents only one reflection per rotation. But in the de Jong-Bouman and in the precession method, the two reflections are, by the nature of the methods, designed to record at the same point on the photographic film. Thus each spot receives two contributions from the same plane in the course of one rotation. This may be allowed for in absolute measurement considerations by assigning $m = 2$ for these methods. Since the factor is applied to all reflections without descrimination, the matter may be neglected except in any attempt to make actual use of the absolute value of the constant K in (86) and (91) of Chapter 3.

Oscillating-crystal method. In this method (unless an ill-chosen oscillation range is employed so that the effects of the next paragraph occur), only one plane reflects to a particular spot on the film, so $m = 1$. On the other hand, since the crystal does not undergo a complete rotation, the ratio of reflection time to dead time is increased, and the time factor, t, of (91) of Chapter 3, must be replaced by the factor $t \times 360°$/oscillation angle.

Rotating-crystal method. In this method, any two planes which have identical spacings, d, and identical slopes with respect to the rotation axis† must necessarily reflect to the same spot on the film. This condition is fulfilled by reflections which are related by a symmetry element containing the rotation axis. Specificially,

1. An n-fold axis parallel to the rotation axis causes n symmetrically related reciprocal-lattice points to fulfill the condition. Consequently, it induces a multiplicity factor of $m_1 = n$.
2. A reflection plane containing the rotation axis causes pairs of points symmetrically related by it to fulfill the condition. For points not on the plane, therefore, it induces a multiplicity factor of $m_2 = 2$. *But for points on the plane, there is no companion point, so* $m_2 = 1$.
3. An inversion center merely causes pairs of planes which reflect to the upper and lower halves of the film to be equivalent. Since the upper and lower records are distinct, the center merely induces a trivial multiplicity factor of $m_3 = 1$.

The resultant multiplicity factor is the product of the contributions of all symmetry elements contained in the rotation axis, namely

$$\begin{aligned} m &= m_1\, m_2\, m_3 \\ &= m_1\, m_2, \end{aligned} \tag{106}$$

† M. J. Buerger. *X-ray crystallography.* (John Wiley and Sons, New York, 1942) Fig. 75, page 135.

since m_3 is unity. For diffraction effects, an inversion center is effectively present in the crystal.† It will be developed more fully in the next section that this has an effect on the multiplicity factor which can be described by saying that the multiplicity factor is that due to the symmetry, not of the crystal, but of its diffraction effects. The symmetries of the diffraction effects of the several crystal classes are listed in Table 3. As a consequence of this, the multiplicity factor of the rotating-crystal method is $2n$ or n for a general reflection (where n is the period of the axis of the diffraction class of the crystal which has been placed parallel to the rotation axis) according as the axis is or is not in a reflection plane of the diffraction class. The multiplicity factor for this general reflection is

Table 3
Multiplicity factors and coincidences for the powder method

Crystal system	Symmetry of crystal class	Symmetry of diffraction effects	Symmetry of lattice	Additional position equivalence in lattice	Reflection type	Multiplicity factor	Additional coincidences due to symmetry of lattice
Triclinic	1, $\bar{1}$	$\bar{1}$	$\bar{1}$		hkl	2	
Monoclinic	2	$\frac{2}{m}$	$\frac{2}{m}$		hkl	4	
	m				$h0l$ (second setting)	2	
	$\frac{2}{m}$				$0k0$	2	
Orthorhombic	222	$\frac{2}{m}\frac{2}{m}\frac{2}{m}$	$\frac{2}{m}\frac{2}{m}\frac{2}{m}$		hkl	8	
	$mm2$				$hk0$, $h0l$, $0kl$	4	
	$\frac{2}{m}\frac{2}{m}\frac{2}{m}$				$h00$, $0k0$, $00l$	2	
Tetragonal	4	$\frac{4}{m}$	$\frac{4}{m}\frac{2}{m}\frac{2}{m}$	hkl, khl	hkl	8	hkl, khl
	$\bar{4}$				hhl	8	
	$\frac{4}{m}$				$h0l$	8	
					$hk0$	4	$hk0$, $kh0$
					$hh0$	4	
					$h00$	4	
					$00l$	2	
	422	$\frac{4}{m}\frac{2}{m}\frac{2}{m}$			hkl	16	
	$4mm$				hhl	8	
	$\bar{4}2m$				$h0l$	8	
	$\frac{4}{m}\frac{2}{m}\frac{2}{m}$				$hk0$	8	
					$hh0$	4	
					$h00$	4	
					$00l$	2	

† M. J. Buerger. *X-ray crystallography.* (John Wiley and Sons, New York, 1942) 56–58.

Table 3 (***Continued***)

Crystal system	Symmetry of crystal class	Symmetry of diffraction effects	Symmetry of lattice	Additional position equivalence in lattice	Reflection type	Multiplicity factor	Additional coincidences due to symmetry of lattice
Hexagonal	3 $\bar{3}$	$\bar{3}$	$\frac{6}{m}\frac{2}{m}\frac{2}{m}$	$hkil$, $khil$ $\overline{hki}l$, $\overline{khi}l$	$hkil$	6	$hkil$, $khil$, $\overline{hk}il$, $khil$
					$hh\overline{2h}l$	6	$hh\overline{2h}l$ $\bar{h}\bar{h}2hl$
					$h0\bar{h}l$	6	$h0\bar{h}l$, $0h\bar{h}l$
					$hki0$	6	$hki0$, $khi0$
					$hh\overline{2h}0$	6	
					$h0\bar{h}0$	6	
					$000l$	2	
	32 $3m$ $\bar{3}\frac{2}{m}$	$\bar{3}\frac{2}{m}$		$hkil$ $\overline{hki}l$	$hkil$	12	$hkil$ $khil$
					$hh\overline{2h}l$	12	
					$h0\bar{h}l$	6	$h0\bar{h}l$ $\bar{h}0hl$
					$hki0$	12	
					$hh\overline{2h}0$		
					$h0\bar{h}0$	6	
					$000l$	2	
	6 $\bar{6}$ $\frac{6}{m}$	$\frac{6}{m}$		$hkil$, $khil$	$hkil$	12	$hkil$, $khil$
					$hh\overline{2h}l$	12	
					$h0\bar{h}l$	12	
					$hki0$	6	$hki0$, $khi0$
					$hh\overline{2h}0$	6	
					$h0\bar{h}0$	6	
					$000l$	2	
	622 $6mm$ $\bar{6}m2$ $\frac{6}{m}\frac{2}{m}\frac{2}{m}$	$\frac{6}{m}\frac{2}{m}\frac{2}{m}$			$hkil$	24	
					$hh\overline{2h}l$	12	
					$h0\bar{h}l$	12	
					$hki0$	12	
					$hh\overline{2h}0$	6	
					$h0\bar{h}0$	6	
					$000l$	2	
Isometric	23 $\frac{2}{m}\bar{3}$	$\frac{2}{m}\bar{3}$	$\frac{4}{m}\bar{3}\frac{2}{m}$	hkl, khl	hkl	24	hkl, khl
					hhl	24	
					$hk0$	12	$hk0$, $kh0$
					$hh0$	12	
					hhh	8	
					$h00$	6	
	432 $\bar{4}3m$ $\frac{4}{m}\bar{3}\frac{2}{m}$	$\frac{4}{m}\bar{3}\frac{2}{m}$			hkl	48	
					hhl	24	
					$hk0$	24	
					$hh0$	12	
					hhh	8	
					$h00$	6	

reduced by a factor of 2 is the reciprocal-lattice point, P_{hkl}, lies on the aforementioned plane.

Powder method. All symmetrically equivalent planes in the crystal have identical d's. They consequently define powder-diffraction cones having identical half-opening angles 2θ. Therefore, in the powder method, all of the n equivalent planes of the same form hkl contribute equally to the energy arriving at a single Debye ring. But even more planes than this have this property. According to Friedel's law,[†] although the phases of pairs of reflections hkl and $\overline{hkl}$ are not the same, their intensities are identical. Consequently all planes of the forms $((hkl))$ and $((\overline{hkl}))$ (if these forms are distinct, as they are when the crystal class lacks an inversion center) not only reflect along the same cone and contribute to the same Debye ring, but they contribute equal energies. As a consequence of this, the true multiplicity factor of the powder method for a reflection hkl is the same as the number of planes in the form hkl for the diffraction class of the crystal.

The specific multiplicity factors applicable to the powder method for the several types of reflections for the various crystal classes are listed in Table 3. The multiplicity is complicated by the fact that reflections from several unrelated planes may be recorded in the same powder line. This effect is discussed below.

Reflection coincidences

The powder method, and to a somewhat lesser extent the rotating-crystal method, are severely handicapped and rendered complex in interpretation by the unfortunate circumstance that, for many crystal classes, reflections which are not even symmetrically related, as discussed in the last section, are recorded in such a way as to render them inseparable. There are several different causes that give rise to this kind of situation and these are discussed briefly below.

Coincidence due to superior lattice symmetry. Each of the 14 Bravais space lattices (and therefore their reciprocal lattices) has higher symmetry than any of the crystal classes consistent with it except its "holohedral" class, as shown in Table 3. In each of the non-holohedral classes, therefore, the positions of the reciprocal-lattice points have a higher symmetry than that of the crystal class. Consequently, in the reciprocal lattice there occur sets of points which, though related in position, do not necessarily have identical reflection energies. If their positions in the reciprocal lattice are related by an inversion center of the lattice, then the points do have identical intensities, as discussed in

[†] M. J. Buerger. *X-ray crystallography.* (John Wiley and Sons, New York, 1942) 56–58.

the last section. But if their positions are related by another symmetry element of the lattice, the points do not have identical intensities. The only other symmetry which the reciprocal lattice can have can be defined by certain "vertical" planes. These have the effect of relating the positions of points *hkl* and *khl*. These points, then, although they have identical origin distances, do not have identical reflection intensities unless the diffraction class of the crystal also contains the "vertical" plane. If it does not, then the unequal diffraction intensities of *hkl* and *khl* are recorded at the same point on the same Debye ring of the powder photograph, and, in most practical orientations, at the same spot on the rotating-crystal photograph. The coincidences, especially as they affect the powder method, are listed for the various crystal classes in Table 3.

This difficulty does not appear in any of the moving-film methods because the points *hkl* and *khl* record at separate points on the film.

Coincidence due to identical reciprocal net distance. In square and hexagonal nets, the distance of a point *hk* from the origin is given by $\xi = a^* \sqrt{h^2+k^2}$ and $\xi = a^* \sqrt{h^2+k^2+hk}$ respectively. Several sets of *hk* sometimes have the same distance ξ.[†] All members of a set of points on the same level having identical ξ values reflect to the same spot on rotating-crystal photographs made with the rotation axis perpendicular to the square or hexagonal net.

All members of such a set of reflections also record on the same Debye ring on powder photographs. In three-dimensional isometric lattices, the origin distance is given by $\sigma = a^* \sqrt{h^2+k^2+l^2}$. Several sets of the three indices *hkl* may define identical distances σ.[§] All members of such a set with identical σ reflect to the same Debye ring, and consequently such reflections are unresolved by the powder method.

Coincidence due to accidentally equal origin distance. In addition to the identical coincidences noted in the last section, coincidences also occur if the values of ξ are accidentally equal for two or more reciprocal-lattice points on the same level. Coincidences occur in the powder method if the value of σ is accidentally the same for two or more reciprocal-lattice points.

Coincidences from twinned crystals. While coincident reflections constitute the bane of the rotating-crystal and powder methods of recording reflections, there is one circumstance in which coincidences affect even moving-film methods of recording. This occurs if the supposed single crystal is in reality a twin or twinned aggregate. In many instances the most uncertain part of the whole crystal-structure analysis

[†] M. J. Buerger. *X-ray crystallography.* (John Wiley and Sons, New York, 1942) 163.

[§] *Internationale Tabellen zur Bestimmung von Kristallstrukturen*, Vol. 2. (Gebruder Borntraeger, Berlin, 1935), 474–501.

lies in the uncertainty whether the crystal used was really a single crystal and not a twin.

Ordinarily a trained crystallographer has sufficient background and intuition to reject a crystal which is twinned, whereas many beginners, or those with little real acquaintances with actual crystals, make the fatal error of beginning the whole crystal-structure analysis with a crystal naïvely supposed to be a single crystal but in reality a twin. But twins even intrude themselves upon the seasoned crystallographer who must be constantly on the lookout for them. Indeed, certain species of crystals have been known only in twin form and these have passed as single crystals until they have yielded to crystal-structure analysis in spite of the twinning.

The lattices of the two individuals of a twin are related by a symmetry element which is a rational plane or rational row (occasionally merely a single point) common to the two lattices. Because of this, the points of the reciprocal lattice of one individual may fall on or near the points of the reciprocal lattice of the other. There are several general categories of twins. In non-merohedral twins, the points of these two reciprocal lattices coincide only on the symmetry element (through the origin) which relates them as twins. In this case, the twin is distinguishable as such by the double-lattice character of the diffraction record. But in merohedral twins and pseudo-merohedral twins, the points of one reciprocal lattice coincide precisely with those of the twin, and in this case, the twin character of the crystal may not be discovered from the diffraction record. In such instances, even the space group may be erroneously determined, since the missing points of one lattice (space-group extinctions) may be filled in by the points present in the twin lattice, and therefore the extinction may go undetected.

In merohedral twins, points like hkl and khl are superposed, and consequently the crystal gives a reflection record in which every diffraction spot (except those for which $h = k$) consists of such a pair. In pseudo-merohedral twins, similar coincidences also occur, but the diffraction record may even deceive the investigator into believing that the crystal is hexagonal, for example, when it is, in fact, a twinned aggregate of orthorhombic individuals. In such instances, the indexing is meaningless.

It goes without saying that every effort should be made to test a crystal for twinning before it is used as a basis for a crystal-structure analysis. To this end, it is worthwhile to examine all candidate crystals optically for evidences of twinning, particularly if the crystal species is transparent, in which case optical examination is easily made with the aid of the polarizing microscope. But optical tests are powerless to detect twins of pure merohedry. For crystals in which there is reason to suspect this kind of twinning, chemical etching may be used to detect its presence or absence.

If the use of a twin for purposes of crystal-structure analysis is unavoidable, it should be remembered that the intensity of a spot conforms to (94). It is also suggested that the reader study the section on twinning in Chapter 5.

Literature

[1] P. Debye and P. Scherrer. *Atombau.* Physik. Z. (1918) 474–483, especially 481–482.

[2] Heinrich Ott. *Das Gitter des Aluminiumnitrids (AlN).* Z. Physik. **22** (1924) 201–214, especially footnote 1, page 212.

[3] H. Hoffmann and H. Mark. *Das Gitter der Oxalsäure.* Z. physik. Chem. **111** (1924) 321–356, especially 334–337.

[4] E. G. Cox and W. F. B. Shaw. *Correction factors in the photographic measurement of x-ray intensities in crystal analysis.* Proc. Roy. Soc. (London) (A) **127** (1930) 71–88, especially 73–77.

[5] J. Bouman and W. F. de Jong. *Die Intensitäten der Punkte einer photographierten reziproken Netzebene.* Physica **5** (1938) 817–832.

[6] George Tunell. *The rotation factor for equi-inclination Weissenberg photographs.* Am. Mineralogist **24** (1939) 448–451.

[7] M. J. Buerger. *The photography of interatomic distance vectors and of crystal patterns.* Proc. Nat. Acad. Sci. U. S. **25** (1939) 383–388, especially 385.

[8] M. J. Buerger. *The correction of x-ray diffraction intensities for Lorentz and polarization factors.* Proc. Nat. Acad. Sci. U. S. **26** (1940) 637–642.

[9] B. E. Warren and I. Fankuchen. *A simplified correction factor for equi-inclination Weissenberg patterns.* Rev. Sci. Instr. **12** (1941) 90–91.

[10] Chia-Si Lu. *Reciprocal Lorentz-polarization factor charts for equi-inclination Weissenberg photographs.* Rev. Sci. Instr. **14** (1943) 331–335.

[11] M. J. Buerger. *The photography of the reciprocal lattice.* Am. Soc. for X-ray and Electron Diffraction, Monograph No. 1 (1944).

[12] A. D. Booth. *A method for calculating reciprocal spacings for x-ray reflections from a monoclinic crystal.* J. Sci. Instr. **22** (1945) 74.

[13] M. J. Buerger and Gilbert E. Klein. *Correction of x-ray diffraction intensities for Lorentz and polarization factors.* J. Appl. Phys. **16** (1945) 408–418.

[14] M. J. Buerger and Gilbert E. Klein. *Correction of diffraction amplitudes for Lorentz and polarization factors.* J. Appl. Phys. **17** (1946) 285–306.

[15] W. L. Bond. *An alignment chart giving the polarization correction of equi-inclination Weissenberg photographic intensities.* J. Appl. Phys. **19** (1948) 82–83.

[16] W. Cochran. *The correction of x-ray intensities for polarization and Lorentz factors.* J. Sci. Instr. **25** (1948) 253–254.

[17] G. H. Goldschmidt and G. J. Pitt. *The correction of x-ray intensities for Lorentz-polarization and rotation factors.* J. Sci. Instr. **25** (1948) 397–398.

[18] G. Kaan and W. F. Cole. *The measurement and correction of intensities from single-crystal x-ray photographs.* Acta Cryst. **2** (1949) 38–43.

[19] Howard T. Evans, Jr., S. G. Tilden, and Douglas P. Adams. *New techniques applied to the Buerger precession camera for x-ray diffraction studies.* Rev. Sci. Instr. **20** (1949) 155–159.

[20] C. J. B. Clews and W. Cochran. *The structures of pyrimidines and purines. III. An x-ray investigation of hydrogen bonding in aminopyrimidines.* Acta Cryst. **2** (1949) 46–57, especially 48.

[21] Hans Gunther Heide. *Ein modifizierter Lorentzfaktor für Drehkristallverfahren.* Acta Cryst. **4** (1951) 29–34.

[22] Hans Gunther Heide. *Zum Lorentzfaktor für Drehkristallverfahren.* Ann. Physik (6) **8** (1951) 240–245.

[23] F. H. Herbstein. *A simple method of applying the rotation factor correction in equi-inclination Weissenberg photographs.* Acta Cryst. **4** (1951) 185.

[24] Jürg Waser. *The Lorentz factor for the Buerger precession method.* Rev. Sci. Instr. **8** (1951) 563–566.

[25] Gopinath Kartha. *Correction charts for Lorentz and polarization factors in anti-equi-inclination photographs.* Acta Cryst. **5** (1952) 549–550.

[26] Robinson D. Burbank. *Upper level precession photography and the Lorentz-polarization correction. Part I.* Rev. Sci. Instr. **23** (1952) 321–327.

[27] H. J. Grenville-Wells and S. C. Abrahams. *Upper level precession photography and the Lorentz-polarization correction. Part II.* Rev. Sci. Instr. **23** (1952) 328–331.

[28] Gopinath Kartha. *Charts for obtaining geometrical corrections directly from anti-equi-inclination photographs.* Rev. Sci. Instr. **24** (1953) 871.

[29] Masao Atoji and William N. Lipscomb. *Lorentz polarization factors for precession angles of 10°, 15°, and 21°.* Acta Cryst. **7** (1954) 595–596.

[30] Masao Atoji and William N. Lipscomb. *Lorentz polarization factors for precession angles of 10°, 15°, and 21°: correction.* Acta Cryst. **8** (1955) 364.

[31] Kurt Boström. *A graphical method for the calculation of* $|F|^2$ *and* $|F|$ *from equi-inclination Weissenberg photographs.* Acta Cryst. **10** (1957) 477–479.

8

Some physical factors affecting intensities

In the last chapter, a discussion was given of some factors of simple geometrical origin for which allowance must be made in transforming the measured energies of spectra to $|F|^2$'s. In addition there are some further factors which arise in a less simple fashion. These are due to "extinction," absorption, and the temperature motion of the atoms. In general it is less easy to allow for these effects than those of the last chapter, and it is not uncommon to assume that some of the effects are so small that they can be neglected. These factors are discussed in this chapter.

"Extinction"

In the derivation of the relation between intensity and $|F|^2$, it was assumed that the crystal in question consisted of a very small volume element dV. This assumption not only eliminated any concern for absorption, but incidentally eliminated the complication usually called "extinction." This feature, now to be considered briefly, is a phenomenon which results in the attenuation of the primary beam of x-radiation when the crystal is in diffracting position, and so reduces the intensity of the diffracted beam. The effect is well known theoretically, but is very difficult to allow for experimentally because it depends on the physical perfection of the crystal. For this reason it is customarily ignored in many current crystal-structure analysis procedures.

Darwin, who first theoretically investigated the effects of "extinction," subdivided the general attenuation of intensity due to this cause into two categories,[5] *primary "extinction"* and *secondary "extinction."* An excellent résumé of the subject is given by Lonsdale.[9]

Primary "extinction." One of the interesting aspects of the geometry of x-ray reflection is that the ray which is reflected at the Bragg angle by a set of planes is directed at just the correct angle to be reflected back again into the direct beam by other planes of the same stack, as shown in Fig. 1. Thus, each plane of the crystal redirects a small fraction, q, of the reflected rays reaching it back into the primary beam. Now, although it was not explicitly pointed out in Chapter 3, radiation scattered by an electron differs in phase from that of the primary beam

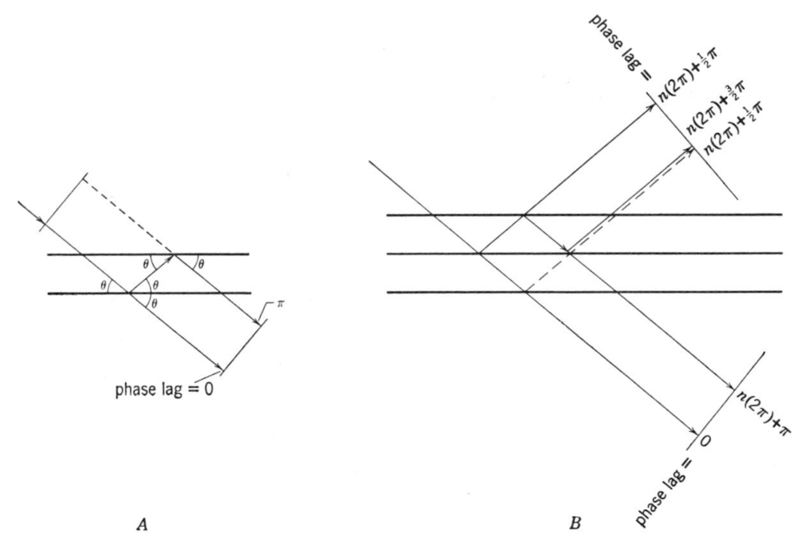

Fig. 1.

by $\pi/2$. It follows that the rays which are twice reflected differ in phase from the primary beam by π, Fig. 1*A* (plus the phase lag due to Bragg reflection, which is always a multiple of 2π) and consequently are exactly out of phase with it. For this reason, the twice-reflected rays, instead of strengthening the primary beam, actually weaken it by destructive interference. As the primary beam penetrates the crystal further and further, it is continuously attenuated in this manner.

This same situation occurs for any two rays which have been reflected n and $n-2$ times respectively. In particular, the thrice-reflected rays and the once-reflected rays are both traveling in the direction of the Bragg reflection, Fig. 1*B*. Since these differ in phase by π, the thrice-reflected rays weaken the reflected beam.

When this weakening phenomenon continues through a large series of planes, the primary beam becomes so attenuated that it is effectively "extinguished." Darwin, who first investigated the effect theoretically, gave the name *primary "extinction"*† to the phenomenon.[5]

It is evident that if a crystal is composed entirely of a single sequence of planes in perfect array, "extinction" imposes complications beyond the simple conditions taken into account in the derivation of the integrated intensity given in (80) of Chapter 3. Darwin also derived a formula for the integrated intensity for these more complex conditions, i.e., for an *ideally perfect crystal.* This formula differs markedly from (80) of Chapter 3, which was derived for a volume element dV so small that neither absorption nor extinction occurred to an appreciable extent. These two formulae are shown below for comparison:

Small volume element:

$$\frac{E\omega}{I} = \left(\frac{Ne^2}{mc^2}\right)^2 \lambda^3 \frac{1 + \cos^2 2\theta}{2 \sin 2\theta} |F|^2 \, dV, \tag{1}$$

Perfect crystal:

$$\frac{E\omega}{I} = \frac{8}{3\pi} \frac{Ne^2}{mc^2} \lambda^2 \frac{1 + |\cos 2\theta|}{2 \sin 2\theta} |F|. \tag{2}$$

It will be observed that, among other differences, the perfect-crystal formula involves the terms $(Ne^2/mc^2)|F|$ to the first power, while the small volume element involves this set of factors to the second power. The perfect-crystal formula thus gives much smaller values of the integrated reflection.

Early measurements of integrated reflections gave values very much in excess of those expected from a perfect crystal by (2). The measurements showed that most crystals had integrated reflections which lay between (1) and (2), with the majority nearer to (1). To account for this, it was necessary to assume that most crystals are not perfect crystals, although they might approximate perfect crystals over small volumes. In examining crystals to gain a clue for the nature of their imperfection, Darwin noted that they showed crinkled surfaces. Since this kind of

† The use of the word "extinction" for this phenomena, and for the phenomenon described in the next section as secondary "extinction," is unfortunate. In both cases the beam becomes attenuated as it traverses the crystal, but in no case is utterly extinguished. There is another and more exact use of the word *extinction* in x-ray crystallography, namely in *space-group extinction,* in which phenomenon the diffracted beam is utterly extinguished independently of the length of the path in the crystal. In this book the word "extinction" is placed in quotes when referring to the phenomena described in this chapter, to emphasize this distinction. It would be better to use the words *primary* and *secondary attenuation* instead of primary and secondary "extinction" for these phenomena.

imperfection provided a difficult model of an imperfect crystal to handle mathematically, Darwin invented an artificial imperfection strictly for ease of mathematical handling. This model, called an *ideally imperfect crystal,* was composed of a *mosaic* of blocks so small that within each one primary "extinction" did not occur to an appreciable amount.

It is an interesting commentary on those who have extended the idea of mosaic theory that they evidently do not know that Darwin invented the model as a matter of mathematical convenience. It is now recognized that crystals do not conform to Darwin's simple jumble of mosaic blocks, yet they do nevertheless contain imperfections which serve the same purpose so far as "extinction" is concerned. For example, it is known from observation that the body of a crystal is branched from the nucleus outward and that the detailed structures of the branches do not register where they touch. This observed imperfection is known as *lineage structure.*[†][§] The adjacent regions of a crystal displaying lineage structure are separated from one another by gaps which are not integral translations, and the adjacent regions are almost, but not quite, parallel. Recent developments in the theory and observation of a type of imperfection in crystal structures known as *dislocations* (about which there is now a vast literature) further confirm the fact that a megascopic crystal is rarely, if ever, so perfect that a single lattice could be drawn through the entire structure. Rather, the crystal is made up of small, nearly perfect regions which are continuous with each other structurally, yet contain regions of disregistry.

In any case, most crystals are imperfect to some extent, and one can imagine an *ideally imperfect crystal* in which the perfect regions are so small that primary "extinction" is inappreciable. In such a crystal, the small perfect regions reflect as units, and since the units are not related to one another in position or orientation by translational repetition, the radiation twice reflected from the planes of one region is not in phase with the radiation twice-reflected from the region farther along the path of the direct beam. Accordingly, the direct beam is not appreciably attenuated by primary "extinction." Furthermore, since the regions below are not necessarily parallel to those above them, these regions do not, in general, reflect at the same time. The range of orientations of the regions has the effect of drawing out the otherwise sharp Bragg reflection so that it extends over an appreciable range of angular orientations of the crystal.

Crystals, even of the same species, differ with regard to the size of the regions and with regard to the range of orientations of the regions of per-

[†] M. J. Buerger. *The significance of "block structure" in crystals.* Am. Mineralogist **17** (1932) 177–191.

[§] M. J. Buerger. *The lineage structure of crystals.* Z. Krist. (A) **89** (1934) 195–220.

fection. Certain crystals, like some diamonds, some calcite, and some quartz, are fairly perfect crystals, i.e., their regions of perfection are large. Others, such as most NaCl crystals, are highly imperfect. Thus, crystals range in "extinction" effects all the way from perfect to ideally imperfect, but fortunately for the crystal-structure analyst, most crystals are nearer to the ideally imperfect model.

Table 1

Some values of the function $\dfrac{\tanh sq}{sq}$

sq	$\tanh sq$	$\dfrac{\tanh sq}{sq}$
0.1	0.0997	0.997
.2	.1974	.987
.3	.2913	.971
.4	.3799	.950
.5	.4621	.924
.6	.5370	.895
.7	.6044	.863
.8	.6640	.830
.9	.7163	.796
1.0	.7616	.762
1.5	.9052	.603
2.0	.9640	.482
2.5	.9866	.395
3.0	.9951	.332
3.5	.9982	.285
4.0	.9993	.250
4.5	.9998	.222
5.0	.9999	.200

Darwin[5] is also responsible for an approximation by which allowance can be made for the primary "extinction" in reflections from crystals whose units are larger than those permissible in the ideally imperfect case. If the "mosaic" unit of the crystal contains s planes in perfect sequence, then the integrated reflection is

$$\frac{E\omega}{I} = QV\frac{\tanh sq}{sq}. \tag{3}$$

In other words, the factor $(\tanh sq)/sq$ corrects the intensity for primary "extinction." Some values of this factor for the argument sq are given in Table 1. In this equation Q is given by (82) of Chapter 3, and V is the volume of the crystal and corresponds to dV of (81) of Chapter 3, except that now it refers to the whole volume of the crystal, since this is the sum of the dV's of the mosaic blocks. The quantity q is the reflecting efficiency of the individual plane and was given by (56) of Chapter 3.

Table 2

Values of the primary "extinction" correction, $\frac{\tanh sq}{sq}$, for orders of the reflection $h00$ of NaCl
(After Lonsdale[9])

Reflection	d	q	Depth of block, D					
			10^{-5} cm.		10^{-4} cm.		10^{-3} cm.	
			$s = \frac{D}{d}$	$\frac{\tanh sq}{sq}$	$s = \frac{D}{d}$	$\frac{\tanh sq}{sq}$	$s = \frac{D}{d}$	$\frac{\tanh sq}{sq}$
200	2.814 Å	2.05×10^{-4}	3.55×10^{2}	1.00	3.55×10^{3}	0.85	3.55×10^{4}	0.14
400	1.407	0.30_7	7.10	1.00	7.10	0.98	7.10	0.46
600	0.938	0.08_1	10.65	1.00	10.65	1.00	10.65	0.81
800	0.703	0.02_6	14.20	1.00	14.20	1.00	14.20	0.90
10·0·0	0.563	0.00_9	17.75	1.00	17.75	1.00	17.75	0.99

To appreciate how primary "extinction" affects particular situations, one can transform the expression for q given in (56) of Chapter 3, which was

$$q = N\, d \frac{\lambda}{\sin \theta} \frac{e^2}{mc^2} p^{1/2} F \tag{4}$$

by substituting from Bragg's law,

$$n\lambda = 2d \sin \theta, \tag{5}$$

$$\frac{\lambda}{\sin \theta} = \frac{2d}{n}. \tag{6}$$

Making the substitution, one obtains

$$q = N \frac{2d^2}{n} \frac{e^2}{mc^2} p^{1/2} F. \tag{7}$$

From this it is evident that primary "extinction," as indicated by the correction for it given by $(\tanh sq)/sq$, is greater for

large structure factors, F,
large spacings, d,
small orders, n.

Since (7) does not involve λ, primary "extinction" is independent of wavelength.

In Tables 2 and 3, quoted from Lonsdale,[9] the specific effect of primary "extinction" in some reflections for NaCl and diamond are given for

Table 3

Values of the primary "extinction" correction, $\frac{\tanh sq}{sq}$, for some reflections of diamond

(After Lonsdale[9])

Reflection hkl	d	q	Depth of block, D			
			10^{-4} cm.		10^{-3} cm.	
			$s = \frac{D}{d}$	$\frac{\tanh sq}{sq}$	$s = \frac{D}{d}$	$\frac{\tanh sq}{sq}$
111	2.05	9.84×10^{-5}	4.87×10^{3}	0.93	4.87×10^{4}	0.21
220	1.26	2.85	7.95	0.98_2	7.95	0.45
113	0.89	1.22	9.32	0.99_5	9.32	0.71
004	0.92	1.09	11.24	0.99_7	11.24	0.68
331	0.15	0.65	12.25	0.99_8	12.25	0.83

various thicknesses D. These tables make it evident that primary "extinction" becomes appreciable for crystal thicknesses greater than about 10^{-4} cm., and that at dimensions of 10^{-3} cm. it is already extreme. In an exaggerated case, obviously the only part of the volume of a fairly perfect crystal which is effective in reflecting x-rays is a relatively thin shell or skin of the order of 10^{-3} cm. thick.

The actual use of Darwin's factor, $(\tanh sq)/sq$, for correcting for "extinction" effects has been criticized by Weiss[14] and others.[15–17] Darwin's derivation of (3) was based upon an infinite plate with the reflecting planes parallel to its surface. Neither condition is approached in diffraction by powder samples. Rather the scattering regions are of approximately equal dimensions and the planes are at arbitrary angles. The average path length is the same for all planes, and not proportional to $1/\sin\theta$. If the particles are taken as spheres of diameter D', then Ekstein's[12] result can be applied in the form

$$\frac{I_{\text{ext}}}{I} = 1 - \tfrac{7}{16}\left(\frac{e^2}{mc^2}|F|pD'N\right)^2. \tag{8}$$

According to Ekstein, this is close to what Zachariasen[8] gives:

$$\frac{I_{\text{ext}}}{I} = \frac{\tanh A}{A} \approx 1 - \tfrac{1}{2}\left(\frac{e^2}{mc^2}|F|pD'N\lambda\right)^2,$$

in which A is the expression in parentheses and

$$D' = \frac{t_0}{|\gamma_0\,\gamma_H|^{1/2}},$$

where t_0 = plate thickness,
γ_0, γ_H = direction cosines of the normal to the plate for direct and diffracted rays.

In current crystal-structure analysis, corrections are not ordinarily made for primary-"extinction" effects because of the difficulty of making the correction. Two means are commonly used to avoid the correction. One is to attempt to eliminate the effect. Two methods have been employed to this end. One method is to crush the crystal to a powder so that the grain size is of the order of 10^{-4} cm. and then employ the powder method of determining intensities. Of course this method suffers all the disadvantages of the powder method, discussed in Chapter 7. Another method is to dip the single crystal into liquid air. This procedure apparently produces such internal strains that the crystal effectively becomes a mosaic. Unfortunately this method does not induce imperfection in all crystals. It has been found to work well on certain organic crystals but has failed on inorganic crystals such as diamond and calcite. The presence or absence of primary "extinction" can be detected by a divergent-beam x-ray technique perfected by Lonsdale.[9]

In the event that primary "extinction" has not been eliminated, it has the effect of reducing the apparent values of the measured intensities for spectra having large $|F|$'s and large d's. This is most annoying during the process of refinement. It is customary, therefore, to omit F's suspected of having large "extinction" effects when computing the residual factor R (Chapter 22). It is also customary to substitute in Fourier syntheses the computed values of the $|F|$'s for the observed values, where "extinction" effects are suspected of being large.

Secondary "extinction." When an x-ray beam is incident upon a crystal whose orientation is such that the Bragg condition is not satisfied for any set of planes, the primary beam, as it penetrates the crystal, is reduced in intensity due to ordinary absorption. But if the Bragg condition is satisfied for any set of planes, there is an additional attenuation of the primary beam even if there is no primary "extinction," or if the primary "extinction" is allowed for. The reason for this is that diversion of the energy from the primary beam by the reflection weakens it. Therefore the intensity of the primary beam received by a plane is equal to that of the original beam less that which has been reflected by the planes preceding it, less any absorption in the path to it. The reduction of the intensity of the primary beam due to this kind of previous reflection was called by Darwin[5] *secondary "extinction."*

Since secondary "extinction" behaves a good deal like ordinary absorption, it is customary to consider it as adding to the ordinary linear absorption coefficient. If the ordinary absorption coefficient is μ, then an absorption coefficient containing an allowance for secondary "extinction" is of the form

$$\mu' = \mu + gQ, \tag{9}$$

where Q is the energy removed in the reflected beam, and g is a constant characteristic of the crystal but independent of λ. When both primary and secondary "extinction" occur, it is necessary to use a more complicated form,

$$\mu' = \mu + gQ - g'Q^2 \cdots, \tag{10}$$

where g' depends on the degree of perfection and other features. In (9) and (10), Q is the actual energy of the reflection corrected for primary "extinction," specifically,

$$Q = Q_0 \frac{\tanh sq}{sq}. \tag{11}$$

As in the case of pure absorption (discussed later in this chapter), the intensity of the reflection tends to decrease with the path, t, due to the factor concerned with reduction of the primary beam, yet, on the other hand, the reflection tends to increase with path since the number of planes contributing to the reflection increases with path. Specifically, the proportion of the energy of the primary beam reflected is

$$\rho = Qe^{-\mu' t}\, t. \tag{12}$$

With these two opposing tendencies, a maximum reflection occurs for a particular value of t. This can be found by differentiating (12) and setting the result equal to zero:

$$\begin{aligned} \frac{d\rho}{dt} &= Q(e^{-\mu' t} - t\mu' e^{-\mu' t}), \\ &= Qe^{-\mu' t}(1 - \mu' t). \end{aligned} \tag{13}$$

If one sets

$$1 - \mu' t = 0,$$

$$t_{\text{opt}} = \frac{1}{\mu'}. \tag{14}$$

This t corresponding to a maximum reflection provides a method of measuring μ'. The proportion reflected at this thickness is found by substituting (14) into (12):

$$\begin{aligned} \rho_{\text{opt}} &= Qe^{-\mu'(1/\mu')} \frac{1}{\mu'} \\ &= \frac{1}{e} \frac{Q}{\mu'}. \end{aligned} \tag{15}$$

To measure μ', one measures ρ for various crystal thicknesses t_0, for which the path t is

$$t = \frac{t_0}{\cos\theta}. \tag{16}$$

When ρ is plotted against t_0 a maximum is found corresponding to (14) for which the value of the coefficient is given by

$$\mu' = \frac{\cos\theta}{t_0}. \tag{17}$$

Although this method of determining μ' has been applied by physicists in investigating "extinction," it is of little help to the crystal-structure analyst, who ordinarily must deal with a tiny single crystal. It can only be applied where relatively large slabs can be cut from single crystals.

Secondary "extinction," and methods of attempting to correct for it, have been discussed by Weiss,[14] Lang,[15] Williamson and Smallman,[16] Vand,[17] Gatineau and Mering,[18] and Chandrasekhar.[19]

Absorption

X-radiation is absorbed by matter. The extent to which this occurs in the tiny crystal samples used in crystal-structure analysis is much greater than ordinarily appreciated, and the effects of absorption can be ignored only for rough work. On the other hand, the amount of absorption which takes place during the transmission of x-radiation through the crystal while a particular reflection is being produced is very tedious to calculate since this is a function of the shape of the crystal sample and the relation of the direct and diffracted beams to this shape. While it is ordinarily out of the question to spend the time making an exact allowance for absorption, it is often possible to approximate the allowance.

Elementary absorption theory. After an x-ray beam of intensity I penetrates an elementary slab of crystal whose thickness is dt, the intensity is reduced by an amount dI. The decrease, $-dI$, is proportional to the original intensity I, to the thickness dt, and to the specific absorption μ (known as the *linear-absorption coefficient*,† expressed in cm.$^{-1}$ units). Consequently,

$$-dI = \mu I\, dt; \tag{18}$$

† This coefficient applies to the absorption of a particular *monochromatic* radiation by a particular material. If the radiation is polychromatic the various wavelengths are differently absorbed, the less penetrating becoming more attenuated with depth. Such a system does not have an absorption coefficient which is independent of depth.

or

$$-\frac{dI}{I} = \mu\, dt. \tag{19}$$

If the intensity of the original x-ray beam is designated I_0 and that of the beam which has traversed a thickness t is designated by I, integration of (19) gives

$$-\log I_0 + \log I = -\mu t, \tag{20}$$

or

$$\log \frac{I}{I_0} = -\mu t. \tag{21}$$

This can be expressed in exponential form as

$$\frac{I}{I_0} = e^{-\mu t}, \tag{22}$$

or

$$I = I_0 e^{-\mu t}. \tag{23}$$

Therefore, after the radiation passes through a thickness t, the original beam is reduced by a factor $e^{-\mu t}$.

Calculation of linear-absorption coefficients. The linear-absorption coefficient for any crystal, required for (23), can be computed from a knowledge of its chemical composition, its density, and a table of *mass-absorption coefficients* of the elements. For minerals the first two data are given for each mineral in Dana's *Textbook of mineralogy.* A table of mass-absorption coefficients of the elements is given in *International tables for x-ray crystallography,* Vol. 3. The mass-absorption coefficients are functions not only of the elements but also of the wavelength of x-radiation employed.

The linear-absorption coefficient, μ, in (23) is computed from the mass-absorption coefficient, μ/ρ, of the crystal, by means of the relation

$$\mu = G\left(\frac{\mu}{\rho}\right), \tag{24}$$

where G is the density of the crystal. The mass absorption for the crystal is computed from the individual mass-absorption coefficients of the elements $A, B, C \cdots$ in the crystal by means of the relation

$$\mu = G\left\{p_A\left(\frac{\mu}{\rho}\right)_A + p_B\left(\frac{\mu}{\rho}\right)_B + p_C\left(\frac{\mu}{\rho}\right)_C \cdots\right\}, \tag{25}$$

where p_A is the fraction by weight of element A in compound $ABC \cdots$. An example of the computation of (25) is given in Table 4. The calculation incidentally brings out the fact that the absorption coefficients may

Table 4
Example of the computation of the linear absorption coefficient for wollastonite, $CaSiO_3$, $G = 2.72$ g./cc.

Atom	Atomic weight	Total weight of atom species in compound	Fraction of total weight p
Ca	40	40	34.5%
Si	28	28	24.1
3O	16	48	41.4
		116	100.0%

For CuKα radiation:

Atom	$\left(\frac{\mu}{\rho}\right)$†	$p\left(\frac{\mu}{\rho}\right)$
Ca	172 g.$^{-1}$	59.4 g.$^{-1}$
Si	60.3	14.5
O	12.7	5.3
$\sum p\left(\frac{\mu}{\rho}\right)$ for compound:		79.2 g.$^{-1}$

$$\mu = G\sum p\left(\frac{\mu}{\rho}\right) = 2.72 \times 79.2 = 215 \text{ cm.}^{-1}$$

For MoKα radiation:

Atom	$\left(\frac{\mu}{\rho}\right)$†	$p\left(\frac{\mu}{\rho}\right)$
Ca	19.8 g.$^{-1}$	6.8 g.$^{-1}$
Si	6.7	1.6
O	1.5	0.6
$\sum p\left(\frac{\mu}{\rho}\right)$ for compound:		9.0 g.$^{-1}$

$$\mu = G\sum p\left(\frac{\mu}{\rho}\right) = 2.72 \times 9.0 = 24 \text{ cm.}^{-1}$$

† From *International tables for x-ray crystallography.*

be reduced by as much as an order of magnitude by using Mo rather than Cu radiation.

The transmission factor. When a crystal diffracts x-rays, the primary beam is absorbed as it enters the crystal and the diffracted beam is absorbed as it leaves the crystal. As a result of absorption, therefore, the diffracted beam is weaker than if there were no absorption. It is convenient to define as the *transmission factor* for a reflection the ratio of the intensity which is diffracted to the intensity which would be diffracted if there were no absorption. Let K be the fraction of the intensity of the direct beam diffracted by a crystal in a particular spectrum; then the transmission factor for that spectrum is

$$T = \frac{KI}{KI_0}. \tag{26}$$

The value of the diffracted intensity is reduced due to absorption according to (23). But if the crystal is not a plate, the various path lengths are unequal, so that (26) must be integrated over the various paths in the volume of the crystal. Thus the value of (26) becomes

$$\begin{aligned} T &= \frac{\int^v KI\,dV}{\int^v KI_0\,dV} \\ &= \frac{\int^v KI_0\,e^{-\mu t}\,dV}{\int^v KI_0\,dV} \\ &= \frac{\int^v e^{-\mu t}\,dV}{V}. \end{aligned} \tag{27}$$

The meaning of (27) is illustrated in Fig. 2. The volume of the crystal is represented by the outline in Fig. 2. The quantity $e^{-\mu t}$ must be integrated over the irradiated volume of the crystal. Note that the total path t is composed of a part t_1 from the boundary of the crystal along the primary-beam direction to the element of volume dV, plus a part t_2 from dV along the diffracted-beam direction to the boundary of the crystal. This integration depends not only on the glancing angle, but also on the detailed shape of the crystal. It is possible to prepare tables of (27) only for a limited number of simple shapes.

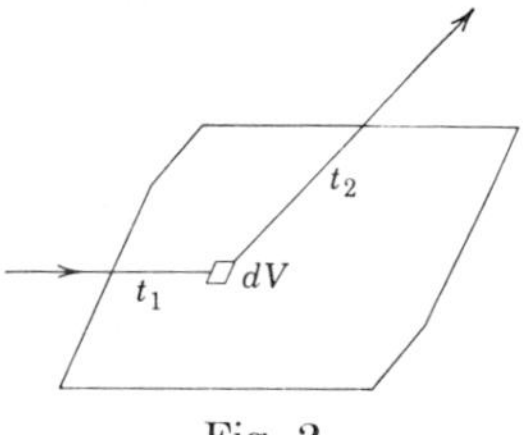

Fig. 2.

Cylindrical specimens. One of the simple shapes for which (27) can be tabulated is a cylinder. The tabulated results are not only applicable to correcting powder-photograph data for absorption, but can also be applied to intensities derived from a single crystal provided that the crystal has been shaped to cylindrical form by one of the methods mentioned in Chapter 5.

The transmission factor for a cylindrical specimen can be expressed by taking the element of volume dV as prismatic, and equal to the product of an element of area dA perpendicular to the cylinder, and length L equal to the length of the irradiated cylinder. Thus (27) becomes

$$T = \frac{\int^{v} e^{-\mu t}\, L\, dA}{LA}$$

$$= \frac{1}{\pi R^2} \int^{A} e^{-\mu t}\, dA, \tag{28}$$

where R is the radius of the cylinder. The integral cannot be evaluated, but it can be solved graphically. This was first done by Claassen[21] and

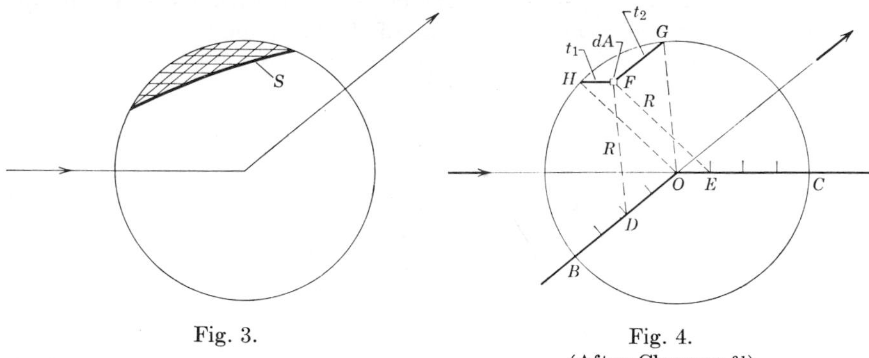

Fig. 3.

Fig. 4. (After Claassen.[21])

later by Bradley.[22] Claassen's method consists of dividing the area of the circular cross-section, Fig. 3, into strips of area S, within which the sum of the lengths of incident and diffracted rays lies between some selected limits t and $t+\Delta t$. It is convenient to specify the length of the path t in terms of a fraction x of the radius of the cylinder, so that

$$t = xR. \tag{29}$$

In this way (28) can be written as a summation:

$$T = \frac{1}{\pi R^2} \sum e^{-\mu x R}\, S. \tag{30}$$

Claassen presented an ingenious method of locating strips of constant xR, which is illustrated in Fig. 4. Radii are drawn from the center and in directions opposite to the primary and diffracted rays. These radii are divided into lengths equal to some specific submultiple of R (in Fig. 4 they are divided into $\frac{1}{4}R$). Suppose one wishes to map the locus of points in such a way that $t_1+t_2 = \frac{3}{4}R$. A distance DOE is selected along BOC equal to $\frac{3}{4}R$. There are a succession of ways of selecting a distance along BOC equal to $\frac{3}{4}R$, a particular one of which is illustrated in Fig. 4. From D and E lines parallel to radii OG and OH are drawn;

at their intersection, F, is the required dA such that $t_1 + t_2 = \frac{3}{4}R$. That this is so can be seen by noting that $DFGO$ is a parallelogram, so that $t_2 = DO$. Similarly $t_1 = OE$. If DO is reduced by one division, and OE increased by one division, and the construction repeated, then another point is located within S of Fig. 3.

The magnitude of T in (30) can be evaluated by measuring the areas S with a planimeter, multiplying the value found by the value of $e^{-\mu x R}$, and summing over all areas S. The value of T as found by Bradley[22] for several values of the glancing angle θ is tabulated against μR in Table 5 and Table 6. To make use of these data, it is necessary to plot a curve of T versus θ (or $\sin \theta$) so that the correction can be applied to reflections having any value of θ.

Cylindrical specimens, upper levels.[41] Figure 5 shows a cross-section of a crystal ground to circular cross-section. The transmission factor for the zero-level reflection with the beam normal to the cylinder axis can be written

$$T = \frac{1}{V} \int^{v} e^{-\mu(t_1+t_2)}\, dV. \qquad (31)$$

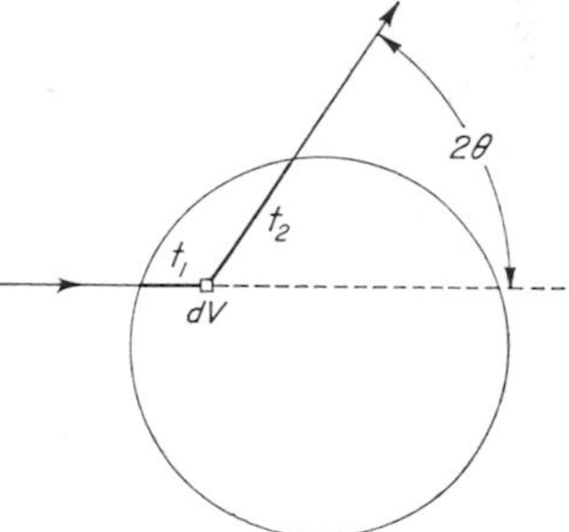

Fig. 5.
(After Buerger and Niizeki.[41])

In the general case, Fig. 6, the primary beam makes an angle $\tilde{\mu}$, and the diffracted

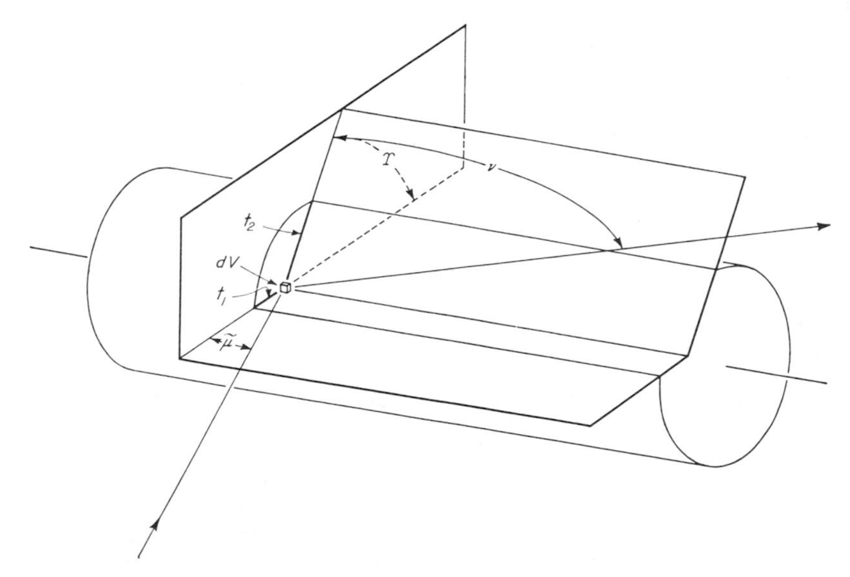

Fig. 6.
(After Buerger and Niizeki.[41])

Table 5
Transmission factors for cylindrical samples ($\mu R < 5$)
(After Bradley[22])

μR	θ						
	0°	$22\frac{1}{2}$°	35°	45°	55°	$67\frac{1}{2}$°	90°
0.0	1.0000	1.0000	1.0000	1.0000	1.0000	1.0000	1.0000
0.1	0.8470	0.8480	—	0.849	—	0.850	0.851
0.2	.7120	.7160	—	.719	—	.724	.729
0.3	.6000	.6060	—	.614	—	.627	.635
0.4	.5000	.5170	—	.527	—	.545	.556
0.5	.4350	.4420	—	.458	—	.478	.490
0.6	.3690	.3780	—	.398	—	.423	.436
0.7	.3140	.3240	—	.348	—	.378	.393
0.8	.2680	.2786	—	.305	—	.337	.356
0.9	.2300	.2410	—	.274	—	.305	.324
1.0	.1977	.2005	—	.242	—	.2785	.295
1.1	.1698	.1828	—	.2170	—	.2550	.2715
1.2	.1459	.1600	—	.1954	—	.2350	.2510
1.3	.1256	.1403	—	.1770	—	.2170	.2335
1.4	.1084	.1233	—	.1611	—	.2010	.2180
1.5	.0038	.1091	—	.1469	—	.1866	.2050
1.6	.0811	.0973	—	.1352	—	.1746	.1932
1.7	.0710	.0871	—	.1217	—	.1641	.1824
1.8	.0615	.0780	—	.1154	—	.1542	.1730
1.9	.0537	.0702	—	.1074	—	.1459	.1644
2.0	.0471	.0635	—	.1005	—	.1384	.1567
2.1	.0416	.0579	—	.0944	—	.1315	.1493
2.2	.0367	.0531	—	.0889	—	.1250	.1426
2.3	.0324	.0486	—	.0838	—	.1189	.1365
2.4	.02865	.0447	—	.0791	—	.1135	.1309
2.5	.0255	.0412	—	.0750	—	.1086	.1256
2.6	.0227	.0382	—	.0711	—	.1010	.1211
2.7	.0202	.0355	—	.0675	—	.0998	.1167
2.8	.01803	.0330	—	.0541	—	.0962	.1127
2.9	.01607	.0308	—	.0610	—	.0925	.1089
3.0	.01436	.02885	0.0443	.0582	0.0724	.0389	.1054
3.1	.01288	.02705	.0122	.0558	.0697	.0857	.1021
3.2	.01159	.0255	.0102	.0535	.0672	.0830	.0990
3.3	.01049	.02415	.0384	.0514	.0648	.0804	.0961
3.4	.00055	.0229	.0368	.0195	.0626	.0778	.0933
3.5	.00871	.0217	.03525	.0477	.0605	.0755	.0906
3.6	.00796	.0206	.03385	.0460	.0585	.0733	.0881
3.7	.00729	.01968	.03255	.0444	.0566	.0711	.0858
3.8	.00670	.01875	.0314	.0429	.0549	.0592	.0836
3.9	.00617	.01787	.0303	.0415	.0533	.0673	.0845
4.0	.00568	.01706	.02925	.0402	.0547	.0353	.0791

Table 5 (*Continued*)
Transmission factors for cylindrical samples ($\mu R < 5$)

μR	θ						
	0°	22½°	35°	45°	55°	67½°	90°
4.1	0.00525	0.01629	0.02825	0.0389	0.0502	0.0635	0.0774
4.2	.00488	.01563	.0273	.0377	.0488	.0618	.0755
4.3	.00453	.01500	.02645	.0366	.0475	.0602	.0738
4.4	.00420	.01445	.02555	.0356	.0462	.0578	.0721
4.5	.00391	.01390	.02485	.0347	.0450	.0573	.0705
4.6	.00364	.01343	.0241	.0638	.0439	.0561	.0589
4.7	.00340	.01300	.0234	.0329	.0428	.0547	.0675
4.8	.00316	.01259	.02275	.0321	.0418	.0535	.0661
4.9	.002945	.01222	.0221	.0313	.0408	.0525	.0647
5.0	.002755	.01189	.02155	.0305	.0399	.0514	.0635

Table 6
Transmission factors for cylindrical samples ($\mu R > 5$)
(After Bradley[22])

μR	θ						
	0°	22½°	35°	45°	55°	67½°	90°
5.0	0.0434	0.1871	0.3393	0.4809	0.628	0.810	1.00
5.5	.0350	.1778	.3295	.4705	.620	.804	1.00
6.0	.0291	.1701	.3210	.4625	.613	.799	1.00
6.5	.0244	.1640	.3140	.4560	.607	.795	1.00
7.0	.0212	.1589	.3080	.4505	.602	.792	1.00
7.5	.0183	.1546	.3030	.4450	.598	.789	1.00
8.0	.0161	.1509	.2985	.4405	.594	.786	1.00
9.0	.0126	.1445	.2910	.4335	.587	.781	1.00
10.0	.0102	.1398	.2855	.4275	.582	.777	1.00
11.0	.0084	.1359	.2810	.4225	.578	.774	1.00
12.0	.0069	.1326	.2770	.4185	.574	.772	1.00
13.0	.0059	.1300	.2740	.4150	.571	.770	1.00
14.0	.0050	.1278	.2715	.4125	.569	.768	1.00
15.0	.0044	.1259	.2690	.4100	.567	.767	1.00
20.0	.0025	.1193	.2605	.4020	.559	.762	1.00
25.0	.0016	.1155	.2560	.3970	.554	.758	1.00
30.0	.0011	.1129	.2525	.3935	.551	.756	1.00
40.0	.0006	.1097	.2485	.3895	.548	.753	1.00
50.0	.0001	.1079	.2460	.3870	.545	.751	1.00
75.0	.0002	.1054	.2430	.3840	.542	.749	1.00
100	.0001	.1042	.2410	.3820	.540	.748	1.00
200	.0000	.1018	.2390	.3800	.538	.746	1.00
∞	.0000	.1007	.2365	.3770	.536	.744	1.00

ray makes an angle ν, with the plane normal to the cylinder axis. The direction of the diffracted beam is more completely defined by the cylindrical direction coordinates ν (the angular component in a plane parallel to the cylinder axis) and Υ (the angular component in a plane normal to the cylinder axis). Both $\bar{\mu}$ and ν are setting coordinates for any method involving a rotating crystal. Υ is a setting coordinate of the counter diffractometer (Chapter 6), and also is the coordinate on the Weissenberg film normal to the center line of the film. Now, if one compares Fig. 6 for a value of Υ equal to 2θ in Fig. 5, it is evident that, for the general case, the path t_1 is replaced by $t_1/\cos \bar{\mu}$, and the path t_2 is replaced by the path $t_2/\cos \nu$. Therefore the transmission factor has a form similar to (31), specifically,

$$T = \frac{1}{V} \int^{v} e^{-(\mu t_1/\cos\bar{\mu} + \mu t_2/\cos\nu)}\, dV. \tag{32}$$

This has a simple solution when $\nu = -\bar{\mu}$ (equi-inclination, general level) or when $\nu = +\bar{\mu}$ (anti-equi-inclination, zero level only). In these cases

$$\begin{aligned} T &= \frac{1}{V} \int^{v} e^{-(\mu t_1 + \mu t_2)/\cos\nu}\, dV \\ &= \frac{1}{V} \int^{v} e^{-\mu(t_1 + t_2)/\cos\nu}\, dV. \end{aligned} \tag{33}$$

This is exactly the same as (31) except that every ray path is increased by the factor $1/\cos \nu$. To correct the upper-level equi-inclination reflection for absorption, therefore, one applies the same correction that would be applied at the same value of Υ for the zero level (where $\Upsilon = 2\theta$), except that the correction should be looked up under the value $\mu R/\cos \nu$, not μR.

This analysis neglects an end effect (for which there is less absorption) for the ends of the cylinder in Fig. 6. This end effect is negligible if the length-to-diameter ratio of the cylinder is large. (If the absolute length of the cylinder is so large that the ends are not in the x-ray beam, then there is no end effect, but there must be a correction for the volume intercepted by the beam. This is proportional to $1/\cos \nu$, so that the "integrated intensity" must be corrected by $\cos \nu$ if the x-ray beam does not bathe the full length of the cylinder.)

To illustrate the importance of making a correction for absorption in data taken from upper levels, an example is given here of the computation of the correction and its application: Three-dimensional data were obtained from a small crystal of wollastonite. The average diameter was

Table 7

Calculation of correction for absorption for rod-shaped crystal of wollastonite (μ_l = 215 cm.$^{-1}$ for Cu$K\alpha$; radius, R = 0.0038 cm.)

(After Buerger and Niizoki[41])

Υ	Transmission factor, T									
	level:	0	1	2	3	4	5	6	7	8
	ν:	0°	6°03′	12°10′	18°25′	24°55′	31°47′	39°12′	47°31′	57°26′
	cos ν:	1	0.994	0.978	0.949	0.907	0.850	0.775	0.675	0.538
	$\mu_1 R/\cos\nu$:	0.81	0.815	0.828	0.853	0.893	0.954	1.045	1.20	1.505
0°		0.264	0.262	0.257	0.248	0.233	0.213	0.185	0.146	0.088
45°		0.274	0.272	0.267	0.258	0.243	0.224	0.198	0.160	0.103
90°		0.302	0.300	0.295	0.287	0.273	0.255	0.230	0.195	0.141
135°		0.334	0.332	0.328	0.320	0.307	0.291	0.268	0.235	0.181
180°		0.353	0.351	0.347	0.339	0.326	0.308	0.285	0.251	0.199

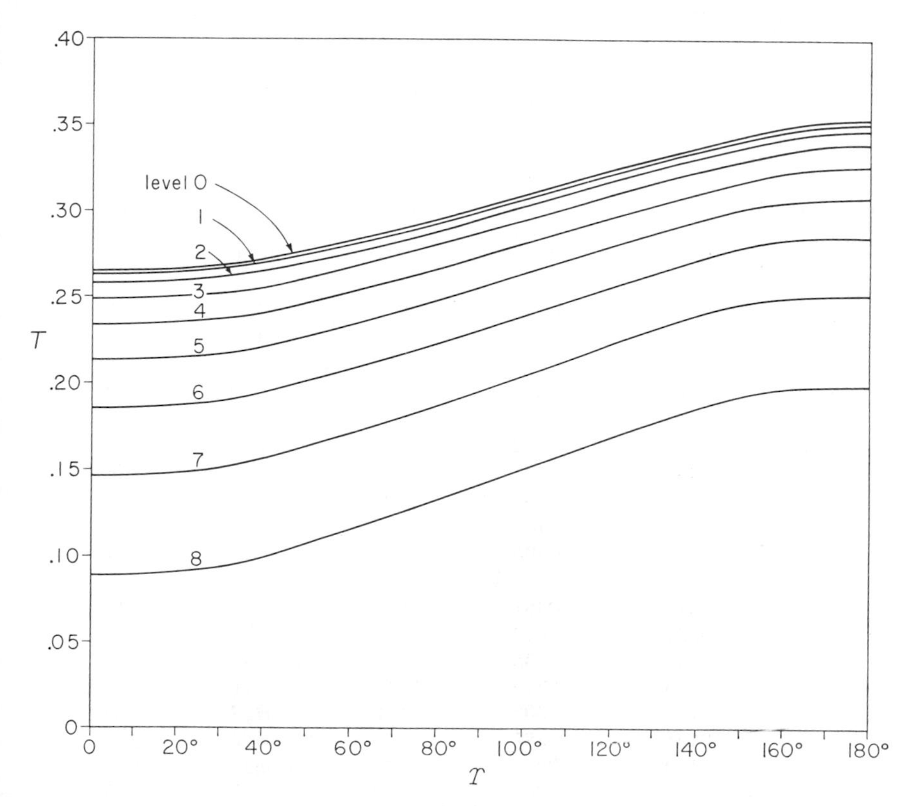

Fig. 7.
(After Buerger and Niizeki.[41])

Table 8
Transmission factors for spherical samples ($\mu R < 5$)
(From Evans and Ekstein[33])

μR	θ				
	0°	22½°	45°	67½°	90°
0.0	1.000	1.000	1.000	1.000	1.000
0.1	0.862	0.862	0.863	0.869	0.872
0.2	.742	.742	.743	.748	.753
0.3	.646	.646	.647	.655	.661
0.4	.560	.560	.567	.580	.580
.05	.489	.490	.502	.520	.531
.06	.422	.424	.441	.462	.476
0.7	.368	.372	.391	.415	.432
0.8	.321	.326	.348	.374	.393
0.9	.281	.287	.313	.340	.359
1.0	.245	.252	.281	.310	.330
1.1	.245	.224	.256	.285	.306
1.2	.189	.200	.233	.263	.286
1.3	.167	.179	.214	.245	.267
1.4	.147	.159	.196	.228	.250
1.5	.131	.143	.180	.215	.236
1.6	.115	.128	.166	.201	.222
1.7	.102	.116	.154	.190	.210
1.8	.0910	.105	.143	.179	.199
1.9	.0814	.0953	.134	.169	.189
2.0	.0731	.0874	.125	.161	.181
2.1	.0653	.0797	.117	.152	.173
2.2	.0585	.0729	.109	.145	.165
2.3	.0528	.0673	.103	.138	.159
2.4	.0476	.0521	.0977	.132	.152
2.5	.0130	.0575	.0925	.126	.147
2.6	.0388	.0532	.0877	.121	.111
2.7	.0352	.0196	.0835	.116	.136
2.8	.0321	.0463	.0796	.112	.131
2.9	.0290	.0433	.0760	.108	.126
3.0	.0267	.0405	.0726	.101	.121
3.1	.0244	.0381	.0695	.100	.118
3.2	.0224	.0359	.0666	.0971	.114
3.3	.0205	.0338	.0638	.0939	.111
3.4	.0189	.0319	.0614	.0911	.108
3.5	.0174	.0302	.0590	.0883	.105
3.6	.0161	.0287	.0569	.0587	.102
3.7	.0149	.0273	.0549	.0833	.0991
3.8	.0138	.0260	.0530	.0809	.0967
3.9	.0128	.0247	.0512	.0786	.0041
4.0	.0119	.0235	.0494	.0765	.0918

Table 8 (*Continued*)

μR	θ				
	0°	$22\frac{1}{2}$°	45°	$67\frac{1}{2}$°	90°
4.1	.0111	.0224	.0481	.0744	.0895
4.2	.0103	.0214	.0466	.0725	.0873
4.3	.00960	.0205	.0457	.0705	.0852
4.4	.00899	.0197	.0439	.0688	.0833
4.5	.00844	.0189	.0127	.0672	.0815
4.6	.00787	.0182	.0415	.0656	.0797
4.7	.00738	.0175	.0403	.0641	.0780
4.8	.00693	.0169	.0393	.0627	.0765
4.9	.00650	.0163	.0383	.0612	.0749
5.0	.00613	.0157	.0373	.0600	.0734

Table 9
Transmission factors for spherical samples ($\mu R > 5$)
(From Evans and Ekstein[33])

μR	θ				
	0°	$22\frac{1}{2}$°	45°	$67\frac{1}{2}$°	90°
5.0	0.0835	0.214	0.508	0.817	1.00
5.5	.0687	.200	.495	.811	1.00
6.0	.0575	.189	.484	.806	1.00
6.5	.0489	.181	.475	.802	1.00
7.0	.0423	.173	.468	.799	1.00
7.5	.0367	.167	.461	.795	1.00
8.0	.0326	.161	.455	.793	1.00
9.0	.0256	.152	.445	.788	1.00
10.0	.0207	.146	.438	.783	1.00
11.0	.0172	.141	.433	.780	1.00
12.0	.0144	.137	.426	.777	1.00
13.0	.0122	.134	.422	.775	1.00
14.0	.0106	.131	.418	.773	1.00
15.0	0.00923	.128	.416	.771	1.00
20.0	.00513	.119	.402	.769	1.00
25.0	.00317	.114	.397	.767	1.00
30.0	.00227	.112	.392	.765	1.00
40.0	.00124	.107	.386	.763	1.00
50.0	.00078	.105	.382	.761	1.00
75.0	.00034	.102	.378	.759	1.00
100.0	.00019	.102	.375	.758	1.00
200.0	.00000	.100	.372	.756	1.00
	.00000	.098	.369	.755	1.00

0.0075 cm. The linear-absorption coefficient, calculated as shown in Table 4, is $\mu = 215$ cm.$^{-1}$ for Cu$K\alpha$. Thus $\mu R = 0.81$. Table 7 shows the computation of the transmission factor. For each level there is derived a value of $\mu R/\cos \nu$. For each of these values, the transmission factor is found by interpolation from the corresponding value of μR in standard tables. To make actual use of these sample values of the transmission, they should be plotted and connected by curves as shown in Fig. 7. Then the transmission factor T for any reflection on any level can be read when the value of Υ for the reflection is known. Since this is a Weissenberg coordinate, and also a setting coordinate for the single-crystal counter diffractometer, the value is known for each reflection.

Figure 7 brings out the importance of making appropriate corrections for absorption in upper-level intensity data. The transmission factors for the higher levels differ so widely from those of the zero level that only a poor residual factor† R can be expected if the zero-level correction is applied to all levels. The crystal in the example is about as small as can be handled conveniently, yet the transmission factor falls in the range 8% to 35% for Cu$K\alpha$ radiation. For Mo$K\alpha$, the value μR is of the order of only 10% of that for Cu$K\alpha$, and the corresponding transmissions are in the range 90% to 100%.

Spherical specimens. The correction for absorption for spherical specimens is comparatively simple since the transmission factor for a given sample is a function of θ only. Bond has shown how crystals of spherical form can be prepared (Chapter 5). The chief disadvantage in using a spherical shape is that the crystal clothed in this spherical shape must be oriented crystallographically before any experimental work can be undertaken. With all external evidences of crystallographic directions removed, this is ordinarily a very trying task.

Evans and Ekstein[33] have tabulated the transmission factors for spherical crystals. Their results are listed in Tables 8 and 9.

Rod-mounted powder specimens. It is common practice to make a powder sample by sticking the powder to the outside of a glass rod or hair. This combination has a more complicated transmission factor than a powder sample composed of a solid sample of the crystalline material only. In the rod-mounted case, the transmission factor is not only a function of μR and θ, but of R_{rod}/R and μ_{rod} also. The transmission factor for such samples has been tabulated by Møller and Jensen.[31]

Prismatic specimens. The correction for absorption for crystals with prismatic cross-section has been treated by many authors. The general

† See Chapter 22.

nature of the problem is illustrated in a solution due to Albrecht.[26] To find the transmission factor, (27) must be evaluated for all paths $t = t_1 + t_2$ in Fig. 8. Albrecht divides the volume of the crystal into n

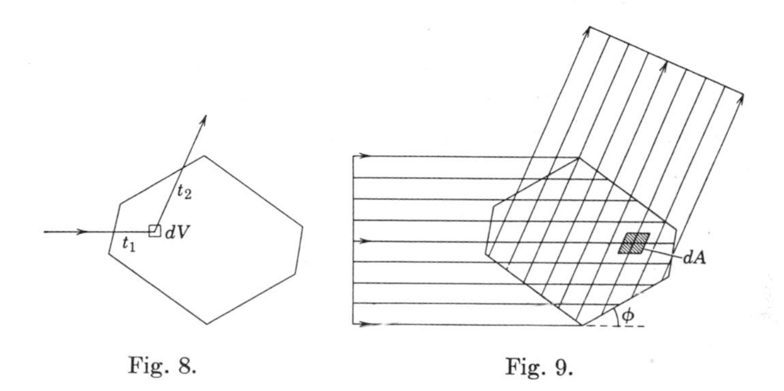

Fig. 8. Fig. 9.

equal prisms whose length L is the length of the crystal, and whose base ΔA is determined by the mesh of a grid whose directions are parallel to the primary and diffracted rays, Fig. 9. Then (27) can be a approximated by the summation

$$T = \frac{\sum e^{-\mu(t_1+t_2)} L\Delta A}{nL\Delta A} \tag{34}$$

$$= \frac{1}{n} \sum e^{-\mu(t_1+t_2)} \tag{35}$$

To determine T, one merely measures with a ruler on a scaled-up drawing each of the paths, t_1+t_2, to and from an intersection in Fig. 9. For each path the quantity $e^{-\mu t}$ is looked up in tables, the sum of these recorded, and the total divided by n, the number of samples into which the cross-section of the specimen was divided. This evaluation of T takes a different form for each reflection because the angle ϕ, Fig. 9, which the crystal makes with the primary beam, is different for each reflection. Rogers and Moffett[37] use a scale graduated into $e^{-\mu t}$ for facilitating the computation.

Evans[32] showed that it is possible to divide the cross-section of a crystal into triangular and parallelogram-shaped areas within which (27) can be integrated. He lists the values of these integrals for five area types.

High-speed digital computing has been recommended[38] for computing transmission factors. Probably this is the only sensible way of carrying out the tedious problem more exactly.

Remarks on correction for absorption. The correction for absorption is an exasperating one. Unlike the correction for "extinction," which cannot be assessed in advance of the experimental work, the theory of the absorption correction is exactly known, and the correction can be computed for a crystal of any given shape, but the computation is so tedious for polyhedral crystals that it is carried out infrequently indeed. To carry it out for the thousand or so reflections required for a three-dimensional analysis is nearly out of the question. For this reason it is advisable to carry out experimental work with a crystal ground to cylindrical or spherical shape.

The correction for absorption is such a large one that it is better to apply an approximate correction than none at all. For example, if a needle-shaped crystal is used it is better to correct reflections for all levels by treating the absorption as if it were from a cylindrical specimen whose diameter is the mean cross-section of the needle. Similarly if the crystal is an approximately equi-dimensional lump, it is better to apply an absorption correction based upon a sphere whose diameter is the mean diameter of the lump.

Mapping transmission in reciprocal space. It was discovered early by Wells[24] that Weissenberg photographs made by plate-shaped crystals rotated about an axis parallel to the plate, or by a needle-shaped crystal rotated about an axis normal to the needle, contain two diagonal lines on which absorption is extreme. The theory of this effect was generalized by Buerger[25] for crystals of any external shape. The effect was rediscovered by MacGillavry and Vos[27] about five years later. While the theory of the effect has been developed for planes of any slope[25] with regard to the rotation axis, the general nature of the effect is illustrated here for planes parallel to the rotation axis only.

Weissenberg photographs. A case which is fundamental to the development of more complicated cases is that of a crystal having one plane surface in the zone of rotation. For the sake of discussion, the extension of the single surface in question is assumed to be indefinite. Figure 10 is a diagrammatic representation of this case. The incoming x-ray beam strikes the plane surface of the crystal, *pp*, at its intersection with the rotation axis, *O*, Fig. 10*c*. The planes of the crystal reflect the incoming beam, producing reflected beams having reflection directions, Υ, over the entire range of $\Upsilon = 0°$ to $\Upsilon = 360°$. For the present discussion, however, the planes may be sharply separated into two categories:

1. Planes, like *bb*, which reflect to a spot, *B*, on the film, which is within the arc $C\bar{E}G$. Such reflections may be termed transmitted reflections because they can only be recorded after either the incident x-ray beam or the reflected beam itself has been transmitted through the body of the crystal. If the crystal has any appreciable absorption, such reflections are most severely reduced in intensity.

2. Planes, like *ff*, which reflect to a spot, *F*, on the film, which is within the arc *CEG*. Such reflections may be termed surface reflections because neither the incident x-ray beam nor the reflected beam need penetrate any appreciable body of the crystal to produce them.

It is plain from Fig. 10 that the arcs of surface reflections and transmitted reflections each occupy an ϒ range of 180°. These arcs are dependent upon the surface, *pp*, for their delimination, since their boundaries are fixed by the points *C* and *G*, the extension of the plane *pp* to the actual film. The arcs therefore travel around the film with the same velocity as the crystal rotation.

Suppose, now, that at the zero azimuth of crystal rotation, the crystal surface has a rotation coordinate ω_p (Fig. 10*a*) and has not yet attained parallelism with the x-ray beam. The surface, *pp*, is in the x-ray shadow of the body of the crystal and so no surface reflections are possible and only transmitted reflections can be recorded on the Weissenberg photograph. A rotation of the crystal through the angle ω_p brings the plane parallel with the x-ray beam (Fig. 10*b*) and permits surface reflections for the first time over the 180° range

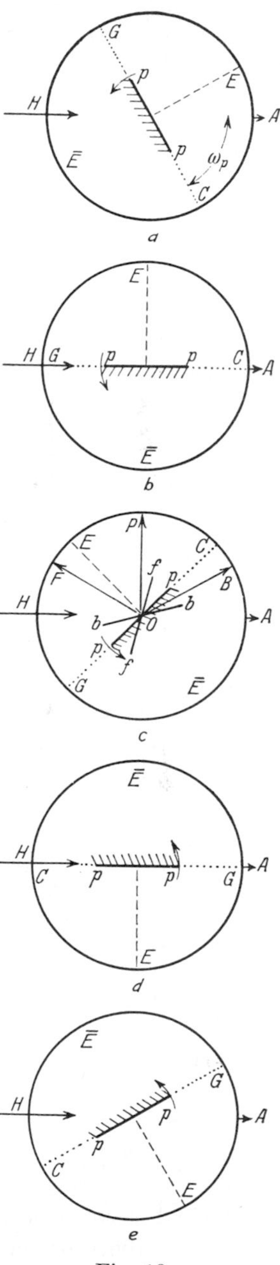

Fig. 10.
(After Buerger.[25])

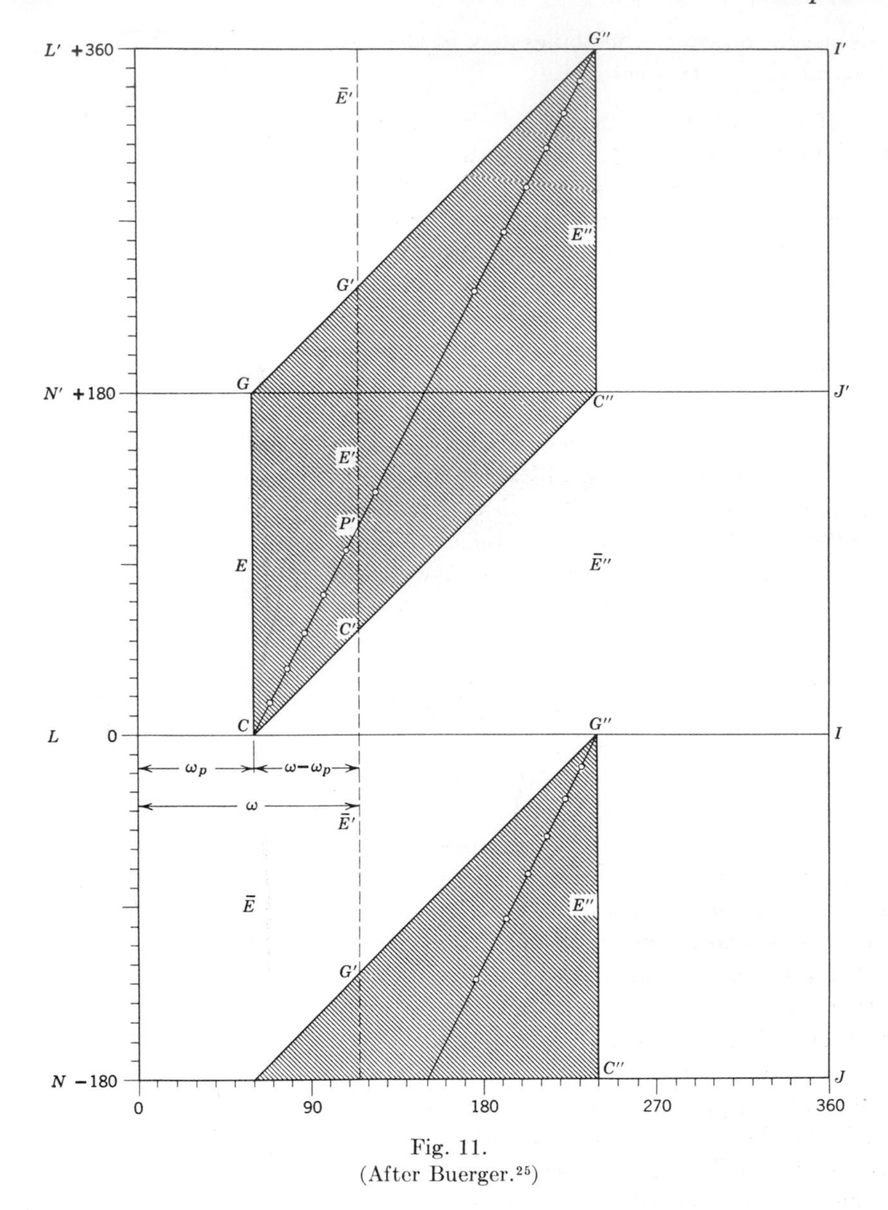

Fig. 11.
(After Buerger.[25])

CEG, which records on the Weissenberg projection as the line *CG*, Fig. 11. The Weissenberg projections illustrated here are drawn with a 180° overlap in Υ to illustrate both the usual Weissenberg film appearance, *NLN′J′IJ*, where the values of Υ range from $-180°$ to $+180°$, and also the back-reflection range *LN′L′I′J′I*, in which the Υ range is from 0° to 360°. The latter range is more convenient for derivations of Weissenberg-projection properties.

As the crystal continues to rotate, the 180° arc *CEG* traverses the film circumference at the speed of crystal rotation and the arc successively occupies the positions *CEG*, *C′E′G′* and *C″E″C″*, Fig. 11, corresponding with the cases represented in Fig. 10*b*, *c*, and *d*, after which no more surface reflections are possible because the surface is in the x-ray shadow of the body of the crystal, Fig. 10*e*. Since the line *CC′C″* represents a plot of the position of crystal rotation, ω, against film translation, which in turn is equal to the crystal rotation, ω, it is a straight line of slope 1. The reflection from a surface travels at twice the angular speed as the surface; therefore, the line of reflections *CP′G″*, is twice the height of *CC′C″* at any point, and consequently has a slope of 2.

From the foregoing discussion it is obvious that the distribution on the Weissenberg projection of surface reflections from any given crystal surface is independent of the extent of the surface and of the possible existence of other surfaces.

The Weissenberg projection of the entire crystal surface is the composite projection of the surface reflection areas of all its separate faces. Any composite case not illustrated may be derived from a knowledge of the two-circle coordinate angles, ρ and ω, of the surfaces which are actually in the x-ray beam.

Consider a crystal whose habit is such that there are *n* natural faces in the zone of the rotation axis. These faces have angular coordinates in the rotation zone of $\omega_1, \omega_2, \omega_3 \cdots \omega_n$. As the rotation of the crystal brings each of these faces to parallelism with the x-ray beam, its arc of surface reflections begins to sweep out a parallelogram-shaped area identical with that just discussed and illustrated in Fig. 11. Where two parallelograms overlap, this common area contains surface reflections from both of the corresponding faces. Two important actual cases commonly arise:

1. A tabular crystal with negligible edge faces, having the rotation axis in the zone of tabular development (Fig. 12).
2. A crystal of prismatic development in the zone of the rotation axis (Fig. 13).

A particular variety of 2 is a crystal bounded by pairs of pinacoids in the zone of the rotation axis. The Weissenberg projection of this case is the same as 2, except that the angular separations of the wedge points are specialized for symmetrical crystals.

It is apparent from the above discussion that in the particular habit discussed by Wells,[24] namely the thin crystal plate (Fig. 12), the diagonals of extreme absorption (those bounding the parallelograms) are only a special case of a much more general absorption situation. All reflections in entire unshaded areas of Fig. 12 are subject to important absorption. This is least in the center of an unshaded area, where the transmission is

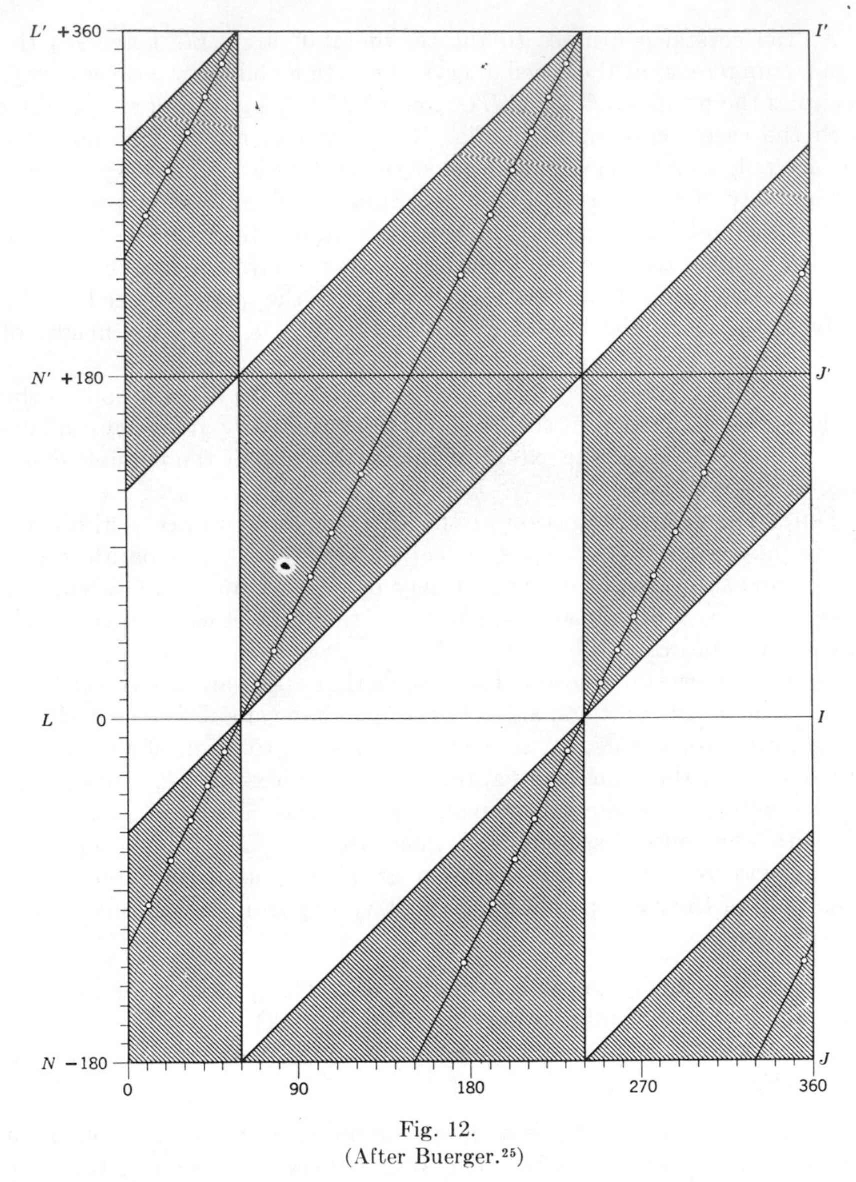

Fig. 12.
(After Buerger.[25])

directly through the plate, and becomes progressively greater toward the diagonals of the parallelogram, as the path of transmission increases. It is possible to prepare a chart for this area giving the correction for absorption in Weissenberg photographs. A similar correction chart for the comparatively small corrections of the shaded areas of surface reflections may also be prepared.

In the more general case of equatorial photographs from crystals with

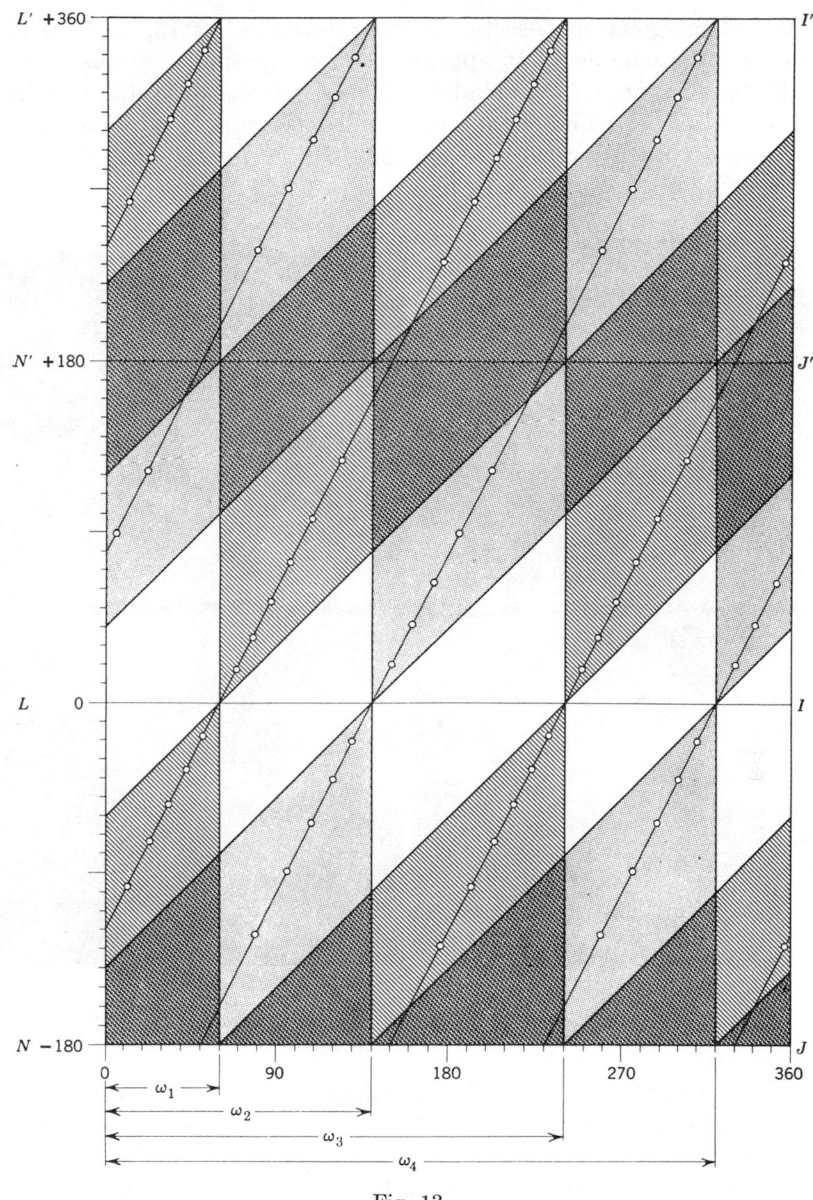

Fig. 13.
(After Buerger.[25])

completed polygonal cross-section such as shown in Fig. 13, the correction factor is more complex. It appears that composite diagrams such as Fig. 13, based upon the actual habit of the experimental crystal, may be of possible aid in determining the appropriate absorption correction factor to be applied for any reflection, for the diagram reveals at a glance the character of the individual contributions to the reflection. With the

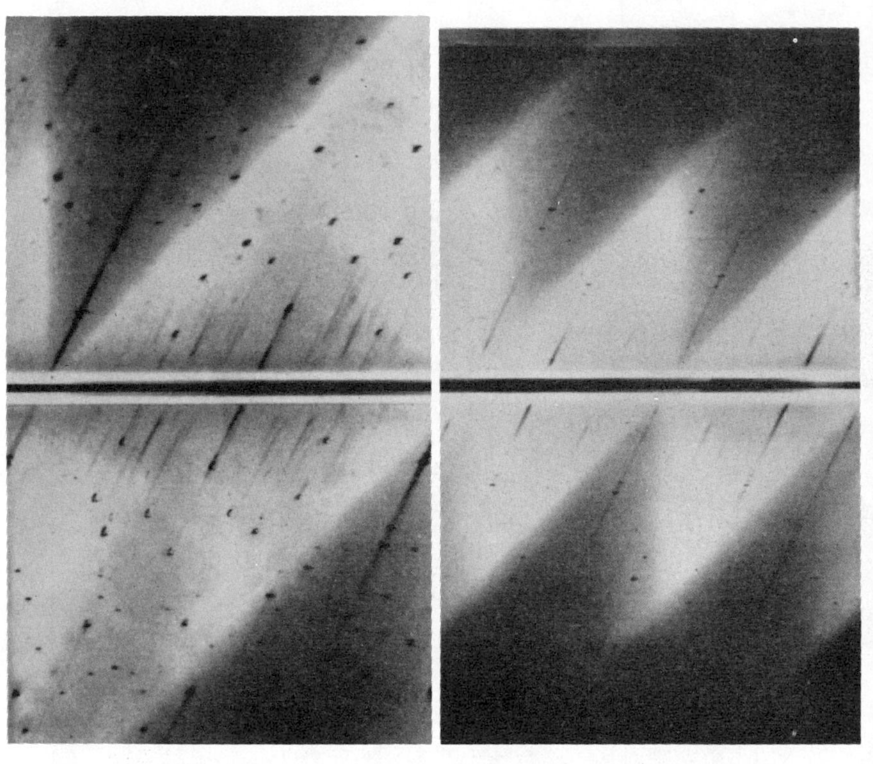

Fig. 14.
(After Buerger.[25])

Fig. 15.
(After Buerger.[25])

aid of the diagram it is possible to quickly decide when the use of a certain reflection for parameter study is dangerous due to high absorption correction.

It is interesting to observe that in the *n*-layer equi-inclination photographs the contributions from a plane surface of slope $\rho = 0$ to all reflections, regardless of position on the Weissenberg photograph, have been transmitted through substantially identical paths and therefore have identical absorption corrections. The transmission factor may therefore be eliminated from any intensity study of the *n*-layers made with this technique, provided that a crystal is employed which has a natural or artificial termination normal to the rotation axis which is of dimensions large enough to receive the entire x-ray beam.

The areas derived as surface-reflection areas can experimentally be made to appear on Weissenberg films as *background patterns.* Several of these are illustrated in Figs. 14 and 15. The blackened areas represent the distribution of soft, incoherently scattered radiation. The scattering can be made especially intense by employing radiation just below the absorption edge of one of the elements present in the crystal. Extreme

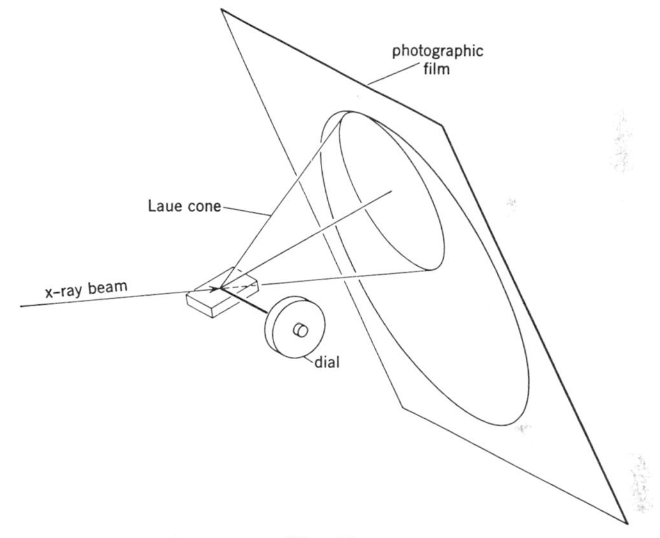

Fig. 16.

absorption and scattering of the incident ray occur in the immediate region of the surface of incidence. The scattered radiation reaches the film in the hemispherical solid angle above the crystal surface, while the film in the other hemisphere is in the shadow of the crystal body and is therefore effectively shielded from scattered radiation. The geometry of this effect is thus the same as that of surface reflections. The amount of background blackening produced by any one crystal face is proportional to the area of the face included in the x-ray beam.

MacGillavry and Vos[27] suggest using the blackening of the background pattern as a measure of the transmission factor in that neighborhood. This is based upon the supposition that the absorption coefficient for the incoherently scattered radiation is the same as that for the coherently diffracted radiation.

Several authors[29, 32, 34] have advocated contouring the map of the Weissenberg film in values of the transmission factor as computed by one of the various methods. When this is done, discontinuities appear which correspond to outlines of the areas on the background pattern.

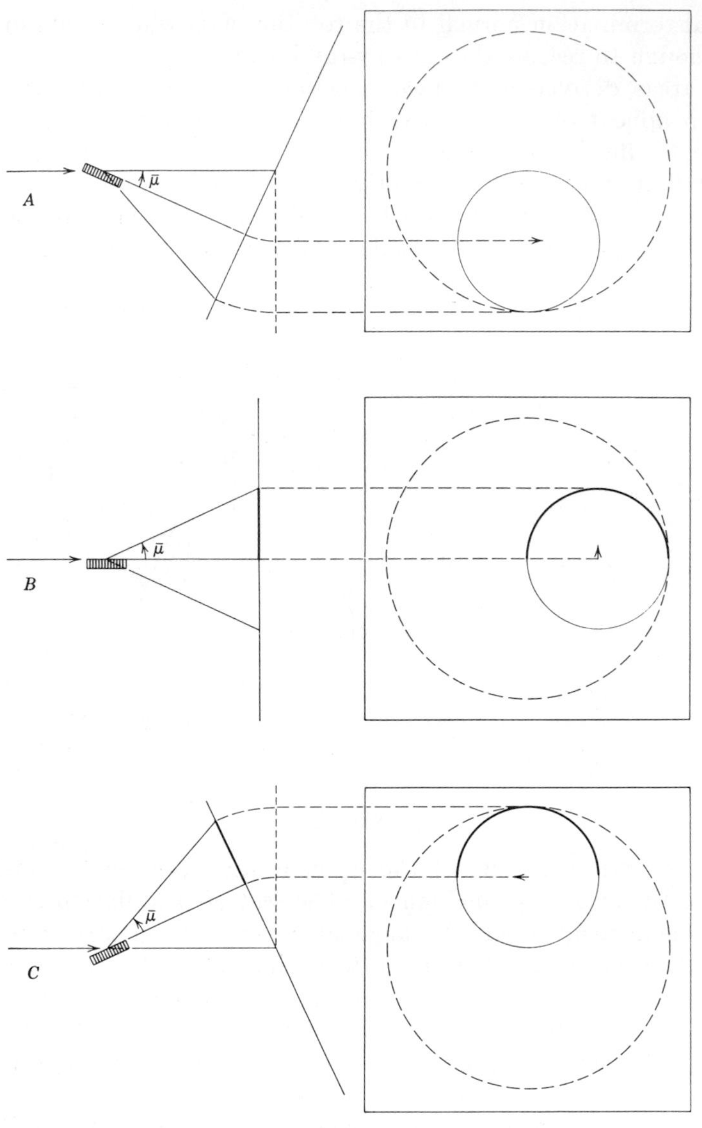

Fig. 17.

Precession photographs. Takéuchi[39] has investigated some of the properties of surface-reflection fields for precession photographs. These and some additional properties are noted below.

The general scheme of the precession method is shown in Fig. 16. To see how the surface-reflection fields are derived, consider a crystal which has a single surface plane. First, let this plane be set so that it contains both the dial axis and the axis of the Laue cone, Fig. 16. As the

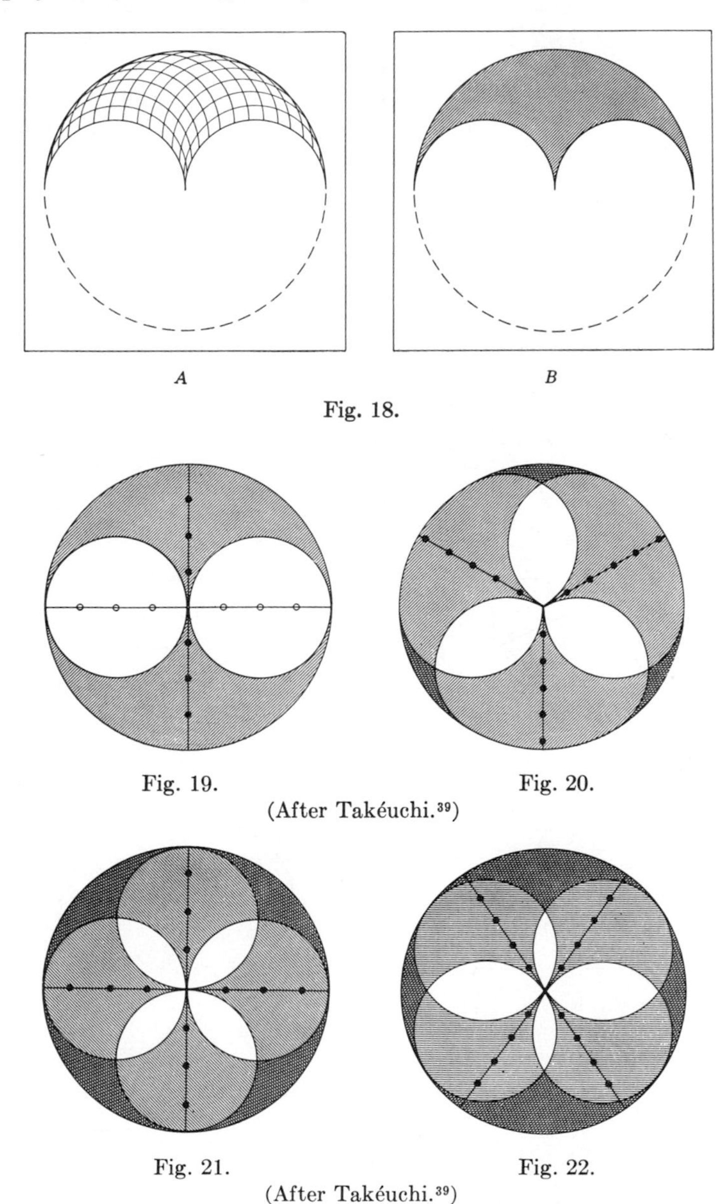

Fig. 18.

Fig. 19. Fig. 20.

(After Takéuchi.[39])

Fig. 21. Fig. 22.

(After Takéuchi.[39])

precessing motion progresses, the Laue cone precesses around the direct beam. When the axis of the cone is directed downward, as in Fig. 17*A*, the surface of the crystal is in the shadow of the direct beam, and there are no surface reflections. When the axis of the cone is just horizontal, Fig. 17*B*, surface reflections become possible in the upper half of the cone. The rays for these surface reflections intersect the photographic film in a semicircular arc, drawn heavy in Fig. 17*B*. The surface

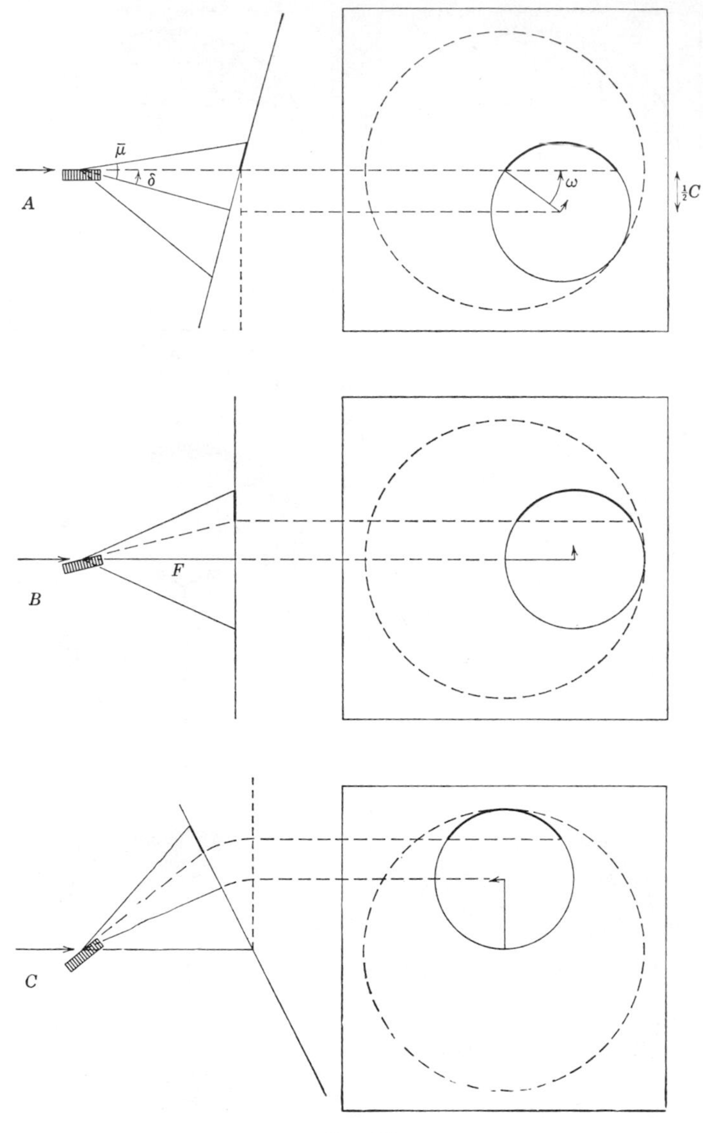

Fig. 23.

reflections continue to record along this semicircular arc as the precessing motion progresses, Fig. 17C. A number of these arc loci are shown in Fig. 18A. These sweep out the shaded field of Fig. 18B, which is the surface-reflection field of this surface plane of the crystal.

The surface-reflection fields for any crystal whose habit is composed of plane faces set parallel to the axis of the cone can be derived by combining fields like the one shown in Fig. 18B, one for each face of the crystal.

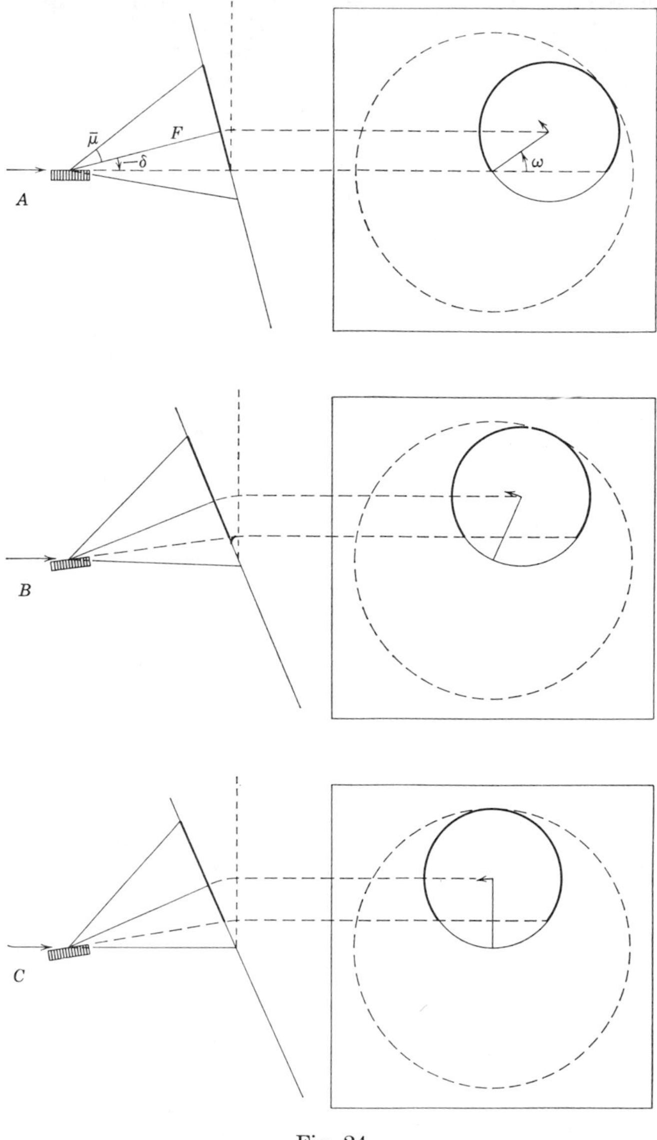

Fig. 24.

For example, the surface-reflection fields for a plate-shaped crystal are given in Fig. 19, for a trigonal prism in Fig. 20, for a tetragonal prism in Fig. 21, and for a crystal bounded by a more general prism in Fig. 22.

If the crystal plane is not parallel to the Laue cone, but makes an angle $+\delta$ with it, Fig. 23, then before the axis of the cone becomes horizontal, the surface becomes parallel with the x-ray beam, and surface

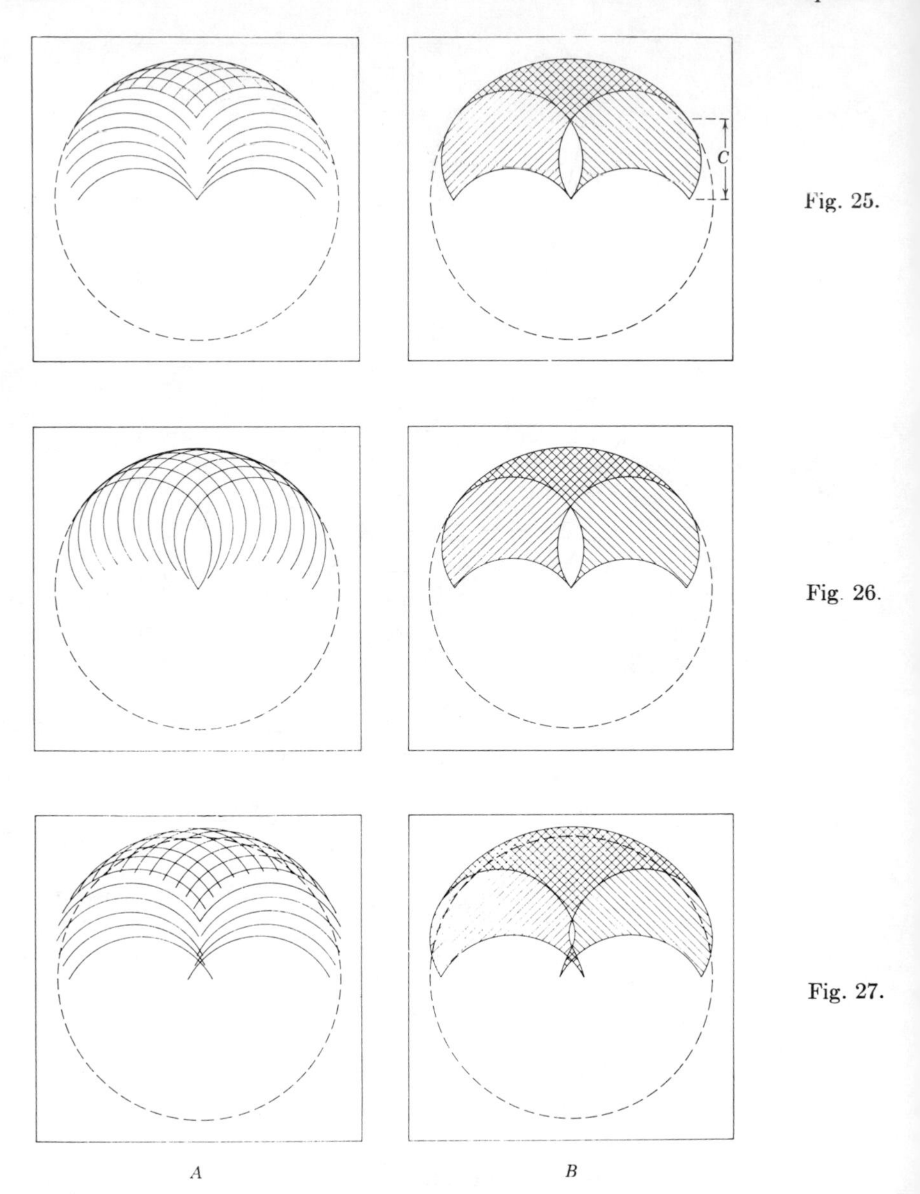

Fig. 25.

Fig. 26.

Fig. 27.

reflections occur along the part of the cone which is above a horizontal plane, Fig. 23*A*. A precession motion through an angle ω is necessary to make the cone axis horizontal. As the precession motion proceeds, Figs. 23*C*, the same arc remains uncovered, and this sweeps out the areas noted in Fig. 25. This results in a field in which surface reflections occur for both right and left sides, and another pair of fields in which surface reflections occur for only one side. The slope, δ, of the plane with respect to the axis of the cone can be computed from the distance C of the cusp from the origin. This cusp occurs at twice the displacement of that circle to first permit surface reflections, Fig. 23A. From Fig. 23A, this gives $\cos \delta = C/2F$. If the plane makes an angle $-\delta$ with the cone axis, Fig. 24, then it reaches parallelism with the direct beam only after the precession motion has proceeded for an angle ω beyond the condition when the cone axis is horizontal. Then surface reflections occur over an arc greater than a semicircle, producing the fields shown in Fig. 26.

Figure 27 illustrates the surface-reflection field for an upper-level photograph for a plane making a positive angle with the cone axis. The case is somewhat similar to that for the zero level, shown in Fig. 25.

The temperature factor

In discussing the intensity of the x-ray diffraction spectra it has been tacitly assumed that the crystal structure is a static one, i.e., that it can be thought of as a periodic pattern of stationary atoms. Temperature modifies this situation since it requires every atom to undergo thermal motion. The general nature of the modification can be appreciated by noting that the effect of the thermal motion is to make the electrons of each atom sweep out a larger average volume than they would occupy if the atom were at rest. This causes the effective f curves of the atoms to fall off more rapidly with $(\sin \theta)/\lambda$ than for the same atoms at rest.

It is a very complex matter to make an accurate allowance for thermal motion. Each atom undergoes a motion such that its electron density is smeared over a small anisotropic volume, usually regarded as a triaxial ellipsoid in the general case. Each non-equivalent atom not only has a different ellipsoid, but the ellipsoids are differently oriented. Obviously the consideration of how such a complex motion affects the measured intensities is extremely difficult, and for discussions of it the reader is referred to original literature[48–77] and to Chapter 22.

The Debye-Waller correction. Waller[53] showed that a fair approximation to the effect of thermal motion on the intensity of x-ray reflection can be made for isometric structures containing only one kind of atom. In this simple case

$$f = f_0 \, e^{-(B \sin^2 \theta)/\lambda^2}. \tag{36}$$

The quantity B is called the temperature coefficient. Its value is given by

$$B = 8\pi^2 \, \overline{u_\perp{}^2}, \tag{37}$$

where $\overline{u_\perp{}^2}$ is the mean square displacement normal to the reflecting plane of the atoms from their mean position. The value of B can also be evaluated in terms of the Debye characteristic temperature:

$$B = \frac{6h^2 T}{mk\Theta^2} Q\left(\frac{\Theta}{T}\right), \tag{38}$$

where
m = the mass of the atom,
h = Planck's constant,
k = Boltzmann's constant,
T = absolute temperature,
Θ = Debye characteristic temperature,
$Q\left(\frac{\Theta}{T}\right)$ = a quantization factor which has been tabulated and does not differ appreciably from unity unless $\frac{\Theta}{T} < 1$.

These relations involve quantities which are ordinarily unknown to the crystal-structure analyst and so are usually of little help in making a temperature correction.

While it has long been recognized that the simple form of temperature correction given in (36) cannot be applied in more complicated cases, it is nevertheless true that it can often be used as an approximation. This type of approximation assumes that, in thermal motion, each atom of the crystal undergoes the same average isometric motion. When this assumption is made the correction $e^{-(B \sin^2 \theta)/\lambda^2}$ can be used for each f in the summation for F, so that

$${}^TF = {}^0Fe^{-(B \sin^2 \theta)/\lambda^2} \tag{39}$$

and

$$|{}^TF|^2 = |{}^0F|^2 e^{-(2B \sin^2 \theta)/\lambda^2} \tag{40}$$

where
${}^0F$ = the structure factor at 0°K.
${}^TF$ = the structure factor at the temperature of the experiment.

If logarithms are taken of both sides of (39) there results

$$\ln\left(\frac{{}^TF}{{}^0F}\right) = -B \frac{\sin^2 \theta}{\lambda^2}. \tag{41}$$

This has the form

$$y = kx, \tag{42}$$

where

$$y = \ln\left(\frac{{}^{T}F}{{}^{0}F}\right),$$

$$x = \frac{\sin^2\theta}{\lambda^2},$$

so that (41) should plot as a straight line of slope

$$k = -B. \tag{43}$$

Some data by Thewlis[64] on tetragonal β-uranium are shown plotted in this way in Fig. 28.

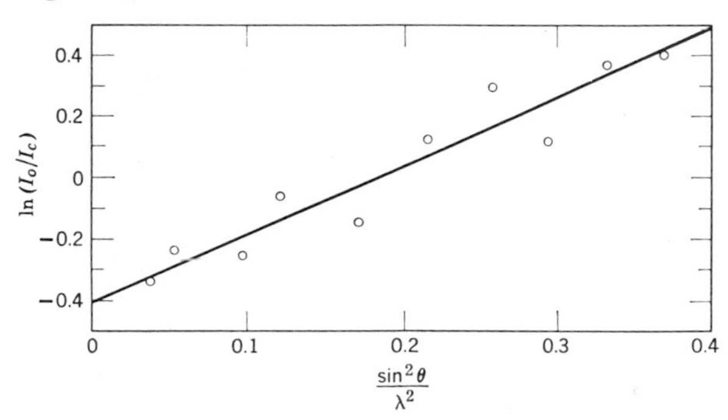

Fig. 28.
(After Thewlis.[64])

Unfortunately this method of determining B requires that the structure be solved, whereas B is desired in advance of the solution. The value of B can fortunately be determined by another method, which is discussed below.

The determination of *B* and the absolute scale. The temperature parameter, B, can be found, and at the same time the set of F's can be placed upon an absolute basis, by a method first presented by Wilson[79] and later rediscovered by Harker.[80] The theory of this is as follows: From (4) of Chapter 10, the general value of the structure factor is

$$F = \sum_j f_j\, e^{i2\pi(hx_j+ky_j+lz_j)}. \tag{44}$$

The square of the absolute value of F is found by multiplying it by its complex conjugate:

$$|F|^2 = F\tilde{F}$$

$$= \left(\sum_i f_i e^{i2\pi(hx_i+ky_i+lz_i)}\right)\left(\sum_j f_j e^{-i2\pi(hx_j+ky_j+lz_j)}\right) \tag{45}$$

$$= \sum_i \sum_j f_i\, f_j e^{i2\pi(h[x_i-x_j]+k[y_i-y_j]+l[z_i-z_j])}. \tag{46}$$

This can be separated into two parts, according as $i = j$ and $i \neq j$:

$$|F|^2 = \sum_j f_j^2 + \sum_i \sum_{\substack{j \\ i \neq j}} f_i f_j \, e^{i2\pi(h[x_i - x_j] + k[y_i - y_j] + l[z_i - z_j])}. \tag{47}$$

If either the average or the sum is taken over all *hkl*, the last term tends to zero since it contains as many positive as negative components, so that

$$\overline{|F|^2} = \overline{\sum_j f_j^2}. \tag{48}$$

Now, consider $|F_{\text{obs}}|^2$ and the absolute value of $|F|^2$. The $|F_{\text{obs}}|^2$ is usually known on an arbitrary scale, so that

$$|F_{\text{obs}}|^2 = K|F|^2. \tag{49}$$

Accordingly

$$\overline{|F_{\text{obs}}|^2} = K\overline{|F|^2}. \tag{50}$$

But the right side of (50) is known from (48). Thus

$$\overline{|F_{\text{obs}}|^2} = K \overline{\sum_j f_j^2}.$$

From this the scale factor is

$$K = \frac{\overline{|F_{\text{obs}}|^2}}{\overline{\sum_j f_j^2}}. \tag{51}$$

But the f's are the true scattering powers under the conditions of observation. If the Debye-Waller temperature correction is assumed, then according to (36)

$$f^2 = {}^0f^2 \, e^{-(2B \sin^2\theta)/\lambda} \tag{52}$$

Accordingly, (51) may be written

$$K = \frac{\overline{|F_{\text{obs}}|^2}}{\overline{\sum_j {}^0f_j^2} \, e^{-(2B \sin^2\theta)/\lambda^2}}. \tag{53}$$

From this it follows that

$$\frac{\overline{|F_{\text{obs}}|^2}}{\overline{\sum_j {}^0f_j^2}} = Ke^{-(2B \sin^2\theta)/\lambda^2}. \tag{54}$$

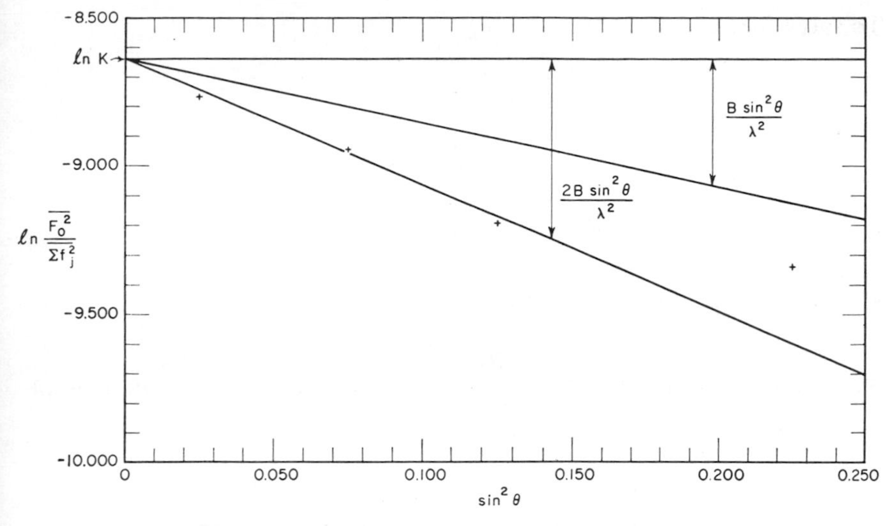

Fig. 29.

If this is plotted against $\sin^2 \theta$ or $\sin^2 \theta/\lambda$, then, as

$$\theta \to 0,$$

$$\frac{\overline{|F_{\text{obs}}|^2}}{\sum_j {}^0f_j^2} \to K. \tag{55}$$

This provides a way of finding coefficient K, which is necessary to place the $|F_{\text{obs}}|^2$'s on absolute scale.

If logarithms are taken of both sides of (54) there results

$$\ln\left(\frac{\overline{|F_{\text{obs}}|^2}}{\sum_j {}^0f_j^2}\right) = \ln K - 2B\,\frac{\sin^2 \theta}{\lambda^2} \tag{56}$$

This is shown plotted in Fig. 29. To solve (56) for B, it is convenient to rearrange it to

$$\ln\left(\frac{\overline{|F_{\text{obs}}|^2}}{\sum_j {}^0f_j^2}\right) - \ln K = \frac{-2B}{\lambda^2}\sin^2 \theta. \tag{57}$$

This has the form

$$y = Ax. \tag{58}$$

To solve for A,

$$A = \frac{y}{x} = \frac{y_2 - y_1}{x_2 - x_1}$$

so that

$$\frac{-2B}{\lambda^2} = \frac{\ln\left(\dfrac{|F_{\text{obs}}|^2}{\sum {}^0f_j^{\,2}}\right) - \ln K}{\sin^2\theta}. \tag{59}$$

Note that natural logarithms are used.

Table 10
Suggested zones for Wilson's method for use with a precession photograph

Zone number	$\sin^2\theta$ range	$\sin^2\theta$ middle	$\sin\theta$ range
1	0 – 0.050	0.025	0 – 0.223
2	0.050 – 0.100	.075	0.223 – 0.316
3	.100 – 0.150	.125	.316 – 0.387
4	.150 – 0.200	.175	.387 – 0.447
5	.200 – 0.250	.225	.447 – 0.500

Table 11
Suggested zones for Wilson's method for use with methods involving a rotating crystal

Zone number	$\sin^2\theta$ range	$\sin^2\theta$ middle	$\sin\theta$ range
1	0 – 0.200	0.100	0 – 0.447
2	0.200 – 0.400	.300	0.447 – 0.632
3	.400 – 0.600	.500	.632 – 0.775
4	.600 – 0.800	.700	.775 – 0.894
5	.800 – 1.000	.900	.894 – 1.000

To make use of this, divide the reciprocal lattice into a limited number of equal $\sin^2\theta$ zones within which it is assumed that an appropriate averaging takes place to warrant the reduction of (47) to (48). Appropriate zones are suggested in Table 10 for precession photographs and in Table 11 for methods involving a rotating crystal. Then

a. Plot all $|F_{\text{obs}}|^2$ on the reciprocal lattice.

b. Find the average for each zone, counting zero intensities and extinguished reflections as points present but having zero intensity.

If only a fraction of the reciprocal lattice is used because of symmetry, remember to count border points at half value. Omit $|F_{000}|^2$.

c. Determine $\overline{\sum_j {}^0f_j^{\,2}}$ for each of the five zones.

d. Prepare a table of $\overline{|F|^2} / \overline{\sum {}^0f_j^{\,2}}$ for each of the five zones.

e. Make a plot† similar to Fig. 29.

Literature

"Extinction"

[1] C. G. Darwin. *The theory of x-ray reflexion.* Phil. Mag. **27** (1914) 675–690.

[2] W. H. Bragg. *The intensity of reflexion of x-rays by crystals.* Phil. Mag. **27** (1914) 881–899.

[3] W. H. Bragg. *An x-ray absorption band.* Nature **93** (1914) 31–32.

[4] W. L. Bragg, R. W. James, and C. H. Bosanquet. *The intensity of reflexion of x-rays by rock salt.* Phil. Mag. (6) **42** (1921) 1–17.

[5] C. G. Darwin. *The reflexion of x-rays from imperfect crystals.* Phil. Mag. **43** (1922) 800–829.

[6] Ivar Waller. *Zur Theorie der Röntgenreflexion.* Ann. Physik. **79** (1926) 261–272.

[7] Vera Daniel and H. Lipson. *The dissociation of an alloy of copper, iron, and nickel. Further x-ray work.* Proc. Roy. Soc. (London) (A) **182** (1944) 378–387.

[8] William H. Zachariasen. *Theory of x-ray diffraction in crystals.* (John Wiley and Sons, New York, 1945).

[9] Kathleen Lonsdale. *Extinction in x-ray crystallography.* Mineral. Mag. **28** (1947) 14–25.

[10] W. A. Wooster and G. L. Macdonald. *Crystalline texture and the determination of structure amplitudes.* Acta Cryst. **1** (1948) 49–54.

[11] G. E. Bacon and R. D. Lowde. *Secondary extinction and neutron crystallography.* Acta Cryst. **1** (1948) 303–314.

[12] H. Ekstein. *Multiple elastic scattering and radiation damping. I.* Phys. Rev. **83** (1951) 721–729.

[13] W. H. Hall and G. K. Williamson. *The diffraction pattern of cold worked metals: I. The nature of extinction.* Proc. Phys. Soc. (London) (B) **64** (1951) 937–946.

[14] R. J. Weiss. *Extinction effects in powders.* Proc. Phys. Soc. (London) (B) **65** (1952) 553–555.

† If no correction has been made for absorption, the slope of the straight line contains a contribution from absorption, and the value of B determined from it is too small. This is because absorption decreases exponentially with increasing $\sin\theta$, and for a spherical specimen is a function of $\sin\theta$ only. Thus the transmission factor is often falsely included in the factor designated the temperature factor.

[15] A. R. Lang. *Extinction in x-ray diffraction patterns of powders.* Proc. Phys. Soc. (B) **66** (1953) 1003–1008.

[16] G. K. Williamson and R. E. Smallman. *X-ray extinction and the effect of cold work on integrated intensities.* Proc. Phys. Soc. (London) (B) **68** (1955) 577–585.

[17] Vladimir Vand. *Methods for the correction of x-ray intensities for primary and secondary extinction in crystal structure analysis.* J. Appl. Phys. **26** (1955) 1191–1194.

[18] L. Gatineau and J. Mering. *Une méthode de correction des effets d'extinction affectant les intensités des rayons X réfléchis par un cristal unique.* Compt. rend. **242** (1956) 2018–2020.

[19] S. Chandrasekhar. *A first-order correction for extinction in crystals.* Acta Cryst. **9** (1956) 954–956.

[20] Walter C. Hamilton. *The effect of crystal shape and setting on secondary extinction.* Acta Cryst. **10** (1957) 629–634.

Absorption

[21] A. Claassen. *The calculation of absorption in x-ray powder-photographs and the scattering power of tungsten.* Phil. Mag. (7) **9** (1930) 57–65.

[22] A. J. Bradley. *The absorption factor for the powder and rotating-crystal methods of x-ray crystal analysis.* Proc. Phys. Soc. (London) **47** (1935) 879–899.

[23] Otis P. Hendershot. *Absorption factor for the rotating crystal method of crystal analysis.* Rev. Sci. Instr. **8** (1937) 324–326.

[24] A. F. Wells. *A note on absorption and Weissenberg photographs.* Z. Krist. (A) **96** (1937) 451–453.

[25] M. J. Buerger. *X-ray surface reflection fields and their application to absorption corrections and to background patterns.* Z. Krist. (A) **99** (1938) 189–204.

[26] Gustav Albrecht. *The absorption factor in crystal spectroscopy.* Rev. Sci. Instr. **10** (1939) 221–222.

[27] C. H. MacGillavry and H. J. Vos. *Anomale Untergrundschwärzung in Weissenberg Diagrammen.* Z. Krist. (A) **105** (1943) 257–267.

[28] M. van Driel and C. H. MacGillavry. *The crystal structure of phosphorus pentabromide.* Rec. trav. chim. **62** (1943) 167–171.

[29] R. Gwynne Howells. *A graphical method of estimating absorption factors for single crystals.* Acta Cryst. **3** (1950) 366–369.

[30] D. Grdenić. *A note on the calculation of the absorption factors for single crystals of high absorbing power.* Acta Cryst. **5** (1952) 283–284.

[31] Eva Møller and E. Jensen. *The absorption factor for rod-mounted specimens in the powder method of x-ray analysis.* Acta Cryst. **5** (1952) 345–348.

[32] Howard T. Evans, Jr. *X-ray absorption corrections for single crystals.* J. Appl. Phys. **23** (1952) 663–668.

[33] H. T. Evans, Jr. and Miriam G. Ekstein. *Tables of absorption factors for spherical crystals.* Acta Cryst. **5** (1952) 540–542.

[34] Allan Zalkin and D. H. Templeton. *The crystal structure of CeB_4, ThB_4, and UB_4.* Acta Cryst. **6** (1953) 269–272, especially 270.

[35] N. Joel, R. Vera, and I. Garaycochea. *A method for the estimation of transmission factors in crystals of uniform cross section.* Acta Cryst. **6** (1953) 465–468.

[36] I. Garaycochea, R. Muñoz, and O. Wittke. *Sobre la posibilidad de interpolar en la estimacion de los factores de transmision* T_{hk0} *de cristales.* P. Dep. Crist. Mineral, **1** (1954) 81–86.
[37] D. Rogers and R. H. Moffett. *A graphical aid for the rapid evaluation of absorption corrections by Albrecht's method.* Acta Cryst. **9** (1956) 1037–1938.
[38] William R. Busing and Henri A. Levy. *High-speed computation of the absorption correction for single crystal diffraction measurements.* Acta Cryst. **10** (1957) 180–182.
[39] Yoshio Takéuchi. *X-ray surface reflection fields on precession photographs.* Mineral. J. (Japan) **2** (1957) 162–168.
[40] A. R. B. Skertchly. *Calculation of the macro absorption factor for a cylindrical specimen irradiated with a fine beam.* Acta Cryst. **10** (1957) 535.
[41] M. J. Buerger and N. Niizeki. *The correction for absorption for rod-shaped single crystals.* Am. Mineralogist **43** (1958) 726–731.

Anomolous absorption

[42] G. Borrmann. *Über extinktions diagramme von Quarz.* Physik. Z. **42** (1941) 157–162.
[43] M. von Laue. *Die Absorption der Röntgenstrahlen in Kristallen im Interferenzfall.* Acta Cryst. **2** (1949) 106–113.
[44] G. Borrmann. *Die Absorption von Röntgenstrahlen im Fall der Interferenz.* Z. Phys. **127** (1950) 297–323.
[45] H. N. Campbell. *X-ray absorption in a crystal set at the Bragg angle.* J. Appl. Phys. **22** (1951) 1139–1142.
[46] W. H. Zachariasen. *On the anomalous transparency of thick crystals to x-rays.* Proc. Nat. Acad. Sci. U. S. **38** (1952) 378–382.
[47] C. Barrère. *Contribution a l'étude de l'absorption des rayons X au voisinage de l'angle de Bragg.* Bull. soc. franç. minéral. et crist. **80** (1957) 344–377.

The temperature factor

[48] P. Debije. *Über den Einfluss der Wärmebewegung auf die Interferenzerscheinungen bei Röntgenstrahlen.* Verhandl. deut. phyik. Ges. **15** (1913) 678–689.
[49] P. Debije. *Über die Intensitätsverteilung in den mit Röntgenstrahlen erzeugten Interferenzbildern.* Verhandl. deut. physik. Ges. **15** (1913) 738–752.
[50] M. v. Laue. *Über den Temperatureinfluss bei den Interferenzerscheinungen an Röntgenstrahlen.* Ann. Physik **42** (1913) 1561–1571.
[51] P. Debye. *Interferenz von Röntgenstrahlen und Wärmebewegung.* Ann. Physik **43** (1914) 49–95.
[52] M. v. Laue. *Der Einfluss der Temperatur auf die Röntgenstrahlinterferenzen.* Ann. Physik **81** (1926) 877–905.
[53] Ivar Waller. *Die Einwirkung der Wärmebewegung der Kristallatome auf Intensität, Lage and Schärfe der Röntgenspektrallinien.* Ann. Physik **83** (1927) 153–183.
[54] R. W. James and Elsie M. Firth. *An x-ray study of the heat motions of the atoms in a rock-salt crystal.* Proc. Roy. Soc. (London) (A) **117** (1927) 62–87.
[55] Ivar Waller and R. W. James. *On the temperature factors of x-ray reflexion for sodium and chlorine in the rock-salt crystal.* Proc. Roy. Soc. (London) (A) **117** (1927) 214–223.

[56] R. W. James and G. W. Brindley. *A quantitative study of the reflexion of x-rays by sylvine.* Proc. Roy. Soc. (London) (A) **121** (1928) 155–171.

[57] Lindsay Helmholz. *The crystal structure of silver phosphate.* J. Chem. Phys. **4** (1936) 316–322, especially 319.

[58] E. W. Hughes. *The crystal structure of melamine.* J. Am. Chem. Soc. **63** (1941) 1737–1752, especially 1743–1744.

[59] Kathleen Lonsdale. *Experimental study of x-ray scattering in relation to crystal dynamics.* Phys. Soc. Repts. Progr. in Phys. **9** (1943) 256–293.

[60] E. A. Owen and R. W. Williams. *The effect of temperature on the intensity of x-ray reflexion.* Proc. Roy. Soc. (London) (A) **188** (1947) 509–521.

[61] Kathleen Lonsdale. *Vibration amplitudes of atoms in cubic crystals.* Acta Cryst. **1** (1948) 142–149.

[62] William J. Dulmage and William N. Lipscomb. *The crystal structures of hydrogen cyanide, HCN.* Acta Cryst. **4** (1951) 330–334, especially 331.

[63] A. I. Snow and R. E. Rundle. *The structure of dimethylberyllium.* Acta Cryst. **4** (1951) 348–352, especially 349.

[64] J. Thewlis. *An x-ray powder study of β-uranium.* Acta Cryst. **5** (1952) 790–794, especially 791.

[65] D. R. Holmes. *Anisotropic temperature factors in* hkl *structure-factor calculations.* Acta Cryst. **6** (1953) 301–302.

[66] H. J. Grenville-Wells. *A nomogram for evaluating the temperature factor.* Acta Cryst. **6** (1953) 665.

[67] W. Cochran. *The effect of anisotropic thermal vibration on the atomic scattering factor.* Acta Cryst. **7** (1954) 503–504.

[68] J. S. Rollett and David R. Davies. *The calculation of structure factors for centrosymmetric monoclinic systems with anisotropic atomic vibration.* Acta Cryst. **8** (1955) 125–128.

[69] David R. Davies and J. J. Blum. *The crystal structure of parabanic acid.* Acta Cryst. **8** (1955) 129–136, especially 131–133.

[70] J. S. Rollett. *The crystal structure of phenyl-propiolic acid.* Acta Cryst. **8** (1955) 487–494, especially 491.

[71] Jürg Waser. *The anisotropic temperature factor in triclinic coordinates.* Acta Cryst. **8** (1955) 731.

[72] Henri A. Levy. *Symmetry relations among coefficients of the anisotropic temperature factor.* Acta Cryst. **9** (1956) 679.

[73] Kathleen Lonsdale and H. Judith Grenville-Wells. *Anisotropic temperature factors in cubic crystals.* Nature **177** (1956) 986–987.

[74] R. E. Gilbert and K. Lonsdale. *Anisotropic temperature vibrations in crystals. I. Direct measurements of Debye factors for urea.* Acta Cryst. **9** (1956) 697–709.

[75] H. J. Grenville-Wells. *Anisotropic temperature vibrations in crystals. II. The effect of changes in atomic scattering factors and temperature parameters on the accuracy of the determination of the structure of urea.* Acta Cryst. **9** (1956) 709–721.

[76] D. W. J. Cruickshank. *The determination of the anisotropic thermal* motion of atoms in crystals. Acta Cryst. **9** (1956) 747–753.

[77] Yoshio Sasada and Isamu Nitta. *A refinement of the crystal structure of tropolone hydrochloride.* Bull. Chem. Soc. Japan **30** (1957) 62–68, especially 63–64.
[78] Arthur Paskin. *A reformulation of the temperature dependence of the Debye characteristic temperature and its effect on Debye-Waller theory.* Acta Cryst. **10** (1957) 667–669.

Determination of *B* and absolute scale

[79] A. J. C. Wilson. *Determination of absolute from relative x-ray intensity data.* Nature **150** (1942) 152.
[80] David Harker. *Absolute intensity scale for crystal diffraction data.* Am. Mineralogist **33** (1948) 764–765.
[81] David Harker. *The meaning of the average of* $|F|^2$ *for large values of the interplanar spacing.* Acta Cryst. **6** (1953) 731–736.
[82] Gopinath Kartha. *A new method of calculating the scale factor in structure analysis.* Acta Cryst. **6** (1953) 817–820.
[83] D. Rogers. *An improved method of scaling intensities.* Acta Cryst. **7** (1954) 628.
[84] Jan Krogh-Moe. *A method of converting experimental x-ray intensities to an absolute scale.* Acta Cryst. **9** (1956) 951–953.
[85] Beatrice S. Magdoff, F. H. C. Crick and V. Luzzati. *The three-dimensional Patterson function of ribonuclease II.* Acta Cryst. **9** (1956) 156–162.
[86] N. Norman. *The Fourier transform method of normalizing intensities.* Acta Cryst. **10** (1957) 370–373.

9

The number of atoms in the unit cell and their distribution among the equipoints

If the phases of the F's as well as their amplitudes could be determined, it would be possible to proceed directly to a Fourier synthesis of the crystal structure without making any use of chemical information about the crystal. In most instances, however, the phases of the F's are not known, and it is desirable to proceed as far as possible in locating some or all of the atoms by making use of the chemical formula of the substance, the size of its cell, and the distribution of symmetry in the cell.

By making use of such information (as detailed in this chapter) it is possible to determine the exact number of atoms in each unit cell, and to limit the number of ways these atoms can be distributed with regard to the symmetry elements of the cell. It is usually not possible to determine uniquely how the atoms are distributed with regard to the symmetry elements, for there are often several possible distributions. But fortunately, the number of such distributions is always limited, even though it may be large in some instances. In simple structures a knowledge of this distribution may partly or wholly determine the structure; otherwise it may lead to a suggestion as to the general nature of the correct structure.

The unit-cell content

General principles. If the unit-cell dimensions and the empirical chemical formula of a crystal are known, the chemical content of the cell can be determined. This possibility arises because the unit cell and formula can be combined to give a theoretical density, and the actual density can be measured experimentally.

Let M be the mass of the collection of atoms constituting a unit of the chemical formula, and let N be the number of such formula units per unit cell. The volume, V, of the cell is known, of course, since the cell dimensions have been measured by x-ray diffraction means. Then the density, G, of the cell, in terms of these data, is evidently

$$G = \frac{\text{cell mass}}{\text{cell volume}} = \frac{NM}{V}. \tag{1}$$

Now, the density of the crystal can also be measured[†] by standard experimental procedures, for example, by the picnometer method. Recently, the Berman torsion density balance[4] has come into considerable use for the measurement of the densities of a few small fragments of a crystal, and is recommended for determining densities required for crystal-structure investigations. When the density of the crystal has been learned by one of these experimental methods, the only unknown remaining in (1) is N, the number of formula units per cell. Consequently, (1) can be solved for N. The usefulness of this information will be discussed in a later place in this chapter.

Density is customarily expressed in grams per cubic centimeter. Therefore, to use (1), the mass of the chemical formula unit must be expressed in grams. The standard units used by chemists to compare masses or weights of various chemical-formula units are the atomic weights, which are based on H $\approx$ 1, O = 16. To convert this unit to grams, it should be multiplied by the factor [§,5] 1.660×10^{-24}. Furthermore, the volume of the cell, V, is derived from the cell edges, a, b, and c, ordinarily expressed in Ångström units, which are 10^{-8} cm. In cgs units, therefore, the volume is ordinarily derived in 10^{-24} cm.^{3} Taking into account these conversion factors, relation (1) should be expressed as follows:

$$G\left(\frac{\text{grams}}{\text{cc.}}\right) = \frac{N \times M\,(\text{atomic-weight units}) \times 1.660 \times 10^{-24}}{V(\text{Å}^3) \times 10^{-24}}. \tag{2}$$

† Measured densities are often too low due to voids in the crystal.

§ If the atomic weight of oxygen is defined as 16, then one mole of oxygen gas, O_2, weighs $2 \times 16 = 32$ grams. The number of molecules in a mole of gas is Avogadro's number, 6.023×10^{23}. The mass of each oxygen atom is thus

$$m = \tfrac{1}{2}\,\frac{32}{6.023 \times 10^{23}} = 2.6565 \times 10^{-23} \text{ grams.}$$

The factor which converts atomic weight to mass in grams is therefore

$$\frac{2.6565 \times 10^{-23}}{16} = 1.660 \times 10^{-24}\,\frac{\text{grams}}{\text{atomic-weight unit}}.$$

Determination of the chemical formula. Under ordinary circumstances, the composition of the crystal corresponds with an ideal Daltonian formula. In these instances, (1) is used chiefly as a solution for N, namely

$$N = \frac{GV}{M}. \tag{3}$$

Unfortunately, in many crystals, deviations from an ideal formula occur because of solid solution.

Solid solution may occur in crystals in any of three ways:[2]

a. If the ideal formula of the crystal is represented by AB, then a certain amount of an element C may be present in the formula in addition to AB. The formula of the crystal is then ABC_p, where p is some irrational fraction, usually small. This is known as *addition solid solution.* In special cases, C may be additional amounts of A or B.

b. If the ideal formula of the crystal is represented by AB, then some fraction of A or B may be omitted. In this case the actual formula would be, say, AB_{1-p}, where p is a fraction, usually small. This is known as *omission solid solution.*

c. If the ideal formula of the crystal is represented by AB, then the place of some B may be taken by C. The formula of the crystal is then $A\left|\begin{matrix}B_{1-p}\\C_p\end{matrix}\right|$. This is known as *substitution solid solution.* Perhaps this type of solid solution is the most common. It is almost universal in crystals of minerals, especially in those minerals which were formed at high temperatures where a certain amount of disorder[3] in the B position is tolerated.

When solid solution occurs, the chemical formula, obtained by attempting to reduce the chemical analysis of the crystal, may appear to be irrational, and in any case not ideal. The analysis of the crystal may not suggest any obvious formula at all. When this condition obtains, then relation (1) has another use: It can be used to find the chemical mass of the unit cell, namely NM:

$$NM = GV. \tag{4}$$

In order to express NM in chemical formula-weight units, which are most convenient for the purpose, the following scheme of units should be observed:

$$NM \text{ (in atomic-weight units)} = \frac{G(\text{g./cc.}) \times V(\text{Å}^3) \times 10^{-24}}{1.660 \times 10^{-24}}. \tag{5}$$

To make use of (4) only a knowledge of the density and cell dimensions is required. When NM is found, it can be multiplied by the weight percentages of each atom as given by the chemical analysis, thus yielding the mass of each atom in the cell. This can be easily converted into the number of each kind of atom per unit cell. Now, this should be an integer, for the space-group operations generate an integral number of motifs from an original motif. Therefore the motif formula, which is the true chemical formula in simplest form, must be a submultiple of the cell formula, which was found by the aid of (4). Furthermore, the multiplicity in question must be one which is a possible multiplicity produced by the operations of the space group.

If one or several atoms should occur in the cell to an extent which does not conform to the integral concept just discussed, it must be concluded that such atoms are involved in deviations from the ideal formula by reason of solid solution. In such cases, a judicious juggling of the amount of atoms which do not occur in proper integral amounts (in accordance with the possible schemes of solid solution discussed above) will usually reveal the nature of the ideal formula of the crystal.

As a simple example of the use of such crystallographic methods in deciding the chemical formula, consider the case of the mineral löllingite.[1] This is an iron arsenide in which the iron and arsenic are not usually present in simple Daltonian amounts. The chemical analysis of a particular löllingite from Franklin Furnace, New Jersey, is shown in the first column of Table 1. The second column of the table gives the atomic

Table 1
Reduction of analysis of Franklin löllingite

	1 Weight %	2 Atomic weight	3 Wt. % / Atomic wt.	4 Atomic %	5 Ratio of atoms
Fe	29.40	55.84	0.530	36.1	1
As	69.80	74.93	0.933	63.3	1.78 (As + S)
S	0.21	32.06	0.007	0.5	
$\sum$	99.41		1.470	99.9	

weights of elements present; the third gives the relative proportions of the atoms in the crystal. In the fourth column this is reduced to percentages, and in the last column the ratio of iron to arsenic is derived. This shows that the empirical formula of this particular löllingite is essentially $FeAs_{1.78}$ which is an irrational formula. Such non-Daltonian formulae are the rule for löllingite, and the ratio of iron to arsenic varies inlöllingite crystals from different mineral localities.

The meaning of this irrational formula can be determined by finding the mass of atoms in the unit cell. This can be done by applying relation

(5). To do this the cell dimensions are required. Löllingite is an orthorhombic mineral with the following cell:

$$a = 5.26\text{Å},$$
$$b = 5.93,$$
$$c = 2.86,$$
$$V = 89.2\ \text{Å}^3.$$

The density, found experimentally, is 7.53. If relation (5) is used, the mass associated with this cell is found to be

$$NM = \frac{7.53 \times 89.2 \times 10^{-24}}{1.660 \times 10^{-24}} = 404 \text{ atomic-weight units.}$$

This total chemical mass may now be apportioned to iron, arsenic, and sulfur by multiplying it successively by the weight percentages of these elements as they are known from the chemical analysis to be present in the crystal. This computation is carried out in columns 1 and 2 of Table 2. By dividing the total mass of each type of atom in the cell

Table 2
Computation of cell content of Franklin löllingite

	1 Weight %	2 Mass per cell (atomic-weight units)	3 Mass of one atom (atomic-weight units)	4 Number of atoms per cell
Fe	29.40	119	55.84	2.13
As	69.80	282	74.93	3.77
S	0.21	0.9	32.06	0.03
$\sum$	99.4	402		5.93

(column 2) by the atomic mass of one such atom (column 3), the absolute number of atoms of each kind in the cell is obtained (column 4). This shows that the average unit cell contains atoms which could be represented by the empirical formula $Fe_{2.13}As_{3.77}S_{0.03}$. In this formula, note that the cell has Fe atoms to the extent of 0.13 in excess of the integral number 2. This is very close (within the limits of error of the chemical analysis, which does not add to quite 100%) to the number of atoms by which As + S fails to attain the integral number 4. Such a correlation very strongly suggests that, on the average, 0.13 atoms of Fe per cell are present in the As position of the structure, so that the formula of this particular crystal could be written

$$Fe_{2.00}\left|\begin{array}{r} As_{3.77} \\ Fe_{0.13} \\ S_{0.03} \\ \hline 3.93 \end{array}\right|.$$

Thus, there is a good preliminary argument in favor of the view that the excess Fe atoms substitute for some of the As, and that the ideal cell formula of löllingite should be Fe_2As_4. But there is also a confirmatory argument, and that is that this formula is an exact multiple of $FeAs_2$, the multiplicity factor being 2. This is a permissible multiplicity of the space group $P\,2_1/n\;2_1/n\;2/m$, to which löllingite conforms. As a consequence it can be said that ideal löllingite has 2 formula weights of $FeAs_2$ per cell, and that usually Fe proxies for As, atom for atom, at least up to a certain small fraction of the As atoms.

An example of the power of this method in an extreme case, even with rather poor density data, is afforded by the crystal tourmaline, which is a common mineral silicate. In tourmaline there is always such extensive solid solution that the very formula was in doubt until x-ray methods were applied to it. But some aspects of the formula are plain after a treatment of the cell data and chemical analysis by the plan used for treating the löllingite data. Tourmaline belongs to space group $R3m$. A white tourmaline from De Kalb, New York, for which both analysis and density are available, has the following dimensions,[†] referred to a (triple) hexagonal cell:

$$A = 15.95 \text{ Å},$$

$$C = 7.24,$$

$$V = \begin{cases} 1595 \quad \text{Å}^3 \text{ for the triple hexagonal cell,} \\ 532 \quad \text{Å}^3 \text{ for the primitive rhombohedral cell.} \end{cases}$$

The analysis and density of this material are given by Pennfield and Foote.[§] Unfortunately, the density is said to vary between 3.06 and 3.12. The lower value of this is taken for present purposes (and this even appears to be slightly too high, judging by the final results). Then, applying (5), one finds the mass associated with the primitive rhombohedral cell to be

$$NM = \frac{3.06 \times 532 \times 10^{-24}}{1.660 \times 10^{-24}} = 980 \text{ atomic-weight units.}$$

[†] Gabrielle E. Hamburger and M. J. Buerger. *The structure of tourmaline.* Am. Mineralogist **33** (1948) 532–540.

[§] S. L. Pennfield and H. W. Foote. *Ueber die chemische Zusammensetzung des Tourmaline.* Z. Krist. **31** (1899) 332.

Table 3 gives the analysis of this tourmaline in the first column. When the 980 chemical units are multiplied by these weight percentages, the weights of each of the oxides present in the cell are obtained, column 2. Dividing these results by the chemical weights of the oxides, column 3, gives the number of "molecules" of each of the oxides in the cell, column

Table 3
Computation of cell content of the De Kalb tourmaline

	1	2	3	4	5	6
	Weight %	Mass per cell (atomic-weight units)	"Molecular weight" of oxide	Number of "molecules" of oxide	Number of metal atoms per cell	Number of oxygen atoms per cell
SiO_2	36.72	360	60.06	5.94	5.94	11.98
TiO_2	0.05	0.5	79.90	0.006	0.01	0.01
B_2O_3	10.81	106	69.64	1.53	3.05	4.58
Al_2O_3	29.68	291	101.94	2.85	5.70	8.55
FeO	0.22	2.2	71.84	0.03	0.03	0.03
MgO	14.92	146	40.32	3.62	3.62	3.62
CaO	3.49	34.2	56.08	0.61	0.61	0.61
Na_2O	1.26	12.35	61.97	0.20	0.40	0.20
K_2O	0.05	0.5	94.20	0.005	0.01	0.01
H_2O	2.98	29.2	18	1.62	3.24	1.62
F	0.93	9.1	19	0.48	0.48	
$\sum$	101.11					31.21
less O equivalent to F	0.39					
	100.72					

4. This number is decomposed into the metal, column 5, and its oxygen content, column 6. The total number of oxygen atoms in the cell is the sum of the figures in this column.

The space group $R3m$, referred to a primitive rhombohedral cell, has 1-fold, 3-fold, and 6-fold equipoints. It is therefore necessary to cause sets of numbers from the second last column of Table 3 to conform to the integers 1, 3, and 6. One way of doing this is

$Na_{0.40}$			$Al_{5.52}$	$Si_{5.94}$	$O_{31.21}$	$H_{3.24}$
$Ca_{0.61}$	$Mg_{3.05}$	$B_{3.05}$	$Mg_{0.57}$			
				$Al_{0.18}$	$F_{0.48}$	
$K_{0.01}$			$Fe_{0.03}$			
1.02			6.12	6.12	31.69	

It will be noticed that these sums are about 2% in excess of the numbers 1, 3, 6, and 31. This suggests that the idealized formula is probably

$$NaMg_3B_3Al_6Si_6O_{27}(OH)_4.$$

Indeed this formula is confirmed by the results of a complete crystal-structure analysis. The 2% excess of the summation probably represents a 2% error in the experimental values entering the density-volume product in the computation above. The elements in the columns of the above formula are isomorphous substitutions, or, to put it another way, there is disorder among the atoms in each column over the equipoint of the column.

Distribution of atoms among the equipoints

Special positions of a space group. If an atom lies in a general position with respect to the symmetry elements of a crystal, the symmetry operations require that the cell contain m atoms, related to one another by the m operations of the cell of the space group. If the atom should lie on a symmetry element, this number may be reduced. The reduction occurs in all cases where the symmetry operation of the element does not contain a translation component, as shown below. It occurs, therefore, when the atom lies on an inversion center, on a reflection plane, or on a rotation axis. These locations are designated *special positions* of the space group. Elements whose operations do involve translation components, namely glide planes and screw axes,† are not special positions, for reasons which will be evident presently.

The reason for the reduction can be appreciated by imagining an atom in the general position to migrate toward an inversion center, a reflection plane, or a rotation axis. When the atom reaches the symmetry element, it can migrate no further without becoming coincident with its $n-1$ partners related by the n operations of the particular symmetry element. Therefore, if the atom comes to be located on the symmetry element, the n atoms related by the symmetry element become 1 atom. The value of n is 2 for an inversion center and for a reflection plane, but is equal to n for an n-fold rotation axis.

If the multiplicity of the general position of the space group is m, and the atom lies on a rotation axis of period n, then the number of atoms in the cell is $(1/n)m$. Of course, by the theory of groups, n is a factor of m, so $(1/n)m$ is an integer. Suppose that the space group contains two intersecting special positions of multiplicity n_1 and n_2. The combined multiplicity of the pair of symmetry elements is their product, $n_1 n_2$.

† Except when the order n and the pitch subscript contain a common factor, which occurs in the screws 4_2, 6_2, 6_3, and 6_4.

If an atom lies on this intersection, it suffers a reduction compared to its number in the general position, by a factor $1/n_1 n_2$. Again, if the multiplicity of the general position of the space group is m, then the number of atoms for the combined special position is $m/n_1 n_2$.

There is no reduction of multiplicity if the atom comes to lie on a screw axis[†] or a glide plane. This follows from the above discussion, because the atom, migrating from the general position and arriving at the screw axis or glide plane, does not coincide with its symmetry equivalents. These are distant from it by amounts which are multiples of the translation component of the symmetry element.

When an atom occupies a special position, it must conform to the symmetry of that equipoint. This restricts the possible locations of atoms of known coordination. For example, Si is known only with tetrahedral coordination with respect to oxygen. It can therefore never be placed upon an inversion center, but may be placed on a 2-fold axis, on a plane or at a $\bar{4}$ equipoint. On the other hand, Al can have either tetrahedral or octahedral coordination. If it occupies an inversion center its coordination must be octahedral.

Rank of the equipoint. An *equipoint* is a collection of points related to one point by the symmetry operations of the space group. The representative point of this collection can be in the general position, or it can be in some special position, usually at an inversion center, on a reflection plane, or on a rotation axis. Thus, the position of an atom in the cell may be described in general terms, and without necessarily involving the variable parameters required to locate the representative atom, by describing the nature of the equipoint which the atom occupies. The number of points per unit cell in an equipoint is defined as its *rank*.

For example, consider the space group $P\,2/m$. This group contains 2-fold rotation axes and, at right angles to them, reflection planes. The points of intersection of these two kinds of symmetry elements are locations of inversion centers. The cell of the space group contains only four operations, namely 1, 2, m, and $\bar{1}$. If the atom lies in the general position, the cell must contain a total of four symmetrically equivalent atoms, one for each of the four space-group operations of the cell. The atom is then said to occupy an equipoint of rank 4. If it lies on either a 2-fold rotation axis or a reflection plane, it occupies an equipoint of rank 2 in either case, while if it occupies the intersection of the 2-fold axis and the reflection plane, it occupies an equipoint of rank 1.

Sets of similar equipoints. The symmetry elements of a space group occur in several similar, but, in general, symmetrically unequivalent, sets. This situation arises as a consequence of products of the

[†] Except when the order n and the pitch subscript contain a common factor, which occurs in the screws 4_2, 6_2, 6_3, and 6_4.

operations of the symmetry elements with the three translations of the space group. Unless some of the operations become equivalent because they become related by other operations of the group, a primitive cell may be expected to have the following number of sets of symmetry elements which constitute special positions:

	8 sets of inversion centers,
	2 sets of parallel reflection planes,
	4 sets of parallel 2-fold rotation axes,
	3 " " " 3-fold " ",
	2 " " " 4-fold " ",
but only	1 set of parallel 6-fold rotation axes.

This means that equipoints also occur in such sets.

In the more complicated space groups, it is usual to have several of the sets of similar symmetry elements related to one another by some other operations of the space group. For example, four of the eight sets of inversion centers may become related to the other four by means of 2-fold screws; there are then only four distinct sets of equipoints on inversion centers. Or, again, the two sets of parallel reflection planes may be made equivalent to one another by a set of 2-fold screws parallel to the planes, and half-way between the two sets. In this case, there is only one special position on reflection planes.

Equipoints are customarily labeled by means of a number followed by a letter. The number designates the rank of the equipoint, and the letter is a somewhat arbitrary way of distinguishing the several equipoints of the space group. The order of the lettering for equipoints proceeds in the order of increasing degrees of freedom. For example, in space group $P\,2/m$, cited above, both rotation axes and reflection planes have rank 2. The first letters of the alphabet are therefore used to designate the equipoints of rank 2 on 2-fold axes, since an atom on the axis has only one degree of freedom, while subsequent letters of the alphabet are used to designate the equipoints of rank 2 on the reflection planes, since an atom on the plane has two degrees of freedom. These, and other equipoint designations of space group $P\,2/m$ are given in Table 4.

Distribution of atoms on the equipoints. The first part of this chapter provided the method of determining how many atoms occur in a unit cell. Of course, this number must fit the numbers which can be provided by the equipoints. Ordinarily there are several ways in which the number of atoms may be distributed among the equipoint numbers.

To make this clear by way of an example, the distribution of atoms

Table 4
Characteristics of equipoints in space group $P\,2/m$ (second setting)

Position	Rank	Designation	Degrees of freedom	Variable parameters
Inversion center at 000	1	$1a$	none	none
Inversion center at $0\frac{1}{2}0$	1	$1b$	"	"
Inversion center at $00\frac{1}{2}$	1	$1c$	"	"
Inversion center at $\frac{1}{2}00$	1	$1d$	"	"
Inversion center at $\frac{1}{2}\frac{1}{2}0$	1	$1e$	"	"
Inversion center at $0\frac{1}{2}\frac{1}{2}$	1	$1f$	"	"
Inversion center at $\frac{1}{2}0\frac{1}{2}$	1	$1g$	"	"
Inversion center at $\frac{1}{2}\frac{1}{2}\frac{1}{2}$	1	$1h$	"	"
2-fold rotation axes at $0y0$; $0\bar{y}0$	2	$2i$	1	y
2-fold rotation axes at $\frac{1}{2}y0$; $\frac{1}{2}\bar{y}0$	2	$2j$	1	y
2-fold rotation axes at $0y\frac{1}{2}$; $0\bar{y}\frac{1}{2}$	2	$2k$	1	y
2-fold rotation axes at $\frac{1}{2}y\frac{1}{2}$; $\frac{1}{2}\bar{y}\frac{1}{2}$	2	$2l$	1	y
Reflection planes at $x0z$; $\bar{x}0\bar{z}$	2	$2m$	2	x, z
Reflection planes at $x\frac{1}{2}z$; $\bar{x}\frac{1}{2}\bar{z}$	2	$2n$	2	x, z
General position at xyz; $x\bar{y}z$; $\bar{x}y\bar{z}$; $\bar{x}\bar{y}\bar{z}$	4	$4o$	3	x, y, z

Table 5
Equipoints of the space group $P\,2_1/n\,2_1/n\,2/m$

Position	Designation	Coordinates
Inversion centers	$2a$	000; $\frac{1}{2}\frac{1}{2}\frac{1}{2}$
Inversion centers	$2b$	$00\frac{1}{2}$; $\frac{1}{2}\frac{1}{2}0$
Inversion centers	$2c$	$0\frac{1}{2}0$; $\frac{1}{2}0\frac{1}{2}$
Inversion centers	$2d$	$\frac{1}{2}00$; $0\frac{1}{2}\frac{1}{2}$
2-fold rotation axes	$4e$	$00z$; $00\bar{z}$; $\frac{1}{2}, \frac{1}{2}, \frac{1}{2}+z$; $\frac{1}{2}, \frac{1}{2}, \frac{1}{2}-z$
2-fold rotation axes	$4f$	$0\frac{1}{2}z$; $0\frac{1}{2}\bar{z}$; $\frac{1}{2}, 0, \frac{1}{2}+z$; $\frac{1}{2}, 0, \frac{1}{2}-z$
Reflection planes	$4g$	$xy0$; $\bar{x}\bar{y}0$; $\frac{1}{2}-x, \frac{1}{2}+y, \frac{1}{2}$; $\frac{1}{2}+x, \frac{1}{2}-y, \frac{1}{2}$
General position	$8h$	xyz; $\bar{x}\bar{y}z$; $\frac{1}{2}-x, \frac{1}{2}+y, \frac{1}{2}+z$; $\frac{1}{2}+x, \frac{1}{2}-y, \frac{1}{2}+z$; $\bar{x}\bar{y}\bar{z}$; $xy\bar{z}$; $\frac{1}{2}+x, \frac{1}{2}-y, \frac{1}{2}-z$; $\frac{1}{2}-x, \frac{1}{2}+y, \frac{1}{2}-z$

among the equipoints for the case of marcasite† may be cited. Marcasite is orthorhombic FeS_2. The cell contains 2 FeS_2 so that 2 Fe and 4 S must be distributed among the equipoints. The space group, $P\,2_1/n\,2_1/n\,2/m$, has the equipoints shown in Table 5.

† M. J. Buerger. *The crystal structure of marcasite.* Am. Mineralogist **16** (1931) 361–395.

It should first be noted that the two Fe atoms can only be distributed among the 2-fold equipoints (i.e., equipoints of rank 2). Thus, the Fe can be in either

$$2a,\ 2b,\ 2c,\ \text{or}\ 2d.$$

Since none of these equipoints has any degree of freedom, whichever one is chosen as the position for Fe cannot be used again as a possible site for S.

The 4 S atoms must now be distributed over equipoints whose ranks sum to 4. This can be done either by placing the S atoms in one position of rank 4, or in two positions of rank 2. Therefore, the 4 S atoms may be distributed in the following possible specific ways:

$$\begin{array}{l} 4e, \\ 4f, \\ 4g, \\ \left.\begin{array}{l} 2a+2b, \\ 2a+2c, \\ 2a+2d, \\ 2b+2c, \\ 2b+2d, \\ 2c+2d, \end{array}\right\} \text{but excluding any combination if Fe is distributed in one of these equipoints.} \end{array}$$

When the several possible distributions of Fe among the equipoints are combined with the above distributions for S, a total number of distributions results as shown in Table 6.

Many of the structures suggested by Table 6 are really duplicates, which become identical with one another on change of origin. That this might be so is suggested by noting that, in Table 5, some of the equipoints obviously can be transformed into others by the addition of, say, $00\frac{1}{2}$ to the coordinates of the set. Now, it is generally true that two similar sets of equipoints can be transformed into one another by the addition of their components of separation. Another way of describing this is to say that similar equipoints can be transformed into one another by a transfer of origin. It follows that certain *combinations* of equipoints are equivalent by transfer of origin.

Since this situation is of general occurrence, it will be followed through in some detail for the example of marcasite, under discussion. Figure 1 shows the locations of the special positions in $P\ 2_1/n\ 2_1/n\ 2/m$. Observe that the four most specialized equipoints, namely a, b, c, and d, can be transformed into one another by adding half a translation in one of the three axial directions. Analytically, this can be demonstrated by adding $00\frac{1}{2}$, $0\frac{1}{2}0$, or $\frac{1}{2}00$ to the coordinates of the equipoints $2a$, $2b$, $2c$, and $2d$ in Table 5. Furthermore, noting that a variable plus a constant is

Table 6
Number of ways of distributing 2 Fe and 4 S among the equipoints of space group $P\,2_1/n\,2_1/n\,2/m$

Combination number	Fe positions	S positions	Characteristics of structures
1	$2a$	$2b+2c$	
2	$2a$	$2b+2d$	
3	$2a$	$2c+2d$	
4	$2b$	$2a+2c$	
5	$2b$	$2a+2d$	
6	$2b$	$2c+2d$	12 symmetry-fixed
7	$2c$	$2a+2b$	structures; sulfur atoms
8	$2c$	$2a+2d$	on inversion centers
9	$2c$	$2b+2d$	
10	$2d$	$2a+2b$	
11	$2d$	$2a+2c$	
12	$2d$	$2b+2c$	
13	$2a$	$4e$	
14	$2a$	$4f$	8 one-parameter struc-
15	$2b$	$4e$	tures; sulfur atoms on
16	$2b$	$4f$	2-fold rotation axes
17	$2c$	$4e$	
18	$2c$	$4f$	z parameter of sulfur
19	$2d$	$4e$	atoms to be determined
20	$2d$	$4f$	
			4 two-parameter struc-
21	$2a$	$4g$	tures; sulfur atoms on
22	$2b$	$4g$	reflection planes
23	$2c$	$4g$	
24	$2d$	$4g$	x and y parameters of
			sulfur atoms to be
			determined

simply another variable, and specifically that

$$x \pm \tfrac{1}{2} = x',$$

$$y \pm \tfrac{1}{2} = y',$$

$$z \pm \tfrac{1}{2} = z',$$

one finds that the additions just mentioned also carry the equipoints on the 2-fold axes into one another and carry the equipoints on reflection planes into one another. This is made even more obvious by imagining shifts of amounts $00\tfrac{1}{2}$, $0\tfrac{1}{2}0$, or $\tfrac{1}{2}00$ to occur in the location of the origin in Fig. 1. To examine systematically what happens with such additions, the transformations of the various individual equipoints of space group

Table 7
Transformations between equipoints with change of origin

Original Equipoints	Equipoint designation when origin is transferred to		
	$00\frac{1}{2}$	$0\frac{1}{2}0$	$\frac{1}{2}00$
a	*b*	*c*	*d*
b	*a*	*d*	*c*
c	*d*	*a*	*b*
d	*c*	*b*	*a*
e	*e*	*f*	*f*
f	*f*	*e*	*e*
g	*g*	*g*	*g*
h	*h*	*h*	*h*

Table 8
Transformation of equipoint combinations with change of origin

Original combination	Equipoint combination when origin is transferred to		
	$00\frac{1}{2}$	$0\frac{1}{2}0$	$\frac{1}{2}00$
a bc	*b ad*	*c da* = *c ad*	*d cb* = *d bc*
a bd	*b ac*	*c db* = *c bd*	*d ca* = *d ac*
a cd	*b dc*	*c ab*	*d ba* = *d ab*
b ac			
b ad			
b cd			
c ab			
c ad			
c bd			
d ab			
d ac			
d bc			
a e	*b e*	*c f*	*d f*
a f	*b f*	*c e*	*d e*
b e			
b f			
c e			
c f			
d e			
d f			
a g	*b g*	*c g*	*d g*
b g			
c g			
d g			

$P\,2_1/n\,2_1/n\,2/m$ into one another are tabulated in Table 7. What the same transformations do to the combinations of Table 6 can then be systematically found by transforming each of the equipoint letters of Table 6 according to the scheme of Table 7. This is carried out in Table 8. Actually, the transformation need not be applied blindly to all equipoints if the following rule is adopted: Each set of letters representing an equipoint combination should be transformed by the several additions

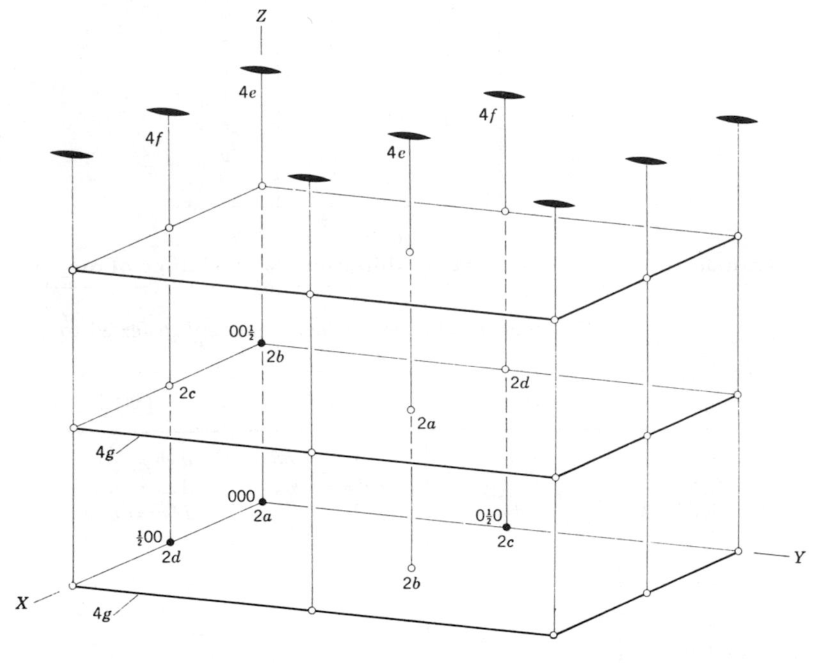

Fig. 1.

unless its combination has already appeared in an earlier transformation. In making the identification, it should be noted that the letters to the right of the space represent the equipoints of the sulfur atoms; when there are two of these, they can be interchanged without changing the structure. When these manipulations have been carried out, it becomes evident that the six combinations *a bc*, *a bd*, *a cd*, *a e*, *a f*, and *a g* represent the only distinctly different ones.

This lengthy analytical procedure of eliminating duplications of equipoint combinations can often be avoided and the distinct combinations can be made intuitively, especially when the geometry is sufficiently simple to sketch, as shown in Fig. 2. In this illustration, only one representative location of each equipoint is shown. Again writing the Fe equipoint first, and the S equipoints after the space, it is obvious that *a bc*, *a bd*, and *a dc* are distinct, and that no new types of combinations would result by shifting any of these combinations in the direction of either of the three axes, since this would merely interchange labels. Furthermore, it is obvious that $a\ e$ is distinct from $a\ f$. In the first case, the sulfur is on the same axis as the inversion center containing the iron; in the second it is on the other axis. Finally, the more general position, g, can be combined with any center, a, b, c, or d, with the same resulting structure.

With the possible distributions of the atoms over the equipoints known, it is often possible (as discussed in detail in Chapter 12) to eliminate certain of them on the basis of fairly simple intensity considerations. The

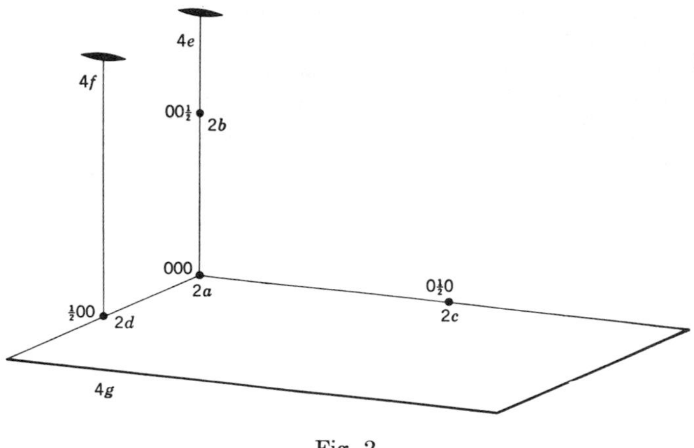

Fig. 2.

correct crystal structure then lies among the uneliminated equipoint distributions.

It may not always be feasible to enumerate the total number of ways the atoms may be distributed among the equipoints when this number is very high. This situation often occurs if the number of atoms to be distributed for some one atomic species is as great or greater than the rank of the general equipoint of the space group. For example, consider the possibility of distributing four atoms in space group $P\,2/m$, the equipoints of which are listed in Table 4. Equipoints occur in the space group for ranks 4, 2, and 1. The four atoms to be distributed can be arranged, therefore, among the following rank sums:

$$4,$$

$$2+2,$$

$$2+1+1,$$

$$1+1+1+1.$$

When an attempt is made to substitute all the combinations of specific equipoints sets in these combinations of rank, the number runs very high. To further combine these possibilities for one atom with the possibilities for another atom becomes very tedious indeed.

Literature

[1] M. J. Buerger. *The crystal structure of löllingite, $FeAs_2$.* Z. Krist. (A) **82** (1932) 165–187.

[2] M. J. Buerger. *The pyrite-marcasite relation.* Am. Mineralogist **19** (1934) 37–61, especially 53–58.

[3] M. J. Buerger. *The temperature-structure-composition behavior of certain crystals.* Proc. Nat. Acad. Sci. U. S. **20** (1934) 444–453.

[4] Harry Berman. *A torsion microbalance for the determination of specific gravities of minerals.* Am. Mineralogist **24** (1939) 434–440.

[5] William G. Schlecht. *Calculation of density for x-ray data.* Am. Mineralogist **29** (1944) 108–110.

10

The structure factor

In Chapter 2 it was shown that the wave scattered by a crystal can be characterized by a quantity F_{hkl}. This is a complex quantity whose magnitude is the amplitude of the scattered wave and whose direction in the complex plane is determined by the phase of the scattered wave. From the point of view of Chapter 3, F_{hkl} is a factor which enters into the expression for the intensity of the scattered wave. Since this factor depends on the arrangement of matter in the specific crystal under discussion, that is, on its crystal structure, the factor is commonly called the *structure factor*. In this chapter the structure factor is discussed in some detail with special attention to how it may be computed for any proposed crystal structure.

Atomic scattering factor

In Chapter 3 it was shown that the scattering unit is the electron. If the amplitude of the wavelet scattered by an electron is taken as the unit amplitude, then the amplitude scattered by a cell of the crystal can be written as the sum of the unit amplitudes of the wavelets, properly phased, which are scattered by the individual electrons in the cell. While this is a possible form in which the structure factor could be written, it is not a very practical one. An obvious objection to using the structure factor in this form is that it requires a summation over a very large number of electrons, and furthermore the positions of these electrons are not closely known.

A more convenient way of arranging the structure factor is to recognize that each chemical atom has a specific number of electrons associated with it in space, the specific form of the distribution of the electrons about each atom being known from atomic-structure theory. Thus it should be possible to group together terms in the structure factor so that each group of terms is associated with one atom. This treatment has advantages, for it turns out that it is unnecessary to know the location of each electron of an atom; one need only know the general distribution of the electrons with respect to the center of the atom. Each group of terms in this condensed structure factor represents the scattering by an atom in the cell.

The scattering power of an atom, designated f (from "form" factor), is expressed in terms of the scattering power of a single, free electron.

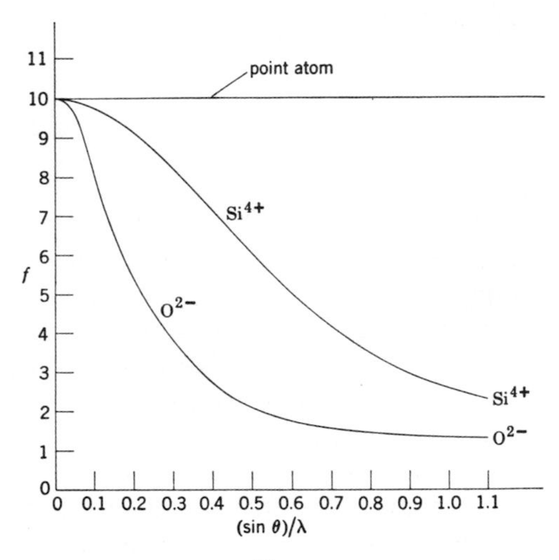

Fig. 1

The maximum scattering by an atom is therefore equal to Z, its atomic number. This occurs at $(\sin \theta)/\lambda = 0$. If the Z electrons of an atom were concentrated at a single point, there would be no destructive interference between the wavelets scattered by them, and there would be no variation of f with $(\sin \theta)/\lambda$, as indicated by the upper line in Fig. 1. But since the electrons of actual atoms are distributed through the volume of the atom, destructive interference sets in, as outlined elsewhere.† The larger the volume of an atom, the greater is the fall-off of f with $(\sin \theta)/\lambda$. If isoelectronic ions are compared, this effect can be readily noticed. A common example is Si^{4+} and O^{2-}, both of which have 10 electrons. In Si^{4+} the electrons are concentrated in a small volume, whereas in O^{2-} they are spread out over a large volume. The corresponding scattering curves have the shapes given by the middle and lower lines respectively of Fig. 1.

The scattering factors for the various chemical atoms are difficult to calculate, but are available both in the original literature[1–24] and in the

† M. J. Buerger. *X-ray crystallography.* (John Wiley and Sons, 1942) 51–53.

International tables for x-ray crystallography,[25] vol. 3. In most work it is assumed that the electron distribution in an atom has spherical symmetry. The environment of an atom, its bonding to its neighbors, and its thermal motion often cause its electron distribution to deviate from spherical symmetry. This causes the reflection amplitudes computed by using spherically symmetrical f values to depart from those observed, especially at small glancing angles.

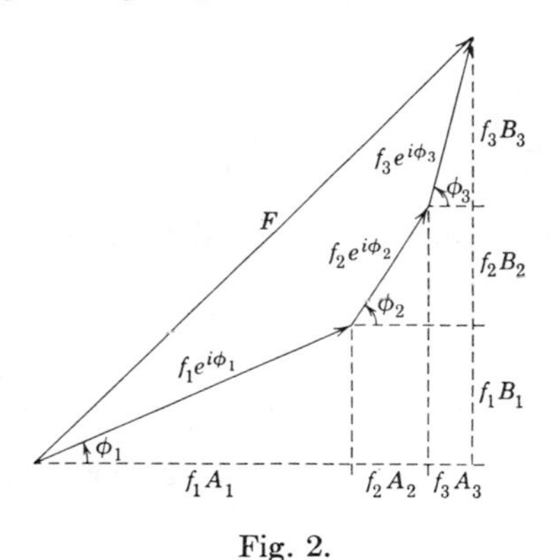

Fig. 2.

Some forms of the structure factor

Exponential form of the structure factor. The composition of the complex quantity F_{hkl} is shown in Fig. 2. Fundamentally it is a summation similar to (6) and (7) of Chapter 2, specifically

$$F_{hkl} = f_1 \, e^{i\phi_1} + f_2 \, e^{i\phi_2} + f_3 \, ^{i\phi_3} + \cdots f_J \, e^{i\phi_J} \tag{1}$$

$$= \sum_j f_j \, e^{i\phi_j}, \tag{2}$$

the summation being taken over the J atoms of the unit cell. According to (43) of Chapter 2, the phase angle, ϕ_j, can be expressed as a function of the fractional coordinates x_j, y_j, and z_j of the atoms of the cell:

$$\phi_j = 2\pi(hx_j + ky_j + lz_j), \tag{3}$$

so that the structure factor can also be written

$$F_{hkl} = f_1 \, e^{i2\pi(hx_1+ky_1+lz_1)} + f_2 \, e^{i2\pi(hx_2+ky_2+lz_2)} + \cdots f_J \, e^{i2\pi(hx_J+ky_J+lz_J)}$$

$$= \sum_j f_j \, e^{i2\pi(kx_j+hy_j+lz_j)}. \tag{4}$$

Component form of the structure factor. For purposes of computing structure factors it is usually convenient to express the structure factors in terms of its real and imaginary components. Reference to Fig 2 shows that

$$F_{hkl} = |F_{hkl}| \cos \phi + i|F_{hkl}| \sin \phi \tag{5}$$

$$\begin{aligned} &= (f_1 \cos \phi_1 + f_2 \cos \phi_2 + f_3 \cos \phi_3 + \cdots) \\ &+ i(f_1 \sin \phi_1 + f_2 \sin \phi_2 + f_3 \sin \phi_3 + \cdots) \end{aligned} \tag{6}$$

$$= \Big(\sum_j f_j \cos \phi_j\Big) + i\Big(\sum_j f_j \sin \phi_j\Big). \tag{7}$$

Again, (6) and (7) can be rewritten as explicit functions of the coordinates of the atoms in the cell by substituting in them the value of ϕ given by (3):

$$\begin{aligned} F_{hkl} = &\{f_1 \cos 2\pi(hx_1+ky_1+lz_1) + f_2 \cos 2\pi(hx_2+ky_2+lz_2) + \cdots\} \\ &+ i\,\{f_1 \sin 2\pi(hx_1+ky_1+lz_1) + f_2 \cos 2\pi(hx_2+ky_2+lz_2) + \cdots\} \end{aligned} \tag{8}$$

$$\begin{aligned} = &\Big\{\sum_j f_j \cos 2\pi(hx_j+ky_j+lz_j)\Big\} \\ &+ i\Big\{\sum_j f_j \sin 2\pi(hx_j+ky_j+lz_j)\Big\}. \end{aligned} \tag{9}$$

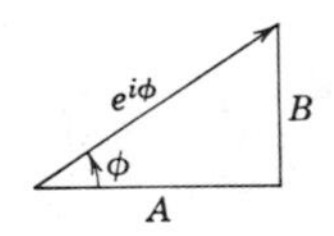

Fig. 3.

As an abbreviation (Fig. 3) it is sometimes convenient to substitute

$$A = \cos \phi, \tag{10}$$

$$B = \sin \phi \tag{11}$$

in (6) and (7), which then become

$$\begin{aligned} F_{hkl} = &(f_1 A_1 + f_2 A_2 + f_3 A_3 + \cdots) \\ &+ i(f_1 B_1 + f_2 B_2 + f_3 B_3 + \cdots) \end{aligned} \tag{12}$$

$$= \Big(\sum_j f_j A_j\Big) + i\Big(\sum_j f_j B_j\Big). \tag{13}$$

This is also illustrated in Fig. 2.

From (10), (11), and (3), it follows that

$$\begin{aligned} A &= \cos 2\pi(hx+ky+lz), \\ B &= \sin 2\pi(hx+ky+lz). \end{aligned} \tag{14}$$

The structure amplitude

The magnitude of the structure factor may be called the *structure amplitude*, $|F_{hkl}|$. Figure 2 shows that this magnitude is

$$|F_{hkl}| = \sqrt{(f_1 A_1 + f_2 A_2 + f_3 A_3 + \cdots)^2 + (f_1 B_1 + f_2 B_2 + f_3 B_3 + \cdots)^2} \tag{15}$$

$$= \sqrt{\Big(\sum_j f_j A_j\Big)^2 + \Big(\sum_j f_j B_j\Big)^2}. \tag{16}$$

By substituting from (14), this can be expressed as a function of the coordinates of the atoms in the cell:

$$|F_{hkl}| = \sqrt{\Big(\sum_j f_j \cos 2\pi(hx_j+ky_j+lz_j)\Big)^2 + \Big(\sum_j f_j \sin 2\pi(hx_j+ky_j+lz_j)\Big)^2}. \tag{17}$$

Simplifications due to symmetry

Simplification due to center of symmetry. An important simplification of (16) and (17) occurs if the crystal class contains a symmetry center, *and if* the origin of coordinates, to which the positions of the atoms are referred, is taken as a symmetry center of the space group. Under these conditions, an atom having coordinates xyz is always accompanied by an equivalent atom† at $\bar{x}\bar{y}\bar{z}$. The phases of the waves contributed by these two atoms are:

$$\phi_{xyz} = 2\pi(hx+ky+lz) \tag{18}$$

$$\phi_{\bar{x}\bar{y}\bar{z}} = 2\pi(-hx-ky-lz) \tag{19}$$

$$= -2\pi(hx+ky+lz)$$

$$= -\phi_{xyz}. \tag{20}$$

The waves contributed by the atoms of the cell therefore always occur in pairs having phases of equal magnitudes but opposite signs. Figure 4 shows that under these circumstances the imaginary B components of the amplitudes of the centrosymmetrically related atoms cancel one

† Except that, if an atom occupies the symmetry center, it is its own centrosymmetrical mate. The discussion following remains valid if the atom on the center is regarded as two half atoms.

another, and (17) reduces to

$$F = \sum_{j=1}^{J} f_j \cos 2\pi(hx_j + ky_j + lz_j) \tag{21}$$

$$= \sum_{j=1}^{J} f_j A_j. \tag{22}$$

Since these j's are alike in pairs, this can also be simplified to

$$F = 2\sum_{j=1}^{J/2} f_j A_j. \tag{23}$$

Thus the use of a symmetry center as origin eliminates the B components.

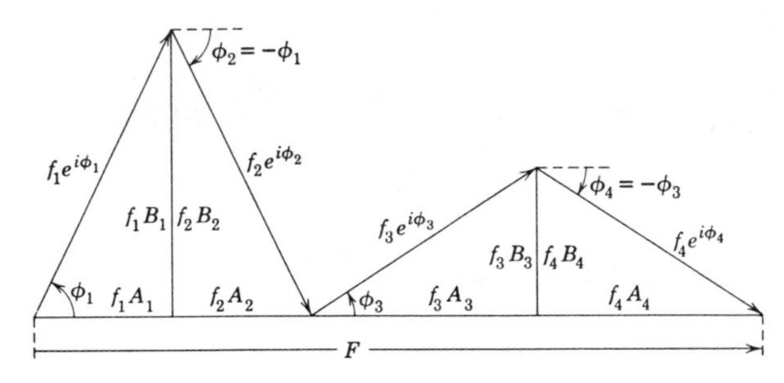

Fig. 4.

A compact analytical proof of this is as follows: Since all the atoms of the cell occur in pairs with coordinates $x_j\, y_j\, z_j$ and $\bar{x}_j\, \bar{y}_j\, \bar{z}_j$, (4) can be rewritten for this case by splitting the summation into two summations, each containing one half of the centrosymmetrical atom pairs:

$$F_{hkl} = \sum_{j=1}^{J/2} f_j\, e^{i2\pi(hx_j+ky_j+lz_j)} + \sum_{j=1}^{J/2} f_j\, e^{-i2\pi(hx_j+ky_j+lz_j)}. \tag{24}$$

This and other expressions can be rendered more compact by making use of the relations

$$e^{i\phi} + e^{-i\phi} = 2\cos\phi, \tag{25}$$

$$e^{i\phi} - e^{-i\phi} = i2\sin\phi. \tag{26}$$

If (25) is substituted into (24) there results

$$F_{hkl} = 2\sum_{j=1}^{J/2} f_j \cos 2\pi(hx_j + ky_j + lz_j). \tag{27}$$

The symmetry factor. In all except the simplest cases, unit cells contain rather large numbers of atoms, and computations involving the summations indicated in (17) would become quite tedious. It is fortunate, therefore, that symmetry conditions permit simplification of such summations, for, in all except the unsymmetrical space group, $P1$, an atom in the general position is accompanied by other atoms having symmetrically related coordinates. Thus there are groups of terms in the summations of (17) which not only have common values of f_j (because the several symmetrically related atoms are identical), but also have related coordinates. The latter feature permits consolidating the trigonometric parts of (17) by the rules of trigonometric combination. Alternatively, terms of (4) may be combined by rules of exponential combination.

The *symmetry factor*, S, can be defined as the wave scattered by a set of unit scattering particles, such as stationary electrons occupying an equipoint, ordinarily the general equipoint of the space group. Suppose the symmetry of the space group relates a set of m points in the general position. Then the symmetry factor is the wave scattered by a set of unit scatterers located on these m points. This has a form like (4) except that $f_j = 1$ and $j = 1, 2 \cdots m$. Specifically,

$$S = e^{i\phi_1} + e^{i\phi_2} + e^{i\phi_3} \cdots e^{i\phi_m} \tag{28}$$

$$= \sum_{j=1}^{m} e^{i\phi_j} \tag{29}$$

$$= \sum_{j=1}^{m} e^{i2\pi(hx_j+ky_j+lz_j)}. \tag{30}$$

The scattering by a single set of equivalent atoms, say set I, in the general position is then simply $f_{\text{I}} S_{\text{I}}$. All the atoms in the cell consist of the several equivalent sets I, II, III $\cdots p$. Therefore (4) can be rewritten in terms of p sets of equivalent atoms and their symmetry factors, namely:

(4): $$F_{hkl} = \sum_j f_j \, e^{i2\pi(hx_j+ky_j+lz_j)}$$

$$= f_{\text{I}} S_{\text{I}} + f_{\text{II}} S_{\text{II}} + f_{\text{III}} S_{\text{III}} + \cdots f_p S_p \tag{31}$$

$$= \sum_{j=\text{I}}^{p} f_j S_j. \tag{32}$$

In the next section specific examples of symmetry factors are discussed.

Illustrations of derivation of symmetry factors

Space group *Pmm2* using trigonometric notation. The derivation of symmetry factors may be illustrated for the space group *Pmm2*,

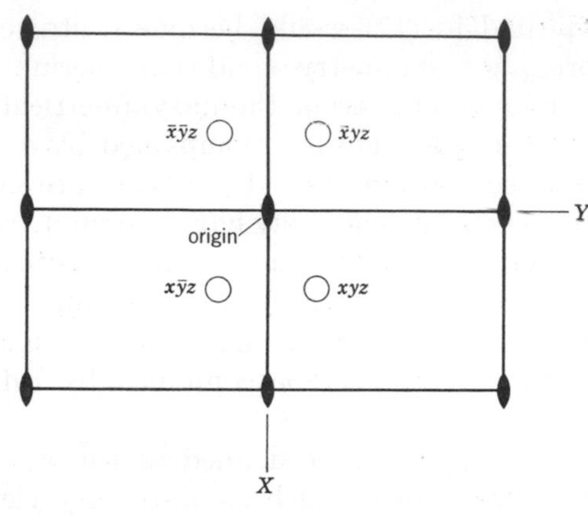

Fig. 5.

Fig. 5. An atom in the general position of this space group is accompanied by three other atoms. If the coordinates of the first atom are xyz, referred to a point on the 2-fold axis as origin, the coordinates of the four atoms related by the symmetry of this space group (Fig. 5) are

$$xyz \quad \bar{x}yz \quad \bar{x}\bar{y}z \quad x\bar{y}z.$$

The symmetry factor for this quadruplet of atoms is composed of a real component, A, and an imaginary component, B. The real component is as follows:

$$A = A_1 + A_2 + A_3 + A_4 \tag{33}$$

$$= \cos\phi_1 + \cos\phi_2 + \cos\phi_3 + \cos\phi_4 \tag{34}$$

$$= \cos 2\pi(hx+ky+lz) + \cos 2\pi(-hx+ky+lz)$$
$$+ \cos 2\pi(-hx-ky+lz) + \cos 2\pi(hx-ky+lz). \tag{35}$$

Applying the trigonometric identity,

$$\cos\alpha + \cos\beta = 2\cos\tfrac{1}{2}(\alpha+\beta)\cos\tfrac{1}{2}(\alpha-\beta), \tag{36}$$

one finds that the sum reduces to

$$A = 2\cos 2\pi(ky+lz)\cos 2\pi hx + 2\cos 2\pi(-ky+lz)\cos 2\pi(-hx)$$
$$= 2\cos 2\pi(ky+lz)\cos 2\pi hx + 2\cos 2\pi(ky-lz)\cos 2\pi hx$$
$$= 2\cos 2\pi hx\{\cos 2\pi(ky+lz) + \cos 2\pi(ky-lz)\}. \tag{37}$$

A further reduction can now be made by applying the trigonometric identity,

$$\cos(\alpha+\beta) + \cos(\alpha-\beta) = 2\cos\alpha\cos\beta. \tag{38}$$

The expression then reduces immediately to

$$A = 4 \cos 2\pi hx \cos 2\pi ky \cos 2\pi lz. \tag{39}$$

In a similar manner, the imaginary component of the symmetry factor can be reduced to

$$B = 4 \cos 2\pi hx \cos 2\pi ky \sin 2\pi lz. \tag{40}$$

Space group *Pmm*2 using exponential notation. It is often easier to derive symmetry factors by starting with exponential notation rather than trigonometric notation. This is because the exponential terms are more compact and can be combined by the simpler laws of exponential combination. The symmetry factor for *Pmm*2, which was derived in the last section using trigonometric notation, can also be derived using exponential notation as follows:

$$(30): \qquad S = \sum_{j=1}^{m} e^{i2\pi(hx_j+ky_j+lz_j)}$$

$$= e^{i2\pi(hx+ky+lz)} + e^{i2\pi(-hx-ky+lz)}$$

$$+ e^{i2\pi(hk-ky+lz)} + e^{i2\pi(-hx+hk+lz)} \tag{41}$$

$$= e^{i2\pi lz}\{e^{i2\pi(hx+ky)} + e^{i2\pi(-hx-ky)}$$

$$+ e^{i2\pi(hx-ky)} + e^{i2\pi(-hx+ky)}\}. \tag{42}$$

Relation (25) can now be applied to pairs of terms in the braces of (42), giving

$$S = e^{i2\pi lz}\{2 \cos 2\pi(hx+ky) + 2 \cos 2\pi(hx-ky)\}. \tag{43}$$

By applying (38), this becomes

$$S = e^{i2\pi lz}(4 \cos 2\pi hx \cos 2\pi ky). \tag{44}$$

The first factor can be resolved into real and imaginary components by using relation (5) of Chapter 2:

$$S = (\cos 2\pi lz + i \sin 2\pi lz)\, 4 \cos 2\pi kx \cos 2\pi ky. \tag{45}$$

To write down the real and imaginary parts of (45), one separates the terms not containing i and those containing i, into A and B parts, thus:

$$\left.\begin{aligned} A &= 4 \cos 2\pi hx \cos 2\pi ky \cos 2\pi lz, \\ B &= 4 \cos 2\pi hx \cos 2\pi ky \sin 2\pi lz. \end{aligned}\right. \tag{46}$$

Space group *Pmc*2₁. In order to illustrate the effect of a glide plane on the structure factor, the structure factor for $Pmc2_1$ is derived in this

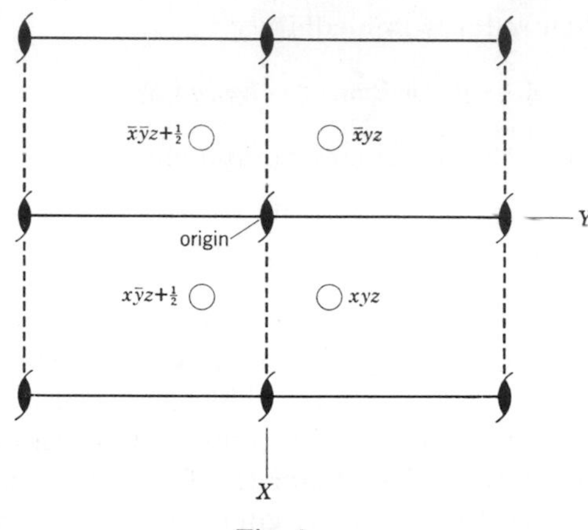

Fig. 6.

section. This can be compared step by step with the derivation of the structure factor for $Pmm2$ which was given in the last section.

Space group $Pmc2_1$ is shown in Fig. 6. The coordinates of the general position shown in Fig. 6 are substituted into (30):

$$(30): \qquad S = \sum_{j=1}^{m} e^{i2\pi(hx_j+ky_j+lz_j)}$$

$$= e^{i2\pi(hx+ky+lz)} + e^{i2\pi(-hx-ky+l[z+\frac{1}{2}])}$$
$$+ e^{i2\pi(-hx+ky+lz)} + e^{i2\pi(hx-ky+l[z+\frac{1}{2}])} \tag{47}$$

$$= e^{i2\pi(hx+ky+lz)} + e^{i\pi l}\, e^{i2\pi(-hx-ky+lz)}$$
$$+ e^{i2\pi(-hx+ky+lz)} + e^{i\pi l}\, e^{i2\pi(hx-ky+lz)} \tag{48}$$

$$= e^{i2\pi lz}\{e^{i2\pi(hx+ky)} + e^{-i2\pi(hx+ky)}\, e^{i\pi l}$$
$$+ e^{i2\pi(-hx+ky)} + e^{i2\pi(+hx-ky)}\, e^{i\pi l}\}. \tag{49}$$

Since $e^{i\phi}$ behaves as an operator which rotates a vector in the complex plane through the angle ϕ, $e^{i\pi}$ rotates through 180° and therefore turns a vector into its negative. Similarly, $e^{i2\pi}$ leaves a vector unmoved. Therefore

$$e^{i\pi l} = \begin{Bmatrix} +1 \text{ when } l \text{ is even} \\ -1 \text{ when } l \text{ is odd} \end{Bmatrix}. \tag{50}$$

Thus (50) has a different value according as l is even or odd. Depending on whether this causes the second and fourth terms in the braces of (49) to become positive or negative, the terms in braces consolidate according to the relations (25) and (26), thus:

When l is even,

$$S = e^{i2\pi lz}\{2 \cos 2\pi(hx+ky) + 2 \cos 2\pi(-hx+ky)\}. \tag{51}$$

By applying (38), this becomes

$$S = e^{i2\pi lz}\, 2\{2 \cos 2\pi hx \cos 2\pi ky\}. \tag{52}$$

The first factor can now be separated into real and imaginary parts according to (5) of Chapter 2:

$$S = (\cos 2\pi lz + i \sin 2\pi lz)\, 4 \cos 2\pi hx \cos 2\pi ky. \tag{53}$$

When the parts not containing i and containing i are separated, the real and imaginary parts may be separated as

$$\begin{aligned} A &= 4 \cos 2\pi hx \cos 2\pi ky \cos 2\pi lz, \\ B &= 4 \cos 2\pi hx \cos 2\pi ky \sin 2\pi lz. \end{aligned} \tag{54}$$

When l is odd,

$$S = e^{i2\pi lz}\{i2 \sin 2\pi(hx+ky) + i2 \sin 2\pi(-hx+ky)\}. \tag{55}$$

Pairs of terms can be consolidated by applying the trigonometric identity

$$\sin(\alpha+\beta) + \sin(\alpha-\beta) = 2 \sin\alpha \cos\beta. \tag{56}$$

If this is applied to (55), it reduces to

$$S = e^{i2\pi lz}\, 2i\{2 \cos 2\pi hx \sin 2\pi ky\}. \tag{57}$$

The first factor is now separated into real and imaginary parts according to (5) of Chapter 2:

$$S = (\cos 2\pi lz + i \sin 2\pi lz) 4i(\cos 2\pi hx \sin 2\pi ky) \tag{58}$$

$$= 4(i \cos 2\pi lz + i^2 \sin 2\pi lz)(\cos 2\pi hx \sin 2\pi ky). \tag{59}$$

Since $i^2 = -1$, this is equivalent to

$$S = 4(i \cos 2\pi lz - \sin 2\pi lz)(\cos 2\pi hx \sin 2\pi ky). \tag{60}$$

The real and imaginary parts of this are

$$\begin{aligned} A &= -4 \cos 2\pi hx \sin 2\pi ky \sin 2\pi lz, \\ B &= 4 \cos 2\pi hx \sin 2\pi ky \cos 2\pi lz. \end{aligned} \tag{61}$$

The various forms of the symmetry factor for $Pmc2_1$ to use under various conditions are thus:

When $l = 2n$ (i.e., l is even),

$$(54): \quad \begin{aligned} A &= 4 \cos 2\pi hx \cos 2\pi ky \cos 2\pi lz, \\ B &= 4 \cos 2\pi hx \cos 2\pi ky \sin 2\pi lz; \end{aligned}$$

When $l = 2n+1$ (*i.e.*, l *is odd*),

$$(61): \quad \begin{aligned} A &= -4 \cos 2\pi hx \sin 2\pi ky \sin 2\pi lz, \\ B &= 4 \cos 2\pi hx \sin 2\pi ky \cos 2\pi lz. \end{aligned}$$

The symmetry factor necessarily contains the extinction rules for the space group. For $Pmc2_1$, the b glide plane and [001] screw axis cause extinctions of $h0l$ when l is odd. The symmetry factor for $h0l$ is obtained by setting $k = 0$ in the above tabulation, which then reduces to:

$h0l$, *l even*,

$$A = 4 \cos 2\pi hx \cdot 1 \cdot \cos 2\pi lz,$$

$$B = 4 \cos 2\pi hx \cdot 1 \cdot \sin 2\pi lz;$$

$h0l$, *l odd*,

$$A = -4 \cos 2\pi hx \cdot 0 \cdot \sin 2\pi lz = 0,$$

$$B = 4 \cos 2\pi hx \cdot 0 \cdot \cos 2\pi lz = 0.$$

The general forms of the symmetry factors for all space groups are available in tables.[26–29]

Change of origin

The form of the expression for the symmetry factor depends on the coordinates of the equivalent points, and these, in turn, depend on the origin chosen. If a highly symmetrical position is chosen as origin, the coordinates of the equivalent points display a symmetrical form and the symmetry factor assumes a correspondingly simple form. It is accordingly customary to choose the origin of coordinates on a symmetry element. If the symmetry group contains inversion centers, the origin should be chosen on a center since, as pointed out in a previous section, this choice eliminates the B component of the symmetry factor.

The general effect on the symmetry factor of a change of origin is illustrated in Fig. 7. This discussion is given for three dimension but, for clearness, the illustration shows only two dimensions. The contribution of point P at coordinates xyz to the symmetry factor is

$$s = e^{i2\pi(hx+ky+lz)}. \tag{62}$$

If the origin is shifted by an amount $\Delta x\ \Delta y\ \Delta z$ with respect to the original origin, the new coordinates of point P are $x-\Delta x$, $y-\Delta y$, $z-\Delta z$. The contribution of the point P to the symmetry factor is now

$$s' = e^{i2\pi(h[x-\Delta x]+k[y-\Delta y]+l[z-\Delta z])} \tag{63}$$

$$= e^{-i2\pi(h\Delta x+k\Delta y+l\Delta z)}\, e^{i2\pi(hx+ky+lz)} \tag{64}$$

$$= e^{-i2\pi(h\Delta x+k\Delta y+l\Delta z)}\, s. \tag{65}$$

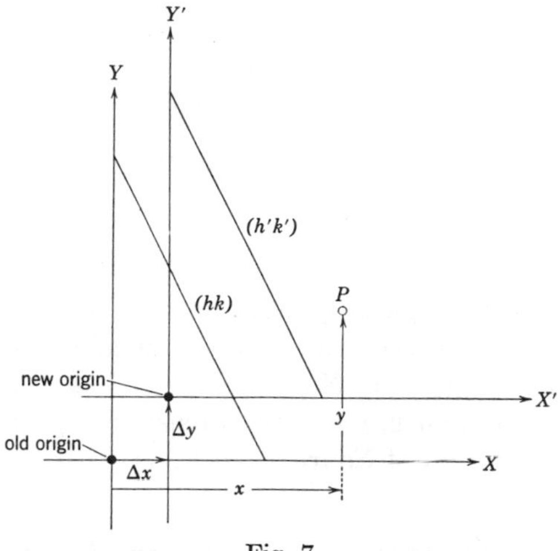

Fig. 7.

The symmetry factor is the sum of the contributions of the m points equivalent to P, i.e.,

$$S = \sum_{j=1}^{m} s_j \tag{66}$$

The symmetry factor for the new origin is

$$\begin{aligned} S' &= \sum_{j=1}^{m} s_j' \\ &= \sum_{j=1}^{m} e^{-i2\pi(h\Delta x+k\Delta y+l\Delta z)}\, s_j \\ &= e^{-i2\pi(h\Delta x+k\Delta y+l\Delta z)} \sum_{j=1}^{m} s_j \\ &= e^{-i2\pi(h\Delta x+k\Delta y+l\Delta z)}\, S. \end{aligned} \tag{67}$$

Thus the symmetry factor for an origin shifted to $\Delta x\ \Delta y\ \Delta z$ can be derived from the original symmetry factor by multiplying the latter by the factor $e^{-i2\pi(h\Delta x+k\Delta y+l\Delta z)}$.

Phase of the structure factor

Figure 2 shows that the phase angle of F_{hkl} can be computed as follows:

$$\tan \phi_{hkl} = \frac{\sum_j f_j B_j}{\sum_j f_j A_j}, \tag{68}$$

$$\cos \phi_{hkl} = \frac{\sum_j f_j A_j}{|F_{hkl}|}, \tag{69}$$

$$\sin \phi_{hkl} = \frac{\sum_j f_j B_j}{|F_{hkl}|}. \tag{70}$$

When $B = 0$, the phase angle ϕ_{hkl} degenerates to 0 or π. This amounts to assigning a sign to the structure factor; specifically when $\phi_{hkl} = 0$, F_{hkl} is $+$, and when $\phi_{hkl} = \pi$, F_{hkl} is $-$.

This phase information is required whenever a proposed structure is to be refined with the aid of Fourier synthesis.

Computation of structure factors

In the course of a crystal-structure analysis, structure factors must be computed many times. There are several ways of doing this: with the aid of a high-speed digital computer, by ordinary hand calculations, or with the aid of an analogue computer. The strategies of computing differ somewhat with the method used to perform the computations.

High-speed digital computation. Little has been written about performing structure-factor computations with the aid of high-speed digital computers.[30–37] To make use of such a computer it is usually necessary to prepare a "program," and feed this to the machine in the form, say, of a punched tape. This program provides the machine with instructions as to how to operate on the data, which is fed to the machine, say, in the form of another punched tape. The computation is accomplished at a very rapid rate. For example, the computer, "Whirlwind," computes a complete set of structure factors like F_{hk0} (with h and k running from zero to 12), for a crystal without symmetry other than centrosymmetry, in about $\frac{3}{4}$ minute, for 9 atoms in the asymmetric unit. The cost of such a set of computations is of the order of four dollars.

With such speed available, it is probably not worthwhile in most cases to attempt to shorten the computation by bulking together symmetrically related atoms with the aid of the symmetry factor. If one wished to use the symmetry factor it would require writing a new program for each one of the 230 space groups. In the absence of a library of programs for the 230 space groups for a particular computer, a practical procedure is to have four sets of general programs available to take care of centrosymmetrical and non-centrosymmetrical cases in two and three dimensions. These are specifically as follows:

Centrosymmetrical:

Two dimensional,

$$F_{hk0} = 2 \sum_{j=1}^{J/2} f_{j,hk0} \cos 2\pi(hx_j+ky_j); \tag{71}$$

Three dimensional,

$$F_{hkl} = 2 \sum_{j=1}^{J/2} f_{j,hkl} \cos 2\pi(hx_j+ky_j+lz_j); \tag{72}$$

Non-centrosymmetrical:

Two dimensional,

$$A_{hk0} = \sum_j f_{j,hk0} \cos 2\pi(hx_j+ky_j); \tag{73}$$

$$B_{hk0} = \sum f_{j,hk0} \sin 2\pi(hx_j+ky_j); \tag{74}$$

$$|F_{hk0}| = \sqrt{A^2_{hk0} + B^2_{hk0}}; \tag{75}$$

$$\tan \phi_{hk0} = \frac{B_{hk0}}{A_{hk0}}; \tag{76}$$

Three dimensional,

$$A_{hkl} = \sum_j f_{j,hkl} \cos 2\pi(hx_j+ky_j+lz_j); \tag{77}$$

$$B_{hkl} = \sum_j f_{j,hkl} \sin 2\pi(hx_j+ky_j+lz_j); \tag{78}$$

$$|F_{hkl}| = \sqrt{A_{hkl}^2 + B_{hkl}^2}; \tag{79}$$

$$\tan \phi_{hkl} = \frac{B_{hkl}}{A_{hkl}}. \tag{80}$$

Parts of these programs have features in common and so can be handled by similar programs with the same "subroutines." The answer given for centrosymmetrical crystals is an ordered list of the numerical values of F_{hkl}, each characterized by a sign + or −. The answer required for non-centrosymmetrical crystals is an ordered list of the four quantities A_{hkl}, B_{hkl}, $|F_{hkl}|$, and ϕ_{hkl}.

The computation of structure factors by a high-speed computor has certain disadvantages. Ordinarily the only part of the computation which shows is the answers. In the event that the computed structure factors do not check those observed, it is not evident how the computed structure factors can be improved, since the steps in the computation are not shown.† Such computations are, however, well suited to the normal refinement process of structure-factor computations followed by Fourier synthesis.

Hand computation. Hand computation is so laborious that if this method of structure-factor computation must be used, every advantage should be taken of means of condensing the labor. For this reason full advantage should be taken of the symmetry factor for each space group, and some form of filing the computations should be devised. An example of the use of such a form is given later.

Reduction of computation by using the symmetry factor. It was shown in (32) that the wave scattered in spectrum *hkl* can be expressed as a function of the symmetry factor:

(32):
$$F_{hkl} = \sum_{j=\mathrm{I}}^{p} f_j S_j.$$

This can be expressed in terms of real and imaginary components. For this purpose each symmetry factor is regarded as decomposed into its real and imaginary parts:

$$S_j = A_j + iB_j, \tag{81}$$

so that (32) is decomposed into

$$F_{hkl} = \sum f_j A_j + i \sum f_j B_j. \tag{82}$$

The absolute value of a vector in the complex plane is equal to the square root of the sum of the squares of its real and imaginary components, so

$$|F_{hkl}| = \sqrt{\left(\sum_j f_j A_j\right)^2 + \left(\sum_j f_j B_j\right)^2}. \tag{83}$$

The advantage of using the symmetry factor in computing the intensities of spectra is seen by comparing (83) with (15). *Each* of the j terms in (83) represents several terms in (15). Thus, in space group $Pmm2$, discussed in a foregoing section, there are four equivalent points in the general position of the space group. Therefore each $f_j A_j$ in (83) corresponds to four terms in (12), for example,

$$f_{\mathrm{I}} A_{\mathrm{I}} = f_1 A_1 + f_2 A_2 + f_3 A_3 + f_4 A_4. \tag{84}$$

Since all the atoms in an equivalent set are alike, they all have the same f's, so that

$$f_{\mathrm{I}} A_{\mathrm{I}} = f_{\mathrm{I}}(A_1 + A_2 + A_3 + A_4). \tag{85}$$

† The program can be arranged so that the intermediate steps *are* printed out, but this slows down the computations and increases their cost considerably.

Thus the several terms in (12) are replaced by

$$\begin{aligned} f_1 A_1 + f_2 A_2 + f_3 A_3 + f_4 A_4 &= f_{\mathrm{I}} A_{\mathrm{I}} \\ f_5 A_5 + f_6 A_6 + f_7 A_7 + f_8 A_8 &= f_{\mathrm{II}} A_{\mathrm{II}} \\ f_9 A_9 + f_{10} A_{10} + f_{11} A_{11} + f_{12} A_{12} &= f_{\mathrm{III}} A_{\mathrm{III}} \\ \cdot \quad \cdot \quad \cdot \quad \cdot \quad & \quad \cdot \\ \cdot \quad \cdot \quad \cdot \quad \cdot \quad & \quad \cdot \, , \end{aligned} \tag{86}$$

where I, II, etc., refer to atom species I, atom species II, etc. Relation (15) therefore consolidates to

$$|F| = \sqrt{(f_{\mathrm{I}} A_{\mathrm{I}} + f_{\mathrm{II}} A_{\mathrm{II}} + \cdots)^2 + (f_{\mathrm{I}} B_{\mathrm{I}} + f_{\mathrm{II}} B_{\mathrm{II}} + \cdots)^2}, \tag{87}$$

where A_{I} and B_{I} are the real and imaginary components respectively of the symmetry factor for atoms I, A_{II} and B_{II} are the real and imaginary components respectively of the symmetry factor of atom II, etc.

Thus the use of symmetry factors commonly reduces the amount of labor required in computing diffraction intensities. The symmetry factors for the 230 space groups are listed in several tables.[26–29] In some cases of high symmetry and few atoms per cell in specialized locations, it may be easier to compute the intensities without using symmetry-factor consolidation of terms. An example of this is found in Chapter 12.

Each of the symmetry-factor components, A and B, is made up of two parts, a numerical coefficient and a group of trigonometric terms. In the variation of the parameters x, y, and z, the symmetry factor can attain a maximum possible value equal to the numerical coefficient. This is equal to the total number of atoms, m, which are equivalent by space-group symmetry. If an atom of the crystal structure should occupy a special position rather than the general position of the space group, as assumed in the above derivation, then the symmetry factor has the same trigonometric form, but the numerical part is reduced to equal the correct multiplicity, m, of the special position.

Arrangement of computations. The trigonometric part of the symmetry factor for many space groups is a product of $\cos 2\pi hx$ and $\sin 2\pi hx$ terms. Tables of these functions for h ranging from 0 through 30, and x ranging from 0 to 1 in steps of 0.001 are available.[38] These tables are arranged so that a straight edge can be laid vertically along the page at any trial value of the atomic parameter x, and then both cosine and sine values for all values of h appear in the column next to the ruler. This arrangement facilitates copying onto the computation sheets as explained later.

The following suggestions are given for arranging computations based upon the tables. From the discussion of the symmetry factor, it is evident that only the trigonometric part of the factor is a function of the parameters. It is, therefore, desirable to split off this trigonometric part from the multiplicity part. Let A and B represent the real and imaginary components of the symmetry factor, respectively, and let A' and B' represent their purely trigonometric parts. Then if m is the multiplicity factor, these two are related by

$$\begin{aligned} A &= mA', \\ B &= mB'. \end{aligned} \tag{88}$$

Expressing the symmetry factor in this form, the square of the structure factor can be written

$$|F_{hkl}|^2 = \{f_{\text{I}}\, m_{\text{I}}\, A_{\text{I}}' + f_{\text{II}}\, m_{\text{II}}\, A_{\text{II}}' + \cdots\}^2 \\ + \{f_{\text{I}}\, m_{\text{I}}\, B_{\text{I}}' + f_{\text{II}}\, m_{\text{II}}\, B_{\text{II}}' + \cdots\}^2. \tag{89}$$

In order to segregate the A', B' parts of (89) from the rest, the following rearrangement is suggested as a computing form:

$$\begin{aligned} |F_{hkl}|^2 = \{&[f_{\text{I}}\, m_{\text{I}}]_{hkl}\, A_{\text{I}}' & + \{&[f_{\text{I}}\, m_{\text{I}}]_{hkl}\, B_{\text{I}}' \\ +\, &[f_{\text{II}}\, m_{\text{II}}]_{hkl}\, A_{\text{II}}' & +\, &[f_{\text{II}}\, m_{\text{II}}]_{hkl}\, B_{\text{II}}' \\ +\, &\cdots\cdots\}^2 & +\, &\cdots\cdots\}^2. \end{aligned} \tag{90}$$

The terms in brackets remain constant for any changes of parameter and consequently may be permanently set down upon the computation sheet. The terms A_{I}', B_{I}', A_{II}', B_{II}', etc., are functions of the trial parameters only, and these terms alone need be changed in the course of the computations.

To illustrate the application of this computing form, an example is drawn from a series of $0kl$ structure-factor computations for the centrosymmetrical crystal valentinite, Sb_2O_3. The space group of this crystal is $P\,2_1/c\,2_1/c\,2/n$. Choosing an origin at a symmetry center, the symmetry factor for an atom in the general position of this space group is:

$$A = 8 \cos 2\pi\left(hx+\frac{l-h}{4}\right) \cos 2\pi\left(ky+\frac{k-l}{4}\right) \cos 2\pi\left(lz+\frac{h-k}{4}\right),$$

$$B = 0.$$

For $0kl$ reflections, this reduces to

$$\left.\begin{aligned} A &= 8 \cos 2\pi ky \cos 2\pi lz, \\ B &= 0, \end{aligned}\right\} \quad \text{for } k \text{ even, } l \text{ even,}$$

and

$$\left.\begin{aligned} A &= -8 \sin 2\pi ky \sin 2\pi lz, \\ B &= 0, \end{aligned}\right\} \quad \text{for } k \text{ odd, } l \text{ even.}$$

From this it is evident that it is desirable to place on one set of computation sheets the computations for reflections having k even, and on another those for k odd. The accompanying illustration is for several reflections of the series having k even.

The unit cell of valentinite contains 8 Sb and 12 O atoms. It is first necessary to ascertain in what manner these atoms are distributed among the various equipoints of the space group. In the case of valentinite, it is possible (as outlined in Chapter 12) to unequivocally ascertain that the eight antimony atoms occupy the general position, $8e$, that eight of the oxygen atoms (designated O_2) occupy the general position, $8e$, and that the remaining four oxygens (designated O_1) occupy the special position, $4c$, on the rotation axis. For valentinite, therefore, the subscripts I, II, and III of (90) refer to Sb, O_2, and O_1, for which m is 8, 8, and 4 respectively.

Before starting the computations, it is necessary to derive from the unit-cell dimensions the values of $(\sin\theta)/\lambda$ for the several $0kl$ reflections. For each such value there corresponds a definite value of f_{Sb} and f_O, which can be read from curves plotted from appropriate data.[3,25] From these values, column 4 of the form, Table 1, can be filled in, and no further change is made in this column during the course of the computations, even if the parameters are changed and new values entered on the same computation sheets.

After the first trial parameters have been decided upon and inserted in columns 2 and 3, *Numerical structure factor tables*[38] provides the necessary values for filling columns 5 and 6. Note that when z_{Sb}, for example, has been decided upon, a single column of *Numerical structure factor tables* provides all values of $\cos 2\pi lz$ (as well as $\sin 2\pi lz$, to be used for the computation sheets involving reflections $0kl$ having k odd). This means that the values for columns 5 and 6 can be filled in for the entire computation form by simply finding in the tables the six columns corresponding with the y and z values of columns 2 and 3. The filling in of columns 5 and 6 of the form is thus reduced to copying six lists of numbers from the tables.

Values in columns 5 and 6 have definite signs, $+$ or $-$, which must be carefully observed. The products of columns 4, 5, and 6 are entered in column 7 or 8, according to sign. The two columns are summed separately, the lesser subtracted from the greater, and the result is the F_{0kl} which is sought. This may be squared and multiplied by Lp to give I, if desired.

Adjustment of parameters. Suppose, now, that after one completes such a set of computations for a series of reflections, it turns out that the calculated structure factors are in general harmony with the observed, but that there are minor disagreements. The general agreement constitutes a strong presumption that the general features of the trial structure are correct, and the minor disagreement indicates that the atoms in the trial structure are not quite in their correct positions. It remains to refine the coordinates of the atoms. Study of the computations for the intensities showing disagreement usually suggests the way the parameters may be varied to improve the agreement. New computations can

Table 1
Example of computation of F_{0kl} for valentinite

1	2	3	4	5	6	7	8
Reflection m, atom	Parameters		mf	Symmetry factor/m		F	
	y	z		$\cos 2\pi ky$	$\cos 2\pi lz$	+	−
042							
8 Sb	0.128	0.179	300.8	−0.997	−0.628	188	
8 O_2	0.058	0.861	37.6	0.113	−0.175		1
4 O_2	0.250	0.029	18.8	1	0.934	18	
						206	1
						− 1	
						205	
044							
8 Sb			235.2	−0.997	−0.212	48	
8 O_2			20.8	0.113	−0.113		2
4 O_1			10.4	1	0.746	8	
						56	2
						− 2	
						54	
046							
8 Sb			191.2	−0.997	0.894		170
8 O_2			14.4	0.113	0.504	1	
4 O_1			7.2	1	0.460	3	
						4	170
							− 4
							166

then be placed on the original sheets by striking out the old values and writing above them the new values of parameter and symmetry factor. When several changes of this sort are contemplated, it is an advantage to use a different colored pencil for each new parameter set with its accompanying computations.

Attention should be called to the fact that, between the values of about 0 and 0.5, the cosine and sine vary rapidly and in an approximately linear manner with x and y, whereas they vary most slowly in the region of unity. If, therefore, two reflections, $h_1\,k_1\,0$ and $h_2\,k_2\,0$, show computed values out of the observed order, they cannot be brought into order by slight shifts of any parameters for which the corresponding value in column 5 or 6 is near 1. Only changes of parameter affecting values in these columns which are in the region 0 to 0.5 are of much aid in improving the values of computed intensities.

Aids to hand computing. A number of devices have been described[38–46] for aiding the computation of structure factors by hand.

Analogue computation. Several mechanical devices for computing structure factors have been prepared.[47–65] Many of these suffer from a common defect. The computing device gives $\cos 2\pi(hx+ky+lz)$ or $\sin 2\pi(hx+ky+lz)$, say, if the coordinates x, y, and z for *one* atom are supplied to it. One later performs the summation over all atoms in the cell. Such methods fail to take advantage of the reduction of computation due to the consolidation of phase information in the symmetry factor.

Literature

Atomic scattering factors

[1] R. W. James and G. W. Brindley. *Some numerical calculations of atomic scattering factors.* Phil. Mag. (7) **12** (1931) 81–112.

[2] Linus Pauling and J. Sherman. *Screening constants for many-electron atoms. The calculation and interpretation of x-ray term values, and the calculation of atomic scattering factors.* Z. Krist. **81** (1932) 1–29.

[3] C. Hermann. *Internationale Tabellen zur Bestimmung von Kristallstrukturen,* Vol. 2. (Gebrüder Borntraeger, Berlin, 1935) 571–578.

[4] H. Viervoll and O. Ögrim. *An extended table of atomic scattering factors.* Acta Cryst. **2** (1949) 277–279.

[5] R. McWeeny. *X-ray scattering by aggregates of bonded atoms. I. Analytical approximations in single-atom scattering.* Acta Cryst. **4** (1951) 513–519.

[6] R. McWeeny. *X-ray scattering by aggregates of bonded atoms. II. The effect of the bonds: with an application to H_2.* Acta. Cryst. **5** (1952) 463–468.

[7] R. McWeeny. *X-ray scattering by aggregates of bonded atoms. III. The bond scattering factor: simple methods of approximation in the general case.* Acta Cryst. **6** (1953) 631–637.

[8] W. G. Henry. *The atomic scattering factors of Au^+* and *Hg^{++}.* Acta Cryst. **7** (1954) 138–139.

[9] R. McWeeny. *X-ray scattering by aggregates of bonded atoms. IV. Application to the carbon atom.* Acta Cryst. **7** (1954) 180–186.

[10] M. M. Qurashi. *On the completion and extension of the table of atomic scattering factors published by Viervoll and Ögrim.* Acta Cryst. **7** (1954) 310–312.

[11] Jean A. Hoerni and James A. Ibers. *Some calculations of atomic form factors.* Acta Cryst. **7** (1954) 744–746.

[12] J. Berghuis, IJbertha M. Haanappel, M. Potters, B. O. Loopstra, Caroline H. MacGillavry. and A. L. Veenendaal. *New calculations of atomic scattering factors.* Acta Cryst. **8** (1955) 478–483.

[13] James A. Ibers. *New atomic form factors for beryllium and boron.* Acta Cryst. **10** (1957) 86.

[14] V. Vand, P. F. Eiland, and R. Pepinsky. *Analytical representation of atomic scattering factors.* Acta Cryst. **10** (1957) 303–306.

[15] L. H. Thomas and K. Umeda. *Atomic scattering factors calculated from the TFD atomic model.* J. Chem. Phys. **26** (1957) 293–303.

[16] Y. Tomiie and C. H. Stam. *Calculation of atomic scattering factors using Slater wave functions: sodium and calcium.* Acta Cryst. **11** (1958) 126–127.

[17] H. C. Freeman and J. E. W. L. Smith. *A polynomial approximation to atomic scattering factor curves.* Acta Cryst. **11** (1958) 819–822.

[18] Edgar L. Eichhorn. *Atomic scattering factors for wolfram.* Acta Cryst. **11** (1958) 824.

[19] J. Gillis. *An application of electronic computing to x-ray crystallography.* Acta Cryst. **11** (1958) 833–834.

[20] A. L. Veenendaal, Caroline H. MacGillavry, B. Stam, M. L. Potters, and Marlene J. H. Römgens. *X-ray scattering factors of* Al^{2+}, $Al,^{3+}$ Mn^{2+}, *Fe,* Zr^{4+}, Au^{+}, and U^{6+}. Acta Cryst. **12** (1959) 242–246.

[21] A. J. Freeman. *Atomic scattering factors for spherical and aspherical charge distributions.* Acta Cryst. **12** (1959) 261–271.

[22] A. J. Freeman and J. H. Wood. *The atomic scattering factor for iron.* Acta Cryst. **12** (1959) 271–273.

[23] James A. Ibers. *Remarks on "improved" atomic form factors.* Acta Cryst. **12** (1959) 347.

[24] J. B. Forsyth and M. Wells. *On an analytic approximation to the atomic scattering factor.* Acta Cryst. **12** (1959) 412–415.

[25] G. D. Rieck and Caroline H. MacGillavry and *International tables for x-ray crystallography.* Vol. 3, *Physical and chemical tables.* (The Kynoch Press, Birmingham, England, 1960)

Symmetry factors

[26] C. Hermann. *Internationale Tabellen zur Bestimmung von Kristallstrukturen,* Vol. 1. (Gebrüder Borntraeger, Berlin, 1935) 94–377.

[27] Kathleen Lonsdale. *Simplified structure factor and electron density formulae for the 230 space groups of mathematical crystallography.* (G. Bell and Sons, London, 1936)

[28] K. Lonsdale. *Errata in Lonsdale's structure factor tables.* Acta Cryst. **3** (1950) 162–163.

[29] Norman F. M. Henry and Kathleen Lonsdale. *International tables for x-ray crystallography,* Vol. 1. (Kynoch Press, Birmingham, 1952) 353–525.

High-speed digital computation

[30] P. A. Shaffer, Jr., Verner Schomaker, and Linus Pauling. *The use of punched cards in molecular structure determinations. I. Crystal structure calculations.* J. Chem. Phys. **14** (1946) 648–658, especially 656–657.

[31] Jerry Donohue and Verner Schomaker. *The use of punched cards in molecular structure determination. III. Structure-factor calculations of x-ray crystallography.* Acta Cryst. **2** (1949) 344–347.
[32] M. D. Grems and J. S. Kasper. *An improved punched-card method for crystal structure-factor calculations.* Acta Cryst. **2** (1949) 347–351.
[33] F. R. Ahmed and D. W. J. Cruickshank. *Crystallography calculations on the Manchester University electronic digital computer (Mark II).* Acta Cryst. **6** (1953) 765–769.
[34] F. Fowweather. *The use of general programmes for crystallographic calculations on the Manchester University electronic digital computer (Mark II).* Acta Cryst. **8** (1955) 633–637.
[35] R. A. Sparks, R. J. Prosen, F. H. Kruse, and K. N. Trueblood. *Crystallographic calculations on the high-speed digital computer SWAC.* Acta Cryst. **9** (1956) 350–358.
[36] Theo Hahn. *Kristallographische Rechnungen mittels schneller elektronischer Zeffernrechenmaschinen.* Neues Jahrb. Mineral. **5** (1957) 112–116.
[37] H. C. Freeman. *Crystallographic calculations on the Silliac electronic digital computer. II. Structure factors.* Australian J. Chem. **11** (1958) 99–103.

Hand computation

[38] M. J. Buerger. *Numerical structure factor tables.* Geol. Soc. Am. Spec. Paper **33** (1941, reprinted 1953).
[39] C. A. Beevers and H. Lipson. *The use of Fourier strips for calculating structure factors.* Acta Cryst. **5** (1952) 673–675.
[40] Leroy Alexander. *A simple but versatile strip technique for calculating structure factors.* Acta Cryst. **6** (1953) 727–731.
[41] E. W. Radoslovich and Helen D. Megaw. *Calculations of geometrical structure factors for space groups of low symmetry. I.* Acta Cryst. **8** (1955) 95–98.
[42] E. Stanley. *A rapid numerical method of calculating structure factors.* Acta Cryst. **8** (1955) 122.
[43] H. Lipson and P. R. Pinnock. *A stencil method of computing structure factors.* Acta Cryst. **8** (1955) 172–174.
[44] E. W. Radoslovich. *Calculation of geometrical structure factors for space groups of low symmetry. II.* Acta. Cryst. **8** (1955) 456–460.
[45] S. Ramaseshan and R. V. G. Sundara Rao. *Some techniques for structure factor calculations.* J. Indian Inst. Sci. **39** (1957) 34–40.
[46] E. W. Radoslovich. *Calculation of geometrical structure factors for space groups of low symmetry. III.* Acta Cryst. **12** (1959) 11–13.

Analogue computation

[47] R. C. Evans and H. S. Peiser. *A machine for the computation of structure factors.* Proc. Phys. Soc. **54** (1942) 457–462.
[48] V. Vand. *A simple device for calculating x-ray structure factors.* J. Sci. Instr. **25** (1948) 352.
[49] A. D. Booth. *Fourier technique in x-ray organic structure analysis.* (Cambridge University Press, Cambridge, 1948)

[50] V. Vand. *A mechanical calculating machine for x-ray structure factors.* Nature **163** (1949) 169–171.
[51] H. Lipson and C. A. Taylor. *A photoelectric device for the evaluation of structure factors.* Acta Cryst. **2** (1949) 130.
[52] G. Hägg. *Two aids for the calculation of crystal structure factors.* Acta Cryst. **3** (1950) 315–316.
[53] V. Vand. *A mechanical x-ray structure-factor calculating machine.* J. Sci. Instr. **27** (1950) 257–261.
[54] G. M. J. Schmidt. *A circular slide rule for structure-factor calculations.* Acta Cryst. **4** (1951) 186.
[55] M. M. Woolfson. *A photoelectric structure-factor machine.* Acta Cryst. **4** (1951) 250–253.
[56] D. Sayre. *The calculation of structure factors by Fourier summation.* Acta Cryst. **4** (1951) 362–367.
[57] F. J. Llewellyn. *A mechanical-electrical unit for calculating structure amplitudes.* J. Sci. Inst. **28** (1951) 229–230.
[58] V. Vand. *A simple mechanical structure-factor computing aid.* Acta Cryst. **5** (1952) 390.
[59] W. L. Bragg. *A device for calculating structure factors.* Acta Cryst. **5** (1952) 474–475.
[60] V. Vand. *A Fourier electron-density balance.* J. Sci. Instr. **29** (1952) 118–121.
[61] D. C. Phillips. *A device for calculating complex x-ray structure factors.* J. Sci. Instr. **29** (1952) 299.
[62] E. Stanley. *A structure-factor calculator.* J. Sci. Instr. **29** (1952) 334–335.
[63] N. T. van der Walt. *Electromechanical crystal structure factor computer.* Rev. Sci. Instr. **27** (1956) 750–756.
[64] K. A. Morley and C. A. Taylor. *A variable structure-factor graph for use in crystal-structure determination.* J. Sci. Instr. **34** (1957) 54–58.
[65] W. Schaffer. *Mechanical structure-factor computer.* J. Sci. Instr. **36** (1959) 75–77.

11

Plane projections in structure-factor calculations

In a trial-and-error investigation of a crystal structure, one attempts to find locations of atoms such that the computed structure factors (or intensities) of the trial structure match those actually observed. This ultimately requires the three coordinates of each atom to be determined. In the initial stages of the investigation, however, it is a great convenience to reduce the number of variables in the problem by confining attention to two coordinates of the atoms at a time. This amounts to working with the projection of the structure on a plane. Only the reflections from the planes parallel to the direction of projection are required for this part of the investigation. Thus, suppose that the x and y coordinates of the atoms are first sought. A projection on which z has no component is required, namely a projection along c and onto a plane at right angles[†] to c. To study such a projection, only the special class of reflection, $hk0$, is required. The intensities of these can be obtained from a single zero-level Weissenberg or precession photograph.

The symmetry of the projection must be one of those possible for a diperiodic pattern in a plane. Consequently, it corresponds to one of the 17 plane groups. An intelligent appreciation of the problem, therefore, calls for an understanding of the relations between the symmetries of space groups and their projections as plane groups.

Relations between plane groups and projections of space groups

There are 230 space groups but only 17 plane groups. Each space group projects along any rational direction as a plane group. The most

[†] Note that, for oblique crystals, this plane is not generally (001), but is a section of the cell normal to the c axis.

Table 1
Projections of symmetry elements

Symmetry element	Projected symmetry element	
	Projected parallel to axis or plane	Projected perpendicular to axis or plane
translation, T		t = proj. of T
rotation axis, n	rotation axis, n	
screw axis, n_m	n	
2		m
2_1		g
$3, 3_1, 3_2$		1
$4, 4_2$		m
$4_1, 4_3$		g
$6, 6_2, 6_4$		m
$6_1, 6_3, 6_5$		g
improper rotation, $\bar{n}$		
$n = 4N$	n	m
$n = 4N \pm 1$	$2n$	2
$n = 4N \pm 2$	$\frac{1}{2}n$	m
m	m	1
c	m	t = proj. $c/2$
a	g (glide component = proj. a)	t = proj. $a/2$
b	g (glide component = proj. b)	t = proj. $b/2$
n	g (glide component = proj. n)	t = proj. of glide
d	g (glide component = proj. g)	t = proj. of glide

useful projections are those along the directions of axial symmetry elements. One important reason for this is that the projection of a space group onto a plane normal to an axial symmetry element is a plane group of related axial symmetry. The plane group corresponding to any space group can be found by applying the relation between the space symmetries and their plane projections, which are shown in Table 1.

Attention is particularly directed to the last entry in the table. The patterns produced by certain symmetry elements having a translation component parallel to the plane of projection degenerate to patterns with submultiple periods when projected. Specifically, if the space cell contains a submultiple "centering" translation, t, Fig. 1, this translation projects as a primitive translation of the plane pattern because its component, $t_{\perp}$, normal to the plane of projection vanishes. Similarly, if the space group contains a glide plane parallel with the plane of projec-

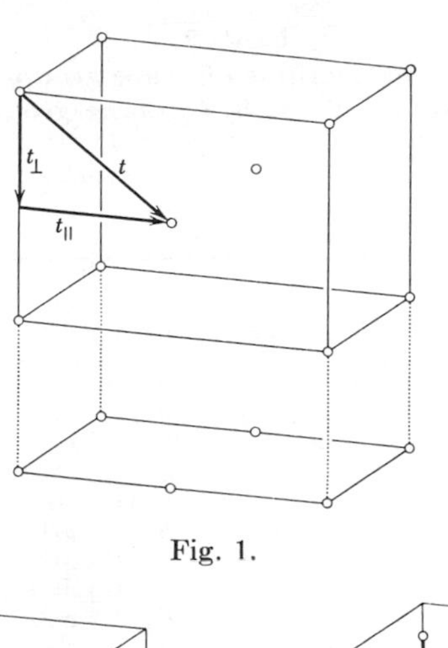

Fig. 1.

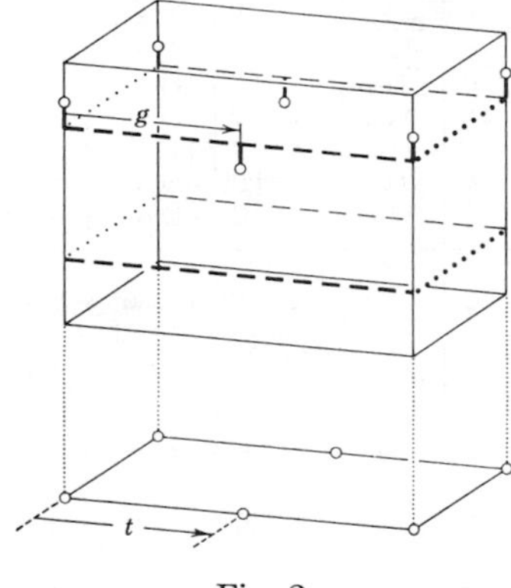

Fig. 2.

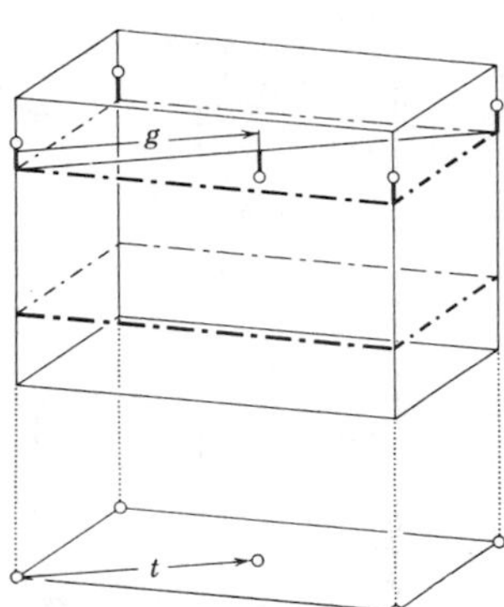

Fig. 3.

tion, Fig. 2, the projected pattern has a translation equal to the glide component of this glide. For example, suppose that the glide is axial as shown in the illustration, then the projected cell has an edge half that of the space cell in the direction of the glide. If the glide is diagonal, Fig. 3, the projected cell is centered due to this glide.

The plane groups resulting from the projections of the triclinic, monoclinic and orthorhombic space groups on planes perpendicular to several important rational directions are shown in Table 2. This table also shows the axial transformations which occur with the projection.

Symmetry factors of the plane groups

Since every space group projects along a rational direction as a plane group, it follows that the symmetry factor for the reflections from the planes in a zone reduces to the symmetry factor for the plane group of

Table 2
Projections of space groups
(Space-group cell: A, B, C; plane-group cell: a, b)

Space group†	Projection ‖ [001]	a	b	Projection ‖ [100]	a	b	Projection ‖ [010]	a	b	Projection ‖ [110]	a	b
$P1$	$p1$	A	B	$p1$	B	C	$p1$	A	C	$p1$	$\frac{1}{2}(A+B)$	C
$P\bar{1}$	$p2$	A	B	$p2$	B	C	$p2$	A	C	$p2$	$\frac{1}{2}(A+B)$	C
$P2$	$p2$	A	B	$p1m1$	B	C	$p1m1$	A	C	$p1m1$	$\frac{1}{2}(A+B)$	C
$P2_1$	$p2$	A	B	$p1g1$	B	C	$p1g1$	A	C	$p1g1$	$\frac{1}{2}(A+B)$	C
$I2$	$p2$	$\frac{1}{2}(A+B)$	$\frac{1}{2}(-A+B)$	$c1m1$	B	C	$c1m1$	A	C	$p1m1$	$\frac{1}{2}(A+B)$	$\frac{1}{2}C$
$A2$	$p2$	A	$\frac{1}{2}B$	$c1m1$	B	C	$p1m1$	B	$\frac{1}{2}C$	$c1m1$	$\frac{1}{2}(A+B)$	C
$B2$	$p2$	$\frac{1}{2}A$	B	$p1m1$	B	$\frac{1}{2}C$	$c1m1$	A	C	$c1m1$	$\frac{1}{2}(A+B)$	C
Pm	$p1$	A	B	$p11m$	B	C	$p11m$	A	C	$p11m$	$\frac{1}{2}(A+B)$	C
Pa	$p1$	$\frac{1}{2}A$	B	$p11m$	B	C	$p11g$	A	C	$p11g$	$\frac{1}{2}(A+B)$	C
Pb	$p1$	A	$\frac{1}{2}B$	$p11g$	B	C	$p11m$	A	C	$p11g$	$\frac{1}{2}(A+B)$	C
Pn	$p1$	$\frac{1}{2}(A+B)$	$\frac{1}{2}(-A+B)$	$p11g$	B	C	$p11g$	A	C	$p11m$	$\frac{1}{2}(A+B)$	C
Im	$p1$	$\frac{1}{2}(A+B)$	$\frac{1}{2}(-A+B)$	$c11m$	B	C	$c11m$	A	C	$p11m$	$\frac{1}{2}(A+B)$	$\frac{1}{2}C$
Am	$p1$	$\frac{1}{2}A$	B	$c11m$	B	C	$p11m$	A	$\frac{1}{2}C$	$c11m$	$\frac{1}{2}(A+B)$	C
Bm	$p1$	A	$\frac{1}{2}B$	$p11m$	B	$\frac{1}{2}C$	$c11m$	A	C	$c11m$	$\frac{1}{2}(A+B)$	C
Ia	$p1$	$\frac{1}{2}(A+B)$	$\frac{1}{2}(-A+B)$	$c11m$	B	C	$c11m$	A	C	$p11m$	$\frac{1}{2}(A+B)$	$\frac{1}{2}C$
Aa	$p1$	$\frac{1}{2}A$	$\frac{1}{2}B$	$c11m$	B	C	$p11g$	A	$\frac{1}{2}C$	$c11m$	$\frac{1}{2}(A+B)$	C
Bb	$p1$	$\frac{1}{2}A$	$\frac{1}{2}B$	$p11g$	B	$\frac{1}{2}C$	$c11m$	A	C	$c11m$	$\frac{1}{2}(A+B)$	C
$P\frac{2}{m}$	$p2$	A	B	$p2mm$	B	C	$p2mm$	A	B	$p2mm$	$\frac{1}{2}(A+B)$	C
$P\frac{2}{a}$	$p2$	$\frac{1}{2}A$	B	$p2mm$	B	C	$p2mg$	A	B	$p2mg$	$\frac{1}{2}(A+B)$	C
$P\frac{2}{b}$	$p2$	A	$\frac{1}{2}B$	$p2mg$	B	C	$p2mm$	A	C	$p2mg$	$\frac{1}{2}(A+B)$	C
$P\frac{2}{n}$	$p2$	$\frac{1}{2}(A+B)$	$\frac{1}{2}(-A+B)$	$p2mg$	$\frac{1}{2}B$	C	$p2mg$	$\frac{1}{2}A$	C	$p2mm$	$\frac{1}{2}(A+B)$	C
$P\frac{2_1}{m}$	$p2$	A	B	$p2gm$	B	C	$p2gm$	A	B	$p2gm$	$\frac{1}{2}(A+B)$	C
$P\frac{2_1}{a}$	$p2$	$\frac{1}{2}A$	B	$p2gm$	B	C	$p2gg$	A	B	$p2gg$	$\frac{1}{2}(A+B)$	C
$P\frac{2_1}{b}$	$p2$	A	$\frac{1}{2}B$	$p2gg$	B	C	$p2gm$	A	C	$p2gg$	$\frac{1}{2}(A+B)$	C
$P\frac{2_1}{n}$	$p2$	$\frac{1}{2}(A+B)$	$\frac{1}{2}(-A+B)$									
$I\frac{2}{m}$	$p2$	$\frac{1}{2}(A+B)$	$\frac{1}{2}(-A+B)$	$c2mm$	B	C	$c2mm$	A	B	$p2mm$	$\frac{1}{2}(A+B)$	$\frac{1}{2}C$
$A\frac{2}{m}$	$p2$	A	$\frac{1}{2}B$	$c2mm$	B	C	$p2mm$	A	$\frac{1}{2}C$	$c2mm$	$\frac{1}{2}(A+B)$	C
$B\frac{2}{m}$	$p2$	$\frac{1}{2}A$	B	$p2mm$	B	$\frac{1}{2}C$	$c2mm$	A	B	$c2mm$	$\frac{1}{2}(A+B)$	C
$I\frac{2}{a}$	$p2$	$\frac{1}{2}(A+B)$	$\frac{1}{2}(-A+B)$	$c2mm$	B	C	$c2mm$	A	B	$p2mm$	$\frac{1}{2}(A+B)$	$\frac{1}{2}C$
$A\frac{2}{a}$	$p2$	$\frac{1}{2}A$	$\frac{1}{2}B$	$c2mm$	B	C	$p2mg$	A	C	$c2mm$	$\frac{1}{2}(A+B)$	C
$B\frac{2}{b}$	$p2$	$\frac{1}{2}A$	$\frac{1}{2}B$	$p2mg$	B	$\frac{1}{2}C$	$c2mm$	A	C	$c2mm$	$\frac{1}{2}(A+B)$	C

† The sequence of space groups and alternative settings is the same as given in M. J. Buerger, *Elementary crystallography* (John Wiley and Sons, New York, 1956) 442–446.

§ In M. J. Buerger, *Elementary crystallography* (John Wiley and Sons, New York, 1956) there are errors in the drawings of these space groups, which appear on pages 337, 338, and 344.

Table 2 (*Continued*)

Space group†	Projection ‖ [001]			Projection ‖ [100]			Projection ‖ [010]			Projection ‖ [110]		
		a	*b*		*a*	*b*		*a*	*b*		*a*	*b*
$P222$	*p2mm*	A	B	*p2mm*	B	C	*p2mm*	A	C			
$P222_1$	*p2mm*	A	B	*p2gm*	B	C	*p2gm*	A	C			
$P2_12_12$	*p2gg*	A	B	*p2mg*	B	C	*p2mg*	A	C			
$P2_12_12_1$	*p2gg*	A	B	*p2gg*	B	C	*p2gg*	A	C			
$C222$	*c2mm*	A	B	*p2mm*	$\frac{1}{2}B$	C	*p2mm*	$\frac{1}{2}A$	C			
$C222_1$	*c2mm*	A	B	*p2gm*	$\frac{1}{2}B$	C	*p2gm*	$\frac{1}{2}A$	C			
$I222$	*c2mm*	A	B	*c2mm*	B	C	*c2mm*	A	C			
$I2_12_12_1$	*c2mm*	A	B	*c2mm*	B	C	*c2mm*	A	C			
$F222$	*p2mm*	$\frac{1}{2}A$	$\frac{1}{2}B$	*p2mm*	$\frac{1}{2}B$	$\frac{1}{2}C$	*p2mm*	$\frac{1}{2}A$	$\frac{1}{2}C$			
$Pmm2$	*p2mm*	A	B	*p1m1*	B	C	*p1m1*	A	C			
$Pcc2$	*p2mm*	A	B	*p1m1*	B	$\frac{1}{2}C$	*p1m1*	A	$\frac{1}{2}C$			
$Pbm2$	*p2gm*	A	B	*p1m1*	$\frac{1}{2}B$	C	*p1m1*	A	C			
$Pnc2$	*p2gm*	A	B	*c1m1*	B	C	*p1m1*	A	$\frac{1}{2}C$			
$Pba2$	*p2gg*	A	B	*p1m1*	$\frac{1}{2}B$	C	*p1m1*	$\frac{1}{2}A$	C			
$Pnn2$	*p2gg*	A	B	*c1m1*	B	C	*c1m1*	A	C			
$Pmc2_1$	*p2mm*	A	B	*p1g1*	B	C	*p1m1*	A	$\frac{1}{2}C$			
$Pbc2_1$	*p2gm*	A	B	*p1g1*	$\frac{1}{2}B$	C	*p1m1*	A	$\frac{1}{2}C$			
$Pnm2_1$	*p2gm*	A	B	*c1m1*	B	C	*p1g1*	A	C			
$Pbn2_1$	*p2gg*	A	B	*p1g1*	$\frac{1}{2}B$	C	*c1m1*	A	C			
$Imm2$	*c2mm*	A	B	*c1m1*	B	C	*c1m1*	A	C			
$Iba2$	*c2mm*	A	B	*p1m1*	$\frac{1}{2}B$	$\frac{1}{2}C$	*p1m1*	$\frac{1}{2}A$	$\frac{1}{2}C$			
$Ibm2$	*c2mm*	A	B	*p1m1*	$\frac{1}{2}B$	$\frac{1}{2}C$	*c1m1*	A	C			
$Cmm2$	*c2mm*	A	B	*p1m1*	$\frac{1}{2}B$	C	*p1m1*	$\frac{1}{2}A$	C			
$Ccc2$	*c2mm*	A	B	*p1m1*	$\frac{1}{2}B$	$\frac{1}{2}C$	*p1m1*	$\frac{1}{2}A$	$\frac{1}{2}C$			
$Cmc2_1$	*c2mm*	A	B	*p1g1*	$\frac{1}{2}B$	C	*p1m1*	$\frac{1}{2}A$	$\frac{1}{2}C$			
$Amm2$	*p2mm*	A	$\frac{1}{2}B$	*c1m1*	B	C	*c1m1*	A	C			
$Abm2$	*p2mm*	A	$\frac{1}{2}B$	*p1m1*	$\frac{1}{2}B$	$\frac{1}{2}C$	*p1m1*	$\frac{1}{2}A$	$\frac{1}{2}C$			
$Ama2$	*p2mg*	A	$\frac{1}{2}B$	*c1m1*	B	$\frac{1}{2}C$	*p1m1*	$\frac{1}{2}A$	C			
$Aba2$	*p2mg*	$\frac{1}{2}A$	$\frac{1}{2}B$	*p1m1*	$\frac{1}{2}B$	$\frac{1}{2}C$	*p1m1*	$\frac{1}{2}A$	$\frac{1}{2}C$			
$Fmm2$	*p2mm*	$\frac{1}{2}A$	$\frac{1}{2}B$	*p1m1*	$\frac{1}{2}B$	$\frac{1}{2}C$	*p1m1*	$\frac{1}{2}A$	$\frac{1}{2}C$			
$Fdd2$	*p2gg*	$\frac{1}{2}A$	$\frac{1}{2}B$	*c1m1*	$\frac{1}{2}B$	$\frac{1}{2}C$	*c1m1*	$\frac{1}{2}A$	$\frac{1}{2}C$			
$P\frac{2}{m}\frac{2}{m}\frac{2}{m}$	*p2mm*	A	B	*p2mm*	B	C	*p2mm*	A	C			
$P\frac{2}{n}\frac{2}{n}\frac{2}{n}$	*c2mm*	A	B	*c2mm*	B	C	*c2mm*	A	C			
$P\frac{2}{c}\frac{2}{c}\frac{2}{m}$	*p2mm*	A	B	*p2mm*	B	$\frac{1}{2}C$	*p2mm*	A	$\frac{1}{2}C$			
$P\frac{2}{b}\frac{2}{a}\frac{2}{n}$	*c2mm*	A	B	*p2mm*	$\frac{1}{2}B$	C	*p2mm*	$\frac{1}{2}A$	C			
$P\frac{2}{m}\frac{2}{c}\frac{2_1}{m}$	*p2mm*	A	B	*p2gm*	B	C	*p2mm*	A	$\frac{1}{2}C$			
$P\frac{2}{n}\frac{2}{a}\frac{2_1}{n}$	*c2mm*	A	B	*c2mm*	B	C	*p2gm*	$\frac{1}{2}A$	C			
$P\frac{2}{m}\frac{2}{n}\frac{2_1}{a}$	*p2mm*	$\frac{1}{2}A$	B	*p2gm*	B	C	*c2mm*	A	C			
$P\frac{2}{b}\frac{2}{c}\frac{2_1}{b}$	*p2mm*	A	$\frac{1}{2}B$	*p2gm*	$\frac{1}{2}B$	C	*p2mm*	A	$\frac{1}{2}C$			
$P\frac{2_1}{b}\frac{2_1}{a}\frac{2}{m}$	*p2gg*	A	B	*p2mm*	$\frac{1}{2}B$	C	*p2mm*	$\frac{1}{2}A$	C			
§ $P\frac{2_1}{c}\frac{2_1}{c}\frac{2}{n}$	*c2mm*	A	B	*p2mg*	B	$\frac{1}{2}C$	*p2mg*	A	$\frac{1}{2}C$			
§ $P\frac{2_1}{b}\frac{2_1}{m}\frac{2}{a}$	*p2gm*	$\frac{1}{2}A$	B	*p2mm*	B	$\frac{1}{2}C$	*p2mg*	A	C			

Table 2 (*Continued*)

Space group†	Projection ‖ [001]			Projection ‖ [100]			Projection ‖ [010]			Projection ‖ [110]	
		a	*b*		*a*	*b*		*a*	*b*	*a*	*b*
$P\frac{2_1}{n}\frac{2_1}{n}\frac{2}{m}$	*p2gg*	A	B	*c2mm*	B	C	*c2mm*	A	C		
$P\frac{2_1}{m}\frac{2_1}{m}\frac{2}{n}$	*c2mm*	A	B	*p2mg*	B	C	*p2mg*	A	C		
$P\frac{2_1}{n}\frac{2_1}{c}\frac{2}{a}$	*p2gm*	$\frac{1}{2}A$	B	*c2mm*	B	C	*p2mg*	A	$\frac{1}{2}C$		
$P\frac{2_1}{b}\frac{2_1}{c}\frac{2_1}{a}$	*p2mg*	$\frac{1}{2}A$	B	*p2gm*	$\frac{1}{2}B$	C	*p2mg*	A	$\frac{1}{2}C$		
$P\frac{2_1}{n}\frac{2_1}{m}\frac{2_1}{a}$	*p2gm*	$\frac{1}{2}A$	B	*c2mm*	B	C	*p2gg*	A	C		
$I\frac{2}{m}\frac{2}{m}\frac{2}{m}$	*c2mm*	A	B	*c2mm*	B	C	*c2mm*	A	C		
$I\frac{2}{b}\frac{2}{a}\frac{2}{m}$	*c2mm*	A	B	*p2mm*	$\frac{1}{2}B$	$\frac{1}{2}C$	*p2mm*	$\frac{1}{2}A$	$\frac{1}{2}C$		
§ $I\frac{2_1}{b}\frac{2_1}{c}\frac{2_1}{a}$	*p2mm*	$\frac{1}{2}A$	$\frac{1}{2}B$	*p2mm*	$\frac{1}{2}B$	$\frac{1}{2}C$	*p2mm*	$\frac{1}{2}A$	$\frac{1}{2}C$		
$I\frac{2_1}{m}\frac{2_1}{m}\frac{2_1}{a}$	*p2mm*	$\frac{1}{2}A$	$\frac{1}{2}B$	*c2mm*	B	C	*c2mm*	A	C		
$C\frac{2}{m}\frac{2}{m}\frac{2}{m}$	*c2mm*	A	B	*p2mm*	$\frac{1}{2}B$	C	*p2mm*	$\frac{1}{2}A$	C		
$C\frac{2}{c}\frac{2}{c}\frac{2}{m}$	*c2mm*	A	B	*p2mm*	$\frac{1}{2}B$	$\frac{1}{2}C$	*p2mm*	$\frac{1}{2}A$	$\frac{1}{2}C$		
$C\frac{2}{m}\frac{2}{m}\frac{2}{a}$	*p2mm*	$\frac{1}{2}A$	$\frac{1}{2}B$	*p2mm*	$\frac{1}{2}B$	C	*p2mm*	$\frac{1}{2}A$	C		
$C\frac{2}{c}\frac{2}{c}\frac{2}{a}$	*p2mm*	$\frac{1}{2}A$	$\frac{1}{2}B$	*p2mm*	$\frac{1}{2}B$	$\frac{1}{2}C$	*p2mm*	$\frac{1}{2}A$	$\frac{1}{2}C$		
$C\frac{2}{m}\frac{2}{c}\frac{2_1}{m}$	*c2mm*	A	B	*p2gm*	$\frac{1}{2}B$	C	*p2mm*	$\frac{1}{2}A$	$\frac{1}{2}C$		
$C\frac{2}{m}\frac{2}{c}\frac{2_1}{a}$	*p2mm*	$\frac{1}{2}A$	$\frac{1}{2}B$	*p2gm*	$\frac{1}{2}B$	C	*p2mm*	$\frac{1}{2}A$	$\frac{1}{2}C$		
$F\frac{2}{m}\frac{2}{m}\frac{2}{m}$	*p2mm*	$\frac{1}{2}A$	$\frac{1}{2}B$	*p2mm*	$\frac{1}{2}B$	$\frac{1}{2}C$	*p2mm*	$\frac{1}{2}A$	$\frac{1}{2}C$		
$F\frac{2}{d}\frac{2}{d}\frac{2}{d}$	*c2mm*	$\frac{1}{2}A$	$\frac{1}{2}B$	*c2mm*	$\frac{1}{2}B$	$\frac{1}{2}C$	*c2mm*	$\frac{1}{2}A$	$\frac{1}{2}C$		

that projection. As a consequence, it is obvious that there are only 17 sets of these zonal symmetry factors, one set for each plane group. These are listed for the various plane groups in Table 3.

In the trial-and-error computations involved in attempting to find a structure which reproduces the observed intensities for reflections from planes in a zone, the symmetry factor to be used must correspond to one of the comparatively simple cases listed in Table 3. To find the correct one, Table 2 should first be consulted. This identifies the correct plane group for the projection, or zone, desired. Table 3 then provides the symmetry factor for the plane group, and hence for the zone. Particular attention should be paid to possible changes in indices which are required in case the space group projects as a pattern with submultiple axes.

In the last chapter it was seen that the symmetry factor, in general, has both a real component, A, and an imaginary component, B. The B component vanishes if the symmetry includes inversion centers *and* if the origin is taken at an inversion center. This requires that an atom at xyz be accompanied by an equivalent atom at $\bar{x}\bar{y}\bar{z}$. In two dimensions this degenerates to equivalence of points at xy and $\bar{x}\bar{y}$. The symmetry element relating these points is usually represented by the symbol of a 2-fold axis (or 2-fold "rotor"). Both a three-dimensional 2-fold axis and a 2-fold screw project in the direction of the axis as a 2-fold rotor in the plane-group projection. Similarly, any even-fold axis, including a screw, projects as the corresponding even-fold rotor. If such locations exist in the plane group, and if one of them is chosen as the origin, the B term of the symmetry factor vanishes, even if the space group from which the projection is made lacks inversion centers. Thus, by making appropriate use of centric projections, computations for non-centric space groups can be reduced to a minimum. Each of the symmetry factors for the plane groups listed in Table 3 is referred to a centric origin if the plane group permits.

The successful solution of a projected structure using a centric plane group, conversely, must not be interpreted as necessarily implying inversion centers in the space group. The space group *may* have inversion centers, but it may alternatively have 2-fold axes or 2-fold screws in locations corresponding to centers in the projection.

Symmetry-factor maps

Since symmetry factors for zones involve only two variables, say x and y, they can be represented for any selected values of h and k on a two-dimensional plot. This was first suggested by Bragg and Lipson.[1,2] Although there are 17 plane groups, only 10 kinds of diagrams for any selected reflection, hk, are needed to cover these groups. This small number results because (*a*) some of the symmetry-factor forms are duplicated in several different plane groups, and (*b*) the sine and cosine of an angle can be obtained from the same graph by a shift of origin. These maps for representative values of h and k are shown in Figs. 4–15.

For each symmetry-factor graph type, a separate map ought to be available for each combination of the indices hk. Actually, certain other combinations can be had by properly shifting the map. Thus, turning the map of hk upside down provides $h\bar{k}$. Since this is a useful saving, the map should be drawn on transparent paper. Furthermore, the maps for

Table 3
Symmetry factors for the plane projections

Symmetry	Origin at	Conditions	A	B
$p1$	1		$\cos 2\pi(hx+ky)$	$\sin 2\pi(hx+ky)$
$p2$	2		$2 \cos 2\pi(hx+ky)$	0
$p1m1$	m		$2 \cos 2\pi hx \cos 2\pi ky$	$2 \cos 2\pi hx \sin 2\pi ky$
$p1g1$	g	$k = 2n$	$2 \cos 2\pi hx \cos 2\pi ky$	$2 \cos 2\pi hx \sin 2\pi ky$
		$k = 2n+1$	$-2 \sin 2\pi hx \sin 2\pi ky$	$2 \sin 2\pi hx \cos 2\pi ky$
$c1m1$	m	$h+k = 2n$	$4 \cos 2\pi hx \cos 2\pi ky$	$4 \cos 2\pi hx \sin 2\pi ky$
		$h+k = 2n+1$	0	0
$p2mm$	$2mm$		$4 \cos 2\pi hx \cos 2\pi ky$	0
$p2mg$	2	$h = 2n$	$4 \cos 2\pi hx \cos 2\pi ky$	0
		$h = 2n+1$	$-4 \sin 2\pi hx \sin 2\pi ky$	0
$p2gg$	2	$h+k = 2n$	$4 \cos 2\pi hx \cos 2\pi ky$	0
		$h+k = 2n+1$	$-4 \sin 2\pi hx \sin 2\pi ky$	0
$c2mm$	$2mm$	$h+k = 2n$	$8 \cos 2\pi hx \cos 2\pi ky$	0
		$h+k = 2n+1$	0	0

$p4$	4		$2\{\cos 2\pi(hx+ky) + \cos 2\pi(kx-hy)\}$ $= 4\cos\pi\{(h-k)x + (h+k)y\}$ $\times\cos\pi\{(h+k)x - (h-k)y\}.$	0
$p4mm$	$4mm$		$4(\cos 2\pi hx \cos 2\pi ky$ $+\cos 2\pi kx \cos 2\pi hy)$	0
$p4gm$	4	$h+k = 2n$	$4(\cos 2\pi hx \cos 2\pi ky$ $+\cos 2\pi kx \cos 2\pi hy)$	0
		$h+k = 2n+1$	$-4(\sin 2\pi hx \sin 2\pi ky$ $-\sin 2\pi kx \sin 2\pi hy)$	0
$p3$	3		$\cos 2\pi(hx+ky)$ $+\cos 2\pi(kx+iy)$ $+\cos 2\pi(ix+hy)$	$\sin 2\pi(hx+ky)$ $+\sin 2\pi(kx+iy)$ $+\sin 2\pi(ix+hy)$
$p3m1$	$3m1$		$2\{\cos\pi h(x+y)\cos\pi(k-i)(x-y)$ $+\cos\pi k(x+y)\cos\pi(i-h)(x-y)$ $+\cos\pi i(x+y)\cos\pi(h-k)(x-y)\}$	$2\{\cos\pi h(x+y)\sin\pi(k-i)(x-y)$ $+\cos\pi k(x+y)\sin\pi(i-h)(x-y)$ $+\cos\pi i(x+y)\sin\pi(h-k)(x-y)\}$
$p31m$	$31m$		$2\{\cos\pi h(x+y)\cos\pi(k-i)(x-y)$ $+\cos\pi k(x+y)\cos\pi(i-h)(x-y)$ $+\cos\pi i(x+y)\cos\pi(h-k)(x-y)\}$	$2\{\sin\pi h(x+y)\cos\pi(k-i)(x-y)$ $+\sin\pi k(x+y)\cos\pi(i-h)(x-y)$ $+\sin\pi i(x+y)\cos\pi(h-k)(x-y)\}$
$p6$	6		$2\{\cos 2\pi(hx+ky)$ $+\cos 2\pi(kx+iy)$ $+\cos 2\pi(ix+hy)\}$	0
$p6mm$	$6mm$		$4\{\cos\pi h(x+y)\cos\pi(k-i)(x-y)$ $+\cos\pi k(x+y)\cos\pi(i-h)(x-y)$ $+\cos\pi i(x+y)\cos\pi(h-k)(x-y)$	0

$h\bar{k}$ and $k\bar{h}$ can be obtained by turning the above hk and kh maps through 90°.

In some of the actual uses of the maps, it is not necessary to have numerical values. If the positive and negative fields alone are outlined, with the divisions marked (i.e., the zero level), and the peaks and depressions marked, the map is useful in showing how a proposed shift of an

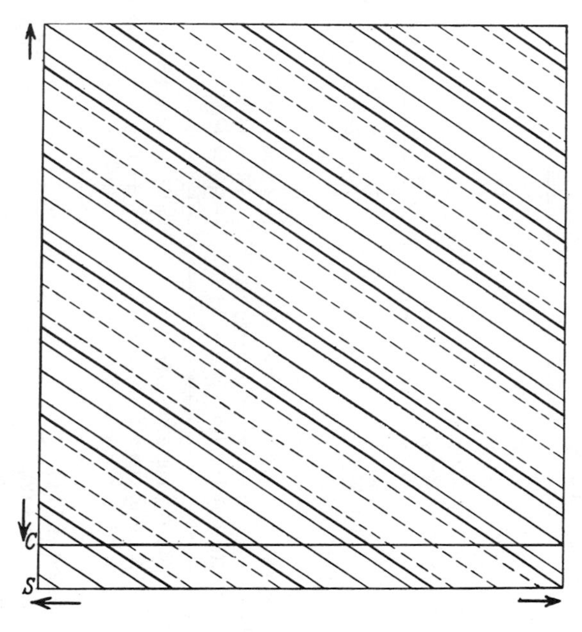

Fig. 4.
(for $p1$ and $p2$)

Variation, over the cell, of

$$\cos 2\pi(hx+ky)$$
$$\text{and } \sin 2\pi(hx+ky)$$

for hk = 23. C and S are the origins for the cosine and sine functions. The contour intervals are 0.5.

(After Bragg and Lipson.[2])

atom changes the relative magnitude of the intensity of the reflection. For all except the plane groups mentioned in the next paragraph, such outline graphs are quickly prepared for all indices, hk.

With the exceptions of the five plane groups mentioned below, the graph for any index is merely a homogeneous deformation of the graph for the particular index hk = 11, repeated as required to cover the cell. For such plane groups, the outline graphs can be very quickly prepared. But this simple geometrical correspondence does not hold for the trigonal, tetragonal, and hexagonal plane groups containing reflection symmetry. These are the plane groups $p4mm$, $p4gm$, $p3m1$, $p31m$, and $p6mm$. For

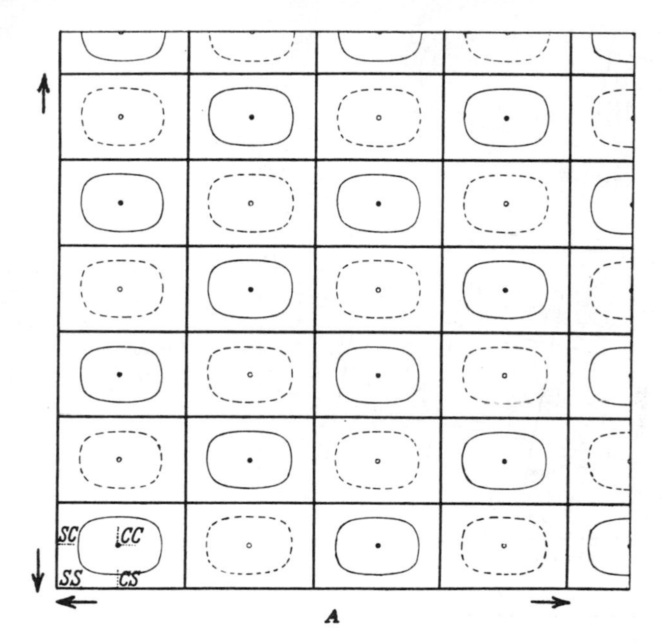

Fig. 5*A*.
(for *p*1*m*1, *p*1*g*1, *c*1*m*1, *p*2*mm*, *p*2*mg*, *p*2*gg*, *c*2*mm*)

Variation, over the cell, of

$$\begin{Bmatrix}\cos\\ \sin\end{Bmatrix} 2\pi hx \begin{Bmatrix}\cos\\ \sin\end{Bmatrix} 2\pi ky$$

for $hk = 23$. The origins of the combinations are indicated by *CC*, *SC*, *CS*, and *SS*. (After Bragg and Lipson.[2])

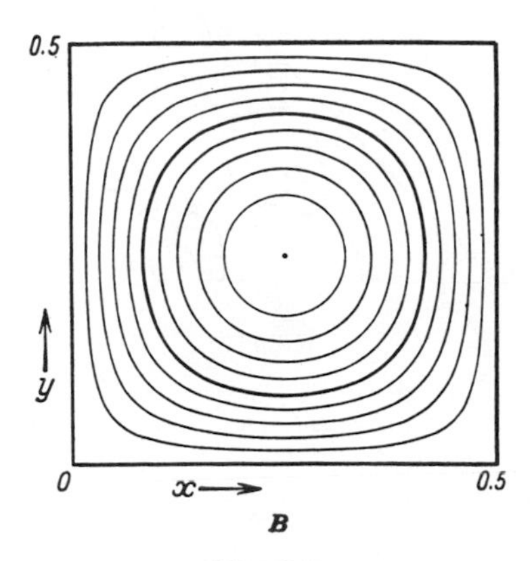

Fig. 5*B*.
Detail of the contouring of Fig. 5*A* for $\sin 2\pi 1x \sin 2\pi 1y$.
(After Bragg and Lipson.[2])

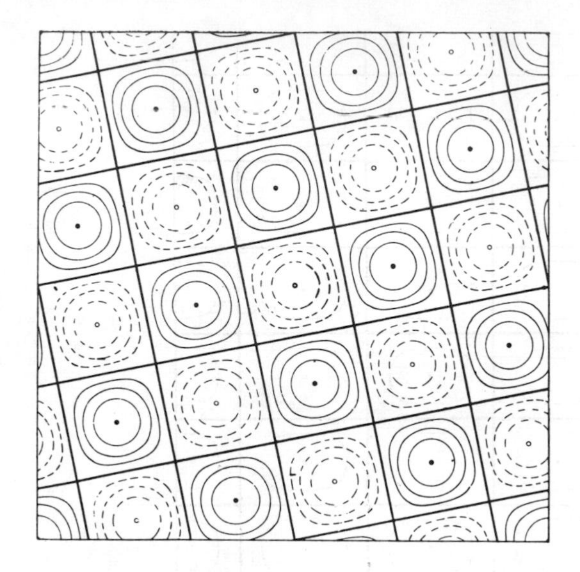

Fig. 6.
(for $p4$)

Variation, over the cell, of
$\cos 2\pi(hx+ky)$
$+ \cos 2\pi(kx-hy)$
for $hk = 23$.
(After Bragg and Lipson.[2])

Fig. 7.
(for $p4mm$ and $p4gm$)

Variation, over the cell, of
$\cos 2\pi hx \cos 2\pi ky$
$+ \cos 2\pi kx \cos 2\pi hy$
for $hk = 23$.
(After Bragg and Lipson.[2])

Fig. 8.
(for $p4mg$)

Variation, over the cell, of

$$\sin 2\pi hx \sin 2\pi ky - \sin 2\pi kx \sin 2\pi hy$$

for $hk = 23$.

(After Bragg and Lipson.[2])

computing reflections for the last three cases, Beevers and Lipson[3] have prepared sheets giving the numerical values of the symmetry factor for the following reflections:

$10\bar{1}0$			
$20\bar{2}0$	$21\bar{3}0$		
$30\bar{3}0$	$31\bar{4}0$	$32\bar{5}0$	
$40\bar{4}0$	$41\bar{5}0$	$42\bar{6}0$	$43\bar{7}0$
$50\bar{5}0$	$51\bar{6}0$	$52\bar{7}0$	
$60\bar{6}0$			

These sheets give the symmetry factor at all points x, y, at intervals of $\frac{1}{60}$ of the cell edge, for a representative triangle of the symmetrical cell. With the aid of these sheets, actual structure-factor computations can be carried out in just those cases where the structure factors would be most tedious to compute by straightforward methods.

Use of symmetry-factor graphs

The advantage of using symmetry-factor graphs is that one avoids having to carry out the arithmetic operations involved in forming the symmetry factor before he can see what effect a shift of atoms causes in

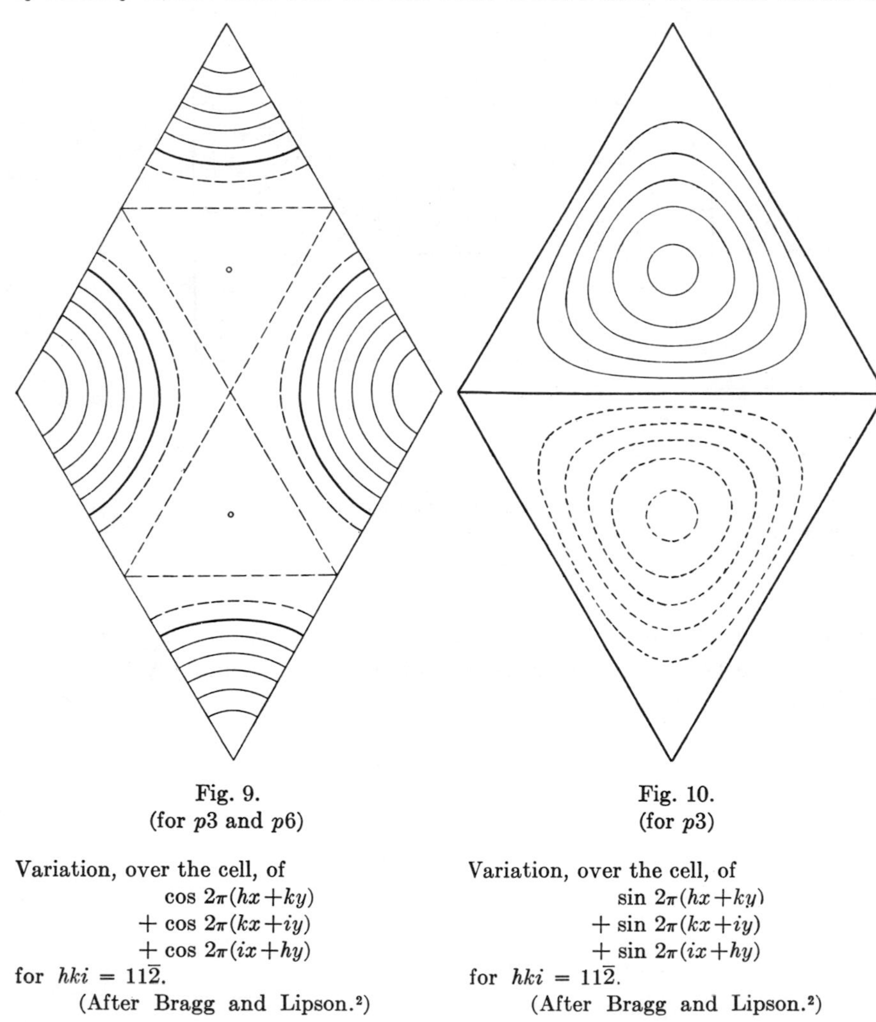

Fig. 9.
(for $p3$ and $p6$)

Variation, over the cell, of

$$\cos 2\pi(hx+ky) + \cos 2\pi(kx+iy) + \cos 2\pi(ix+hy)$$

for $hki = 11\bar{2}$.
(After Bragg and Lipson.[2])

Fig. 10.
(for $p3$)

Variation, over the cell, of

$$\sin 2\pi(hx+ky) + \sin 2\pi(kx+iy) + \sin 2\pi(ix+hy)$$

for $hki = 11\bar{2}$.
(After Bragg and Lipson.[2])

the intensity of the reflection. In this way, the gap between cause and effect is greatly reduced.

In practice, one plots the positions of the atoms of the proposed structure on paper, and then lays over this plot the transparent symmetry-factor graphs for the various values of hk. A rough value of the F_{hk} of each reflection can then be quickly obtained by multiplying the value of the symmetry factor for each atom by its f and summing the results. (If the projection is non-centrosymmetrical, both A and B parts must be determined in this way; then after the summation, the computation $|F|^2 = A^2 + B^2$ must be performed.) By simplifying the initial stages of

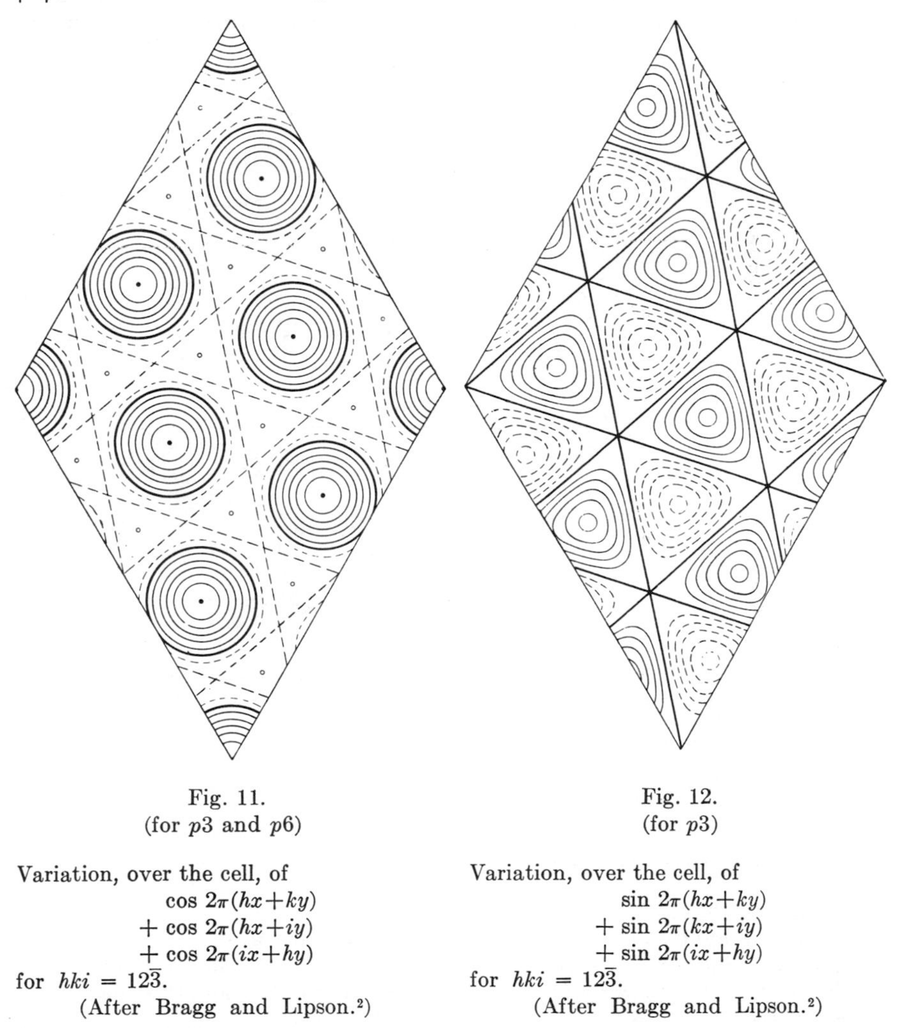

Fig. 11.
(for $p3$ and $p6$)

Variation, over the cell, of
$$\cos 2\pi(hx+ky) + \cos 2\pi(hx+iy) + \cos 2\pi(ix+hy)$$
for $hki = 12\bar{3}$.
(After Bragg and Lipson.[2])

Fig. 12.
(for $p3$)

Variation, over the cell, of
$$\sin 2\pi(hx+ky) + \sin 2\pi(kx+iy) + \sin 2\pi(ix+hy)$$
for $hki = 12\bar{3}$.
(After Bragg and Lipson.[2])

the work in this manner, the symmetry-factor graphs serve a very useful purpose since one quickly perceives whether or not the trial structure has any merit in explaining the observed intensities. If not, the structure can be rejected, and a new one tested as quickly.

In this connection it should be observed that if the F's of the low index reflections compare more or less favorably with the observed F's, it is an indication that the trial structure is probably approximately correct, even if there is lack of agreement between observed and computed F's for reflections of higher order. Such a structure can usually be improved by shifting the atoms in such a way as to cause the computed values of the F's for the higher indices to approach the observed values. It is in the process of this improvement that the symmetry-factor graphs have a great

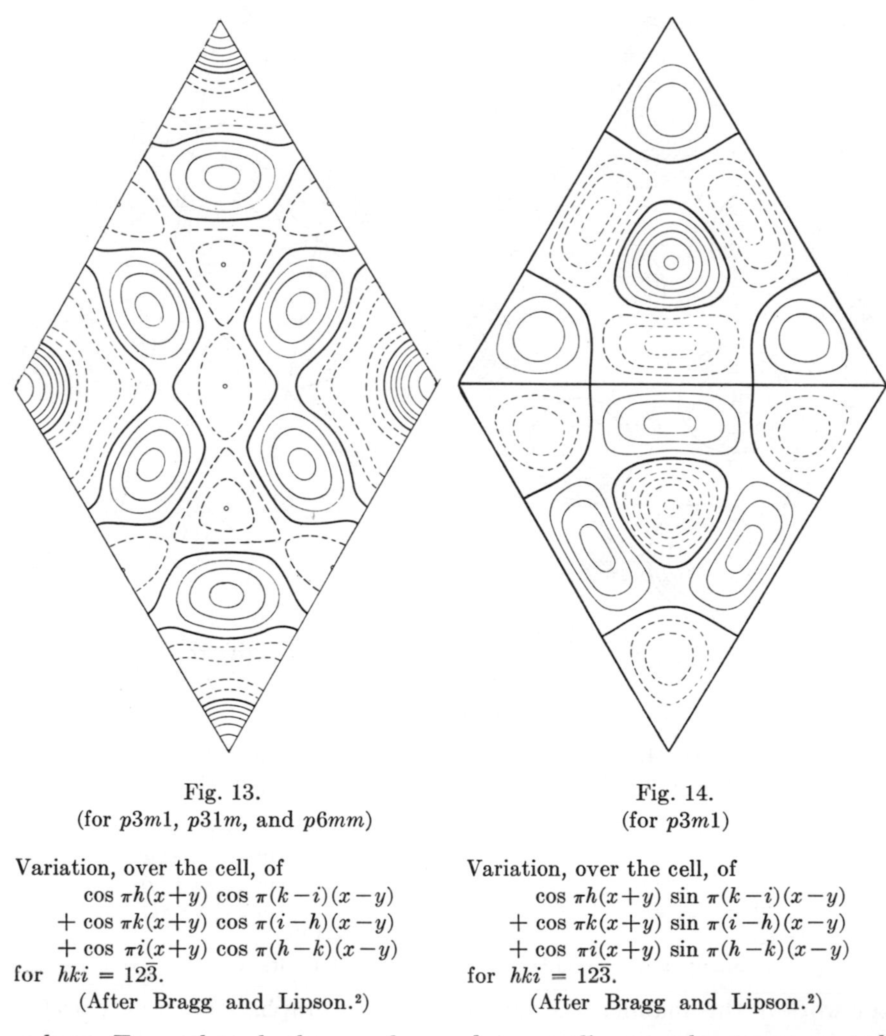

Fig. 13.
(for $p3m1$, $p31m$, and $p6mm$)

Variation, over the cell, of

$$\cos \pi h(x+y) \cos \pi(k-i)(x-y) + \cos \pi k(x+y) \cos \pi(i-h)(x-y) + \cos \pi i(x+y) \cos \pi(h-k)(x-y)$$

for $hki = 12\bar{3}$.
(After Bragg and Lipson.[2])

Fig. 14.
(for $p3m1$)

Variation, over the cell, of

$$\cos \pi h(x+y) \sin \pi(k-i)(x-y) + \cos \pi k(x+y) \sin \pi(i-h)(x-y) + \cos \pi i(x+y) \sin \pi(h-k)(x-y)$$

for $hki = 12\bar{3}$.
(After Bragg and Lipson.[2])

value. Even though the graphs used are outline graphs, as suggested above, a glance at the positions of the representative atoms on the graph of a particular reflection hk immediately shows which particular atoms are in positions to which the reflection is sensitive. The graph also furnishes a guide as to which way the shift should be made to increase or decrease the computed intensity. The sensitivity, or lack of sensitivity, with location is simply a generalization to two dimensions of the strategy mentioned in Chapter 10: In general, the graphs are found to have hills and depressions. The contributions of atoms located on crests and troughs are insensitive to slight shifts since the gradients at these places

Fig. 15.
(for $p31m$)

Variation, over the cell, of

$$\sin \pi h(x+y) \cos \pi h(x-y) + \sin \pi k(x+y) \cos \pi k(x-y) + \sin \pi i(x+y) \cos \pi i(x-y)$$

for $hki = 12\bar{3}$.
(After Bragg and Lipson.[2])

are small. On the other hand, the contributions of atoms at the greatest values of the gradient are most sensitive to changes in position, and a shift in the direction of the steepest gradient produces the greatest effect. If several atoms can be shifted to produce the desired effect on a particular reflection, the shift should be limited to those which are on insensitive positions of those other reflections whose intensities have about the correct value.

Graphical computation of structure factors

Hoppe and Camerer[4] have developed a graphical method of computing structure factors which has a certain similarity to the quantitative use of the simplest Bragg and Lipson charts. The general theory of this method is as follows:

The contribution of an atom j to a reflection $hk0$ is

$$^{j}F = f_j \cos 2\pi(hx_j+ky_j). \tag{1}$$

The map of this function against x and y is essentially the structure-factor graph shown in Fig. 4. The same situation can be described from another point of view. An atom makes its maximum contribution to a reflection $hk0$ when it is located on one of the set of planes $hk0$. When the atom is not on any plane of this set, it makes an inferior contribution dependent on the distance of the atom from the nearest plane. Specifically, let the distance between planes $hk0$ be d, and let the atom be located at a distance s from the plane. Then the contribution of the atom to the reflection can also be expressed in the form

$$^{j}F = f_j \cos 2\pi \frac{s}{d}. \tag{2}$$

Of course, this is equivalent to (1). Thus, if one plots the atom in relation to the plane, its contribution can be found by measuring its distance from the plane and applying (2).

To make practical use of this relation, one prepares a map showing the locations of all the atoms in the projection of the structure. (In the example just given, the projection is on a plane normal to the c axis of the crystal.) He also makes maps of each set of lines hk in this projection, employing the same scale used for the map of the structure. For each such hk map, there is attached a plot of the trigonometric part of (2) for the particular value of d for the plane $hk0$. (Alternatively, the same information may be contained in a chart showing the variation of the cosine in (2) with s.) To find the contribution of the atom to the structure factor for $hk0$, one lays the chart on the map of the structure, and for each atom measures (as with a pair of dividers) the distance of that atom from the nearest plane, and then applies this distance to the abscissa of the cosine plot. The ordinate gives $\cos 2\pi(s/d)$. This value multiplied by the f of the atom gives it contribution, and the algebraic sum of such contributions for all the individual atoms of the structure gives the structure factor for the plane $hk0$.

This method can be extended to the graphical computation of structure factors for the general plane hkl very simply. The symmetry factor for hkl is $\cos 2\pi(hx+ky+lz)$. But this still corresponds to the simple form $\cos 2\pi(s/d)$. Thus, if the origin of the chart or curve is displaced by an amount corresponding to lz, the same mapping and procedure can be used to find the structure factors of hkl.

Finally, for non-centrosymmetrical projections the imaginary part of the symmetry factor, which is $\sin 2\pi(hx+ky)$ for $hk0$ reflections, or $\sin 2\pi(hx+ky+lz)$ for the more general reflections hkl, can be determined with the aid of the charts discussed above by displacing the origin by an amount corresponding to $\pi/2$.

Related graphical devices for evaluating the trigonometric part of structure factors have been suggested by Grenville-Wells[5, 6] and Morley and Taylor.[7]

Literature

[1] W. L. Bragg. *Structure-factor graphs for crystal analysis.* Nature **138** (1936) 362–364.

[2] W. L. Bragg and H. Lipson. *The employment of contoured graphs of structure-factor in crystal analysis.* Z. Krist. (A) **95** (1936) 323–337.

[3] C. A. Beevers and H. Lipson. *On the evaluation of some hexagonal structure factors.* Proc. Phys. Soc. **50** (1938) 275–282. (Copies of the figure fields described in this paper were made available by the authors to those who could use them.)

[4] Walter Hoppe and L. Camerer. *Ein graphische Verfahren zur Berechnung von Structurfaktoren für die Röntgenanalyse von Kristallen.* Z. Naturforsch., **4a** (1949) 637–639.

[5] H. J. Grenville-Wells. *A graphical method of evaluating trigonometric functions used in crystal-structure analysis.* Parts I and II. J. Appl. Phys. **25** (1954) 485–490.

[6] H. J. Grenville-Wells. *A new graphical principle for the evaluation of Fourier transforms.* Acta Cryst. **8** (1955) 737.

[7] K. A. Morley and C. A. Taylor. *A variable structure-factor graph for use in crystal-structure determination.* J. Sci. Instr. **34** (1957) 54–58.

[8] C. A. Taylor and F. A. Underwood. *An optical method for producing structure-factor graphs.* Acta Cryst. **12** (1959) 336–339.

12

Examples of the determination of some simple crystal structures

There are certain classes of crystal structures which can be determined by very simple means and without recourse to advanced methods. A structure is amenable to such comparatively simple attack if it falls into one of the following categories:

1. All the atoms of the structure are fixed by symmetry.
2. The number of variable parameters determining the locations of the atoms is small, and they can be separated and studied one at a time.
3. The number of variable parameters determining the locations of the atoms is small, and although they cannot be studied one at a time, the crystal contains one set of sufficiently heavy atoms whose parameters *can* be so studied, in spite of the presence of variables due to lighter atoms. Furthermore, the parameters of the lighter atoms are limited in number.

In this chapter these three cases are illustrated by outlines of the determination of the structures of NaCl, of marcasite (FeS_2), and of valentinite (Sb_2O_3).

Determination of the structure of NaCl, an example of a symmetry-fixed structure

The sodium chloride structure type provides an example of a symmetry-fixed structure. That the structure is indeed symmetry fixed is a matter which becomes evident when the structure analysis has proceeded only a comparatively short distance, specifically when one has determined the number of formula weights per unit cell and looks up the available equipoints of the space groups in question.

Because of the high symmetry of crystals of NaCl, it is also possible to perform a structure analysis with the use of a single rotating-crystal photograph, and even a powder photograph is adequate. But because of the limited symmetry information which can be derived from rotating-crystal photographs,† the analysis takes a somewhat more devious initial course than would be necessary if one of the moving-film methods were used. If the crystal has symmetry other than isometric, tetragonal, or hexagonal, the rotating-crystal method would be a hopeless way of recording the reflections, not only because symmetry information would be almost lacking, but also because the reflections could not be indexed with certainty.

Since the structure of NaCl is determinable by the rotating-crystal method due to its high symmetry, it incidentally offers an example of the procedure to be used when this method of recording reflections is chosen for a structure analysis.

Unit cell. Since it is well established that NaCl is isometric, a single rotating-crystal photograph, with rotation about an a axis, is all that is necessary to establish the dimensions of the unit cell as well as to provide the intensities of the reflections for the structure analysis. The rotation photograph quickly shows that the edge of the cubic cell is $a = 5.63\text{Å}$. The density of NaCl is found in standard reference works to be 2.16 g./cc. The formula weight of NaCl is $23.00 + 35.46 = 58.46$ atomic-weight units. Applying relation (3) of Chapter 9, the number of formula weights per unit cell is

$$N = \frac{G \times V \times 10^{-24}}{M \times 1.660 \times 10^{-24}}$$

$$= \frac{2.16 \times 5.63^3 \times 10^{-24}}{58.46 \times 1.660 \times 10^{-24}} = 3.97 \approx 4.$$

It is usual that the number of formula weights per unit cell does not come out exactly an integer. This failure, in the case of simple, chemically pure compounds such as NaCl, may be attributed to two causes. In the first place the experimentally determined density tends to be too low due to voids in the crystal, and in the second place there may be an error in the experimental determination of the volume of the cell. In this particular case, since V is simply a^3, any fractional error in the determination of the cell edge turns up as three times the fractional error in V, and hence in N.

Space group. The rotating-crystal photograph must first be indexed

† M. J. Buerger. *X-ray crystallography.* (John Wiley and Sons, New York, 1942) 165.

by one of the methods discussed elsewhere.† The indexing by the graphical procedure is very illuminating with respect to certain features of the rotating-crystal method, and is illustrated in Fig. 1. In following through this process it becomes obvious that *hkl* and *khl* record on the same spot on the film, and therefore that it is not possible to tell whether these are equivalent in intensity or not. For this reason it is not possible to determine the presence or absence of "vertical planes," and

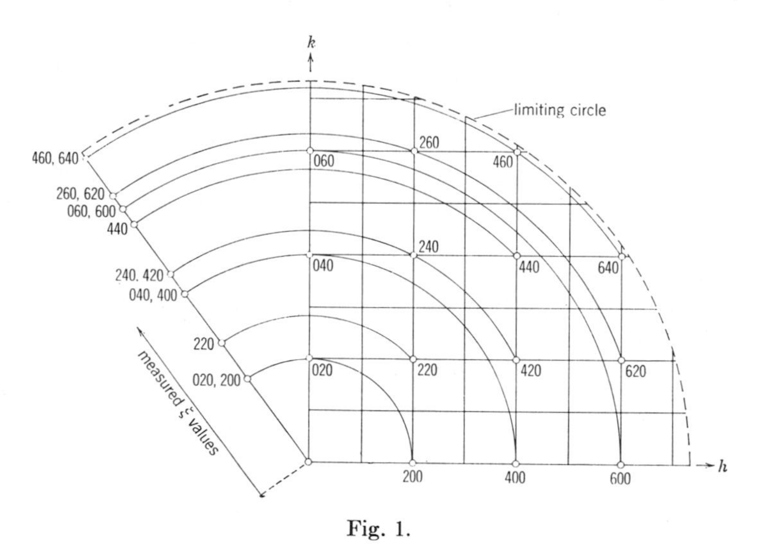

Fig. 1.

consequently it is not possible to distinguish between certain crystal classes and their space groups.

It is possible, however, to determine the lattice of the crystal, since the same lattices apply to all crystal classes of this crystal system. This is done by arranging the observed reflections in a systematic list, as shown in Table 1. It is then evident that the reflections *hkl* are present when *h*, *k*, and *l* are all even or all odd, and absent when *h*, *k*, and *l* are mixed, indicating that the structure is based on a face-centered lattice.

Since there is no symmetry information whatever on the rotating-crystal photograph, no crystal class of the isometric system is excluded by the x-ray observations, and all five classes of the system must be considered in any rigorous deduction of the structure. The possible space groups in these classes which are based upon a face-centered lattice are listed in Table 2. In each crystal class the several possible space groups based upon the same lattice contain isogonal symmetry elements which differ by a characteristic translation component. When this translation

† M. J. Buerger. *X-ray crystallography.* (John Wiley and Sons, New York, 1942) 157–165.

component is other than zero, missing reflections occur in the diffraction record. Now, a further study of the reflection list, Table 1, makes it plain that there are no systematic absences other than the ones required by the face-centering of the lattice, namely that *hkl* is absent unless *h*, *k*, and *l* are all even or all odd. From this it can be concluded that the

Table 1
Catalog of reflections recorded on rotating-crystal photograph of NaCl (CuKα radiation)

				—			
				440	—	640	
			—	—	—	—	
*hk*0:		220	—	420	—	620	—
	—	—	—	—	—	—	—
	—	200	—	400	—	600	—
					551 (711)		
				—	—		
*hk*1:			331	—	531	—	
		—	—	—	—	—	
	111	—	311	—	511	—	711 (551)
	—	—	—	—	—	—	—
				442	—		
			—	—	—		
*hk*2:		222	—	422	—	622	
		—	—	—	—	—	
	—	202	—	402	—	602	
				—			
			333	—			
*hk*3:		—	—	—	—		
		—	313	—	513		
		—	—	—	—		

translation components of all symmetry elements of the space group of NaCl are zero, and this limits the possible space groups to the five contained in the boxes of Table 2. Note that this elimination has been based on reflections which actually appear in the diffraction record and hence is a positive elimination of space groups which do not fit the observed diffraction.

Distribution of atoms among the equipoints. The unit cell of NaCl must accommodate 4 formula weights, as discovered above, and therefore must accommodate 4 Na and 4 Cl atoms. An examination of the equipoint lists[†] reveals that none of the uneliminated space groups contains equipoints of rank lower than 4. (That this is the lower limit of equipoint rank could have been predicted without consulting tables by noting that, in general, the lattice points themselves have the lowest equipoint rank; that in this particular case there are 4 face-centered lattice points per unit cell, and that since NaCl is based on a face-centered

Table 2
Isometric space groups based on a face-centered lattice

Crystal class	Space group
T, 23	$\boxed{F23}$
T_h, $\frac{2}{m}\bar{3}$	$\boxed{F\frac{2}{m}\bar{3}}$
	$F\frac{2}{d}\bar{3}$
T_d, $\bar{4}3m$	$\boxed{F\bar{4}3m}$
	$F\bar{4}3c$
O, 432	$\boxed{F43}$
	$F4_13$
O_h, $\frac{4}{m}\bar{3}\frac{2}{m}$	$\boxed{F\frac{4}{m}\bar{3}\frac{2}{m}}$
	$F\frac{4}{m}\bar{3}\frac{2}{c}$
	$F\frac{4_1}{d}\bar{3}\frac{2}{m}$
	$F\frac{4_1}{d}\bar{3}\frac{2}{c}$

lattice, none of the space groups in question could have any equipoint rank lower than 4.) Consequently, the obvious distribution of the 4 Na atoms is on an equipoint of rank 4 and that of the 4 Cl atoms is on another equipoint of rank 4. The 4-fold equipoints are all symmetry fixed and therefore without any degrees of freedom.

There is another less obvious but possible distribution, which, though it does not apply in this particular case as explained below, should be

[†] Norman F. M. Henry and Kathleen Lonsdale. *International tables for x-ray crystallography*. Vol. 1. (Kynoch Press, Birmingham, England, 1952)

borne in mind in considering distributions of the atoms over the available equipoints. Posnjak and Barth[†][§] have shown that, in some crystals, two atoms of different species may occupy a single equipoint by being distributed at random on that equipoint. (In this particular case, this would correspond to Na and Cl being distributed at random over an equipoint of rank 8.) This is called a *variate-atom equipoint.* This occurs because of disorder among the two atoms in question, a situation which can sometimes develop if the crystal is grown at (or subsequently exposed to) high temperatures.[¶] This does not apply to the crystal of

Table 3
4-fold equipoints of possible space groups of NaCl

Space group	4-fold equipoints	Diffraction symbol
$F23$	4*a*, 4*b*, 4*c*, 4*d*	$m3F$ - -
$F\frac{2}{m}\bar{3}$	4*a*, 4*b*	
$F\bar{4}3m$	4*a*, 4*b*, 4*c*, 4*d*	$m3mF$ - - -
$F43$	4*a*, 4*b*	
$F\frac{4}{m}\bar{3}\frac{2}{m}$	4*a*, 4*b*	

NaCl under consideration, so that the possibility can be safely ignored. The possibility of variate-atom equipoints, however, should be considered in assigning atoms to equipoints in any structure analysis.

The possible 4-fold equipoints in the uneliminated space groups are listed in Table 3. The various ways they can be combined two at a time, corresponding to one for Na, another for Cl, are shown in Table 4. It will be observed that there are six different combinations. Two of these define halite structures, with different origins. The other four define a pair of sphalerite structures with different origins, and another pair which can be derived from them by inversion in the origin; for convenience these are termed sphalerite and "inverse" sphalerite. [They are not enantiomorphic since they have symmetry planes parallel to (110).] With appropriate choice of origin, all combinations can be represented by halite, sphalerite, and inverse-sphalerite structures. Since structures related by an inversion give identical diffraction intensities, it is sufficient to consider only one possible sphalerite structure. Omitting

† Tom F. W. Barth and E. Posnjak. *The spinel structure: An example of variate atom equipoints.* J. Wash. Acad. Sci. **21** (1931) 255–258.

§ E. Posnjak and Tom F. W. Barth. *A new type of crystal fine-structure: Lithium ferrite ($Li_2O{\cdot}Fe_2O_3$).* Phys. Rev. **38** (1931) 2234–2239.

¶ M. J. Buerger. *The temperature-structure-composition behavior of certain crystals.* Proc. Nat. Acad. Sci. U. S. **20** (1934) 444–453.

the possibility of variate-atom equipoints, then, NaCl can have only one of two possible structures, the halite structure or the sphalerite structure. Both of these structures are symmetry fixed, and the problem of structure determination is reduced to deciding between the two possibilities.

Incidentally, the space groups possible for NaCl have now been reduced to two, namely $F\bar{4}3m$ for the sphalerite structure or $F\,4/m\,\bar{3}\,2/m$ for for the halite structure. The reason for this is as follows: The sphalerite structure can be represented as made up of equipoints $4a+4c$, and while these both occur in two space groups, as shown in Table 3, the symmetry

Table 4
Possible equipoint combinations for NaCl

Combination	Coordinates of origin point of each equipoint		Structure type
$4a+4b$	000	$\frac{1}{2}\frac{1}{2}\frac{1}{2}$	halite (1)
$4a+4c$	000	$\frac{1}{4}\frac{1}{4}\frac{1}{4}$	sphalerite (1)
$4a+4d$	000	$\frac{3}{4}\frac{3}{4}\frac{3}{4}$	"inverse" sphalerite (2)
$4b+4c$	$\frac{1}{2}\frac{1}{2}\frac{1}{2}$	$\frac{1}{4}\frac{1}{4}\frac{1}{4}$	"inverse" sphalerite (3)
$4b+4d$	$\frac{1}{2}\frac{1}{2}\frac{1}{2}$	$\frac{3}{4}\frac{3}{4}\frac{3}{4}$	sphalerite (4)
$4c+4d$	$\frac{1}{4}\frac{1}{4}\frac{1}{4}$	$\frac{3}{4}\frac{3}{4}\frac{3}{4}$	halite (2)

of the combination is that of the highest symmetry, namely $F\bar{4}3m$. Similarly, the halite structure can be represented as made up of equipoints $4a+4b$. While these occur in all five space groups, as shown in Table 3, the symmetry of the combination is that of the highest symmetry, namely $F\,4/m\,\bar{3}\,2/m$. Table 3 shows that the two uneliminated space groups, $F\bar{4}3m$ and $F\,4/m\,\bar{3}\,2/m$ both belong to the same diffraction symbol. The other diffraction symbol, along with its two space groups, could have been eliminated at the very outset if a moving-film method of recording had provided the necessary diffraction-symmetry information. It is this failure to eliminate space groups which are capable of elimination by simple diffraction means, that is a particularly inconvenient feature of structure determination by the rotating-crystal and powder methods.

Distinguishing between the two possible structures. The Na and Cl atoms for the two possible structures are located in the following positions:

$$\text{halite structure}\begin{cases}\text{Na: } 4a & 000 \quad 0\tfrac{1}{2}\tfrac{1}{2} \quad \tfrac{1}{2}0\tfrac{1}{2} \quad \tfrac{1}{2}\tfrac{1}{2}0,\\ \text{Cl: } 4b & \tfrac{1}{2}\tfrac{1}{2}\tfrac{1}{2} \quad \tfrac{1}{2}00 \quad 0\tfrac{1}{2}0 \quad 00\tfrac{1}{2};\end{cases} \tag{1}$$

$$\text{sphalerite structure}\begin{cases}\text{Na: } 4a & 000 \quad 0\tfrac{1}{2}\tfrac{1}{2} \quad \tfrac{1}{2}0\tfrac{1}{2} \quad \tfrac{1}{2}\tfrac{1}{2}0,\\ \text{Cl: } 4c & \tfrac{1}{4}\tfrac{1}{4}\tfrac{1}{4} \quad \tfrac{1}{4}\tfrac{3}{4}\tfrac{3}{4} \quad \tfrac{3}{4}\tfrac{1}{4}\tfrac{3}{4} \quad \tfrac{3}{4}\tfrac{3}{4}\tfrac{1}{4}.\end{cases} \tag{2}$$

It is plain from Chapter 2 that the test of a structure is that the intensities computed for it are the same as the observed intensities. It follows that the incorrect structure can be eliminated by showing that it gives computed intensities not in harmony with the observed intensities. Actually, it is unnecessary to carry out the computation for all reflections; usually the comparison can be made for some set of reflections which are easily computed. In this instance, the easiest comparison is indicated by the following considerations: The sphalerite structure lacks a center of symmetry so that computations of the intensities of its general spectra

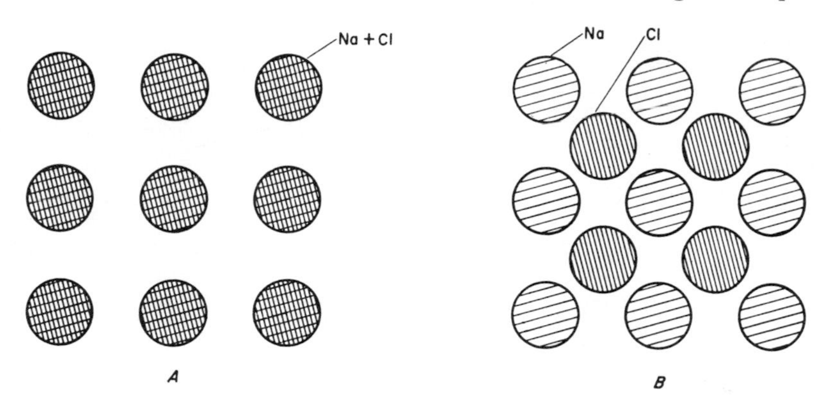

Fig. 2.

Projection of halite-type structure parallel to a_3.

Projection of sphalerite-type structure parallel to a_3.

require separate computations of the real and imaginary parts, then combining these along the lines indicated in Chapter 10. In noncentrosymmetrical structures, however, it is often possible to find special projections of the structure which are centrosymmetrical. This is true of the sphalerite structure, whose projection on (001) is centrosymmetrical, as illustrated in Fig. 2*B*. This means that the imaginary part of the structure factor vanishes for the *hk*0 reflections.

The amplitude of the reflection is given by (4) of Chapter 10, namely

$$F_{hk0} = \sum_j f_j \, e^{i2\pi(hx_j+ky_j)}. \tag{3}$$

The j atoms include 4 of Na and 4 of Cl, so that (3) can be expanded to

$$F_{hk0} = \left\{ f_{\mathrm{Na}} \sum e^{i2\pi(hx+ky)} \right\} + \left\{ f_{\mathrm{Cl}} \sum e^{i2\pi(hx+ky)} \right\}. \tag{4}$$

Here the summation is over the four appropriate coordinates of (1) or (2). Since the projection is centrosymmetrical, the exponentials become

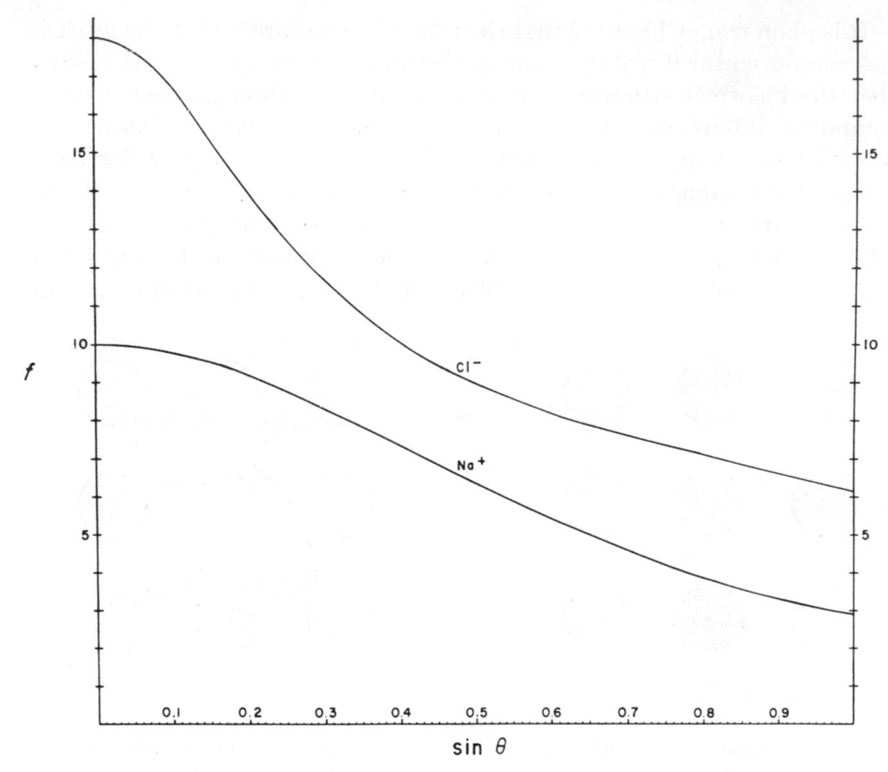

Fig. 3.

cosines, and (4) takes the more simple form

$$F_{hk0} = \left\{f_{\mathrm{Na}} \sum \cos 2\pi(hx+ky)\right\} + \left\{f_{\mathrm{Cl}} \sum \cos 2\pi(hx+ky)\right\}. \tag{5}$$

When the actual coordinates of (1) are substituted for the x's and y's in (5), this can be reduced to a very limited number of types, namely

Halite structure:

$$F_{hk0} = 4(f_{\mathrm{Na}}+f_{\mathrm{Cl}}), \qquad h+k = 2n \quad (n \text{ is an integer}). \tag{6}$$

Sphalerite structure:

$$F_{hk0} = 4(f_{\mathrm{Na}}+f_{\mathrm{Cl}}), \qquad h+k = 4n, \tag{7}$$

$$F_{hk0} = 4(f_{\mathrm{Na}}-f_{\mathrm{Cl}}), \qquad h+k = 4n+2. \tag{8}$$

The f values are functions of $\sin \theta/\lambda$ (see Chapter 10), and numerical values can be found in suitable tables.† It is convenient to plot such numerical values for the radiation actually employed in the diffraction experiment, as shown in Fig. 3. The appropriate values for each reflection hkl can then be found as a function of $\sin \theta_{hkl}$.

The computations indicated in (6), (7), and (8) are so simple that they can be performed in tabular fashion as shown in Table 5. The first

Table 5
Comparison of $hk0$ reflections for two models of NaCl
(reflections in order of increasing distance from layer origin)

1	2	3	4	5		6	7	8	
Reflection	$\sin \theta$	f_{Na}	f_{Cl}	$\frac{F}{4}$		m	$L \cdot p$	$I = mLpF^2$	
				Halite type	Sphalerite type			Halite type	Sphalerite type
200	0.274	8.6	12.2	20.8	3.6	4	1.6	44,000	1,300
220	0.388	7.5	10.2	17.7	17.7	4	1.0	20,000	20,000
400	0.548	5.9	8.6	14.5	14.5	4	0.63	8,500	8,500
420	0.613	5.3	8.1	13.4	2.8	8	0.55	13,000	550
440	0.775	4.0	7.2	11.2	11.2	4	0.53	4,300	4,300
600	0.822	3.7	7.0	10.7	3.3	4	0.60	4,400	420
620	0.866	3.5	6.8	10.3	10.3	8	0.71	9,600	9,600
640	0.989	3.0	6.2	9.2	3.2	8	2.7	29,000	3,500

column of Table 5 lists the observed reflections of the type $hk0$ (i.e., the reflections on the zero layer of the rotating-crystal photograph). The second column is the value of $\sin \theta_{hkl}$ computed from

$$\sin \theta = \frac{\sigma}{2}$$

$$= \frac{1}{2} \frac{\lambda}{a} (h^2+k^2+l^2)^{1/2}, \tag{9}$$

or derived more simply and with sufficient accuracy for the purpose by reading the cylindrical coordinates of the reciprocal lattice with the aid of a Bernal chart§ and using the relationship

$$\sin \theta = \tfrac{1}{2}\sigma$$

$$= \tfrac{1}{2} \sqrt{\xi^2+\zeta^2}. \tag{10}$$

† G. D. Rieck and Caroline H. MacGillavry. *International tables for x-ray crystallography.* Vol. 3. (Kynoch Press, Birmingham, England, 1960)

§ M. J. Buerger. *X-ray crystallography.* (John Wiley and Sons, New York, 1942) 145.

For these $hk0$ reflections $\zeta = 0$, and this degenerates to simply

$$\sin \theta_{hk0} = \frac{\xi}{2}. \tag{11}$$

Columns 3 and 4 are the values of f for Na and Cl as taken from the plot of Fig. 3. The next columns are the sum or difference of the f's as required by (6), (7), or (8). The result is now in the form of $F/4$. The remaining columns convert this into intensity by squaring and applying multiplicity, Lorentz, and polarization factors.

The multiplicity factor, m, is easily determined by noting the total number of points which cut the circle of reflection at the same place in Fig. 1. The corrections for the Lorentz and polarization factors are most readily obtained by laying the Kaan and Cole chart directly over the rotation photograph as described in Chapter 7.

A comparison of the zero layer of the rotating-crystal photograph shows that the blackening of the sequence of spots is definitely not consistent with that in the sphalerite column of Table 5, but is in good agreement with that in the halite column. This then provides the criterion for deciding that the structure of NaCl is the halite type. Incidentally, its space group, from the earlier discussion, must be the one in Table 3 of highest symmetry containing the equipoints $4a + 4b$, specifically $F\,4/m\,\bar{3}\,2/m$.

Confirmation of the structure determination. To fully confirm a structure determination requires proof that all observed intensities are in harmony with those computed from the proposed structure. In this instance, the intensities of all reflections hkl which are within range of recording on the rotating-crystal photograph must be computed, using the coordinates given in (1).

The amplitude of the general reflection hkl is given by (21) of Chapter 10. Since the structure is centrosymmetrical, this takes the more simple form (provided a center is used as origin)

$$F_{hkl} = \sum_j f_j \cos 2\pi(hx_j + ky_j + lz_j), \tag{12}$$

where the summation is taken over all j atoms of the cell. Since the atoms are of two kinds, the f's can be taken outside the summation sign, giving specifically, for NaCl,

$$F_{hkl} = \left\{ f_{\mathrm{Na}} \sum \cos 2\pi(hx + ky + lz) \right\} + \left\{ f_{\mathrm{Cl}} \sum \cos 2\pi(hx + ky + lz) \right\}, \tag{13}$$

where the summation is now limited to the coordinates x, y, and z of the four atoms whose specific values of x, y, and z are listed in (1).

There are two different ways to proceed at this point. If the structure is a complicated one, with atoms in general positions, one can find in suitable references† the general expression for the trigonometric terms inside the summation, which is the symmetry factor for the space group. In cases of high symmetry like this one, such general symmetry factors are highly complicated. For the space group of NaCl, namely $F\,4/m\,\overline{3}\,2/m$, it is a complicated trigonometric expression about four lines long. To utilize this general expression for a very few atoms in very special positions is much more trouble than it is worth. An alternative way of proceeding is to simply substitute in (13) the specific values of x, y, and z for the four atoms each of Na and Cl given by (1).

Actually, in very simple structures such as this one, where the atoms occupy an equipoint whose rank is identical with the number of lattice points per cell, one can be sure that the structure composed of n kinds of atoms can be decomposed into n lattice arrays of atoms, in this case specifically two. Each lattice array scatters in phase for all reflections permitted by the lattice, and the only problem is to take account of the interaction between the two lattice arrays, which amounts to compounding their scattered amplitudes with proper phase relations. The phase difference in the scattering [see (65) of Chapter 10] is $e^{-i2\pi(h\Delta x+k\Delta y+l\Delta z)}$ where Δx, Δy, Δz, are the differences in coordinates of representative lattice points. From (1), taking the representative point on the Na array as 000, and the representative point on the Cl array as $\frac{1}{2}\frac{1}{2}\frac{1}{2}$, this phase difference is

$$\begin{aligned} e^{\Delta\phi} &= e^{-i2\pi(h\cdot\frac{1}{2}+k\cdot\frac{1}{2}+l\cdot\frac{1}{2})} \\ &= e^{-i\pi(h+k+l)} \\ &= \begin{matrix} +1 & \text{for } h+k+l \text{ even} \\ -1 & \text{for } h+k+l \text{ odd.} \end{matrix} \end{aligned} \tag{14}$$

By either the long-hand computation of the last part of the previous paragraph, or by the reasoning of this paragraph, it is evident that for all permissible reflections of the structure (which are limited to reflections for which h, k, and l are all even or all odd, due to the face-centered mode of the lattice), the Na and Cl atoms scatter in phase for reflections having $h+k+l$ even, but scatter in opposite phase for reflections having $h+k+l$ odd. This can be expressed analytically as the specific solution of (13)

† Norman F. M. Henry and Kathleen Lonsdale. *International tables for x-ray crystallography.* Vol. 1. (Kynoch Press, Birmingham, England, 1952)

in the form

$$F_{hkl} = 4f_{Na} + 4f_{Cl}, \qquad h+k+l \text{ even},$$

$$F_{hkl} = 4f_{Na} - 4f_{Cl}, \qquad h+k+l \text{ odd}, \tag{15}$$

the whole set of reflections being limited by the lattice to h, k, and l either all even or all odd.

The computations are so simple that they can be carried out again in tabular form, as shown in Table 6. The reflections are arranged by layer lines for convenience in making the comparison with the photograph.

Table 6
Computed intensities for NaCl structure
(reflections in order of increasing distance from layer origin)

1	2	3	4	5	6	7	8
Reflection	$\sin\theta$	f_{Na}	f_{Cl}	$\frac{F}{4}$	Multiplicity factor m	Lorentz-polarization factor $L{\cdot}p$	Intensity $I = mLpF^2$
200	0.274	8.6	12.2	20.8	4	1.6	44,000
220	0.388	7.5	10.2	17.7	4	1.0	20,000
400	0.548	5.9	8.6	14.5	4	0.63	8,500
420	0.613	5.3	8.1	13.4	8	0.55	13,000
440	0.775	4.0	7.2	11.2	4	0.53	4,300
600	0.822	3.7	7.0	10.7	4	0.60	4,400
620	0.866	3.5	6.8	10.3	8	0.71	9,600
640	0.989	3.0	6.2	9.2	8	2.7	29,000
111	0.232	8.9	13.1	4.2	4	1.7	1,900
311	0.455	6.8	9.4	2.6	8	0.89	770
331	0.597	5.4	8.2	2.8	4	0.59	300
511	0.713	4.4	7.5	3.1	8	0.52	640
531	0.810	3.8	7.1	3.3	8	0.60	840
{551, 711}	0.976	3.0	6.3	3.3	{4, 8}	2.7	5,600
202	0.388	7.5	10.2	17.7	4	1.6	32,000
222	0.474	6.6	9.2	15.8	4	1.0	16,000
402	0.613	5.3	8.1	13.4	4	0.67	7,700
422	0.671	4.8	7.7	12.5	8	0.61	12,000
442	0.822	3.7	7.0	10.7	4	0.62	4,500
602	0.866	3.5	6.8	10.3	4	0.91	6,200
622	0.909	3.3	6.6	9.9	8	1.3	16,000
313	0.597	5.4	8.2	2.8	8	1.0	1,000
333	0.713	4.4	7.5	3.1	4	0.86	530
513	0.810	3.8	7.1	3.3	8	1.1	1,500

Sin θ can be found as indicated by either (9) or (10). The remainder of the computation is similar to that already described for Table 5.

A comparison of the computed reflection intensities, as tabulated in Table 6, with the blackening of the spots on the film, shows that there is good agreement between computed and observed intensities, so that the structure deduced may be regarded as confirmed.

Determination of structures with separable parameters, illustrated by the structure of marcasite, FeS_2

In case certain sets of spectra are functions of only one parameter, this parameter may be said to be separable from the others. In this case it is not necessary to try to find the parameter blindly by trial and error, for the intensities of this set of spectra are functions of this one parameter alone. Thus, one can be solved for in terms of the other. The most illuminating way of carrying out the solution is by graphical means. If the intensity data are normalized to absolute values, the solution can be carried out analytically also, as shown later.

One of the convenient features of the determination of a crystal structure having separable parameters is that it can be carried out with a minimum of effort expended on the determination of experimental intensities. All that is required in the way of intensity data is a list of the observed intensities on a relative basis arranged in order of decreasing intensity.

A good example of the use of this method is afforded by the determination of the structure of the mineral marcasite,[1] FeS_2. All marcasite parameters can be fixed by using reflections from prism-zone planes alone, so that all reflections required for the parameter determination can be found on zero-level photographs. An ideal way of recording zero-level reflections is by means of precession photographs.† These are accordingly assumed to be used in the following example.

Unit cell, space group, and equipoints. Marcasite is orthorhombic and conforms to space group $P\,2_1/n\,2_1/n\,2/m$, D_{2h}^{12}, Fig. 4. Its unit cell has the following dimensions:[2]

$$a = 4.45\ \text{Å},$$

$$b = 5.40,$$

$$c = 3.38.$$

Using relation (3) of Chapter 9, it is found that the unit cell contains 2 FeS_2.

† M. J. Buerger. *The photography of the reciprocal lattice.* Am. Soc. for X-Ray and Electron Diffraction, Monograph No. 1 (1944).

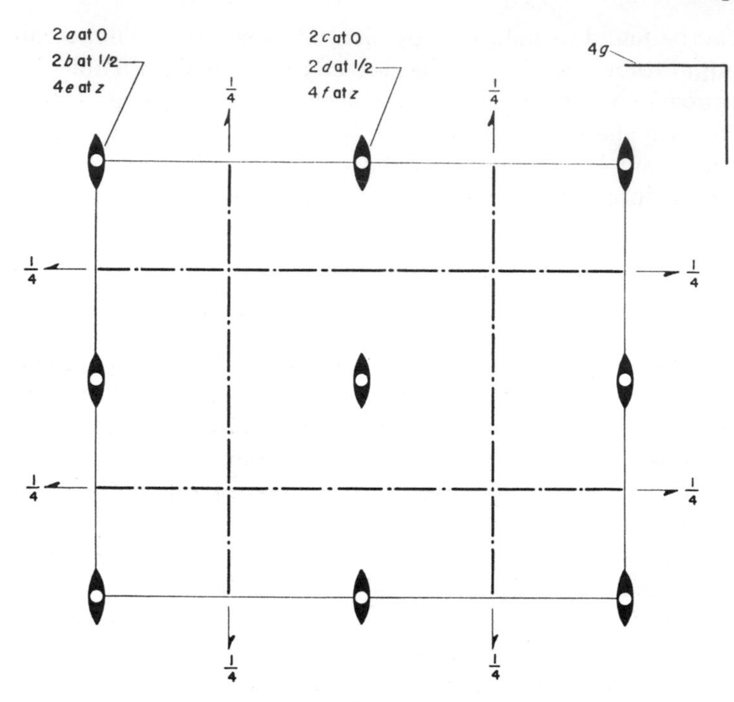

Fig. 4.
Equipoints of $P2_1/n\ 2_1/n\ 2/m$.

A discussion of the possible distribution of the 2 Fe and 4 S atoms among the equipoints of $P\ 2_1/n\ 2_1/n\ 2/m$ was given in Chapter 9. The conclusions of this discussion are now succinctly presented in Table 7. This shows that the several atoms can be distributed in a manner consistent with the symmetry in six different ways among the equipoints. The first step in the structure determination will be to eliminate five of these combinations on the basis of intensity considerations.

Table 7
Possible distributions of atoms in marcasite

Symmetry-determined structures		Structures with one variable parameter, z_S		Structure with two variable parameters, x_S and y_S	
Fe	S	Fe	S	Fe	S
$2a$	$2b+2c$	$2a$	$4e$	$2a$	$4g$
$2a$	$2b+2d$	$2a$	$4f$		
$2a$	$2c+2d$				

Intensities. According to (4) of Chapter 10, the amplitude of a reflection is given by

$$F_{hkl} = \sum_j f_j \, e^{i2\pi(hx_j+ky_j+lz_j)}. \tag{16}$$

In this case, the different atoms j comprise 2 atoms of Fe and 4 atoms of S. For each of these kinds of atoms, the f_j is constant and can be taken

Table 8

Symmetry factor for the equipoints of space group $P\frac{2_1}{n}\frac{2_1}{n}\frac{2}{m}$

Equipoint	Symmetry factor	
	$h+k+l$ even	$h+k+l$ odd
2*a*	2	0
2*b*	$(-1)^l\ 2$	0
2*c*	$(-1)^k\ 2$	0
2*d*	$(-1)^h\ 2$	0
4*e*	$4 \cos 2\pi lz$	0
4*f*	$(-1)^h\ 4 \cos 2\pi lz$	0
4*g*	$4 \cos 2\pi hx \cos 2\pi ky$	$-4 \sin 2\pi hx \sin \pi ky$

outside the summation sign, so that for marcasite, (16) has the specific form

$$F_{hkl} = f_{\text{Fe}} \sum e^{i2\pi(hx+ky+lz)} + f_{\text{S}} \sum e^{i2\pi(hx+ky+lz)}. \tag{17}$$

The term within the summation sign is simply the appropriate symmetry factor or factors, as discussed in Chapter 10. The form of the symmetry factor for space group $P\,2_1/n\,2_1/n\,2/m$, is

$$S = m \cos 2\pi hx \cos 2\pi ky \cos 2\pi lz \quad \text{for } h+k+l \text{ even}, \tag{18}$$

and

$$S = -m \sin 2\pi hx \sin 2\pi ky \cos 2\pi lz \quad \text{for } h+k+l \text{ odd}, \tag{19}$$

where m is the rank of the equipoint, and x, y, and z are the coordinates of one of its atoms. The coordinates of the equipoints involved in the combinations of Table 7 are given in Table 5 of Chapter 9. By substituting these values of x, y, and z into (18) and (19), it is found that the symmetry factor assumes the several simple values shown in Table 8. The appropriate values of f_{Fe} and f_{S} can be found in standard references.[†]

[†] G. D. Rieck and Caroline H. MacGillavry. *International tables for x-ray crystallography.* Vol. 3. (Kynoch Press, Birmingham, England, 1960)

The variation of these values for sin θ for the Mo$K\alpha$ radiation used in making the precession photographs is shown graphically in Fig. 5.

Elimination of incorrect equipoint combinations. It will be observed that the Fe is in symmetry-fixed positions in all six equipoint combinations given in Table 7. The only variables involved are z_S for the structures $2a\ 4e$ and $2a\ 4f$, and both x_S and y_S for the structure $2a\ 4g$. Therefore three of the structures have all spectra symmetry fixed, two have the $hk0$ spectra symmetry fixed, and the last has only the $00l$

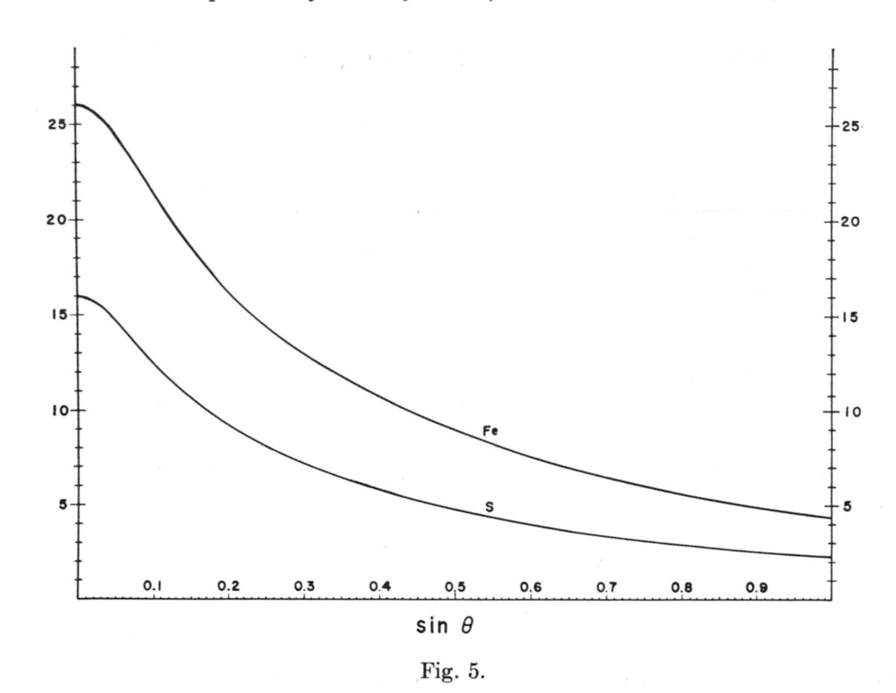

Fig. 5.

spectra symmetry fixed. The intensities of the symmetry-fixed spectra of all six structures could be computed and compared with the observed intensities with the object of rejecting any structures which do not give intensities which conform with the observed ones. Actually it is sufficient to compute merely the intensities of the special reflections $h00$, $0k0$, and $00l$. This has the advantage of considerably reducing the labor, since these classes of spectra are present for the space group $P\,2_1/n\,2_1/n\,2/m$ only when the indices are even, and Table 8 shows that the amplitude (17) reduces to the very simple form

$$F = 2f_{Fe} + 4f_S \tag{20}$$

for these particular symmetry-fixed reflections. The computation of these F's and their conversion into intensities with the aid of

$$I = LpF^2 \tag{21}$$

are shown in Table 9.

Table 9
Intensities computed for pinacoid reflections of possible marcasite structures, compared with observed intensities

Reflection	$\sin\theta$	f_{Fe}	f_S	F	$\frac{1}{Lp}$	Lp	Computed intensity, $I = LpF^2$: Arrangements *a bc*, *a bd*, *a cd*	Arrangements *a e*, *a f*	Arrangements *a g*	Observed intensity
200	0.160	18.2	10.1	76.8	0.53	1.88	11100	11100	variable	absent
400	0.320	12.5	7.3	54.2	0.92	1.09	3200	3200	with	medium
600	0.480	9.2	5.4	39.0	0.50	1.82	2800	2800	x	medium
020	0.132	19.8	11.0	85.6	0.34	3.9	29000	29000	variable	strong
040	0.264	14.0	8.1	60.4	0.71	1.41	5100	5100	with	absent
060	0.395	10.8	6.3	46.8	0.99	1.01	2200	2200	y	weak
002	0.211	15.8	9.0	67.6	0.67	1.49	6800	variable	6800	strong
004	0.422	10.3	6.1	45.0	0.85	1.18	2400	with z	2400	medium-strong

An estimate of the relative intensities of $h00$, $0k0$, and $00l$ taken from the three axial precession photographs is listed in the farthest-right column of Table 9. It will be observed that the only structure which provides intensities not in actual conflict with the observed intensities is arrangement *a g*. This definitely eliminates all the other equipoint combinations.

Determination of sulfur parameters. As a result of the discussion just given, the only uneliminated structure is the one which can be described as follows:

$$\begin{aligned} &\text{Fe in } 2a \text{ (inversion centers) at } 000;\ \tfrac{1}{2}\tfrac{1}{2}\tfrac{1}{2}. \\ &\text{S in } 4g \text{ (reflection planes) at } xy0;\ \tfrac{1}{2}+x,\ \tfrac{1}{2}-y,\ \tfrac{1}{2}; \\ &\qquad\qquad\qquad\qquad\qquad\quad \bar{x}\bar{y}0;\ \tfrac{1}{2}-x,\ \tfrac{1}{2}+y,\ \tfrac{1}{2}. \end{aligned} \tag{22}$$

It remains to determine the unknown sulfur parameters x and y. This can be readily done by making use of all reflections which do not involve two variables at a time. Recalling that the factor within the summation sign of (17) is merely the symmetry factor given in Table 8, the specific value of (17) for the combination $2a$ $4g$ is had by substituting

Table 10
Data for plotting the variation of $I_{h0l}^{1/2}$ as a function of x, Fig. 6

Reflection	$\sin\theta$	f_{Fe}	f_S	S_{Fe}	Precession $\frac{1}{Lp}$, c^* horizontal	$\sqrt{Lp}$	$f_{Fe}\,S_{Fe}\sqrt{Lp}$	$f_S\,S_S\sqrt{Lp}$
002	0.210	15.8	9.0	2	0.68	1.21	38	43
004	.420	10.3	6.0	2	.88	1.07	22	26
101	.133	19.7	11.0	2	.40	1.58	62	$70\cos 2\pi x$
103	.328	12.3	7.2	2	.92	1.04	26	$30\cos 2\pi x$
200	.160	18.2	10.1	2	.42	1.54	56	$62\cos 2\pi 2x$
202	.268	13.8	8.0	2	.77	1.14	31	$36\cos 2\pi 2x$
204	.450	9.7	5.7	2	.78	1.13	22	$26\cos 2\pi 2x$
301	.261	14.0	8.1	2	.72	1.18	33	$38\cos 2\pi 3x$
303	.398	10.8	6.3	2	.95	1.03	22	$26\cos 2\pi 3x$
400	.320	12.5	7.3	2	.86	1.08	27	$30\cos 2\pi 4x$
402	.385	10.8	6.4	2	.97	1.02	22	$26\cos 2\pi 4x$
501	.414	10.4	6.1	2	.98	1.01	21	$25\cos 2\pi 5x$
600	.480	9.2	5.4	2	.70	1.19	22	$26\cos 2\pi 6x$

in (17) the appropriate values for Fe in $2a$ and S in $4g$ provided by Table 8. This gives

$$F_{hkl} = 2f_{Fe} + 4f_S \cos 2\pi hx \cos 2\pi ky, \qquad h+k+l \text{ even}, \tag{23}$$

$$F_{hkl} = \qquad -4f_S \sin 2\pi hx \sin 2\pi ky, \qquad h+k+l \text{ odd}. \tag{24}$$

Since the reflections not involving x are $0kl$, and the reflections not involving y are $h0l$, and since the glide planes of $P\,2_1/n\,2_1/n\,2/m$ permit only those classes of reflections to be present for which $k+l$ and $h+l$ are even, respectively, the amplitudes involving only one variable are

$$F_{h0l} = 2f_{Fe} + 4f_S \cos 2\pi hx, \tag{25}$$

and

$$F_{0kl} = 2f_{Fe} + 4f_S \cos 2\pi ky. \tag{26}$$

The intensities can be computed by simply squaring F and multiplying it by the Lorentz and polarization factors as indicated in (21). This gives the intensities in units based upon the amplitude scattered by a single electron.

The most illuminating way of determining the values of x is to plot (25) as a function of x in order to see where all observed intensities are

Table 11
Data for plotting the variation of $I_{0kl}^{1/2}$ as a function of y, Fig. 7

Reflection	$\sin\theta$	f_{Fe}	f_S	S_{Fe}	Precession $\frac{1}{Lp}$, c^* horizontal	$\sqrt{Lp}$	$f_{Fe}\,S_{Fe}\sqrt{Lp}$	$f_S\,S_S\sqrt{Lp}$
002	0.210	15.8	9.0	2	0.68	1.21	38	43
004	.420	10.3	6.0	2	.88	1.07	22	26
011	.125	20.1	11.2	2	.39	1.60	64	72 cos $2\pi y$
013	.321	12.4	7.2	2	.91	1.05	26	30 cos $2\pi y$
020	.131	19.8	11.0	2	.34	1.71	68	75 cos $2\pi 2y$
022	.249	14.4	8.3	2	.74	1.16	33	39 cos $2\pi 2y$
024	.441	9.9	5.8	2	.81	1.11	22	26 cos $2\pi 2y$
031	.222	15.3	8.9	2	.62	1.27	39	45 cos $2\pi 3y$
033	.372	11.3	6.6	2	.95	1.03	23	27 cos $2\pi 3y$
040	.262	14.0	8.1	2	.70	1.19	33	39 cos $2\pi 4y$
042	.336	12.1	7.1	2	.91	1.05	25	30 cos $2\pi 4y$
044	.495	9.0	5.3	2	.25	2.00	36	42 cos $2\pi 4y$
051	.344	11.9	6.9	2	.92	1.04	25	29 cos $2\pi 5y$
053	.457	9.8	5.7	2	.80	1.12	22	26 cos $2\pi 5y$
060	.393	10.9	6.4	2	.99	1.01	22	26 cos $2\pi 6y$
062	.446	9.8	5.7	2	.90	1.05	20	24 cos $2\pi 6y$
071	.469	9.4	5.5	2	.80	1.12	21	25 cos $2\pi 7y$

satisfied. The plot of the variation of intensities in the form of

$$\sqrt{I} = \sqrt{Lp}\,|F| \tag{27}$$

of all $h0l$ spectra as a function of x and which can be recorded on the 30° precession photograph is shown in Fig. 6. This plot is readily prepared from the data in Table 10. A corresponding plot for the variation of the intensities of the $0kl$ spectra with y is shown in Fig. 7, which is prepared from the data in Table 11. In Table 12 are listed the observed relative intensities. An examination of Figs. 6 and 7 shows that these intensity

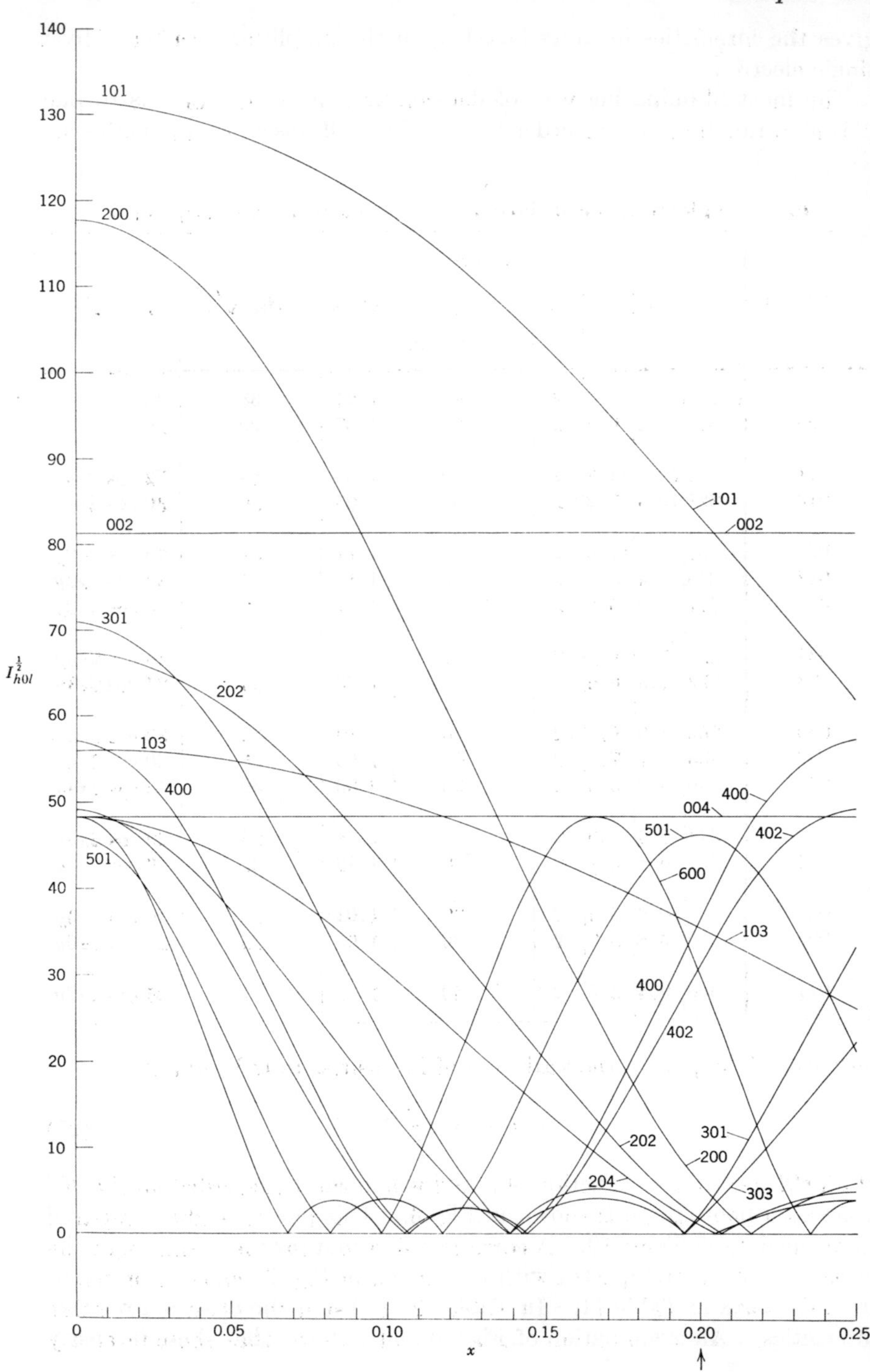

Fig. 6.
Variation of $I^{1/2}_{h0l}$ with sulfur parameter x; marcasite, Mo$K\alpha$.

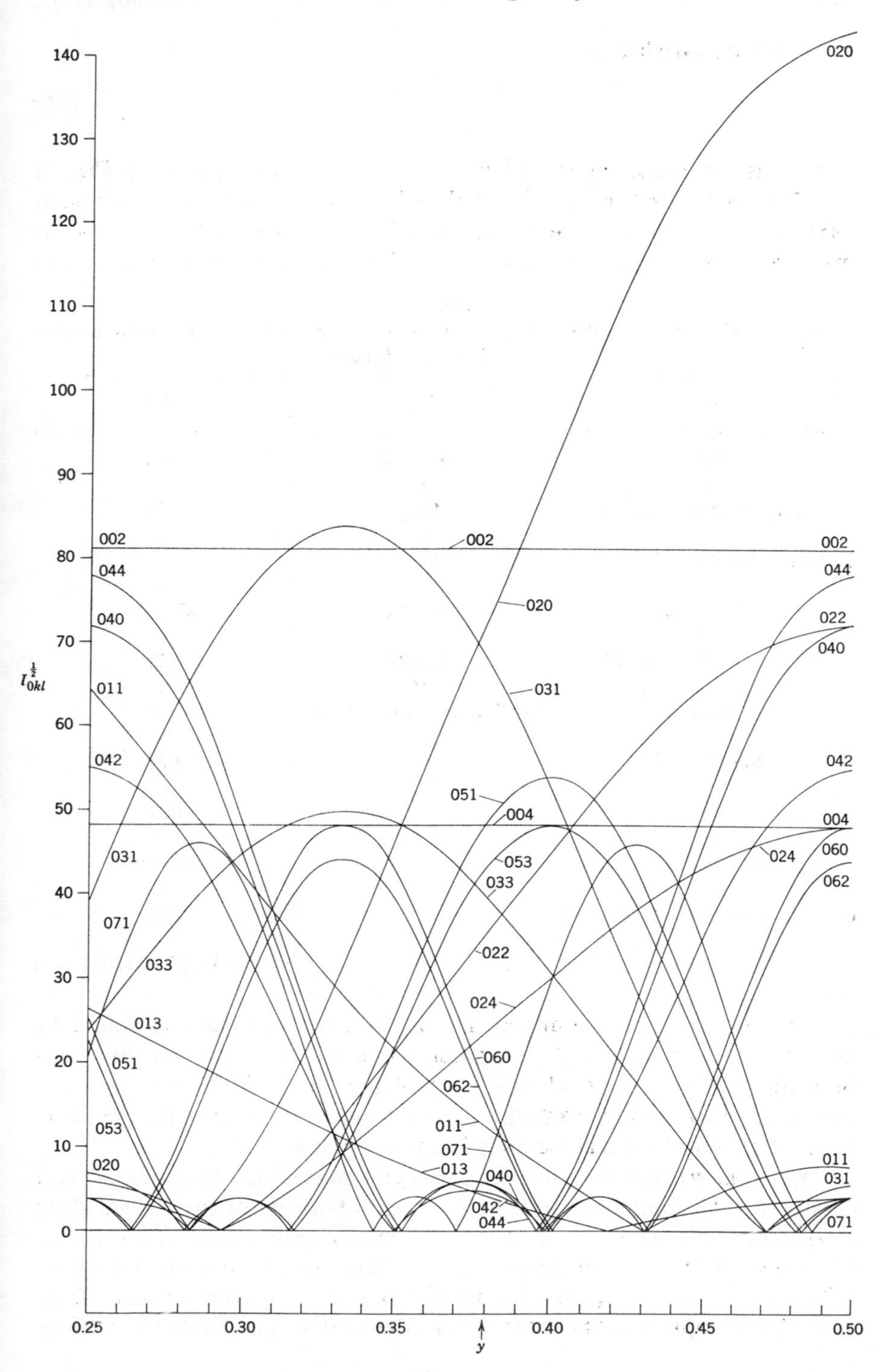

Fig. 7.

Variation of $I_{0kl}^{1/2}$ with sulfur parameter y; marcasite, Mo$K\alpha$.

relations are satisfied at

$$x = 0.20, \quad y = 0.38. \tag{28}$$

In this connection it should be observed that the graphs of Figs. 6 and 7 are computed from (27), which is based on the amplitudes given in (25) and (26). These are cosine functions. Since the cosine is symmetrical about 0 and π, there are two solutions symmetrically related to

Table 12
Observed relative intensities on marcasite, zero-level precession photographs using Mo$K\alpha$ radiation

$hk0$	$h0l$	$0kl$
020	101, 002	002
V	V	V
120, 110, 310, 240, 130	004	020
V	V	V
530, 620, 220, 400, 600, 150, 160	501	031
V	V	V
610, 420, 260	103, 400	051, 004
V	V	V
320, 210, 410, 430, 060	402, 600	053
V	V	V
230, 170	200, 202, 204, 301, 303 = 0	033
V		V
450, 360, 270		022
V		V
440		024
V		V
250, 330, 200		060, 062
V		V
040, 140, 340, 350, 510, 520, 540 = 0		011
		V
		013, 040, 042, 044, 071 = 0

$x = 0$ and $x = \frac{1}{2}$, and similarly for y. When these pairs of solutions have been found for x and y respectively, a decision between them can be made on the basis of reflections involving sines. Table 8 shows that any of the reflections for which $h+k+l$ is odd can be used for this purpose. Some of these will be used incidentally now.

A reasonably extensive check on the correctness of the structure derived may be had by computing all the observed intensities of the special class of spectra not used so far in the structure determination, namely, the $hk0$ reflections. The amplitudes of these are computed with the aid of (23) and (24) according to whether the sum $h+k$ is even or odd. The intensities computed from these amplitudes are listed in Table 13. When

Table 13

Computation of intensities of $hk0$ reflections for marcasite with sulfur parameters $\begin{Bmatrix} x = 0.20 \\ y = 0.38 \end{Bmatrix}$, Mo$K\alpha$ radiation

Reflection	$\sin\theta$	f_{Fe}	f_S	$\frac{S_{Fe}}{2}$		$\frac{S_S}{4}$		$F = f_{Fe} S_{Fe} + f_S S_S$	$\frac{1}{Lp}$	$\sqrt{Lp}$	$\sqrt{I} = \sqrt{Lp}\, F$
010											
020	0.131	19.8	11.0	1	1	1	0.038	41.3	0.34	1.9	78.5
030											
040	.263	14.0	8.1	1	1	1	−.997	−4.2	.71	1.41	−5.9
050											
060	.394	10.8	6.3	1	1	1	−.113	−18.7	.99	1.01	−18.8
070											
100											
110	.103	21.3	12.1	1	1	0.309	−.720	31.8	.32	1.7	54.1
120	.153	18.5	10.3	0	0	.951	−.999	40.0	.43	1.5	60.0
130	.212	15.7	9.0	1	1	.309	.666	38.8	.58	1.31	50.8
140	.274	13.7	8.0	0	0	.951	−.075	2.3	.75	1.15	2.6
150	.337	12.1	7.0	1	1	.309	.771	30.9	.91	1.05	32.4
160	.401	10.7	6.3	0	0	.951	.994	−13.8	.98	1.01	−13.9
170	.465	9.4	5.5	1	1	.309	−.608	14.3	.82	1.10	15.7
200	.160	18.2	10.1	1	1	−.809	1	3.7	.53	1.37	5.1
210	.178	17.1	9.7	0	0	.588	.694	−15.4	.56	1.33	−20.5
220	.207	15.9	9.1	1	1	−.809	.038	30.4	.62	1.27	38.6
230	.253	14.3	8.2	0	0	.588	.746	−14.4	.72	1.18	−16.9
240	.307	12.8	7.5	1	1	−.809	−.997	49.8	.85	1.08	53.8
250	.364	11.5	6.7	0	0	.588	−.637	10.0	.96	1.02	10.2
260	.424	10.2	6.0	1	1	.809	−.113	22.6	.98	1.01	22.8
270	.486	9.1	5.3	0	0	.588	−.794	9.9	.55	1.35	13.4
300											
310	.249	14.4	8.3	1	1	−.809	−.720	48.2	.77	1.14	54.9
320	.274	13.7	8.0	0	0	−.588	−.999	−18.8	.81	1.11	−20.8
330	.310	12.7	7.4	1	1	−.809	.666	9.4	.88	1.07	10.1
340	.356	11.7	6.8	0	0	−.588	−.075	−1.2	.94	1.03	−1.2
350	.406	10.6	6.2	1	1	−.809	.771	5.7	.97	1.01	5.8
360	.461	9.5	5.6	0	0	−.588	.994	13.1	.80	1.12	14.7
400	.320	12.5	7.3	1	1	.309	1	34.0	.92	1.04	35.4
410	.327	12.3	7.2	0	0	−.951	.694	18.8	.92	1.04	19.6
420	.346	11.9	6.9	1	1	.309	.038	23.3	.93	1.04	24.2
430	.375	11.2	6.6	0	0	−.951	.746	18.7	.95	1.02	19.1
440	.414	10.4	6.1	1	1	.309	−.997	13.3	.93	1.04	13.8
450	.458	9.6	5.6	0	0	−.951	−.637	−13.6	.76	1.15	−15.6

Table 13 (*Continued*)

Reflection	$\sin\theta$	f_{Fe}	f_S	$\frac{S_{Fe}}{2}$		$\frac{S_S}{4}$		$F = f_{Fe} S_{Fe} + f_S S_S$	$\frac{1}{Lp}$	$\sqrt{Lp}$	$\sqrt{I} = \sqrt{Lp}\,F$
500											
510	.405	10.6	6.2	1	1	1	−.720	3.3	.92	1.04	3.4
520	.421	10.3	6.0	0	0	0	−.999	0	.88	1.07	0
530	.445	9.8	5.7	1	1	1	.666	34.8	.80	1.12	39.0
540	.478	9.2	5.4	0	0	0	.075	0	.55	1.35	0
600	.481	9.2	5.4	1	1	.309	1	25.1	.50	1.35	32.7
610	.485	9.1	5.4	0	0	.951	.694	−14.3	.34	2.0	−28.6
620	.498	8.9	5.3	1	1	.309	.038	18.0	.20	2.2	39.6

checked against the c-axis precession photograph the agreement is quite satisfactory. Incidentally, this check serves to confirm the particular parameter pair given in (28) above, and eliminates combinations like $x = -0.20$, $y = 0.38$, which give discordant intensities for $hk0$, $h+k$ odd.

The marcasite structure resulting from this analysis is shown in Fig. 8.

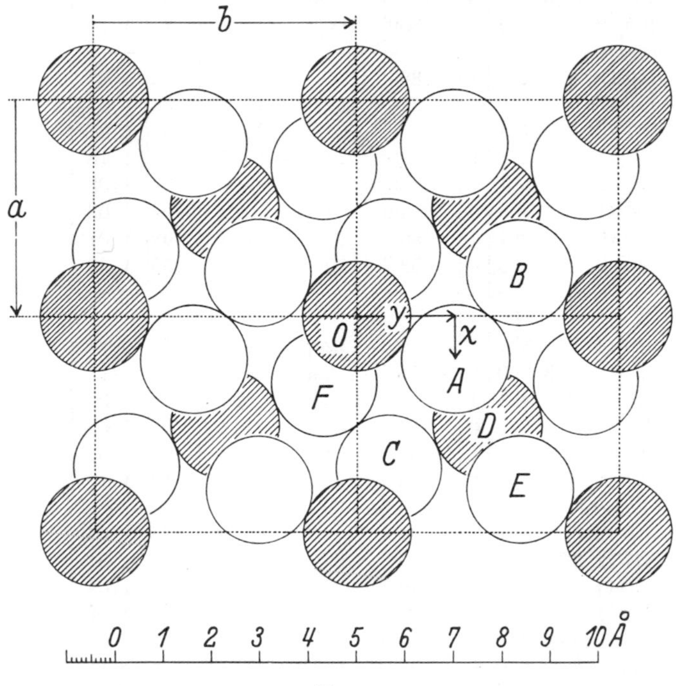

Fig. 8.

Analytical parameter determination. The parameters can also be determined by an analytical procedure provided that the absolute intensities of some spectra are known. Those intensities which are too weak to record are convenient to consider in this connection, since they have absolute values substantially zero. These may be used in the following way.

If a systematic catalog of all observed reflections is examined, it is observed that those of type $5kl$ are missing for $h+k+l$ odd. Out of 10 reflections which could have occurred, not one was strong enough to observe. The amplitude of this type of reflection is given by (24). These reflections have no contribution from the Fe atom, and the reflection is due entirely to the sulfurs. It is evident from (24) that for the expression to vanish, $\sin 2\pi 5x \sin 2\pi ky$ must vanish for all values of k. This requires that $\sin 2\pi 5x = 0$, which occurs for $2\pi 5x = 0, \pi, 2\pi \cdots n\pi$. The possible solutions are $x = n/10$. The solution $x = 2/10$ corresponds with the solution $x = 0.20$ determined graphically.

It is also observed that reflections of the type $h4l$ are missing for $h+k+l$ odd. Again applying (24), it is apparent that for the expression to vanish, $\sin 2\pi hx \sin 2\pi 4y$ must vanish for all values of h. This requires that $\sin 2\pi 4y = 0$, which occurs for $2\pi 4y = 0, \pi, 2\pi \cdots n\pi$. The possible solutions are $y = n/8$. The solution $y = 3/8$ corresponds with the solution $y = 0.38$ determined graphically.

Determination of structures with heavy atoms which have separable parameters, as illustrated by the structure of valentinite,[3] Sb_2O_3

The compound Sb_2O_3 has both an isometric form, *senarmontite*, and an orthorhombic form, *valentinite*. Valentinite offers a good example of the kind of moderately complex structure which can be determined by straightforward elementary means. Even though the structure is somewhat complex, its solution is feasible because it contains a single set of symmetry-equivalent heavy atoms. In such instances, the heavy atoms so dominate the intensities that they can be located without regard to the other atoms. Once located, their coordinates also fix enough diffraction phases so that the rest of the atoms can be located.

The determination of the structure of valentinite, as given here, is based upon a set of Weissenberg photographs made with Cu$K\alpha$ radiation, and supplemented by certain zero-level Weissenberg photographs made with the Mo$K\alpha$ radiation. The shorter Mo$K\alpha$ wavelength enables one to obtain many additional reflections which are important in fixing the parameters of the antimony atoms very exactly.

Unit cell and space group. The diffraction effects as recorded on the Weissenberg photographs place valentinite in the diffraction group *mmmPccn*. This uniquely fixes the space group as $P\,2_1/c\,2_1/c\,2/n$. The cell has the following dimensions:

$$a = 4.93\ \text{Å},\quad b = 12.48,\quad c = 5.43.$$

The density of the crystals as given in standard mineralogical references is 5.76 g./cc. Using these data, relation (3) of Chapter 9 shows that the cell contains 4 Sb_2O_3.

Table 14

Equipoints of space group $P\,\frac{2_1}{c}\,\frac{2_1}{c}\,\frac{2}{n}$

Designation	Location	Coordinates			
$4a$	inversion centers	000;	$\frac{1}{2}\frac{1}{2}0$;	$0\frac{1}{2}\frac{1}{2}$;	$\frac{1}{2}0\frac{1}{2}$.
$4b$	" "	$00\frac{1}{2}$;	$\frac{1}{2}\frac{1}{2}\frac{1}{2}$;	$0\frac{1}{2}0$;	$\frac{1}{2}00$.
$4c$	2-fold axes	$\frac{1}{4}\frac{1}{4}z$;	$\frac{3}{4}\frac{3}{4}\bar{z}$;	$\frac{1}{4}, \frac{1}{4}, \frac{3}{4}+z$;	$\frac{3}{4}, \frac{3}{4}, \frac{1}{2}-z$.
$4d$	" "	$\frac{1}{4}\frac{3}{4}z$;	$\frac{3}{4}\frac{1}{4}\bar{z}$;	$\frac{1}{4}, \frac{3}{4}, \frac{1}{2}+z$;	$\frac{1}{4}, \frac{3}{4}, \frac{1}{2}-z$.
$8e$	general position	xyz;	$\frac{1}{2}-x, \frac{1}{2}-y, z$;	$\frac{1}{2}+x, \bar{y}, \frac{1}{2}-z$;	$\bar{x}, \frac{1}{2}+y, \frac{1}{2}-z$;
		$\bar{x}\bar{y}\bar{z}$;	$\frac{1}{2}+x, \frac{1}{2}+y, \bar{z}$;	$\frac{1}{2}-x, y, \frac{1}{2}+z$;	$x, \frac{1}{2}-y, \frac{1}{2}+z$.

Table 15

Possible ways of accommodating 8 Sb and 12 O in space group $P\,\frac{2_1}{c}\,\frac{2_1}{c}\,\frac{2}{n}$

8 Sb in equipoints	12 O in equipoints
1. $4a+4b$	1. $4a+4b+4c$
2. $4a+4c$	2. $4a+4b+4d$
3. $4a+4d$	3. $4a+4c+4c$
4. $4b+4c$	4. $4a+4c+4d$
5. $4b+4d$	5. $4a+4d+4d$
6. $4c+4c$	6. $4b+4c+4c$
7. $4c+4d$	7. $4b+4c+4d$
8. $4d+4d$	8. $4b+4d+4d$
9. $8e$	9. $8e+4a$
	10. $8e+4b$
	11. $8e+4c$
	12. $8e+4d$

Distribution of atoms among the equipoints. Space group $P\,2_1/c\,2_1/c\,2/n$ has equipoints of ranks 4 and 8 only. They are listed in Table 14 and illustrated in Fig. 9. The possible ways of accommodating the 8 Sb and the 12 O atoms on the equipoints are shown in Table 15.

Note that not all ways of combining sets of equipoints, individually suitable to 8 Sb and 12 O, are permissible combinations. This is because positions $4a$ and $4b$ are without any degrees of freedom and can consequently only be used for one set of atoms. They may thus not be used more than once in any combination. Actually, it will be unnecessary to

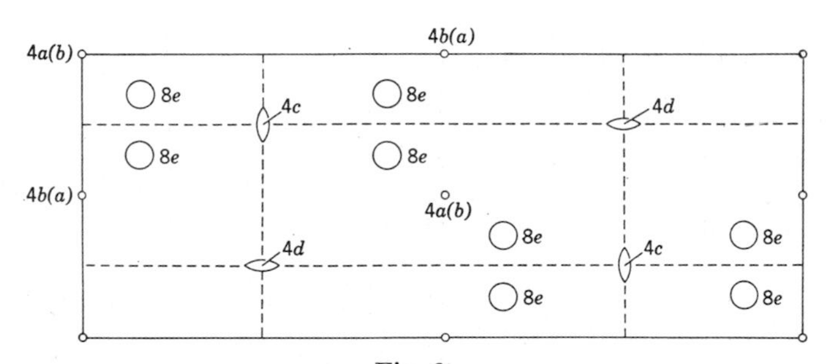

Fig. 9.
Equipoints of $P2_1/c\,2_1/c\,2/n$.

form these combinations in advance, for the scheme of analysis to be followed first locates the Sb atoms, thus eliminating, as inapplicable to valentinite, eight of the nine possible ways of distributing the Sb atoms on the equipoints shown in Table 14.

Reflection intensities. The amplitude of a reflection is given by (45) of Chapter 2, namely

$$F_{hkl} = \sum_{j=1}^{J} f_j\, e^{i2\pi(hx_j+ky_j+lz_j)}. \tag{29}$$

The J atoms of the cell comprise the 8 Sb atoms and the 12 O atoms. Thus, (29) can be split into two terms, and in each of these an f can be taken outside the summation sign. The exponential remaining within the summation is simply the sum of the symmetry factors, S, of Chapter 10, pertaining to the equipoint combinations. Since it is not known at this stage how the atoms are distributed among the equipoints, this can be allowed for by letting the sum of the symmetry factors be represented by ΣS for each atom. If these considerations are taken into account, (29) can be expressed as

$$F_{hkl} = f_{\mathrm{Sb}} \sum S_{\mathrm{Sb}} + f_{\mathrm{O}} \sum S_{\mathrm{O}}. \tag{30}$$

The symmetry factor, S, for particular classes of reflections for space group $P\,2_1/c\,2_1/c\,2/n$ are listed in Table 16.

Table 16

Symmetry factors for space group $P\,\frac{2_1}{c}\,\frac{2_1}{c}\,\frac{2}{n}$

Reflection	Conditions	Symmetry factor
hkl	$\begin{Bmatrix} h+k \text{ even} \\ h+l \text{ even} \\ k+l \text{ even} \end{Bmatrix} \Rightarrow \begin{Bmatrix} h \text{ even} \\ k \text{ even} \\ l \text{ even} \end{Bmatrix}$ or $\begin{Bmatrix} h \text{ odd} \\ k \text{ odd} \\ l \text{ odd} \end{Bmatrix}$	$8 \cos 2\pi hx \cos 2\pi ky \cos 2\pi lz$
	$\begin{Bmatrix} h+k \text{ odd} \\ h+l \text{ odd} \\ k+l \text{ even} \end{Bmatrix} \Rightarrow \begin{Bmatrix} h \text{ odd} \\ k \text{ even} \\ l \text{ even} \end{Bmatrix}$ or $\begin{Bmatrix} h \text{ even} \\ k \text{ odd} \\ l \text{ odd} \end{Bmatrix}$	$-8 \sin 2\pi hx \cos 2\pi ky \sin 2\pi lz$
	$\begin{Bmatrix} h+k \text{ odd} \\ h+l \text{ odd} \\ k+l \text{ odd} \end{Bmatrix} \Rightarrow \begin{Bmatrix} h \text{ even} \\ k \text{ odd} \\ l \text{ even} \end{Bmatrix}$ or $\begin{Bmatrix} h \text{ odd} \\ k \text{ even} \\ l \text{ odd} \end{Bmatrix}$	$-8 \cos 2\pi hx \sin 2\pi ky \sin 2\pi lz$
	$\begin{Bmatrix} h+k \text{ even} \\ h+l \text{ odd} \\ k+l \text{ odd} \end{Bmatrix} \Rightarrow \begin{Bmatrix} h \text{ even} \\ k \text{ even} \\ l \text{ odd} \end{Bmatrix}$ or $\begin{Bmatrix} h \text{ odd} \\ k \text{ odd} \\ l \text{ even} \end{Bmatrix}$	$-8 \sin 2\pi hx \sin 2\pi ky \cos 2\pi lz$
$hk0$	h even, k even	$8 \cos 2\pi hx \cos 2\pi ky$
	h odd, k odd	$-8 \sin 2\pi hx \sin 2\pi ky$
$h0l$	h even, l even	$8 \cos 2\pi hx \cos 2\pi lz$
	h odd, l even	$-8 \sin 2\pi hx \sin 2\pi lz$
$0kl$	k even, l even	$8 \cos 2\pi ky \cos 2\pi lz$
	k odd, l even	$-8 \sin 2\pi ky \sin 2\pi lz$
$h00$	h even	$8 \cos 2\pi hx$
$0k0$	k even	$8 \cos 2\pi ky$
$00l$	l even	$8 \cos 2\pi lz$

The computation of intensities may be carried out by the relation

$$I_{hkl} = LpF^2_{hkl}. \tag{31}$$

This expresses intensities in units based upon the amplitude scattered by one electron. In qualitative structure determinations, it is more convenient to compare intensities with the use of the relation

$$\sqrt{I_{hkl}} = \sqrt{Lp}\,|F_{hkl}|. \tag{32}$$

This has the advantage that it involves fewer computing operations and also that the graphical representation of intensity as a function of parameter reduces to drawing cosine curves. Furthermore, (32) is as good as (31) for comparison of relative intensities, since, if a set of reflections is arranged in order of decreasing I, it is also arranged in order of decreasing $\sqrt{I}$. It is evident, therefore, that most of the required intensity computations can be made with the aid of (32). Substituting from (30) into (32) gives the complete computation

$$\sqrt{I_{hkl}} = \sqrt{Lp}\left| f_{\text{Sb}} \sum S_{\text{Sb}} + f_{\text{O}} \sum S_{\text{O}} \right|. \tag{33}$$

In much of the following preliminary part of the analysis, advantage is taken of the fact that the scattering power of Sb is very great in comparison with that of oxygen. The ratio of scattering powers at $\sin\theta = 0$ is the same as the ratio of atomic numbers, namely $f_{\text{Sb}}/f_{\text{O}} = Z_{\text{Sb}}/Z_{\text{O}} = 51/8$. This means that the Sb atoms dominate the resultant amplitude in both magnitude and phase. For these preliminary purposes, (30) and (33) may accordingly be approximated by

(30): $$F_{hkl} \approx f_{\text{Sb}} \sum S_{\text{Sb}}. \tag{34}$$

(33): $$\sqrt{I_{hkl}} \approx \sqrt{Lp}\, f_{\text{Sb}} \left| \sum S_{\text{Sb}} \right|. \tag{35}$$

Elimination of incorrect antimony equipoint combinations. Certain possible combinations of antimony equipoints listed in Table 15 can be easily eliminated by very simple intensity considerations. For the present purpose it is sufficient to compare relative observed pinacoid reflection intensities with those to be expected with the Sb atoms occupying the several possible equipoint combinations. The expected intensities can be approximated closely enough for the purpose by means of the amplitudes due to the antimony atoms alone, as just explained. The amplitudes for the sequence of orders of reflection of $h00$, $0k0$, and $00l$ are listed in Table 17. The actual intensities must follow the essential features of the several series listed in this table because of the relatively great scattering power of the Sb atom compared with that of the O atom. The actual locations of the O atoms affect the amplitudes indicated, at most, by giving rise to weak reflections where the amplitudes of Table 17 indicate an absent reflection, and by very slightly disturbing the relative intensities involved in regular declines indicated by the designation $8f, 8f, 8f \cdots$, etc.

Table 17 indicates that normal intensity declines for the even-order reflections of $h00$ and $0k0$, either with or without alternate absences, characterize all structures containing Sb occupying the special equipoints. The photographs display an intensity series comparable with one of these regularities, namely, an approximately regular decline with alternate absences, but only for one series, namely, $0k0$; other series are distinctly irregular. Accordingly the Sb atoms can only occupy the general position $8e$. The meaning of the regular decline of even order of $0k0$ with alternate absences is obviously that the Sb parameters are specialized; specifically these atoms are arranged in sheets parallel to (010) which are spaced approximately one quarter of the b translation apart. This would give the y parameter of Sb a value of approximately $\frac{1}{8}$.

Table 17
Amplitude patterns to be expected from various Sb equipoint combinations

Sb equipoint combination	Regularities in amplitude series for even-order pinacoid reflections		
	$h00$	$0k0$	$00l$
$4a + 4b$	$8f,\ 8f,\ 8f,\ 8f \cdots$	$8f,\ 8f,\ 8f,\ 8f \cdots$	$8f,\ 8f,\ 8f,\ 8f \cdots$
$4a + 4c$	$8f,\ 0,\ 8f,\ 0 \cdots$	$8f,\ 0,\ 8f,\ 0 \cdots$	irregular
$4a + 4d$	$8f,\ 0,\ 8f,\ 0 \cdots$	$8f,\ 0,\ 8f,\ 0 \cdots$	irregular
$4b + 4c$	$8f,\ 0,\ 8f,\ 0 \cdots$	$8f,\ 0,\ 8f,\ 0 \cdots$	irregular
$4b + 4d$	$8f,\ 0,\ 8f,\ 0 \cdots$	$8f,\ 0,\ 8f,\ 0 \cdots$	irregular
$4c + 4c$	$\overline{8f},\ 8f,\ \overline{8f},\ 8f \cdots$	$\overline{8f},\ 8f,\ \overline{8f},\ 8f \cdots$	irregular
$4c + 4d$	$\overline{8f},\ 8f,\ \overline{8f},\ 8f \cdots$	$\overline{8f},\ 8f,\ \overline{8f},\ 8f \cdots$	irregular
$4d + 4d$	$\overline{8f},\ 8f,\ \overline{8f},\ 8f \cdots$	$\overline{8f},\ 8f,\ \overline{8f},\ 8f \cdots$	irregular
$8e$	irregular	irregular	irregular

Antimony parameters. In the discussion of reflection intensities it was pointed out that antimony has such a large scattering power compared with oxygen that it dominates the spectra. This fact is now utilized to locate the antimony atoms regardless of what locations the oxygen atoms occupy. To see that this is possible, return to (33). This can now be given a more specific form since it is known that the Sb atoms occupy the position $8e$. For pinacoid reflections like $h00$, Table 16 shows that the symmetry factor reduces to a cosine function. Thus (33) takes the specific and simple form

$$\sqrt{I_{h00}} = \sqrt{Lp}\,\left| 8f_{\mathrm{Sb}} \cos 2\pi hx + f_{\mathrm{O}} \sum S_{\mathrm{O}} \right|, \tag{36}$$

$$\sqrt{I_{0k0}} = \sqrt{Lp}\,\left| 8f_{\mathrm{Sb}} \cos 2\pi ky + f_{\mathrm{O}} \sum S_{\mathrm{O}} \right|, \tag{37}$$

$$\sqrt{I_{00l}} = \sqrt{Lp}\,\left| 8f_{\mathrm{Sb}} \cos 2\pi lz + f_{\mathrm{O}} \sum S_{\mathrm{O}} \right|. \tag{38}$$

Now the maximum possible value of the oxygen contribution to the amplitude, namely $f_{\mathrm{O}}\,\Sigma S_{\mathrm{O}}$, is obtained when the 12 oxygen atoms scatter in phase. Even at a value of $\sin\theta = 0$, this is limited to $12 \times 8 = 96$

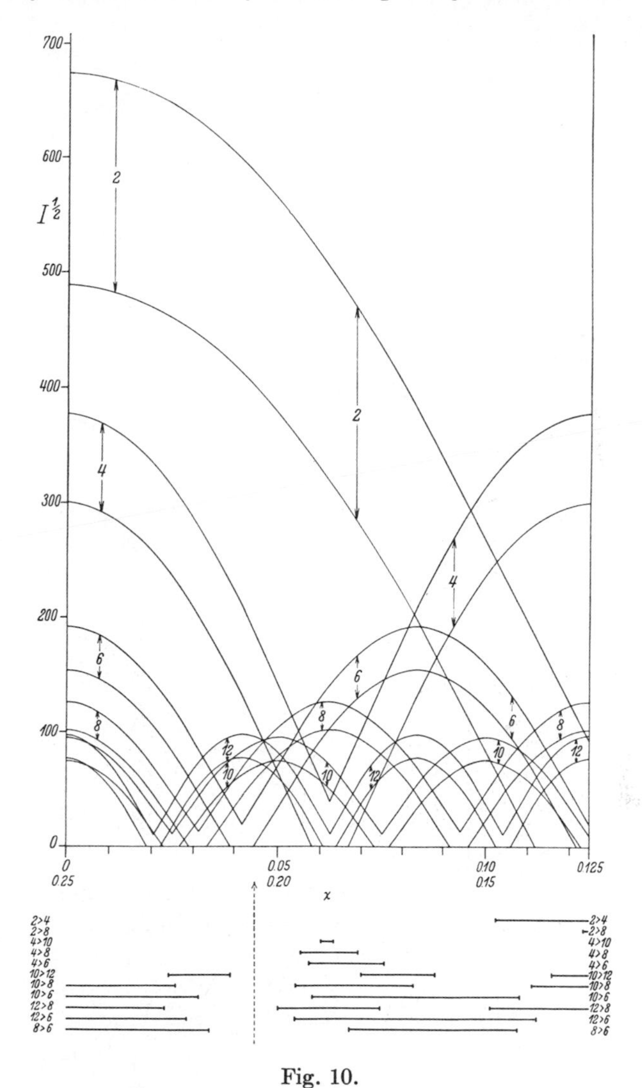

Fig. 10.
Variation of $I^{\frac{1}{2}}_{h00}$ with antimony x parameter; valentinite, Cu$K\alpha$.
(After Buerger and Hendricks.[5])

electron units. The corresponding maximum value of the antimony contribution is then $8 \times 51 = 408$ electron units. From this it is evident that the resultant amplitude of any reflection is dominated by antimony and could be represented by that due to antimony alone, plus or minus an uncertainty equal to the contribution due to oxygen, which has a maximum value for any reflection hkl of $12 f_{\mathrm{O},hkl}$. Using this value of the uncertainty, functions (36), (37), and (38) are plotted against x, y,

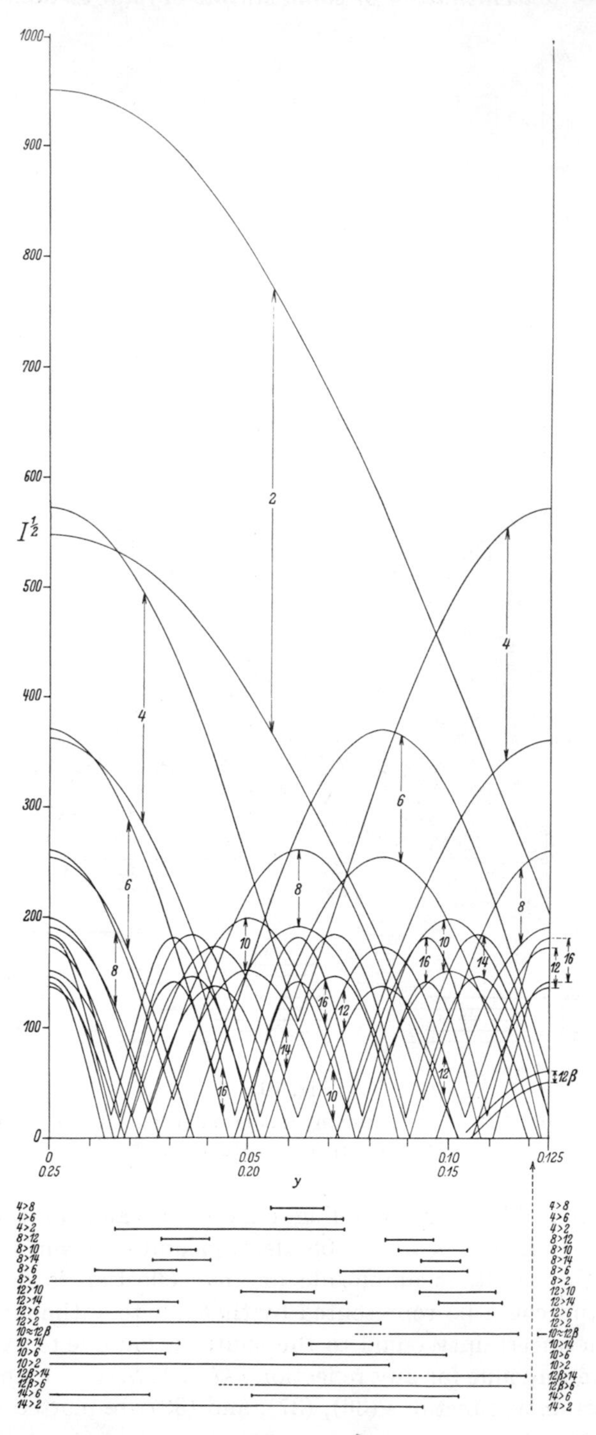

Fig. 11.

Variation of $I_{0k0}^{1/2}$ with antimony y parameter; valentinite, Cu$K\alpha$. (After Buerger and Hendricks.[5])

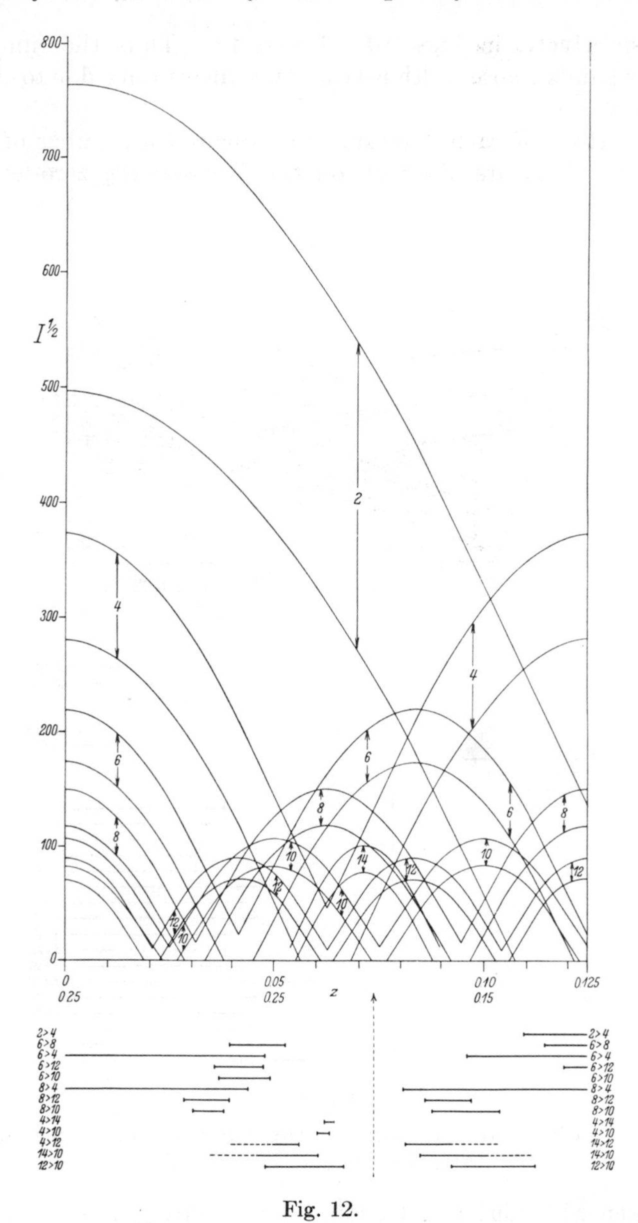

Fig. 12.

Variation of $I_{00l}^{1/2}$ with antimony z parameter; valentinite, Cu$K\alpha$. (After Buerger and Hendricks.[5])

and z, respectively, in Figs. 10, 11 and 12. Thus the functions are pictured as bands whose width is twice this uncertainty due to the oxygen contribution.

At the bottom of each diagram are indicated a number of intensity relationships which are observed on the Weissenberg zero-level photo-

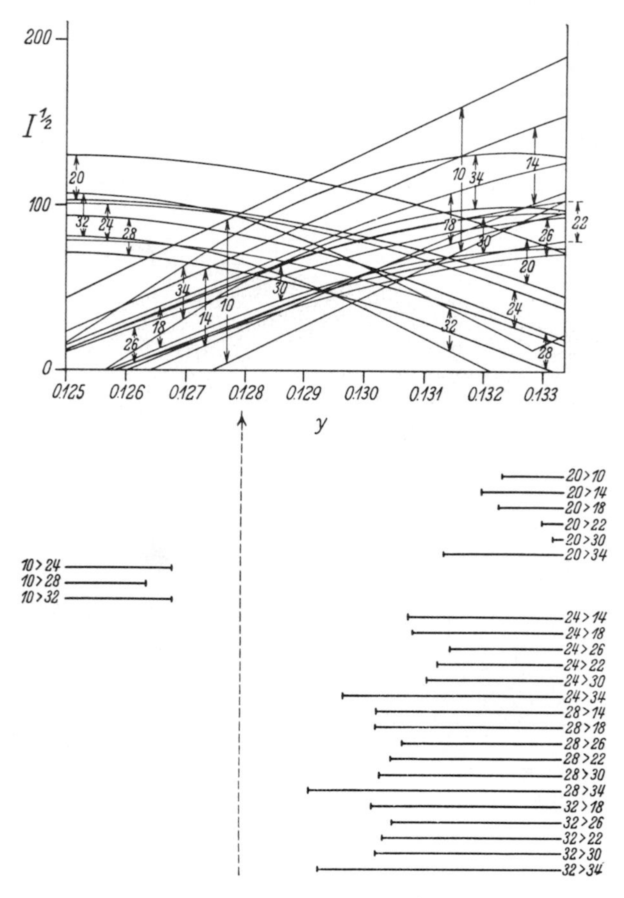

Fig. 13.
Variation of $I^{1/2}_{0k0}$ with antimony z parameter for high values of k; valentinite, Mo$K\alpha$.
(After Buerger and Hendricks.[5])

graphs taken with Cu$K\alpha$ radiation. The notation $2 > 4$, for example, means that the second order of the reflection ($h00$ in Fig. 10, $0k0$ in Fig. 11, and $00l$ in Fig. 12) is observed to be more intense than the fourth order. Next to each such relation a line is drawn which indicates the region of the two curves in question where this could not theoretically hold regardless of the oxygen contribution. Such regions may be regarded as eliminated values of the Sb coordinate.

Because of the great length of the b axis, namely, 12.48 Å, it is possible to record 34 orders of reflection from (010) with Mo$K\alpha$ radiation. Since the amplitudes of the high-order reflections fluctuate very rapidly with parameter, it becomes possible to refine the y parameter of Sb to a very accurate value indeed, irrespective of the positions of the oxygen atoms. This refinement is carried out graphically in Fig. 13.

Since the space group provides for halved spacings of all the pinacoid planes, only even orders of pinacoid reflections appear. The absence of odd-order reflections gives two solutions to each parameter value. This corresponds with the fact that (36), (37), and (38) are cosine functions. With the eliminations shown in Figs. 10, 11, 12, and 13, these approximate solutions are

$$\begin{aligned} x &= 0.045 \text{ or } 0.205, \\ y &= 0.122 \text{ or } 0.128, \\ z &= 0.074 \text{ or } 0.176. \end{aligned} \tag{39}$$

The incorrect value of each of these pairs can be eliminated by a consideration of more general spectra with odd values of h, k, and l, respectively, and therefore involving sine functions.

Table 16 shows that the symmetry factor for $hk0$ reflections for odd values of h is $-8 \sin 2\pi hx \sin 2\pi ky$. When h is 5, this is a maximum at $x = 0.050$ and a minimum at $x = 0.200$. Since the $5k0$ reflections form one of the strongest festoons on the Weissenberg c-axis zero-level photograph, it may be safely concluded that the correct value of x is approximately 0.045, not 0.205.

To eliminate the incorrect value of y, the structure factor may be examined for $h{\cdot}29{\cdot}0$ and $h{\cdot}31{\cdot}0$. For these the trigonometric part of the symmetry factor has the specific values

$$\text{at } y = 0.122, \quad \begin{cases} \text{for } h{\cdot}29{\cdot}0, & \sin(29 \times 43.8°) = \sin 1270° \\ & \quad = \sin(-190°) \ \therefore \text{ almost zero,} \\ \text{for } h{\cdot}31{\cdot}0, & \sin(31 \times 43.8°) = \sin 1358° \\ & \quad = \sin(-82°) \quad \therefore \text{ almost a maximum,} \end{cases}$$

$$\text{at } y = 0.128, \quad \begin{cases} \text{for } h{\cdot}29{\cdot}0, & \sin(29 \times 46.2°) = \sin 1340° \\ & \quad = \sin(-90°) \quad \therefore \text{ almost a maximum,} \\ \text{for } h{\cdot}31{\cdot}0, & \sin(31 \times 46.2°) = \sin 1432° \\ & \quad = \sin(-8°) \quad \therefore \text{ almost zero.} \end{cases}$$

The series $h{\cdot}29{\cdot}0$ is strong and the series $h{\cdot}31{\cdot}0$ is absent on Weissenberg c-axis zero-level photographs. The correct value of the y parameter is accordingly the one near $y = 0.128$.

Unfortunately, the elimination of incorrect values of the z parameter cannot be made in a similar manner by using $h0l$ and $0kl$ spectra because the reflections for which l is odd are extinguished by the (100) and (010) glide planes which both have translation components $c/2$. Recourse may be had, however, to the general reflections, and the same method may be used. The two possible values of the z parameter lie in the neighborhood of $0.083 = \frac{1}{12}$ and $0.167 = \frac{1}{6}$. For these regions, the $hk3$ give important information because

$$\text{at } z = \tfrac{1}{12}, \qquad \begin{cases} \cos(3 \times 30°) = \cos 90° = 0, \\ \sin(3 \times 30°) = \sin 90° = \text{a maximum}; \end{cases}$$

$$\text{at } z = \tfrac{1}{6}, \qquad \begin{cases} \cos(3 \times 60°) = \cos 180° = \text{a maximum}, \\ \sin(3 \times 60°) = \sin 180° = 0. \end{cases}$$

A general survey of $hk3$ spectra shows that those referable to symmetry factors (Table 16) involving $\cos 2\pi lz$ are usually strong, while those referable to symme ry factors involving $\sin 2\pi lz$ are either weak or absent. This definitely places z in the region of $\frac{1}{6}$ rather than $\frac{1}{12}$, and retains 0.176 as the correct alternative.

The correct approximate antimony parameters may now be tabulated:

$$x = 0.045, \quad y = 0.128, \quad z = 0.176. \tag{40}$$

A slight refinement of these values may have to be made after the oxygen atoms are located.

Method of location of oxygen atoms.[†] The Sb atoms were located with the aid of the intensity bands illustrated in Figs. 10 through 13. For all the information taken into account, the calculated and observed intensities are in complete harmony for the Sb parameters listed in the last section. There are, however, other intensity features which have not as yet been considered, and which depend upon the location of the oxygen atoms. These will now receive consideration.

The plan of attack in the location of the oxygen positions is as follows: First of all, the y coordinates of the Sb atoms can be located with great precision because so many orders of $0k0$ are available (17 even orders, from 2 to 34). The band width, as already explained, represents the

[†] From this point on, the discussion represents what *can* be done by these primitive methods and not what *would* be done in a present-day structure analysis. The oxygens are more easily located by using Fourier synthesis, basing the signs of the F's on the antimony-atom locations just established. This heavy-atom method is discussed in Chapter 19.

intensity uncertainty due to the unknown oxygen positions. In order to have the calculated intensities of the $0k0$ spectra arranged in the observed relative order, the calculated intensities must be within certain very definite regions of the oxygen uncertainty bands. In other words, the phase (positive or negative) and even an estimate of the absolute value of the oxygen contribution may be obtained for some spectra, especially where the maximum oxygen contribution is large, as in reflections of small $\sin\theta$ value. Certain equipoint combinations can be discarded on the grounds that they could not furnish the required oxygen contributions, and the y parameters of the correct oxygen equipoint combination may be determined independently of every other variable except the Sb y parameter, which, as already mentioned, is very certainly and accurately established.

With the Sb x and y parameters known, the oxygen equipoints and y parameters established, the oxygen x parameters may be determined by a study of the relative intensities of the $hk0$ spectra. Many comparisons, such as 800 with 840, 10·0·0 with 10·4·0, 12·0·0 with 12·4·0, etc., may be made which are almost independent of the antimony parameters, so the small uncertainty of the Sb x coordinate is no bar to the accurate determination of the oxygen x parameters. Definite phase and amplitude requirements within the oxygen uncertainty bands then locate the oxygen atoms quite accurately.

At this stage, the projection of the valentinite structure on (001) becomes accurately established and the rough structural plan is fairly obvious. Only a few trial calculations need be made in physically possible oxygen positions for the correct structural alternative.

Elimination of certain incorrect oxygen equipoint combinations and determination of oxygen *y* parameters. The Sb y parameter may be taken, from Figs. 11 and 13, as $y = 0.128$ with very little uncertainty. Referring, now, to the photographs taken with copper radiation, it is striking that 020 and 060, which have a considerable possible oxygen contribution, are absent and exceedingly weak respectively. The absolute values of the amplitudes of these two reflections must, therefore, be approximately zero. Figure 11 shows that the oxygen amplitude contributions required to give this net value of zero at antimony $y = 0.128$ must be about $+30$ for 020 and -38 for 060. The further relationships $0{\cdot}10{\cdot}0 = 0{\cdot}12{\cdot}0\beta \geqq 0{\cdot}14{\cdot}0$ give the added requirements that the oxygen amplitude contribution to 0·10·0 be negative and rather extreme, and to 0·14·0, approximately zero. These conditions are summarized in Table 18.

The oxygen atoms may be in any of the eight equipoint combinations listed in Table 15. The contribution of the oxygens to the $0k0$ reflec-

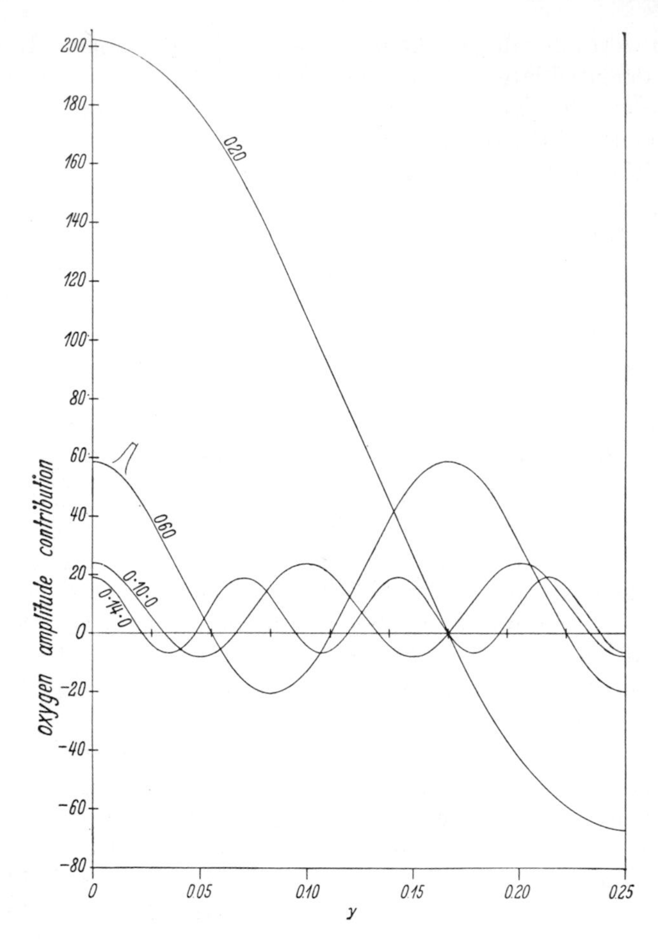

Fig. 14.
Oxygen contribution to amplitudes of $0k0$ as a function of y for the equipoint combination of $8e+4a$ or $8e+4b$; valentinite, Cu$K\alpha$.
(After Buerger and Hendricks.[5])

tions, according to (33) and Table 16, is

$$\Delta \sqrt{I_{0k0}} = \sqrt{Lp}\, fm \cos 2\pi ky, \tag{41}$$

where m is the rank of the equipoint. For the several equipoints this reduces to

$$8e: \quad \Delta \sqrt{I_{0k0}} = \sqrt{Lp}\, f8 \cos 2\pi ky, \tag{42}$$

$$4a \text{ or } 4b: \quad \Delta \sqrt{I_{0k0}} = \sqrt{Lp}\, f4, \tag{43}$$

$$4c \text{ or } 4d: \quad \Delta \sqrt{I_{0k0}} = \sqrt{Lp}\, f4(-1)^k. \tag{44}$$

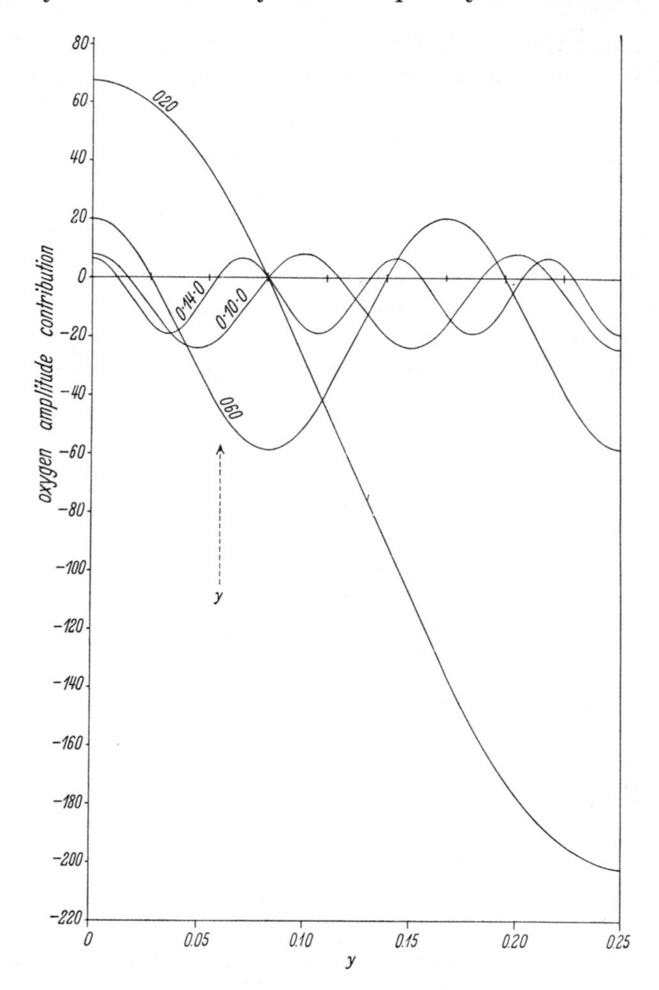

Fig. 15.
Oxygen contribution to amplitudes of $0k0$ as a function of y for the equipoint combination $8e+4c$ or $8e+4d$; valentinite, Cu$K\alpha$.
(After Buerger and Hendricks.[5])

Table 18
Intensity relations for $0k0$ reflections useful in fixing oxygen y coordinates of valentinite

Intensity condition	Oxygen contribution to amplitude of reflection, required at Sb $\theta_b = 46.2°$	
020 absent	020	+30
060 very faint	060	−38
$0{\cdot}10{\cdot}0 = 0{\cdot}12{\cdot}0\ \beta$	$0{\cdot}10{\cdot}0$	−16, (−) maximum
$0{\cdot}14{\cdot}0 \leq 0{\cdot}12{\cdot}0\beta$	$0{\cdot}14{\cdot}0$	≈0

Combinations 1 through 8, Table 15, involving equipoints of the special positions, give rise to amplitudes which are invariable, and equal to

$$\Delta \sqrt{I_{0k0}} = \sqrt{Lp}\, fN, \tag{45}$$

where N is 12, 8, 4, or -4. None of these combinations has the possibility of making the oxygen contribution to the amplitude of 020 equal to $+30$. Hence all combinations involving only special positions, namely combinations 1 through 8, Table 15, are eliminated.

The other equipoint combinations of Table 15, namely 9 through 12, have structure factors containing (42) and hence involve the variable

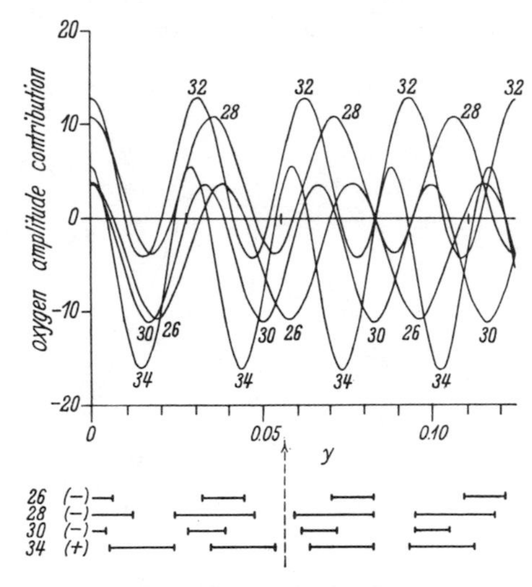

Fig. 16.
Oxygen contribution to amplitudes of $0k0$ as a function of oxygen parameter y_{I} for high values of k; valentinite, Mo$K\alpha$.
(After Buerger and Hendricks.[5])

Table 19
Phase requirements of certain $0k0$ reflections of valentinite

Reflection	Phase of oxygen contribution to amplitude of reflection required at $y_{\mathrm{Sb}} = 0.128_3$
0·26·0	(−)
0·28·0	(−)
0·30·0	(−)
0·32·0	uncertain
0·34·0	(+)

$\cos 2\pi ky$. These give more flexible intensity possibilities. The variations of oxygen amplitude with y for combinations 9, $8e+4a$, and 10, $8e+4b$, are the same, and are shown in Fig. 14. It can be seen that no value of y gives the required oxygen contributions, and therefore these combinations may be eliminated.

The variations of oxygen amplitude contribution with y are the same for combinations 11, $8e+4c$, and 12, $8e+4d$, and are shown in Fig. 15. Both of these combinations give an excellent agreement with the required contributions for $y = 0.06_1$. This is the parameter of the 8 oxygen atoms in the general position, the other 4 oxygens being on one of the two possible rotation axes, $4c$ or $4d$. The decision between these is arrived at in the determination of the x parameters of the oxygen atoms, where the more general reflections $hk0$ are considered.

The above discussion applies, as already noted, to reflections present on photographs taken with copper radiation. This solution for the parameter may be confirmed by a study of the more sensitive $0k0$ reflections of high order present on photographs taken with molybdenum radiation. The oxygen phase requirements are given in Table 19. Figure 16 shows that these conditions are simultaneously satisfied in a narrow region centering about $y = 0.056$.

Determination of oxygen *x* parameter. The oxygen and antimony y parameters are now both accurately fixed, and the antimony x parameter is rather closely known. The reflections $hk0$ are therefore defined except for the uncertainty of the oxygen x parameter, which may, accordingly, be sought. The correct parameter will reproduce observed intensity relations among the $hk0$ spectra. Table 16 shows that the intensity calculations for $hk0$ spectra take the following form:

$\left.\begin{matrix} h \text{ even} \\ k \text{ even} \end{matrix}\right\}$:

$$\begin{aligned}\sqrt{I_{hk0}} = \sqrt{Lp}\,\{&f_{\text{Sb}} \cdot 8 \cos 2\pi hx \cos 2\pi(k \cdot 0.128) \\ &+ f_{\text{O}} \cdot 8 \cos 2\pi hx \cos 2\pi(k \cdot 0.056) \\ &+ f_{\text{O}} \cdot 4 \cos 2\pi h \cdot \frac{n}{4} \cos 2\pi \tfrac{1}{2}\},\end{aligned} \tag{46}$$

$\left.\begin{matrix} h \text{ odd} \\ k \text{ odd} \end{matrix}\right\}$:

$$\begin{aligned}\sqrt{I_{hk0}} = \sqrt{Lp}\,\{&f_{\text{Sb}} \cdot 8 \sin 2\pi hx \sin 2\pi(k \cdot 0.128) \\ &+ f_{\text{O}} \cdot 8 \sin 2\pi hx \sin 2\pi(k \cdot 0.056) \\ &+ f_{\text{O}} \cdot 4 \sin 2\pi h \cdot \frac{n}{4} \sin 2\pi k \tfrac{1}{4}\},\end{aligned} \tag{47}$$

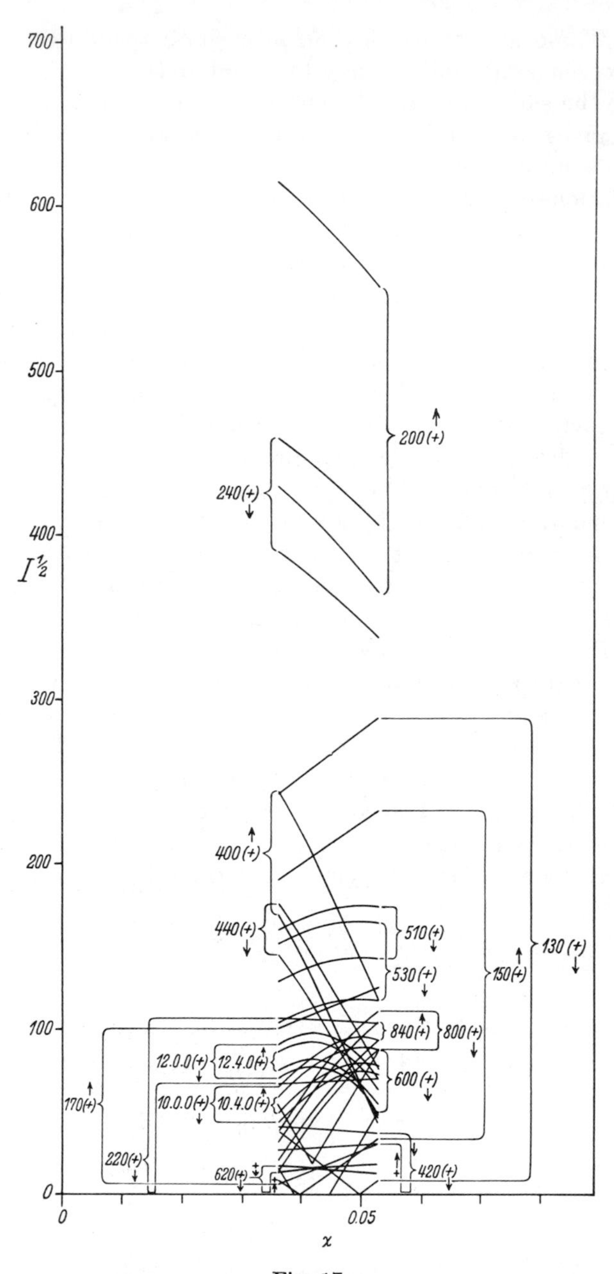

Fig. 17.
Variation of $I_{hk0}^{1/2}$ with antimony x parameter; valentinite, Cu$K\alpha$. (After Buerger and Hendricks.[5])

where $n = 1$ or 3, depending on whether the 4-fold oxygens occupy equipoint $4c$ or $4d$ respectively. The antimony x parameter is purposely left in undetermined form; it will appear that if the general region of the antimony x parameter is known, the x parameter of oxygen can be fixed independently of it.

In the foregoing section, it was shown that the 4-fold oxygen must be on either of the two possible rotation axes $4c$ or $4d$, located at $y = \frac{1}{4}$

Table 20

Intensity relations for *hk*0 reflections and conditions they impose on oxygen contributions of valentinite

Intensity condition	Oxygen amplitude contribution requirement
$130 > \left\{\begin{matrix}8.00\\10.00\\12.00\end{matrix}\right\}$	130 cannot be (+) max.
150 absent 170 ≈ 800	150 extremely (−) 170 nearly neutral
200 ≦ 240 220 absent	200 must be (−) 220 must be very close to +20
400 ≦ 400 420 > 600 (absent)	400 must be (−) 420 must be extreme, preferably (−)
620 > 600 (absent)	$\left\{\begin{matrix}\text{600 must be (+)}\\\text{620 must be extreme, preferably (+) max.}\end{matrix}\right.$
(10·0·0 > 10·4·0) > (840 > 800) > (12·4·0 > 12·0·0)	−(10·4·0 + 10·0·0) > (800 + 840) > (12·0·0 + 12·4·0)
$(510 > 530) > \left\{\begin{matrix}(840 > 800)\\(10{\cdot}0{\cdot}0 > 10{\cdot}4{\cdot}0)\\(12{\cdot}4{\cdot}0 > 12{\cdot}0{\cdot}0)\end{matrix}\right.$	$(530 - 510) > \left\{\begin{matrix}(800 + 840)\\(10{\cdot}4{\cdot}0 + 10{\cdot}0{\cdot}0)\\(12{\cdot}0{\cdot}0 + 12{\cdot}4{\cdot}0)\end{matrix}\right.$

and $y = \frac{3}{4}$ respectively. This is allowed for by an alternative in the last term of the above calculation forms. One of these two alternatives may now be definitely eliminated. The reflection 150 has an exceptionally high oxygen variation due to its small $\sin\theta$ value (See Fig. 17). The absolute value of this reflection is observed to be zero on molybdenum-radiation films (although it appears very faintly on copper-radiation photographs). Figure 17 shows that the value zero can only be attained by nearly a maximum negative total oxygen contribution to the amplitude of this reflection. If the 4-fold oxygens are on equipoint $4d$, with

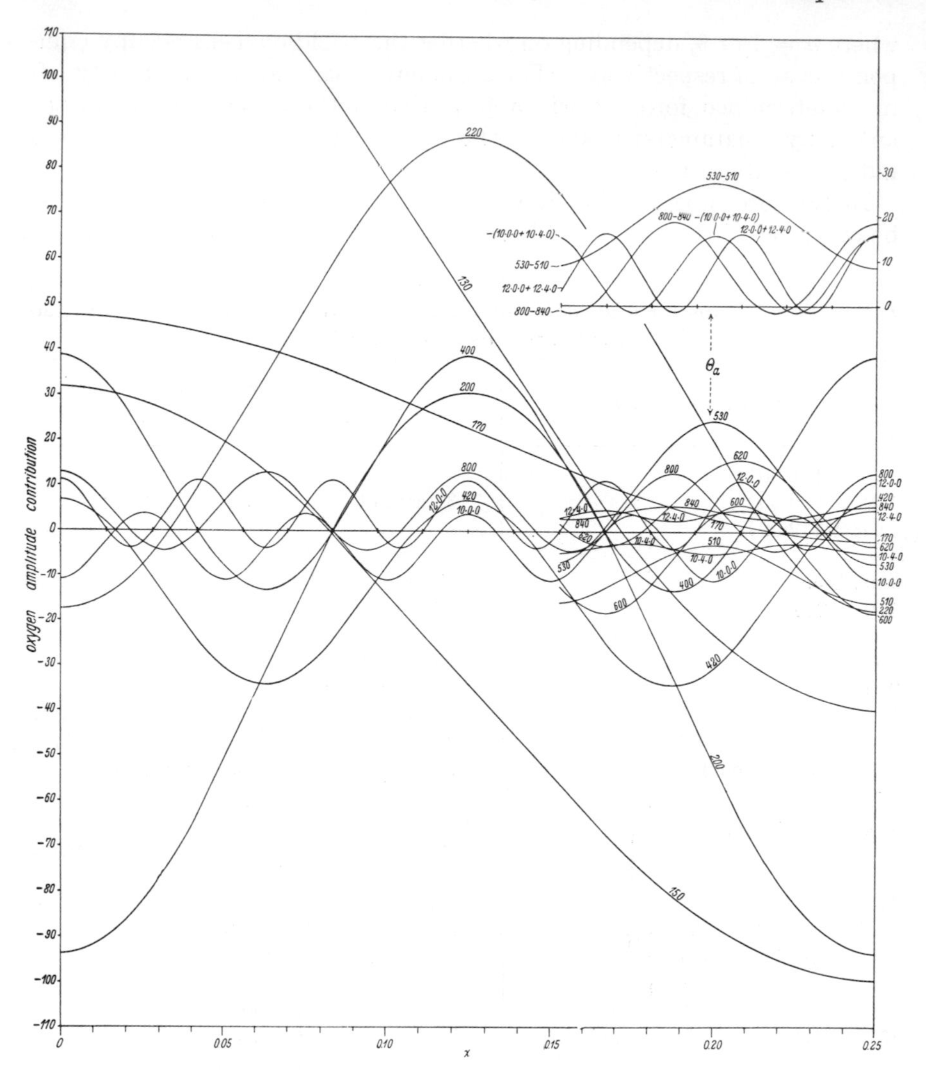

Fig. 18.
Variation of oxygen contribution to amplitudes of $h00$ as a function of oxygen parameter x_{I}; valentinite, Cu$K\alpha$.
(After Buerger and Hendricks.[5])

$y = \frac{3}{4}$, then their individual amplitude contribution is positive, and the minimum possible net intensity of 150 would be a little less than that of 080, which is of intermediate intensity. Since the reflection 150 is absent, this eliminates equipoint $4d$ for the 4-fold oxygens, and determines the oxygen equipoint combination as $8e+4c$; i.e., the oxygens are in the general positions and on the 2-fold rotation axes through $\frac{1}{4}\frac{1}{4}0$.

The necessary oxygen amplitude contributions for the several reflections $hk0$ may be determined from a study of Fig. 17 together with the c-axis Weissenberg photograph made with molybdenum radiation. The conditions and the requirements they impose upon the oxygen contributions are given in Table 20. The variation of the oxygen amplitude contributions of a number of the important reflections with the variation of the 8-fold oxygen parameter, x, is shown in Fig. 18. Study of Fig. 18 for the conditions listed in Table 20 shows that they are satisfied at the unique parameter, $x = 0.147$. A change in this parameter by $\Delta x = 0.014$ completely spoils the intensity relation $(10{\cdot}0{\cdot}0 > 10{\cdot}4{\cdot}0) > (840 > 800) > (12{\cdot}4{\cdot}0 > 12{\cdot}0{\cdot}0)$, so the parameter $x = 0.147$ may be considered accurate to about ± 0.007.

Table 21
Oxygen contribution to $h00$ reflection of valentinite

Reflection	Net oxygen amplitude contribution at oxygen $x = 0.147$
200	-49
400	-9
600	$+3$
800	$+8\frac{1}{2}$
10·0·0	-11
12·0·0	$+4\frac{1}{2}$

With the correct oxygen equipoint combination known, and the 8-fold oxygen x parameter accurately determined, the oxygen uncertainty in the determination of the antimony x parameter is removed. The net oxygen contribution at oxygen $x = 0.147$ may be found from Fig. 18 for $h00$ reflections used in the determination of the antimony x parameter (see the original antimony x parameter determination, Fig. 10). These are listed in Table 21. The critical region of Fig. 10 replotted with these oxygen phase allowances then permits a refining of the antimony x to $x = 0.444 \pm 0.003$.

Physically likely oxygen z parameters. The correct equipoint combination is now known, and the x and y parameters of all the atoms are unequivocally and accurately determined purely with aid of intensity criteria. The projection of the valentinite structure on (001) is therefore unequivocally and accurately determined. It is illustrated diagrammatically in Fig. 19. The only parameter normal to this projection known at this stage of the investigation is the antimony parameter, $z = 0.176 \pm 0.007$. A study of the projection together with this antimony parameter shows that the only physically acceptable interpretation of these data is that the 8-fold oxygen atoms are between pairs of antimony atoms [between (100) glide equivalents], and form chains with these extending down the c axis. Since the antimony positions are known, the oxygen elevations are fixed if the oxygen atoms are to occur equally spaced between antimony atoms, the only reasonable physical interpretation. If this is the case, the 8-fold oxygen atoms have a z parameter of either 0.007 or -0.142. Now, this 8-fold oxygen is in the immediate region of a symmetry center, so that a centrosymmetrically

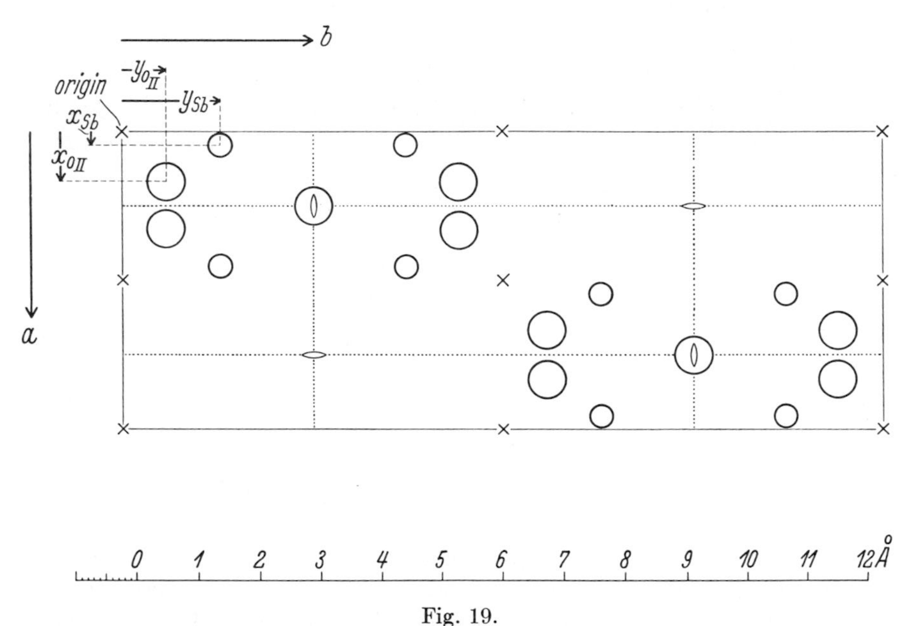

Fig. 19.
Structure of valentinite projected parallel to c. Small circles are antimony atom locations, large circles oxygen atom locations.
(After Buerger and Hendricks.[5])

equivalent oxygen atom is nearby. These are separated by about 2.04 Å for the first possible parameter and 2.54 Å for the second possible parameter. The first distance is closer than known oxygen-oxygen separations, while the second is approximately the standard separation. The 8-fold oxygen parameter, $z = -0.142$ may therefore be accepted as the physically likely one.

Complete sets of antimony and 8-fold oxygen parameters are now available. They define an antimony-oxygen distance of 2.00 Å, which is in good agreement with the electron-pair bond radius sum, 2.02 Å. If this antimony-oxygen separation continues to hold for the distance between the antimony atoms and the 4-fold oxygens, then the latter must have a z parameter of either -0.174 or 0.026. In the former case, the 4-fold and 8-fold oxygens occur on almost exactly the same vertical levels and have a spacing of about 2.46 Å; in the latter case they occur spaced almost exactly half-way between one another and have a spacing of about 2.61 Å. The first possible parameter gives too close an oxygen-oxygen separation without any obvious physical or chemical justification for it; consequently the second parameter, $z = 0.026$, may be accepted as the more likely one.

It should be stated at this point that the oxygen parameters, z_I and z_{II}, suggested by the above physical discussion, are tied directly to the antimony z parameters by dimensional considerations. The space-group symmetry is such that if this antimony z parameter is increased, the oxygen parameters above mentioned are increased by the same amount, Δ, without in any other way changing the discussion. This leaves all interatomic distances mentioned the same, with the exception of the separation between the centrosymmetrical 8-fold oxygen atoms. For the antimony z parameter used, this gives an oxygen-oxygen distance of 2.54 Å, as already mentioned. The antimony z parameter limits have already been defined as ± 0.007. This allowance permits the centrosymmetrical oxygen separation to rise as high as 2.59 Å or fall as low as 2.50 Å.

The parameters of the valentinite structure can now be listed as follows:

	x	y	z
Sb	0.044	0.128	$0.176+\Delta$
O_I	$\frac{1}{4}$	$\frac{1}{4}$	$0.026+\Delta.$
O_{II}	0.147	0.058	$-0.142+\Delta$

(48)

Final parameters. To check the structure suggested by the physical requirements discussed in the last section, the intensities of a set of representative reflections involving the z parameters can be calculated. For this purpose the following $0kl$ reflections are used:

002	004	006	008	0·0·10	0·0·12	0·0·14
012	014	016	018			
022	024	026	028			
032	034	036	038			
042	044	046	048			
052	054	056				
062	064	066				
	074	076				
	084	086				

The $0k4$ and $0k6$ reflection sequences are carried somewhat farther than the other reflections because some of them are quite sensitive to the vertical parameters to be tested. The intensity order calculated with the parameters listed in the foregoing section proved to be in comparatively satisfactory agreement with the observed intensity order, but there is a very distinct improvement made by adding a small correction as allowed for in (48). An increase of $\Delta z = 0.003$ gives an excellent agreement between calculated and observed intensities for this group of $0kl$

Table 22
Calculated and observed intensities of important $0kl$ reflections (arrows indicate observed positions not corresponding with calculated order)

Reflection	$\sqrt{I}$ calculated	Reflection	$\sqrt{I}$ calculated
002	370.3	036	62.8
012	340.4	062	48.2
042	330.6	076	47.8
052	313.8	056	47.0
054	230.7	0·0·12	43.5
014	217.1	018	41.5
032	207.7	084	33.8
034	198.7	024	31.5
006	188.3	022	28.4
046	163.3	→ →	
074	141.4	038	24.6
086	140.6	0·0·10	17.9
048	117.7	066	13.4
008	112.9	028	10.1
0·0·14	91.7	026	5.3
004	86.6	064	1.5
044	83.2		
016	80.0		
→			

reflections, as shown in Table 22. Three inconsequential discrepancies remain: The relatively lower visually estimated intensity of 0·0·14 is obviously due to the fact that it is resolved into an $\alpha_1+\alpha_2$ doublet. [This is the case also for 0·0·10 (absent) and 0·0·12 (very weak), but these reflections are already so weak that the further weakening due to resolution into doublets makes very little difference.] The only other discrepanices in observed intensity order are possible slightly high positions of 062 and 084, as indicated. With the three exceptions noted, the observed and calculated intensity orders are in perfect agreement.

The final parameters, with due account given to the correction, are as follows:

	x	y	z	
Sb	0.044	0.128	0.179	
O_I	$\frac{1}{4}$	$\frac{1}{4}$	0.029.	(49)
O_{II}	0.147	0.058	-0.139	

Literature

[1] M. J. Buerger. *The crystal structure of marcasite.* Am. Mineralogist **16** (1931) 361–395.

[2] M. J. Buerger. *Interatomic distances in marcasite and notes on the bonding in crystals of löllingite, arsenopyrite, and marcasite types.* Z. Krist. (A) **97** (1937) 504–513.

[3] M. J. Buerger and Sterling B. Hendricks. *The crystal structure of valentinite (orthorhombic Sb_2O_3).* Z. Krist. (A) **98** (1937) 1–30.

13

Fourier synthesis, general theory

It is well known that a periodic function (within reasonable limitations) can be represented by an appropriate sum of cosine and sine terms known as a *Fourier series.* Since a crystal is periodic, its electron density can be neatly represented by such a series. Extensive use is made of the Fourier representation of the electron density in crystal-structure analysis.

The general form of Fourier series

The most general form of a Fourier series involves a sum of weighted exponentials of positive and negative multiples of an angle ϕ:

$$\begin{aligned}\rho_{(\phi)} &= K_0\, e^{i0\phi} + K_1\, e^{i1\phi} + K_2\, e^{i2\phi} + K_3\, e^{i3\phi} + \cdots \\ &\qquad + K_{-1}\, e^{i(-1)\phi} + K_{-2}\, e^{i(-2)\phi} + K_{-3}\, e^{i(-3)\phi} + \cdots \qquad (1)\\ &= \sum_{n=-\infty}^{\infty} K_n\, e^{in\phi} \qquad (2)\end{aligned}$$

The coefficients K_n are the Fourier coefficients; in the general case these are complex numbers.

The Fourier series can also be expressed as a sum of weighted cosines and sines of positive and negative multiples of an angle. To make this transformation, let K_n be expressed as the sum of its real and imaginary components,

$$K_n = R_n + iI_n, \qquad (3)$$

and let the exponential be represented by its real and imaginary parts, according to Euler's relation:

$$e^{in\phi} = \cos n\phi + i \sin n\phi. \qquad (4)$$

Then the product in (2) becomes

$$K_n e^{in\phi} = (R_n + iI_n)(\cos n\phi + i \sin n\phi) \tag{5}$$

$$= R_n \cos n\phi + iR_n \sin n\phi + iI_n \cos n\phi - I_n \sin n\phi \tag{6}$$

$$= (R_n \cos n\phi - I_n \sin n\phi) + i(R_n \sin n\phi + I_n \cos n\phi). \tag{7}$$

Thus, the general trigonometric form of (2) is

$$\begin{aligned} \rho_{(\phi)} = & \sum_{n=-\infty}^{\infty} R_n \cos n\phi \\ & - \sum_{n=-\infty}^{\infty} I_n \sin n\phi \\ & + i \sum_{n=-\infty}^{\infty} R_n \sin n\phi \\ & + i \sum_{n=-\infty}^{\infty} I_n \cos n\phi. \end{aligned} \tag{8}$$

If the function $\rho_{(\phi)}$, to be represented by the Fourier series, is real, then the imaginary contributions in (8) must cancel one another. Since

$$\sin(-n\phi) = -\sin n\phi, \tag{9}$$

then provided

$$R_{-n} = R_n, \tag{10}$$

it follows that

$$R_{-n} \sin(-n\phi) = -R_n \sin n\phi, \tag{11}$$

so that the terms in the third line of (8) cancel in pairs. Similarly, since

$$\cos(-n\phi) = \cos n\phi, \tag{12}$$

then, provided

$$I_{-n} = -I_n, \tag{13}$$

it follows that

$$I_{-n} \cos(-n\phi) = -I_n \cos n\phi, \tag{14}$$

so that the terms in the fourth line of (8) cancel in pairs. Thus when the function to be represented by the Fourier summation is real, (8) reduces to

$$\begin{aligned} \rho_{(\phi)} = & \sum_{n=-\infty}^{\infty} R_n \cos n\phi \\ & - \sum_{n=-\infty}^{\infty} I_n \sin n\phi. \end{aligned} \tag{15}$$

There is a compact formulation of the consequences of $\rho_{(\phi)}$ being real. According to (3), K_n can be expressed in terms of its real and imaginary components. The comparison of K_n with K_{-n}, obtained from (10) and (13), is

$$\begin{aligned} K_n &= R_n + iI_n, \\ K_{-n} &= R_n - iI_n. \end{aligned} \tag{16}$$

This corresponds to an experimental observation known as *Friedel's law*, which is discussed further in Chapters 15 and 20.

It should be observed that the sign of the second summation in (15) is negative. This is a consequence of defining (2) in terms of a positive angle. If it is defined in terms of a negative angle, (4) becomes

$$e^{-in\phi} = \cos n\phi - i \sin n\phi. \tag{17}$$

Then the corresponding forms of (2), (8), and (15) are

(2):
$$\rho_{(\phi)} = \sum_{n=-\infty}^{\infty} K_n e^{-in\phi}; \tag{18}$$

(8):
$$\begin{aligned} \rho_{(\phi)} = & \sum_{n=-\infty}^{\infty} R_n \cos n\phi \\ & + \sum_{n=-\infty}^{\infty} I_n \sin n\phi \\ & - i \sum_{n=-\infty}^{\infty} R_n \sin n\phi \\ & + i \sum_{n=-\infty}^{\infty} I_n \cos n\phi; \end{aligned} \tag{19}$$

(15) (ρ real):
$$\begin{aligned} \rho_{(\phi)} = & \sum_{n=-\infty}^{\infty} R_n \cos n\phi \\ & + \sum_{n=-\infty}^{\infty} I_n \sin n\phi. \end{aligned} \tag{20}$$

Forms (18), (19), and (20) are preferred in crystal-structure analysis provided (45) of Chapter 2 is written with the exponential of a positive angle. The reason for this is pointed out in Chapter 15.

One-dimensional synthesis

Angular, absolute, and fractional coordinates. Some of the properties of Fourier synthesis as applied to crystal-structure analysis can be studied without incurring the complications of three dimensions by considering a corresponding one-dimensional series. In the summation given in (2) and (18) the electron density is given for a point expressed as a phase location, ϕ, of the phase cycle 0 to 2π. In some applications it is desirable to express this location in terms of an absolute linear measure X along the pattern of the function. The function repeats at translation interval a, and the distance along the interval is proportional to the phase ϕ. This provides the proportion

$$\frac{\phi}{2\pi} = \frac{X}{a}. \tag{21}$$

Therefore, the value

$$\phi = 2\pi \frac{X}{a} \tag{22}$$

can be substituted into (18) to give

$$\rho_{(X)} = \sum_{n=-\infty}^{\infty} K_n \, e^{-i2\pi nX/a}. \tag{23}$$

In other applications, it simplifies the notation if the location along the cycle is expressed in terms of a fraction x, of the full cycle. This amounts to normalizing the cycle to unity, so that the proportion can be written

$$\frac{\phi}{2\pi} = \frac{x}{1}. \tag{24}$$

This permits the value

$$\phi = 2\pi x \tag{25}$$

to be substituted into (18) to give a third form

$$\rho_{(x)} = \sum_{n=-\infty}^{\infty} K_n \, e^{-i2\pi nx}. \tag{26}$$

If (21) and (24) are combined, the relation between the three sets of coordinates is found to be

$$\frac{\phi}{2\pi} = \frac{X}{a} = x. \tag{27}$$

The angular, absolute, and fractional coordinates are compared graphically in Fig. 1.

Evaluation of the Fourier coefficients. Relations (18), (23), and (26) express the electron density at a desired point in the period in terms of a set of unknown Fourier coefficients, K_n. The values of these coefficients can be determined from the following analysis.

The amplitude of the hth interference maximum resulting from diffraction by a one-dimensional crystal can be expressed, in terms of the scattering powers, f_j, of its atoms located at points X_j, as

$$F_h = \sum_j f_j \, e^{i2\pi h X_j/a} \tag{28}$$

If, instead of assuming discrete atoms at locations X_j, one assumes that the structure is composed of an electron density, $\rho_{(X)}$, which varies continuously along the period 0 to a, the resulting amplitude of the scattered

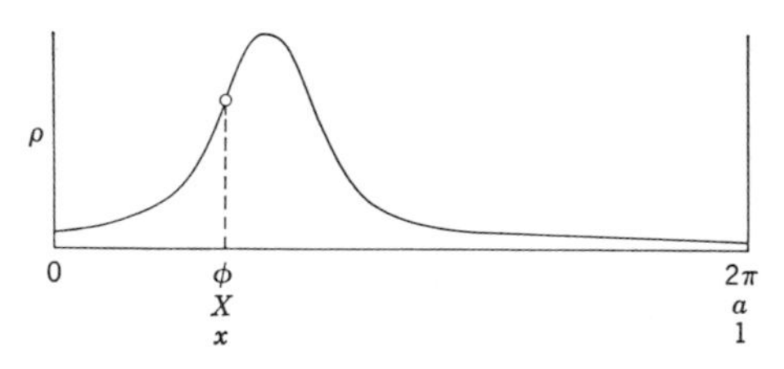

Fig. 1.

waves can be expressed by replacing the summation over the several atoms in (28) by an integration over a range 0 to a:

$$F_h = \int_0^a \rho_{(X)} \, e^{i2\pi hX/a} \, dX. \tag{29}$$

Into this expression can be substituted the Fourier representation of the electron density given in (23):

$$F_h = \int_0^a \left(\sum_{n=-\infty}^{\infty} K_n \, e^{-i2\pi nX/a} \right) e^{i2\pi hX/a} \, dX. \tag{30}$$

Since an integral of a sum of terms is the sum of the individual integrals, this can be rewritten

$$\begin{aligned} F_h &= \sum_{n=-\infty}^{\infty} \left(\int_0^a K_n \, e^{-i2\pi nX/a} \, e^{i2\pi hX/a} \, dX \right) \\ &= \sum_{n=-\infty}^{\infty} \left(\int_0^a K_n \, e^{i2\pi(h-n)X/a} \, dX \right). \end{aligned} \tag{31}$$

Each of the integrals in parentheses is characterized by two integers,

h (the order of the reflection) and n (the Fourier coefficient). That integral characterized by $h = n$ has the following value:

$$\begin{aligned}\int_0^a K_n\, e^{i2\pi(h-n)X/a}\, dX &= \int_0^a K_n\, e^{i2\pi(n-n)X/a}\, dX \\ &= \int_0^a K_n\, e^0\, dX \\ &= K_n \int_0^a dX \\ &= K_n\, a. \end{aligned} \tag{32}$$

On the other hand, those integrals characterized by $h \neq n$ vanish:

$$\begin{aligned}\int_0^a K_n\, e^{i2\pi(h-n)X/a}\, dX &= \frac{a}{i(h-n)2\pi}\, e^{i2\pi(h-n)X/a} \Big[_0^a \\ &= \frac{a}{i(h-n)2\pi} \left(e^{i2\pi(h-n)1} - e^{i2\pi(h-n)0}\right) \\ &= \frac{a}{i(h-n)2\pi}(1-1) \\ &= 0. \end{aligned} \tag{33}$$

Since the only non-vanishing integral has the value given in (32), the value of (31) is

$$F_h = \sum_{n=-\infty}^{\infty} aK_{n=h}. \tag{34}$$

This merely means that for each h,

$$K_n = \frac{1}{a} F_h. \tag{35}$$

This integration has an interesting graphical interpretation. The integral is a sum of a sequence of infinitesimal vectors in the complex plane, each differing from its neighbor by a phase angle $2\pi(h-n)X/a$. Since h is an integer and n is an integer, $(h-n)$ is an integer. When this difference equals 1, the phase of each infinitesimal vector differs from its neighbor by $2\pi X/a$, and the entire series (integrating X from zero to a) turns through a complete circuit; when this difference equals 2, the series turns through two complete circuits, etc. In each case the end of the circuit is the origin, and the resultant of the series is zero. But when the difference $(h-n)$ is zero, all the little vectors have the same phase angle and accordingly they lie in a straight line. In this case their resultant is their arithmetic sum.

If the value of K_n in (35) is substituted in (23), the specific form of Fourier synthesis used in crystal-structure analysis is obtained:

$$\rho_{(X)} = \frac{1}{a} \sum_{h=-\infty}^{\infty} F_h \, e^{-i2\pi hX/a}. \tag{36}$$

The significance of (35) is that for any grating of periodically varying scattering power, the hth Fourier component of the density scatters in phase only in the hth-order reflection. For all other orders the various regions of its scattering power are completely out of phase, and thus each order except the hth order destroys itself.

The term F_0 has a special significance. It can be evaluated from (29) for $h=0$:

$$F_0 = \int_0^a \rho_{(X)} \, e^0 \, dX$$
$$= Z. \tag{37}$$

Here Z is the arithmetic sum of all the scattering matter in the period 0 to a. For crystals, this corresponds to the number of electrons in the interval. In effect, this says that in the direction of the direct beam, all parts of the grating scatter in phase, and, therefore, the zero-order maximum is the sum of the distributed scattering power of the grating.

Another view of Fourier synthesis

In the beginning of this chapter, it was assumed that the reader was familiar with the fact that a periodic function (within reasonable limitation) can be represented by a Fourier series. The proof of this is given in many books on mathematics, but their treatments share the same formal approach and require a good deal of supplementary background. An alternative approach is given below. This has several advantages: 1, only a limited background is required; 2, the general reason why a Fourier series can represent a periodic function is obvious from the outset, and 3, some of the reasons why the series is so useful in crystal-structure analysis are brought out explicitly.

Representation of a periodically repeated discrete value. The following trigonometric series has some interesting properties:

$$s = \cos 0\psi + \cos 1\psi + \cos 2\psi + \cos 3\psi \cdots + \cos H\psi$$
$$+ \cos(-1\psi) + \cos(-2\psi) + \cos(-3\psi) \cdots + \cos(-H\psi)$$
$$= \sum_{h=-H}^{H} \cos h\psi. \tag{38}$$

This series can be represented geometrically as the projection of a set of unit vectors each of whose directions differs from its neighbor's by the same angle ψ, Fig. 2. It can be seen that the various projections have various values and that these are both positive and negative. If ψ has a general value, these projections tend to annul one another, possibly leaving a small residue as the sum, s, in (38). But the situation is altogether different if ψ is $0, 2\pi, 4\pi \cdots m2\pi$. In these cases all vectors

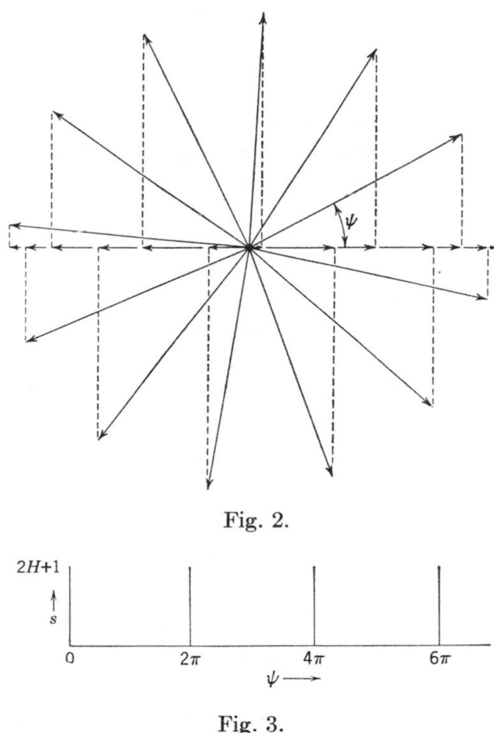

Fig. 2.

Fig. 3.

point in the same direction, and their sum is their arithmetic sum, namely $2H+1$.

In order to see how (38) may be useful, ignore, temporarily, the small residue noted above. Then, if the sum, s, in (38) is plotted against ψ, the ideal result is shown in Fig. 3. When $\psi = 0, 2\pi, 4\pi \cdots m2\pi$, the sum is $2H+1$, but for other values of ψ it is zero. Thus (38) has the property of representing a magnitude recurring periodically at interval 2π. Another description is that (38) has the property of representing a periodically recurring point whose weight is given by this magnitude.

This gives a rough qualitative appreciation of the nature of the sum in (38), but a more quantitative appreciation of it is desirable. Series (38), as it stands, is not easy to treat quantitatively. Fortunately, as will be shown, it is equivalent to the following exponential series, which can be readily evaluated:

$$S = e^{i0\psi} + e^{i1\psi} + e^{i2\psi} + e^{i3\psi} \cdots + e^{iH\psi}$$
$$+ e^{-i1\psi} + e^{-i2\psi} + e^{-i3\psi} \cdots + e^{-iH\psi}$$
$$= \sum_{h=-H}^{H} e^{-ih\psi}. \tag{39}$$

That (38) and (39) are equivalent can be seen as follows: The first term in both (38) and (39), namely, $\cos 0\psi$ and $e^{i0\psi}$, is equal to unity. The values of the two paired terms in (38) are equal; each pair is therefore equal to $2 \cos n\psi$. Each such pair is equal to the corresponding pair in (39) by virtue of the relation

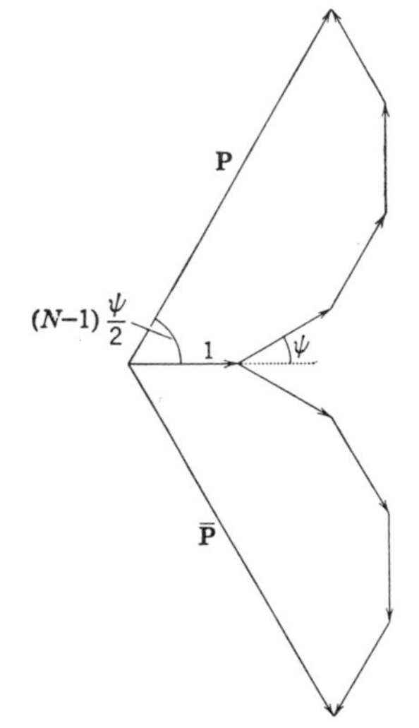

Fig. 4.

$$e^{ix} + e^{-ix} = 2 \cos x. \tag{40}$$

Thus (38) and (39) are equal.

The representation of (39) on the Argand diagram is shown in Fig. 4. This shows two sequences of equal vectors, each of unit length. In one sequence each vector is inclined to its neighbors by a positive angle ψ, in the other by a corresponding negative angle. The resultant of the two sequences is the required sum, S, except that the initial vector occurs twice. The quantities $\mathbf{P}$ and $\bar{\mathbf{P}}$ are complex. The required sum can be expressed as

$$S = \mathbf{P} + \bar{\mathbf{P}} - 1. \tag{41}$$

The complex quantity $\mathbf{P}$ was evaluated in (18) of Chapter 2 as

$$\mathbf{P} = \frac{\sin \dfrac{N\psi}{2}}{\sin \dfrac{\psi}{2}} e^{i(N-1)\psi/2}. \tag{42}$$

In deriving this in Chapter 2 it was convenient to sum n from 0 to $N-1$. In (39) it is desired to sum h from $-H$ through 0 to H. This requires that $H = N-1$ so that N in (42) must be changed to $H+1$. The desired form of (42) is therefore

$$\mathbf{P} = \frac{\sin\,(H+1)\,\dfrac{\psi}{2}}{\sin \dfrac{\psi}{2}} e^{iH\psi/2}. \tag{43}$$

If this value of (43) is substituted into (41), it can be reduced as follows:

$$S = \frac{\sin (H+1)\dfrac{\psi}{2}}{\sin \dfrac{\psi}{2}} e^{iH\psi/2} + \frac{\sin (H+1)\dfrac{\psi}{2}}{\sin \dfrac{\psi}{2}} e^{-iH\psi/2} - 1 \tag{44}$$

$$= 2\frac{\sin (H+1)\dfrac{\psi}{2}}{\sin \dfrac{\psi}{2}} \cos H\frac{\psi}{2} - 1. \tag{45}$$

This expression can be rendered more compact by making use of the trigonometric identity

$$\sin \alpha \cos \beta = \tfrac{1}{2} \sin (\alpha+\beta) + \tfrac{1}{2} \sin (\alpha-\beta). \tag{46}$$

If this is applied to the term $\sin (H+1)\dfrac{\psi}{2} \cos H\dfrac{\psi}{2}$, (45) may be reduced as follows:

$$S = \frac{\sin (H+1)\dfrac{\psi}{2}}{\sin \dfrac{\psi}{2}} \cos H\frac{\psi}{2} - 1$$

$$= 2\frac{\tfrac{1}{2}\sin\left[(H+1)\dfrac{\psi}{2} + H\dfrac{\psi}{2}\right] + \tfrac{1}{2}\sin\left[(H+1)\dfrac{\psi}{2} - H\dfrac{\psi}{2}\right]}{\sin \dfrac{\psi}{2}} - 1$$

$$= \frac{\sin (2H+1)\dfrac{\psi}{2} + \sin\dfrac{\psi}{2}}{\sin \dfrac{\psi}{2}} - 1$$

$$= \frac{\sin (2H+1)\dfrac{\psi}{2}}{\sin \dfrac{\psi}{2}} + 1 - 1$$

$$= \frac{\sin (2H+1)\dfrac{\psi}{2}}{\sin \dfrac{\psi}{2}}. \tag{47}$$

This is an interesting and important function. Its behavior is indicated in Fig. 5. The function has a narrow maximum at $\psi = 0, 2\pi, 4\pi \cdots$, and a small background ripple elsewhere. The somewhat similar function $\sin N(\psi/2)/\sin(\psi/2)$ was discussed in Chapter 2, and it was noted there that this function has a peak at 2π when N is odd, but an inverted peak at 2π when N is even. In (47) the place of the integer, N, is taken by $2H+1$, which is always odd, so that the peak is always a positive peak.

When $H = \infty$, (47) perfectly reproduces a periodically recurring vertical straight line. This shows that the function does have the behavior

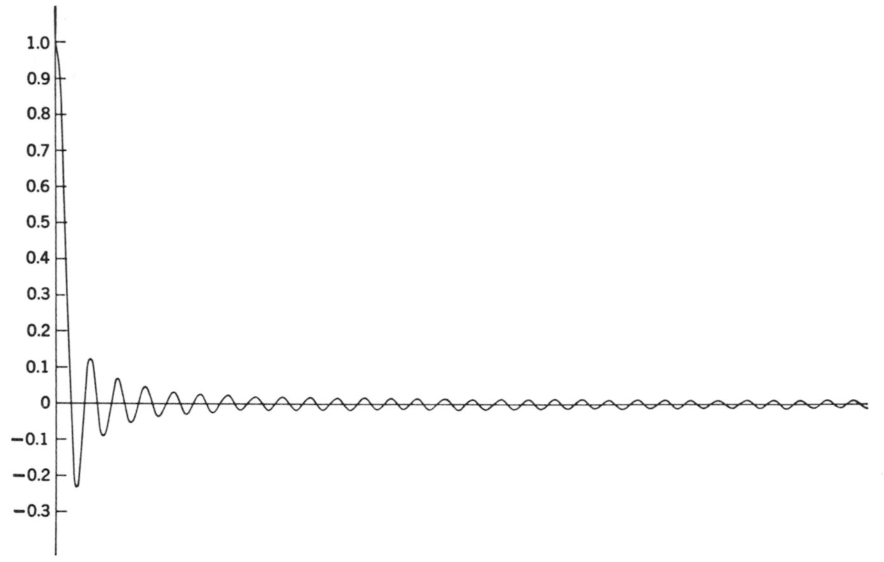

Fig. 5.

suggested by this idealization in Fig. 3, provided that the number of terms, $2H+1$, in the summation is infinite. For a finite number of terms, there are two kinds of departure from ideal behavior: In the first place, the vertical line of Fig. 3 is blurred into a narrow "peak", Fig. 6, and secondly, it is surrounded by a region in which there is a ripple in the background. The ripple is of such a nature as to surround the point first by a shallow negative region, after which it oscillates with decreasing amplitude about zero. This effect is known as the *series termination effect,* since it is caused by using a series which, for practical reasons, is terminated short of an infinite number of terms. This imperfection in representation decreases as the number of terms H increases.

The height of the peak is always $2H+1$. If one wishes to represent a recurring unit magnitude, the sum in (38) and (39) must be divided by $2H+1$. Similarly, if one wishes to represent a recurring magnitude of weight w, the sum must be multiplied by $w/(2H+1)$. This more generalized series is

$$S = \frac{w}{2H+1} \sum_{h=-H}^{H} e^{-ih\psi}. \tag{48}$$

Such a summation is an elementary but very fundamental kind of Fourier series.

Representation of a periodic sequence of several discrete values. The function given in (48) can be generalized in several ways. In the

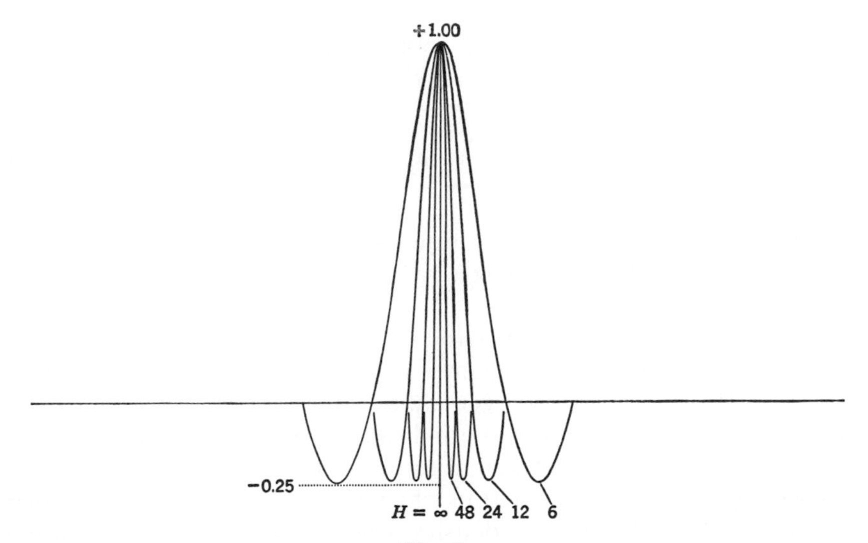

Fig. 6.

The shape of a vertical line as represented by a Fourier series with various numbers of Fourier terms.

first place, the peak can be shifted from $\psi = 0$ to $\psi = \phi$ by the following device: One wishes the exponentials to have the value e^0 when $\psi = \phi$. This occurs if the exponentials are given the form $e^{ih(\psi-\phi)}$. The more general form of (48) is, therefore,

$$S = \frac{w}{2H+1} \sum_{h=-H}^{H} e^{-ih(\psi-\phi)}. \tag{49}$$

This general form represents a sequence of peaks, all of height w at locations $\psi = \phi,\ 2\pi+\phi,\ 4\pi+\phi \cdots$.

It is obvious that if a single peak can be represented by the single summation of (49), several peaks can be represented by as many summations. Suppose one wishes to represent peaks $p_1,\ p_2,\ p_3 \cdots$ of weights $w_1, w_2, w_3 \cdots$ at locations $\phi_1, \phi_2, \phi_3 \cdots$. The Fourier representation of this sequence of points is

$$S = S_1 + S_2 + \cdots$$

$$= \frac{1}{2H+1} \left\{ w_1 \sum_{h=-H}^{H} e^{-ih(\psi-\phi_1)} + w_2 \sum_{h=-H}^{H} e^{-ih(\psi-\phi_2)} \cdots \right\} \quad (50)$$

$$= \frac{1}{2H+1} \sum_{h=-H}^{H} e^{-ih\psi} \sum_{j=1}^{J} w_j \, e^{ih\phi_j}. \quad (51)$$

It is convenient to express (51) in terms of a linear instead of angular coordinate. This can be done by making use of (27):

$$S = \frac{1}{2H+1} \sum_{h=-H}^{H} e^{-i2\pi hX/a} \sum_{j=1}^{J} w_j \, e^{i2\pi hX_j/a}. \quad (52)$$

Relation (52) could be used, for example, to represent a periodic sequence of point atoms of atomic numbers $w_1, w_2, w_3 \cdots$ occupying locations $X_1, X_2, X_3 \cdots$.

Representation of a continuously variable periodic function. Summation (52) provides a sequence of J discrete values, periodically repeated. It can be readily generalized to provide a continuously variable function which repeats periodically. In making this change, it is desirable to transform the weighting to a density. This can be done by selecting small ranges ΔX within which the weighting can be regarded as uniform. In such a range a density ρ can be defined as the weighting w_j in a range ΔX_j, or

$$w_j = \rho_j \, \Delta X_j, \quad (53)$$

so that (52) can be written

$$S = \frac{1}{2H+1} \sum_{h=-H}^{H} e^{-i2\pi hX/a} \sum_{j=1}^{J} \rho_j \, \Delta X_j \, e^{i2\pi hX_j/a}. \quad (54)$$

Now, in the limit

$$\rho_j \, \Delta X_j \xrightarrow[\Delta X \to 0]{\lim} \rho_{(X)} \, dX, \quad (55)$$

and (54) becomes

$$S = \frac{1}{2H+1} \sum_{h=-H}^{H} e^{-i2\pi hX/a} \int_0^a \rho_{(X)} \, e^{i2\pi hX/a} \, dX. \quad (56)$$

Comparison with (29) shows that the integral in (56) is the same as F_h, so that (56) can be more compactly written

$$S = \frac{1}{2H+1} \sum_{h=-H}^{H} F_h \, e^{-i2\pi hX/a}. \tag{57}$$

The term $1/(2H+1)$ is the reciprocal of the total number of terms in the summation. This is a constant for any summation. Thus (57) can be expressed

$$S = C \sum_{h=-H}^{H} F_h \, e^{-i2\pi hX/a}. \tag{58}$$

This has the same form as (36), which is the Fourier representation of the electron density.

Some characteristics of Fourier syntheses. The summation in (58) was devised to represent any periodic density function. Suppose that this density is concentrated in discrete atoms. Then F_h, the complex amplitude scattered by the period, can be expressed in terms of the scattering powers, $f_{h,j}$ of the J atoms of the cell:

$$\begin{aligned} F_h &= f_{h,1} \, e^{i2\pi hx_1} + f_{h,2} \, e^{i2\pi hx_2} \cdots \\ &= \sum_{j=1}^{J} f_{h,j} \, e^{i2\pi hx_j}. \end{aligned} \tag{59}$$

Thus (58) can be expressed as

$$S = C \sum_{h=-H}^{H} e^{-i2\pi hx} \sum_{j=1}^{J} f_{h,j} \, e^{i2\pi hx_j}. \tag{60}$$

This has the same form as (52), which holds for a set of J discrete points. The sharp points having weights w of (52) are replaced in (60) by the "broad points," or atoms, whose scattering powers are f's. The parallelism between series (52) representing sharp points and series (60) representing broad points permits the properties of the latter, which is the actual case, to be interpreted in terms of the former, which is the simple synthetic case.

The simple case was devised by adding together, in (50), summations, each of which represented a single point. Thus (60) can be decomposed into a sum of separate Fourier series, each representing one atom of the period (cell). Each such individual synthesis is similar to (49) in that, at the location of the atom, all phases are zero. The fact that all Fourier waves are in phase at this point is the reason for the appearance of a peak in the individual synthesis. When the phase of a Fourier wave is zero at an atom location, its phase at the origin of the period (cell) is fixed. It is this phase, the ϕ of (49) and the $2\pi hx$ of (59), which provides the phase character of the Fourier coefficient of the individual synthesis.

For simplicity, now, consider the important case of a centrosymmetrical structure with the origin chosen at a symmetry center. Then (as noted in the next section) the phases mentioned above can be only 0 or π. Then each Fourier wave of the individual synthesis has its crest (in the sense of a positive value) at the atom location, and centrosymmetry requires this now to have either a crest (for phase 0) or a trough (for phase π) at the origin. When several individual syntheses are added, the same Fourier wave may have a crest at the origin for one atom, and a trough at the origin for another atom. The phase of the composite wave depends on the algebraic sum of these individual crests and troughs. If the atom locations are known in advance, the phase of the composite wave at the origin can be determined. Usually the atom locations are unknown, so the phase remains unknown.

Now, since each Fourier wave must have a crest or trough exactly at the origin, it cannot be expected that a crest will occur exactly at an arbitrary atom location. In general a particular Fourier wave may have a general region, either positive or negative, at the atom location. But the sum of all the waves of different h must give a net positive residue at each atom location in order that a peak representing the atom may appear there. But of those special waves whose wavelengths permit them to have a crest or trough very close to the atom location, there must be many which have crests. These waves, especially, operate to more sharply define the center of the peak for the atom. This is especially true of Fourier waves of high index.

Phase determination using intense high-index reflections. The last feature of Fourier synthesis is the basis of the *method of intense high-index-reflections* for the determination of the signs of certain F's. The method is illustrated in Fig. 7. When certain high-index reflections are very intense it is assumed that they are of the type which corresponds to Fourier waves that have crests at most, or all, of the atom locations.[1] That they have crests at many such locations is the reason why the wave has a large amplitude. An example is shown in Fig. 7. The crystals of $C_6(CH_3)_6$, though triclinic, have a cell whose (001) section is almost exactly hexagonal. There are four very strong reflections, 001, 340, $4\bar{7}0$, and $\bar{7}30$. The strong 001 reflection assures that all atoms lie close to the (001) planes. The other strong reflections are from planes parallel to the c axis. They are outlined in Fig. 7. The atoms must also lie close to these planes. If the molecule (whose shape is assumed from the chemical formula) is shifted around the projection, it is found that its atoms fit the intersections of these planes neatly as shown in Fig. 7. With one molecule per cell, the origin is the only permissible location, and the only unknown is the orientation of the molecule. In general, any of the Fourier waves for 340, $4\bar{7}0$, and $\bar{7}30$ might have been either + or − at the origin. Had any been negative, its crest would have occurred half way between the lines of Fig. 7, and then the molecules could not have been fitted to the intersections.

This method of phase determination and atom location has been used chiefly with a certain class of organic compounds, and especially by Robertson and his school.[2,3] It is especially useful in centrosymmetrical

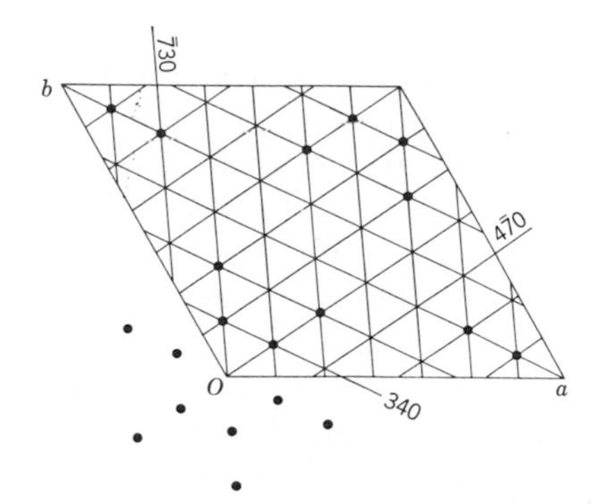

Fig. 7.
Relation between the locations of the carbon atoms in $C_6(CH_3)_6$ and the crests of large-amplitude reflections.
(After Lonsdale.[1])

cases when one dimension of the cell is small, and when the known shape of the molecule is planar and centrosymmetrical. If it is assumed that the number of molecules per cell is such that the molecule may occupy a symmetry center of the space group, the problem reduces to finding the orientation of the molecule. This must be such that the atoms of the molecules lie close to near-intersections of crests or troughs of Fourier waves corresponding to intense reflections.

Convergence of successive Fourier syntheses

One of the most important uses of Fourier syntheses in crystal-structure analysis is to improve the knowledge of the locations of atoms of a structure for which the approximate locations are already known. This matter will be discussed in more detail in Chapter 22. The general reasons why this is so may be anticipated by considering a centrosymmetrical crystal referred to an origin taken at a center. Under these circumstances (as will be shown in Chapter 15), the quantity F_h, in (28) and (59), reduces to

$$F_h = f_{h,1} \cos 2\pi h x_1 + f_{h,2} \cos 2\pi h x_2 \cdots$$
$$= \sum_{j=1}^{J} f_{h,j} \cos 2\pi h x_j. \tag{61}$$

These are merely positive or negative numbers, $\pm|F_h|$. Furthermore, the exponential in (58) reduces to a cosine, so that

$$S = C \sum_{h=-H}^{H} \pm|F_h| \cos 2\pi h x. \tag{62}$$

Each Fourier component is therefore a cosine wave of amplitude $|F_h|$, which is either positive or negative at the origin.

Now, if the approximate locations of the atoms are known, the several real quantities F_h can be computed from (61). Due to the only-approximate knowledge of the atom location, the computed quantities may not be very accurate, yet they all have correct signs, except for those whose magnitudes are close to zero. For these particular F's, the near balance of terms in (61) may permit the sum to be thrown to the wrong side of zero. If the calculated signs are attributed to the experimentally measured F_h's, (omitting any with doubtful signs) the correct electron density (62) can be computed. This computed electron density shows all atoms in essentially their correct positions except for slight shifts due to omitted small F_h's with doubtful signs[§]. If a new computation of F_h in (61) is now based upon the improved locations of atoms revealed by this electron-density map, the F_h's which were omitted from the first electron-density synthesis because of doubtful signs can be added to the series. The new series yields atom locations further improved. In this way *a sequence of successive Fourier series converges to the correct atom locations.*

A sequence of successive Fourier syntheses converges to the correct atom locations even if a considerable number of F_h's with incorrect signs are included in the first synthesis, provided that the F's with incorrect signs have relatively small numerical values.[†] This is because if the sign of one of the many cosine waves which build up the appearance of an atom is reversed, it weakens the appearance of the atom but does not annihilate it. For this reason, a limited number of F's with incorrect signs can be tolerated in a Fourier synthesis provided these are not F's of large numerical values.

[†] An exception occurs for crystals which can be regarded as having substructures. For such crystals, the correctness of the structure depends upon F's which are often quite small.

[§] And due to F_h's omitted beyond $h = H$; this gives rise to the series-termination error, noted more fully in Chapter 22.

If a crystal is composed of both heavy and light atoms, the amplitudes of most F's are dominated by the heavy atoms. For such crystals, it is often possible to determine the signs of the F's as computed with (61) by using terms for the heavier atoms only. The Fourier synthesis based upon these F's then shows both heavy and light atoms. This is the basis of the *heavy-atom method* discussed in detail in Chapter 19.

Literature

Examples of the method of intense high-index reflection

[1] Kathleen Lonsdale. *The structure of the benzene ring in* $C_6(CH_3)_6$. Proc. Roy. Soc. (London) (A) **123** (1929) 494–515.

[2] J. Monteath Robertson and J. G. White. *The crystal structure of coronene: A quantitative x-ray investigation.* J. Chem. Soc. (1945) 607–617.

[3] D. M. Donaldson and J. M. Robertson. *The crystal and molecular structure of ovalene: A quantitative x-ray investigation.* Proc. Roy. Soc. (London) (A) **220** (1953) 157–170.

[4] A. Tulinsky and J. G. White. *Rigid-body torsional vibrations in three typical members of a class of benzene derivatives.* Acta Cryst. **11** (1958) 7–14.

14

Forms of Fourier syntheses useful in crystal-structure analysis

In the last chapter the general nature of Fourier synthesis was discussed and some of its properties useful in crystal-structure analysis were demonstrated. In this chapter the particular forms of Fourier syntheses which are actually used in crystal-structure analysis are developed. The most important forms are the three-dimensional synthesis of the electron density, the synthesis of sections through the electron density, and the synthesis of projections of the electron density. There are other less important forms, most of which fall in the category of syntheses of projections of limited portions of a unit cell, and modulated projections.

Three-dimensional synthesis

The representation of a function of one variable by means of Fourier synthesis was discussed in the last chapter. A function of several variables can also be represented by Fourier series. In particular, the electron density as a function of the x, y, and z coordinates of three-dimensional space can be so represented. The relation behind this extension is fundamentally the same as that used by the mathematician Cantor to enumerate the points in a plane in terms of the points along a line.

Any point, P_{xyz}, lies in a line radiating from the origin. Along this line the electron density may be represented by a one-dimensional Fourier series. This Fourier series may be thought of as composed of a fundamental wave and its harmonics. Such a Fourier series synthesizes the electron density along the entire line from the origin. If every point in space is to be covered, then there must be a Fourier series along each of the infinite number of rays in space. This requires, ideally, a triply infinite number of Fourier coefficients. It should be noted that at every point P_{xyz} there occurs not only a summation due to the Fourier waves normal to the ray to P_{xyz}, but also a summation of Fourier waves normal to all other rays.

Correspondence between fundamental Fourier waves and crystal planes. Since the structure of the crystal has the periodicity of its lattice, only those fundamental Fourier waves are permitted which are periodic with the lattice. It is evident that the wavefronts of a

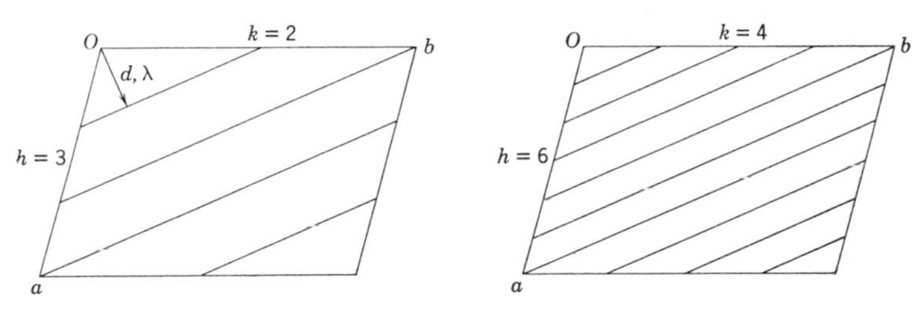

Fig. 1.

fundamental Fourier wave must have the same geometry as the stack of rational planes parallel to the wavefront. It is convenient to associate a Fourier wave with that particular crystal plane having the same direction and spacing. Let the indices of a particular stack of crystal planes be $(h_1\, k_1\, l_1)$, Fig. 1; then the corresponding Fourier wave may be designated by the same set of indices $h_1\, k_1\, l_1$, written without parentheses. The wavelength λ_{hkl} of the Fourier wave is identical with the spacing d_{hkl} of the stack of crystal planes.

If a crystal plane is rational, then its three indices (referred to a primitive cell) can contain no common factor. Consider the contrary proposition, namely a set of indices (nh, nk, nl) containing the common factor n. If these indices were interpreted on a crystallographic basis, they would correspond to the set of planes (hkl), plus $n-1$ more sets of interleaved planes. Since the plane (hkl) contains all the points of the lattice, the additional $n-1$ sets of planes can contain no lattice points, so that the indices nh, nk, nl do not correspond to a set of rational planes. But they do correspond to the nth harmonic of the Fourier wave hkl as shown on the right of Fig. 1. Accordingly, if every combination of all integers h, k, and l is assured for the indices of Fourier waves, then not only are fundamental Fourier waves of every direction included, but all their harmonic wave are included also. To write a Fourier series to represent a three-dimensional function, therefore, every different combination of indices hkl must be represented by a different Fourier coefficient K_n in series (18) of Chapter 13. Thc coefficient is accordingly designated K_{hkl}. In this way, the three-dimensional coefficients K_{hkl} are enumerated against the one-dimensional set K_n. (This enumeration is similar to that used by Cantor to enumerate the points in a plane in terms of the points along a line.)

The phase of a wave at a point in the unit cell. It remains to determine the value of the exponent $n\phi$ in (18) of Chapter 13, which corresponds with the Fourier coefficient K_{hkl}. Figure 2 shows a diagrammatic representation of a Fourier wave hkl of wavelength d_{hkl}. What is the phase of the wave at point P_{xyz}, where the coordinates xyz of the point are fractions of the cell edges a, b, and c? Since the phase of the wave is proportional to the distance along its wave normal, d_{hkl}, the phase at P_{xyz} is the projection of vector $\mathbf{p}$ on d_{hkl}:

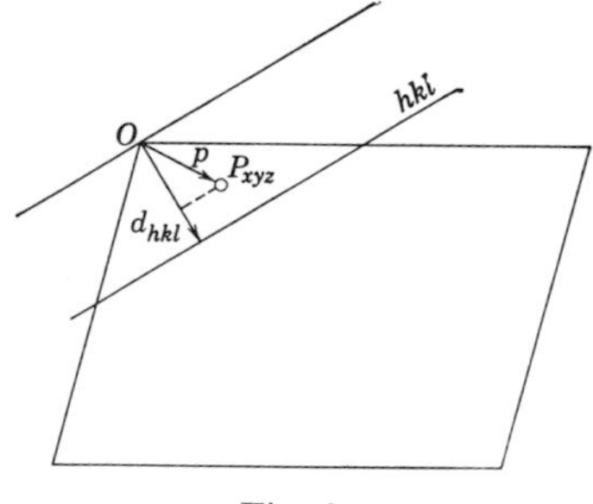

Fig. 2.

$$\frac{\phi}{2\pi} = \frac{\text{proj. } \mathbf{p}_{xyz}}{d_{hkl}}$$

$$= \frac{1}{d_{hkl}} (\text{proj. } \mathbf{p}_{xyz}). \tag{1}$$

It is convenient to express the right of (1) in vector notation. The vector of length $1/d_{hkl}$ and having the direction of d_{hkl} is the vector in reciprocal space from the origin to reciprocal-lattice point hkl; this vector is customarily written $\boldsymbol{\sigma}_{hkl}$. The right of (1) calls for the scalar product of this vector with vector $\mathbf{p}$, so that (1) can be written

$$\frac{\phi}{2\pi} = \boldsymbol{\sigma}_{hkl} \cdot \mathbf{p}_{xyz}. \tag{2}$$

To find the numerical value of this, one can substitute the values of $\boldsymbol{\sigma}_{hkl}$ and $\mathbf{p}_{hkl}$ expressed in terms of their components:

$$\boldsymbol{\sigma}_{hkl} = h\mathbf{a}^* + k\mathbf{b}^* + l\mathbf{c}^*, \tag{3}$$

and

$$\mathbf{p}_{xyz} = x\mathbf{a} + y\mathbf{b} + z\mathbf{c}. \tag{4}$$

The scalar product of these is

$$\boldsymbol{\sigma}_{hkl} \cdot \mathbf{p}_{xyz} = (h\mathbf{a}^* + k\mathbf{b}^* + l\mathbf{c}^*) \cdot (x\mathbf{a} + y\mathbf{b} + z\mathbf{c})$$

$$= hx + ky + lz. \tag{5}$$

If this is substituted in the right of (2), then the value of ϕ is seen to be

$$\phi = 2\pi(hx + ky + lz). \tag{6}$$

Therefore, the terms of the three-dimensional Fourier series corresponding to (18) of Chapter 13 are

$$\rho_{(xyz)} = \sum_h \sum_k^{\infty} \sum_l K_{hkl}\, e^{-i2\pi(hx+ky+lz)}, \tag{7A}$$
$$-\infty$$

or, in absolute coordinates,

$$\rho_{(XYZ)} = \sum_h \sum_k^{\infty} \sum_l K_{hkl}\, e^{-i2\pi(hX/a+kY/b+lZ/c)}. \tag{7B}$$
$$-\infty$$

Evaluation of the Fourier coefficient. Relation (7) expresses the electron density at a desired point (coordinates xyz) in the unit cell in terms of a set of unknown Fourier coefficients K_{hkl}. The values of these coefficients can be determined by an analysis similar to that used in the one-dimensional case of Chapter 13:

The amplitude of the reflection hkl in terms of the scattering powers f_j of the several atoms located at fractional coordinates x_j, y_j, and z_j was given in (45) of Chapter 2:

$$F_{hkl} = \sum_j f_j\, e^{i2\pi(hx_j+ky_j+lz_j)}. \tag{8}$$

If, instead of assuming discrete atoms at locations $x_j\, y_j\, z_j$, one assumes that the structure is composed of an electron density $\rho_{(xyz)}$, which varies continuously over the volume of the cell, then the amplitude of the scattered wave can be expressed by replacing the summation over the several atoms in (8) by an integration over the absolute volume V:

$$F_{hkl} = \int_0^V \rho_{(xyz)}\, e^{i2\pi(hx+ky+lz)}\, dV. \tag{9}$$

Since the volume is a function of a, b, and c, this integration over V can be expressed as a triple integration over a, b, and c. The relation between V and a, b, and c, is

$$V = abc\sqrt{1 - \cos^2\alpha - \cos^2\beta - \cos^2\gamma + 2\cos\alpha\cos\beta\cos\gamma}, \tag{10}$$

where α, β, and γ are the interaxial angles of the unit cell. In a similar way, the elementary unit of volume dV may be expressed in terms of dX, dY, and dZ, as follows:

$$dV = dX\,dY\,dZ\sqrt{1 - \cos^2\alpha - \cos^2\beta - \cos^2\gamma + 2\cos\alpha\cos\beta\cos\gamma}. \tag{11}$$

If (11) is divided by (10), the result is

$$\frac{dV}{V} = \frac{dX\,dY\,dZ}{abc}, \tag{12}$$

so that the change of variables is given by

$$dV = \frac{V}{abc}\,dX\,dY\,dZ. \tag{13}$$

The equations are less bulky if fractional coordinates,

$$\begin{aligned} x &= \frac{X}{a}, \\ y &= \frac{Y}{b}, \\ z &= \frac{Z}{c}, \end{aligned} \tag{14}$$

are used. From these it follows that

$$\begin{aligned} dX &= a\,dx, \\ dY &= b\,dy, \\ dZ &= c\,dz. \end{aligned} \tag{15}$$

If these values are substituted with (13), the relation between differential volume and coordinates becomes

$$\begin{aligned} dV &= \frac{V}{abc}\,a\,dx\,b\,dy\,c\,dz \\ &= V\,dx\,dy\,dz. \end{aligned} \tag{16}$$

When this is substituted into (9), it takes the form

$$F_{hkl} = \int_0^1 \int_0^1 \int_0^1 \rho_{(xyz)}\, e^{i2\pi(hx+ky+lz)}\, V\,dx\,dy\,dz. \tag{17}$$

To evaluate the Fourier coefficient, the value of $\rho_{(xyz)}$ is substituted from (7*A*) into (17):

$$F_{hkl} = \int_0^1 \int_0^1 \int_0^1 \sum_h \sum_k \sum_l K_{hkl}\, e^{-i2\pi(hx+ky+lz)}\, e^{i2\pi(hx+ky+lz)}\, V\,dx\,dy\,dz. \tag{18}$$

For convenience, designate the two sets of *hkl*'s in (18) as $h_1\,k_1\,l_1$ and $h_2\,k_2\,l_2$. With this change, and when the exponentials are consolidated, (18) becomes

$$F_{h_2k_2l_2} = \sum_{h_1} \sum_{k_1} \sum_{l_1} \int_0^1 \int_0^1 \int_0^1 K_{h_1k_1l_1}\, e^{i2\pi[(h_2-h_1)x+(k_2-k_1)y+(l_2-l_1)z]}\, V\,dx\,dy\,dz. \tag{19}$$

Each integral in (19) is characterized by six integers, h_1, k_1, l_1, and

h_2, k_2, l_2. The specific integral characterized by

$$\begin{aligned} h_2 &= h_1, \\ k_2 &= k_1, \\ l_2 &= l_1 \end{aligned} \tag{20}$$

has the following value:

$$\int_0^1 \int_0^1 \int_0^1 K_{hkl}\, e^0\, V\, dx\, dy\, dz = VK_{hkl}. \tag{21}$$

On the other hand, any integral for which one or more of the conditions of (20) do not hold, vanishes. For example, suppose $h_2 \neq h_1$. Then the integral is

$$\int_0^1 \int_0^1 \int_0^1 K_{h_1k_1l_1}\, e^{i2\pi(h_2-h_1)x}\, e^{i2\pi(k_2-k_1)y}\, e^{i2\pi(l_2-l_1)z}\, V\, dx\, dy\, dz. \tag{22}$$

According to (33) of Chapter 13, the integration of dx over the range 0 to 1 vanishes.

Since the only non-vanishing integral of the summation is that for which (20) holds, and whose value is given by (21), the expression in (19) simplifies to

$$F_{h_1k_1l_1} = VK_{h_1k_1l_1}. \tag{23}$$

This means that for each set of indices *hkl*,

$$K_{hkl} = \frac{1}{V} F_{hkl}. \tag{24}$$

Therefore the three-dimensional Fourier synthesis (7*A*) of the electron density can be written as follows in terms of the amplitudes scattered in the reflections *hkl*:

$$\rho_{(xyz)} = \frac{1}{V} \sum_h \sum_{k=-\infty}^{\infty} \sum_l F_{hkl}\, e^{-i2\pi(hx+ky+lz)} \tag{25}$$

The zero term of this Fourier summation is provided by substituting 0 0 0 for *hkl* in (24):

$$\begin{aligned} K_{000} &= \frac{1}{V} F_{000} \\ &= \frac{1}{V} Z, \end{aligned} \tag{26}$$

where Z is the number of electrons in the unit cell.

The synthesis. A Fourier summation such as (25) supplies the value of ρ when the set of coordinates xyz is substituted into the right side of the relation. Theoretically, x, y, and z may be thought of as continuous variables; then (25) supplies the value of the function ρ as it varies over the three-dimensional cell. In actual practice, if one wishes to know the numerical value of ρ, it is necessary to specify a location, such as x_1, y_1, z_1, where this value is required; substitution of the numerical values of x_1, y_1, and z_1 in the right of (25) then provides $\rho_{(x_1,y_1,z_1)}$ as the result of a triple summation over h, k, and l. In other words, the computation implied by (25) is a discrete process which applies to a discrete point in space.

A single value of ρ is ordinarily of no use by itself. In crystal-structure analysis, one usually wishes to know how the value of ρ varies over the cell since maximum values of ρ denote centers of atoms. To find out how ρ varies over the cell, the cell can be imagined to be sampled by points uniformly distributed throughout its volume. This amounts to choosing a sampling interval for x, y, and z. Several such intervals are in common use. For example, the cell edge, a (or b or c) can be sampled at 60 uniformly spaced points by letting x (or y or z) assume the discrete values $x = \frac{0}{60}, \frac{1}{60}, \frac{2}{60} \cdots \frac{59}{60}$. Common sampling intervals are $\frac{1}{60}$, $\frac{1}{120}$, $\frac{1}{50}$, and $\frac{1}{100}$. If the cell has no symmetry, the electron density is different, in general, at every point in the cell, so that the number of different computations required for these intervals is:

Interval	*Total number of sample points per cell*
$\frac{1}{50}$ of cell edge	$50^3 = 125{,}000$
$\frac{1}{60}$ of cell edge	$60^3 = 216{,}000$
$\frac{1}{100}$ of cell edge	$100^3 = 1{,}000{,}000$
$\frac{1}{120}$ of cell edge	$120^3 = 1{,}728{,}000$

The number of sample points required is reduced by a factor equal to the reciprocal of the number of symmetry operations per primitive cell in the space group.

When all these numerical values of the electron density are available, the set of results is not, in itself, readily interpreted. Part of the difficulty is that the three dimensions of space are utilized by the sample points, and there is no further dimension available for representing the value of the function (electron density) as it varies over the three dimensions. This difficulty is customarily avoided by regarding three-dimensional space as represented by a stack of parallel planes, spaced by the amount of the sampling interval not contained in the plane. In other words, the three-dimensional distribution of electron density is regarded as sampled along a sequence of parallel planes. The distribution in each sample plane can be regarded as hills and valleys in the third dimension. For convenience the height of the function in this third dimension is customarily represented by drawing, in the two-dimensional section, contours connecting equal values of the electron density in that section.

Electron-density sections

Not only are sections through the three-dimensional distribution of electron density used as a routine method of studying the total distribution of electron density in space, but, in some problems, it is sufficient to synthesize one such section only, or a limited number of sections. Sections are normally selected parallel to a rational plane. If a section is attempted parallel to an irrational plane, the pattern in it does not repeat because the plane does not contain translations. The simplest sections are parallel to the pinacoids. If the cell is a *reduced cell*† the pinacoids are characterized by having the desirable features of greatest spacing, and meshes of smallest areas.

Although the full three-dimensional summation involves a triple summation, a section requires only a double summation. The forms of these summations for the various cases are derived below.

Pinacoidal sections. Suppose one wishes a section parallel to (001). Since this is parallel to the a and b axes, the x and y coordinates are variables for the synthesis, but z is constant, at some value z_1. The forms of the synthesis can be derived by substituting the constant z_1 for z in (25), and permitting x and y to remain variables. The specific form of (25) is then

$$\rho_{(xyz_1)} = \frac{1}{V} \sum_h \sum_k \sum_l F_{hkl}\, e^{-i2\pi(hx+ky+lz_1)}. \qquad (27)$$

It is convenient to separate this into variable and constant parts:

$$\rho_{(xyz_1)} = \frac{1}{V} \sum_h \sum_k \sum_l F_{hkl}\, e^{-i2\pi(hx+ky)}\, e^{-i2\pi lz_1}$$

$$= \frac{1}{V} \sum_h \sum_k \left\{ \sum_l F_{hkl}\, e^{-i2\pi lz_1} \right\} e^{-i2\pi(hx+ky)}. \qquad (28)$$

The part in parentheses should be performed first, giving a set of terms

$$Q_{hk} \equiv \left\{ \sum_l F_{hkl}\, e^{-i2\pi lz_1} \right\}. \qquad (29)$$

† M. J. Buerger. *Reduced cells.* Z. Krist. **109** (1957) 42–60.

This merely amounts to performing, for each row of the reciprocal lattice having a constant hk, a summation over all the F's in the row, except that every F is weighted by the factor $e^{-i2\pi lz_1}$ before adding. This becomes a Fourier coefficient for the double summation

$$\rho_{(xyz_1)} = \frac{1}{V} \sum_{h}\sum_{k}{}_{-\infty}^{\infty} Q_{hk}\, e^{-i2\pi(hx+ky)}. \tag{30}$$

This summation is particularly easy when there is a symmetry center in the origin and a mirror at $xy0$. The exponentials of (28) then becomes cosines, and (30) becomes

$$\rho_{(xyz_1)} = \frac{2}{V} \sum_{h=-\infty}^{\infty} \sum_{k=-\infty}^{\infty} \left\{ \sum_{l=0}^{\infty}{}' F_{hkl} \cos 2\pi l z_1 \right\} \cos 2\pi(hx+ky), \tag{30A}$$

where the prime indicates that the level $l = 0$ enters at half weight. For sections at $z = 0$ and $z = \frac{1}{2}$, the forms of (29) for (30A) for centrosymmetrical crystals are particularly simple, namely

$$\begin{aligned} z_1 = 0: \qquad Q_{h_1k_1} &= \sum_{l=0}^{\infty}{}' F_{h_1k_1l}, \\ z_1 = \tfrac{1}{2}: \qquad Q_{h_1k_1} &= \sum_{l=0}^{\infty}{}' (-1)^l F_{h_1k_1l}. \end{aligned} \tag{31}$$

An example of the Fourier synthesis of an electron-density section is shown in Fig. 3. In order to show the significant features of several or all the sections of a cell at once it is customary to transfer to one sheet the contours of each atom from that sheet where that atom attains its maximum electron density. An example of this kind of representation is shown in Fig. 4.

General sections. A less simple form is taken by the summation for a section parallel to a more general rational plane $h_1\, k_1\, l_1$. To derive the form for such a section, the relation between the coordinate xyz and the indices $h_1\, k_1\, l_1$ can be used. This relation is†

$$h_1 x + k_1 y + l_1 z = m. \tag{32}$$

This equation ordinarily represents a stack of parallel planes, each plane

† M. J. Buerger. *Elementary crystallography.* (John Wiley and Sons, New York, 1956) 18–21.

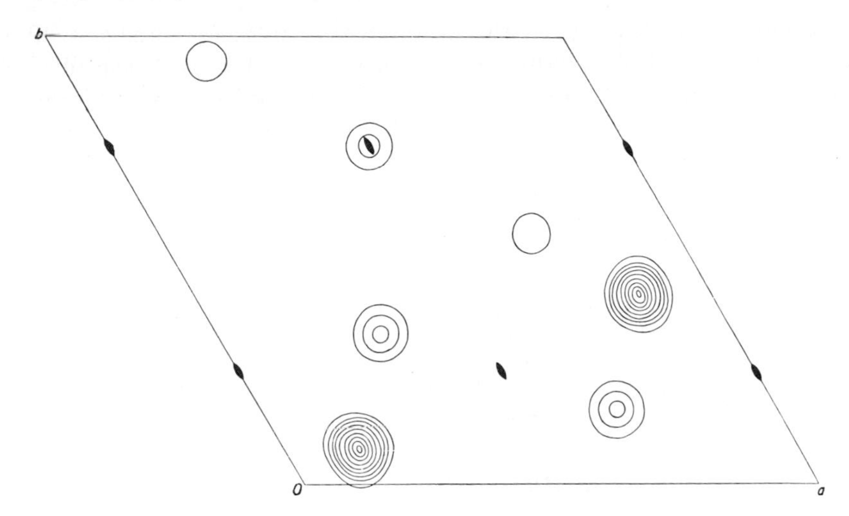

Fig. 3.
Electron-density section $\rho_{(xyz_1)}$; coesite, SiO_2.

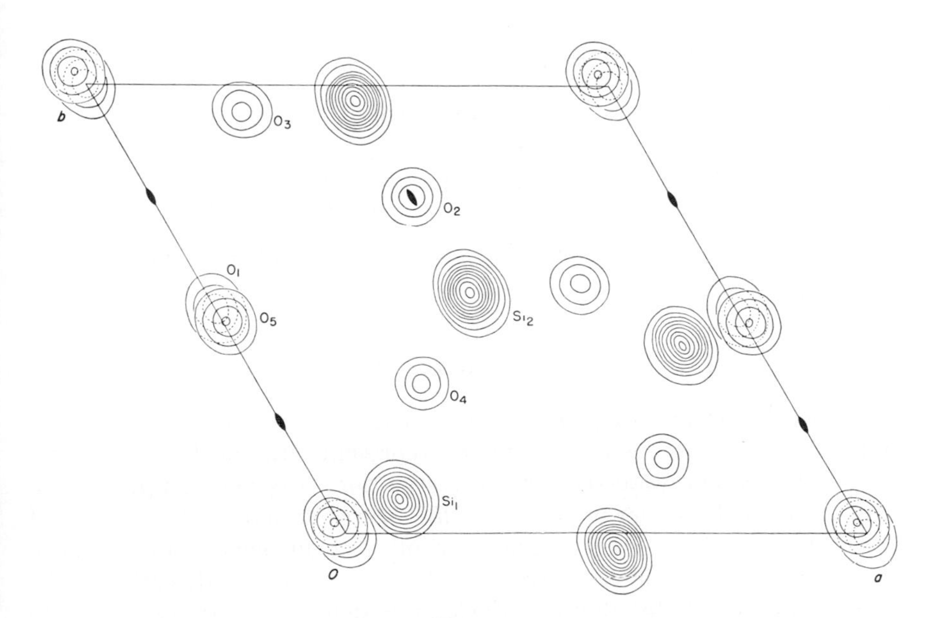

Fig. 4.
Set of electron-density sections $\rho_{(xyz_1)}$, for various levels z_1 containing maxima, projected parallel to c; coesite, SiO_2.

at a distance $m\, d_{h_1k_1l_1}$ from the origin, where m is an integer. When $m = 0$, the plane passes through the origin; when $m = 1$, the plane is the next rational one from the origin, and spaced a distance d_{hkl} from the origin. If one wishes to synthesize the electron density at a distance from the origin equal to a fraction, q, of d_{hkl}, then this fraction should be substituted for m in (32), so that the desired relation for a level q is

$$h_1\, x + k_1\, y + l_1\, z = q \qquad (q = \text{a fraction}). \tag{33}$$

The expression for the electron density is

(25): $$\rho_{(xyz)} = \frac{1}{V} \sum_h \sum_k \sum_l F_{hkl}\, e^{-i2\pi(hx+ky+lz)}.$$

The phase of the exponential is given by

$$(hx+ky+lz) = \frac{\phi}{2\pi}. \tag{34}$$

This can be combined with (33) to eliminate x, y, or z:

(34): $$hx + ky + lz = \frac{\phi}{2\pi}.$$

(33): $$h_1\, x + k_1\, y + l_1\, z = q\,.$$

If z is eliminated, the phase of the exponential is found to be

$$\frac{\phi}{2\pi} = \left\{\left(h - \frac{h_1}{l_1}\, l\right) x + \left(k - \frac{k_1}{l_1}\, l\right) y + q \frac{l}{l_1}\right\}. \tag{35}$$

The form of the synthesis is found by substituting this for the expression in parentheses in (25).

Projections of the electron density

General features of projections. Three-dimensional syntheses are difficult to compute and tedious to represent graphically. For preliminary work in crystal-structure investigations it is common practice to utilize the much simpler *projections of the electron density*. The relation of the full three-dimensional distribution of electron density and its two-dimensional projection is illustrated in Figs. 5 and 6. In this relation the three-dimensional distribution of electron density is *always* projected parallel to a rational crystallographic direction (that is, parallel to a translation).

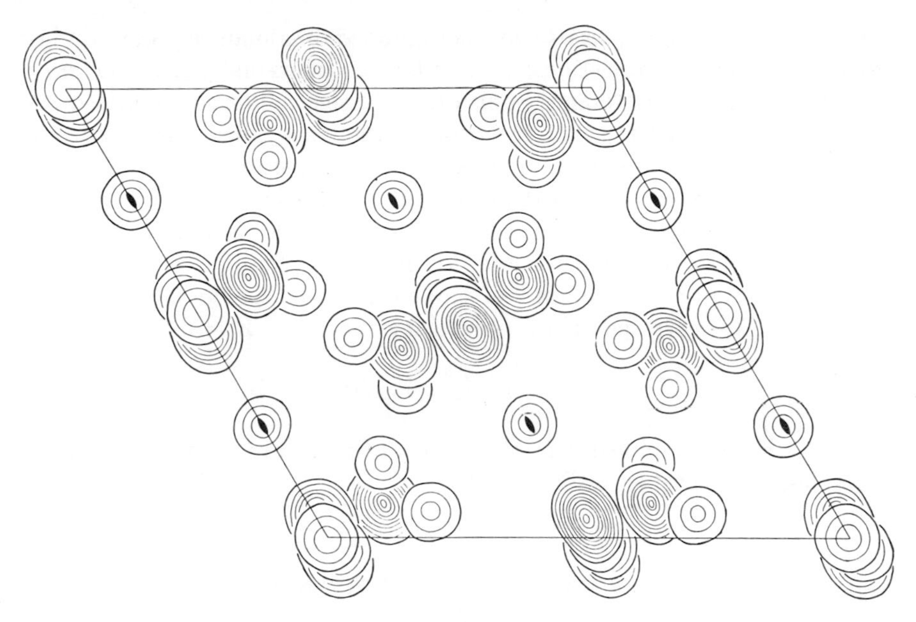

Fig. 5.
Set of all electron-density sections $\rho_{(xyz_1)}$ containing maxima, projected parallel to c. (This is similar to Fig. 4, except that in Fig. 4 a limited number of levels are projected, while in this figure *all* levels containing maxima throughout the cell are projected.)

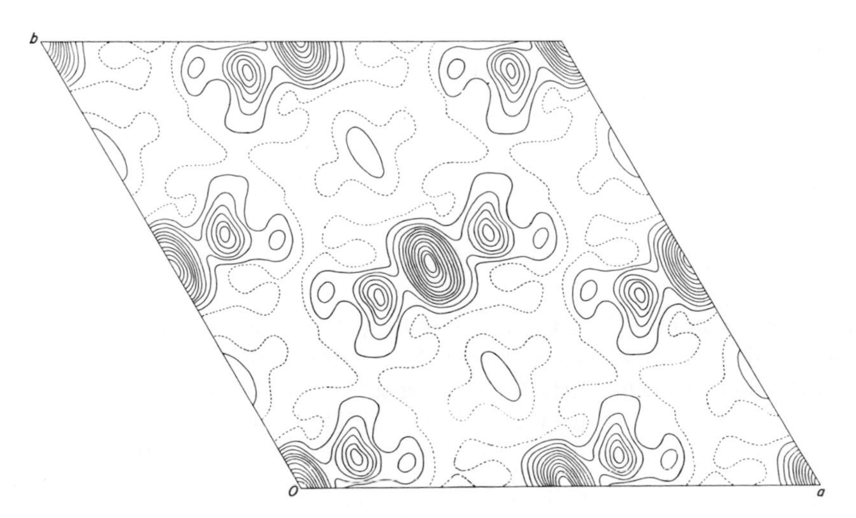

Fig. 6.
Electron-density projection $\rho_{(xy)}$ (projected parallel to c) for coesite, SiO_2, corresponding to Fig. 5.

The projected density is *nearly always*[†] mapped on a plane at right angles to the direction of projection. This plane is irrational unless the projection direction is a symmetry axis of the lattice. The commonest projection directions are those of crystallographic axes, although for special purposes non-axial directions such as [110] are sometimes used.

Deduction from three-dimensional synthesis. A projection of the electron density in a particular direction corresponds mathematically to an integration over the lattice period in that direction. Suppose that the projection in the direction c of the electron density given by (25) is required. This corresponds to integrating (25) from $Z = 0$ to $Z = c$:

$$\rho_{(XY)} = \int_0^c \rho_{(XYZ)}\, dZ. \tag{36}$$

The notation is simplified if fractional coordinates are used. To change variables, note that, according to (15)

$$dZ = c\, dz, \tag{37}$$

and the limits become $z = 0$ to $z = 1$. Thus (36) transforms into

$$\rho_{(xy)} = \int_0^1 \rho_{(xyz)}\, c\, dz, \tag{38}$$

and, substituting from (25) for $\rho_{(xyz)}$, one obtains

$$\rho_{(xy)} = \int_0^1 \frac{1}{V} \sum_h \sum_k \sum_l F_{hkl}\, e^{-i2\pi(hx+ky+lz)}\, c\, dz. \tag{39}$$

Some of the quantities in (39) are functions of z; others are not. If these parts are separated, (39) becomes

$$\rho_{(xy)} = \int_0^1 \frac{c}{V} \sum_h \sum_k \sum_l F_{hkl}\, e^{-i2\pi(hx+ky)}\, e^{-i2\pi lz}\, dz \tag{40}$$

$$= \frac{1}{S} \sum_h \sum_k \sum_l F_{hkl}\, e^{-i2\pi(hx+ky)} \int_0^1 e^{-i2\pi lz}\, dz, \tag{41}$$

where the volume, V, of the cell has been expressed as the length of the c axis times the cross-sectional area, S, at right angles to it. The integral in (41) behaves differently, according as $l = 0$ or $l \neq 0$. When $l = 0$, the integral is unity; otherwise it is zero. Thus the only contributions to the summation over l in (41) are those terms for which $l = 0$, so that (41)

[†] If the plane of projection is not taken normal to the direction of projection, the geometry is distorted; for example, circular atom outlines become elliptical. Such oblique projections should be used only if there is a reason to depart from the normal projection.

Table 1

Forms of Fourier synthesis of the electron-density projection for various projection directions

Direction of projection	Reflections used	Form of synthesis
[100]	$0kl$	$\frac{1}{S}\sum_k\sum_l F_{0kl}\, e^{-i2\pi(ky+lz)}$
[010]	$h0l$	$\frac{1}{S}\sum_h\sum_l F_{h0l}\, e^{-i2\pi(hx+lz)}$
[001]	$hk0$	$\frac{1}{S}\sum_h\sum_k F_{hk0}\, e^{-i2\pi(hx+ky)}$
[110]	$h\bar{h}l$	$\frac{1}{S}\sum_h\sum_l F_{h\bar{h}l}\, e^{-i2\pi(hx+lz)}$
$[1\bar{1}0]$	hhl	$\frac{1}{S}\sum_h\sum_l F_{hhl}\, e^{-i2\pi(hx+lz)}$
[101]	$\bar{h}kh$	$\frac{1}{S}\sum_h\sum_k F_{\bar{h}kh}\, e^{-i2\pi(hx+ky)}$
$[10\bar{1}]$	hkh	$\frac{1}{S}\sum_h\sum_k F_{hkh}\, e^{-i2\pi(hx+ky)}$
[011]	$h\bar{l}l$	$\frac{1}{S}\sum_h\sum_l F_{h\bar{l}l}\, e^{-i2\pi(hx+lz)}$
$[01\bar{1}]$	hll	$\frac{1}{S}\sum_h\sum_l F_{hll}\, e^{-i2\pi(hx+lz)}$
[111]	$hkl,\quad h+k+l=0$ i.e. $hk(\bar{h}+\bar{k})$	$\frac{1}{S}\sum_h\sum_k F_{hk(\bar{h}+\bar{k})}\, e^{-i2\pi(hx+ky)}$
$[\bar{1}11]$	$hkl,\quad -h+k+l=0$ i.e. $hk(h+\bar{k})$	$\frac{1}{S}\sum_h\sum_k F_{hk(h+\bar{k})}\, e^{-i2\pi(hx+ky)}$
$[1\bar{1}1]$	$hkl,\quad h-k+l=0$ i.e. $hk(\bar{h}+k)$	$\frac{1}{S}\sum_h\sum_k F_{hk(\bar{h}+k)}\, e^{-i2\pi(hx+ky)}$
$[11\bar{1}]$	$hkl,\quad h+k-l=0$ i.e. $hk(h+k)$	$\frac{1}{S}\sum_h\sum_k F_{hk(h+k)}\, e^{-i2\pi(hx+ky)}$

reduces to

$$\rho_{(xy)} = \frac{1}{S} \sum_h \sum_k F_{hk0}\, e^{-i2\pi(hx+ky)}, \tag{42A}$$

which, expressed in absolute coordinates, is equivalent to

$$\rho_{(XY)} = \frac{1}{S} \sum_h \sum_k F_{hk0}\, e^{-i2\pi(hX/a+kY/b)}. \tag{42B}$$

An example of the Fourier synthesis of a projection of the electron density is given in Fig. 6.

Direct determination of form of the two-dimensional synthesis. This result could have been expected on less formal grounds. From the discussion given in the initial part of "Three-dimensional synthesis," it is evident that to represent a two-dimensional periodic function by Fourier synthesis requires that the two-dimensional cell be covered by Fourier waves corresponding to the rational lines of the two-dimensional lattice. The Fourier coefficients are therefore K_{hk}. An analysis similar to that given for one dimension and three dimensions shows that

$$K_{hk} = \frac{1}{S} F_{hk}, \tag{43}$$

where S is the area of the two-dimensional cell. The coefficients F_{hk}, which apply strictly for a two-dimensional grating, are obviously F_{hk0} when labeled according to three-dimensional indices. The cell edges, Fig. 7A, correspond to the projections of a and b of the three-dimensional cell. Figure 7B shows the geometry in two dimensions, corresponding to the geometry in three dimensions, as already discussed, which leads to the conclusions that the phase at P_{xy} is given by

$$\phi = \left(\frac{1}{d_{hk}} \text{ proj. } p_{xy}\right) 2\pi \tag{44}$$

$$= (\boldsymbol{\sigma}_{hk} \cdot \mathbf{p}_{xy}) 2\pi \tag{45}$$

$$= (hx+ky) 2\pi. \tag{46}$$

From these considerations (42) could have been deduced without reference to the three-dimensional case.

Projections in non-axial directions. As a result of the discussion of the foregoing sections it becomes an easy matter to write down the form of a Fourier synthesis for a projection in any rational direction. The Fourier waves to use for any projection direction correspond to the planes in the zone of (that is, parallel to) that direction. If the direction is $[uvw]$, the planes corresponding to the required reflections hkl must satisfy the relation for planes in a zone, namely

$$hu+kv+lw = 0. \tag{47}$$

The directions most commonly used, other than the directions of the crystallographic axes, are the several diagonals of the cell. Table 1 lists the forms of the Fourier synthesis for the most commonly used projections, and the reflections whose amplitudes are required for the synthesis.

In planning for an electron-density projection, one should bear in mind that, if the crystal has symmetry, its projection has the projection

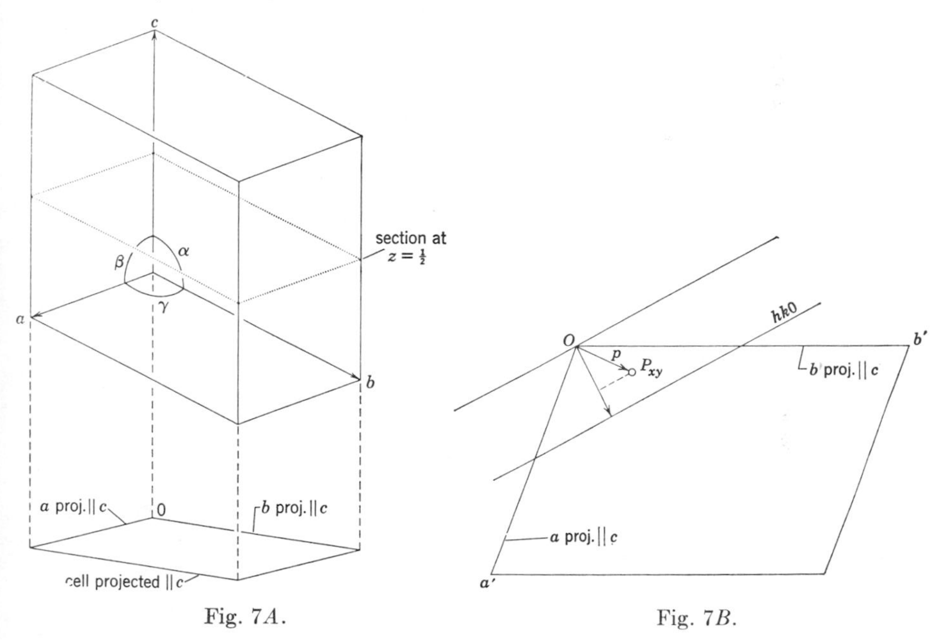

Fig. 7*A*. Fig. 7*B*.

of that symmetry. An example of this is shown in Fig. 8, which represents several projections in different directions of the mineral realgar, As_4S_4. The space group of this crystal is $P\,2_1/n$. This symmetry projects parallel to both [100] and [001] as the plane group $p2gg$. But it projects in the direction [101] as plane group $p2gm$. The importance of the projected symmetry is discussed in Chapter 16.

Stereoscopic pairs of projections. Cowley[18] has pointed out that if two projections of the same structure are prepared for somewhat different directions, the pair of projections has stereoscopic qualities. The pair can therefore be used to determine all three coordinates of the atoms. Stereoscopic pairs of projections have been little used.

Mapping the synthesis. In the discussion of the Fourier synthesis of the three-dimensional electron-density function, it was pointed out that each summation provides a sample of the function at one point in the volume of the cell. Similarly, in any Fourier synthesis of a two-dimensional function, each summation provides a sample of the function at one point in the area of the section or projection.

If the summation is computed by digital methods, it is customary to distribute the sample points uniformly over the area of the cell. Thus the a (or b or c) axis of the cell is divided into N parts, and the sampling is carried out by letting $x = 0(1/N),\ 1(1/N),\ 2(1/N) \cdots (N-1)(1/N)$.

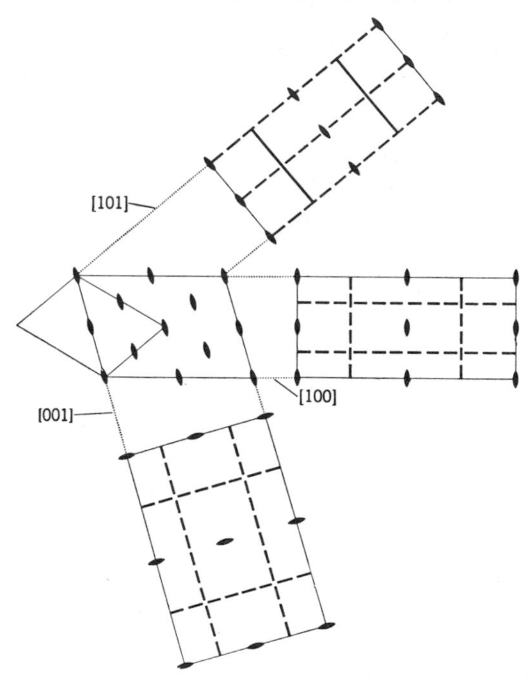

Fig. 8.
Some projections of the space group $P2_1/n$.

If the section or projection contains no symmetry, this calls for the following number of sample points in the cell:

Interval	*Total number of sample points per cell*
$\frac{1}{50}$	$50^2 = 2{,}500$
$\frac{1}{60}$	$60^2 = 3{,}600$
$\frac{1}{100}$	$100^2 = 10{,}000$
$\frac{1}{120}$	$120^2 = 14{,}400$

If the two-dimensional cell contains any symmetry, this number is reduced by a factor equal to the reciprocal of the number of symmetry

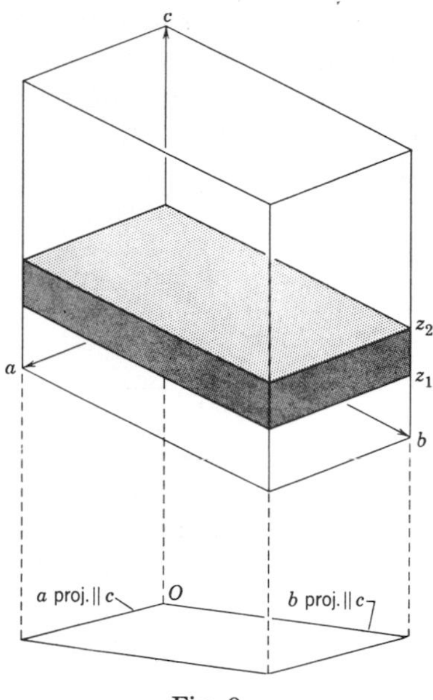

Fig. 9.

operations in the primitive cell. It is convenient to study the distribution of electron density in a section or projection by drawing contours of equal electron density.

It is also possible to use an analogue computor to provide a contoured picture of the distribution of electron density in either a section or projection. The best-known device for accomplishing this is Professor Ray Pepinsky's X-RAC.†

Projections of part of the electron density

There are several ways in which portions of the cell contents can be projected. These are noted below in order of their importance.

Projected slabs. The possibility of devising a Fourier synthesis of a slab of the electron density of a cell was discovered by Verner Schomaker (unpublished, but reported by E. W. Hughes§) and rediscovered by Booth[20] some years later. In deducing the form of the Fourier synthesis of the projection of the electron density from the three-dimensional form, it was pointed out that projection of the entire cell contents corresponds to integration over the period of the projection direction. It follows that

† R. Pepinsky. *An electronic computor for x-ray crystal structure analyses.* J. Appl. Phys. **18** (1947) 601–604.

§ E. W. Hughes. *Recent x-ray and electron-diffraction work at the California Institute of Technology.* Am. Soc. for X-ray and Electron Diffraction, Gibson Island Meeting, July 30, 1941.

the projection of a portion of the cell contents corresponds to integrating over a portion of the period.

Suppose that the projection on a plane normal to c is required of the cell contents lying between two (001) sections at levels z_1 and z_2, Fig. 9. This calls for integrating (25) between the limits z_1 and z_2. Following the pattern of (36) through (38), this integration is

$$\rho_{(xyz)} \Big[_{z_1}^{z_2} = \int_{z_1}^{z_2} \rho_{(xyz)} \, c \, dz. \tag{48}$$

Substituting for $\rho_{(xyz)}$ from (25), this becomes

$$\rho_{(xyz)} \Big[_{z_1}^{z_2} = \int_{z_1}^{z_2} \frac{1}{V} \sum_h \sum_k \sum_l F_{hkl} \, e^{-i2\pi(hx+ky+lz)} \, c \, dz. \tag{49}$$

If the quantities which are not functions of z are separated from those which are, they may be moved to the left of the integral sign, giving

$$\rho_{(xyz)} \Big[_{z_1}^{z_2} = \frac{c}{V} \sum_h \sum_k \sum_l F_{hkl} \, e^{-i2\pi(hx+ky)} \int_{z_1}^{z_2} e^{-i2\pi lz} \, dz \tag{50}$$

$$= \frac{1}{S} \sum_h \sum_k \sum_l F_{hkl} \, e^{-i2\pi(hx+ky)} \left\{ \frac{1}{-i2\pi l} (e^{-i2\pi l z_2} - e^{-i2\pi l z_1}) \right\}. \tag{51}$$

The term in braces can be reduced as follows:

$$\begin{aligned}
\{---\} &= -\frac{1}{i2\pi l} \{e^{-i2\pi l z_2} - e^{-i2\pi l z_1}\} \\
&= -\frac{1}{i2\pi l} \frac{e^{i2\pi l(z_2+z_1)/2}}{e^{i2\pi l(z_2+z_1)/2}} \{e^{-i2\pi l z_2} - e^{-i2\pi l z_1}\} \\
&= -\frac{1}{i2\pi l} \frac{1}{e^{i2\pi l(z_2+z_1)/2}} \{e^{-i2\pi l(z_2-z_1)/2} - e^{-i2\pi l(z_1-z_2)/2}\} \\
&= -\frac{1}{i2\pi l} \frac{1}{e^{i2\pi l(z_2-z_1)/2}} \{e^{-i2\pi l(z_2-z_1)/2} - e^{i2\pi l(z_2-z_1)/2}\} \\
&= -\frac{1}{i2\pi l} e^{-i2\pi l(z_2+z_1)/2} \{-2i \sin [2\pi l(z_2 - z_1)/2]\} \\
&= e^{-i2\pi l(z_2+z_1)/2} \frac{1}{\pi l} \sin \pi l(z_2 - z_1).
\end{aligned} \tag{52}$$

When this is substituted for the term in braces in (51) the following simpler form results:

$$\rho_{(xyz)} \Big[_{z_1}^{z_2} = \frac{1}{S} \sum_h \sum_k \sum_l F_{hkl}\, e^{-i2\pi(hx+ky)}\, e^{-i2\pi l(z_2+z_1)/2} \frac{\sin \pi l(z_2-z_1)}{\pi l} \tag{53}$$

$$= \frac{1}{S} \sum_h \sum_k \sum_l F_{hkl} \frac{\sin \pi l(z_2-z_1)}{\pi l} e^{i2\pi[hx+ky+l(z_2+z_1)/2]}. \tag{54}$$

To make this expression more understandable, note that

$$\frac{z_2+z_1}{2} = z_m, \text{ the mid level of the slab,} \tag{55}$$

$$z_2-z_1 = \Delta z, \text{ the thickness of the slab.} \tag{56}$$

With this simplification, (53) can be rewritten

$$\rho_{(xyz_m\,\Delta z)} = \frac{1}{S} \sum_h \sum_k \sum_l F_{hkl} \frac{\sin \pi l\, \Delta z}{\pi l} e^{-i2\pi(hx+ky+lz_m)}. \tag{57}$$

This reveals that the synthesis has the same exponential form as the synthesis of a section of level z_m, except that the Fourier coefficient is modulated by the term $(\sin \pi l\, \Delta z)/\pi l$. When $\Delta z = 1$, corresponding to the projection of the full cell, this term becomes $(-1)^l$, and the second exponential term in (53) can be given the value $(-1)^l$, $(z_2 = 1, \quad z_1 = 0)$. The synthesis then degenerates into that for an axial projection, (42*A*).

The work of performing the synthesis can be displayed by expressing (53) or (57) in the form

$$\rho_{(xyz)} \Big[_{z_1}^{z_2} = \frac{1}{S} \sum_h \sum_k \sum_l R_{hkl}\, e^{-i2\pi(hx+ky+lz_m)}, \tag{58}$$

where

$$R_{hkl} = F_{hkl} \frac{\sin \pi l\, \Delta z}{\pi l}. \tag{59}$$

This brings out the fact that the synthesis is performed by first finding the coefficients R_{hkl}, after which the remaining work is that for standard synthesis for a section at level z_m.

Another view of the work is to express (57) in terms of a two-dimensional summation:

$$\rho_{(xyz)} \Big[_{z_1}^{z_2} = \frac{1}{S} \sum_h \sum_k Q_{hk}\, e^{-i2\pi(hx+ky)} \tag{60}$$

where

$$Q_{hk} = \sum_l F_{hkl} \left[\frac{\sin \pi l\, \Delta z}{\pi l} e^{-i2\pi l z_m} \right]. \tag{61}$$

Table 2

Some values of $\frac{\sin \pi l \Delta z}{\pi l}$

for use with the Fourier synthesis of projected slabs

l	Δz			
	$\frac{1}{2}$	$\frac{1}{3}$	$\frac{1}{4}$	$\frac{1}{6}$
0	$\frac{1}{2}$	$\frac{1}{3}$	$\frac{1}{4}$	$\frac{1}{6}$
1	.3183	.2756	.2250	.1591
2	0	.1378	.1591	.1378
3	−.1061	0	.0750	.1061
4	0	−.0689	0	.0689
5	.0637	−.0551	−.0450	.0318
6	0	0	−.0531	0
7	−.0455	.0394	−.0321	−.0227
8	0	.0345	0	−.0345
9	.0354	0	.0250	−.0354
10	0	−.0276	.0318	−.0276
11	−.0289	−.0251	.0205	−.0145
12	0	0	0	0
13	.0245	.0212	−.0173	.0122
14	0	.0197	−.0227	.0197
15	−.0212	0	−.0150	.0212
16	0	−.0172	0	.0172
17	.0187	−.0162	.0132	.0094
18	0	0	.0177	0
19	−.0177	.0145	.0118	−.0084
20	0	.0138	0	−.0138

This shows that the simple one-dimensional summations over l in (61), one for each combination hk, should be performed first. When this is done, the remaining work, (60), is exactly that of synthesizing a projection. But, unlike in the projection of the full cell, all F_{hkl}'s, in general, are required, not just a set of F_{hk0}'s.

The form of the summation in (61) is comparatively simple, especially if the values of Δz and z_m are selected with care. In the first place the term in brackets in (61) simply modulates F_{hkl}, and is constant for a given l; i.e., all F's in the same level of the reciprocal lattice are to be multiplied by the same modulating factor. Thus at most only $2L+1$ factors need be computed, where L is the highest level for which data are available. In some choices of Δz and z_m the factors are zero. Some examples are given in Table 2.

To see the behavior of the function $(\sin \pi l \Delta z)/\pi l$, multiply both numerator and denominator by Δz:

$$\frac{\Delta z}{\Delta z} \frac{\sin \pi l \Delta z}{\pi l} = \Delta z \frac{\sin (\pi l \Delta z)}{(\pi l \Delta z)}. \tag{62}$$

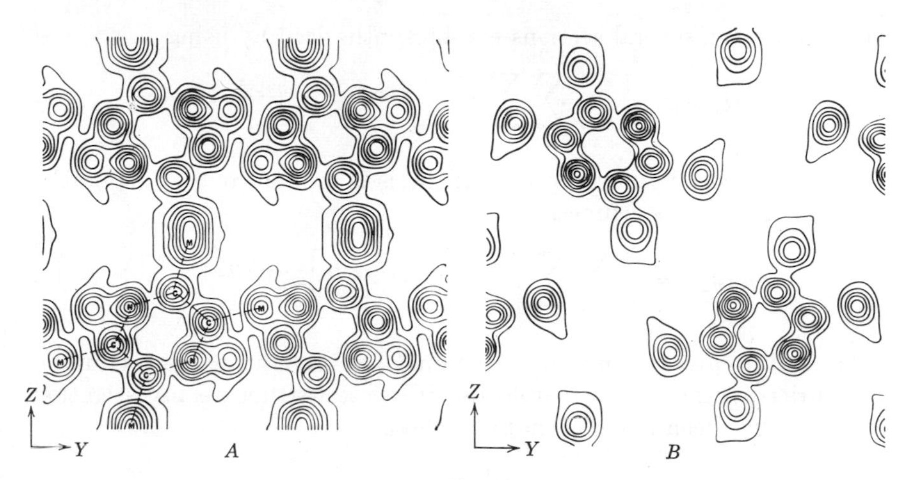

Fig. 10.
Projections of tetramethylpyrazine.
A. Projection $\rho_{(yz)}$ of the full unit cell. *B*. Bounded projection

$$\rho_{(xyz)} \begin{bmatrix} x_2 = \frac{1}{4} \\ x_1 = -\frac{1}{4} \end{bmatrix}$$

(After Cromer, Ihde, and Ritter.[24])

The fraction on the right is the well-known diffraction function $(\sin \phi)/\phi$. Therefore the value of (62) is $\Delta z(\sin \phi/\phi)$, where $\phi = \pi l \,\Delta z$.

Projected slabs are used chiefly in cases where the projection of the entire cell causes superposition or partial superposition of atoms on different levels. Often the projection of half the cell resolves the atoms in such instances. An example is shown in Fig. 10. Other examples are to be found in literature.[22–27]

Projected sections. Booth[20] has also shown that it is possible to synthesize the projections of several sections at the same time. For example, suppose it is desired to synthesize the two sections $\rho_{(xyz_1)}$ and $\rho_{(xyz_2)}$. The multiple synthesis has the Fourier representation

$$\begin{aligned} \rho_{(xyz_1,z_2)} &= \rho_{(xyz_1)} + \rho_{(xyz_2)} \\ &= \frac{1}{V} \sum_h \sum_k \sum_l F_{hkl}\, e^{-i2\pi(hx+ky+lz_1)} \\ &\quad + \frac{1}{V} \sum_h \sum_k \sum_l F_{hkl}\, e^{-i2\pi(hx+ky+lz_2)} \\ &= \frac{1}{V} \sum_h \sum_k \sum_l F_{hkl}\, e^{-i2\pi(hx+ky+l[z_1+z_2])}. \end{aligned} \tag{63}$$

More generally, several sections can be synthesized by using

$$\rho_{(xy\Sigma z)} = \frac{1}{V} \sum_h \sum_k \sum_l F_{hkl}\, e^{-i2\pi(hx+ky+l\Sigma z)}. \tag{64}$$

The two-dimensional form of the synthesis is brought out by separating the portion which is a function of z:

$$\rho_{(xy\Sigma z)} = \frac{1}{V} \sum_h \sum_k \left[\sum_l F_{hkl}\, e^{-i2\pi l\Sigma z} \right] e^{-i2\pi(hx+ky)}. \tag{65}$$

Miscellaneous syntheses. Huggins[21] has brought together a number of Fourier syntheses of possible use in crystal-structure investigation. Little use has been made of many of these.

Modulated projections

Several investigators have devised functions of the electron density which may be called *modulated projections*. They are modulated in the sense that the electron density is modified by a weighting function which varies, for example, with the coordinate z of each atom. These functions have several uses, which will be noted presently.

Generalized projections. Clews and Cochran,[28] Dyer,[29] and Cochran and Dyer[32] have described the most general modulated function, which has been called a *generalized projection*. In this function the electron density is modulated by a factor $e^{i2\pi Lz}$, where L is a constant value of the index l. Before going on, it should be noted that $e^{i2\pi Lz}$ is merely an operator which shifts the phase through the angle $2\pi Lz$, so that, whereas $\rho_{(xyz)}$ is a real, positive quantity, $\rho_{(xyz)}\, e^{i2\pi Lz}$ is a function composed of the same absolute magnitudes, but all of the phases, instead of being zero, are $2\pi Lz$.

The generalized projection is the integral of $\rho_{(xyz)}\, e^{i2\pi Lz}$ over the period of the projection. If the projection is to be made along the c axis, the generalized projection is defined by

$$\rho_{L(xy)} = \int_0^1 \rho_{(xyz)}\, e^{i2\pi Lz}\, c\, dz. \tag{66}$$

If the Fourier expression for $\rho_{(xyz)}$ is substituted from (25) into this, it becomes

$$\rho_{L(xy)} = \int_0^1 \frac{1}{V} \sum_h \sum_k \sum_l F_{hkl}\, e^{-i2\pi(hx+ky+lz)}\, e^{i2\pi Lz}\, c\, dz \tag{67}$$

$$= \frac{c}{V} \sum_h \sum_k \sum_l F_{hkl}\, e^{-i2\pi(hx+ky)} \int_0^1 e^{i2\pi(L-l)z}\, dz. \tag{68}$$

The integral in (68) is 1 when $L = l$, but zero when $L \neq l$. Therefore (68) reduces to

$$\rho_{L(xy)} = \frac{1}{S} \sum_h \sum_k F_{hkL}\, e^{-i2\pi(hx+ky)}. \tag{69}$$

Note that, when $L = 0$, (69), becomes

$$\rho_{0(xy)} = \frac{1}{S} \sum_h \sum_k F_{hk0}\, e^{-i2\pi(hx+ky)} \tag{70}$$

$$= \rho_{(xy)}. \tag{71}$$

That is, when $L = 0$, (69) reduces to an ordinary projection parallel to c. The ordinary projection is synthesized from zero-level F's. In contrast, (69) is synthesized from the F's of the Lth level. Partly for this reason such syntheses are called *generalized projections.*

Generalized projections are useful for several purposes. Fundamentally, a generalized projection, $\rho_{L(xy)}$, may be used to determine the coordinates xy of the atoms in the structure. A separate and independent determination is available from the set of F's of each level of the reciprocal lattice. The projection can also be used to determine the approximate z coordinate of each of the atoms in the structure.

To understand how this is possible, let the electron density due to the J atoms of the crystal structure be resolved into the contributions, ${}^j\rho$, of the individual atoms of the structure. Thus the electron density of the crystal can be expressed as the sum of the electron densities of its individual atoms:

$$\rho_{(xyz)} = {}^1\rho_{(xyz)} + {}^2\rho_{(xyz)} \cdots {}^J\rho_{(xyz)}$$

$$= \sum_{j=1}^{J} {}^j\rho_{(xyz)}. \tag{72}$$

Now consider the electron-density contribution, ${}^j\rho_{(xyz)}$, of the jth atom only. This is

$${}^j\rho_{(xyz)} = \frac{1}{V} \sum_h \sum_k \sum_l {}^jF_{hkl}\, e^{-i2\pi(hx+ky+lz)}, \tag{73}$$

and its generalized projection is

$${}^j\rho_{L(xy)} = \frac{1}{S} \sum_h \sum_k {}^jF_{hkL}\, e^{-i2\pi(hx+ky)}. \tag{74}$$

For $L = 0$, this reduces to the ordinary projection

$$^{j}\rho_{0(xy)} = \frac{1}{S}\sum_{h}\sum_{k} {}^{j}F_{hk0}\, e^{-i2\pi(hx+ky)}. \tag{75}$$

But, for an individual atom,

$$^{j}F_{hkl} = {}^{j}f_{hkl}\, e^{i2\pi(hx_j+ky_j+lz_j)}, \tag{76}$$

so that

$$^{j}F_{hk0} = {}^{j}f_{hk0}\, e^{i2\pi(hx_j+ky_j)} \tag{77}$$

and

$$^{j}F_{hkL} = {}^{j}f_{hkL}\, e^{i2\pi(hx_j+ky_j+Lz_j)} \tag{78}$$

$$= {}^{j}f_{hkL}\, e^{i2\pi(hx_j+ky_j)}\, e^{i2\pi Lz_j}. \tag{79}$$

The relation between $^{j}F_{hkL}$ and $^{j}F_{hk0}$ is

$$\frac{^{j}F_{hkL}}{^{j}F_{hk0}} = \frac{^{j}f_{hkL}\, e^{i2\pi(hx_j+ky_j)}\, e^{i2\pi Lz_j}}{^{j}f_{hk0}\, e^{i2\pi(hx_j+ky_j)}}, \tag{80}$$

therefore,

$$^{j}F_{hkL} = {}^{j}F_{hk0}\,\frac{^{j}f_{hkL}}{^{j}f_{hk0}}\, e^{i2\pi Lz_j}. \tag{81}$$

If this is substituted into (74), it becomes

$$^{j}\rho_{L(xy)} = \frac{1}{S}\sum_{h}\sum_{k} {}^{j}F_{hk0}\,\frac{^{j}f_{hkL}}{^{j}f_{hk0}}\, e^{i2\pi(hx+ky)}\, e^{i2\pi Lz_j}. \tag{82}$$

Now let (82) and (75) be compared:

(75): $$^{j}\rho_{0(xy)} = \frac{1}{S}\sum_{h}\sum_{k} {}^{j}F_{hk0}\, e^{-i2\pi(hx+ky)},$$

(82): $$^{j}\rho_{L(xy)} = \frac{1}{S}\sum_{h}\sum_{k} {}^{j}F_{hk0}\,\frac{^{j}f_{hkL}}{^{j}f_{hk0}}\, e^{-i2\pi(hx+ky)}\, e^{i2\pi Lz_j}.$$

Insofar as the approximation

$$^{j}f_{hkL} \approx {}^{j}f_{hk0} \tag{83}$$

can be said to hold, these two expressions are the same, except for the modulating factor $e^{i2\pi Lz}$, so that

$$^{j}\rho_{0(xy)} \approx {}^{j}\rho_{L(xy)}\, e^{i2\pi Lz_j}. \tag{84}$$

The nature of the approximation will be noted presently, but, to the extent that it is valid, a valuable conclusion can be drawn: *The generalized projection of the* j*th atom is the same as its ordinary projection, except that* *it is modulated by the factor* $e^{i2\pi Lz_j}$. The ordinary projection is everywhere real and positive. The modulating factor $e^{i2\pi Lz_j}$ is merely an operator which changes the phase of the representation of the atom from zero to the angle $2\pi Lz_j$.

Since the modulating factor is, in general, complex, it can be represented by the sum of its real and imaginary components, so that the projection of the jth atom can be expressed as

$$^j\rho_{L(xy)} = {}^j\rho_{(xy)}(\cos 2\pi Lz_j + i \sin 2\pi Lz_j) \tag{85}$$

$$= {}^j\rho_{(xy)} \cos 2\pi Lz_j + i\, {}^j\rho_{(xy)} \sin 2\pi Lz_j \tag{86}$$

$$= {}^jR_{(xy)} + i\, {}^jI_{(xy)}. \tag{87}$$

The representation of this relation in the complex plane is shown in Fig. 11. This provides that

$$\tan 2\pi Lz_j = \frac{I_{(xy)}}{R_{(xy)}}, \tag{88}$$

so that the phase angle of the atom is given by

$$2\pi Lz_j = \tan^{-1} \frac{I_{(xy)}}{R_{(xy)}}. \tag{89}$$

Thus, the z coordinate of the atom can be found from the real and imaginary components of its generalized projection.

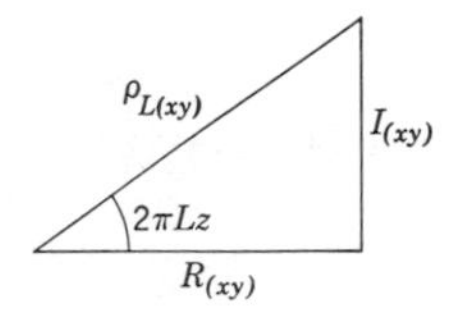

Fig. 11.
Relation between generalized projection and its component projections.

Before relating the generalized projection of an atom to the entire generalized projection, consider briefly the nature of the approximation involved in (84). In the first place, the modulating factor $e^{i2\pi Lz_j}$ in (82) is not a function of h and k; it simply shifts the phase of the summation. Therefore both (75) and (82) have the same Fourier phase factor, $e^{-i2\pi(hx+ky)}$, so that, in any event, both Fourier syntheses determine the same peak location. Secondly, although approximations such as (83) are fairly good when h and k are large, and poorer when h and k are small, nevertheless the whole set of approximations in (82) comprises a very good net approximation because a full set of f_{hkl}'s and f_{hk0}'s are merely different samplings of the same Fourier transform of the atom, as explained in the next chapter. For these reasons, the italicized conclusions noted above are both valid.

This reasoning has been concerned with the synthesis of a particular one of the J atoms of the crystal structure. The Fourier synthesis of the entire structure is composed of the sum of the syntheses of the component atoms, according to (72). It follows that in the generalized projection of the entire structure, (69), *the peak due to each of the component atoms of the structure appears at its correct location* xy, *and that the height of the peak of each atom is the same as that of the zero-level projection, but that it has a phase shifted by* $e^{i2\pi Lz_j}$.

With these properties of the generalized projection established, some details can be considered. Returning to (69), it is evident that, in general, the F's and the exponentials are both complex. To transform (69) to a form suitable for computing, each can be expressed in terms of real and imaginary parts:

$$F_{hkL} = A_{hkL} + iB_{hkL}, \tag{90}$$

and

$$e^{-i2\pi(hx+ky)} = \cos 2\pi(hx+ky) - i \sin 2\pi(hx+ky). \tag{91}$$

If these are substituted with (69), it is expanded to

$$\rho_{L(xy)} = \frac{1}{S}\sum_h \sum_k [A_{hkL}+iB_{hkL}][\cos 2\pi(hx+ky) - i \sin 2\pi(hx+ky)] \tag{92}$$

$$= \frac{1}{S}\sum_h \sum_k [A_{hkL} \cos 2\pi(hx+ky) + B_{hkL} \sin 2\pi(hx+ky)]$$
$$+ i\frac{1}{S}\sum_h \sum_k [-A_{hkL} \sin 2\pi(hx+ky) + B_{hkL} \cos 2\pi(hx+ky)]. \tag{93}$$

This expression is composed of two parts:
a real part:

$$R_{L(xy)} = \frac{1}{S}\sum_h \sum_k [A_{hkL} \cos 2\pi(hx+ky) + B_{hkL} \sin 2\pi(hx+ky)], \tag{94}$$

and an imaginary part:

$$I_{L(xy)} = \frac{1}{S}\sum_h \sum_k [-A_{hkL} \sin 2\pi(hx+ky) + B_{hkL} \cos 2\pi(hx+ky)]. \tag{95}$$

Each of these parts is called a *component* of the generalized projection. Then the full generalized projection, (93), can be abbreviated as

$$\rho_{L(xy)} = R_{L(xy)} + iI_{L(xy)}. \tag{96}$$

The absolute value of the generalized projection can be computed with the aid of

$$\left|\rho_{L(xy)}\right| = [R^2_{L(xy)} + I^2_{L(xy)}]^{1/2}, \tag{97}$$

as suggested by Fig. 11. In contrast to the component syntheses, (94) and (95), this is called a *modulus projection.* The modulus projection has the property that, *ideally it is the same for every level, including zero.*

In ordinary projections, the Fourier synthesis represents electron density which is everywhere positive. In generalized projections this positive density is modulated by the factor $e^{i2\pi Lz}$, and the component syntheses are modulated by the factors $\cos 2\pi Lz$ and $\sin 2\pi Lz$. Therefore an atom, comprising positive electron density, may appear in a component of the generalized projection as its negative equivalent. In this manner a new symmetry operation, a reversal of sign, which is unknown in classical crystallography, appears in the components of the general projection. Whereas an ordinary projection must have one of the 17 plane-group symmetries, the components of a generalized projection may have one of the 46 two-dimensional reversal symmetries.[31]

Some excellent examples of generalized projections are provided by Fridrichsons and Mathieson.[39] They worked with DL-isocryptopleurine methiodide, whose cell data are

space group: $P\,2_1/n$

$$a = 9.95\ \text{Å},$$
$$b = 24.2 \qquad \beta = 112°,$$
$$c = 9.95,$$
$$Z = 4\ C_{25}H_{30}O_3NI \text{ per cell.}$$

Although a generalized projection along b would have avoided the imaginary component, the projection direction a was used in order to resolve the molecule. The ordinary projection $\rho_{(yz)}$ is shown in Fig. 12. The components and modulus of the generalized projection $\rho_{1(yz)}$ are shown in Fig. 13, and of $\rho_{5(yz)}$ are shown in Fig. 14. It is evident that all three projections determine the same atom locations. Fridrichsons and Mathieson added all three in the projection $\sum \frac{1}{2}\rho_0 + \rho_1 + \rho_5$ to obtain Fig. 15.

This brings out the feature that generalized projections usually display enhanced resolving powers over that of ρ_0. This is because atoms on different levels are differently affected by the factor $e^{i2\pi Lz}$, so that there is usually less interference between atoms which are adjacent in projection.

Sinusoidal modulation. A closely related type of modulation has been devised by Zachariasen,[35] who weighted the electron density by the

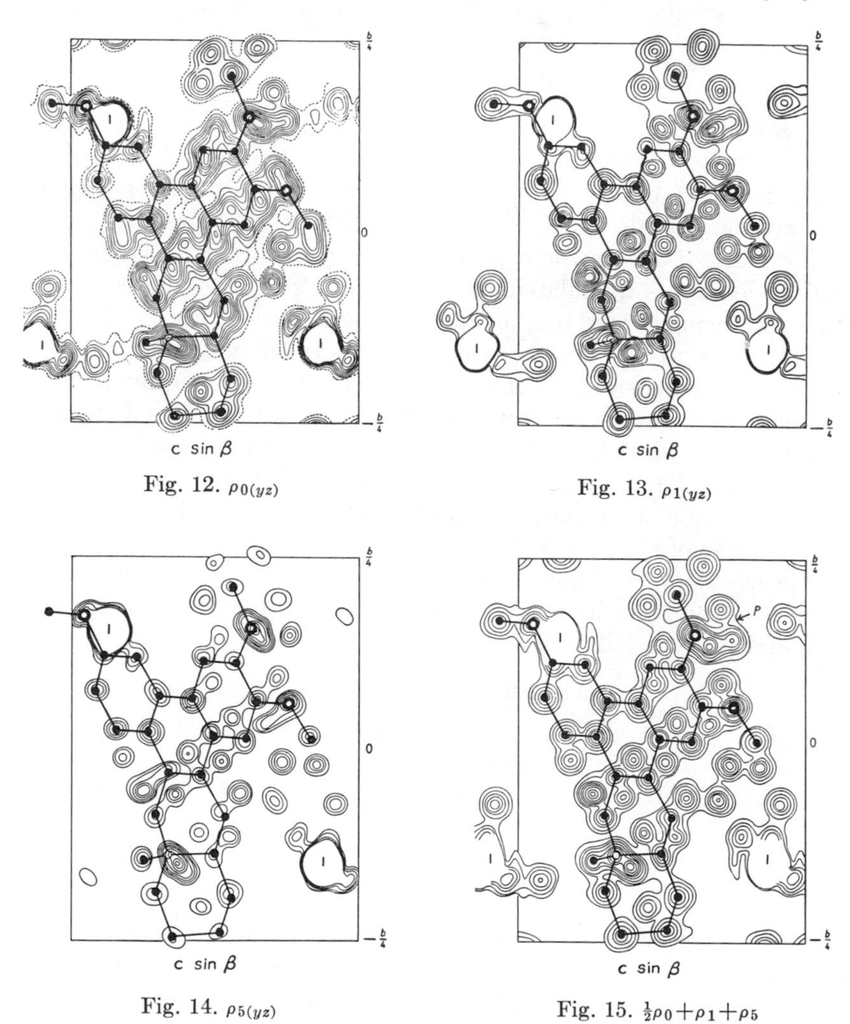

Fig. 12. $\rho_{0(yz)}$

Fig. 13. $\rho_{1(yz)}$

Fig. 14. $\rho_{5(yz)}$

Fig. 15. $\frac{1}{2}\rho_0 + \rho_1 + \rho_5$

Figs. 12–15.
Generalized projections of DL-isocrystopleurine methiodide.
(After Friderichsons and Mathieson[39])

factor $\cos 2\pi z$ and integrated it from $z = 0$ to $z = 1$. This weighting results in a projection which bears an interesting relation to the generalized electron-density projection discussed in the last section. To understand this relation, note that in (66) the function integrated is composed of two parts, a real part $\rho_{(xyz)}$ and a part, $e^{i2\pi Lz}$, which has, in general, a complex component. If the real and complex parts of $e^{i2\pi Lz}$ are separated as $\cos 2\pi Lz + i \sin 2\pi Lz$, (66) can be written

$$\begin{aligned} \rho_{L(xy)} &= c \int_0^1 \rho_{(xyz)} \cos 2\pi Lz \, dz \\ &\quad + ic \int_0^1 \rho_{(xyz)} \sin 2\pi Lz \, dz. \end{aligned} \tag{98}$$

Now, without bothering to go through the details of integrating the two lines of (98), which are the real and imaginary parts, respectively, of the generalized projection, it is evident that they should be the same as the real and imaginary parts of the integrated result. These are given in the first and second lines, respectively, of (93). From this it follows that if the sinusoidal weighting function is $\cos 2\pi Lz$, the result is a function proportional to the real part of the generalized projection, whereas if

Table 3
Sinusoidally modulated projections

Type	Form for centrosymmetrical crystals	Weighting Extremes	
		Lz	Weight
$\int_0^1 \rho_{(xyz)} \cos 2\pi Lz \, c \, dz$	$\frac{1}{S} \sum_h \sum_k F_{hkL} \cos 2\pi(hx+ky)$	0 $\frac{1}{2}$	1 -1
$\int_0^1 \rho_{(xyz)} \sin 2\pi Lz \, c \, dz$	$-\frac{1}{S} \sum_h \sum_k F_{hkL} \sin 2\pi(hx+ky)$	$\frac{1}{4}$ $\frac{3}{4}$	1 -1
$\int_0^1 \rho_{(xyz)}(1 + \cos 2\pi Lz) \, c \, dz$	$\frac{1}{S} \sum_h \sum_k (F_{hk0}+F_{hkL}) \cos 2\pi(hx+ky)$	0 $\frac{1}{2}$	2 0
$\int_0^1 \rho_{(xyz)}(1 - \cos 2\pi Lz) \, c \, dz$	$\frac{1}{S} \sum_h \sum_k (F_{hk0}-F_{hkL}) \cos 2\pi(hx+ky)$	0 $\frac{1}{2}$	0 2
$\int_0^1 \rho_{(xyz)}(1 + \sin 2\pi Lz) \, c \, dz$	$\frac{1}{S} \sum_h \sum_k F_{hk0} \cos 2\pi(hx+ky)$ $-F_{hkL} \sin 2\pi(hx+ky)$	$\frac{1}{4}$ $\frac{3}{4}$	2 0
$\int_0^1 \rho_{(xyz)}(1 - \sin 2\pi Lz) \, c \, dz$	$\frac{1}{S} \sum_h \sum_k F_{hk0} \cos 2\pi(hx+ky)$ $+F_{hkL} \sin 2\pi(hx+ky)$	$\frac{1}{4}$ $\frac{3}{4}$	0 2

the sinusoidal weighting function is $\sin 2\pi Lz$, the result is a function proportional to the imaginary part of the generalized projection. Zachariasen's weighting function[35] is therefore a special case of the real part of (98) in that $L = 1$, and the proportionality factor c is omitted.

Zachariasen[35] and his followers[38, 41] also used an interesting composite weighting function which, unlike the parts of (98), has no negative values. These weighting functions are $(1 + \cos 2\pi z)$ and $(1 + \sin 2\pi z)$. If these

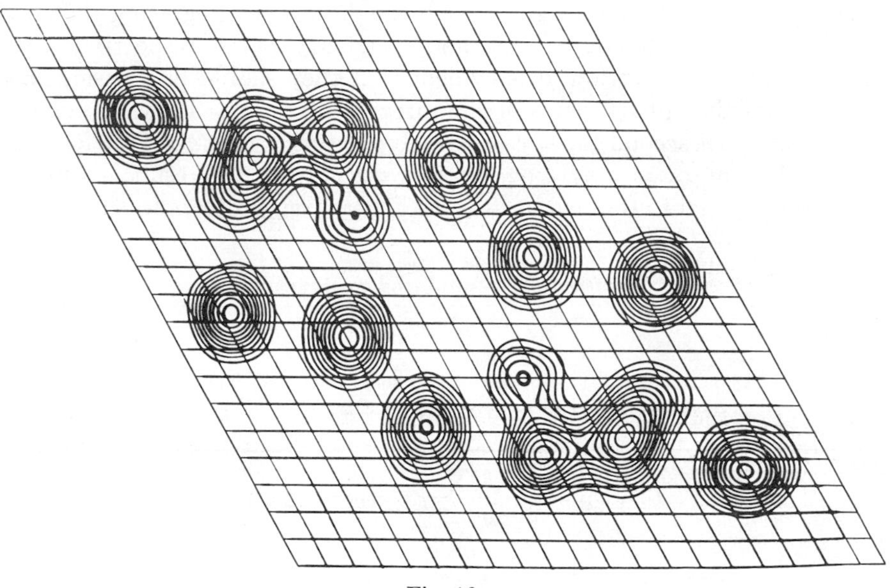

Fig. 16.
Electron-density projection $\rho_{(xy)}$; orthoboric acid. Note lack of resolution. (After Zachariasen.[35])

are generalized to $c(1 + \cos 2\pi Lz)$ and $c(1 + \sin 2\pi Lz)$, their relation to the generalized projection is obvious. For example

$$\int_0^1 \rho_{(xyz)}(1 + \cos 2\pi Lz)c\,dz = c\int_0^1 \rho_{(xyz)}\,dz + c\int_0^1 \rho_{(xyz)} \cos 2\pi Lz\,dz$$
$$= \rho_{(xy)} + R(\rho_{L(xy)}). \tag{99}$$

That is, this function is the sum of the ordinary electron-density projection plus the real part of the Lth-level projection.

The simplicities of these composite functions can be appreciated by evaluating them for centrosymmetrical crystals. In these cases, the exponential summation for the ordinary projection, (70), reduces to a summation of $F_{hk0} \cos 2\pi(hx+ky)$. The reduction for the generalized-projection part can be derived from (93). From what has already been said, the weighting function $\cos 2\pi Lz$ in (99) calls for merely the real part of (93), namely $A_{hkL} \cos 2\pi(hx+ky) + B_{hkl} \sin 2\pi(hx+ky)$. But for centrosymmetrical crystals $B_{hkL} = 0$, and $A_{hkL} = F_{hkL}$, so that the second part of (99) is a summation over $F_{hkL} \cos 2\pi(hx+ky)$. The composite summation for (99) for a centrosymmetrical crystal is therefore over $(F_{hk0} + F_{hkL}) \cos 2\pi(hx+ky)$.

Table 3 lists this and other sinusoidally modulated projections with their forms and properties. The composite projections have the interesting property that they can be used to emphasize atoms at certain

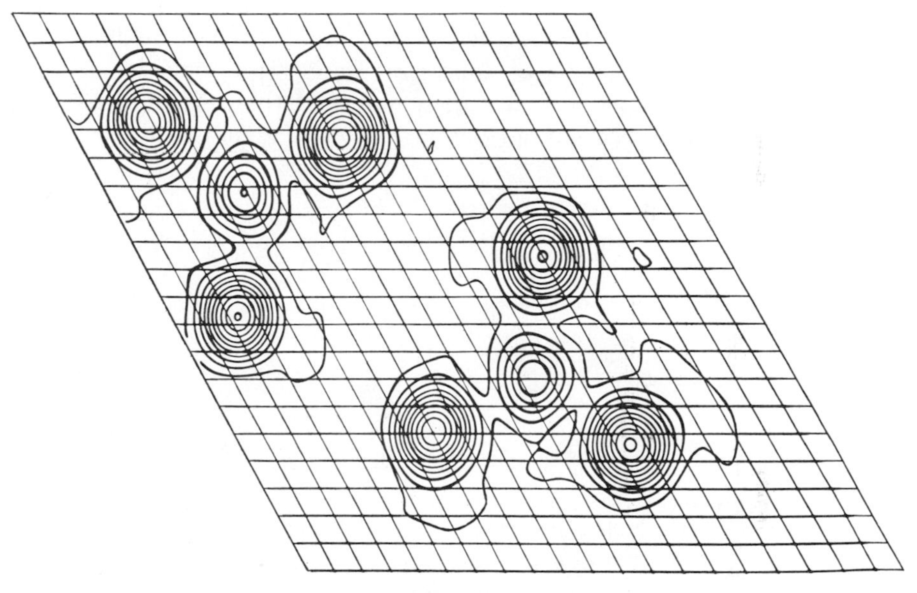

Fig. 17.
Modulated electron-density projection

$$\int_0^1 \rho_{(xyz)}[1 + \sin 2\pi z]\, dz;$$

orthoboric acid. Note improved resolution.
(After Zachariasen.[35])

chosen levels z but suppress atoms at certain other levels. For example, in studying the structure of orthoboric acid, Zachariasen[35] found that the projection of the full cell, Fig. 16, showed lack of resolution between atoms on levels $\frac{1}{4}$ and $\frac{3}{4}$. The atoms in level $\frac{3}{4}$ are eliminated, while those on level $\frac{1}{4}$ are given full weight by using the composite projection $\int_0^1 \rho_{(xyz)}(1 + \sin 2\pi z)\, dz$, Fig. 17.

Elimination of atoms having subperiods

It is an easy matter to set up a Fourier expression to which atoms having a subperiodic arrangement within the cell make no contribution, and are hence eliminated.[50] This subject is treated in outline here, because a fuller treatment† would make use of the subject matter of the next chapter.

Suppose one or more sets of atoms have a period which is $1/n$th of that of the b-axis period. Then this set, considered by itself, diffracts as if the b axis were b/n. These atoms contribute only to reflections $h\,nk\,l$; other atoms contribute to all reflections.

As in the discussion of the generalized projection, the entire electron density of the crystal can be regarded as the sum of the electron densities of its several atoms. Let ${}^s\rho_{(xyz)}$ represent the electron densities of the atoms having the subperiod, and ${}^p\rho_{(xyz)}$ represent the electron densities of the atoms having the normal period. Then the composite electron density is

$$\rho_{(xyz)} = {}^s\rho_{(xyz)} + {}^p\rho_{(xyz)}. \tag{100}$$

If the portion ${}^p\rho_{(xyz)}$ alone is desired, this is given by

$$\begin{aligned} {}^p\rho_{(xyz)} &= \rho_{(xyz)} - {}^s\rho_{(xyz)} \\ &= \frac{1}{V}\sum_h\sum_k\sum_l F_{hkl}\, e^{-i2\pi(hx+ky+lz)} \\ &\qquad - \frac{1}{V}\sum_h\sum_k\sum_l {}^sF_{hkl}\, e^{-i2\pi(hx+ky+lz)} \\ &= \frac{1}{V}\sum_h\sum_k\sum_l (F_{hkl} - {}^sF_{hkl})e^{-i2\pi(hx+ky+lz)}, \end{aligned} \tag{101}$$

where ${}^sF_{hkl}$ is the partial F_{hkl} due to the atoms having the subperiod. This can be rewritten as a sum of two summations, one for $k \neq nm$ and one for $k = nm$, where m is an integer. For $k \neq nm$, ${}^sF_{hkl} = 0$, so that

$$\begin{aligned} {}^p\rho_{(xyz)} &= \frac{1}{V}\sum_h\sum_{k\neq nm}\sum_l F_{hkl}\, e^{-i2\pi(hx+ky+lz)} \\ &\quad + \frac{1}{V}\sum_h\sum_{k=nm}\sum_l (F_{hkl} - {}^sF_{hkl})e^{-i2\pi(hx+ky+lz)}. \end{aligned} \tag{102}$$

Each of these parts represents a fraction of the same summation. The second part gives $1/n$th of the summation; the first part gives $1 - 1/n = (n - 1)/n$th of the summation. Each part is a different sampling of the same Fourier transform (see next chapter), and either part produces a part of the same electron density. The second line in (102) is difficult to evaluate, but the first line is easily evaluated, and it is alone sufficient to

† A fuller treatment is given in M. J. Buerger, *Vector space and its application in crystal-structure investigation.* (John Wiley and Sons, New York, 1959) Chapter 14.

reproduce the desired synthesis. It is called a partial Fourier synthesis, and is given by

$$\partial\rho_{(xyz)} = \frac{1}{V}\sum_{h}\sum_{k\neq nm}\sum_{l} F_{hkl}\, e^{-i2\pi(hx+ky+lz)}. \tag{103}$$

An example of a partial Fourier synthesis is given in Figs. 18 and 19. In the structure of pectolite,[51] $NaHCa_2Si_3O_9$, some of the atoms have a

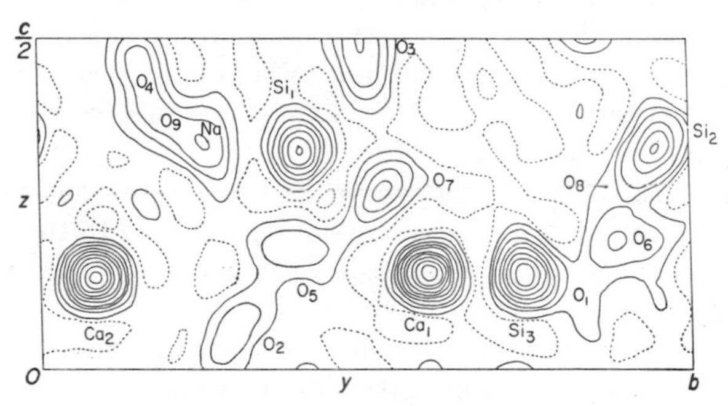

Fig. 18.
Fourier synthesis, $\rho_{(yz)}$; pectolite, $NaHCa_2Si_3O_9$.
(After Buerger.[51])

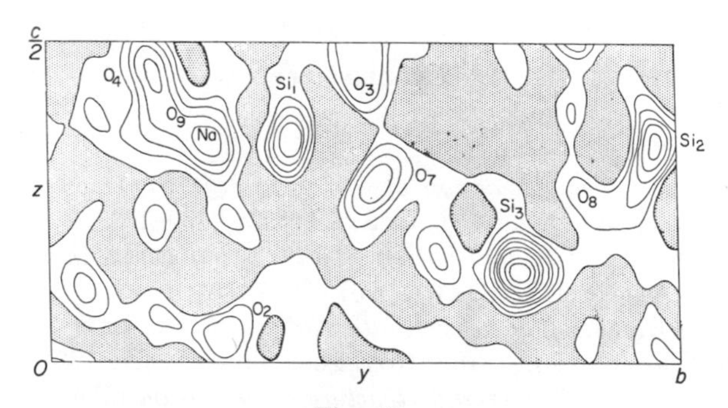

Fig. 19.
Partial Fourier synthesis, $\partial\rho_{(yz)}$; pectolite, $NaHCa_2Si_3O_9$. Note suppression of Ca_1, Ca_2 and O_5, O_6 of Fig. 18.
(After Buerger.[51])

subperiod $b/2$. The electron density projection $\rho_{(yz)}$ is shown in Fig. 18. It can be seen that the following pairs of atoms are separated by $b/2$: Ca_1, Ca_2, O_5, O_6. Those having subperiod $b/2$ contribute only to reflections hkl for k even. They can be eliminated by constructing a Fourier synthesis for which the F's of these spectra are absent. Such syntheses are

$$\partial\rho_{(xyz)} = \frac{1}{V}\sum_{h}\sum_{\substack{k\\ \text{odd}}}\sum_{l} F_{hkl}\, e^{-i2\pi(hx+ky+lz)}. \tag{104}$$

The particular form for the projection $\partial\rho_{(yz)}$ is

$$\partial\rho_{(xy)} = \frac{1}{S}\sum_{\substack{k\\ \text{odd}}}\sum_{l} F_{0kl}\cos 2\pi(hx+ky). \qquad (105)$$

This is shown in Fig. 19. Note that the atom pairs Ca_1 Ca_2 and O_5 O_6 have been eliminated. A peculiarity of this synthesis is that it is composed of positive and negative portions which alternate at an interval $b/2$. In Fig. 19 the negative portion is shaded, since it is a negative repetition of the positive portion. These syntheses evidently have one of the reversal symmetries of Cochran.[31]

Equivalent positive and negative peaks bear an interesting relation to one another. Both are alternative regions permitted to the atoms which do not have the subperiod. The positive regions show where such atoms *are*, the negative regions where they *are not*.

Partial Fourier syntheses can be devised for any situation in which an atom or set of atoms has a subperiod. The synthesis has the normal form except that the F_{hkl}'s are omitted to which the atoms in the substructure contribute.

Literature

General and early papers

[1] William Duane. *The calculation of the x-ray diffracting power at points in a crystal.* Proc. Nat. Acad. Sci. U. S. **11** (1925) 489–493.

[2] R. J. Havighurst. *The distribution of diffracting power in sodium chloride.* Proc. Nat. Acad. Sci. U. S. **11** (1925) 502–507.

[3] W. Lawrence Bragg. *The determination of parameters in crystal structures by means of Fourier series.* Proc. Roy. Soc. (London) (A) **123** (1929) 537–559.

[4] W. H. Zachariasen. *The crystal structure of potassium chlorate.* Z. Krist. **71** (1929) 501–516.

[5] W. L. Bragg and J. West. *A note on the representation of crystal structure by Fourier series.* Phil. Mag. **10** (1930) 823–841.

[6] W. H. Zachariasen. *The crystal lattice of oxalic acid dihydrate, $H_2C_2O_2 \cdot 2H_2O$ and the structure of the oxalate radical.* Z. Krist. **89** (1934) 442–447.

[7] Arthur H. Compton and Samuel K. Allison. *X-rays in theory and experiment.* (D. Van Nostrand Co., New York, 1935) 449–461.

[8] J. Monteath Robertson. *X-ray analysis of the structure of dibenzyl. II. Fourier analysis.* Proc. Roy. Soc. (London) (A) **150** (1935) 348–362.

[9] S. B. Hendricks and M. E. Jefferson. *Electron distribution in $(NH_4)_2C_2O_4 \cdot H_2O$ and the structure of the oxalate group.* J. Chem. Phys. **4** (1936) 102–107.

[10] J. Monteath Robertson and Ida Woodward. *The structure of the carboxyl group. A quantitative investigation of oxalic acid dihydrate by Fourier synthesis of the x-ray crystal data.* J. Chem. Soc. (London) (1936) 1817–1824.

[11] J. Monteath Robertson. *X-ray analysis and application of Fourier series to molecular structures.* Phys. Soc. Repts. Progr. in Phys. **4** (1938) 332–367.
[12] A. D. Booth. *Fourier technique in x-ray organic structure analysis.* (Cambridge University Press, Cambridge, 1948)
[13] W. Cochran. *The Fourier method of crystal-structure analysis.* Acta Cryst. **1** (1948) 138–142.
[14] G. S. Parry and G. J. Pitt. *The derivation of atomic co-ordinates from planar and linear Fourier syntheses.* Acta Cryst. **2** (1949) 145–147.
[15] Werner Nowacki. *Fouriersynthese von Kristallen.* (Birkhäuser, Basel, Switzerland, 1952)
[16] D. M. Burns. *An analytic method of dealing with unresolved peaks in Fourier projections.* Acta Cryst. **8** (1955) 517–518.
[17] P. J. Black. *Oblique projections.* Acta Cryst. **8** (1955) 656–657.
[18] J. M. Cowley. *'Stereoscopic' three-dimensional structure analysis.* Acta Cryst. **9** (1956) 399–401.
[19] Masao Atoji. *Spherical Fourier method.* Acta Cryst. **11** (1958) 827–829.

Limited parts of cell

[20] A. D. Booth. *Two new modifications of the Fourier method of x-ray structure analysis.* Trans. Faraday Soc. **41** (1945) 434–438.
[21] Maurice L. Huggins. *Equations for various types of summations.* Am. Soc. for X-ray and Electron Diffraction. Abstracts, Lake George Meeting (1946) 24–31.
[22] G. A. Jeffrey. *The structure of polyisoprene. VI. An investigation of the molecular structure of dibenzyl by x-ray analysis.* Proc. Roy. Soc. (London) (A) **188** (1947) 222–236.
[23] Walter L. Roth and David Harker. *The crystal structure of octamethylspiro* [5·5] *pentasiloxane: Rotation about the ionic silicon-oxygen bond.* Acta Cryst. **1** (1948) 34–42.
[24] Don T. Cromer, Aaron J. Ihde, and H. L. Ritter. *The crystal structure of tetramethylpyrazine.* J. Am. Chem. Soc. **73** (1951) 5587–5590.
[25] Maja Edstrand. *On the structures of antimony (III) oxidehalides. II. The crystal structure of SbOCl.* Arkiv. Kemi **6** (1953) 89–112.
[26] Maja Edstrand and Nils Ingri. *The crystal structure of double lithium antimony (V) oxide* $LiSbO_3$. Acta Chem. Scand. **8** (1954) 1021–1031.
[27] Yoshiharu Okaya and Ray Pepinsky. *The crystal structure of ammonium acid phthalate.* Acta Cryst. **10** (1957) 324–328.

Modulated projections

[28] C. J. B. Clews and W. Cochran. *The structures of the pyrimidines and purines. III. An x-ray investigation of hydrogen bonding in aminopyramidines.* Acta Cryst. **2** (1949) 46–57, especially 51.
[29] H. B. Dyer. *The crystal structure of cysteylglycine–sodium iodide.* Acta Cryst. **4** (1951) 42–50, especially 44.
[30] R. F. Raeuchle and R. E. Rundle. *The structure of* $TiBe_{12}$. Acta Cryst. **5** (1952) 85–93, especially 89.
[31] W. Cochran. *The symmetry of real periodic two-dimensional functions.* Acta Cryst. **5** (1952) 630–633.

[32] W. Cochran and H. B. Dyer. *Some practical applications of generalized crystal-structure projections.* Acta Cryst. **5** (1952) 634–636.
[33] J. C. Speakman. *The crystal structure of an analogue of nickel phthalocyanine.* Acta Cryst. **6** (1953) 784–791, especially 788–789.
[34] D. C. Phillips. *Atomic resolution in generalized crystal-structure projections.* Acta Cryst. **7** (1954) 221–222.
[35] W. H. Zachariasen. *The precise structure of orthoboric acid.* Acta Cryst. **7** (1954) 305–310, especially 306–307.
[36] K. Eriks and C. H. MacGillavry. *The crystal structure of* $N_2O_5{\cdot}3SO_3$. Acta Cryst. **7** (1954) 430–434, especially 432.
[37] H. Steinfink, B. Post, and I. Fankuchen. *The crystal structure of octamethyl cyclotetrasiloxane.* Acta Cryst. **8** (1955) 420–424, especially 422.
[38] Richard M. Curtis and R. A. Pasternak. *The crystal structure of methylguanidinium nitrate.* Acta Cryst. **8** (1955) 675–681, especially 676.
[39] J. Fridrichsons and A. McL. Mathieson. *A direct determination of molecular structure:* DL-*isocryptopleurine methiodide.* Acta Cryst. **8** (1955) 761–772, especially 763–764.
[40] Noel E. White and C. J. B. Clews. *The crystal and molecular structure of 4,5-diamino-2-chloropyrimidine.* Acta Cryst. **9** (1956) 586–593, especially 587–588.
[41] Yoshikaru Okaya and Ray Pepinsky. *Crystal structure of triaminoguanidinium chloride,* $(NH_2{\cdot}NH)_3C.Cl$. Acta Cryst. **10** (1957) 681–684.
[42] D. June Sutor. *The structure of the pyrimidines and purines. VI. The crystal structure of theophylline.* Acta Cryst. **11** (1958) 83–87.
[43] M. G. Rossmann and H. M. M. Shearer. *Some improvements in the method of generalized projections.* Acta Cryst. **11** (1958) 829–832.

Substructures and superstructures

[44] J. W. Jeffery. *X-ray diffraction by a crystal possessing periodicities within the unit cell.* Proc. Phys. Soc. (A) **44** (1951) 1003–1006.
[45] A. L. Mackay. *A statistical treatment of superlattice reflexions.* Acta Cryst. **6** (1953) 214–215.
[46] M. J. Buerger. *Some relations for crystals with substructures.* Proc. Nat. Acad. Sci. U. S. **40** (1954) 125–128.
[47] P. Vousden. *The determination of pseudosymmetric structures.* Acta Cryst. **7** (1954) 321–322.
[48] Erik von Sydow. *On the structure of the crystal form* A′ *of* n-*pentadecanoic acid.* Acta Cryst. **7** (1954) 529–532.
[49] A. D. Wadsley. *The crystal structure of chalcophanite,* $ZnMn_3O_7{\cdot}3H_2O$. Acta Cryst. **8** (1955) 165–172.
[50] M. J. Buerger. *Partial Fourier syntheses and their application to the solution of certain types of crystal structures.* Proc. Nat. Acad. Sci. U. S. **42** (1956) 776–781.
[51] M. J. Buerger. *The determination of the crystal structure of pectolite,* $Ca_2NaHSi_3O_9$. Z. Krist. **108** (1956) 248–262.
[52] Ronald L. Sass, Rosemary Vidale, and Jerry Donohue. *Interatomic distances and thermal anisotropy in sodium nitrate and calcite.* Acta Cryst. **10** (1957) 567–570.

15

Reciprocal space

In the last two chapters Fourier synthesis and some of its forms useful in crystal-structure analysis have been considered. In the discussion it was seen that the quantity $1/d_{hkl}$ was needed, and this indicated a relation between Fourier synthesis and the reciprocal lattice. In this chapter, this relation is investigated in its own right. It turns out that the kind of transformation involved in Fourier synthesis is a generalization of that involved in transforming a lattice to its reciprocal.

Generalization of the reciprocal lattice

In that part of x-ray crystallography dealing with the investigation of the unit cell and space group of a crystal, the details of the structure can be ignored. In this restricted use of x-ray diffraction, the geometry of the diffraction maxima is important and their intensities are unimportant. Under these circumstances it is sufficient to develop the theme that for a given crystal and its lattice there can be constructed a *reciprocal lattice*, and that to each point of this reciprocal lattice there corresponds an x-ray reflection by that crystal.

For this limited purpose the reciprocal lattice is constructed as follows.† To each stack of planes (hkl), a normal is constructed, thus defining the directions of a set of interplanar spacings. All such directions are assembled so as to radiate from a common origin, and on each a point is placed at a distance $\sigma_{hkl} = 1/d_{hkl}$ from this origin. Thus these points are at origin distances equal to the reciprocals of the interplanar spacings. These points lie on the points of a lattice which is said to be "reciprocal" to the original lattice.

† M. J. Buerger. *X-ray crystallography.* (John Wiley and Sons, New York, 1942) Chapters 6 and 7.

Now, in those branches of crystallography which are concerned only with the geometry of rational planes, the indices h, k, and l of a stack of planes cannot contain a common factor. To see this, consider a stack of planes, referred to a primitive cell, (hkl) where h, k and l do not contain a common factor. Any particular plane of the stack is repeated by all the translations of the lattice to become the other planes of the stack. Therefore every lattice point contains one plane of the stack. To every plane there correspond lattice points, and no plane is without a lattice point. Now consider a plane $(nh\ nk\ nl)$, where n is a common factor. The indices $(nh\ nk\ nl)$ imply a plane having $1/n$th the spacing of (hkl), and hence imply a stack with n times as many planes as (hkl). Every nth plane in the stack $(nh\ nk\ nl)$ contains lattice points, but the remainder do not. Therefore $nh\ nk\ nl$ is not a set of legitimate indices for a rational stack of planes. But for the purposes of x-ray crystallography the nth-order Bragg reflection from hkl behaves precisely as if it were the 1st-order reflection from $(nh\ nk\ nl)$. Although such a plane is fictitious, the designation is accepted in x-ray crystallography in order that the complicating notion of order-of-reflection can be dropped. When $(nh\ nk\ nl)$ is accepted as a possible plane, then the indices hkl contain all possible combinations of all integers. (If such indices are not accepted certain combinations of hkl are missing.)

Thus all values of $d_{nh\ nk\ nl} = (1/n)d_{hkl}$ can be regarded as occurring in the direct lattice, and consequently all values of $n(1/d_{hkl})$ occur in the reciprocal lattice. Therefore, on the line normal to the rational plane (hkl) there occurs a series of points having distances of $n(1/d_{hkl})$ from the origin, i.e. a row of points of interval $1/d_{hkl}$. In effect, then, a set of rational planes of spacing d_{hkl} in direct space is represented in reciprocal space by a row of points of separation $1/d_{hkl}$ and lying on the normal to the set of planes, Fig. 1.

Now, the planes of index (hkl) constitute a periodical sequence. If this periodicity is thought of in most general terms, the repetition can be described as cyclic, and the interval of repetition, namely d, corresponds to 2π. The reciprocal representation of this sequence of planes is a sequence of points which is also cyclically repeated, and the interval $1/d$ represents 2π.

To generalize the notion of reciprocal, consider a single plane, Fig. 2, similar to the plane of Fig. 1. In this generalized case, however, the plane is an individual plane and is not a member of a stack. Let the reciprocal representation of the plane be defined along its normal, OQ, similar to what it was for Fig. 1. Let the reciprocal be characterized by a periodic change along the normal. If this period is expressed in terms of phase, then the phase is zero at the origin, 2π at distance $1/\delta$ from the origin, and $n(2\pi)$ at a distance $n(1/\delta)$ from the origin. This represents a generalization over the reciprocal-lattice case, where, in the corresponding reciprocal, $1/d$, d represented the interplanar spacing, and was the distance from the origin only if the origin was placed at a lattice point. In the generalized case, there are no necessary lattice points,

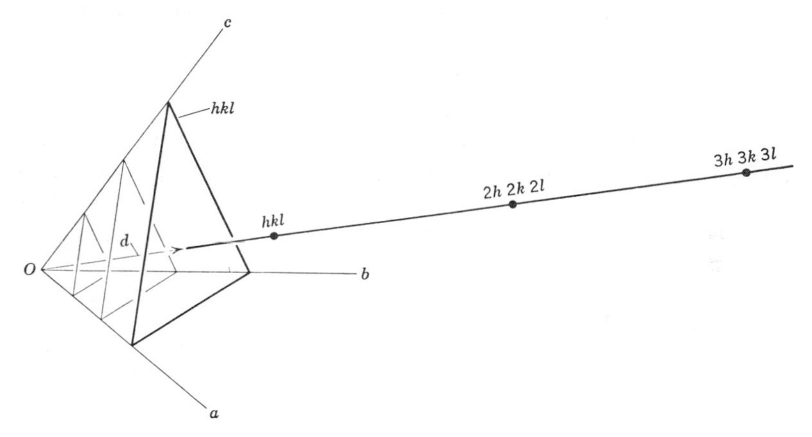

Fig 1.

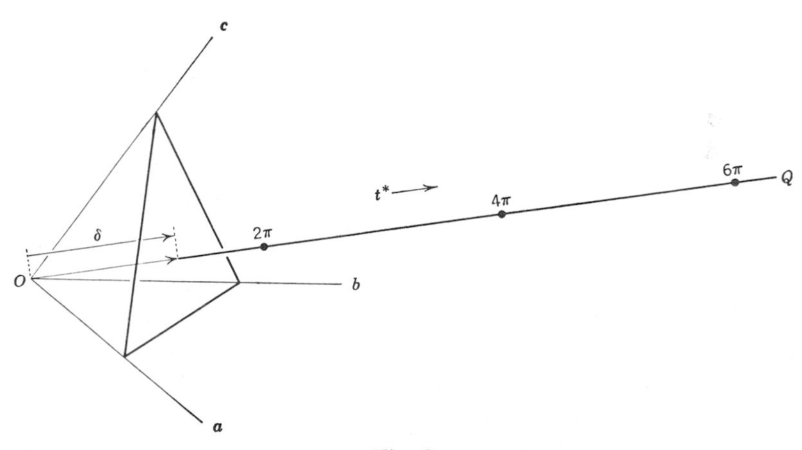

Fig. 2.

and the absolute value of the distance of the plane from the origin determines the distances from the origin where the phase of the reciprocal has values of $n(2\pi)$.

As a second generalization, let the phase be defined at *all* points along the normal, not just at a sequence of discrete points where its value is $n(2\pi)$. Therefore, the reciprocal to a plane is taken as being represented by a continuous phase variation along the normal to the plane. The phase variation is linear along the normal, namely proportional to t^* where t^* is the distance from the origin to the point on the normal being considered. Since the period in reciprocal space is $1/\delta$, the phase at any distance t^* is $2\pi \dfrac{t^*}{1/\delta} = 2\pi\ \delta t^*$. Therefore the product δt^* determines the phase at a point whose origin distance is t^* due to a plane whose origin distance is δ. If one represents the phase component† by the notation of complex numbers, it is $e^{i2\pi\delta t^*}$.

Now, the phasal significance of the idea of the reciprocal, which has just been set up, is quite independent of the idea of the periodicity of a lattice and can, indeed, be applied to any plane whatever. It merely states that the phase along a line in reciprocal space normal to a plane in

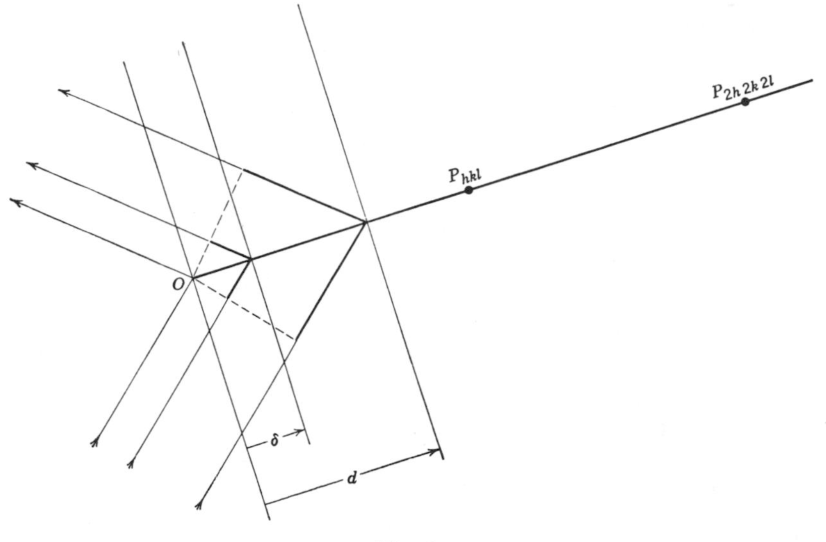

Fig. 3.

direct space is controlled by the distance, δ, of the plane from the origin. The amplitude component of variation along the line in reciprocal space has not yet been mentioned. This must obviously be controlled by some quality of the plane in direct space. Let this amplitude be W, which corresponds with a weighting attributed to the plane. (For x-ray diffraction purposes W is taken equal to the scattering power of the

† Incidentally, this corresponds to the phase scattered by an atom in a cell. The phase difference scattered by planes separated by spacing d_{hkl} is fixed by Bragg's law as $n2\pi$, where n is the order of the reflection. This establishes the sequence of reciprocal-lattice points P_{hkl}, $P_{2h\ 2k\ 2l} \cdots$, Fig. 3. If the origin is taken at O of Fig. 3, then the phase at P_{hkl} due to the plane at a distance d is 2π. The phase at P_{hkl} due to a point on a plane at a smaller distance δ from the origin is therefore $2\pi\delta/d$, and the phase at $P_{nh\ nk\ nl}$ is $2\pi n\delta/d$. Since the distance from the nth reciprocal-lattice point is $t^* = n(1/d)$, this phase can be written $2\pi\delta t^*$, and the phase operator is $e^{i2\pi dt^*}$, which has the form mentioned.

plane, i.e., to the number of electrons in the plane.) It is now possible to define the reciprocal to a plane of weight W. The value of this reciprocal at a point at distance t^* along the normal to the plane is given by

$$R_{t^*} = We^{i2\pi\delta t^*}. \tag{1}$$

The reciprocal of the plane thus behaves like a line normal to the plane along which there is a stationary wave whose amplitude is W and whose wavelength is $1/\delta$, and whose phase increases uniformly with t^*.

In order to display the relation between the reciprocal lattice and this more general kind of reciprocal, the discussion made use of the important geometrical element of the space lattice, namely a plane. But the

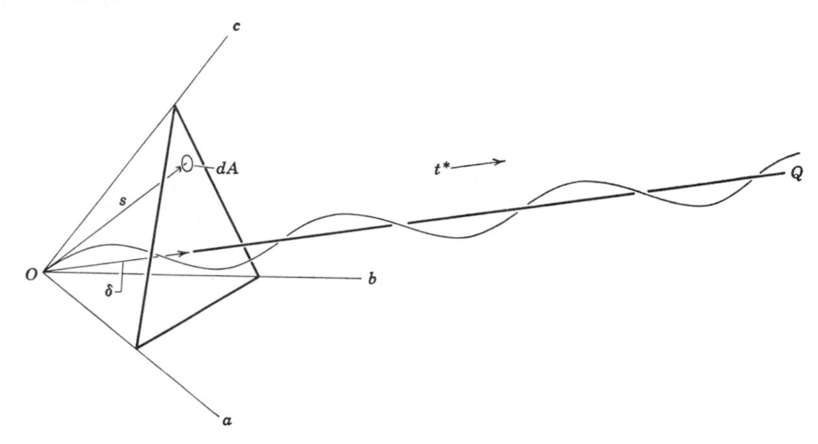

Fig. 4.

generalization is quite independent of the use of planes, as can be seen by the following discussion: All points on the plane have the same phase effect, since this is controlled by the distance, δ, of the plane from the origin. On the other hand, the distribution of weighting in the plane may be irregular (corresponding to the distribution of electron density in the plane), and consequently different regions of the plane may have different effects on the amplitude. Isolate a small area, dA, of the plane, Fig. 4, and attribute to it a density ρ. Then the effect of this small area on the phase at t^* is the same as that of any other part of the plane; namely it is proportional to δt^*. But δ is the projection of the vector $\mathbf{s}$ (which connects the origin with dA) on the normal OQ (Fig. 4). Therefore a neat way of writing the product δt^* is the scalar product of the two vectors $\mathbf{s} \cdot \mathbf{t}^*$. One of these two vectors extends from the origin to the point experiencing the reciprocal. (Since each of the terms of a dot product has equal control on the product, the meaning of the word reciprocal takes on a new significance.) The contribution to the reciprocal at the end of vector $\mathbf{t}^*$ from the infinitesimal area at the end of vector $\mathbf{s}$ can evidently be written

$$dR_{t^*} = dW\, e^{i2\pi\mathbf{s}\cdot\mathbf{t}^*} \qquad (2)$$
$$= \rho_s\, dA\, e^{i2\pi\mathbf{s}\cdot\mathbf{t}^*}$$

By integrating this over the area of the plane, there results

$$R_{t^*} = \int_{\text{area}} \rho_s\, e^{i2\pi\mathbf{s}\cdot\mathbf{t}^*}\, dA. \qquad (3)$$

Finally, this effect, produced at the end of a vector $\mathbf{t}^*$ due to a point at the end of a vector $\mathbf{s}$, can be integrated over a volume, Fig. 5. The

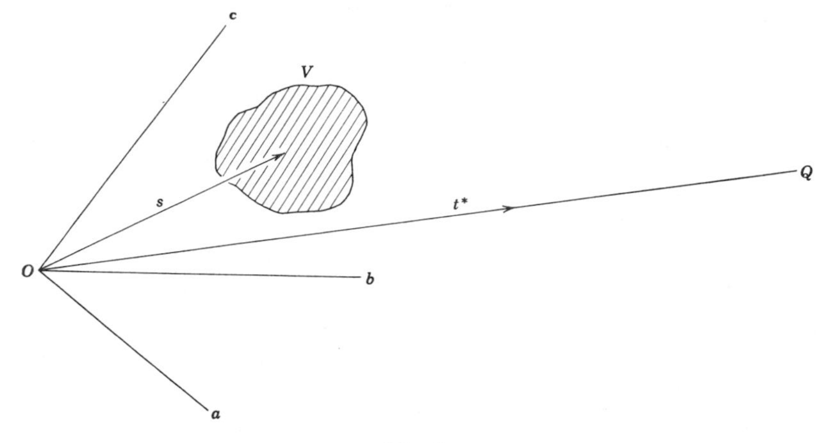

Fig. 5.

reciprocal at the end of $\mathbf{t}^*$ due to points in the volume is obviously

$$R_{t^*} = \int_{\substack{\text{direct}\\ \text{space}}} \rho_s\, e^{i2\pi\mathbf{s}\cdot\mathbf{t}^*}\, dV. \qquad (4)$$

The notion of reciprocal is now divorced from any necessary reference to a plane, or series of planes, or to a periodic structure of any sort. Relation (4) provides the general way by which the effects of all points in the volume in direct space are felt at the end of a vector $\mathbf{t}^*$ in reciprocal space, in respect to both phase and amplitude. The phase aspect is due to the positions of the volume elements dV as well as the position of $\mathbf{t}^*$, while the amplitude aspect is due to weightings, the effects of all elements being added in the complex plane. The function R_{t^*} provides the reciprocal of the distribution of the density, ρ, throughout the direct volume.

Some fundamental properties of reciprocals

The transformation to reciprocal space is identical with the Fourier transformation. The mathematics of the Fourier transformation† are tedious and indirect. The general features of the transformation can, however, be demonstrated on a simplified model.

Elementary direct and inverse transformations. Since all objects of which reciprocals may be required can be regarded as made up of a collection of small elementary volumes, consider the reciprocal of one elementary volume, dV, of density ρ, at the end of vector $\mathbf{s}$ from the origin. Since there is only one elementary volume, the integration has a very simple form. To emphasize the constant nature of $\mathbf{s}$ for the particular elementary volume dV, let it be written as $\mathbf{k}$. For this simple case the integration in (4) reduces to

$$R_{t^*} = \rho_k \, e^{i2\pi \mathbf{k}\cdot\mathbf{t}^*}. \tag{5}$$

If both sides of (5) are multiplied by $e^{-i2\pi\mathbf{k}\cdot\mathbf{t}^*}$ there results

$$R_{t^*} \, e^{-i2\pi\mathbf{k}\cdot\mathbf{t}^*} = \rho_k. \tag{6}$$

This evidently represents the contribution of one point (at the end of vector $\mathbf{t}^*$) in reciprocal space to the transform of this reciprocal to direct space. But the point at the end of the vector $\mathbf{k}$ in direct space receives transform contributions from all points in reciprocal space. The full contribution requires integrating the left side of (6) over all reciprocal space. The full value of the transform at the end of the vector $\mathbf{k}$ is therefore

$$\int_{V^*} R_{t^*} \, e^{-i2\pi\mathbf{k}\cdot\mathbf{t}^*} \, dV^* = \rho_k. \tag{7}$$

Thus, for a simple elementary volume at the end of vector $\mathbf{s}$ in direct space, the following transformations hold:

$$(5): \qquad R_{t^*} = \rho_s \, e^{i2\pi\mathbf{s}\cdot\mathbf{t}^*}, \qquad (T), \tag{5'}$$

$$(7): \qquad \rho_s = \int_{V^*} R_{t^*} \, e^{-i2\pi\mathbf{s}\cdot\mathbf{t}^*} \, dV^*, \qquad (T^{-1}). \tag{7'}$$

For this simple case these are solutions of each other. They can also be regarded as a transformation T from direct to reciprocal space, and

† See, for example:

R. Courant. *Differential and integral calculus.* Vol. 2. (Interscience Publishers, New York, 1936) 318.

E. C. Titchmarsh. *The theory of functions.* (Oxford University Press, Oxford, 1939) 432–435.

E. C. Titchmarsh. *Introduction to the theory of Fourier integrals.* (Clarendon Press, Oxford, 1948) 1–4.

the inverse transformation, T^{-1}, from reciprocal space to direct space, respectively.

Characteristics of the transformation of a point. Before proceeding, it is useful to explore some of the properties of these simple transformations. In the first place, in transformation (5′), the locus in reciprocal space of constant phase occurs for constant $e^{i2\pi\mathbf{s}\cdot\mathbf{t}^*}$. This is constant for constant values of $\mathbf{s}\cdot\mathbf{t}^*$. Since $\mathbf{s}$ is constant in this case by hypothesis, constant phase occurs for constant value of the projection of $\mathbf{t}^*$ on $\mathbf{s}$, that is, along planes in reciprocal space which are normal to

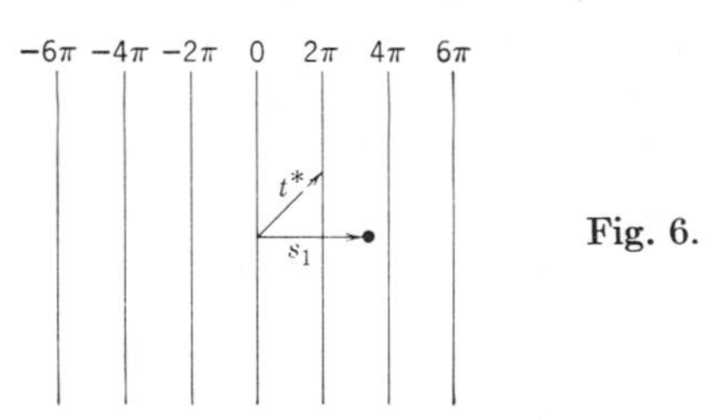

Fig. 6.

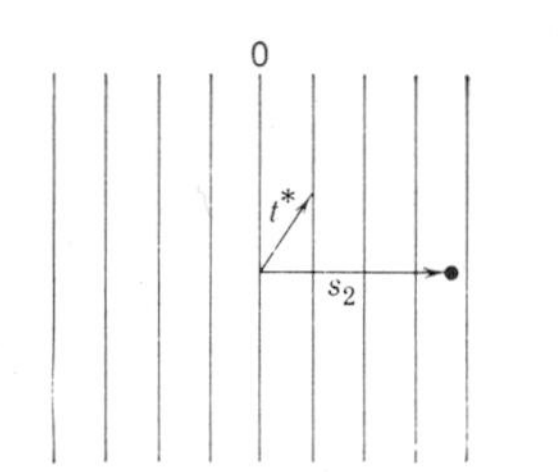

Fig. 7.

vector $\mathbf{s}$ in direct space. The phase is zero when this projection, Fig. 6, is zero, and consequently the plane of zero phase contains the origin.

When the scalar product $\mathbf{s}\cdot\mathbf{t}^*$ is unity, the phase angle $2\pi\mathbf{s}\cdot\mathbf{t}^*$ is 2π. Since $\mathbf{s}$ is constant this requires that proj $\mathbf{t}^* = 1/s$. The origin distance of the plane in reciprocal space where the phase angle is 2π is therefore the reciprocal of the origin distance of the point in direct space. The spacing of the planes in reciprocal space where the phase angle is $0, 2\pi, 4\pi \cdots$ is therefore the reciprocal of the origin distance of the point in direct space. If s is small this spacing is large, Fig. 6, and vice versa, Fig. 7.

It has been seen that by using (7′) the value of ρ at the end of vector $\mathbf{s}$ can be recovered. The question arises, what value is recovered at the ends of other possible vectors $\mathbf{u}$ in direct space. According to (7′) this is

$$\rho_u = \int_{V^*} R_{t^*} e^{-i2\pi\mathbf{u}\cdot\mathbf{t}^*} dV^*. \tag{8}$$

But R_{t^*} is given by (5′), so this can be substituted in (8) to give

$$\rho_u = \int_{V^*} \rho_s \, e^{i2\pi \mathbf{s}\cdot\mathbf{t}^*} \, e^{-i2\pi \mathbf{u}\cdot\mathbf{t}^*} \, dV^*$$

$$= \rho_s \int_{V^*} e^{i2\pi(\mathbf{s}-\mathbf{u})\cdot\mathbf{t}^*} \, dV^*. \tag{9}$$

This integral is zero unless $\mathbf{s} = \mathbf{u}$, in which case (9) reduces to $\rho_u = \rho_s$. Thus, *the transformation for an elementary volume to reciprocal space results in a reciprocal which, when transformed back to direct space, vanishes everywhere except at the original elementary volume, where it reproduces the original density.*

Extension to more general objects. It follows from the last result that to every point in one space there corresponds a wave-like variation of phase in the other space. Reciprocally, to this wave-like variation of phase in one space, there corresponds one and only one point of generally non-vanishing density in the other. Now, any more general object can be regarded as compounded of elementary volumes each characterized by its own density ρ and location $\mathbf{s}$. To each such elementary volume there corresponds a wave-like variation of phase in the other space, having characteristic amplitude ρ and wavelength $1/s$. A compound object in one space is represented by a pattern of superposed waves in the other. In the reverse transformation, each component wave reproduces its own characteristic elementary volume of density ρ at the end of vector $\mathbf{s}$.

This argument shows that the reciprocal transformations (5′) and (7′), demonstrated previously for a single point, hold generally for compound objects. In particular, for a compound object, (5′) must be integrated over direct space, so that it develops into (4). The integration changes the value of the reciprocal at the end of each vector $\mathbf{t}^*$. Nevertheless (7′) remains valid. Thus the transformations from one space to the other, and the reverse transformation are

$$T: \qquad R_{t^*} = \int_{\substack{\text{direct}\\\text{space}}} \rho_s \, e^{i2\pi \mathbf{s}\cdot\mathbf{t}^*} \, dV, \tag{10}$$

$$T^{-1}: \qquad \rho_s = \int_{\substack{\text{reciprocal}\\\text{space}}} R_{t^*} \, e^{-i2\pi \mathbf{s}\cdot\mathbf{t}^*} \, dV^*. \tag{11}$$

These two functions are called *Fourier mates.* Each is a solution in terms of the other. They have rather symmetrical forms. An important difference is that in one the phase of the phase operator is the negative of that of the other.

Repeated transformations. The transformation T involves the phase operator $e^{i2\pi \mathbf{s}\cdot\mathbf{t}^*}$ while the inverse transformation T^{-1} requires $e^{-i2\pi \mathbf{s}\cdot\mathbf{t}^*}$. Suppose that one starts with an original object, then transforms it with the aid of T. It has been seen that if this result is transformed by T^{-1}, the end result is the original object. This sequence of transformations can be expressed as $T^{-1}(T)$. But consider the possibility of using the original transformation a second time. This sequence can be expressed as $T(T)$. The result can be written as follows:

$$T(T) = T(R) = \int_{\substack{\text{reciprocal} \\ \text{space}}} R_{t^*}\, e^{i2\pi \mathbf{s}\cdot\mathbf{t}^*}\, dV^* \tag{12}$$

$$= \int_{\substack{\text{reciprocal} \\ \text{space}}} R_{t^*}\, e^{-i2\pi(-\mathbf{s})\cdot\mathbf{t}^*}\, dV^*. \tag{13}$$

This is similar to (11) but differs from it in that every vector **s** in the original object is replaced by its inverse vector $-\mathbf{s}$. Therefore a transform of

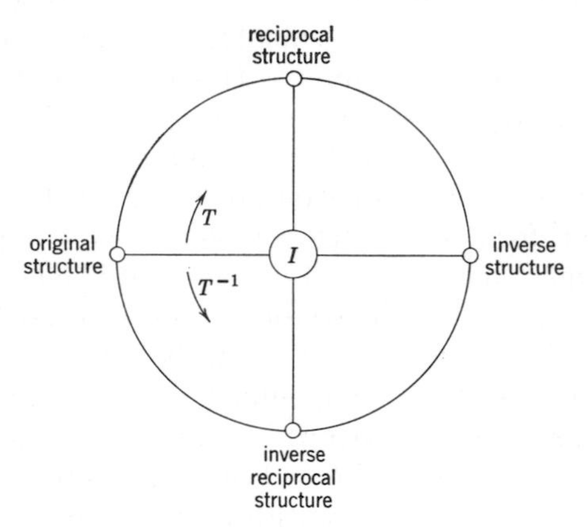

Fig. 8.
Results of repeated Fourier transformations.
(After Ewald.[2])

an object twice repeated is the inverse object. An object and its inverse are identical with each other only if the object is centrosymmetrical.

Ewald[2] has diagramed the results of transformation along the lines shown in Fig. 8. If I represents the operation of inversion, and 1 is the identical operation, then

$$\begin{aligned} TT^{-1} &= 1, \\ T^2 &= I, \\ T^4 &= 1. \end{aligned} \tag{14}$$

These transformations form a group of order 4.

Reciprocity. In the theory of reciprocal lattices, a translation of one lattice and the corresponding spacing of the other are reciprocals:

$$\begin{aligned} t^*_{hkl} \cdot d_{hkl} &= 1, \\ t_{uvw} \cdot d^*_{uvw} &= 1. \end{aligned} \tag{15}$$

An important feature of this transformation is that the star can be placed on either the translation or the spacing. Thus the two lattices are of equal rank and in every way reciprocal.

A similar characteristic applies to generalized reciprocals, that is, to reciprocal structures or Fourier transforms. Thus either structure can be regarded as the original structure, and the other the reciprocal structure. The phase operator of the transformation from one to the other is given by $e^{i2\pi \mathbf{s}\cdot\mathbf{t}}$. Either $\mathbf{s}$ or $\mathbf{t}$ can be regarded as the vector in "direct" space, the other then occurring in reciprocal space. Thus $\mathbf{s}$ and $\mathbf{t}$ can be described as components of sets of reciprocal vectors, such as $\mathbf{a}$, $\mathbf{b}$, $\mathbf{c}$ and $\mathbf{a}^*$, $\mathbf{b}^*$, $\mathbf{c}^*$, respectively.

Reciprocals of some periodic objects

In order to avoid leaving a feeling of abstractness, the reciprocals of some simple objects are discussed in the following sections. Part of the discussion will also show that the generalized definition of "reciprocal" is consistent with the original simple definition of reciprocal lattice.

Reciprocal of a lattice row. Let the origin be taken as one of the points of the lattice row. If the period along the row is $\mathbf{a}$, then the points of the row are defined by the vectors

$$\mathbf{T} = m\mathbf{a} \qquad (m \text{ an integer}). \tag{16}$$

Thus the integration in (10) becomes a summation over all the points in the row. If each point has unit weight the sum is

$$R_{t^*} = \sum_{m=-\infty}^{\infty} e^{i2\pi m \mathbf{a}\cdot\mathbf{t}^*}. \tag{17}$$

This summation of this exponential was given in (47) of Chapter 13. Therefore

$$R_{t^*} = \frac{\sin \pi(2m+1)\mathbf{a}\cdot\mathbf{t}^*}{\sin \pi\mathbf{a}\cdot\mathbf{t}^*}. \tag{18}$$

When m is ∞, this function is zero for all values of t^* except those for which the $\mathbf{a}\cdot\mathbf{t}^*$ is an integer, when the value is infinite. For example, when $\mathbf{a}\cdot\mathbf{t}^* = 0$, a plane through the origin is defined. When $\mathbf{a}\cdot\mathbf{t} = 1$, a plane at distance $t^* = 1/a$ is defined. When $\mathbf{a}\cdot\mathbf{t}^* = 2$, a plane at a distance $t^* = 2(1/a)$ is defined, etc. The reciprocal of a lattice row of period **a** is therefore a stack of planes normal to the row and of spacing $1/a$.

Reciprocal of a plane lattice. Let the origin be taken at a lattice point. Then, if the primitive translations of the net are **a** and **b**, the lattice points are at the ends of vectors

$$\mathbf{T} = m\mathbf{a}+n\mathbf{b} \qquad (m,\ n,\ \text{integers}). \tag{19}$$

The integral in (10) therefore degenerates into the summation

$$R_{t^*} = \sum_m \sum_n {}_{-\infty}^{\infty} e^{i2\pi(m\mathbf{a}+n\mathbf{b})\cdot\mathbf{t}^*} \tag{20}$$

$$= \sum_{-\infty}^{\infty} \sum_{-\infty}^{\infty} e^{i2\pi m\mathbf{a}\cdot\mathbf{t}^*}\, e^{i2\pi n\mathbf{b}\cdot\mathbf{t}^*}$$

$$= \frac{\sin 2\pi(2m+1)\mathbf{a}\cdot\mathbf{t}^*}{\sin \pi\mathbf{a}\cdot\mathbf{t}^*}\ \frac{\sin 2\pi(2n+1)\mathbf{b}\cdot\mathbf{t}^*}{\sin \pi\mathbf{b}\cdot\mathbf{t}^*}. \tag{21}$$

Each of the two terms in (21) behave as the right-hand side of (18). The first factor is zero everywhere except on the planes of a stack normal to **a** whose spacing is $1/a$, where the value is infinite. Similarly the second factor is zero everywhere except on the planes of a stack normal to **b** whose spacing is $1/b$, where the value is infinite. The product of the two factors in (21) is accordingly zero everywhere except at the intersection of the two stacks of planes, where the value is infinite. This locus is an array of parallel lines, each normal to the plane of the original plane lattice, and intersecting the plane of the plane lattice at the points of the reciprocal plane lattice.

Reciprocal of a space lattice. Let the origin be taken at a lattice point. Then, if the primitive translations of the net are **a**, **b** and **c**, the lattice points are at the ends of vectors

$$T = m\mathbf{a}+n\mathbf{b}+p\mathbf{c}. \tag{22}$$

The integral in (10) therefore degenerates into the summation

$$R_{t^*} = \sum_m \sum_n \sum_p {}_{-\infty}^{\infty} e^{i2\pi(m\mathbf{a}+n\mathbf{b}+p\mathbf{c})\cdot\mathbf{t}^*} \tag{23}$$

$$= \sum_{-\infty}^{\infty} \sum_{-\infty}^{\infty} \sum_{-\infty}^{\infty} e^{i2\pi m\mathbf{a}\cdot\mathbf{t}^*}\, e^{i2\pi n\mathbf{b}\cdot\mathbf{t}^*}\, e^{i2\pi p\mathbf{c}\cdot\mathbf{t}^*}$$

$$= \frac{\sin \pi(2m+1)\mathbf{a}\cdot\mathbf{t}^*}{\sin \pi\mathbf{a}\cdot\mathbf{t}^*}\ \frac{\sin \pi(2n+1)\mathbf{b}\cdot\mathbf{t}^*}{\sin \pi\mathbf{b}\cdot\mathbf{t}^*}\ \frac{\sin \pi(2p+1)\mathbf{c}\cdot\mathbf{t}^*}{\sin \pi\mathbf{c}\cdot\mathbf{t}^*}. \tag{24}$$

The first term in (24) is zero everywhere except on the planes of a stack normal to **a** whose spacing is $1/a$; the second term in (24) is zero everywhere except on the planes of a stack normal to **b** whose spacing is $1/b$; the third term in (24) is zero everywhere except on the planes of a stack normal to **c** whose spacing is $1/c$. Along the planes of each stack the value of the corresponding term is infinite.† The three terms of (24) are simultaneously non-zero only when the three sets of planes mutually intersect.† These planes mark out parallelopipeds of altitudes $1/a$, $1/b$, and $1/c$ and the planes mutually intersect only at the vertices. The reciprocal given by (24) is therefore a lattice whose spacings are the reciprocals of the translations of the original lattice. It is therefore the lattice reciprocal to the original lattice.

Reciprocal of a stack of planes. Let the origin be taken in one of the planes. Then the distances of the planes from the origin are given by

$$\mathbf{s} = m\mathbf{d} \qquad (m \text{ an integer}), \tag{25}$$

where **d** is the spacing of planes in the stack. The reciprocal of an individual plane, according to (10) is

$$R_{t^*} = e^{i2\pi \mathbf{d}\cdot\mathbf{t}^*}. \tag{26}$$

The reciprocal of the collection is

$$R_{t^*} = \sum_{m=-\infty}^{\infty} e^{i2\pi m\mathbf{d}\cdot\mathbf{t}^*} \tag{27}$$

$$= \frac{\sin \pi(2m+1)\mathbf{d}\cdot\mathbf{t}^*}{\sin \pi\mathbf{d}\cdot\mathbf{t}^*}. \tag{28}$$

This function is zero for all values of $\mathbf{t}^*$ except those for which $\mathbf{d}\cdot\mathbf{t}^*$ is an integer, when the value is infinite. For example, when $\mathbf{d}\cdot\mathbf{t}^* = 0$, a point at the origin is defined. When $\mathbf{d}\cdot\mathbf{t}^* = 1$, a point at a distance $t^* = 1/d$ is defined, where t^* is taken in the direction of the spacing, d. When $\mathbf{d}\cdot\mathbf{t}^* = 2$, a point at a distance $t^* = 2(1/d)$ is defined, etc. The reciprocal of a stack of planes of spacing d is therefore a row of points of interval $1/d$ normal to the planes.

Reciprocity. The truly reciprocal nature of the Fourier transformation has already been pointed out. Some reciprocals have now been derived for some simple periodic objects, and examples of this reciprocity have been encountered. For example, the reciprocal of a lattice stack is a

† This is true provided m, n, and p are infinite. If they are finite, then a *peak function*, similar to that shown in Fig. 5, Chapter 13, occurs at each lattice point, and between lattice points the function fluctuates near zero. When m, n and p are large integers (as they are for visible crystals) the departure from zero background between lattice points cannot be detected.

Table 1
Reciprocals of some simple periodic objects

Object	Reciprocal
Lattice row (translation a)	Stack of planes (planes perpendicular to row, spacing $d^* = 1/a$)
Plane lattice (translations a and b)	Lattice array of lines (lines perpendicular to plane of plane lattice, spacings of array $1/a$ and $1/b$)
Space lattice (translations a, b, and c)	Space lattice (spacings $1/a$, $1/b$, and $1/c$)

lattice row, and, reciprocally, the reciprocal of a lattice row is a lattice stack. Some other examples are noted in Table 1.

Reciprocals of some more general objects

A reciprocal can be formed for any object whether periodic or not. In this section the reciprocals of some more general objects are considered.

Real and complex reciprocals. Since the formation of the reciprocal involves the phase operator $e^{i2\pi \mathbf{s}\cdot\mathbf{t}^*}$, reciprocals are, in general, complex. But if the object is centrosymmetrical and if the origin is taken at this center, then the imaginary component vanishes. This is readily demonstrated by writing the exponential in its trigonometric equivalent:

$$R_{t^*} = \int_{\substack{\text{direct}\\ \text{space}}} \rho_s \, e^{i2\pi \mathbf{s}\cdot\mathbf{t}^*} \, dV$$

$$= \int_{\substack{\text{direct}\\ \text{space}}} \rho_s \, \{\cos(2\pi \mathbf{s}\cdot\mathbf{t}^*) + i \sin(2\pi \mathbf{s}\cdot\mathbf{t}^*)\} \, dV. \tag{29}$$

If the object is centrosymmetrical, and if the origin is taken at the center, then for every point at the end of vector $\mathbf{s}^*$ there exists an equivalent point at the end of vector $-\mathbf{s}^*$. Since

$$\int_{-x_1}^{x_1} \sin x \, dx = 0, \tag{30}$$

the sine component of (29) vanishes in this case. The reciprocal is then real, and can be computed from

$$R_{t^*} = \int_{\substack{\text{all}\\ \text{direct}\\ \text{space}}} \rho_s \cos(2\pi \mathbf{s}\cdot\mathbf{t}^*) \, dV. \tag{31}$$

Since $\cos(-x) = \cos x$, this can be reduced to

$$R_{t^*} = 2 \int_{\substack{\text{half of}\\ \text{direct}\\ \text{space}}} \rho_s \cos(2\pi\mathbf{s}\cdot\mathbf{t}^*)\, dV. \tag{32}$$

Reciprocal of a line segment. To illustrate, in its simplest form, the derivation of the transform of a centrosymmetrical object, consider the reciprocal of a line in the one-dimensional space defined by the direction of the line. Let the line have length l. To take advantage of the simple form of a centrosymmetrical object, let the origin be taken at

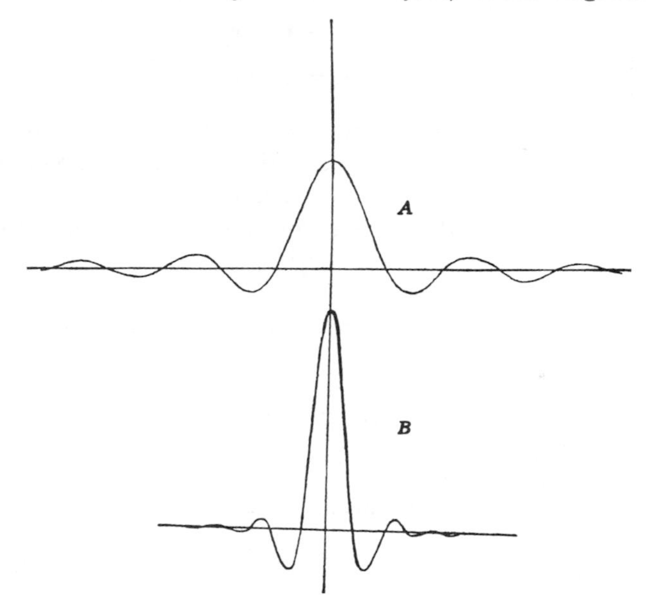

Fig. 9.
Reciprocals of a line segment; *A*, short segment, *B*, long segment. (After Garrido.[11])

the center of the line, so that the coordinates of its ends are $l/2$ and $-l/2$. Let the weighting of the line be uniform and have a value of ρ per unit length. Then (32) has the specific form

$$R_{t^*} = 2 \int_{x=0}^{l/2} \rho_s \cos(2\pi x\cdot t^*)\, dx \tag{33}$$

$$= 2\rho_s \frac{1}{2\pi t^*} \sin 2\pi \frac{l}{2} t^*$$

$$= \rho_s \frac{\sin \pi l t^*}{\pi t^*}. \tag{34}$$

A graph of this function is shown in Fig. 9. The horizontal spread of this curve is inversely proportional to the length of the line.

Reciprocal of a rectangle. A somewhat more complicated example of the derivation of a transform is that of the transform of a rectangle in the two dimensions defined by the plane of the rectangle. Let the length of the rectangle be l, taken in the direction of the X axis, and its width w, taken in the direction of the Y axis. Let the rectangle have a density of ρ per unit area.

To perform the integration, let the vector $\mathbf{s}$ be referred to coordinates $\mathbf{x}\ \mathbf{y}$, and let $\mathbf{t}^*$ be referred to the reciprocal set of coordinates $\mathbf{x}^*\ \mathbf{y}^*$. Then

$$\begin{aligned} \mathbf{s} &= \mathbf{x}+\mathbf{y}, \\ \mathbf{t}^* &= \mathbf{x}^*+\mathbf{y}^*, \\ dV \rightarrow dA &= dx\,dy \end{aligned} \tag{35}$$

To take advantage of the centrosymmetry of the rectangle, let the origin be taken at its center. Then the integration need be carried out for only half the rectangle, say from $x = 0$ to $l/2$ and $y = -w/2$ to $w/2$. Using these limits in (32) and substituting from (35), it is found the reciprocal is

$$R_{x^*y^*z^*} = 2\int_{y=-w/2}^{w/2}\int_{x=0}^{l/2} \rho\cos 2\pi(\mathbf{x}+\mathbf{y})\cdot(\mathbf{x}^*+\mathbf{y}^*+\mathbf{z}^*)\,dx\,dy \tag{36}$$

$$= 2\rho\int_{y=-w/2}^{w/2}\int_{x=0}^{l/2} \cos 2\pi(xx^*+yy^*)\,dx\,dy \tag{37}$$

$$= 2\rho\int_{y=-w/2}^{w/2}\int_{x=0}^{l/2} (\cos 2\pi xx^*\cos 2\pi yy^* - \sin 2\pi xx^*\sin 2\pi yy^*)\,dx\,dy \tag{38}$$

$$= 2\rho\int_{y=-w/2}^{w/2}\left[\frac{\sin 2\pi xx^*}{2\pi x^*}\cos 2\pi yy^* + \frac{\cos 2\pi xx^*}{2\pi x^*}\sin 2\pi yy^*\right]_{x=0}^{l/2} dy \tag{39}$$

$$= 2\rho\left[\frac{\sin 2\pi xx^*}{2\pi x^*}\,\frac{\sin 2\pi yy^*}{2\pi y^*} - \frac{\cos 2\pi xx^*}{2\pi x^*}\,\frac{\cos 2\pi yy^*}{2\pi y^*}\right]_{x=0}^{l/2}\Bigg|_{y=-w/2}^{w/2} \tag{40}$$

$$= 2\rho\left[\left(\frac{\sin 2\pi\frac{1}{2}lx^*}{2\pi x^*}\right)\frac{\sin 2\pi yy^*}{2\pi y^*} - \left(\frac{\sin 2\pi 0x^*}{2\pi x^*}\right)\frac{\sin 2\pi yy^*}{2\pi y^*} - \left(\frac{\cos 2\pi\frac{1}{2}lx^*}{2\pi x^*}\right)\frac{\cos 2\pi yy^*}{2\pi y^*} + \left(\frac{\cos 2\pi 0x^*}{2\pi x^*}\right)\frac{\cos 2\pi yy^*}{2\pi y^*}\right]_{y=-w/2}^{w/2} \tag{41}$$

$$= 2\rho \left[\frac{\sin 2\pi\frac{1}{2}lx^*}{2\pi x^*} \left\{ \frac{\sin 2\pi\frac{1}{2}wy^* - \sin(-2\pi\frac{1}{2}wy^*)}{2\pi y^*} \right\} - \frac{\cos 2\pi\frac{1}{2}lx^*}{2\pi x^*} \left\{ \frac{\cos 2\pi\frac{1}{2}wy^* - \cos(-2\pi\frac{1}{2}wy^*)}{2\pi y^*} \right\} + \frac{1}{2\pi x^*} \left\{ \frac{\cos 2\pi\frac{1}{2}wy^* - \cos(-2\pi\frac{1}{2}wy^*)}{2\pi y^*} \right\} \right] \quad (42)$$

$$= \rho \left(\frac{\sin \pi l x^*}{\pi x^*} \, \frac{\sin \pi w y^*}{\pi y^*} \right). \quad (43)$$

Note that this is a generalization of (34). A graphical representation of the function is shown in Fig. 10. The cross-sections along the X^* and Y^* axes have the same general form as Fig. 9.

Reciprocal of an atom. If a spherically symmetrical atom is placed at the origin, it is evident that its reciprocal is also spherically symmetrical. The value of the reciprocal along any radial line can be found by evaluating (10) along a radial line in reciprocal space. This requires a knowledge of the way the density ρ_s varies in the atom with radial coordinate s. When the function is properly evaluated, it can be represented by a plot of R_{t^*} against the radial coordinate s. When s is expressed in wavelength units it is the σ of x-ray crystallography, whose value is $(2 \sin \theta)/\lambda$. Thus the reciprocal of an atom is its f curve.

Reciprocal of a structure. The reciprocal of a structure can be readily computed by making appropriate substitutions in (10). For this purpose

$$\begin{aligned} \mathbf{s} &= x\mathbf{a}+y\mathbf{b}+z\mathbf{c}, \\ \mathbf{t}^* &= h\mathbf{a}^*+k\mathbf{b}^*+l\mathbf{c}^*, \\ \mathbf{s}\cdot\mathbf{t}^* &= hx+ky+lz, \\ dV &= V dx\, dy\, dz. \end{aligned} \quad (44)$$

where $\mathbf{a}$, $\mathbf{b}$, $\mathbf{c}$ and $\mathbf{a}^*$, $\mathbf{b}^*$, $\mathbf{c}^*$ are reciprocal sets of unit vectors. With these substitutions the reciprocal has the form

$$R_{hkl} = V \int_{-\infty}^{\infty} \int_{-\infty}^{\infty} \int_{-\infty}^{\infty} \rho_{(xyz)}\, e^{i2\pi(hx+ky+lz)}\, dx\, dy\, dz. \quad (45)$$

The reciprocal of a structure is often loosely called its "reciprocal lattice," or its "weighted reciprocal lattice." This is a crude designation. The reciprocal of a lattice is a reciprocal lattice, but the reciprocal of a structure is appropriately called a *reciprocal structure*. The word "lattice" is inappropriate, since it does not have repetition by translation. It may be real or it may be complex.

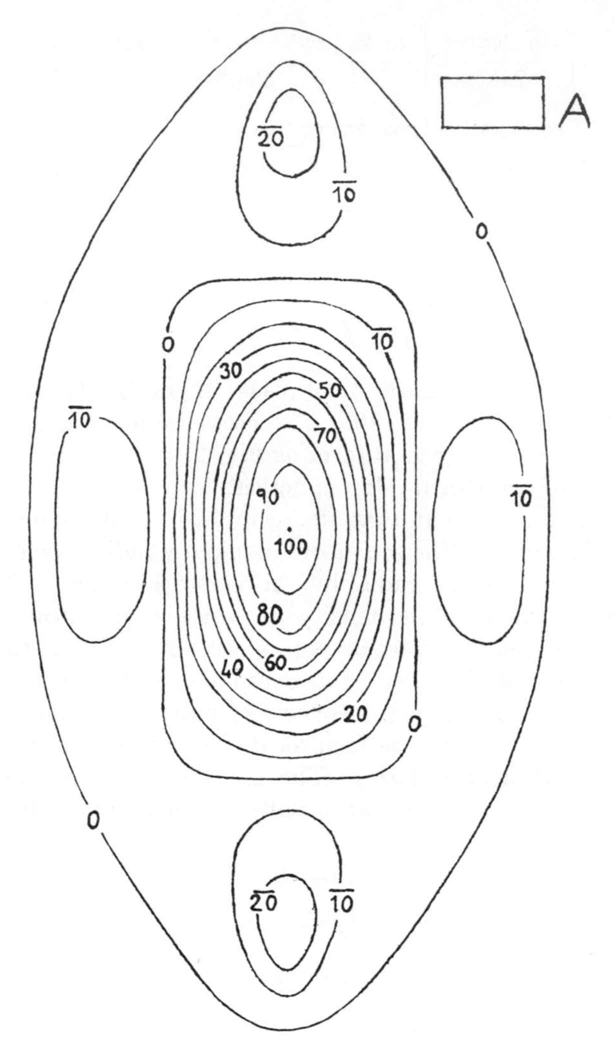

Fig. 10.
Reciprocal of the rectangle shown at A.
(After Garrido.[11])

The discussion of this and the preceding section makes it plain that the mathematics of diffraction are the same as those of forming the reciprocal of the structure.

Combinatorial properties of reciprocals

Reciprocals can be combined with one another in various ways. Some important combinations are discussed on the following page.

Reciprocal of a sum. By definition, the reciprocal is an integral over all space. The same result is obtained if space is subdivided and the integral broken up into regions which together cover all space. This provides an additivity theorem:

Theorem: The reciprocal of a sum is the sum of the reciprocals. This can be expressed symbolically as

$$R_{t^*}(A+B) = R_{t^*}(A) + R_{t^*}(B). \tag{46}$$

This relation is useful in developing the form of the Fourier summation for symmetrical crystals, for example.

Reciprocals of products. Let the product be represented by $\rho_1\rho_2 = P$. The reciprocal of this is

$$\begin{aligned} R_{t^*,P} &= \int_V P e^{i2\pi \mathbf{s}\cdot\mathbf{t}^*}\, dV \\ &= \int_V \rho_1\, \rho_2\, e^{i2\pi \mathbf{s}\cdot\mathbf{t}^*}\, dV. \end{aligned} \tag{47}$$

Each of these ρ's can be represented as the reciprocal of its R, but with the position vector of R ranging over different vectors $\mathbf{t}^*$ which must be integrated over all reciprocal space. Let the position vector of one R be designated as $\mathbf{t}^*$, that of the other as $\mathbf{u}^*$. Then, substituting the reciprocal of ρ_2 in (47) there results

$$R_{t^*,P} = \int_V \rho_1 \left[\int_{V^*} R_{u^*\rho_2}\, e^{-i2\pi \mathbf{s}\cdot\mathbf{u}^*}\, dV^* \right] e^{i2\pi \mathbf{s}\cdot\mathbf{t}^*}\, dV.$$

By changing the order of integration, this becomes

$$R_{t^*,P} = \int_{V^*} R_{u^*\rho_2} \left[\int_V \rho_1\, e^{i2\pi \mathbf{s}\cdot(\mathbf{t}^*-\mathbf{u}^*)}\, dV \right] dV^* \tag{48}$$

$$= \int_{V^*} R_{u^*,\rho_2}\, R_{(t^*-u^*),\rho_1}\, dV^*. \tag{49}$$

This result is called the *convolution* of the functions $R_{u^*\rho_2}$ and $R_{u^*\rho_1}$. *Thus, the transform of a product is the convolution of the transforms.*

On the other hand the product of two transforms can be reduced as follows:

$$R_{t^*,\rho_1}\, R_{t^*,\rho_2} = \int \rho_{1r}\, e^{i2\pi \mathbf{r}\cdot\mathbf{t}^*}\, dV \int \rho_{2,s-r}\, e^{i2\pi(\mathbf{s}-\mathbf{r})\cdot\mathbf{t}^*}\, dV \tag{50}$$

$$= \int \left[\int \rho_{1r}\rho_{2,s-r}\, dV \right] e^{i2\pi \mathbf{s}\cdot\mathbf{t}^*}\, dV \tag{51}$$

$$= R\left[\int \rho_{1r}\rho_{2,s-r}\, dV \right]. \tag{52}$$

This is the reciprocal of the convolution of ρ_{1r} and ρ_{2r}. Thus, *the product of two transforms is the transform of the convolution.* These results show that a product in one space corresponds to a convolution in the other.

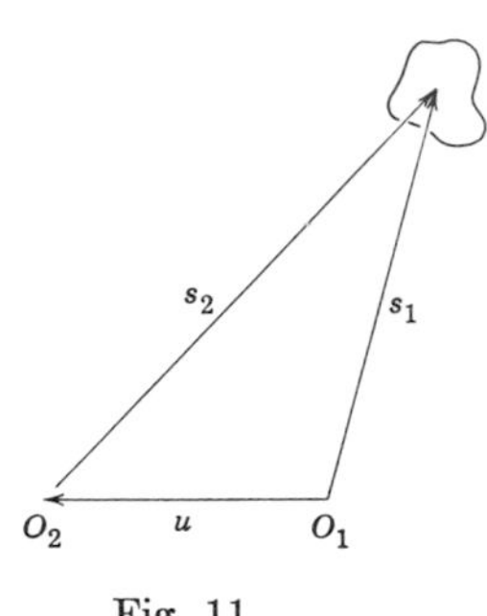

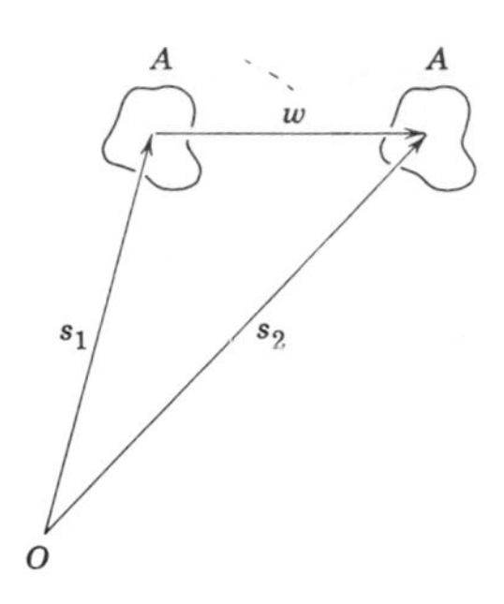

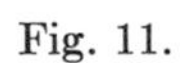
Fig. 11.

Fig. 12.

Change of origin. According to (10), the reciprocal of an object with respect to origin O_1 (Fig. 11) is

$$^1R_{t^*} = \int \rho_{s_1} e^{i2\pi \mathbf{s}_1 \cdot \mathbf{t}^*} dV. \tag{53}$$

Suppose the origin is shifted to O_2. The reciprocal is then

$$^2R_{t^*} = \int \rho_{s_2} e^{i2\pi \mathbf{s}_2 \cdot \mathbf{t}^*} dV. \tag{54}$$

These two reciprocals have the same form, but they involve different vectors $\mathbf{s}_1$ and $\mathbf{s}_2$ in direct space. These are related by the vector relating the two origins, specifically

$$\mathbf{s}_2 = \mathbf{s}_1 - \mathbf{u}. \tag{55}$$

If this substitution is made in (54), there results

$$\begin{aligned} ^2R_{t^*} &= \int \rho_s e^{i2\pi(\mathbf{s}_1 - \mathbf{u}) \cdot \mathbf{t}^*} dV \\ &= \int \rho_s e^{i2\pi \mathbf{s} \cdot \mathbf{t}^*} e^{-i2\pi \mathbf{u} \cdot \mathbf{t}^*} dV \\ &= e^{-i2\pi \mathbf{u} \cdot \mathbf{t}^*} \int \rho_s e^{i2\pi \mathbf{s} \cdot \mathbf{t}^*} dV \\ &= e^{-i2\pi \mathbf{u} \cdot \mathbf{t}^*}\, {}^1R_{t^*}. \end{aligned} \tag{56}$$

This simple analysis shows that at any point in reciprocal space defined by a given vector $\mathbf{t}^*$, a shift of origin defined by vector $\mathbf{u}$ produces a phase shift at $\mathbf{t}^*$ having a phase component $e^{-i2\pi \mathbf{u} \cdot \mathbf{t}^*}$.

For some applications, it is desirable to regard the origin as fixed and the object in direct space shifted by a translation represented by vector $\mathbf{w}$, Fig. 12. A shift of the object by $\mathbf{w}$ has the same geometry as shifting the origin, provided

$$\mathbf{w} = -\mathbf{u}. \tag{57}$$

Therefore, if $\mathbf{w}$ is substituted for $-\mathbf{u}$ in (54), there results

$$ {}^{w}R_{t^*} = e^{i2\pi \mathbf{w}\cdot\mathbf{t}^*} R_{t^*}. \tag{58} $$

This signifies that if the object is shifted in direct space by an amount defined by vector $\mathbf{w}$, all points in reciprocal space experience a phase shift of $2\pi\mathbf{w}\cdot\mathbf{t}^*$.

Reciprocal of a structure. The shifted-object relation, (58), provides a neat tool for formulating the reciprocal of a crystal structure. The reciprocal of the entire structure is the sum of the reciprocals of all the unit cells, each of which is the same except for a shift by translation.

Let ${}^{c}R$ be the reciprocal of the original cell. Then, according to (58), the reciprocal of an individual shifted cell is

$$ {}^{sc}R_{t^*} = {}^{c}R_{t^*}\, e^{i2\pi \mathbf{T}\cdot\mathbf{t}^*}. \tag{59} $$

Now, each lattice translation has the form

$$ \mathbf{T} = m\mathbf{a}+n\mathbf{b}+p\mathbf{c} \qquad (m,\ n,\ p,\ \text{integers}). \tag{60} $$

If this is substituted into (59) it becomes

$$ {}^{sc}R_{t^*} = {}^{c}R_{t^*}\, e^{i2\pi(m\mathbf{a}+n\mathbf{b}+p\mathbf{c})\cdot\mathbf{t}^*}. \tag{61} $$

By reason of the additivity theorem, the reciprocal of the collection of shifted cells is the sum of these reciprocals over m, n, and p:

$$ \begin{aligned} {}^{s}R_{t^*} &= \sum_m \sum_n \sum_p {}^{sc}R_{t^*} \\ &= {}^{c}R_{t^*} \sum_m \sum_n \sum_p e^{i2\pi(m\mathbf{a}+n\mathbf{b}+p\mathbf{c})\cdot\mathbf{t}^*}. \end{aligned} \tag{62} $$

But the summation is exactly that of (23), which is the reciprocal of the lattice. One concludes that the reciprocal of a structure is the product of the reciprocal of one individual cell and the reciprocal of the lattice.

Now the reciprocal of the lattice vanishes everywhere except at the points of the reciprocal lattice, as pointed out in the section on *Reciprocal of a space lattice.* It follows that the reciprocal of a structure vanishes everywhere except at the points of the reciprocal lattice, where it has the value of the reciprocal of the cell times the reciprocal of the lattice. The latter term is infinite if the extension of the crystal lattice is infinite, as pointed out under *Reciprocal of a space lattice.* But if the crystal lattice has a finite extension, the term is finite, and proportional to the total number of cells in the limited lattice, that is, to the volume of the crystal.

Theorem: The reciprocal of a structure is proportional to the reciprocal of the cell as sampled at the reciprocal lattice points.

Solution of molecular structures with the aid of molecular Fourier transforms

The arrangement of molecules in certain molecular crystals can be found with the aid of the Fourier transforms of the molecule. This kind of investigation requires that the structure of the molecule be already known, so the only unknown feature of the structure is the orientation of the molecules with respect to the cell edges plus possibly the distances of the molecules from any symmetry elements of the structure. Such an investigation is easy provided that there is one molecule per cell, or one projected molecule in the two-dimensional cell of the projected structure. This is equivalent to requiring that the molecule be not repeated by any other operation than the lattice translations. The method can be extended to the case of molecules repeated by operations within the cell, but then the simplicity of the method is replaced by complexity.

The general theory of the simple case is as follows: The only unknown feature of the structure is the orientation of the molecule with respect to the cell edges. According to the last section, the transform of the structure is the transform of the cell (that is, of the molecule) sampled at the points of the reciprocal lattice. Since the shape of the molecule and of the cell are known, their transforms, namely the transform of the molecule and the locations of the points of the reciprocal lattice are known. The only unknown feature is the relative orientations of the two. If these are plotted, the problem is resolved into finding a relative orientation whose sampling of the molecule transform at the points of the reciprocal lattice agrees in magnitude with those observed in the diffraction experiment.

This is illustrated in Fig. 13, by the solution of the crystal structure of naphthalene, as given by Knott.[9] Both the reciprocal of the molecule and the reciprocal lattice are shown in the figure. Solving the structure consists merely of rotating the reciprocal lattice (as drawn on transparent paper) over the reciprocal of the molecule until a position is found such that the reciprocal-lattice points cover values on the reciprocal of the molecule equal in absolute value to the observed F values. A good fit was obtained by Knott in the position shown in the figure.

This problem is easy if there is one centrosymmetrical molecule in the cell. Then the transform of the molecule is real. The problem is somewhat more difficult if the molecule is non-centric. It is a much more difficult problem if there are several molecules per cell, related by symmetry elements. In this case the transform of each molecule must be added (in the complex plane if the molecule is non-centric).

Several examples of the use of Fourier transforms in finding the arrange-

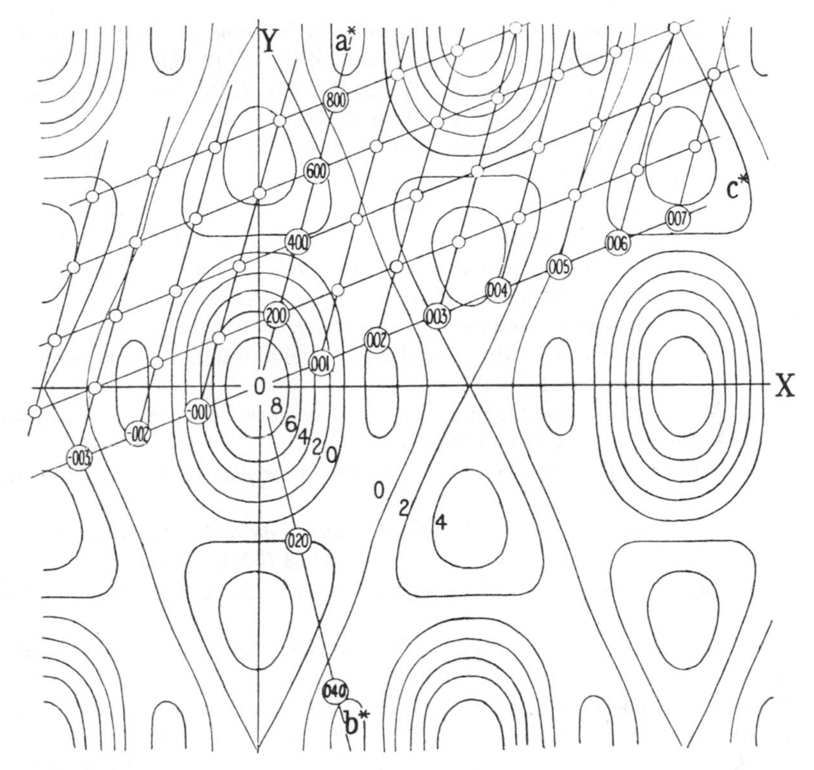

Fig. 13.
Fourier transform of the naphthalene molecule, and the *h*0*l* level of the reciprocal lattice of the naphthalene crystal.
(After Knott.[9])

ment of molecules in crystals are to be found in the following literature list.

Literature

[1] M. A. Bravais. *Mémoire sur les systèmes formés par des points distribués régulièrement sur un plan ou dans l'espace.* J. école polytech. (Paris), *Cahier* 33, *Tome* XIX (1850). (Also translated by Amos J. Shaler as Memoir No. 1, Crystallographic Society of America, 1949.)

[2] P. P. Ewald. *Das "reziproke Gitter" in der Strukturtheorie.* Z. Krist. (A) **56** (1921) 129–156.

[3] J. D. Bernal. *On the interpretation of x-ray, single crystal, rotation photographs.* Proc. Roy. Soc. (London) (A) **113** 117–160, especially 118–120.

[4] A. L. Patterson. *The Gibbs-Ewald reciprocal lattice.* Nature **125** (1930) 238, 447.

[5] Alfred Hettich. *Beiträge zur Methodik der Strukturbestimmung. IV. Ein Eigendiagramm für den Strukturfaktorbeitrag von Atomgruppen, besonders in reziproken Raum.* Z. Krist. (A) **90** (1935) especially 483–492.

[6] P. P. Ewald. *Bemerkungen zur vorstehenden Arbeit von. A. Hettich Abschnitt IV. "Eigendiagramme" von Atomgruppen.* Z. Krist. (A) **90** (1935) 493–494.

[7] A. Charlesby, G. I. Finch, and H. Wilman. *The diffraction of electrons by anthracene.* Proc. Phys. Soc. **51** (1939) 479–528.

[8] P. P. Ewald. *X-ray diffraction by finite and imperfect crystal lattices.* Proc. Phys. Soc. **52** (1940) 167–174.

[9] George Knott. *Molecular structure factors and their application to the solution of the structures of complex organic crystals.* Proc. Phys. Soc. **52** (1940) 229–238.

[10] Dorothy Wrinch. *Fourier transforms and structure factors.* Am. Soc. for X-ray and Electron Diffraction Monograph No. 2 (1946).

[11] Julio Garrido. *El espacio recíproco y su applicacion a algunos problemas de la difusión de los rayos x.* Mem. real acad. cienc. Madrid (Ser. cienc. fís) *Tomo* II, *Memoria* No. 1 (1947).

[12] E. J. W. Whittaker. *Evaluation of Fourier transforms by a Fourier synthesis method.* Acta Cryst. **1** (1948) 165–167.

[13] A. Klug. *The crystal and molecular structure of triphenylene,* $C_{18}H_{12}$. Acta Cryst. **3** (1950) 165–175.

[14] A. Klug. *The application of the Fourier-transform method to the analysis of the structure of triphenylene,* $C_{18}H_{12}$. Acta Cryst. **3** (1950) 176–181.

[15] H. P. Stadler. *The crystal structure of flavanthrone.* Acta Cryst. **6** (1953) 540–542.

[16] Jürg Waser and Verner Schomaker. *The Fourier inversion of diffraction data.* Rev. Mod. Phys. **25** (1953) 671–690.

[17] C. A. Taylor. *A method of determining the relative positions of molecules using Fourier-transform principles.* Acta Cryst. **7** (1959) 757–763.

[18] G. A. Sim, J. Monteath Robertson, and T. H. Goodwin. *The crystal and molecular structure of benzoic acid.* Acta Cryst. **8** (1955) 157–164.

[19] M. Bailey. *The crystal structure of N:N′—diacetylhexamethylenediamine.* Acta Cryst. **8** (1955) 575–578.

[20] V. Vand and R. Pepinsky. *Determination of molecular translations and orientation using modified disagreement functions.* Z. Krist. **108** (1956) 1–14.

[21] K. Doi. *Nouvelle transformation de Fourier pour les analyses de structures désordonnées.* Bull. soc. franç. minéral. crist. **80** (1957) 325–343.

[22] A. M. Cormack. *Fourier transformations in cylindrical co-ordinates.* Acta Cryst. **10** (1957) 354–358.

[23] Hideo Watase, Kenji Osaki, and Isamu Nitta. *Crystal structure of a monoclinic form of naphthazarin.* Bull. Chem. Soc. Japan **30** (1957) 532–536.

[24] M. Bailey. *The crystal structure of* 1:5-*dichloroanthraguinone.* Acta Cryst. **11** (1958) 103–107.

[25] H. Lipson and C. A. Taylor. *Fourier transforms and x-ray diffraction.* (G. Bell and Sons, London, 1958)

[26] C. A. Taylor and K. A. Morley. *An improved method for determining the relative positions of molecules.* Acta Cryst. **12** (1959) 101–105.

[27] James Trotter. *Determination of the positions of molecules in a unit cell.* Acta Cryst. **12** (1959) 339–341.

16

Symmetry in reciprocal space

In the last chapter some general features of reciprocal space were discussed. In this chapter some of the symmetry properties of reciprocal space are considered, and these several symmetry properties are then applied to the reciprocals of crystals. This discussion incidentally provides a convenient approach to the practical problem of taking advantage of the symmetry of the crystal when computing Fourier summations. This application is discussed in the next chapter. Crystallographic symmetry in reciprocal space has been treated by Buerger,[2,3] Nowacki,[4,5] Waser,[6] and Bertaut.[7–10]

Point-group symmetry in reciprocal space

General rotational symmetry. In the last chapter it was shown that the reciprocal of a point has a phase character which varies regularly along a line defined by the origin and the point; this phase is constant along planes at right angles to this line. Consider a general point-group symmetry operation, that is, a proper or improper rotation such as A_α. Such a symmetry operation not only carries the point into a symmetrical point, but it also carries the line from the origin to the point, together with every constant-phase plane, to a corresponding symmetrical location. According to the additivity theorem of the last chapter, the reciprocal of any object is compounded from the properly weighted reciprocals of all its points. The same symmetry operation that carries one point into a symmetrical point necessarily at the same time carries the collection of all points into the collection of their symmetrical points, and therefore carries the entire reciprocal of the set of points into the reciprocal of the symmetrical set. Since every point-group symmetry can be formulated in terms of combinations of groups of rotational operations, proper and improper, the truth of the following theorem is evident:

Theorem 1: If an object conforms to the symmetry of a particular point group, its reciprocal does also.

Several points are worth mentioning in connection with this theorem. In the first place, it is not restricted to crystallographic point groups, but includes non-crystallographic point groups as well. Thus, if the object is symmetrical with respect to a 5-fold axis, so is its reciprocal.

Secondly, the effect on reciprocal space of an individual operation of rotation which requires that operation in the reciprocal could, in another sense, be performed by the reverse rotation in reciprocal space. Thus the reciprocal of a point can be written $\rho e^{i2\pi \mathbf{s}\cdot\mathbf{t}^*}$. A rotation of the vector

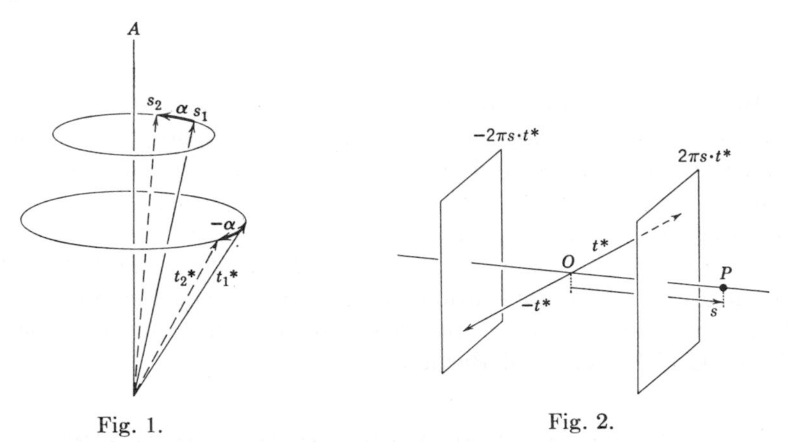

Fig. 1. Fig. 2.

s has exactly the same effect on this reciprocal as the reverse rotation of the vector $\mathbf{t}^*$. For example, vectors **s** and $\mathbf{t}^*$ can be regarded as lying in separate vertical planes containing axis A, Fig. 1. Then the reciprocal is changed in the same way if vector $\mathbf{s}_1$ is given a positive rotation, α, to $\mathbf{s}_2$, or if the vector $\mathbf{t}_1^*$ is given a negative rotation $-\alpha$, to $\mathbf{t}_2^*$. In both cases the angle between the final **s** and **t** is the same; therefore $\mathbf{s}_2\cdot\mathbf{t}_1^* = \mathbf{s}_1\cdot\mathbf{t}_2^*$, and so $e^{i2\pi\mathbf{s}_2\cdot\mathbf{t}_1^*} = e^{i2\pi\mathbf{s}_1\cdot\mathbf{t}_2^*}$. Thus, a rotation of **s** has the same effect on the reciprocal as the corresponding negative rotation of $\mathbf{t}^*$. In this sense, every rotation in direct space is equivalent to an equal but opposite rotation in reciprocal space. For considerations involving the whole symmetry group, however, this fine point is not important, because every group which contains an operation also contains its reverse.

Conjugate symmetry. In addition to conforming to the point symmetry of the object, the complex reciprocal has further symmetry. The nature of this is illustrated for the reciprocal of point P in Fig. 2.

If the weighting at P in direct space is ρ, the weighting all over reciprocal space, due to P, is also ρ, but the phase, which is constant in any plane normal to OP, varies uniformly along OP. If any arbitrary vector $\mathbf{t}^*$ is selected in reciprocal space, the phase angle at the end of the vector is $2\pi\mathbf{s}\cdot\mathbf{t}^*$. If the centrosymmetrical vector $-\mathbf{t}^*$ is selected, the phase angle at its end is $2\pi\mathbf{s}\cdot(-\mathbf{t}^*) = -2\pi\mathbf{s}\cdot\mathbf{t}^*$. Thus the phase angle *increases* on the side of the origin toward P but *decreases* on the side away from P. Therefore at the ends of $\mathbf{t}^*$ and $-\mathbf{t}^*$ the magnitudes are the same, but the phases are reversed.

This symmetry in the reciprocal, demonstrated for a single point at which the density is ρ, can be shown to occur for more general objects. The reciprocal at $\mathbf{t}^*$ for a more general object is given by

$$R_{t^*} = \int_V \rho e^{i2\pi\mathbf{s}\cdot\mathbf{t}^*}\, dV. \tag{1}$$

The reciprocal at the end of the reverse vector, $-\mathbf{t}^*$, is

$$\begin{aligned} R_{-t^*} &= \int_V \rho e^{i2\pi\mathbf{s}\cdot(-\mathbf{t}^*)}\, dV \\ &= \int_V \rho e^{-i2\pi\mathbf{s}\cdot\mathbf{t}^*}\, dV. \end{aligned} \tag{2}$$

The reciprocals at the end of a vector and at the end of the centrosymmetrical vector therefore have exactly the same magnitude (since the integrations are of the same function of ρ and $\mathbf{s}$ over direct space) but the reciprocals have opposite phases. This can be formulated as follows: Let the phase of the reciprocal at the end of the vector $\mathbf{t}^*$ be ϕ. Then the complex reciprocal can be described as

$$R_{t^*} = |R_{t^*}|e^{i\phi}. \tag{3}$$

By using this notation, the conjugate symmetry of reciprocals can be expressed as

$$\begin{aligned} |R_{-t^*}| &= |R_{t^*}|, \\ R_{t^*} &= |R_{t^*}|e^{i\phi}, \\ R_{-t^*} &= |R_{t^*}|e^{-i\phi}. \end{aligned} \tag{4}$$

All reciprocals have conjugate symmetry. It is interesting to see how this can be reconciled with centrosymmetry when this occurs also. Again it is illuminating to study this in the simplified case of the reciprocal of a single point at the end of the vector $\mathbf{s}$, and at which the density is ρ. If there is no further symmetry the reciprocal at the end of vector $\mathbf{t}^*$ is

$$R_{t^*} = \rho e^{i2\pi\mathbf{s}\cdot\mathbf{t}^*}. \tag{5}$$

But if there is also centrosymmetry in the object, then the point at the end of the vector **s** is accompanied by an equivalent point at the end of the vector $-\mathbf{s}$. In this case the reciprocal at the end of vector $\mathbf{t}^*$ is

$$R_{t^*} = \rho e^{i2\pi \mathbf{s}\cdot\mathbf{t}^*} + \rho e^{i2\pi(-\mathbf{s})\cdot\mathbf{t}^*} \tag{6}$$

$$= \rho(e^{i2\pi \mathbf{s}\cdot\mathbf{t}^*} + e^{-i2\pi \mathbf{s}\cdot\mathbf{t}^*})$$

$$= 2\rho \cos 2\pi \mathbf{s}\cdot\mathbf{t}^*. \tag{7}$$

Therefore, when centrosymmetry is present in the object, the reciprocal is everywhere real, so the conjugate of the reciprocal is identical with the reciprocal, but its value fluctuates, and it may be positive or negative depending on the trigonometric term. Furthermore, since the cosine in an even function, the reciprocal has the same value at t^* and $-t^*$.

The symmetries of the reciprocals of crystals

The foregoing treatment of the symmetries of reciprocals has been perfectly general. Since most applications of the theory will be to the reciprocals of crystals, this case is considered in some detail below.

Point-group symmetry. Although the results of the general treatment already given can be applied to the case of crystals, it is interesting to derive the results in a slightly different fashion. First, let the material of the crystal be regarded as contained in a sequence of planes parallel to the rational lattice planes $(h_1\,k_1\,l_1)$. Now, the reciprocal to a sequence of parallel planes vanishes everywhere except along the normal to the planes, so that attention can be limited to this line. Consider a symmetry operation which carries the planes $(h_1\,k_1\,l_1)$ into the planes $(h_2\,k_2\,l_2)$, Fig. 3. The same symmetry operation carries the normal of the planes $(h_1\,k_1\,l_1)$ into the normal of the planes $(h_2\,k_2\,l_2)$. Except for the orientations of the two sets of planes, they are identical. Therefore if a point-group symmetry operation is satisfied by a crystal it is also satisfied by its reciprocal.

Theorem 2: The point-group symmetry of a crystal and its reciprocal are the same.

It is evident that if the reciprocal of a crystal is available, the point-group of the crystal can be determined. It might be supposed, from this, that the point-group symmetry could be ascertained from x-ray diffraction photographs. Unfortunately this is not ordinarily so. The reason may be formulated as follows:

For the crystals of interest here, the reciprocal has non-vanishing values only at reciprocal-lattice points, which are located by coordinates h, k, and l. The reciprocal at such a point is designated F_{hkl}. In general the reciprocal is complex, so that by following the notation of (3), it can be expressed

$$F_{hkl} = |F_{hkl}|e^{i\phi}.$$

Conjugate symmetry then provides relation (4), which, for a crystal, can be expressed

$$\begin{aligned} |F_{hkl}| &= |F_{\bar{h}\bar{k}\bar{l}}|, \\ F_{hkl} &= |F_{hkl}|e^{i\phi}, \\ F_{\bar{h}\bar{k}\bar{l}} &= |F_{hkl}|e^{-i\phi}. \end{aligned} \tag{8}$$

Now, x-ray diffraction experiments provide means of determining the magnitude $|F_{hkl}|$, but no way has been found to measure phases. It follows that F_{hkl} and $F_{\bar{h}\bar{k}\bar{l}}$ appear the same on a diffraction record and cannot be distinguished.† In other words, conjugate symmetry cannot be distinguished from centrosymmetry in the diffraction record. All crystals therefore appear to be centrosymmetrical from diffraction experiments.

This was first pointed out by Friedel,[1] but the false centrosymmetry shown by diffraction records is usually called *Laue symmetry*, probably because Laue photographs were used in earlier days to study the symmetry of the diffraction effects of a crystal. Because of the false centrosymmetry of the diffraction effects, it is not possible to distinguish the 32 crystal classes by their diffraction records. Instead the only 11 centrosymmetrical classes, sometimes called "Laue classes," can be distinguished.§

Space-group symmetry. Point-group symmetries can be formulated in terms of combinations of proper and improper rotations. Space-group symmetries, in addition, involve translation components, that is, changes in **s**, and these can be expected to affect the phases of reciprocals.

Consider how an n-fold screw axis through the origin relates two equivalent stacks of planes. For the sake of definiteness, let the screw axis, A, Fig. 4, be along the crystallographic c axis. The illustration shows two stacks of planes related by an operation of the screw. One set of planes may be indexed (hkl), the other ($h'k'l'$), but both have the common interplanar spacing d_{hkl}. The reciprocal to a stack of planes is a lattice row, normal to the planes through the origin. The separation of points on the row is $1/d_{hkl}$, one point being located at the origin. Let the first point from the origin of the row reciprocal to the stack (hkl) be Q and the first point of the row reciprocal to the stack ($h'k'l'$) be Q'. Both Q and Q' have the same origin distance, namely $1/d_{hkl}$. Since the direction of the normal is unaffected by the displacement of the planes due to the translation component of the axis, it is evident that the rotation com-

† Except when anomalous scattering occurs, as discussed in Chapter 20.

§ See M. J. Buerger. *X-ray crystallography* (John Wiley and Sons, New York, 1942) Table 1, page 57.

ponent of the crystal symmetry element is preserved in the relative locations of the points Q and Q' in reciprocal space.

The relative phases of the two points, however, depend on the relative displacements of the two sets of planes from the origin. In Fig. 4, the phase difference at Q corresponding with the displacement between the origin and the first plane of the set (hkl) is 2π. This difference can be measured by the displacement of the planes along the c axis. Since there are l planes per unit translation along the c axis, the total phase difference per unit translation along the c axis is $2\pi l$. However, the translation component of the screw produces a displacement of planes only a fraction of a translation, namely $1/q$, so that the phase difference due to this translation component is $2\pi l/q$.

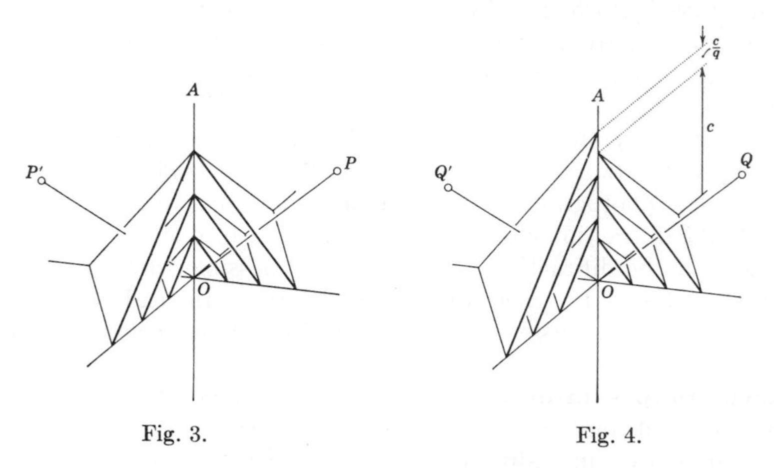

Fig. 3. Fig. 4.

Let the complex value of the reciprocal at the point Q be F_{hkl}. In general, the phase angle of F_{hkl} is not zero, but some value, say ϕ. Thus the phase component of F_{hkl} can be represented in the complex plane by $e^{i\phi}$, as noted in (8). Then the phase component at Q', according to the last paragraph, is $e^{i\phi}\, e^{i2\pi l/q}$. Since the absolute magnitudes at Q and Q' are identical, the reciprocals at these two points must be related as follows:

$$F_{h'k'l'} = F_{hkl}\, e^{i2\pi l/q}, \tag{9}$$

where $1/q$ is the fraction of the c translation comprising the translation component of the symmetry operation. Note that the phase difference vanishes for symmetry elements without a translation component (i.e. $q = 1$), and also for indices such that $l/q =$ an integer.

By permitting the rotation component of the symmetry element to include improper rotations, this result can be extended to include symmetry operations of the second sort. Alternatively, corresponding results can be derived directly for reflections, glide reflections, and roto-inversions. Neither the sequence of planes from the origin nor their distances away from the origin are affected by the improper character of the rotation, so that the only effect on reciprocal space that the nature of the rotation has is with respect to the location of the point.

As a consequence of these considerations, the following statement can be made:

Theorem 3: If a crystal contains a symmetry element through a chosen origin in direct space, its reciprocal has the isogonal, translation-free, symmetry element at the origin, but the points in the fields related by the operations of the symmetry element have phases related by $e^{i2\pi ph/q}$, *where* p *is the power of the operation relating the fields, and* a/q *is the translation component of the operation.*

For diagonal symmetry elements having the general translation component $(a+b+c)/q$, the term $(h+k+l)$ should be substituted for h.

An equivalent statement of this is

Theorem 4: The symmetry of reciprocal space is the same as the isogonal point-group symmetry of direct space, except that every symmetry element containing the origin produces a phase change in equivalent space fields of $e^{i2\pi h/q}$ *where* 1/q *is the translation component of the symmetry operation.*

Note that the phase shift becomes zero for $q = 1$, i.e., for pure reflections and pure rotations.

In case the origin is not taken at the symmetry element, but is displaced by an amount Δx, then, as shown in Fig. 5, the planes *hkl* are displaced together with the origin, while the symmetry-equivalent planes are displaced in the opposite direction. The resulting phase difference between the reciprocals of the symmetry-equivalent planes is easy to evaluate in case the symmetry element is a mirror normal to an axis (*a* in Fig. 5*A*), a 2-fold axis, normal to two axes (*a* and *b*, Fig. 5*B*), or an inversion center. In these cases, if the displacement of the origin is Δx, no phase change occurs in the reciprocal, F_{hkl}, of the planes (*hkl*), since they are carried along with the origin, but the symmetry-equivalent planes are displaced backward by $-2\Delta x$, and they suffer a phase change according to (58) of Chapter 15, which can be represented by

$$e^{i2\pi\mathbf{w}\cdot\mathbf{t}^*} = e^{i2\pi(-2h\Delta x)}. \tag{10}$$

This result is easy to generalize, and a useful form is

Theorem 5: If the origin, at first taken in a symmetry element, is shifted by $\Delta x\ \Delta y\ \Delta z$, *this change induces a phase difference in the reciprocals of the symmetry-equivalent planes of:* $e^{-i2\pi(2h\Delta x)}$: *for a mirror normal to* a;

$e^{-i2\pi(2h\Delta x+2k\Delta y)}$ *for a 2-fold axis normal to* ab; *and* $e^{-i2\pi(2h\Delta x+2k\Delta y+2l\Delta z)}$ *for an inversion center.*

Theorem 5 incidentally provides a tool for predicting the changes in phase relations which accompany any displacement of the origin of the space group.

Reciprocals of space-lattice types. In the last chapter it was shown that if the object in direct space has a periodicity which can be

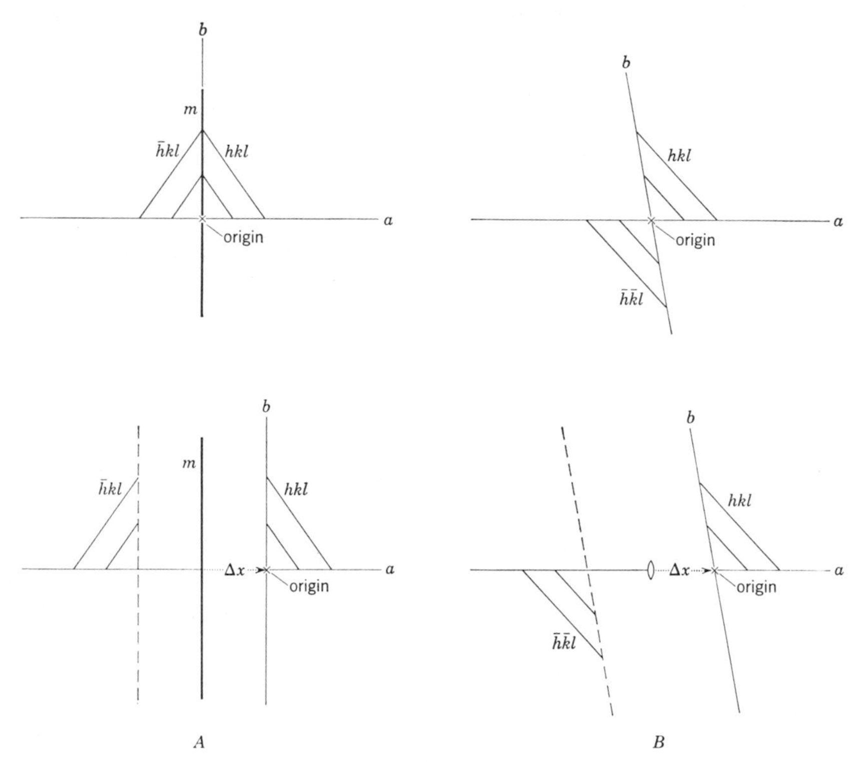

Fig. 5.

A The result of displacing the origin by an amount Δx from an original location on a mirror, m.

B. The result of displacing the origin by an amount Δx from an original location on a 2-fold axis.

described by a space lattice, the lattice periodicity has a characteristic effect upon reciprocal space. Specifically the lattice character limits the non-vanishing regions of the Fourier transform to points of the reciprocal lattice. In other words, if the object has a lattice, the reciprocal of the object has non-zero values only on the points of the reciprocal lattice.

To each space-lattice type there is a unique reciprocal-lattice type.†

† M. J. Buerger. *X-ray crystallography.* (John Wiley and Sons, New York, 1942) 492.

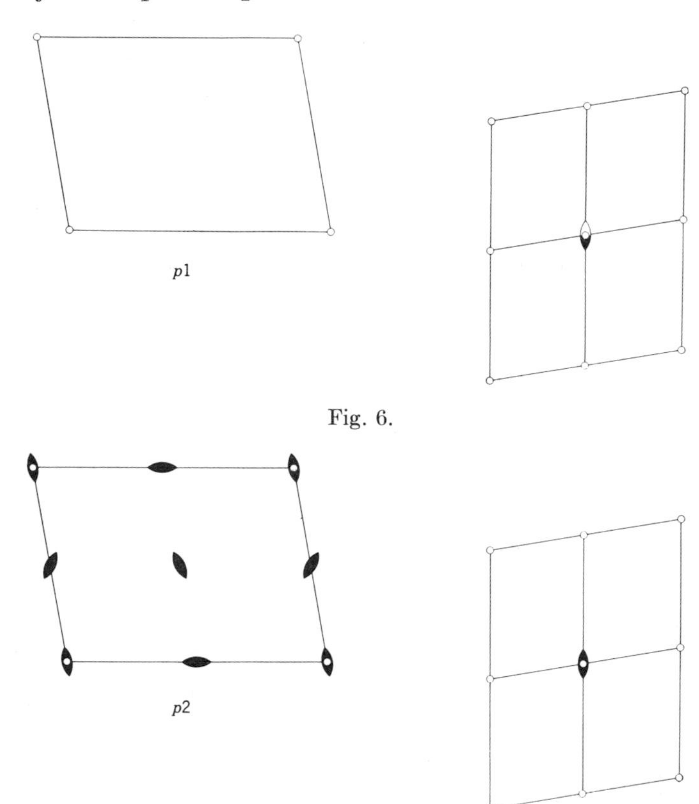

Fig. 6.

Fig. 7.

Fig. 6–22.
The symmetries of the plane groups (left), and the corresponding symmetries in reciprocal space (right). In reciprocal space some unusual symmetry elements are used. A partly blackened rotation axis represents conjugate rotational symmetry, and a wavy line represents conjugate reflectional symmetry. A hatched line shows the direction affecting a phase shift for certain values of the index in that direction or directions.

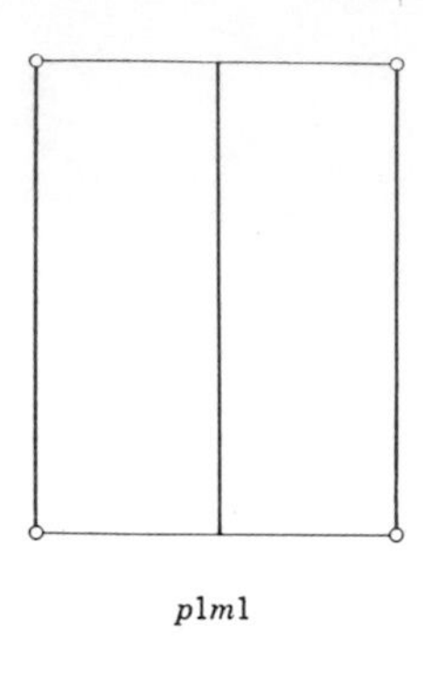

$p1m1$

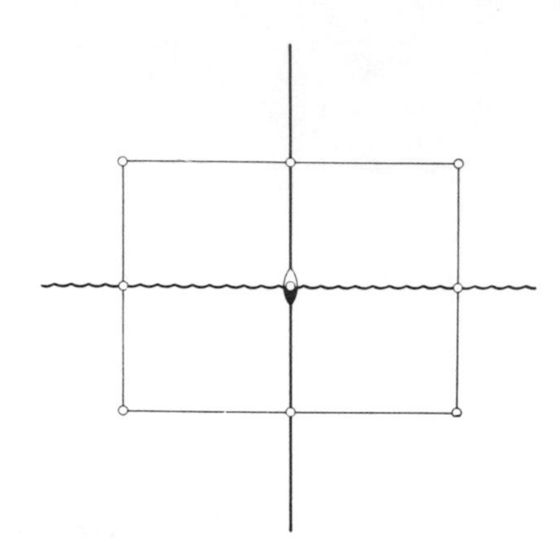

Fig. 8.

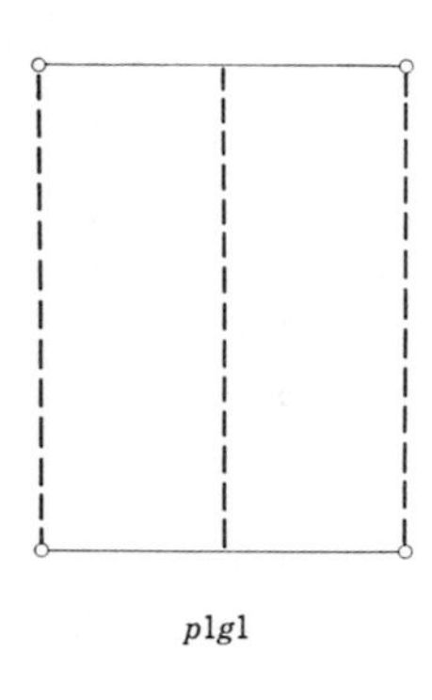

$p1g1$

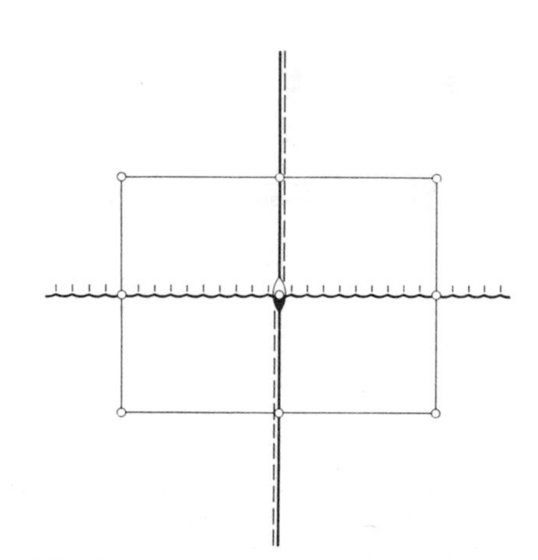

Fig. 9.

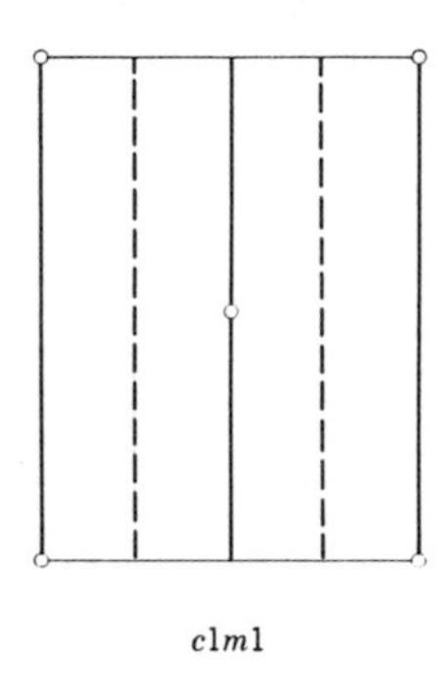

$c1m1$

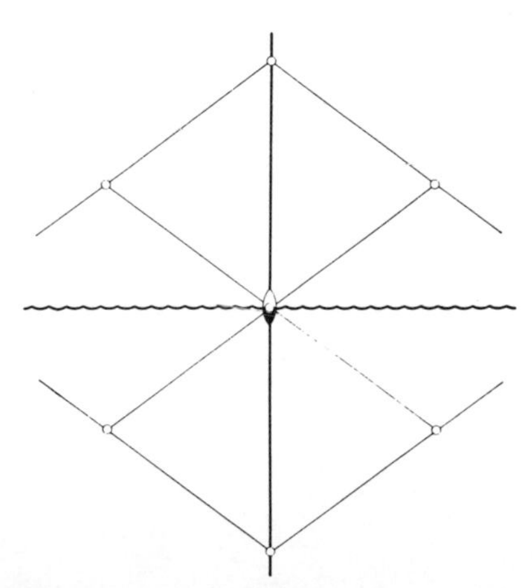

Fig. 10.

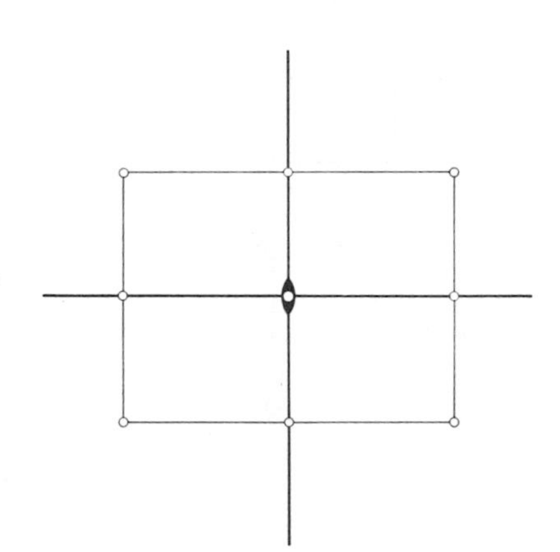

Fig. 11.

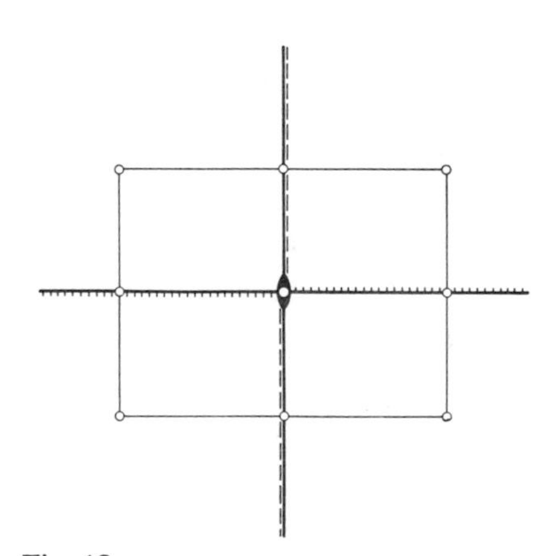

Fig. 12.

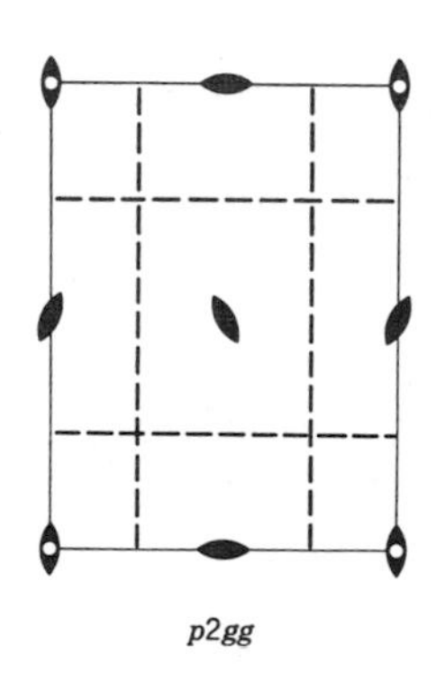

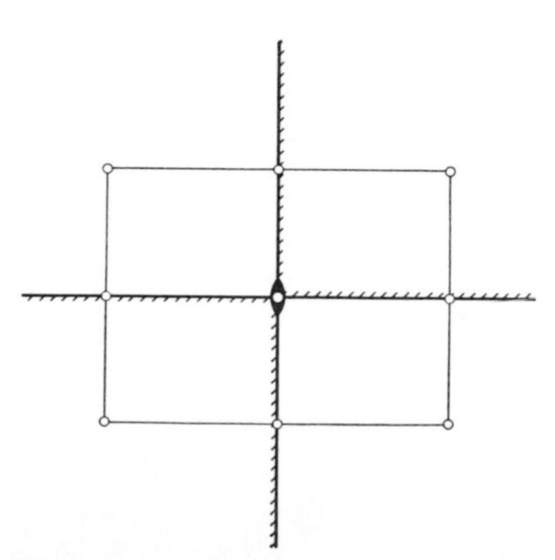

Fig. 13.

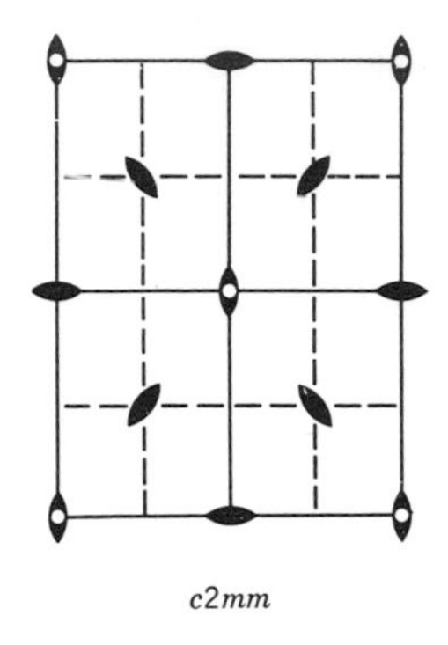

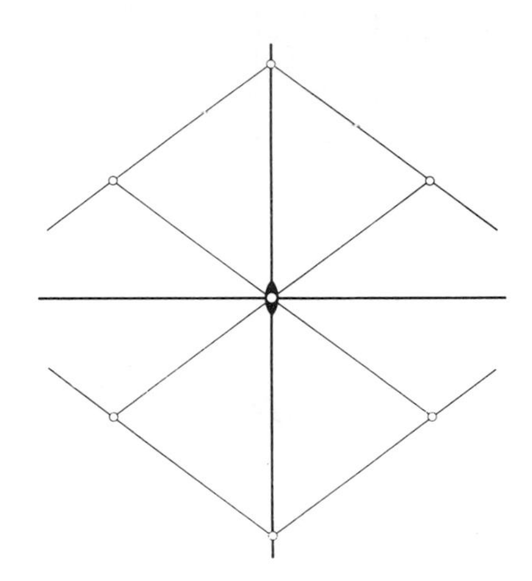

Fig. 14.

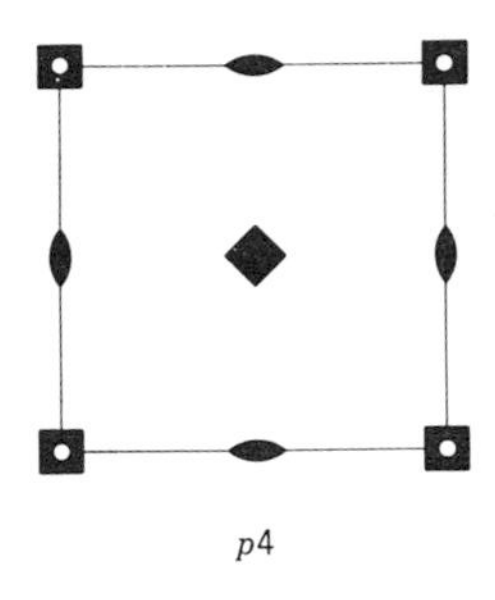

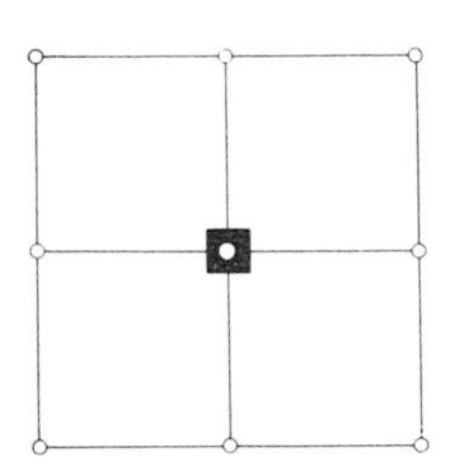

Fig. 15.

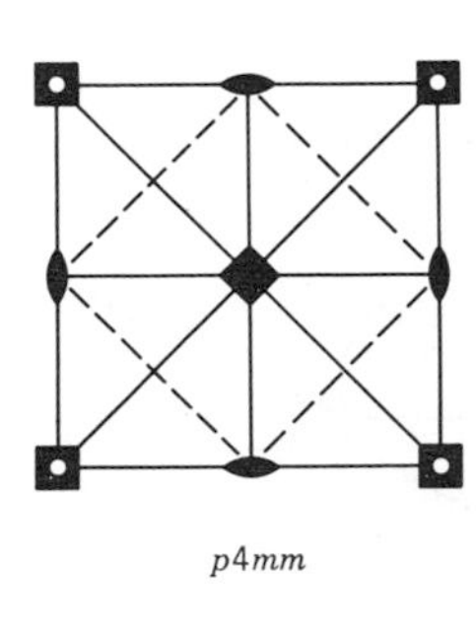

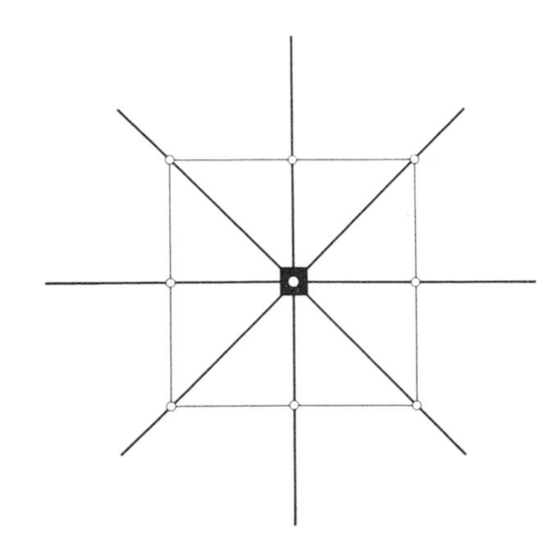

Fig. 16.

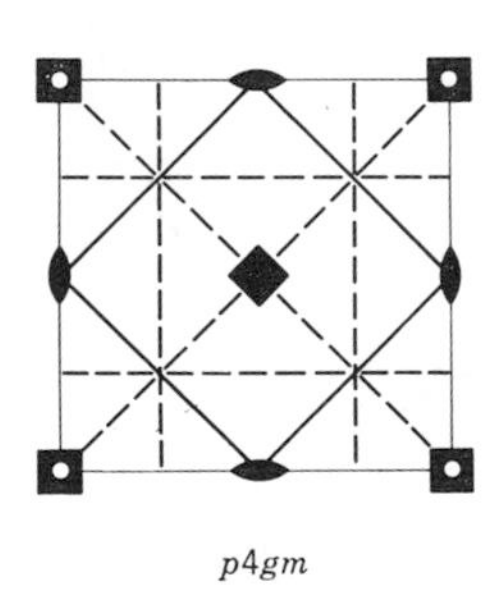

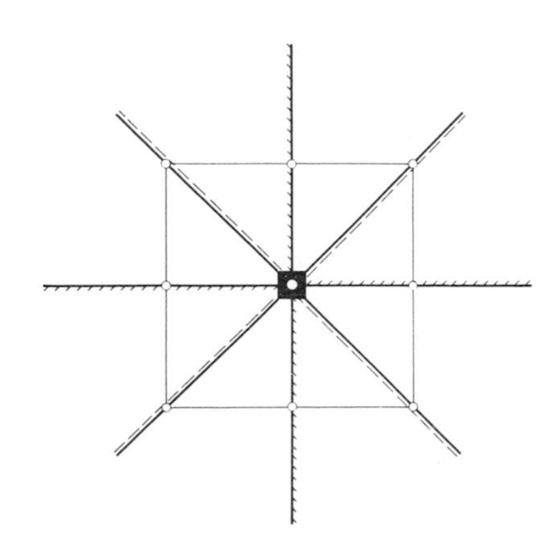

Fig. 17.

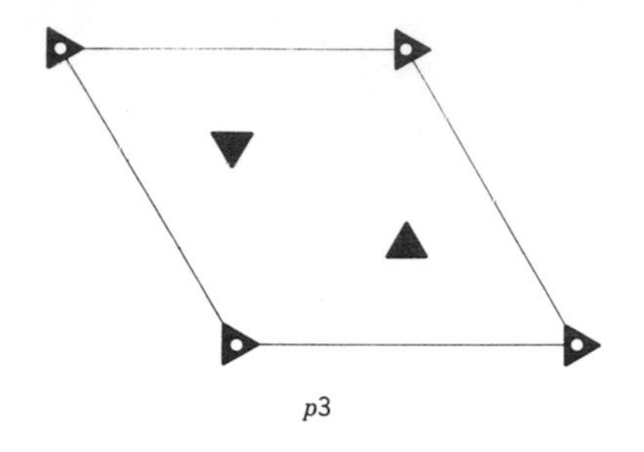

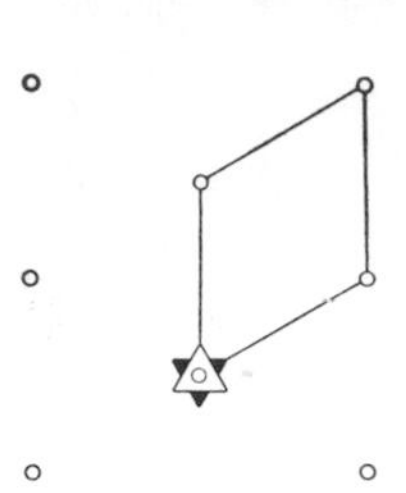

Fig. 18.

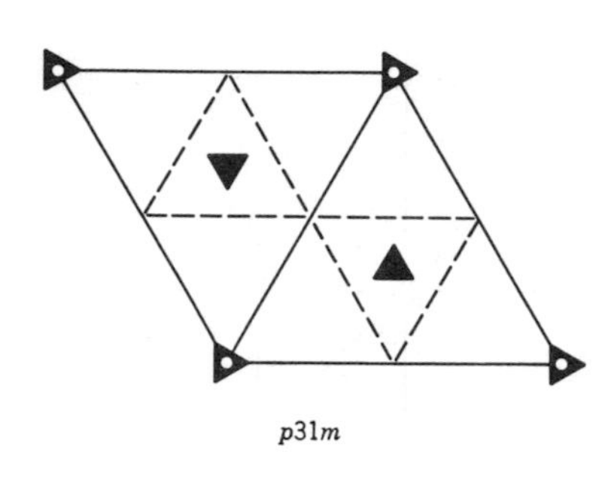

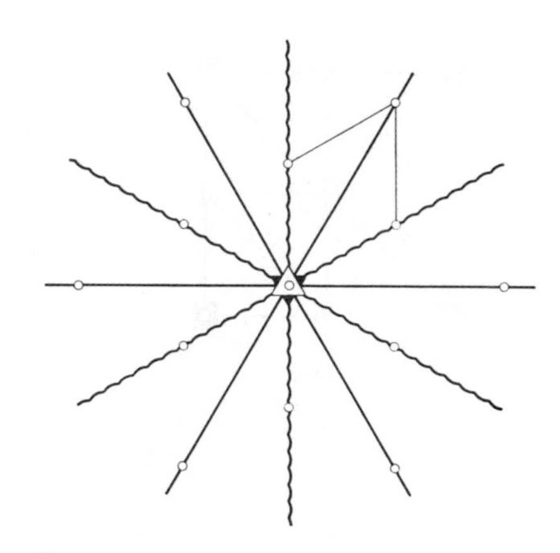

Fig. 19.

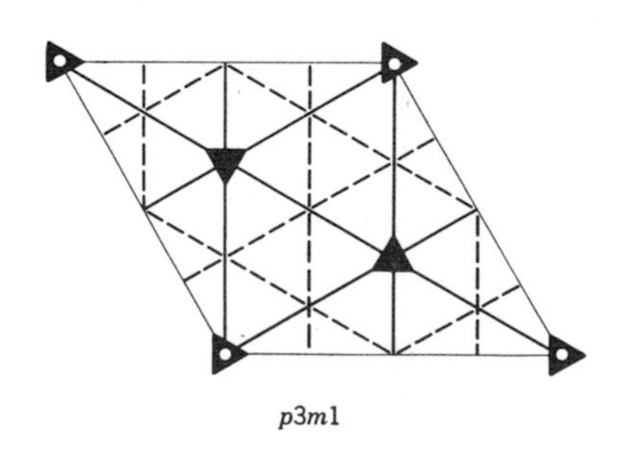

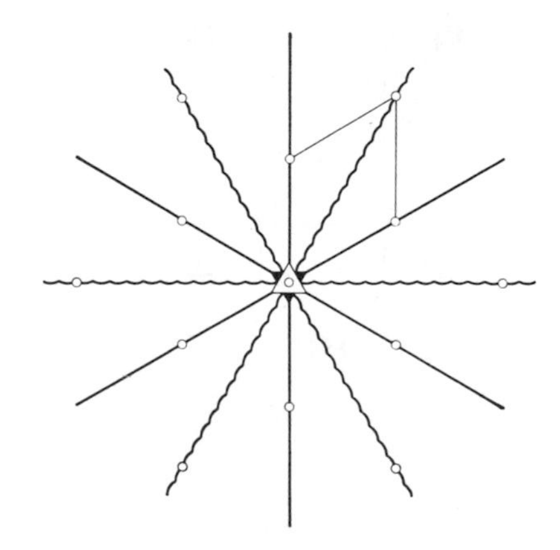

Fig. 20.

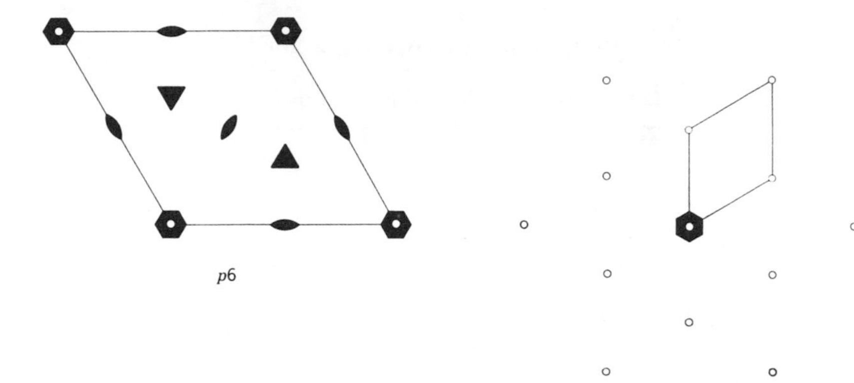

Fig. 21.

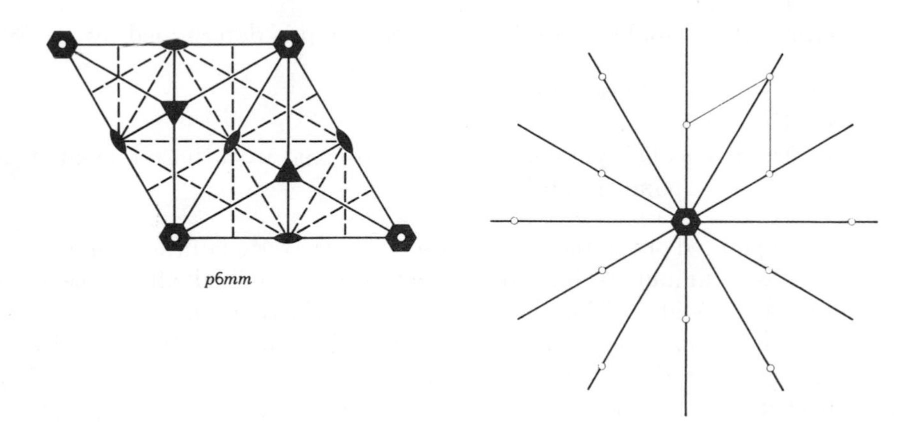

Fig. 22.

Table 1
Reciprocals of lattice types

Lattice type	Reciprocal-lattice type
P	*P*
R	*R*
A	*A*
B	*B*
C	*C*
I	*F*
F	*I*

These are listed in Table 1. Note that each lattice type is its own reciprocal except that body-centered (*I*) and face-centered (*F*) space lattices are each other's reciprocals.

Reciprocals of space groups. A space group is determined by

(*a*) The space-lattice type.
(*b*) The point group.
(*c*) The particular symmetry elements isogonal with the symmentry elements of the point groups.

It has been pointed out in the foregoing sections that each of these characteristics has a unique reciprocal representation provided allowance is made for origin shift. Thus, to each space group there actually corresponds a triply infinite spectrum of reciprocals, depending on the variables x, y, and z, which determine the origin. Nevertheless, the members of this spectrum of reciprocals differ only by relative phases. *For a specific space-group origin, each space group has a unique reciprocal and vice versa.*

As an example of the symmetry of reciprocal space, Figures 6 through 22 show the 17 plane groups on the left, and their reciprocals on the right. In addition to the usual symbols for 2-, 3-, 4-, and 6-fold rotations, and for mirrors and glides, the representations of the plane symmetries in reciprocal space require some new symbols. In these diagrams, axial conjugate symmetry is indicated by a symmetry symbol partly outlined and partly black, reflectional conjugate symmetry by a wavy line, and a reflection accompanied by a phase shift for certain values of an index is indicated by a "one-sided" line having hatching in the direction of that index or combination of indices.

Literature

[1] G. Friedel. *Sur les symétries cristallines que peut révéler la diffraction des rayons Röntgen.* Compt. rend. **157** (1913) 1533–1536.

[2] M. J. Buerger. *Crystallographic symmetry in reciprocal space.* Proc. Nat. Acad. Sci. U. S. **35** (1949) 198–201.

[3] M. J. Buerger. *Fourier summations for symmetrical crystals.* Am. Mineralogist **34** (1949) 771–788.

[4] Werner Nowacki. *Bezichungen zwischen der Symmetrie des Kristall-, Fourier- und Patterson-Raumes.* Schweiz, mineral. petrog. Mitt. **30** (1950) 147–160.

[5] Werner Nowacki. *Fouriersynthese von Kristallen.* (Birkhauser, Basel, Switzerland, 1952).

[6] Jürg Waser. *Symmetry relations between structure factors.* Acta Cryst. **8** (1955) 595.

[7] E. F. Bertaut. *Algèbre des facteurs de structure.* Acta Cryst. **9** (1956) 769–770.

[8] E. F. Bertaut and J. Waser. *Structure factor algebra. II.* Acta Cryst. **10** (1957) 606–607.

[9] E. F. Bertaut. *IV. Algèbre des facteurs de structure.* Acta Cryst. **12** (1959) 541–549.

[10] E. F. Bertaut. *V. Algèbre de facteurs de structure.* Acta Cryst. **12** (1959) 570–574.

17

Application of symmetry to Fourier summations

In the last chapter it was seen that to each symmetry relation which can occur in crystal space there corresponds a kind of symmetry in reciprocal space. This relation between the symmetries of the two spaces provides a tool for simplifying the computation of Fourier summations. To appreciate the significance of this, consider the summations involved in the synthesis of a projection of electron density, $\rho_{(xy)}$. Suppose F_{hk0}'s are available for h ranging from -10 to 10, and for k ranging from -10 to 10. This provides a field of $21 \times 21 = 441$ F_{hk0}'s in reciprocal space. Suppose that the density is required in a cell which is sampled at 60 intervals along a and 60 intervals along b, that is, at $60 \times 60 = 3600$ different points. Without taking advantage of any possible symmetry of the crystal or of its F's in reciprocal space, such a Fourier synthesis would require a summation of 441 F's, properly phased, at 3600 points.

Fortunately the number of summations can be reduced with the aid of symmetry. Even if the projection is without symmetry, conjugate symmetry relates the two halves of reciprocal space. If the projection is symmetrical, the F's of symmetrical segments of reciprocal space are related, and furthermore the summations need be performed for only a fraction of the crystal cell. In this chapter the condensation of the number of terms entering the required summations in Fourier syntheses is considered, but the actual computing methods are reserved for the next chapter.

The application of the additivity theorem in Fourier summations for symmetrical crystals

In the last chapter some of the combinatorial properties of Fourier transforms were noted. An important combinatorial property is additivity. The *additivity theorem* may be restated here in slightly different form.

Theorem 1: The transform of a sum is the sum of the transforms.

Of course this applies either to a transform from direct space to reciprocal space, or the reverse. To emphasize this feature, let $T_{(xyz)}$ represent the transform at xyz in either space, and let M and N represent the object in the other space. Then the additivity theorem, stated above, can be expressed as

$$T_{(xyz)}(M+N) = T_{(xyz)}(M) + T_{(xyz)}(N). \tag{1}$$

In the application required in this chapter, the transformation is from reciprocal space to direct space. Thus $T_{(xyz)}$ refers to the transform at a

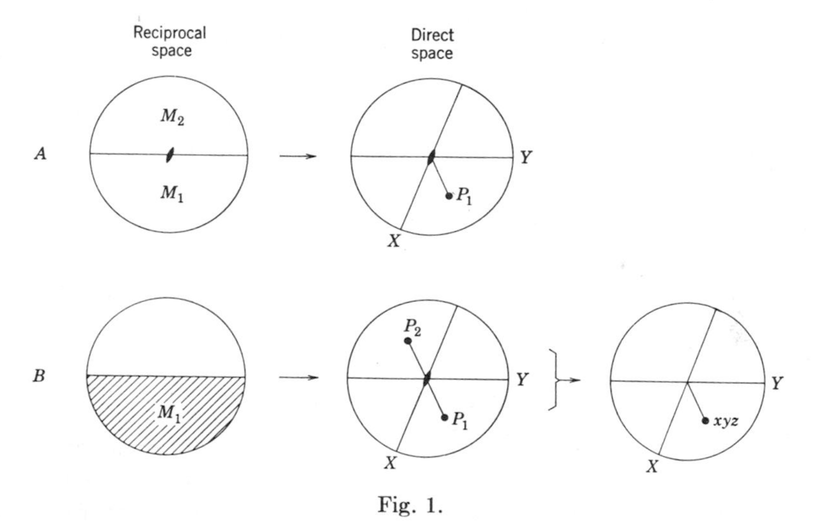

Fig. 1.

point xyz in direct space, and M and N refer to objects in reciprocal space, which are sets of F_{hkl}'s.

Suppose that the transform at xyz in direct space is required for all the objects in reciprocal space. Then reciprocal space can be subdivided into any convenient blocks, and the additivity theorem provides that the transform of all the blocks is the same as the sum of the transforms of the separate blocks. In particular, each block may be an unsymmetrical motif sector of reciprocal space which, when operated upon by the symmetry elements, fills all reciprocal space.

There are two ways of taking advantage of this, Fig. 1. An expression can be set up which gives the transform at P_1 in direct space of the several equivalent segments M_1 and M_2 of reciprocal space, and then symmetry relations of the F's between segments of reciprocal space can be used to reduce the expression. This process is illustrated diagrammatically in Fig. 1*A*. This amounts to performing the addition in reciprocal space. Alternatively, Fig. 1*B*, the transform of one segment M_1 in reciprocal space can be found at the several symmetrical points in direct space P_1 and P_2. The additivity theorem provides

$$T_{(P_1)}(M_1 + M_2) = T_{(P_1)}(M_1) + T_{(P_1)}(M_2). \tag{2}$$

But, because of symmetry,

$$T_{(P_1)}(M_2) = T_{(P_2)}(M_1). \tag{3}$$

That is, the transform of segment M_2 at P_1 is the same, due to symmetry, as the transform of segment M_1 at P_2. Therefore this substitution can be made in (2), giving

$$T_{(P_1)}(M_1 + M_2) = T_{(P_1)}(M_1) + T_{(P_2)}(M_1). \tag{4}$$

This can be formulated in the following general way:

Theorem 2: The transform at a point P *in direct space of a set of equivalent segments which fill reciprocal space is equal to the sum of the transforms at several points equivalent to* P *in direct space of one of the equivalent segments in reciprocal space.*

This theorem can be used to advantage when it is found difficult to write relations between equivalent points in reciprocal space. It has the effect of transferring the work of summation by reducing the number of terms but increasing the number of points at which the summation is made.

The transform of a sector

Exponential form. To apply Theorems 1 and 2, it is convenient to develop the general form for the transform of a motif segment. This can be developed from first principles as follows: The general transformation from reciprocal to direct space, from (7′) of Chapter 15, is

$$\rho_s = \int_{V^*} R_{t^*}\, e^{-i2\pi \mathbf{s}\cdot\mathbf{t}^*}\, dV^*. \tag{5}$$

Let the elementary volume dV^* have edges $d\mathbf{X}^*$, $d\mathbf{Y}^*$, and $d\mathbf{Z}^*$, parallel to the three reciprocal cell axes $\mathbf{a}^*$, $\mathbf{b}^*$, and $\mathbf{c}^*$ respectively. Then

$$\begin{aligned} dV^* &= d\mathbf{X}^*\cdot d\mathbf{Y}^*\times d\mathbf{Z}^* \\ &= \mathbf{a}^*\, dh\cdot\mathbf{b}^*\times dk\cdot\mathbf{c}^*\, dl \\ &= \mathbf{a}^*\cdot\mathbf{b}^*\times\mathbf{c}^*\, dh\, dk\, dl \\ &= V^*\, dh\, dk\, dl \\ &= \frac{1}{V}\, dh\, dk\, dl, \end{aligned} \tag{6}$$

where V and V^* are the volumes of the direct and reciprocal cells, respectively. Let the vectors $\mathbf{s}$ and $\mathbf{t}^*$ also be expressed in terms of the edges of the crystal cell and the reciprocal cell. Then

$$\mathbf{s} = x\mathbf{a}+y\mathbf{b}+z\mathbf{c}, \tag{7}$$

$$\mathbf{t}^* = h\mathbf{a}^*+k\mathbf{b}^*+l\mathbf{c}^*, \tag{8}$$

so that

$$\mathbf{s}\cdot\mathbf{t}^* = hx+ky+lz. \tag{9}$$

If these are substituted in (5) and the variables changed to h, k, and l according to (6), it becomes

$$\rho_{(xyz)} = \frac{1}{V}\iiint F_{hkl}\, e^{-i2\pi(hx+ky+lz)}\, dh\, dk\, dl. \tag{10}$$

For crystals, the F_{hkl}'s are non-vanishing only at points of the reciprocal lattice, where h, k, and l have integral values. Therefore the integration can be replaced by a summation over h, k and l:

$$\rho_{(xyz)} = \frac{1}{V}\sum_h\sum_k\sum_l F_{hkl}\, e^{-i2\pi(hx+ky+lz)}. \tag{11}$$

This has the same form as that for the Fourier summation for the entire crystal, except that the summation limits are not specified. These must be taken in such a way as to include all F_{hkl}'s within the desired segment.

The form corresponding to (11) for the projection of the electron density along the c axis is

$$\rho_{(xy)} = \frac{1}{S}\sum_h\sum_k F_{hk0}\, e^{-i2\pi(hx+ky)}. \tag{12}$$

This form can be deduced by eliminating the variable z from (11), or it can be derived from first principles by repeating steps (5) through (11) for two dimensions. If this is done, S is seen to be the reciprocal of $|\mathbf{a}^*\times\mathbf{b}^*|$, and therefore equal to the area of the cell normal to $\mathbf{c}$. The limits of the summation over h and k are undefined in (12), but these must be taken so as to include all F's in the appropriate sector of the reciprocal lattice.

Trigonometric form. Since (11) involves complex quantities it is convenient for purposes of computation to recast it into trigonometric form. Both F's and exponentials are complex. They can be expressed in terms of real and imaginary components as follows:

$$F_{hkl} = A_{hkl} + iB_{hkl}, \tag{13}$$

$$e^{-i\psi} = \cos\psi - i\sin\psi. \tag{14}$$

If these are substituted into (11), it can be reduced as follows:

$$\rho_{(xyz)} = \frac{1}{V} \sum_h \sum_k \sum_l \{A_{hkl} + iB_{hkl}\} \times \{\cos 2\pi(hx+ky+lz) - i \sin 2\pi(hx+ky+lz)\} \quad (15)$$

$$= \frac{1}{V} \sum_h \sum_k \sum_l \{A_{hkl} \cos 2\pi(hx+ky+lz) - iA_{hkl} \sin 2\pi(hx+ky+lz) + iB_{hkl} \cos 2\pi(hx+ky+lz) + B_{hkl} \sin 2\pi(hx+ky+lz)\}. \quad (16)$$

This shows that, in general, the transform is complex since it contains terms with the imaginary component i. But this complication can be avoided if advantage is taken of conjugate symmetry. Conjugate symmetry provides that

$$F_{hkl} = A_{hkl} + iB_{hkl}, \quad (17)$$

$$F_{\bar{h}\bar{k}\bar{l}} = A_{hkl} - iB_{hkl}. \quad (18)$$

The relation like (17) for $\bar{h}\bar{k}\bar{l}$ is

$$F_{\bar{h}\bar{k}\bar{l}} = A_{\bar{h}\bar{k}\bar{l}} + iB_{\bar{h}\bar{k}\bar{l}}. \quad (19)$$

If (18) and (19) are compared it is evident that

$$A_{\bar{h}\bar{k}\bar{l}} = A_{hkl} \quad (20)$$

$$\text{but} \quad B_{\bar{h}\bar{k}\bar{l}} = -B_{hkl}. \quad (21)$$

Furthermore, a similar relation holds for cosine and sine:

$$\cos 2\pi(\bar{h}x+\bar{k}y+\bar{l}z) = \cos 2\pi(hx+ky+lz), \quad (22)$$

$$\sin 2\pi(\bar{h}x+\bar{k}y+\bar{l}z) = -\sin 2\pi(hx+ky+lz). \quad (23)$$

If a transform such as (16) is added to the transform of the opposite segment, then one transform requires summations over hkl, the other over $\bar{h}\bar{k}\bar{l}$. If these two are combined, (20) and (22) cause the first line of (16) to be doubled, (20) and (23) cause the second line to cancel, (21) and (22) cause the third line to cancel, while (21) and (23) cause the fourth line to double.

As a result, the transform of a pair of opposite segments has the simpler form

$$\rho_{(xyz)} = \frac{2}{V} \sum_h \sum_k \sum_l \{A_{hkl} \cos 2\pi(hx+ky+lz) + B_{hkl} \sin 2\pi(hx+ky+lz)\}. \tag{24}$$

This has no imaginary component. The corresponding two-dimensional form is

$$\rho_{(xy)} = \frac{2}{S} \sum_h \sum_k \{A_{hk0} \cos 2\pi(hx+ky) + B_{hk0} \sin 2\pi(hx+ky)\}. \tag{25}$$

The Fourier summation for any fragment of reciprocal space can be found by the use of (24) or (25) for that fragment plus its opposite fragment provided, of course. that the pair do not include more than all reciprocal space. Furthermore, by the use of Theorem 1, the transform of all reciprocal space can be found by adding the transforms of enough segment pairs so that all reciprocal space is included.

The transforms of two-dimensional sectors

Product form of the summation. The general form for the summation for a two-dimensional sector pair is given by (25). This form involves the trigonometric functions of a sum of angles. For computational purposes it is convenient to change this to a product form. This can be done with the aid of the identities

$$\cos(\alpha+\beta) = \cos\alpha \cos\beta - \sin\alpha \sin\beta, \tag{26}$$

$$\sin(\alpha+\beta) = \sin\alpha \cos\beta + \cos\alpha \sin\beta. \tag{27}$$

When these are substituted into (25) it takes the form

$$\rho_{(xy)} = \frac{2}{S} \sum_h \sum_k \{A_{hk0} \cos 2\pi hx \cos 2\pi ky - A_{hk0} \sin 2\pi hx \sin 2\pi ky + B_{hk0} \sin 2\pi hx \cos 2\pi ky + B_{hk0} \cos 2\pi hx \sin 2\pi ky\}. \tag{28}$$

This is the form for a pair of sectors between which and within which there is no symmetry other than the universal conjugate symmetry of reciprocal space.

Sectors related by symmetry 2. If the sectors of the pair are related by a 2-fold axis, then

$$F_{hk0} = F_{\bar{h}\bar{k}0}. \tag{29}$$

By expressing this in terms of real and imaginary components

$$A_{hk0} + iB_{hk0} = A_{hk0} - iB_{hk0}, \tag{30}$$

it follows that

$$B_{hk0} = 0$$

$$\text{and} \qquad A_{hk0} = F_{hk0}. \tag{31}$$

For this symmetry of the sector pair, therefore, (28) takes the simpler form

$$\rho_{(xy)} = \frac{2}{S} \sum_h \sum_k \{F_{hk0} \cos 2\pi hx \cos 2\pi ky - F_{hk0} \sin 2\pi hx \sin 2\pi ky\}. \tag{32}$$

Transforms for the plane groups

The application of the theory of condensing the number of terms in a Fourier summation may be illustrated by deriving the specific forms of the summation for the 17 plane groups. This is done in the following sections.

Plane group *p*1. Figure 2 shows that reciprocal space for $p1$ can be subdivided into sectors having index fields hk, $\bar{h}k$, $\bar{h}\bar{k}$, and $h\bar{k}$. Let opposite sectors be called a sector pair. Since only conjugate symmetry relates the members of a pair, the summation form for each sector pair is (28). The summation for each sector pair is from 0 to ∞ for both h and k, but the terms for the $\bar{h}k$ field in Fig. 2 require negative h indices. The summations over the two sector pairs can be written

$$\begin{aligned}\rho_{(xy)} = \frac{2}{S} \sum_h{}' \sum_k^{\infty}{}' \{ & A_{hk0} \cos 2\pi hx \cos 2\pi ky + A_{\bar{h}k0} \cos 2\pi hx \cos 2\pi ky \\ & - A_{hk0} \sin 2\pi hx \sin 2\pi ky - A_{\bar{h}k0}(-\sin 2\pi hx) \sin 2\pi ky \\ & + B_{hk0} \sin 2\pi hx \cos 2\pi ky + B_{\bar{h}k0}(-\sin 2\pi hx) \cos 2\pi ky \\ & + B_{hk0} \cos 2\pi hx \sin 2\pi ky + B_{\bar{h}k0} \cos 2\pi hx \sin 2\pi ky\}.\end{aligned} \tag{33}$$

This can be condensed to give

$$\rho_{(xy)} = \frac{2}{S}\sum_{h}{}'\sum_{0}^{\infty}{}'_{k} \{(A_{hk0} + A_{\bar{h}k0}) \cos 2\pi hx \cos 2\pi ky$$

$$-(A_{hk0} - A_{\bar{h}k0}) \sin 2\pi hx \sin 2\pi ky$$

$$+(B_{hk0} - B_{\bar{h}k0}) \sin 2\pi hx \cos 2\pi ky$$

$$+(B_{hk0} + B_{\bar{h}k0}) \cos 2\pi hx \sin 2\pi ky\}. \quad (34)$$

In these and the other summations of this chapter, the following precaution should be noted. The lines dividing the sectors occur through the

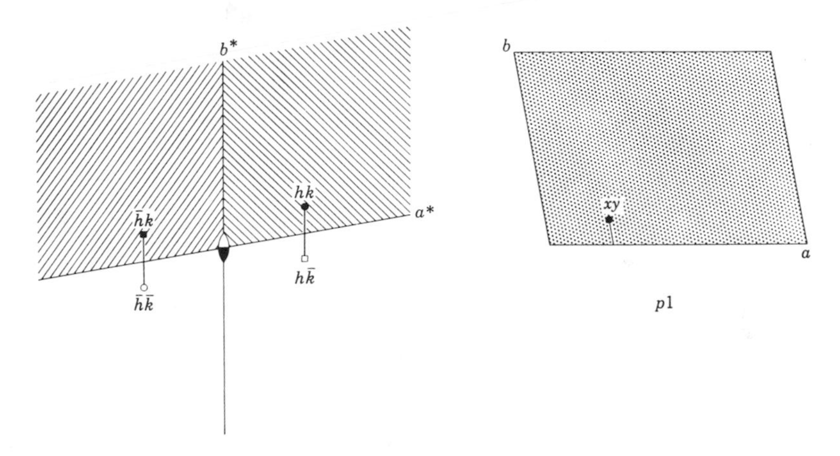

Fig. 2.

points $h00$, $0k0$, $\bar{h}00$, and $0\bar{k}0$. Each of these lines, therefore, belongs to its two adjoining sectors equally. The lines $h00$, $0k0$, $\bar{h}00$, and $0\bar{k}0$ should be counted at only half value for each segment, and the origin point at only quarter value. The special nature of the multiplicity of the edges and point of each sector is indicated by adding a prime to the summation sign.

In Fig. 2 and subsequent illustrations of this chapter, the shaded area in reciprocal space shows the region in which data are required for the Fourier synthesis. The shaded area in direct space shows the area of cell for which the summations must be made. The area is chosen so that the multiplication of the area by the symmetry elements fills the entire cell. Figure 2 shows that data are required from half of reciprocal space and the summation must be made for the full cell in direct space.

Plane group $p2$. Reciprocal space for this symmetry can be divided in the same way as for $p1$, as shown in Fig. 3. In $p2$, however, a 2-fold axis relates the opposite sectors of a pair, and the form of the summation for each sector is given by (32). This is the same as (28) except for condition (31). When these conditions are applied to (34), the B terms vanish and the A's become F's, giving the simpler form

$$\rho_{(xy)} = \frac{2}{S} \sum_{h}{}' \sum_{0\ k}^{\infty}{}' \{(F_{hk0} + F_{\bar{h}k0}) \cos 2\pi hx \cos 2\pi ky - (F_{hk0} - F_{\bar{h}k0}) \sin 2\pi hx \sin 2\pi ky\}. \tag{35}$$

The shading in Fig. 3 indicates that data are required from half of reciprocal space, but the summation need be made for only half the cell in direct space.

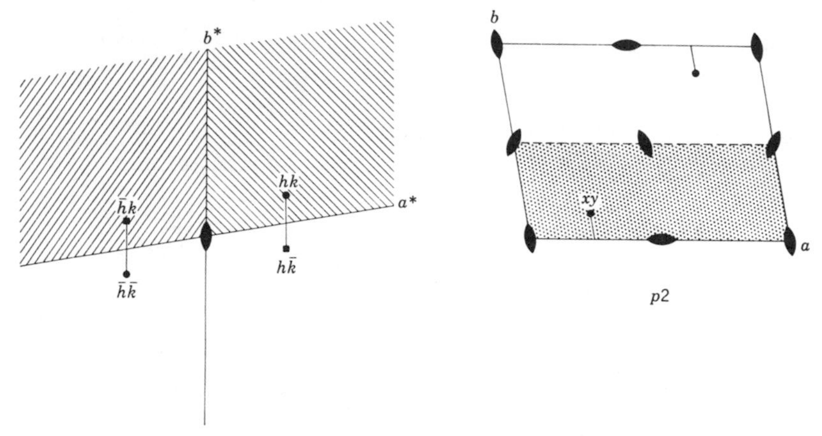

Fig. 3.

Plane groups $p1m1$ and $c1m1$. Plane groups $p1m1$ and $c1m1$ have identical symmetries in reciprocal space as shown in Figs. 4A and 5A respectively. They differ only in the locations of the points where the transform does not vanish. The forms of the summations can be derived by imposing on the forms of $p1$ and $c1$ respectively the special symmetry of the reciprocal space for $p1m1$ and $c1m1$, both of which have a symmetry plane normal to a^*, Fig. 3. This requires that

$$F_{hk0} = F_{\bar{h}k0}, \tag{36}$$

so that $A_{hk0} = A_{\bar{h}k0}$

and $B_{hk0} = B_{\bar{h}k0}$. (37)

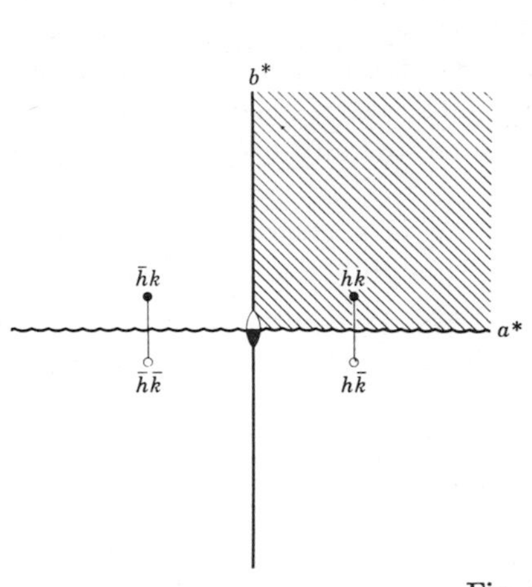

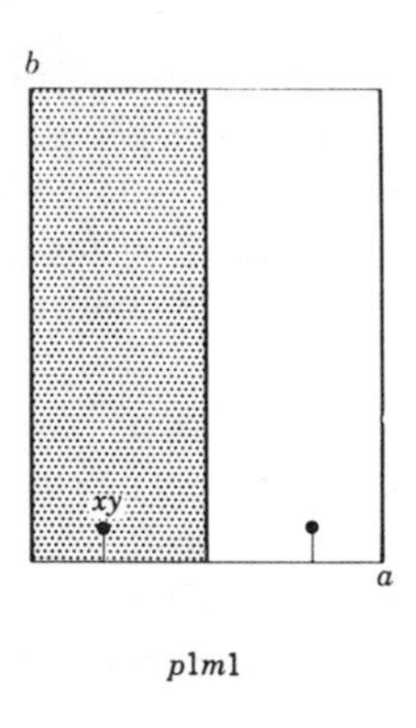

Fig. 4.

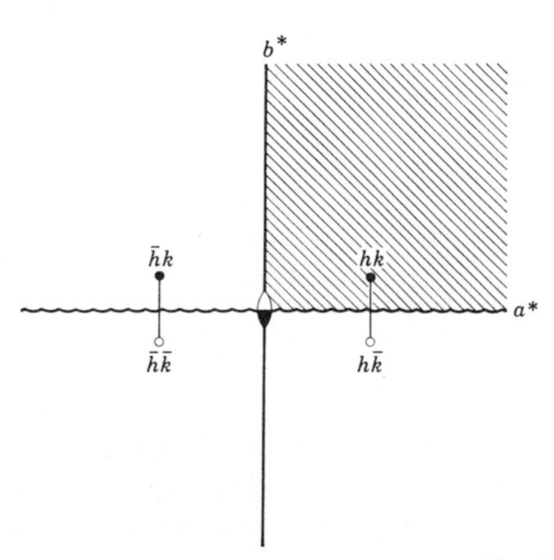

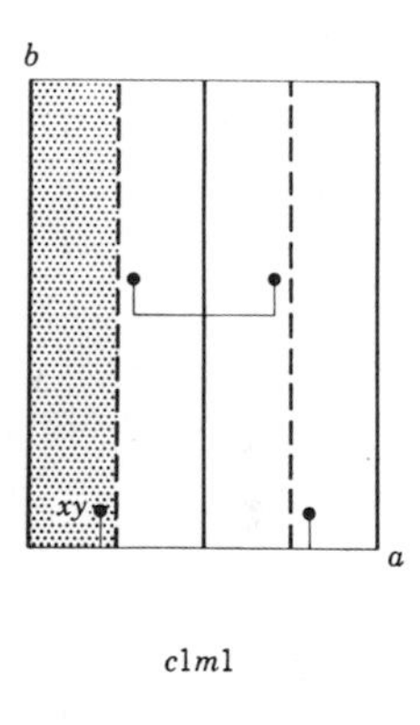

Fig. 5.

If these conditions are applied to (34), the second and third lines vanish, leaving

$$\rho_{(xy)} = \frac{4}{S} \sum_h{}' \sum_{0\,k}^{\infty}{}' \{A_{hk0} \cos 2\pi hx \cos 2\pi ky + B_{hk0} \cos 2\pi hx \sin 2\pi ky\} \tag{38}$$

$$= \frac{4}{S} \sum_{0\,h}^{\infty}{}' \left[\sum_{0\,k}^{\infty}{}' A_{hk0} \cos 2\pi ky + B_{hk0} \sin 2\pi ky \right] \cos 2\pi hx. \tag{39}$$

The form of (39) implies that the summation over k should be performed first. Figures $4A$ and $5A$ show that data are required for only one quarter of reciprocal space. Figure $4B$ shows that the summation just be carried out for a half cell for $p1m1$, while Fig. $5B$ shows it need be carried out for only a quarter cell (equivalent, however, to only half a primitive cell) for $c1m1$.

Plane group *p1g1*. This symmetry is a specialization of $p1$, the specialization consisting of the addition of a glide plane g normal to a with glide component $b/2$. In reciprocal space, relations consequently occur between the F's of the hk and $\bar{h}k$ fields. According to Theorem 4, Chapter 16, this relation is

$$F_{\bar{h}k0} = F_{hk0}\, e^{i2\pi k/2}. \tag{40}$$

It is convenient to divide the reflections into two groups according as k is even or odd:

$$\text{for } k \text{ even:} \quad \begin{aligned} F_{\bar{h}k0} &= F_{hk0}, \\ A_{\bar{h}k0} &= A_{hk0}, \\ B_{\bar{h}k0} &= B_{hk0}; \end{aligned} \tag{41}$$

$$\text{for } k \text{ odd:} \quad \begin{aligned} F_{\bar{h}k0} &= -F_{hk0}, \\ A_{\bar{h}k0} &= -A_{hk0}, \\ B_{\bar{h}k0} &= -B_{hk0}. \end{aligned} \tag{42}$$

If the summation for $p1$, namely (34), is split into two parts, one for k even, the other for k odd, then for k even, conditions (41) apply, and lines 2 and 3 of (34) vanish. For k odd, conditions (42) apply and lines 1 and 4 of (34) vanish. The resulting synthesis for $p1g1$ is

$$\begin{aligned} \rho_{(xy)} = \frac{4}{S} \sum_{h}{}' \sum_{\substack{k \\ \text{even} \\ 0}}^{\infty}{}' & \{A_{hk0} \cos 2\pi hx \cos 2\pi ky \\ & + B_{hk0} \cos 2\pi hx \sin 2\pi ky\} \\ + \frac{4}{S} \sum_{h}{}' \sum_{\substack{k \\ \text{odd} \\ 0}}^{\infty}{}' & \{-A_{hk0} \sin 2\pi hx \sin 2\pi ky \\ & + B_{hk0} \sin 2\pi hx \cos 2\pi ky\} \end{aligned} \tag{43}$$

$$= \frac{4}{S}\left\{\sum_{\substack{0\\h}}^{\infty}{}' \left[\sum_{\substack{0\\k\\\text{even}}}^{\infty}{}' \quad A_{hk0} \cos 2\pi ky + B_{hk0} \sin 2\pi ky \right] \cos 2\pi hx\right.$$

$$\left. + \sum_{\substack{0\\h}}^{\infty}{}' \left[\sum_{\substack{0\\k\\\text{odd}}}^{\infty}{}' - A_{hk0} \sin 2\pi ky + B_{hk0} \cos 2\pi ky \right] \sin 2\pi hx \right\}. \tag{44}$$

Although this looks complicated, it requires no more work than (39). It does require separation of terms for k even and k odd.

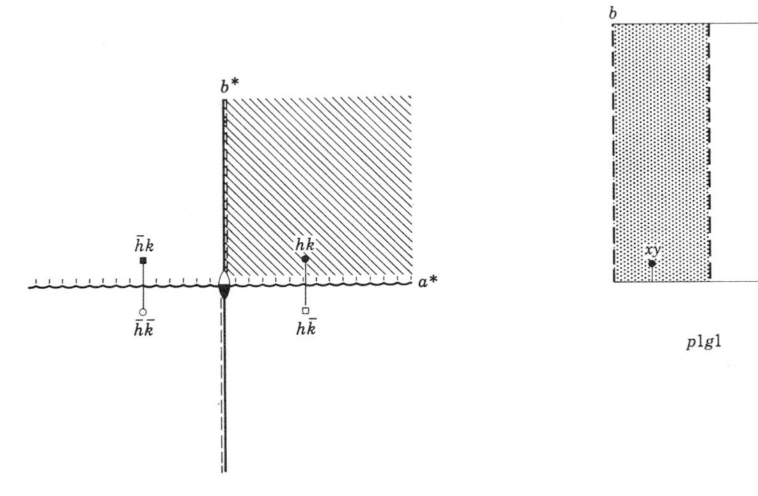

Fig. 6.

The shading in Fig. 6A shows that data are required from a quarter of reciprocal space. Figure 6B shows that the summation is required over a half the direct cell.

There is an alternative form of summation for this plane group which is more convenient for subsequent derivation of the form of $p2gg$. In that derivation it is desirable to have the origin of the plane group chosen half way between neighboring glide planes, that is, with the glide plane displaced $a/4$ from the origin. According to Theorem 5 of Chapter 16, this shift produces a phase change of $e^{i2\pi h/2}$. Accordingly (40) is replaced by

$$\begin{aligned} F_{\bar{h}k0} &= (F_{hk0}\, e^{i2\pi k/2})e^{i2\pi h/2} \\ &= F_{hk0}\, e^{i2\pi(h+k)/2}. \end{aligned} \tag{45}$$

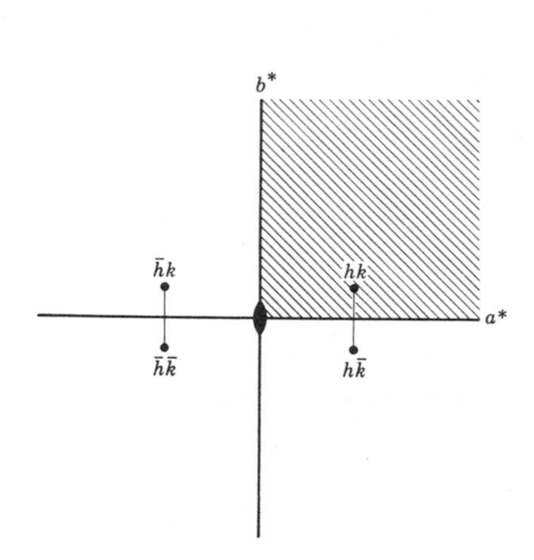

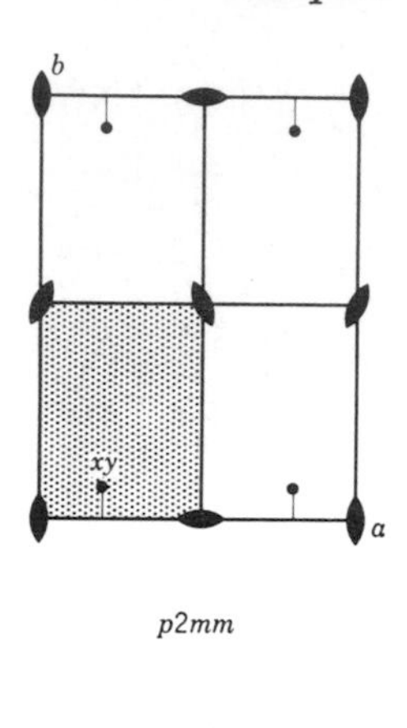

Fig. 7.

Using the same reasoning as before, the summation is split into two groups, according as $h+k$ is even or odd. The resulting summation is

$$\rho_{(xy)} = \frac{4}{S}\Bigg\{\sum_{\substack{0\\h\\h+k\text{ even}}}^{\infty}{}' \Bigg[\sum_{\substack{0\\k}}^{\infty}{}' \quad A_{hk0}\cos 2\pi ky + B_{hk0}\sin 2\pi ky\Bigg]\cos 2\pi hx$$

$$+ \sum_{\substack{0\\h\\h+k\text{ odd}}}^{\infty}{}' \Bigg[\sum_{\substack{0\\k}}^{\infty}{}' - A_{hk0}\sin 2\pi ky + B_{hk0}\cos 2\pi ky\Bigg]\sin 2\pi hx\Bigg\}. \quad (46)$$

Plane groups *p2mm* and *c2mm*. The distribution of symmetry elements in reciprocal space for *p2mm* and *c2mm* is the same, as shown in Figs. 7*A* and 8*A*. Plane group *p2mm* has subgroups *p2* and *p1m1*, while plane group *c2mm* has subgroups *c2* and *c1m1*. The forms from their summations can be derived from the first subgroup by adding *m* perpendicular to *a*, or from the second by adding 2 at the origin. By the first route, conditions (36) are imposed on (35), and by the second route, conditions (29) and (31) are imposed on (39). By either route the following form results:

$$\rho_{(xy)} = \frac{4}{S}\sum_{h}{}' \sum_{\substack{0\\k}}^{\infty}{}' F_{hk0}\cos 2\pi hx \cos 2\pi ky. \quad (47)$$

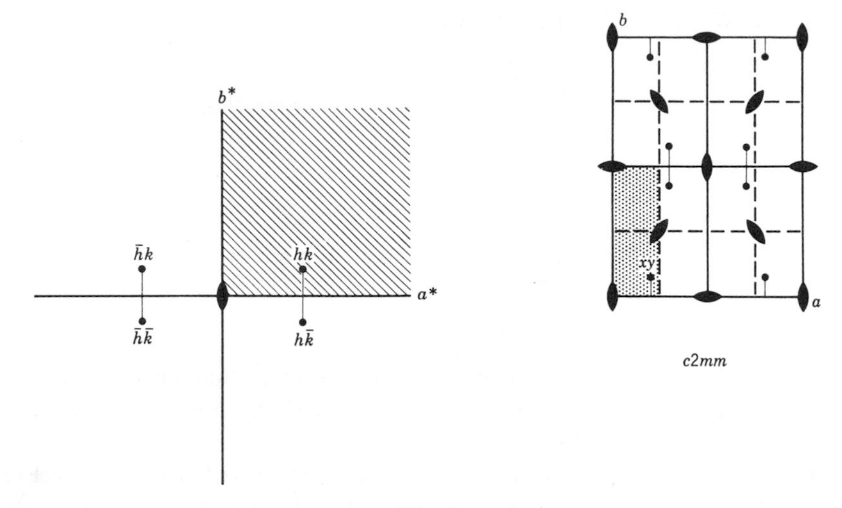

Fig. 8.

Figures 7*A* and 8*A* show that data for this summation are required from a quarter of reciprocal space. Figure 7*B* shows that for *p*2*mm* the summation is required over a quarter cell in direct space, while Fig. 8*B* shows it is required over only an eighth of a cell for *c*2*mm*.

Plane group *p*2*gm*. Plane group *p*2*gm* can be derived from its subgroup *p*1*g*1 adding the operation 2. If the 2-fold axis is taken as the origin, Fig. 9*B*, then this imposes conditions (29) and (31) on (44). This

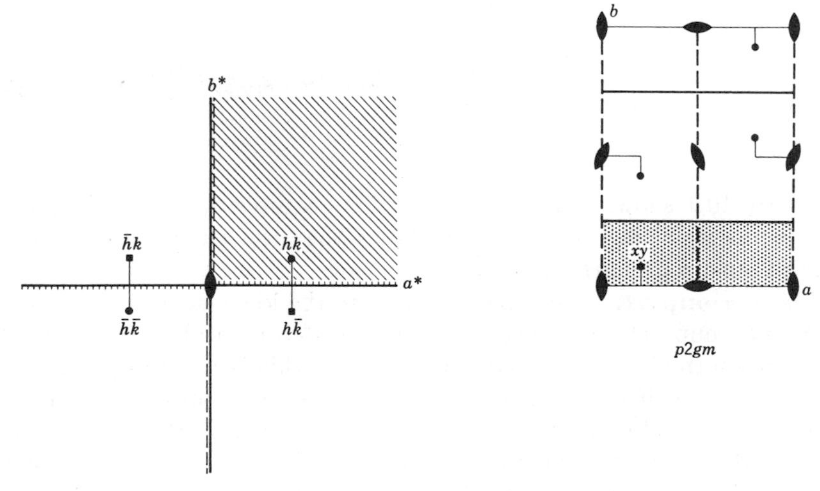

Fig. 9.

causes the B term to vanish, leaving

$$\rho_{(xy)} = \frac{4}{S}\left\{\sum_{h}^{\infty}{}' \sum_{\substack{k \\ 0 \\ k \text{ even}}}{}' F_{hk0} \cos 2\pi hx \cos 2\pi ky - \sum_{h}^{\infty}{}' \sum_{\substack{k \\ 0 \\ k \text{ odd}}}{}' F_{hk0} \sin 2\pi hx \sin 2\pi ky\right\}. \tag{48}$$

Figure 9A shows that data are required from a quarter of reciprocal space for the summation. Figure 9B shows that the summation must be made over a quarter cell in direct space.

Plane group *p*2*gg*. If the origin is placed at the 2-fold axis, Fig. 10A, then the glides are perpendicular to a and b and removed from the origins by $a/4$ and $b/4$. The form of summation for $p1g1$ with the glide removed from the origin by $a/4$ was given in (46). The form for $p2gg$ can be derived from this by imposing the conditions on reciprocal space required by a 2-fold axis at the origin of direct space. These conditions are given in (29) and (31). These cause the vanishing of B terms in (46), leaving

$$\rho_{(xy)} = \frac{4}{S}\left\{\sum_{h}^{\infty}{}' \sum_{\substack{k \\ 0 \\ h+k \text{ even}}}{}' F_{hk0} \cos 2\pi hx \cos 2\pi ky - \sum_{h}^{\infty}{}' \sum_{\substack{k \\ 0 \\ h+k \text{ odd}}}{}' F_{hk0} \sin 2\pi hx \sin 2\pi ky\right\}. \tag{49}$$

Figure 10A shows that this summation requires data from a quarter of reciprocal space. Figure 10B shows that the summation must be carried out for a quarter cell in direct space.

Plane group *p*4. Since $p4$ conforms to the less restrictive symmetry of its subgroup, $p2$, the electron density for $p4$ can also be synthesized by using the form for $p2$, namely (35). When this is done, data must be derived from half of reciprocal space, that is from both the hk and $\bar{h}k$ fields, Fig. 3. But because of the 4-fold symmetry, the motif unit of reciprocal space is contained in a single 90° sector. To take advantage of this, let (35) be written so that the summation over the 180° sector is separated into two summations, one over each of the differently shaded

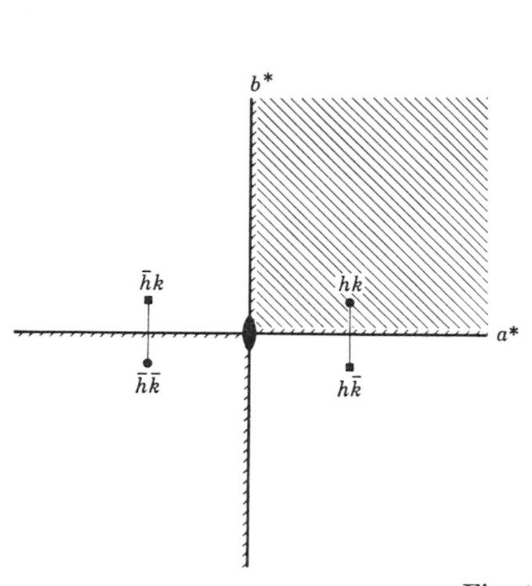

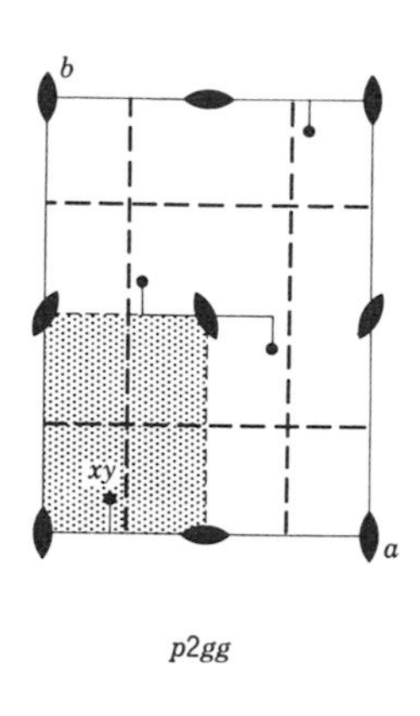

Fig. 10.

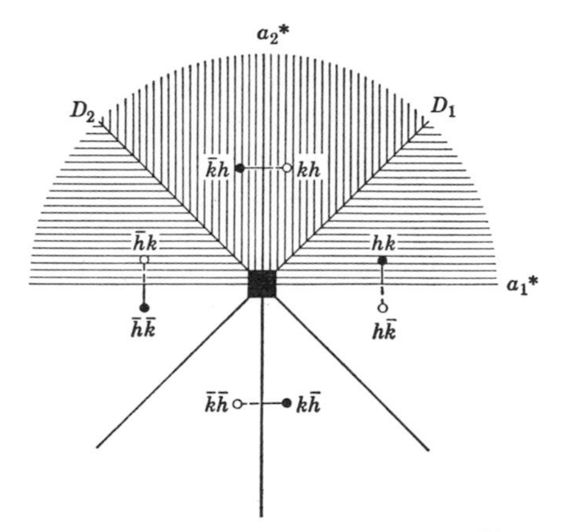

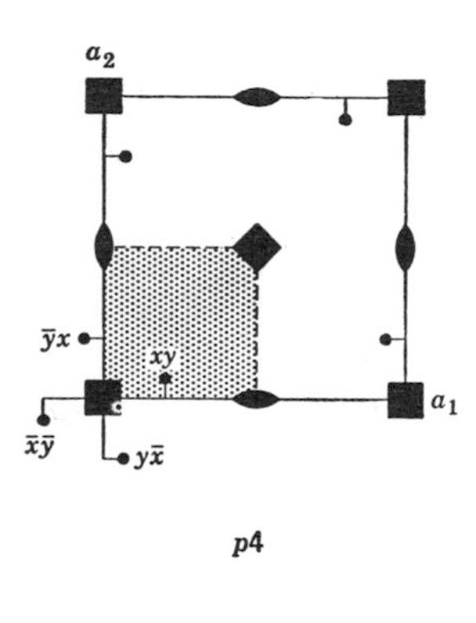

Fig. 11.

regions of Fig. 11*A*. The summation can then be written

$$\rho_{(xy)} = \frac{2}{S} \sum_{\substack{D_1 \\ h}}^{\infty}{}' \sum_{\substack{0 \\ k}}^{D_1}{}' \{(F_{hk0} + F_{\bar{h}k0}) \cos 2\pi hx \cos 2\pi ky - (F_{hk0} - F_{\bar{h}k0}) \sin 2\pi hx \sin 2\pi ky\}$$

$$+ \frac{2}{S} \sum_{\substack{0 \\ h}}^{D_1}{}' \sum_{\substack{D_1 \\ k}}^{\infty}{}' \{(F_{hk0} + F_{\bar{h}k0}) \cos 2\pi hx \cos 2\pi ky - (F_{hk0} - F_{\bar{h}k0}) \sin 2\pi hx \sin 2\pi ky\}. \quad (50)$$

Theorem 2 can now be applied: Because of the 4-fold symmetry of both reciprocal space and direct space, the second part of (50) produces the same effect at xy as the first part does at $\bar{y}x$ (See Fig. 11). Therefore, for the second part of (50) there may be substituted a summation like the first part of (50) except that x becomes $\bar{y}$ and y becomes x. When this is done, (50) can be rewritten as

$$\begin{aligned}\rho_{(xy)} = \frac{2}{S}\sum_{h=D_1}^{\infty}{}' \sum_{k=0}^{D_1}{}' \{ &(F_{hk0} + F_{\bar{h}k0}) \cos 2\pi hx \cos 2\pi ky \\ &- (F_{hk0} - F_{\bar{h}k0}) \sin 2\pi hx \sin 2\pi ky \\ &+ (F_{hk0} + F_{\bar{h}k0}) \cos 2\pi h\bar{y} \cos 2\pi kx \\ &- (F_{hk0} - F_{\bar{h}k0}) \sin 2\pi h\bar{y} \sin 2\pi kx \}. \end{aligned} \tag{51}$$

This can be consolidated to

$$\begin{aligned}\rho_{(xy)} = \frac{2}{S}\sum_{h=D_1}^{\infty}{}' \sum_{k=0}^{D_1}{}' \{ &(F_{hk0} + F_{\bar{h}k0})(\cos 2\pi hx \cos 2\pi ky + \cos 2\pi kx \cos 2\pi hy) \\ &- (F_{hk0} - F_{\bar{h}k0})(\sin 2\pi hx \sin 2\pi ky - \sin 2\pi kx \sin 2\pi hy)\}. \end{aligned} \tag{52}$$

This form of the synthesis uses data from only a single motif unit of reciprocal space, and may therefore be regarded as an elegant form for $p4$. Nevertheless the work of summing (35), from which (51) was developed, is probably no greater.

Plane group *p*4*mm*. Plane group $p4mm$ has $p4$ for a subgroup and can be derived from it by the addition of a mirror normal to a. The consequence of this in reciprocal space, Fig. 12*A*, is

$$F_{\bar{h}k0} = F_{hk0}. \tag{53}$$

If this condition is applied to (35), it reduces to

$$\rho_{(xy)} = \frac{4}{S}\sum_{h}{}' \sum_{k}{}'^{\infty}_{0} F_{hk0} \cos 2\pi hx \cos 2\pi ky. \tag{54}$$

This summation is over the range 0 to ∞ for both h and k, and therefore requires a number of terms equal to the number of points in one quarter of reciprocal space. Figure 12*A* shows that only the data from one-eighth of reciprocal space should be necessary. The less condensed form

(54) resulted because it was based upon the addition of a reflection plane to (35), the form for $p2$.

To limit the summation to a motif unit of reciprocal space, the reflection plane can be used to place restriction (53) on the properly condensed form for $p4$, namely (52). When this is done, the condensed form for symmetry $p4mm$ is obtained, as follows:

$$\rho_{(xy)} = \frac{4}{S} \sum_{h}^{\infty}{}' \sum_{k}^{D}{}' F_{hk0}(\cos 2\pi hx \cos 2\pi ky + \cos 2\pi kx \cos 2\pi hy). \tag{55}$$

(In the printed formula the summation limits are written as $\sum_{D,h}^{\infty}$ and $\sum_{0,k}^{D}$.)

This form is formally satisfactory because it uses only those terms in the shaded eighth of reciprocal space. But the number of summations

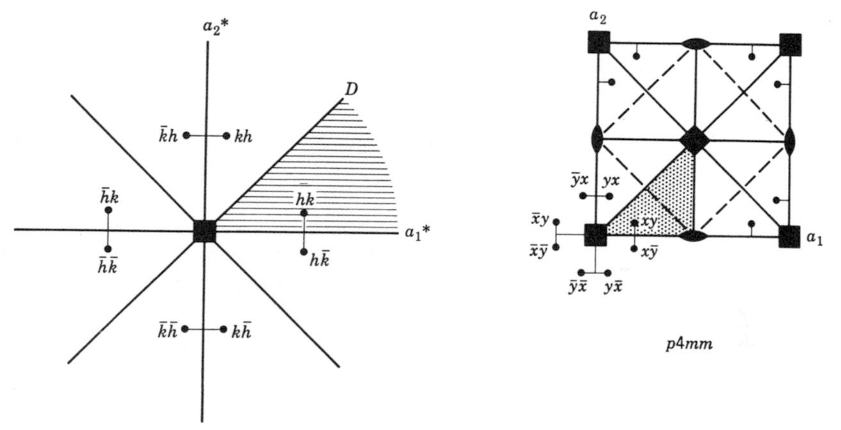

Fig. 12.

is the same as in (54), and (54) has a simpler general form which is preferable.

The data required for summation (54) must be obtained from one-quarter of reciprocal space. The data required for (55), on the other hand, is obtained from one-eighth of reciprocal space. The summations must be carried out, according to Fig. 12*B*, for one-eighth of the cell. (An alternative procedure, making use of Theorem 2, is to perform only the first term of (55), but to carry it out for a quarter cell of Fig. 12*B*, then add the results at xy and yx.)

Plane group *p4gm*. Plane group $p4gm$, Fig. 13*B*, can be derived from its subgroup $p4$ by the addition of a glide plane through the origin, diagonal to the axes, and having glide component $a/2 + b/2$. According to Theorem 4 of Chapter 16, this gives rise to the following relation in reciprocal space, Fig. 13*A*:

$$F_{kh0} = F_{hk0}\, e^{i2\pi(h+k)/2}. \tag{56}$$

It is therefore convenient to distinguish two classes of reflections, for which the conditions are

$$\text{for } h+k \text{ even:} \qquad F_{kh0}(=F_{\bar{h}k0}) = F_{hk0}, \tag{57}$$

$$\text{for } h+k \text{ odd:} \qquad F_{kh0}(=F_{\bar{h}k0}) = -F_{hk0}. \tag{58}$$

As in the cases of $p4$ and $p4mm$, forms for the synthesis of $p4gm$ can be derived which use either a double motif unit, or only a single motif unit of reciprocal space. To derive the form using the double motif unit (a 90° sector in this case), conditions (57) and (58) are applied to (35), giving

$$\rho_{(xy)} = \frac{4}{S}\left\{\sum_{\substack{h \\ h+k \text{ even}}}' \sum_{k=0}^{\infty}{}' F_{hk0} \cos 2\pi hx \cos 2\pi ky - \sum_{\substack{h \\ h+k \text{ odd}}}' \sum_{k=0}^{\infty}{}' F_{hk0} \sin 2\pi hx \sin 2\pi ky\right\}. \tag{59}$$

The form requiring only a single motif unit of reciprocal space may be derived by applying the special conditions (57) and (58) to (52). This gives

$$\rho_{(xy)} = \frac{4}{S}\left\{\sum_{\substack{h=D \\ h+k \text{ even}}}^{\infty}{}' \sum_{k=0}^{D}{}' F_{hk0}(\cos 2\pi hx \cos 2\pi ky + \cos 2\pi kx \cos 2\pi hy) - \sum_{\substack{h=D \\ h+k \text{ odd}}}^{\infty}{}' \sum_{k=0}^{\infty}{}' F_{hk0}(\sin 2\pi hx \sin 2\pi ky - \sin 2\pi kx \sin 2\pi hy)\right\}. \tag{60}$$

This summation makes use of data in one-eighth of reciprocal space, Fig. 13*A*. It is carried out for one-eighth of the cell, Fig. 13*B*.

Plane group *p*3. The only subgroup of $p3$ is the trivial group $p1$. The form for the summation of $p3$ can best be derived as the sum of the summations for the three sector pairs for the fields hk, ki, and ih, Fig. 14*A*. The general form for each sector pair follows (28). If the partial synthesis due to a single sector of field hk is designated $\partial_{hk}\,\rho_{(xy)}$, then the full density is given by

$$\rho_{(xy)} = \partial_{hk}\,\rho_{(xy)} + \partial_{ki}\,\rho_{(xy)} + \partial_{ih}\,\rho_{(xy)}, \tag{61}$$

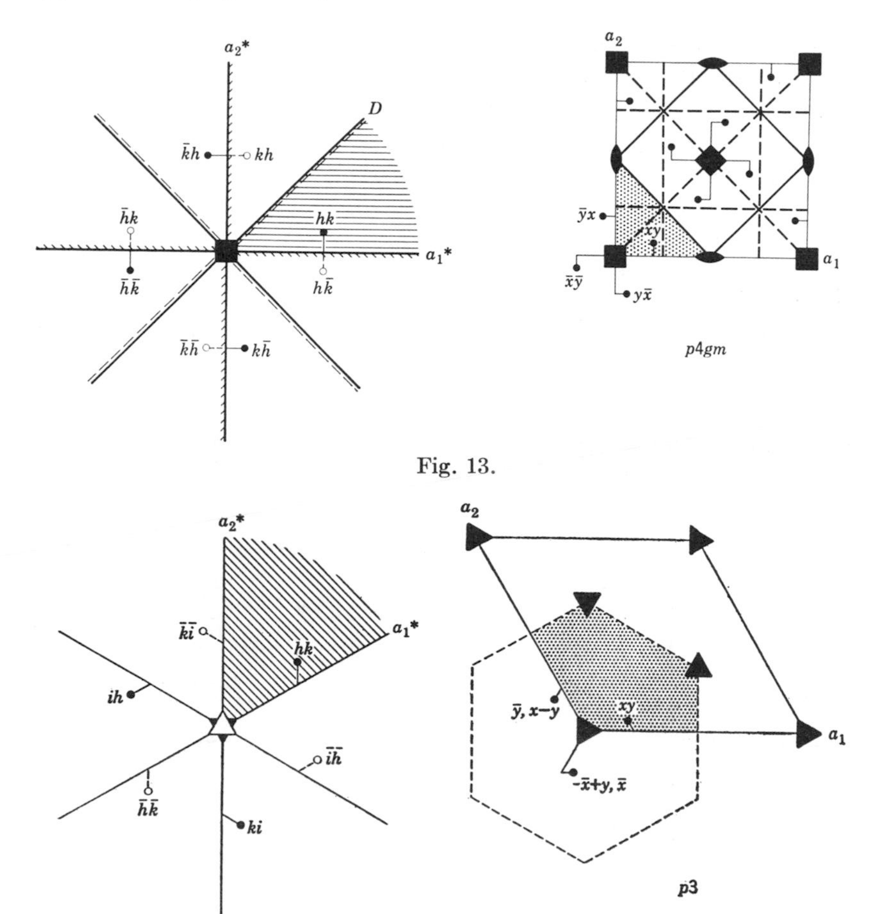

Fig. 13.

Fig. 14.

where, for example, following (28),

$$\partial_{hk}\,\rho_{(xy)} = \frac{2}{S}\sum_{h}^{\infty}{}'\sum_{k}{}' \{A_{hk0}\cos 2\pi hx \cos 2\pi ky - A_{hk0}\sin 2\pi hx \sin 2\pi ky + B_{hk0}\sin 2\pi hx \cos 2\pi ky + B_{hk0}\cos 2\pi hx \sin 2\pi ky.\} \tag{62}$$

Figure 14*A* shows that partial summation (62) requires data from only one-sixth of reciprocal space. But to find the full density, (61), the summation must be performed for three pairs of sectors. According to Fig. 14*B*, the summation must be carried out for one-third of a cell.

Alternatively, Theorem 2 can be used. According to this theorem, only the summation for one sector, (62), need be carried out, but it must be summed over the full cell. The final density at xy is the sum of the partial densities at the three equivalent points shown in Fig. 14B. Using the notation of (61), one obtains

$$\rho_{(xy)} = \partial_{hk}\,\rho_{(xy)} + \partial_{hk}\,\rho_{(\bar{y},\,x-y)} + \partial_{hk}\,\rho_{(\bar{x}+y,\,\bar{x})}. \tag{63}$$

Plane group *p*31*m*. In reciprocal space, Fig. 15A, the symmetry of $p31m$ provides the relation

$$F_{hk0} = F_{kh0}. \tag{64}$$

This relation could have been deduced from (55) for $p4mm$, and it can be used to condense the hk sector of Fig. 15A so that each term of (62) for

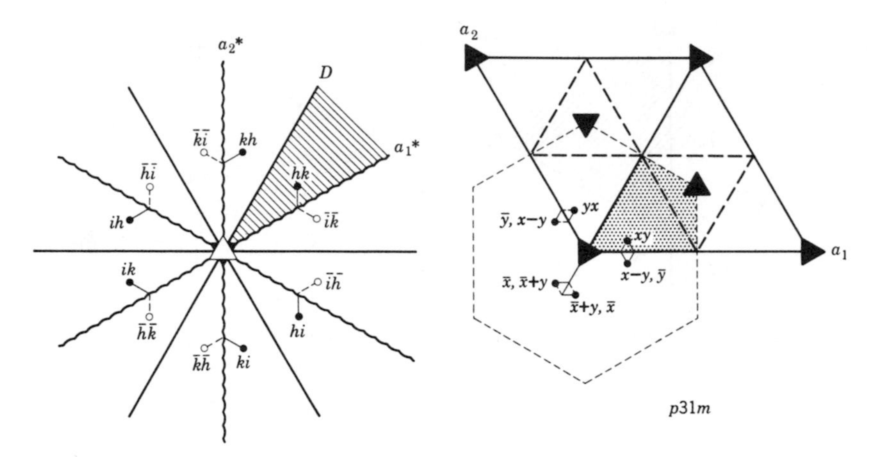

Fig. 15.

$p3$ takes a form like (55). But a much simpler procedure is to make use of Theorem 2 in the following way. The motif sectors of reciprocal space are six pairs such as the one shaded in Fig. 15A. This is bounded by $a_1{}^*$ and the diagonal D. For this sector pair the synthesis follows (28):

$$\begin{aligned}\partial_{hD}\,\rho_{(xy)} = \frac{2}{S}\sum_{h=D}^{\infty}{}' \sum_{k=0}^{D}{}' \{&A_{hk0}\cos 2\pi hx\cos 2\pi ky \\ &- A_{hk0}\sin 2\pi hx\sin 2\pi ky \\ &+ B_{hk0}\sin 2\pi hx\cos 2\pi ky \\ &+ B_{hk0}\cos 2\pi hx\sin 2\pi ky\}.\end{aligned} \tag{65}$$

This partial summation must be carried out over the full area of the cell. Then the full summation is the result of recording at xy the values found for all six equivalent points shown in Fig. 15*B*.

Plane group *p*3*m*1. In reciprocal space, Fig. 16*A*, the symmetry of $p3m1$ provides the relation

$$F_{kh0} = F_{hk0}\, e^{i2\pi(h+k)/2}. \tag{66}$$

Therefore,

$$\text{for } h+k \text{ even:} \qquad F_{kh0} = F_{hk0}; \tag{67}$$

$$\text{for } h+k \text{ odd:} \qquad F_{kh0} = -F_{hk0}. \tag{68}$$

Again these relations enable writing the partial density due to a sector pair like (62) in condensed form. But a more straightforward summation can

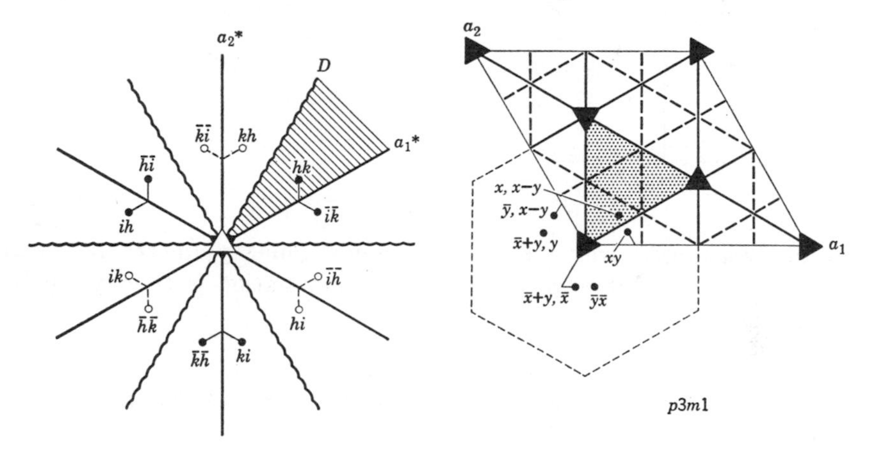

Fig. 16.

be performed by making use of Theorem 2. For this purpose a sector is again chosen between a^* and the diagonal D, Fig. 16*A*. The form of the summation is then the same as (65). This must be summed for all sample points in the full cell. The full density is found by recording at xy the partial densities found at the six equivalent points in Fig. 16*B*.

Plane group *p*6. This symmetry has $p2$ as a subgroup so that the form for a sector pair is given by (32). It would appear that the best way to take advantage of the hexagonal symmetry would be to perform the summation for the field of the hexagonal sector pair of Fig. 17*A*. The summation then has the form

$$\partial_{hk}\, \rho_{(xy)} = \frac{2}{S} \sum_{h}{}' \sum_{k}{}'_{0}^{\infty} \{F_{hk0} \cos 2\pi hx \cos 2\pi ky$$

$$- F_{hk0} \sin 2\pi hx \sin 2\pi ky\}. \tag{69}$$

The number of terms in the summation is the number of F_{hk0}'s in the hexagonal sector. This must be summed over the full hexagonal cell, and, to obtain the full summation, the results at the three points xy; $\bar{y}, x-y$; and $\bar{x}+y, \bar{x}$ must be added and recorded at xy.

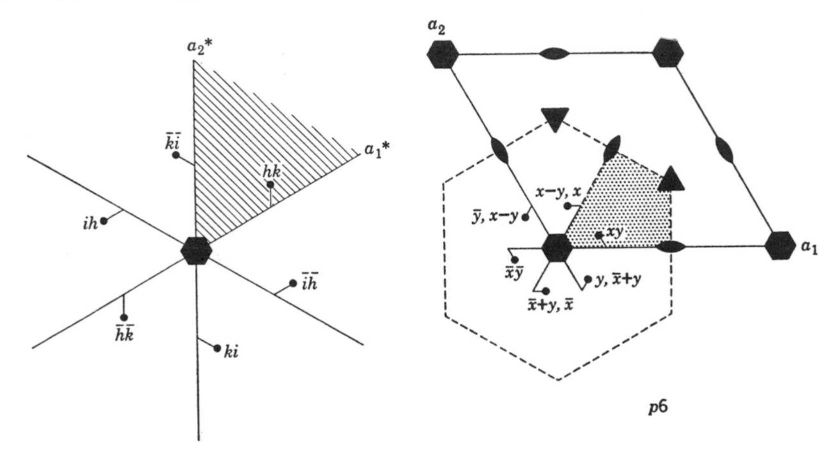

Fig. 17.

Actually a simpler way is to ignore the hexagonal symmetry and treat the symmetry as if it were $p2$. The full summation is then given by (35), namely

$$\rho_{(xy)} = \frac{2}{S} \sum_{h}' \sum_{0\; k}^{\infty}{}' \{(F_{hk0} + F_{\bar{h}k0}) \cos 2\pi hx \cos 2\pi ky - (F_{hk0} - F_{\bar{h}k0}) \sin 2\pi hx \sin 2\pi ky\}. \tag{70}$$

This needs be summed only for one-sixth of the cell, as shaded in Fig. 17*B*. Summation (70) therefore need be made for only a fraction of the number of direct-cell points as that required for (69). In other respects (69) and (70) are equally easy, since both have the same number of summation terms.

Plane group *p6mm*. This symmetry has subgroups $p2$, $p3$, $p1m1$, $p2mm$, $p3m1$, and $p31m$, so that the Fourier summation for it can be performed in a large number of ways. Perhaps the simplest way is to treat it as $p2mm$, chosing a_1* and D_2 as reciprocal axes, Fig. 18*A*. The form of the summation is the same as (47), namely

$$\rho_{(xy)} = \frac{4}{S} \sum_{h}' \sum_{0\; k'}^{\infty}{}' F_{hk'0} \cos 2\pi hx \cos 2\pi k'y. \tag{71}$$

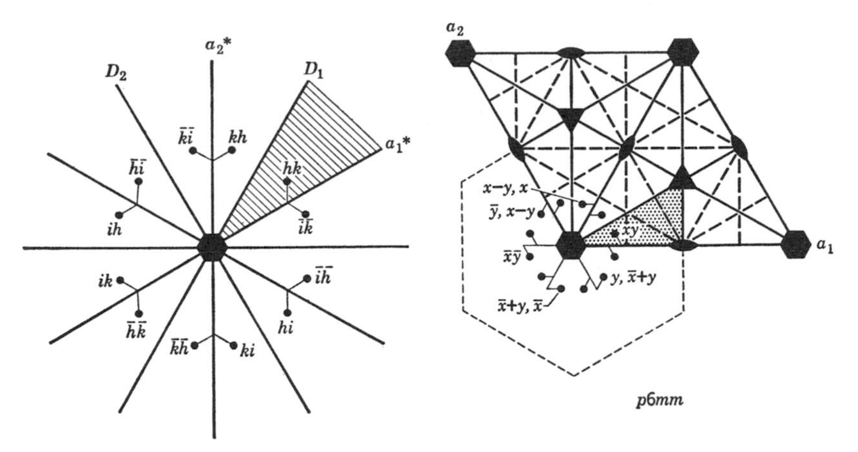

Fig. 18.

The summation need be carried out only for the points within $\frac{1}{12}$ of the unit cell, as shaded in Fig. 18*B*.

Alternatively, full advantage may be taken of the symmetry *p6mm* in reciprocal space. In this case the summation is performed for a hexagonal sector pair. Since *p6mm* provides

$$F_{hk0} = F_{kh0},$$

a summation form analogous to (55) can be written for each sector pair, specifically

$$\partial_{hk}\, \rho_{(xy)} = \frac{2}{S} \sum_{h}^{\infty}{}' \sum_{k}^{D_1}{}' F_{hk0}(\cos 2\pi hx \cos 2\pi ky + \cos 2\pi kx \cos 2\pi hy). \quad (72)$$

with lower limits D_1 (for h) and 0 (for k).

The full summation is given by adding the results at xy; $\bar{y}, x-y$; $\bar{x}+y, \bar{x}$ and recording it at xy.

Three-dimensional summations

Three-dimensional summations, designed for computation one level at a time, can be referred to the pattern established for two-dimensional summations. For sake of clearness, assume that the electron density is desired for a level z_1. Then z_1 is constant for the purposes of the summation.

The general form of the Fourier summation for a pair of opposite segments was given in (24). Since z_1 is constant for the level, it is convenient to separate the trigonometric parts of (24) into variable and

Table 1
Forms of Fourier syntheses for symmetrical projections

Symmetry of projection	Form of synthesis	
$p1$	$\frac{2}{S}\sum_{h}{}' \sum_{k=0}^{\infty}{}' \{(A_{hk0}+A_{\bar{h}k0}) \cos 2\pi hx \cos 2\pi ky$ $- (A_{hk0}-A_{\bar{h}k0}) \sin 2\pi hx \sin 2\pi ky$ $+ (B_{hk0}-B_{\bar{h}k0}) \sin 2\pi hx \cos 2\pi ky$ $+ (B_{hk0}+B_{\bar{h}k0}) \cos 2\pi hx \cos 2\pi ky\}.$	(34)
$p2$	$\frac{2}{S}\sum_{h}{}' \sum_{k=0}^{\infty}{}' \{(F_{hk0}+F_{\bar{h}k0}) \cos 2\pi hx \cos 2\pi ky$ $- (F_{hk0}-F_{\bar{h}k0}) \sin 2\pi hx \sin 2\pi ky\}.$	(35)
$p1m1$, $c1m1$	$\frac{4}{S}\sum_{h=0}^{\infty}{}' \left[\sum_{k=0}^{\infty}{}' A_{hk0} \cos 2\pi ky + B_{hk0} \sin 2\pi ky\right] \cos 2\pi hx.$	(39)
$p1g1$	$\frac{4}{S}\left\{\sum_{h=0}^{\infty}{}' \left[\sum_{k=0,\ k\ \text{even}}^{\infty}{}' A_{hk0} \cos 2\pi ky + B_{hk0} \sin 2\pi ky\right] \cos 2\pi hx.\right.$ $\left.- \sum_{h=0}^{\infty}{}' \left[\sum_{k=0,\ k\ \text{odd}}^{\infty}{}' A_{hk0} \sin 2\pi ky - B_{hk0} \cos 2\pi ky\right] \sin 2\pi hx\right\}.$	(44)

Table 1 (*Continued*)

Symmetry of projection	Form of synthesis	
$p2mm$, $c2mm$	$\frac{4}{S}\sum_{h}{}^{'}\sum_{k}{}^{'}{}_{0}^{\infty} F_{hk0} \cos 2\pi hx \cos 2\pi ky.$	(47)
$p2gm$	$\frac{4}{S}\left\{\sum_{h}{}^{'}\sum_{k}{}^{'}{}_{0}^{\infty} F_{hk0} \cos 2\pi hx \cos 2\pi ky\right.$ (k even)	
	$\left.-\sum_{h}{}^{'}\sum_{k}{}^{'}{}_{0}^{\infty} F_{hk0} \sin 2\pi hx \sin 2\pi ky\right\}.$ (k odd)	(48)
$p2gg$	$\frac{4}{S}\left\{\sum_{h}{}^{'}\sum_{k}{}^{'}{}_{0}^{\infty} F_{hk0} \cos 2\pi hx \cos 2\pi ky\right.$ ($h+k$ even)	
	$\left.-\sum_{h}{}^{'}\sum_{k}{}^{'}{}_{0}^{\infty} F_{hk0} \sin 2\pi hx \sin 2\pi ky\right\}.$ ($h+k$ odd)	(49)

Table 1 (*Continued*)

Symmetry of projection	Form of synthesis
$p4$	$\frac{2}{S}\sum_{h}{}'\sum_{k}{}'_{0}^{\infty} \{(F_{hk0}+F_{\bar{h}k0}) \cos 2\pi hx \cos 2\pi ky - (F_{hk0}-F_{\bar{h}k0}) \sin 2\pi hx \sin 2\pi ky\}.$ (35)
or:	$\frac{2}{S}\sum_{h=D_1}^{\infty}{}'\sum_{k=0}^{D_1}{}' \{(F_{hk0}+F_{\bar{h}k0})(\cos 2\pi hx \cos 2\pi ky + \cos 2\pi kx \cos 2\pi hy) - (F_{hk0}-F_{\bar{h}k0})(\sin 2\pi hx \sin 2\pi ky - \sin 2\pi kx \sin 2\pi hy)\}.$ (52)
$p4mm$	$\frac{4}{S}\sum_{h}{}'\sum_{k}{}'_{0}^{\infty} F_{hk0} \cos 2\pi hx \cos 2\pi ky$ (54)
or:	$\frac{4}{S}\sum_{h=D}^{\infty}{}'\sum_{k=0}^{D}{}' F_{hk0} (\cos 2\pi hx \cos 2\pi ky + \cos 2\pi kx \cos 2\pi hy).$ (55)
$p4gm$	$\frac{4}{S}\left\{\sum_{h}{}'\sum_{k}{}'_{0,\ h+k \text{ even}}^{\infty} F_{hk0} \cos 2\pi hx \cos 2\pi ky - \sum_{h}{}'\sum_{k}{}'_{0,\ h+k \text{ odd}}^{\infty} F_{hk0} \sin 2\pi hx \sin 2\pi ky\right\}$ (59)
or:	$\frac{4}{S}\left\{\sum_{h=D}^{\infty}{}'\sum_{k=0}^{D}{}'_{h+k \text{ even}} F_{hk0} (\cos 2\pi hx \cos 2\pi ky + \cos 2\pi kx \cos 2\pi hy) - \sum_{h=D}^{\infty}{}'\sum_{k=0}^{D}{}'_{h+k \text{ odd}} F_{hk0} (\sin 2\pi hx \sin 2\pi ky - \sin 2\pi kx \sin 2\pi hy)\right\}.$ (60)

Table 1 (*Continued*)

Symmetry of projection	Form of synthesis
$p3$	$$\frac{2}{S}\sum_{h}^{\infty}{}'\sum_{k}{}'\{A_{hk0}\cos 2\pi hx\cos 2\pi ky - A_{hk0}\sin 2\pi hx\sin 2\pi ky + B_{hk0}\sin 2\pi hx\cos 2\pi ky + B_{hk0}\cos 2\pi hx\sin 2\pi ky\}. \quad (62)$$ (lower limit 0) Add the values for xy; $\bar{y}, x-y$; $\bar{x}+y, \bar{x}$ and record the sum at xy.
$p31m$	$$\frac{2}{S}\sum_{D\,h}^{\infty}{}'\sum_{0\,k}^{D}{}'\{A_{hk0}\cos 2\pi hx\cos 2\pi ky - A_{hk0}\sin 2\pi hx\sin 2\pi ky + B_{hk0}\sin 2\pi hx\cos 2\pi ky + B_{hk0}\cos 2\pi hx\sin 2\pi ky\}. \quad (65)$$ D is the reciprocal-cell diagonal along [110]*. Add the values for xy; $\bar{y}, x-y$; $\bar{x}+y, \bar{x}$; yx; $\bar{x}, \bar{x}+y$; $x-y, \bar{y}$ and record the sum at xy.
$p3m1$	Same as for $p31m$, except: Add the values for xy; $\bar{y}, x-y$; $\bar{x}+y, \bar{x}$; $\bar{y}\bar{x}$; $x, x-y$; $\bar{x}+y, y$ and record the sum at xy.
$p6$	$$\frac{2}{S}\sum_{h}^{\infty}{}'\sum_{k}{}'\{(F_{hk0}+F_{\bar{h}k0})\cos 2\pi hx\cos 2\pi ky - (F_{hk0}-F_{\bar{h}k0})\sin 2\pi hx\sin 2\pi ky\}. \quad (70)$$ (lower limit 0) or: $$\frac{2}{S}\sum_{h}^{\infty}{}'\sum_{k}{}'\{F_{hk0}\cos 2\pi hx\cos 2\pi ky - F_{hk0}\sin 2\pi hx\sin 2\pi ky\}. \quad (69)$$ (lower limit 0) Add the values at yx; $\bar{y}, x-y$; $\bar{x}+y, \bar{x}$ and record the sum at xy.

Table 1 (*Continued*)

Symmetry of projection	Form of synthesis
$p6mm$	$\frac{4}{S}\sum_{h=0}^{\infty}{}'\sum_{k'}{}' F_{hk'0}\cos 2\pi hx \cos 2\pi k'y.$ (71) h and k' are referred to orthogonal axes a^* and b'^*
or:	$\frac{2}{S}\sum_{h=D}^{\infty}{}'\sum_{k=0}^{D}{}' F_{hk0}(\cos 2\pi hx \cos 2\pi ky + \cos 2\pi kx \cos 2\pi hy).$ (72) D is the reciprocal-cell diagonal along [110]*. Add the values at xy; $\bar{y}, x-y$; $\bar{x}+y, \bar{x}$ and record the results at xy.

constant parts. This can be done by using the trigonometric identities (26) and (27). When these are applied to (24) it becomes

$$\begin{aligned}\rho_{(xyz_1)} = \frac{2}{V}\sum_h{}'\sum_k{}'\sum_l{}' \{&A_{hkl}\cos 2\pi(hx+ky)\cos 2\pi lz_1 \\ &-A_{hkl}\sin 2\pi(hx+ky)\sin 2\pi lz_1 \\ &+B_{hkl}\sin 2\pi(hx+ky)\cos 2\pi lz_1 \\ &+B_{hkl}\cos 2\pi(hx+ky)\sin 2\pi lz_1\}. \end{aligned} \quad (73)$$

Now, for any selected level z_1 the following parts of (73) are fixed and can be computed in advance of the main computation:

$$\begin{aligned}\sum_l{}' A_{hkl}\cos 2\pi lz_1 &\equiv C_{A,hk}\\ \sum_l{}' A_{hkl}\sin 2\pi lz_1 &\equiv S_{A,hk}\\ \sum_l{}' B_{hkl}\cos 2\pi lz_1 &\equiv C_{B,hk}\\ \sum_l{}' B_{hkl}\sin 2\pi lz_1 &\equiv S_{B,hk}.\end{aligned} \quad (74)$$

Each of these can be computed, and the results inserted into (77), which then takes the simpler form

$$\rho_{(xyz_1)} = \frac{2}{V}\sum_h{}' \sum_k{}' \{C_{A,hk}\cos 2\pi(hx+ky) - S_{A,hk}\sin 2\pi(hx+ky) + C_{B,hk}\sin 2\pi(hx+ky) + S_{B,hk}\cos 2\pi(hx+ky)\} \tag{75}$$

$$= \frac{2}{V}\sum_h{}' \sum_k{}' \{(C_A+S_B)_{hk}\cos 2\pi(hx+ky) + (C_B-S_A)_{hk}\sin 2\pi(hx+ky)\}. \tag{76}$$

By using identities (26) and (27) again, this can be written in product form:

$$\rho_{(xyz_1)} = \frac{2}{V}\sum_h{}' \sum_k{}' \{(C_A+S_B)_{hk}\cos 2\pi hx \cos 2\pi ky - (C_A+S_B)_{hk}\sin 2\pi hx \sin 2\pi ky + (C_B-S_A)_{hk}\sin 2\pi hx \cos 2\pi ky + (C_B-S_A)_{hk}\cos 2\pi hx \sin 2\pi ky\}. \tag{77}$$

This has the same form as that for the general two-dimensional case, (28), except that

$$A_{hk0} \to (C_A+S_B)_{hk}, \quad B_{hk0} \to (C_B-S_A)_{hk}. \tag{78}$$

The detailed forms of the Fourier summation for the 230 space groups are listed by Lonsdale[2] and in the *International tables*.[4]

Literature

[1] C. A. Beevers and H. Lipson. *A rapid method for the summation of two-dimensional Fourier series.* Phil. Mag. (7) **17** (1934) 855–859.

[2] Kathleen Lonsdale. *Simplified structure factor and electron density formulae for the 230 space groups of mathematical crystallography.* (G. Bell and Sons London, 1936) especially 7–8.

[3] M. J. Buerger. *Fourier summations for symmetrical crystals.* Am. Mineralogist **34** (1949) 771–788.

[4] Norman F. M. Henry and Kathleen Lonsdale. *International tables for x-ray crystallography.* Vol. I. (Kynoch Press, Birmingham, England, 1952) 353–525, especially 353–366.

18

Practical methods of summing Fourier series

The last five chapters have been devoted to some of the theoretical aspects of Fourier series. In this chapter some of the practical procedures for performing the actual numerical summations are considered.

The large number of summations which would be required in a straightforward Fourier summation was emphasized in the beginning of the last chapter. That chapter was devoted to showing how the number of summations actually required could be reduced by taking advantage of the symmetry of the crystal. There are also some simple mathematical devices which considerably reduce the labor of summations. These were introduced in the early works of Beevers and Lipson[1–3] as well as Robertson[4] and, in part, have been incorporated in the forms used in the last chapter. The basic mathematical devices are as follows:

a. The trigonometric part of the summation may be recast into product form. This has an advantage in computation since it reduces a two-dimensional summation into products of one-dimensional summations.

b. Trigonometric functions of positive and negative indices may be transformed into functions of positive indices only. This permits a reduction of the summation by bulking together functions of positive and negative angles.

c. Advantage can be taken of the symmetrical or antisymmetrical properties of the trigonometric function about the angles 0, $\frac{1}{4}\cdot 2\pi$, $\frac{1}{2}\cdot 2\pi$, $\frac{3}{4}\cdot 2\pi$, and 2π. If the computation is properly designed, this permits limiting the strictly Fourier part of the summation to the range 0 to $\frac{1}{4}$ of the cell edge.

Sampling of the Fourier transform

A Fourier function, such as the one which provides the value of $\rho_{(xy)}$, does so for continuously variable values of the parameters x and y. But if one is going to compute $\rho_{(xy)}$ by digital methods, he cannot do this for the infinity of values of parameters. As a practical matter it is necessary to select a limited set of samples of x and y at intervals sufficiently close so that values of the function do not vary greatly between sample points. The values of the function for xy positions lying between sampled values can be found by some sort of interpolation. The actual interpolation is commonly accomplished by drawing contours of the function $\rho_{(xy)}$ using as data points the values found at the sampled coordinates xy.

The fineness of the interval chosen between sampled values depends on the amount of detail required in the synthesis. It is desirable that, whatever the interval chosen, the sampling should explicitly include the fractions $\frac{1}{2}$, $\frac{1}{3}$, and $\frac{1}{4}$, since these are commonly equipoint locations. Thus intervals of $\frac{1}{100}$ of the cell edge are not suitable, while intervals of $\frac{1}{120}$, $\frac{1}{60}$, $\frac{1}{48}$, $\frac{1}{36}$, and $\frac{1}{24}$th of the cell edge are normally used. Perhaps the interval of $\frac{1}{60}$ is currently most common.

Use of the symmetrical properties of the trigonometric functions

By making use of the symmetrical and antisymmetrical properties of the cosine and sine, the actual trigonometric part of the summation

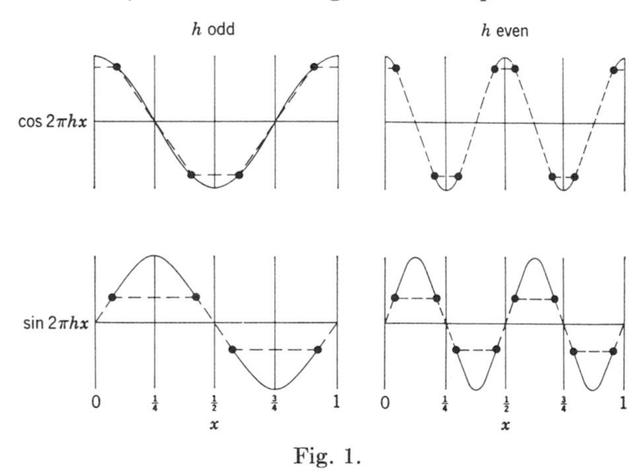

Fig. 1.

involved in a Fourier summation need not be carried beyond $x = \frac{1}{4}$, $y = \frac{1}{4}$. The summation beyond these limits is always related in a simple way to the summation up to these limits. The basis of this is as follows:

In Fig. 1 are shown graphs of the functions $\cos 2\pi hx$ and $\sin 2\pi hx$ for

Table 1
Coordinates of symmetrical and antisymmetrical points for trigonometric functions

	$\cos 2\pi hx$		$\sin 2\pi hx$	
	symmetry point, i.e., no sign change at	antisymmetry point, i.e., sign change at	symmetry point, i.e., no sign change at	antisymmetry point, i.e., sign change at
	$x =$	$x =$	$x =$	$x =$
h even	$0, \frac{1}{4}, \frac{1}{2}, \frac{3}{4}$	$(\frac{1}{8}, \frac{3}{8}, \frac{5}{8}, \frac{7}{8})$	$(\frac{1}{8}, \frac{3}{8}, \frac{5}{8}, \frac{7}{8})$	$0, \frac{1}{4}, \frac{1}{2}, \frac{3}{4}$
h odd	$0, \quad \frac{1}{2}$	$\frac{1}{4}, \quad \frac{3}{4}$	$\frac{1}{4}, \quad \frac{3}{4}$	$0, \quad \frac{1}{2}$

$h = 1$ and for $h = 2$. The behaviors of these functions are typical of their general behaviors for h odd and h even respectively. Note that the symmetry and antisymmetry is distributed about the coordinates shown in Table 1. It is evident that if the value of the trigonometric function has been computed in the range 0 to $\frac{1}{4}$, it has the same absolute value at symmetrically related points in the ranges $\frac{1}{4}$ to $\frac{1}{2}$, $\frac{1}{2}$ to $\frac{3}{4}$, and $\frac{3}{4}$ to 1, although the actual value may involve a sign change. These relations are, more specifically,

$$
\begin{array}{lrcl}
 & \text{Range } x = \tfrac{1}{2} \text{ to } \tfrac{1}{4} & & \text{Range } x = 0 \text{ to } \tfrac{1}{4} \\
h \text{ even:} & \cos 2\pi h(\tfrac{1}{2}-x) & = & \cos 2\pi hx, \\
 & \sin 2\pi h(\tfrac{1}{2}-x) & = & -\sin 2\pi hx; \\
h \text{ odd:} & \cos 2\pi h(\tfrac{1}{2}-x) & = & -\cos 2\pi hx, \\
 & \sin 2\pi h(\tfrac{1}{2}-x) & = & \sin 2\pi hx.
\end{array}
\tag{1}
$$

One can take advantage of these relations by performing the summations for h even and h odd separately. If this is done, then the summations in the several ranges can be found by combining the summations in the 0-to-$\frac{1}{4}$ range in the following manner:

Let

$$
\begin{array}{lll}
C_e = \cos 2\pi hx & \text{when } h \text{ is even,} & x = 0 \text{ to } \tfrac{1}{4}, \\
C_o = \quad \text{"} & \text{" " " odd,} & \quad \text{"} \quad , \\
S_e = \sin 2\pi hx & \text{" " " even,} & \quad \text{"} \quad , \\
S_o = \quad \text{"} & \text{" " " odd} & \quad \text{"} \quad .
\end{array}
$$

Then, for

Range $x = 0$ *to* $\frac{1}{4}$:

$$\left.\begin{array}{l}\text{a cosine summation is}\\ \sum C = \sum (C_e + C_o)\\ \text{a sine summation is}\\ \sum S = \sum (S_e + S_o)\end{array}\right\} \qquad (2)$$

Range $x = \frac{1}{2}$ to $\frac{1}{4}$ (note backward progression of x, also see (1)):

$$\left.\begin{array}{l}\text{a cosine summation is}\\ \sum C = \sum (C_e - C_o)\\ \text{a sine summation is}\\ \sum S = \sum (-S_e + S_o)\end{array}\right\} \qquad (3)$$

To extend the range beyond $x = \frac{1}{2}$, one merely makes use of the universal symmetry of the cosines about either 0 or $\frac{1}{2}$, and the universal antisymmetry of the sines about these points. To make this extension, therefore, all signs of the C's in (2) and (3) are retained and the S's reversed:

Range $x = \frac{1}{2}$ *to* $\frac{3}{4}$:

$$\left.\begin{array}{l}\sum C = \sum (C_e - C_o)\\ \sum S = \sum (S_e - S_o)\end{array}\right\} \qquad (4)$$

Range $x = 1$ *to* $\frac{3}{4}$ (note backward progression of x):

$$\left.\begin{array}{l}\sum C = \sum (C_e + C_o)\\ \sum S = \sum (-S_e - S_o)\end{array}\right\} \qquad (5)$$

A compact representation is given in Table 2. In this table note the symmetry and antisymmetry of the component parts about 0, $\frac{1}{4}$, $\frac{1}{2}$, $\frac{3}{4}$. The second and fourth ranges are designated in reverse order to display this symmetry.

By the use of these relationships a non-centrosymmetrical summation, which involves both cosine and sine components, can be reduced to a quarter of the nominal number of summations required. In any summation where the symmetry of the cell requires that the summation must be carried out beyond x, y, or $z = \frac{1}{4}$, there is a considerable saving of labor in using these relations.

Table 2
Form of the Fourier summations based on summations in the range 0 to $\frac{1}{4}$

Range	Summation combinations
	0
0 to $\frac{1}{4}$	$C_e + C_o + S_e + S_o$
	.......$\frac{1}{4}$
$\frac{1}{2}$ to $\frac{1}{4}$	$C_e - C_o - S_e + S_o$
	.......$\frac{1}{2}$
$\frac{1}{2}$ to $\frac{3}{4}$	$C_e - C_o + S_e - S_o$
	.......$\frac{3}{4}$
1 to $\frac{3}{4}$	$C_e + C_o - S_e - S_o$
	1

Summation of one-dimensional series

There are two popular "strip" methods of summing Fourier series, the Beevers-Lipson[1–3] and Patterson-Tunell methods.[7] Both of these are essentially methods of summing one-dimensional series of the form

$$C = \sum_{h=0}^{H} A_h \cos 2\pi hx \tag{6}$$

and

$$S = \sum_{h=0}^{H} A_h \sin 2\pi hx. \tag{7}$$

It will be shown later how the methods can be adapted to two-dimensional summations. Meanwhile they are treated merely as methods for one-dimensional summations.

The Beevers-Lipson method. A method was devised by Beevers and Lipson[1–3] which relieves the computer of all the preliminary mathematical work in computing (6) and (7), leaving him only the job of performing the additions. The method does this by solving once and for all the parts of (6) and (7) likely to be used again and again. Specifically it provides the values of

$$A_h \cos 2\pi hx \tag{8}$$

and

$$A_h \sin 2\pi hx, \tag{9}$$

for all values of A and h likely to be needed (here A_h is the amplitude). For a particular value of A and h, the values of (8), or (9), are printed on a strip of cardboard for the 16 successive x locations of $\frac{0}{60}, \frac{1}{60}, \frac{2}{60} \cdots \frac{15}{60}$. For example, for $A = 39$ and $h = 4$, the various values of (8) for the parameter x are printed on a strip as follows:

		x =	0	1	2	3	4	5	6	7	8	9	10	11	12	13	14	15	
39	C	4	39	36	26	12	$\bar{4}$	$\overline{19}$	$\overline{32}$	$\overline{38}$	$\overline{38}$	$\overline{32}$	$\overline{19}$	$\bar{4}$	12	26	36	39	(40)

Thus one starts with numerical values for (8) say, for each A and h to be summed. To perform the summation (6), one merely places the $H + 1$ strips so that their successive x values lie in vertical columns, as in Table 4, and adds the numbers in each column. Strips are available for every combination of A and h, with A ranging from 1 to 100 and h ranging from 0 to 30. The negative values of the function are also provided. In some sets of strips these are on separate strips, in others on the backs of the positive-function strips. The values of the function are ordinarily given to two figures, which is ample for most crystallographic work. (Three-figure strips are also available.)

The strips are stored in a special box, which aids in finding the strips and returning them to the proper place. The strips are arranged so that all having the same h values are kept in the same compartment, within which they are filed according to increasing amplitude A.

The use of the strips is illustrated by the following example. It is desired to make a Fourier synthesis $\Sigma A_h \cos 2\pi hx$ using the 18 Fourier coefficients in Table 3. (This is an actual case, specifically the Fourier synthesis of the Harker line $P_{(\frac{1}{2}y\frac{1}{2})}$ of realgar.† From each of the compartments of the cosine-strip box one selects a strip having the A_h value listed in Table 3. These are placed in order, as shown in Table 4, with even and odd values separated. Note that Table 3 shows that A_{16} and A_{17} are zero. They therefore contribute nothing to the summation

Table 3
Data for example of computation of the one-dimensional Fourier synthesis of Table 4

Index, h	Fourier coefficient, A_h	Index, h	Fourier coefficient, A_h
0	34	9	9
1	−2	10	−7
2	−17	11	2
3	−26	12	−3
4	39	13	3
5	11	14	−4
6	1	15	−3
7	−8	16	0
8	1	17	0

† Harker sections are discussed in M. J. Buerger. *Vector space and its application in crystal-structure investigation.* (John Wiley and Sons, New York, 1959) Chapter 7.

Table 4
Example of computation of one-dimensional Fourier synthesis by the Beevers-Lipson method

		$x =$	0	1	2	3	4	5	6	7	8	9	10	11	12	13	14	15	
34	C	0	34	34	34	34	34	34	34	34	34	34	34	34	34	34	34	34	(544)
$\bar{17}$	C	2	$\bar{17}$	$\bar{17}$	$\bar{16}$	$\bar{14}$	$\bar{11}$	$\bar{8}$	$\bar{5}$	$\bar{2}$	2	5	8	11	14	16	17	17	(0)
39	C	4	39	36	26	12	$\bar{4}$	$\bar{19}$	$\bar{32}$	$\bar{38}$	$\bar{38}$	$\bar{32}$	$\bar{19}$	$\bar{4}$	12	26	36	39	(40)
1	C	6	1	1	0	0	$\bar{1}$	$\bar{1}$	$\bar{1}$	0	0	1	1	1	0	0	$\bar{1}$	$\bar{1}$	(0)
1	C	8	1	1	0	$\bar{1}$	$\bar{1}$	0	0	1	1	0	0	$\bar{1}$	$\bar{1}$	0	1	1	(2)
$\bar{7}$	C	10	$\bar{7}$	$\bar{3}$	3	7	3	$\bar{3}$	$\bar{7}$	$\bar{3}$	3	7	3	$\bar{3}$	$\bar{7}$	$\bar{3}$	3	7	(0)
$\bar{3}$	C	12	$\bar{3}$	$\bar{1}$	2	2	$\bar{1}$	$\bar{3}$	$\bar{1}$	2	2	$\bar{1}$	$\bar{3}$	$\bar{1}$	2	2	$\bar{1}$	$\bar{3}$	($\bar{6}$)
$\bar{4}$	C	14	$\bar{4}$	0	4	1	$\bar{4}$	$\bar{2}$	3	3	$\bar{3}$	$\bar{3}$	2	4	$\bar{1}$	$\bar{4}$	0	4	(0)
$\sum C_e$			44	51	53	41	15	$\bar{2}$	$\bar{9}$	$\bar{3}$	1	11	26	41	53	71	89	98	(580)
$\bar{2}$	C	1	$\bar{2}$	$\bar{2}$	$\bar{2}$	$\bar{2}$	$\bar{2}$	$\bar{2}$	$\bar{2}$	$\bar{1}$	$\bar{1}$	$\bar{1}$	$\bar{1}$	$\bar{1}$	$\bar{1}$	0	0	0	($\bar{20}$)
$\bar{26}$	C	3	$\bar{26}$	$\bar{25}$	$\bar{21}$	$\bar{15}$	$\bar{8}$	0	8	15	21	25	26	25	21	15	8	0	(69)
11	C	5	11	10	5	0	$\bar{5}$	$\bar{10}$	$\bar{11}$	$\bar{10}$	$\bar{5}$	0	5	10	11	10	5	0	(26)
$\bar{8}$	C	7	$\bar{8}$	$\bar{6}$	$\bar{1}$	5	8	7	2	$\bar{3}$	$\bar{7}$	$\bar{8}$	$\bar{4}$	2	6	8	5	0	(6)
9	C	9	9	5	$\bar{3}$	$\bar{9}$	$\bar{7}$	0	7	9	3	$\bar{5}$	$\bar{9}$	$\bar{5}$	3	9	7	0	(14)
2	C	11	2	1	$\bar{1}$	$\bar{2}$	0	2	2	0	$\bar{2}$	$\bar{1}$	1	2	1	$\bar{1}$	$\bar{2}$	0	(2)
3	C	13	3	1	$\bar{3}$	$\bar{2}$	2	3	$\bar{1}$	$\bar{3}$	0	3	1	$\bar{2}$	$\bar{2}$	1	3	0	(4)
$\bar{3}$	C	15	$\bar{3}$	0	3	0	$\bar{3}$	0	3	0	$\bar{3}$	0	3	0	$\bar{3}$	0	3	0	(0)
$\sum C_o$			$\bar{14}$	$\bar{16}$	$\bar{23}$	$\bar{25}$	$\bar{15}$	0	8	7	6	13	22	31	36	42	29	0	(101)
		$x =$	0	1	2	3	4	5	6	7	8	9	10	11	12	13	14	15	
$\sum C_e + \sum C_o$			30	35	30	16	0	$\bar{2}$	$\bar{1}$	4	7	24	48	72	89	113	118	98	
$\sum C_e - \sum C_o$			58	67	76	66	30	$\bar{2}$	$\bar{17}$	$\bar{10}$	$\bar{5}$	$\bar{2}$	4	10	17	29	60	98	
		$x =$	30	29	28	27	26	25	24	23	22	21	20	19	18	17	16	15	

and are omitted from Table 4. The strips have now been arranged in two blocks corresponding to C_e and C_o of (2) and (3), with corresponding x values in columns. When a column is summed the result is $\Sigma A_h \cos 2\pi hx$ for that value of x. Accordingly the numbers in all columns are summed for each block. The results are ΣC_e and ΣC_o. According to (2) and (3) the Fourier synthesis for the range $x = 0$ to $\frac{15}{60}$ is obtained by adding ΣC_o to ΣC_e, while the range from $\frac{30}{60}$ to $\frac{15}{60}$ is obtained by subtracting ΣC_o from ΣC_e. The results of this sum and difference appear at the bottom of the table. Since this is a cosine synthesis, it is symmetrical about $\frac{1}{2}$, so this completes the summation.

The right side of Table 4 shows a series of numbers in parentheses. Each number is the sum of the numbers on that strip. When this column is added the result should equal the sum of the bottom row of sums, and so serves as a check on the correctness of the synthesis.

The result of such a synthesis is usually plotted, as in Fig. 2. In the event that this one-dimensional synthesis is part of a two-dimensional one, only the numerical result at the bottom of the columns is recorded, no record being kept of the other figures in the two banks of Table 4.

When one wishes to include an amplitude A_h which exceeds the 0-to-100 range of available strips, one can substitute any two (or more) A_h

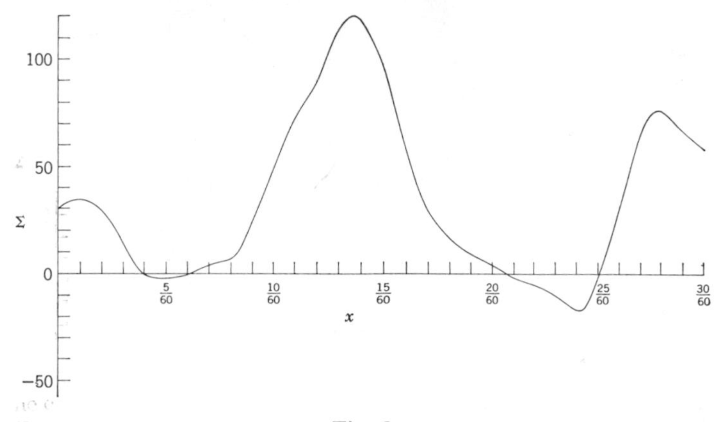

Fig. 2.
Plot of the example of the one-dimensional Fourier summation carried out in Table 4; realgar, $P_{(\frac{1}{2}y\frac{1}{2})}$.

strips of the same h such that the sum of their amplitudes equals the desired A_h.

The Patterson-Tunell method. An alternative method of summing one-dimensional Fourier series using strips was devised by Patterson and Tunell.[7] The theoretical backgrounds for the Beevers-Lipson and Patterson-Tunell methods are identical. They differ chiefly in the number of kinds of strips needed.

In the Beevers and Lipson method, separate strips were required for $A \cos 2\pi 1x$, $A \cos 2\pi 2x$, $A \cos 2\pi 3x$, etc. In the Patterson-Tunell method, it is recognized that the numbers which appear on the strip $A \cos 2\pi 2x$ are the same as the alternate numbers on the strip $A \cos 2\pi 1x$, and that the numbers on strip $A \cos 2\pi 3x$ consist of every third number on the strip $A \cos 2\pi 1x$, etc. Consequently, one could obtain all these required values of $A \cos 2\pi hx$, for a specific value of x, by arranging a mask which would screen all but the nth number (where $n = hx$) to obtain $A \cos 2\pi 1x$, another to screen all but the $2n$th number for $A \cos 2\pi 2x$, a third to screen all but the $3n$th number for $A \cos 2\pi 3x$, etc. Thus, only one kind of strip is needed, namely $A \cos 2\pi 1x$, for each of the different amplitudes, A. In practice one has a collection of strips with the amplitude, A, ranging from 1 to ± 300. White strips indicate positive A, colored strips are used to indicate negative A.

To perform the summation, a strip is selected from a storage box (Fig. 3) for each value of A_h. These are laid on a board, Fig. 4, with the strip for $h = 0$ in the bottom position, the strip for $h = 1$ in the next

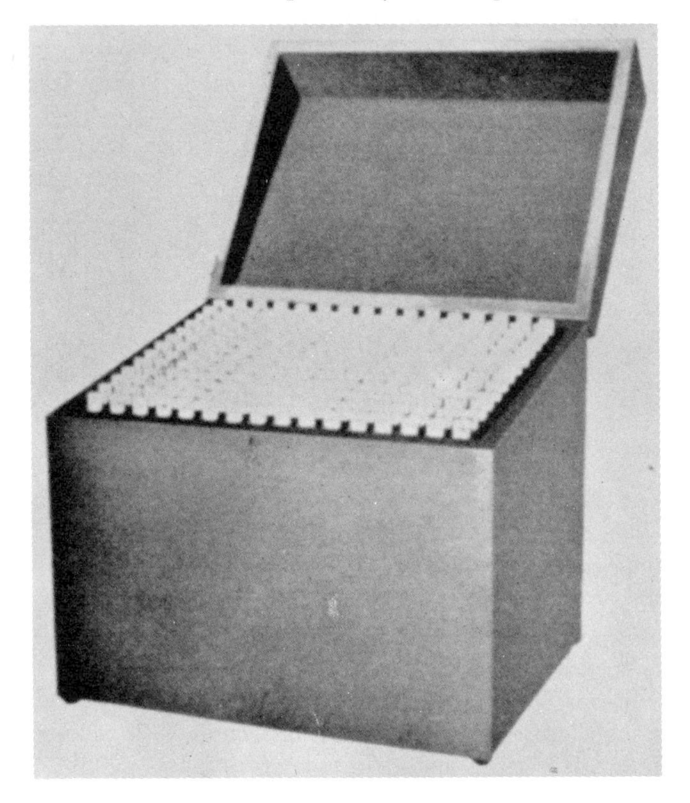

Fig. 3.
Box for storing Patterson-Tunell strips.
(After Patterson and Tunell.[7])

position above it, the strip for $h = 2$ in the next higher position, etc. until all values of h are represented and are in order from bottom to top.

To perform the summation for the point $x = 0$, h even, the value of $n = hx$ is 0. Therefore the appropriate mask would permit one to see the 0th number of each strip. This cosine mask is shown in the upper left of Fig. 5. To perform the summation for $x = 1$, h even, the value of n is $n = h \cdot 1 = h$. The appropriate mask (Fig. 5, upper row, next to left end) has openings to display the 0th number for strip $h = 0$, the 2nd number for strip $h = 2$, the 4th number for strip $h = 4$, etc. All the numbers visible are added, with the precaution that colored numbers are negative, and a colored ring about the hole in the mask indicates that the sign of the number must be reversed (i.e., corresponds with a negative value of the trigonometric function).

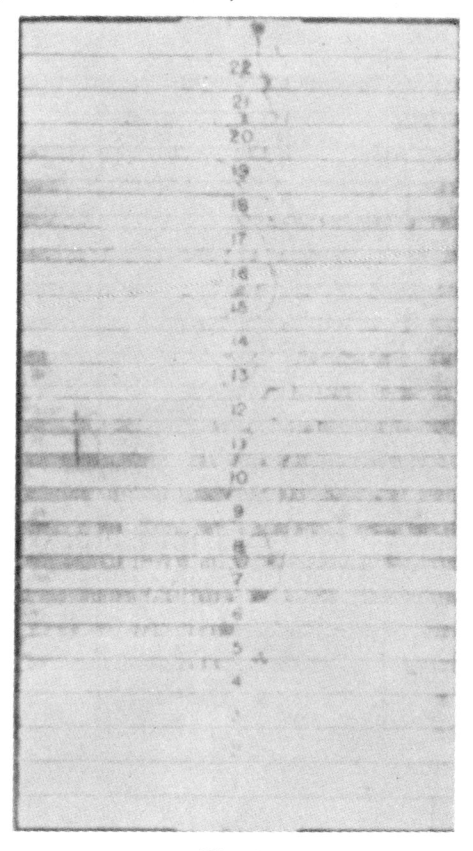

Fig. 4.
Board for arranging Patterson-Tunell strips. A strip corresponding to A_h is laid in the slot numbered h, for each number.
(After Patterson and Tunell.[7])

The advantage of the Patterson-Tunell method is that a much smaller collection of different kinds of strips is employed, since the strips are functions of A only, and not of both A and h, as in the Beevers-Lipson method. Disadvantages include the requirement for the large paraphernalia of the mask and special board, and the consequent large amount of space required for their use, and the possibility of making an error in judging sign, especially when the mask sign reverses a negative sign of a strip.

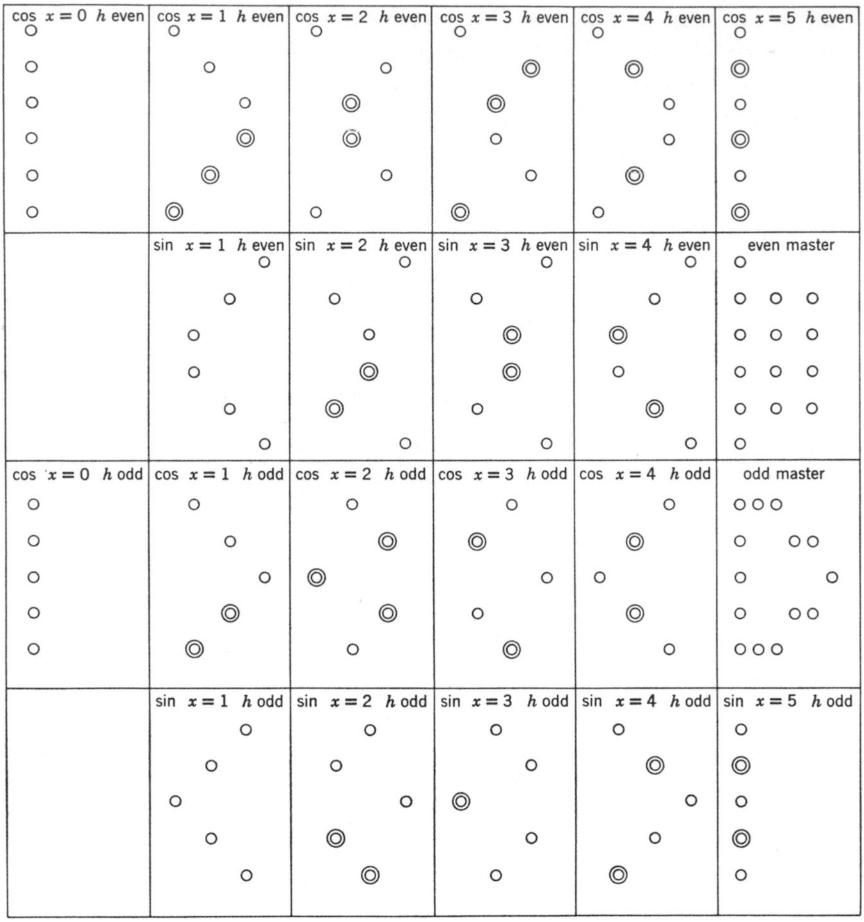

cos $x = 0$ h even	cos $x = 1$ h even	cos $x = 2$ h even	cos $x = 3$ h even	cos $x = 4$ h even	cos $x = 5$ h even
	sin $x = 1$ h even	sin $x = 2$ h even	sin $x = 3$ h even	sin $x = 4$ h even	even master
cos $x = 0$ h odd	cos $x = 1$ h odd	cos $x = 2$ h odd	cos $x = 3$ h odd	cos $x = 4$ h odd	odd master
	sin $x = 1$ h odd	sin $x = 2$ h odd	sin $x = 3$ h odd	sin $x = 4$ h odd	sin $x = 5$ h odd

Fig. 5.

Set of Patterson-Tunell masks for sampling at $\frac{1}{20}$th of the cell periods. Each rectangle is a separate sheet with circular holes through which a certain number can be seen on each of several strips, which are arranged on a board like Fig. 4. The circle about certain circular holes in the mask requires that the sign of the number seen be reversed.

(After Patterson and Tunell.[7])

Adaptation of one-dimensional methods to two-dimensional Fourier syntheses

Beevers and Lipson[1] pointed out that if the trigonometric part of a two-dimensional Fourier synthesis is expressed in product form, the summation can be regarded as successive sets of one-dimensional series. The form of a general two-dimensional Fourier synthesis is given in (34) of Chapter 17. The forms of the synthesis for various two-dimensional symmetries are also given in product form in Chapter 17.

In order to illustrate the adaptation of one-dimensional methods to two-dimensional Fourier syntheses, the simplest product form will be used as an example. This is the form for symmetry *p2mm*, which is given in (47) of Chapter 17. Omitting the scale factor, this is

$$\rho_{(xy)} = \sum_{h=0}^{H}{}' \sum_{k=0}^{K}{}' F_{hk0} \cos 2\pi hx \cos 2\pi ky. \tag{10}$$

The prime marks after the Σ''s indicate that the border terms of the representative segment of the coefficients (i.e., the border of the representative section of the F's in reciprocal space) are to enter the synthesis at half their true value to allow for their occurring on a line symmetry. One can regard (10) as summed over a k first, leaving h constant. To express this it can be written

$$\begin{aligned}\rho_{(xy)} = \quad & \Big\{\Big[\sum_k F_{0k0} \cos 2\pi ky\Big] \cos 2\pi 0x \\ + \quad & \Big[\sum_k F_{1k0} \cos 2\pi ky\Big] \cos 2\pi 1x \\ + \quad & \Big[\sum_k F_{2k0} \cos 2\pi ky\Big] \cos 2\pi 2x \\ & \cdots \\ + \quad & \Big[\sum_k F_{Hk0} \cos 2\pi ky\Big] \cos 2\pi Hx\Big\},\end{aligned} \tag{11}$$

or, more compactly,

$$\rho_{(xy)} = \sum_h \Big[\sum_k F_{hk0} \cos 2\pi ky\Big]_{h \text{ constant}} \cos 2\pi hx. \tag{12}$$

This can be written

$$\rho_{(xy)} = \sum_k K_h \cos 2\pi hx, \tag{13}$$

where

$$K_h = \Big[\sum_h F_{hk0} \cos 2\pi ky\Big]_h. \tag{14}$$

Thus one separates the F_{hk0} coefficients into groups with h constant, and sums the group with respect to k. This gives a set of numerical coefficients K_h, (14), to replace the terms which occur in the brackets of (11) and (12). This process is usually described by saying that one sums first over k (h being regarded as constant), then over h. It is immaterial, mathematically, which index is summed first, although, as will be seen, there may be practical advantages to summing one or the other first.

Table 5
Data for example of computation of two-dimensional Fourier summation
(Data for Patterson synthesis $P_{(xy)}$, realgar)

k \ h	0	1	2	3	4	5	6	7
17	0	0	0					
16	2	0	0					
15	0	3	0	6	0			
14	14	2	2	0	0			
13	0	2	14	0	5			
12	4	5	3	11	0	0		
11	0	5	16	0	9	3	0	
10	4	5	0	0	0	0	0	
9	0	3	15	15	0	9	0	
8	118	5	0	2	20	0	0	
7	0	1	11	3	0	3	0	0
6	146	24	2	24	14	2	3	1
5	0	40	0	0	18	46	0	0
4	0	0	1	44	2	0	0	0
3	0	0	2	2	2	9	5	0
2	66	0	3	3	14	0	2	2
1	0	16	37	6	35	17	0	3
0	604	40	0	24	138	0	0	0

An example of the computation of a Fourier summation for this symmetry is $P_{(xy)}$ of realgar. (This kind of Fourier summation gives the Patterson synthesis, discussed elsewhere,† rather than the electron density. For Fourier coefficients it uses $|F|^2$'s instead of F's.) The Fourier coefficients for this synthesis are those in Table 5. The first step in starting the synthesis is to take account of the meaning of the primes on the Σ''s of (10). These required that the borders of the asymmetrical set of Fourier coefficients be counted at half value, and that the origin point be counted at one-quarter value. The borders and origin of the $|F|^2$'s in Table 5 are therefore altered to produce the lower-left block of Table 6 and Fig. 6.

† M. J. Buerger. *Vector space and its application in crystal-structure investigation.* (John Wiley and Sons, New York, 1959)

It will be evident shortly that the second summation is most tedious. One therefore arranges to sum over the largest number of terms first and the smallest number second. Referring to Table 6, it is plain that in this example one should sum over k first and h second. Accordingly each *vertical* column of the $|F_{hk0}|^2$ table is treated as the set of coefficients of a one-dimensional synthesis and summed over k to produce the values for different y locations shown in the upper left table. The summation over terms with k even and k odd are performed separately and the results are recorded separately. To facilitate expansion the ΣC_e terms are placed above the corresponding ΣC_o terms.

The first summation (over k) is now complete except for expansion to the y ranges of 0 to $\frac{15}{60}$ and $\frac{30}{60}$ to $\frac{15}{60}$. This is done using the upper parts

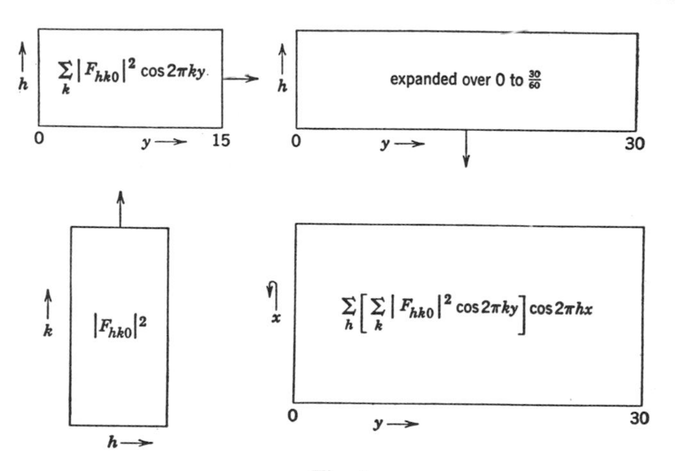

Fig. 6.
Sequence of operations in Table 6.

of (2) and (3). The results of the expansion are recorded in the upper right block of Fig. 6. Each vertical column in this table is now a set of Fourier coefficients with y constant. These coefficients are to be summed over h to produce values at various points x. Each vertical column is therefore handled as a one-dimensional summation, even and odd terms being separately summed and recorded. The results are recorded in the block in the lower right of Fig. 6.

It still remains to expand this block over x ranges from 0 to $\frac{15}{60}$ and $\frac{30}{60}$ to $\frac{15}{60}$. It is most convenient to record the results of this expansion directly on a scaled projection of the crystal cell (a scale of 1 Å = 2 in. is usually appropriate). A good way of preparing the projection is to make a tracing of the cell, and spot the points represented by the xy sampling. Since many Fourier syntheses are usually made using this same plan, copies of this original should be made by some reproduction process like

Table 6

h	0	1	2	3	4	5	6	7	8	9	10	11	12	13	14	15
7	3 3	3 3	2 3	2 3	0 3	0 3	0 2	0 2	0 2	0 2	0 2	0 1	$\bar{2}$ 1	$\bar{2}$ 1	$\bar{3}$ 0	$\bar{3}$ 0
6	5 5	4 5	3 4	1 3	$\bar{1}$ 2	$\bar{2}$ 0	$\bar{1}$ $\bar{2}$	$\bar{1}$ $\bar{3}$	1 $\bar{4}$	1 $\bar{5}$	2 $\bar{3}$	1 $\bar{5}$	$\bar{1}$ $\bar{4}$	$\bar{3}$ $\bar{3}$	$\bar{4}$ $\bar{2}$	$\bar{5}$ 0
5	2 87	2 74	1 42	$\bar{1}$ 7	$\bar{2}$ $\bar{14}$	$\bar{2}$ $\bar{25}$	$\bar{2}$ $\bar{27}$	$\bar{1}$ $\bar{23}$	1 $\bar{16}$	2 $\bar{3}$	2 17	2 35	1 46	$\bar{1}$ 43	$\bar{2}$ 24	$\bar{2}$ 0
4	119 59	109 48	85 24	61 13	47 17	51 18	66 6	82 $\bar{5}$	88 $\bar{4}$	80 13	65 25	51 29	47 23	51 15	59 8	63 0
3	96 32	78 19	42 $\bar{3}$	10 $\bar{9}$	$\bar{9}$ $\bar{3}$	$\bar{22}$ 2	$\bar{38}$ 9	$\bar{45}$ 18	$\bar{31}$ 16	$\bar{2}$ $\bar{4}$	22 $\bar{18}$	25 $\bar{10}$	20 9	22 11	34 4	42 0
2	11 95	7 66	1 10	$\bar{2}$ $\bar{7}$	3 19	4 48	$\bar{3}$ 47	$\bar{5}$ 27	$\bar{1}$ 21	3 25	2 22	$\bar{1}$ 8	$\bar{2}$ $\bar{1}$	$\bar{3}$ 4	$\bar{3}$ 7	$\bar{3}$ 0
1	61 70	46 56	18 27	$\bar{1}$ 5	$\bar{2}$ $\bar{5}$	2 $\bar{16}$	8 $\bar{25}$	15 $\bar{23}$	27 $\bar{9}$	40 7	44 25	36 44	25 48	12 37	4 17	$\bar{1}$ 0
0	298	255	160	76	43	57	107	177	231	231	189	147	126	124	127	128
y	30	29	28	27	26	25	24	23	22	21	20	19	18	17	16	15

h \ y	0	1	2	3	4	5	6	7	8	9	10	11	12	13	14	15	16	17	18	19	20	21	22	23	24	25	26	27	28	29	30
7	6	6	5	5	3	3	2	2	2	2	2	1	$\bar{1}$	$\bar{1}$	$\bar{3}$	$\bar{3}$	$\bar{3}$	$\bar{3}$	$\bar{3}$	$\bar{1}$	$\bar{2}$	$\bar{2}$	$\bar{2}$	$\bar{2}$	$\bar{2}$	$\bar{3}$	$\bar{3}$	$\bar{1}$	$\bar{1}$	0	0
6	10	9	7	4	1	$\bar{2}$	$\bar{3}$	$\bar{4}$	$\bar{3}$	$\bar{4}$	$\bar{3}$	$\bar{4}$	$\bar{5}$	$\bar{6}$	$\bar{6}$	$\bar{5}$	$\bar{2}$	0	3	6	7	7	5	2	1	$\bar{2}$	$\bar{3}$	$\bar{2}$	$\bar{1}$	$\bar{1}$	0
5	89	76	43	6	$\bar{16}$	$\bar{27}$	$\bar{29}$	$\bar{24}$	$\bar{15}$	$\bar{1}$	19	37	47	42	22	$\bar{2}$	$\bar{26}$	$\bar{44}$	$\bar{45}$	$\bar{33}$	$\bar{15}$	5	17	22	25	23	12	$\bar{8}$	$\bar{41}$	$\bar{72}$	$\bar{85}$
4	178	157	109	74	64	69	72	77	84	93	90	80	70	66	67	63	51	36	24	22	40	67	92	87	60	33	30	48	61	61	60
3	128	97	39	1	$\bar{12}$	$\bar{20}$	$\bar{29}$	$\bar{27}$	$\bar{15}$	$\bar{6}$	4	15	29	33	38	42	30	11	11	35	40	2	$\bar{47}$	$\bar{63}$	$\bar{47}$	$\bar{20}$	$\bar{6}$	19	45	59	64
2	106	73	11	$\bar{9}$	22	52	44	22	20	28	24	7	$\bar{3}$	1	4	$\bar{3}$	$\bar{10}$	$\bar{7}$	$\bar{1}$	$\bar{9}$	$\bar{20}$	$\bar{22}$	$\bar{22}$	$\bar{32}$	$\bar{50}$	$\bar{44}$	$\bar{16}$	5	$\bar{9}$	$\bar{59}$	$\bar{84}$
1	131	102	45	4	$\bar{7}$	$\bar{14}$	$\bar{17}$	$\bar{8}$	18	47	69	80	73	49	21	$\bar{1}$	$\bar{13}$	$\bar{23}$	$\bar{23}$	$\bar{8}$	19	33	36	38	17	14	3	$\bar{6}$	$\bar{9}$	$\bar{10}$	$\bar{9}$
0	298	255	160	76	43	57	107	177	231	231	189	147	126	124	127	128	127	124	126	147	189	231	231	177	107	57	43	76	160	255	298

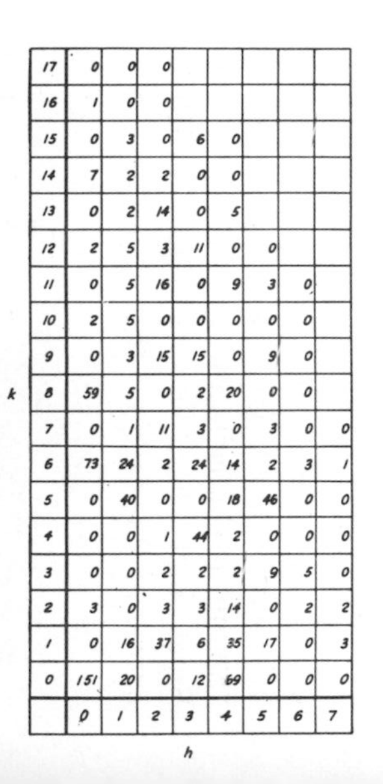

k \ h	0	1	2	3	4	5	6	7
17	0	0	0					
16	1	0	0					
15	0	3	0	6	0			
14	7	2	2	0	0			
13	0	2	14	0	5			
12	2	5	3	11	0	0		
11	0	5	16	0	9	3	0	
10	2	5	0	0	0	0	0	
9	0	3	15	15	0	9	0	
8	59	5	0	2	20	0	0	
7	0	1	11	3	0	3	0	0
6	73	24	2	24	14	2	3	1
5	0	40	0	0	18	46	0	0
4	0	0	1	44	2	0	0	0
3	0	0	2	2	2	9	5	0
2	3	0	3	3	14	0	2	2
1	0	16	37	6	35	17	0	3
0	151	20	0	12	69	0	0	0

x		0	1	2	3	4	5	6	7	8	9	10	11	12	13	14	15
15	15	360 0	330 0	251 0	155 0	84 0	76 0	138 0	236 0	298 0	300 0	258 0	224 0	204 0	195 0	196 0	199 0
16	14	349 14	320 15	243 12	150 0	78 $\bar{7}$	71 $\bar{11}$	132 $\bar{8}$	228 $\bar{6}$	290 $\bar{2}$	292 6	250 15	216 20	197 24	188 17	189 3	193 $\bar{12}$
17	13	317 23	290 24	221 18	133 0	66 $\bar{11}$	56 $\bar{17}$	116 $\bar{14}$	210 $\bar{9}$	270 $\bar{2}$	268 11	228 26	196 39	178 40	169 28	170 4	175 $\bar{24}$
18	12	270 20	248 25	187 21	107 2	45 $\bar{10}$	35 $\bar{17}$	92 $\bar{13}$	182 $\bar{6}$	240 1	236 17	197 35	165 49	148 48	141 31	143 $\bar{1}$	147 $\bar{34}$
19	11	216 7	197 14	148 17	77 5	22 $\bar{7}$	13 $\bar{11}$	68 $\bar{4}$	151 2	207 8	199 24	162 40	131 51	117 43	111 25	112 $\bar{7}$	119 $\bar{41}$
20	10	166 $\bar{15}$	150 $\bar{5}$	107 7	47 6	1 2	$\bar{5}$ 1	46 8	124 12	176 17	167 31	129 41	99 43	88 31	85 12	85 $\bar{19}$	93 $\bar{45}$
21	9	129 $\bar{39}$	112 $\bar{26}$	75 $\bar{6}$	22 6	$\bar{15}$ 10	$\bar{17}$ 14	33 20	105 23	155 27	144 36	107 39	77 34	66 14	66 $\bar{3}$	67 $\bar{27}$	74 $\bar{44}$
22	8	116 $\bar{55}$	96 $\bar{43}$	54 $\bar{19}$	6 4	$\bar{22}$ 16	$\bar{16}$ 24	31 28	99 31	146 34	136 38	97 35	67 25	56 1	57 $\bar{16}$	59 $\bar{31}$	64 $\bar{37}$
23	7	132 $\bar{53}$	106 $\bar{45}$	52 $\bar{25}$	2 $\bar{1}$	$\bar{18}$ 17	$\bar{4}$ 26	43 30	105 32	152 36	144 41	105 34	71 18	60 $\bar{4}$	61 $\bar{19}$	63 $\bar{26}$	68 $\bar{25}$
24	6	179 $\bar{25}$	144 $\bar{25}$	69 $\bar{21}$	10 $\bar{5}$	$\bar{3}$ 13	19 21	65 23	125 25	171 34	168 40	125 35	87 23	72 3	76 $\bar{12}$	79 $\bar{16}$	80 $\bar{11}$
25	5	252 31	204 17	105 $\bar{2}$	31 $\bar{6}$	21 5	51 8	96 8	154 12	202 27	203 40	159 42	115 36	94 23	97 7	101 2	99 4
26	4	342 110	281 79	150 26	59 $\bar{4}$	50 $\bar{5}$	87 $\bar{8}$	130 $\bar{13}$	187 $\bar{5}$	237 17	243 39	198 52	147 59	121 53	123 35	128 23	123 16
27	3	436 196	360 150	201 63	91 2	75 $\bar{16}$	121 $\bar{27}$	166 $\bar{34}$	220 $\bar{25}$	274 7	284 40	237 67	179 84	148 87	147 67	153 44	147 26
28	2	517 277	430 217	245 99	119 9	106 $\bar{25}$	150 $\bar{44}$	194 $\bar{54}$	248 $\bar{42}$	304 $\bar{2}$	318 41	270 80	206 108	168 118	167 96	174 63	165 32
29	1	573 333	476 263	277 123	138 14	124 $\bar{30}$	169 $\bar{54}$	214 $\bar{69}$	266 $\bar{54}$	326 $\bar{8}$	340 41	292 90	224 127	183 141	180 115	187 74	179 35
30	0	592 354	494 281	287 132	145 16	130 $\bar{32}$	176 $\bar{58}$	220 $\bar{73}$	272 $\bar{57}$	332 $\bar{10}$	348 42	300 93	230 133	188 148	185 123	192 78	183 36

x		16	17	18	19	20	21	22	23	24	25	26	27	28	29	30
15	15	190 0	167 0	148 0	172 0	242 0	313 0	340 0	294 0	216 0	136 0	92 0	121 0	187 0	376 0	442 0
16	14	186 $\bar{21}$	164 $\bar{25}$	147 $\bar{25}$	171 $\bar{27}$	240 $\bar{17}$	308 5	333 28	285 35	210 30	132 21	88 10	117 $\bar{10}$	182 $\bar{34}$	370 $\bar{55}$	435 $\bar{63}$
17	13	171 $\bar{41}$	154 $\bar{46}$	142 $\bar{47}$	168 $\bar{51}$	232 $\bar{31}$	294 12	311 52	263 66	193 56	120 38	79 18	104 $\bar{18}$	165 $\bar{63}$	350 $\bar{99}$	415 $\bar{114}$
18	12	150 $\bar{52}$	141 $\bar{58}$	135 $\bar{59}$	163 $\bar{62}$	219 $\bar{39}$	272 15	279 68	231 87	166 70	102 45	64 20	86 $\bar{19}$	142 $\bar{79}$	322 $\bar{123}$	385 $\bar{140}$
19	11	127 $\bar{56}$	125 $\bar{56}$	126 $\bar{57}$	156 $\bar{65}$	204 $\bar{43}$	245 15	240 75	191 94	135 74	81 46	49 18	66 $\bar{27}$	115 $\bar{83}$	287 $\bar{122}$	348 $\bar{139}$
20	10	104 $\bar{45}$	110 $\bar{51}$	117 $\bar{47}$	146 $\bar{55}$	186 $\bar{39}$	215 15	201 72	151 92	103 66	61 37	33 12	48 $\bar{26}$	89 $\bar{69}$	254 $\bar{100}$	310 $\bar{110}$
21	9	87 $\bar{40}$	97 $\bar{27}$	109 $\bar{27}$	137 $\bar{39}$	169 $\bar{29}$	190 15	168 64	119 80	74 53	42 24	22 5	33 $\bar{23}$	69 $\bar{49}$	223 $\bar{62}$	275 66
22	8	77 $\bar{23}$	90 $\bar{5}$	104 $\bar{5}$	128 $\bar{18}$	150 $\bar{13}$	169 16	145 52	96 63	53 35	29 10	15 $\bar{2}$	27 $\bar{16}$	57 $\bar{23}$	201 $\bar{19}$	248 $\bar{16}$
23	7	77 $\bar{6}$	88 14	102 15	122 2	146 2	161 19	137 39	88 45	43 18	21 1	13 $\bar{5}$	31 $\bar{8}$	55 3	189 20	230 29
24	6	85 7	93 23	105 24	121 16	145 19	164 22	146 28	95 29	42 5	18 $\bar{5}$	16 $\bar{7}$	41 $\bar{3}$	65 20	189 46	223 58
25	5	98 15	102 21	111 22	126 23	152 31	179 27	169 18	115 16	51 $\bar{5}$	21 $\bar{5}$	23 $\bar{4}$	56 3	83 29	196 53	226 82
26	4	117 13	115 1	120 1	134 21	166 39	203 31	202 12	145 7	67 $\bar{9}$	27 $\bar{2}$	31 $\bar{2}$	76 6	105 27	211 45	236 64
27	3	136 8	129 $\bar{14}$	131 $\bar{14}$	145 14	183 43	232 33	239 7	177 0	86 $\bar{11}$	32 3	40 1	96 6	128 18	226 25	249 29
28	2	151 $\bar{2}$	142 $\bar{35}$	142 $\bar{35}$	156 4	200 43	258 36	275 5	207 $\bar{8}$	101 $\bar{9}$	38 10	47 4	112 5	149 7	242 2	261 1
29	1	162 $\bar{9}$	150 $\bar{53}$	149 $\bar{54}$	163 $\bar{5}$	212 43	276 38	297 5	227 $\bar{4}$	114 $\bar{7}$	42 13	52 5	123 4	162 $\bar{3}$	252 $\bar{16}$	271 $\bar{22}$
30	0	166 $\bar{12}$	153 $\bar{59}$	152 $\bar{60}$	166 $\bar{7}$	216 42	283 38	306 4	234 $\bar{5}$	118 $\bar{7}$	44 14	54 6	127 4	167 $\bar{6}$	256 $\bar{23}$	274 $\bar{30}$

y

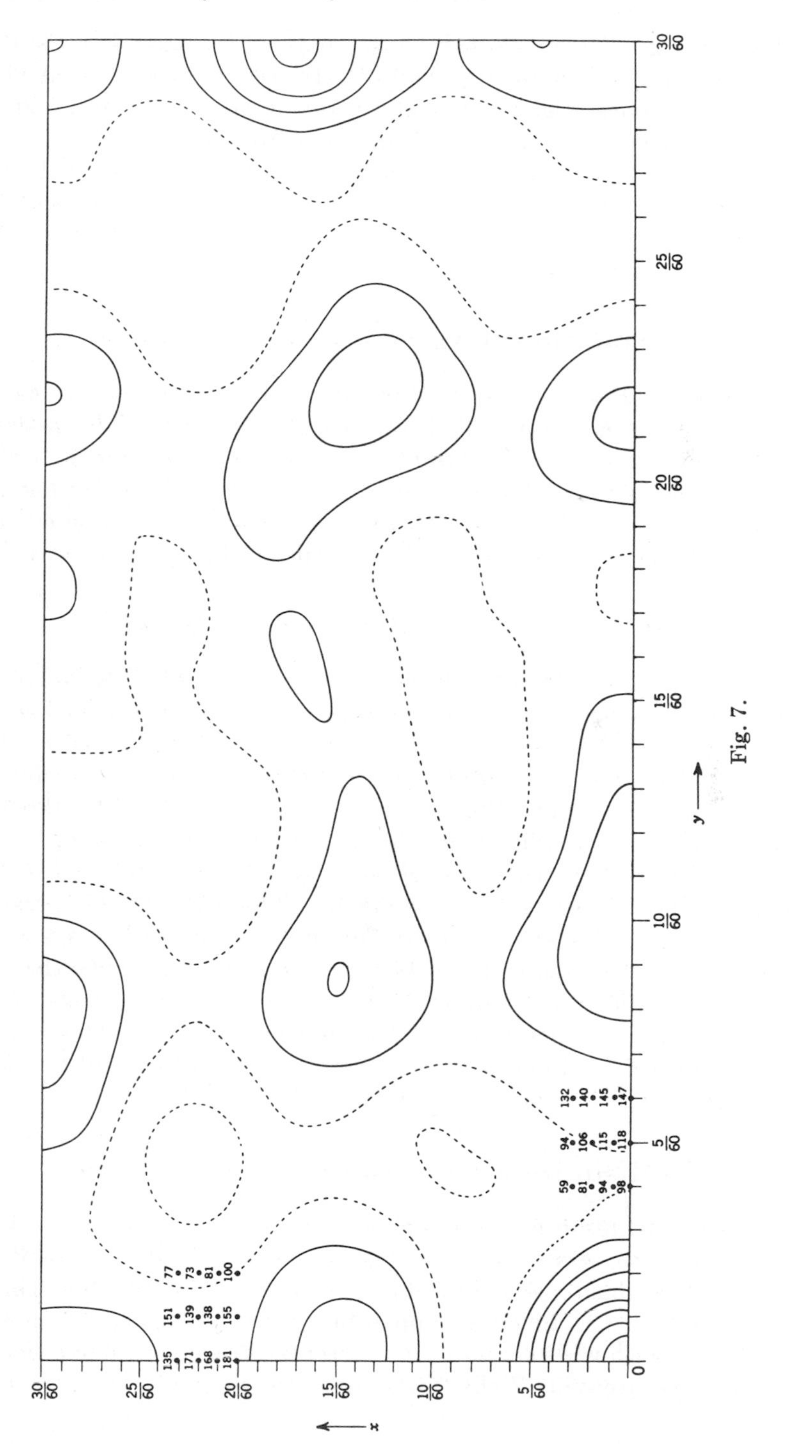

x
y
0
5/60
10/60
15/60
20/60
25/60
30/60
135 171 168 181
151 139 138 155
77 73 81 100
59 81 94 98
94 106 115 118
132 140 145 147

Fig. 7.

ozalid printing. The numerical values obtained by expanding the lower right block in Table 6 are recorded directly on this scaled print of the xy sample locations, Fig. 7. A piece of tracing material is then laid over this number field and contours are drawn on the tracing sheet.

The example just discussed was computed using two-figure Patterson-Tunell strips. If Beevers-Lipson strips were used, it would be found that the results are slightly different due to the different rounding-off errors in the two kinds of strips.

Adaptation to three-dimensional series

The general Beevers-Lipson type of Fourier computation can be adapted to the computation of three-dimensional series. This matter is discussed by Goodwin and Hardy.[6] One has two choices of procedure. He may compute one-dimensional syntheses along lines having each sampled xy value, with z variable (say). Alternatively he may compute two-dimensional syntheses at constant z values with x and y variable.

Outline for less symmetrical projections

In order to facilitate the discussion of Fourier syntheses for various symmetries it is desirable to treat operations such as shown in Fig. 6 in outline. An outline corresponding to Fig. 6 is shown in Table 7. Outlines for other symmetries are made up of similar parts but usually contain more individual operations. These may be illustrated for three less simple projections, specifically for the symmetries $p2gg$, $p2$, and $p1$. The forms of the Fourier syntheses for these symmetries are found in (49), (35), and (34) of Chapter 17, respectively. The sequences of operations for these symmetries are shown in Tables 8, 9, and 10A, respectively. In these tables it is assumed that the range of k is greater than that of h, so that summations are performed first over k. If the range of h is greater than k, the roles of h and k should be reversed. This is easily arranged for Tables 8 and 9, which are symmetrical in h and k. Table 10B is the alternative form of the difficult Table 10A.

Direct summation of two-dimensional series

The Beevers-Lipson method of summation breaks down a two-dimensional series into sets of one-dimensional series. It is also possible to sum two-dimensional series in a direct manner. This was first pointed out by Robertson.[4] Since his apparatus was cumbersome this method did not meet with wide acceptance. Recently it has been independently discovered by Grenville-Wells,[19] who made the method quite practical.

Table 7

Sequence of operations in Fourier synthesis for symmetry *p2mm*

$$\sum_h{}' \sum_k{}' F_{hk0} \cos 2\pi hx \cos 2\pi ky$$

$$F_{hk0}$$

$$\downarrow$$

$$\sum_k{}' F_{hk0} \cos 2\pi ky$$

$$\downarrow$$

expand over $y = 0$ to $\frac{1}{2}$

$$\downarrow$$

$$\sum_h{}' \left[\sum_k{}' F_{hk0} \cos 2\pi ky\right] \cos 2\pi hx$$

$$\downarrow$$

expand over $x = 0$ to $\frac{1}{2}$

$$\downarrow$$

$$\sum_h{}' \sum_k{}' F_{hk0} \cos 2\pi hx \cos 2\pi ky$$

Table 8

Sequence of operations in Fourier synthesis for symmetry *p2gg*

$$\sum_{\substack{h \quad k \\ h+k \text{ even}}}{}' \sum{}' F_{hk0} \cos 2\pi hx \cos 2\pi ky - \sum_{\substack{h \quad k \\ h+k \text{ odd}}}{}' \sum{}' F_{hk0} \sin 2\pi hx \sin 2\pi ky$$

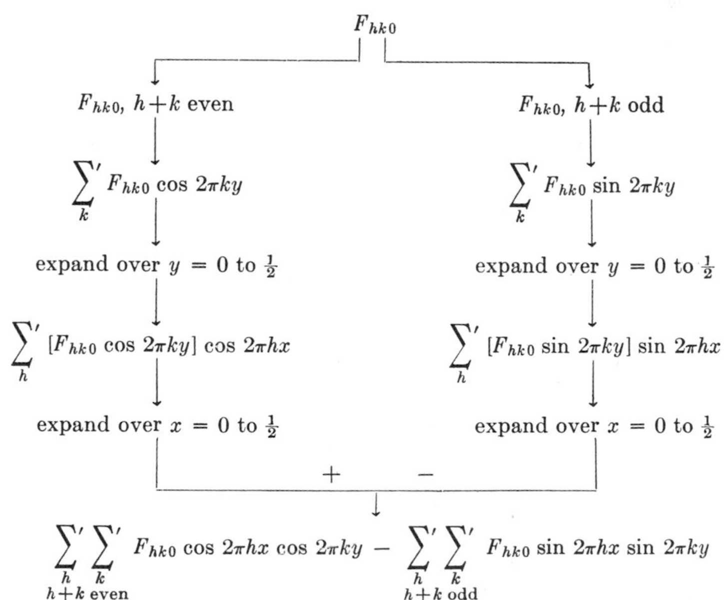

In summing two-dimensional series directly, one wishes to sum the Fourier series in the form (omitting a trivial scale constant)

$$\rho_{(xy)} = \sum_h{}' \sum_k{}' F_{hk0} \cos 2\pi(hx+ky). \tag{15}$$

This can be broken into a set of subseries, each with k constant, thus:

$$\begin{aligned} \rho_{(xy)} = \sum_h{}' \{ & F_{h00} \cos 2\pi(hx+0y) \\ + \; & F_{h10} \cos 2\pi(hx+1y) \\ + \; & F_{h20} \cos 2\pi(hx+2y) \\ & \cdot \\ & \cdot \\ + \; & F_{hK0} \cos 2\pi(hx+Ky) \}. \end{aligned} \tag{16}$$

Table 9
Sequence of operations in Fourier synthesis for symmetry $p2$

$$\sum_h{}' \sum_k{}' (F_{hk0}+F_{\bar{h}k0}) \cos 2\pi hx \cos 2\pi ky - \sum_h{}' \sum_k{}' (F_{hk0}-F_{\bar{h}k0}) \sin 2\pi hx \sin 2\pi ky$$

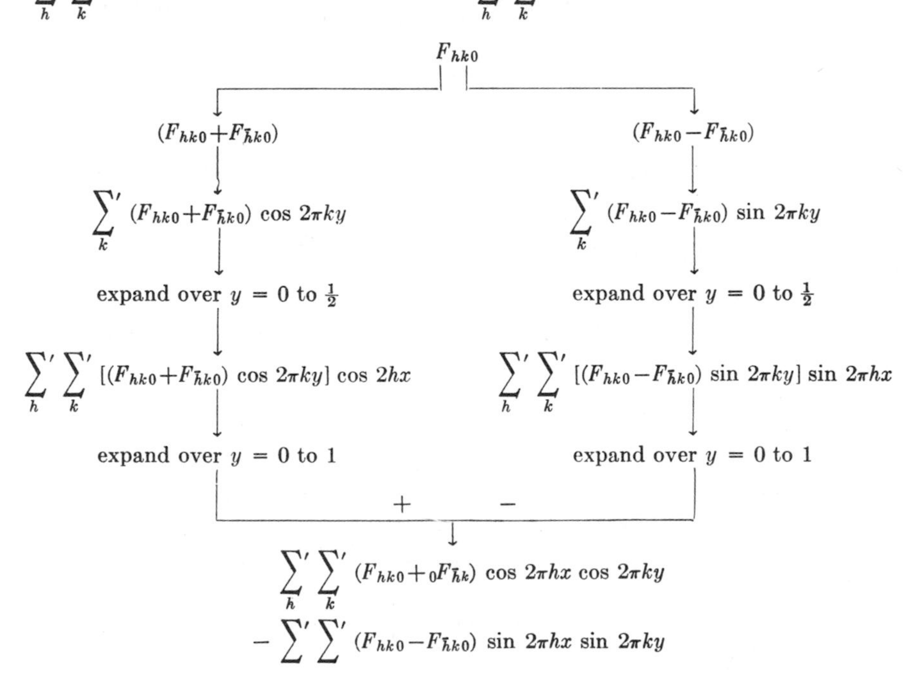

Each such subseries can be summed (as will be shown) on a set of strips for various values of x and y. The sum for the several subseries is the entire summation.

To see how this can be arranged, let

$$\phi = 2\pi(hx+ky). \tag{17}$$

Then each subseries is of the form

$$\sum_h{}' F_{hk_10} \cos \phi_{k_1} \tag{18}$$

The angle $2\pi\phi_{k_1}$ is made up of two components, $2\pi hx$ and $2\pi ky$. With k held constant, the contributions of these two components to ϕ are diagrammatically shown in Fig. 8*A*. Since k is constant, the contribution

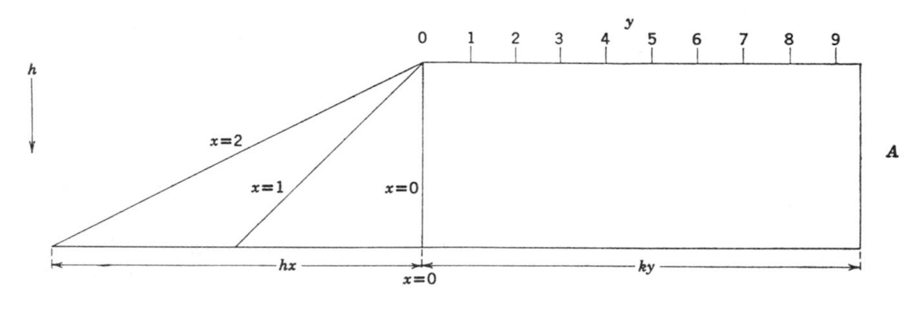

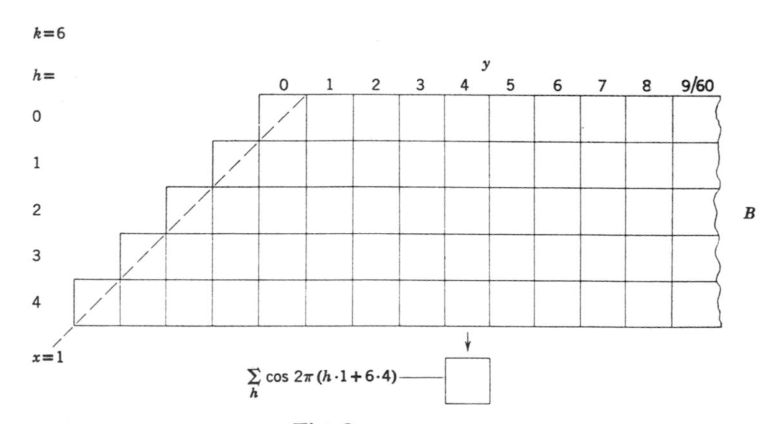

Fig. 8.
Scheme of the Grenville-Wells method of summing two-dimensional Fourier series.

of $2\pi ky$ varies uniformly with y. Since both h and x are variable, the contribution of $2\pi hx$ increases uniformly with h if x is held constant, and is doubled if x is doubled, etc. The cosine of ϕ can be found by laying a Patterson-Tunell-type strip with its left-hand side at the given x line,

Fig. 8*B*. The contribution to the point xy of $F_{hk0} \cos \phi$ (h and k being constant) can therefore be found by putting the left side of the strip at the proper value of x for the proper level of h and reading the $F \cos \phi$ under the proper value of y. Accordingly, if several strips for the various values of h are laid with the left end at the proper x value, their sum in the proper y column gives the contribution at xy of all F_{hk_10}. Their contributions at other y values are found by adding different columns. These contributions at other x's are given by moving the left end of the

Table 10*A*

Sequence of operations in Fourier synthesis for symmetry *p*1, *K* > *H*

$$\sum_h{}' \sum_k{}' \{(A_{hk0}+A_{\bar{h}k0}) \cos 2\pi ky + (B_{hk0}+B_{\bar{h}k0}) \sin 2\pi ky\} \cos 2\pi hx$$

$$+ \sum_h{}' \sum_k{}' \{(B_{hk0}-B_{\bar{h}k0}) \cos 2\pi ky - (A_{hk0}-A_{\bar{h}k0}) \sin 2\pi ky\} \sin 2\pi hx$$

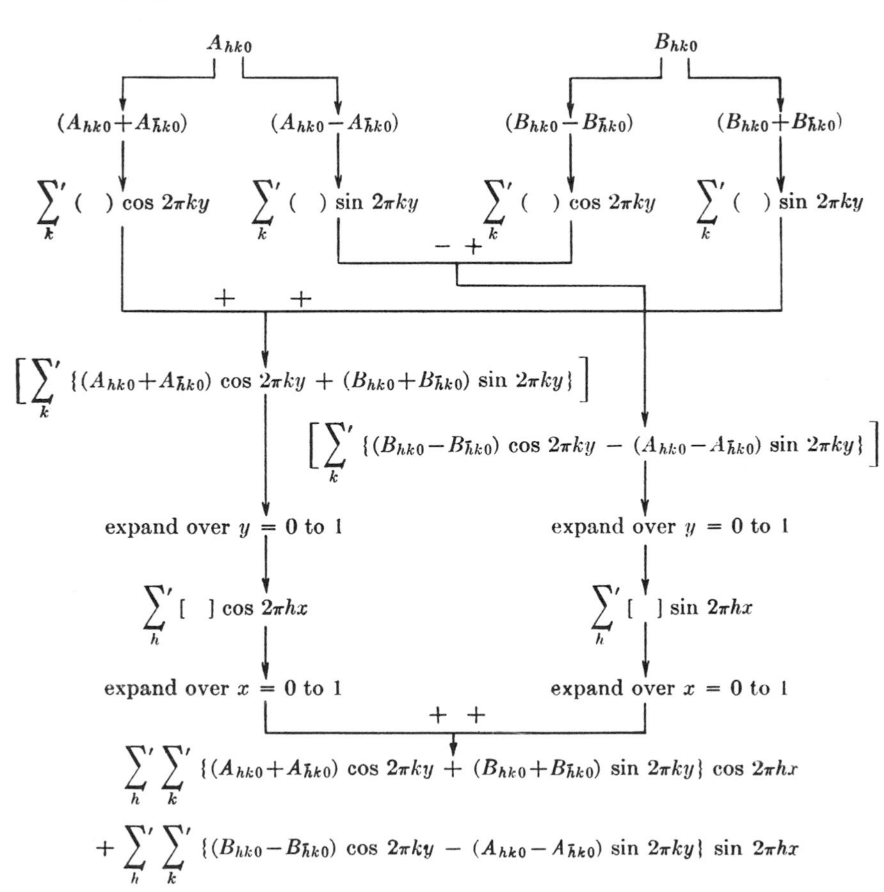

strips to another diagonal x line. The contribution of other F_{hk0}'s for different k's can be similarly determined by setting up other sets of strips for different k values.

In the Grenville-Wells method, a card based upon Fig. 8 is supplied for each value of k. On each card one sets strips for all F_{hk0} values, k being constant for the card. The sum of all the numbers in y=constant columns for all strips gives $\rho_{(xy)} = \Sigma F_{hk0} \cos 2\pi(hx+ky)$ for one value of the pair xy. When sine summations are necessary, $\Sigma F_{hk0} \sin 2\pi(hx+ky)$ can also be summed by setting the strips to a different origin.

Table 10B

Sequence of operations in Fourier synthesis for symmetry p1, $H > K$

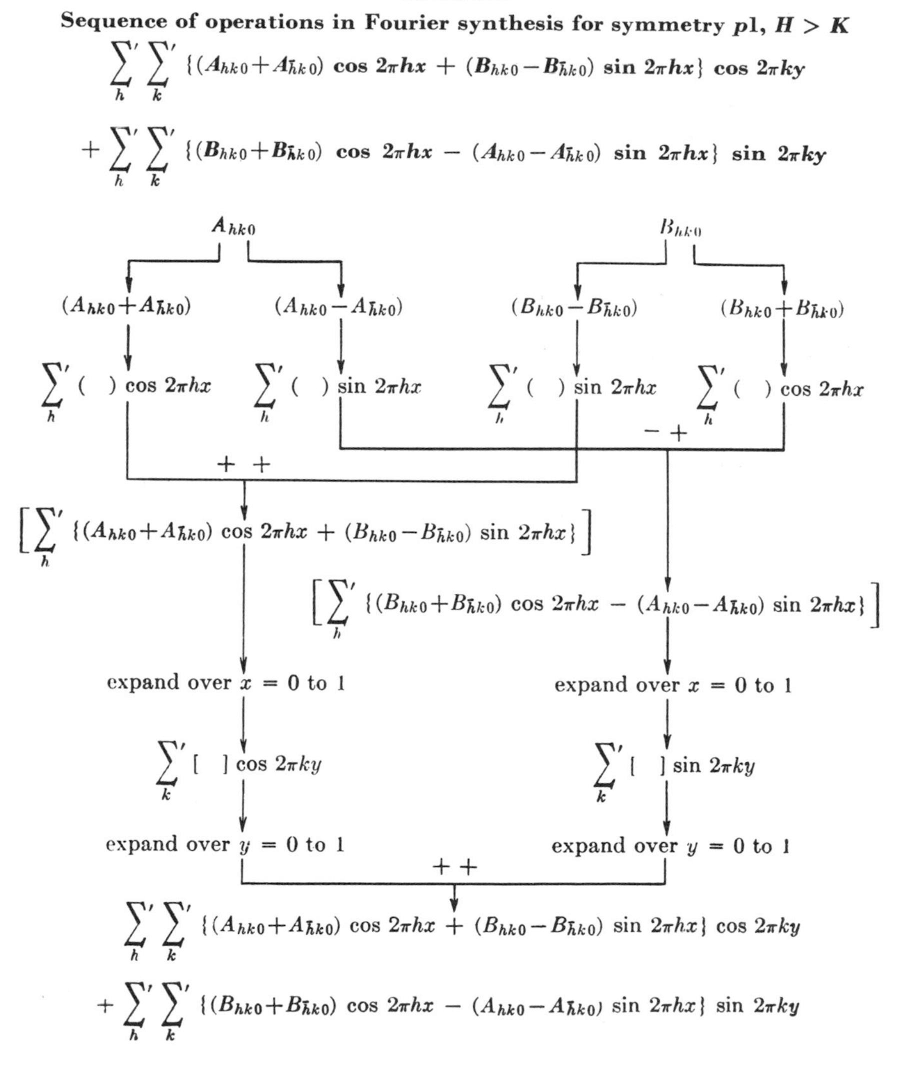

This method has the advantage that the synthesis can be independently performed for one point xy without going through the first entire summation over k, say, as required in the Beevers-Lipson method.

Phase-table method

W. de Beauclair[14] has presented values of $\cos 2\pi(hx+ky)$ for a field of x and y divided into 48ths of a cycle, for different values of h and k. To use the table for computing (15) one must multiply each F_{hk0} by the appropriate trigonometric value, and sum results for a particular xy. This method has not come into general use.

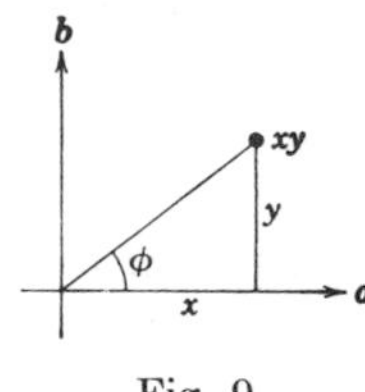

Fig. 9.

The Lukesh semipolar method

Lukesh devised a novel method of Fourier synthesis,[8, 9] Starting with

(15): $$\rho_{(xy)} = \sum_h \sum_k F_{hk0} \cos 2\pi(hx+ky),$$

one substitutes, (Fig. 9),

$$y = x \tan \phi, \tag{19}$$

so that (15) becomes

$$\rho_{(xy)} = \sum_h \sum_k F_{hk0} \cos 2\pi(hx+kx \tan \phi) \tag{20}$$

$$= \sum_h \sum_k F_{hk0} \cos 2\pi x(h+Ck), \tag{21}$$

$$\text{where} \qquad C = \tan \phi. \tag{22}$$

To apply the system one uses an unusual polar net of sample locations. This method has not come into general use.

The high-index line synthesis

The relations between the direct and reciprocal lattices suggests an interesting way of performing a two-dimensional synthesis by means of a one-dimensional summation. Consider a set of rational planes in the reciprocal lattice, Fig. 10*A*. The transform of this set of planes lies along a line at right angles to the planes. In other words if the $(uv)^*$ reciprocal-lattice planes are used in a Fourier transform, the line $[uv]$ in the direct lattice results. This line is periodic with the period of the lattice translation $\mathbf{t}_{uv} = u\mathbf{a}+v\mathbf{b}$.

Now the interesting thing about this synthesis is that the period of the line $[uv]$ is not complete until it crosses several cells, Fig. 10*B*. Each cell is crossed in a different region. If all traversed cells are translated to the same location, the cell is crossed by a uniformly spaced set of lines, Fig. 10*B*. This is the line $[uv]$ modulo Γ, where Γ is the group of lattice translations. Indeed this complete set of lines $[uv]$ which crosses one cell is simply translated fragments of the reciprocal to the complete set of planes $(uv)^*$ crossing the reciprocal cell. If a plane $(u1)^*$ is chosen, Fig. 11*A*, the b^* translation repeats the first plane as successive planes. In direct space, Fig. 11*B*, successive applications of the $-a$ translation restore successive segments of the line $[u1]$ to the first cell.

This Fourier transformation can be readily carried out by using the standard Fourier relation, (11), Chapter 15, but expressed for a point at a distance s from the origin in direct space:

$$\rho_s = \int_{V^*} R_{t^*}\, e^{-i2\pi \mathbf{s}\cdot\mathbf{t}^*}\, dV^*. \tag{23}$$

In the case of points on a lattice, this relation reduces to a summation for points of weight F_n confined to discrete planes at distances $\mathbf{t}^* = n\mathbf{d}^*$ from the origin. The direct-lattice distance, $\mathbf{s}$, can be expressed $x\mathbf{t}$, i.e., as a fraction x of the period of $\mathbf{t}_{uv}$. Thus (23) becomes

$$\rho_{(x)} = \sum F_n\, e^{-i2\pi x\mathbf{t}\cdot n\mathbf{d}_n^*}. \tag{24}$$

Since

$$d_n^* = \frac{1}{t_n}, \tag{25}$$

this reduces to

$$\rho_{(x)} = \sum F_n\, e^{-i2\pi nx}. \tag{26}$$

For centrosymmetrical crystals this has the simpler form

$$\rho_{(x)} = \sum F_n \cos 2\pi nx. \tag{27}$$

Here n indicates the nth reciprocal-lattice plane from the origin and x is the fractional coordinate along a period of the line $[uv]$.

The weighting of the nth $(uv)^*$ plane consists of all the F_{hk0}'s on that plane, i.e., the F_n's. The F's on the same plane can be distinguished by relations among their indices. Thus, if the planes are $(uv)^*$, the intercepts of the first rational plane from the origin are v, u. Figures 10*A* and 11*A* show that indices of *F*'s lying on the same plane have indices *h-mv*, *k+mu*, where *m* is an integer. The particular plane on which an F_{hk0} belongs could be found by reducing the index in this way to the index of a point on a plane separated from the origin plane by *n* translations. This would show that the F_{hk0} is on the *n*th plane from the origin.

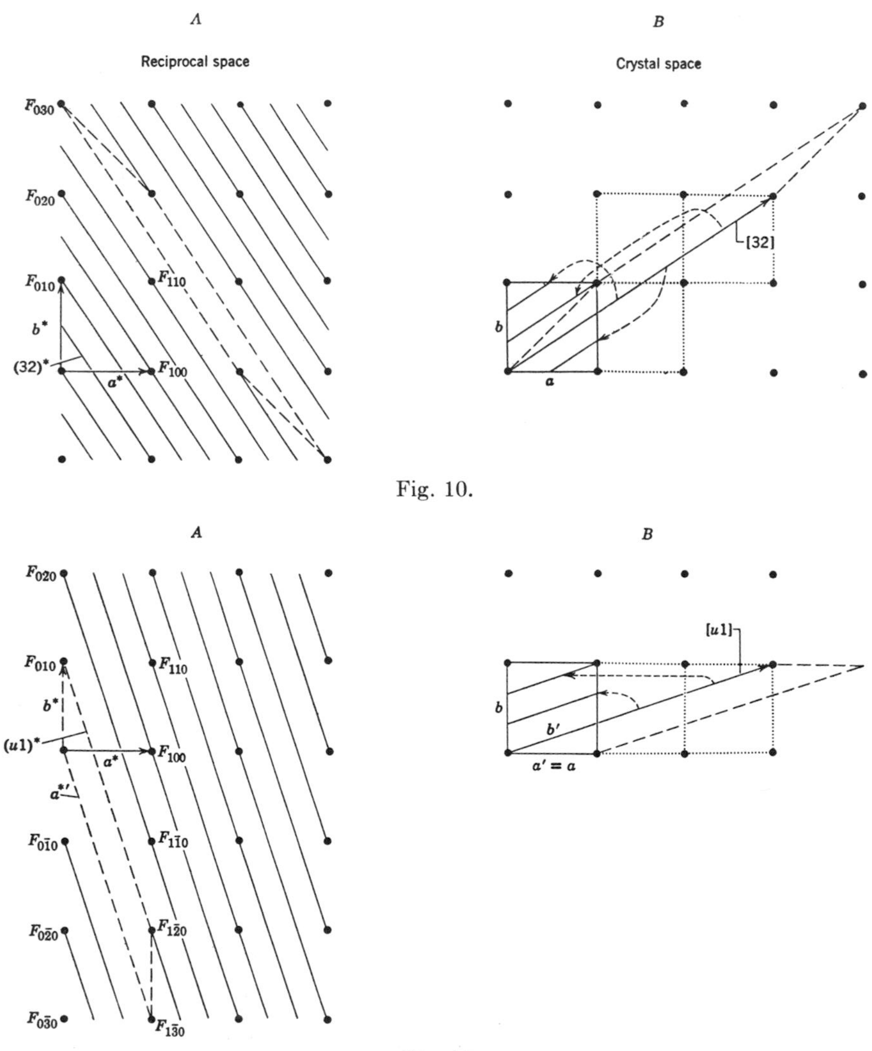

Fig. 10.

Fig. 11.

When a plane $(u1)$ is used, Fig. 11*A*, the indices of *F*'s on the same plane have the form $h-m$, $k+mu$. The index of an *F* on the *n*th plane from the origin could be reduced by this relation to $0n0$.

A more general way of finding the plane in the reciprocal lattice to which each F_{hk0} belongs is to make use of an index transformation.

Thus the high-index line in the direct lattice may be taken as a cell edge, and the high-index plane in the reciprocal lattice may be taken as the corresponding pinacoid. Figure 11A shows that the required direct lattice transformation is as follows:

New, oblique cell from original cell:

$$\begin{aligned} a' &= a, \\ b' &= ua + b \end{aligned} \qquad \text{matrix} \begin{Vmatrix} 1 & 0 \\ u & 1 \end{Vmatrix}. \tag{28}$$

The index transformation (i.e., reciprocal-lattice transformation) has the same matrix.† Therefore the index transformation is

$$\begin{aligned} h' &= h, \\ k' &= uh + k. \end{aligned} \tag{29}$$

For example, the particular cell transformation of Fig. 11B is

$$\begin{aligned} a' &= a, \\ b' &= 3a + b \end{aligned} \qquad \text{matrix} \begin{Vmatrix} 1 & 0 \\ 3 & 1 \end{Vmatrix} \tag{30}$$

Thus, the point $hk = 11$ of Fig. 11A transforms as follows:

$$\begin{aligned} h' &= 1h + 0k = 1{\cdot}1 + 0{\cdot}1 = 1, \\ k' &= 3h + 1k = 3{\cdot}1 + 1{\cdot}1 = 4. \end{aligned} \tag{31}$$

Thus $\cdot 11 \cdot^* \to \cdot 14 \cdot^*$ and $F_{110} \to F_{140}$; i.e., F_n of equations (26) and (27) is F_4, since F_{140} is on a "plane" removed from the origin pinacoid of the reciprocal cell by 4 b^* translations.

To make practical use of this property of a high-index line synthesis, one must select a line which traverses the cell at sufficiently closely spaced intervals. For simplicity, confine attention to the special kinds of high-index lines $[u1]$, Fig. 11B. Such a line must cross u cells to attain a lattice point, so that the line, modulo Γ, crosses the origin cell u times. If one would accept a Beevers-Lipson sampling of the cell edges in 60 parts, then a line [60, 1] would accomplish an equivalently close sampling. The sampling interval along the line should be such that one cell crossing, i.e., $\frac{1}{60}$th of the period of the line [60, 1], would be divided into 60 parts. The entire line synthesis would thus be summed at 3600 points.

Evidently the computation for a completely asymmetric cell by the high-index line method would require the same number of summation points as the Beevers-Lipson summation. The number of sample points for which summations must be made is reduced for symmetrical crystals

† M. J. Buerger. *X-ray crystallography.* (John Wiley and Sons, New York, 1942) 12, 13.

in both methods. The advantages of the high-index line method are that only one type of operation is used, that only cosine summations are necessary unless an inversion center is missing, and that the summation for any individual point can be undertaken without preliminary computation. The disadvantage is that no "strips" are commercially available for the method.

This method is well adapted to high-speed digital computation and can be readily extended to routine full three-dimensional summations. It has not come into common use. It has been exploited to a limited extent by Rose and Rimsky[13] and by Donnay and Donnay.[15]

The circle theory of Fourier synthesis

Grenville-Wells[18] pointed out that the contribution of a reflection $hk0$ to a Fourier synthesis at point xy is related to a circle controlled by xy. Let the plane and point be as shown in Fig. 12*A*. For simplicity, let

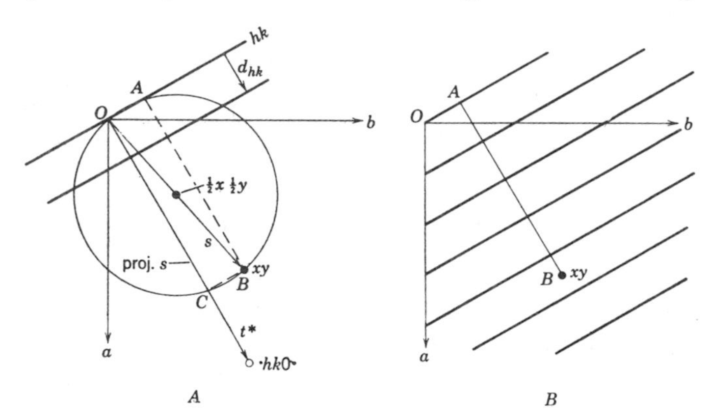

Fig. 12.

the crystal be centrosymmetrical. Then the contribution of F_{hk0} to point xy is $F_{hk0} \cos 2\pi(hx+ky)$. The phase part of this is given by reading the Bragg-Lipson chart,† symmetry $p2$, Fig. 12*B*, for the length and direction of AB. This is the same as the length and direction of OC. Thus the Fourier contribution at xy of any plane hk is controlled by the intersection of a circle whose center is at $\frac{1}{2}x\ \frac{1}{2}y$ with the normal to the plane. Unfortunately this is not of much help since the resulting value must be multiplied by F_{hk0} before entering into a summation.

A somewhat more direct interpretation of this same relation can be appreciated by using reciprocal space. The contribution of an F_{hk0} is

$$^{hk0}\rho_{(xy)} = F_{hk0}\, e^{-i2\pi \mathbf{s}.\mathbf{t}^*_{hk0}}. \tag{32}$$

† Chapter 11, Fig. 4.

The phase part of this is dependent on the projection of the direct-lattice vector **s** on the vector $\mathbf{t}^*_{hk0}$ which runs to the reciprocal-lattice point $\cdot hk0\cdot$. Unfortunately, this phase factor must be multiplied by F_{hk0} before it enters the summation.

Mechanical methods of Fourier synthesis

Many mechanical and graphical methods of computing Fourier syntheses have been prepared and some of these methods are frequently used. These methods are so numerous that they would require a separate volume for adequate discussion. Some devices for computing Fourier syntheses are mentioned in the literature list.

High-speed digital computing

With the development of high-speed digital computors, more and more Fourier syntheses are being carried out by this method. While strip methods may require several days for the computation of the summations for the desired number of sample points of an electron-density projection, high-speed digital computors can do the same job in, say, four minutes. In fact the greatest time is taken by the human elements of entering the data into the computor and copying the results into the appropriately scaled outline of the cell.

The advent of high-speed computing methods marks the vanishing of one of the important barriers in crystal-structure analysis. Few three-dimensional syntheses would be undertaken without the aid of high-speed computors. The literature list will furnish a guide to application of the various methods to Fourier syntheses.

Literature

[1] C. A. Beevers and H. Lipson. *A rapid method for the summation of a two-dimensional Fourier series.* Phil. Mag. (7) **17** (1934) 855–859.

[2] C. A. Beevers and H. Lipson. *A numerical method for two-dimensional Fourier synthesis.* Nature **137** (1936) 825–826.

[3] H. Lipson and C. A. Beevers. *An improved numerical method of two-dimensional Fourier synthesis for crystals.* Proc. Phys. Soc. (London) **48** (1936) 772–780.

[4] J. Monteath Robertson. *Numerical and mechanical methods in double Fourier synthesis.* Phil. Mag. (4) **21** (1936) 176–187.

[5] J. Monteath Robertson. *Calculation of structure factors and summation of Fourier series in crystal analysis: non-centrosymmetrical projections.* Nature **138** (1936) 683–684.

[6] T. H. Goodwin and R. Hardy. *The computation of three-dimensional Fourier syntheses in crystal structure analysis.* Phil. Mag. (7) **25** (1938) 1096–1104.

[7] A. L. Patterson and George Tunell. *A method for the summation of the Fourier series used in the x-ray analysis of crystal structures.* Am. Mineralogist **27** (1942) 655–679.

[8] Joseph S. Lukesh. *A semi-polar form of Fourier series and its use in crystal structure analysis.* J. Appl. Phys. **18** (1947) 321–326.

[9] Joseph S. Lukesh. *Crystal pattern synthesis by an approximate summation of Fourier series.* J. Appl. Phys. **18** (1947) 493–498.

[10] J. Monteath Robertson. *Three-figure cosine factors for Fourier analysis and synthesis. Part I—A modified stencil method.* J. Sci. Instr. **25** (1948) 28–30.

[11] J. Monteath Robertson. *Three-figure cosine factors for Fourier analysis and synthesis. Part II—A mechanical sorting device.* J. Sci. Instr. **25** (1948) 216–218.

[12] W. Cochran. *A critical examination of the Beevers-Lipson method of Fourier series summation.* Acta Cryst. **1** (1948) 54–56.

[13] A. J. Rose and A. Rimsky. *Méthode de calcul pour la sommation des séries de Fourier. Coordonnées polaires.* Bull. soc. franç. minéral. et crist. (1949) 305–318.

[14] W. de Beauclair and U. Sinogowitz. *Untersuchengen über die Fouriersynthese der Ladungsverteilung in Kristallen. Band I. Verfahren und Geräte zur mehrdimensionalen Fouriersynthese. Band II. Phasenfaktorentafel zur kristallographischen zweidimensionalen Fouriersynthese in Punkten eines Achtundvierzigstel-Netzes.* (Akademic-Verlag, Berlin, 1949).

[15] J. D. H. Donnay and Gabrielle Hamburger Donnay. *The one-dimensional crystal. I. General.* Acta Cryst. **2** (1949) 366–369. *II. A graphical method for computing structure factors.* Acta Cryst. **2** (1949) 370–371.

[16] C. A. Beevers. *Fourier strips at a 3° interval.* Acta Cryst. **5** (1952) 670–673.

[17] Fernando Huerta y José M.ª Casals. *Un método grafico para sumar series de Fourier.* Anales real soc. españ. fís. y quím. **48** (1952) 238–243.

[18] H. J. Grenville-Wells. *A graphical method of evaluating trigonometric functions used in crystal-structure analysis. Parts I and II.* J. Appl. Phys. **25** (1954) 485–490.

[19] H. J. Grenville-Wells. *Moving-strip Fourier analyzer.* Rev. Sci. Instr. **25** (1954) 1156–1161.

[20] H. J. Grenville-Wells. *A new graphical principle for the evaluation of Fourier transforms.* Acta Cryst. **8** (1955) 737.

[21] P. Fellgett. *On numerical Fourier transformation, with special reference to Lipson-Beevers strips.* J. Sci. Instr. **35** (1958) 257–258.

Miscellaneous machine methods

[22] W. Lawrence Bragg. *An optical method of representing the results of x-ray analysis.* Z. Krist. **70** (1929) 475–492.

[23] W. Lawrence Bragg. *Die Untersuchung der Atomanordnung mittels Röntgenstrahlen.* Metallwirtschaft **9** (1930) 461–465.

[24] J. Monteath Robertson. *A simple harmonic continuous calculating machine.* Phil. Mag. (7) **13** (1932) 413–419.

[25] Alfred Hettich. *Beiträge zur Methodik der Strukturbestimmung. V. Zur*

photographischen Summierung von Fouriergliedern. Z. Krist. (A) **90** (1935) especially 488–492.

[26] C. A. Beevers. *A machine for the rapid summation of Fourier series.* Proc. Phys. Soc. **51** (1939) 660–667.

[27] Louis R. Maxwell. *An electrical method for compounding sine functions.* Rev. Sci. Instr. **11** (1940) 47–54.

[28] Maurice L. Huggins. *Photographic Fourier syntheses.* J. Am. Chem. Soc. **63** (1941) 66–69.

[29] Douglas Macewan and C. A. Beevers. *A machine for the rapid summation of Fourier series.* J. Sci. Instr. **19** (1942) 150–156.

[30] Maurice L. Huggins. *Cooperation between the sciences.* Am. Scientist **31** (1943) 338–345.

[31] G. Hägg and T. Laurent. *A machine for the summation of Fourier series.* J. Sci. Instr. **23** (1946) 155–158.

[32] Dan McLachlan, Jr. and E. F. Champaygne. *A machine for the application of sand in making Fourier projections of crystal structures.* J. Appl. Phys. **17** (1946) 1006–1014.

[33] R. Pepinsky. *An electronic computer for x-ray crystal structure analyses.* J. Appl. Phys. **18** (1947) 601–604.

[34] Roscoe H. Wooley and Dan McLachlan, Jr. *A multiple projector for the Huggins masks.* (Presented in part as a paper before the Cornell meeting of the American Society for X-ray and Electron Diffraction, June 19, 1949. Unpublished.)

[35] H. Shimizu, P. J. Elsey, and D. McLachlan, Jr. *A machine for synthesizing two-dimensional Fourier series in the determination of crystal structures.* Rev. Sci. Instr. **21** (1950) 779–783.

[36] J. L. Amorós. *Empleo de métodos mecanicos en el calculo de la sintesis tridimensional del Co_2Al_9.* Anales real soc. españ. fís. y quím. (A) **47** (1951) 1–12.

[37] Gérard von Eller. *Sur un nouvel appareil pour le développement par voie optique des séries de Fourier à plusiers dimensions.* Comptes rend. **232** (1951) 1122–1124.

[38] Barton Howell, Carl J. Christensen, and Dan McLachlan, Jr. *A method for making accurately reproducible masks of the Huggins type.* Nature **168** (1951) 282–284.

[39] C. A. Beevers and J. Monteath Robertson. *A multi-component integrator for Fourier analysis and structure factor calculator.* Computing Methods and the Phase Problem in X-ray Crystal Analysis (X-Ray Crystal Analysis Laboratory, The Pennsylvania State College, 1952) 119–129.

[40] I. W. Ramsay, H. Lipson, and D. Rogers. *An electrical analogue machine for Fourier synthesis.* Computing Methods and the Phase Problem in X-Ray Crystal Analysis (X-Ray Crystal Analysis Laboratory, The Pennsylvania State College, 1952) 130–131.

[41] Ray Pepinsky. *X-RAC and X-FAC: Electric analogue computers for x-ray analysis.* Computing Methods and the Phase Problem in X-Ray Crystal Analysis (X-Ray Crystal Analysis Laboratory, The Pennsylvania State College, 1952) 167–390.

[42] A. J. Rose. *Machine á calculer permettant la détermination de fonctions*

périodiques et leur introduction dans des calculs. Application a la sommation des séries de Fourier et au calcul des facteurs de structure en cristallographie. Journal des recherches **7,** 1–6.

[43] V. Vand. *A Fourier electron-density balance.* J. Sci. Instr. **29** (1952) 118–121.

[44] Leonid V. Azaroff. *A one-dimensional Fourier analog computer.* Rev. Sci. Instr. **25** (1954) 471–477.

[45] G. Suryan. *An analogue computer for double Fourier series summation for x-ray crystal-structure analysis.* Acta Cryst. **10** (1957) 82–83.

[46] V. Frank. *A Fourier analogue-computer of the Hägg-Laurent type.* J. Sci. Instr. **34** (1957) 210–211.

[47] Leonid V. Azároff. *Semiautomatic, two-dimensional Fourier analog computer.* Rev. Sci. Instr. **29** (1958) 317–318.

High-speed digital computing

[48] P. A. Shaffer, Jr., Verner Schomaker, and Linus Pauling. *The use of punched cards in molecular structure determinations. I. Crystal structure calculations.* J. Chem. Phys. **14** (1946) 648–658.

[49] P. A. Shaffer, Jr., Verner Schomaker, and Linus Pauling. *The use of punched cards in molecular structure determinations. II. Electron diffraction calculations.* J. Chem. Phys. **14** (1946) 659–664.

[50] W. Nowacki. *Die Verwendung von Lochkartenmaschinen zur Fourier- und Patterson-Synthese von Kristallen.* Chimia. **2** 12 (1948).

[51] M. L. Hodgson, C. J. B. Clews, and W. Cochran. *A punched-card modification of the Beevers-Lipson method of Fourier synthesis.* Acta Cryst. **2** (1949) 113–116.

[52] E. G. Cox and G. A. Jeffrey. *The use of 'Hollerith' computing equipment in crystal-structure analysis.* Acta Cryst. **2** (1949) 341–343.

[53] E. G. Cox, L. Gross, and G. A. Jeffrey. *A Hollerith technique for computing three-dimensional differential Fourier syntheses in x-ray crystal-structure analyses.* Acta Cryst. **2** (1949) 351–355.

[54] D. M. S. Greenhalgh and G. A. Jeffrey. *A new punched-card method of Fourier synthesis.* Acta Cryst. **3** (1950) 311–312.

[55] J. M. Bennett and J. C. Kendrew. *The computation of Fourier syntheses with a digital electronic calculating machine.* Acta Cryst. **5** (1952) 109–116.

[56] N. Kitz and B. Marchington. *A method of Fourier synthesis using a standard Hollerith Senior rolling total tabulator.* Acta Cryst. **6** (1953) 325–326.

[57] S. W. Mayer and K. N. Trueblood. *Three-dimensional Fourier summations on a high-speed digital computer.* Acta Cryst. **6** (1953) 427.

[58] T. R. Thompson, D. T. Caminer, L. Fantl, W. B. Wright, and G. S. D. King. *The use of a high-speed automatic calculator in the refinement stages of crystal-structure determinations.* Acta Cryst. **7** (1954) 260–269.

[59] J. Monteath Robertson. *Some properties of Fourier strips, with applications to the digital computer.* Acta Cryst. **8** (1955) 286–288.

[60] V. Timbrell. *A calculator for numerical Fourier synthesis.* J. Sci. Instr. **35** (1958) 313–318.

[61] F. R. Ahmed and W. H. Barnes. *Generalized programmes for crystallographic computations.* Acta Cryst. **11** (1958) 669–671.

19

Phase determination for structures having a set of heavy or replaceable atoms

In order to prepare a Fourier synthesis of the electron density it is necessary to know the phases of the Fourier coefficients. These cannot be determined experimentally. Nevertheless, there are some special cases in which some or all of the phases can be learned from other data. In these instances the electron density, or an approximation to it, can be found by Fourier synthesis. Usually this reveals the locations of the atoms in the structure in sufficient detail so that the remaining phases can be computed.

There are two important situations in which the phases can be determined. If the crystal contains a heavy atom whose scattering power dominates the intensities, it may control some or all of the phases. If the location of the atom in the cell is known, then these phases can be determined. The possibility of phase determination also arises if two isomorphous crystals are available in which there is one replaceable atom. It is then often possible to determine the phase scattered by the replaceable atom, and this information can be used in conjunction with the changes in intensity of the reflections for the two crystals to determine the phases of the reflections. In this chapter these methods are discussed.

Phase determination for crystals having a set of heavy atoms

Possibilities of locating the heavy atom. There are at least two ways in which the location of a heavy atom can be found without further knowledge about the crystal structure. One way of doing this arises whenever the number of heavy atoms per cell is so small that they must occupy one or more sets of equipoints without degrees of freedom, or, at least, without degrees of freedom in projection (See Chapter 9). A second way of finding the location of the heavy atom is with the aid of Patterson or Harker syntheses.†

Centrosymmetrical case. When the heavy atom has been located by one of the above-mentioned means, the contribution of this atom to each of the reflections F_{hkl} can be computed. Let the contribution of the heavy atom to this wave be ${}^{H}F_{hkl}$, and the contribution of the rest of the structure be ${}^{R}F_{hkl}$. The general relation between these complex quantities is

$$F_{hkl} = {}^{H}F_{hkl} + {}^{R}F_{hkl}. \tag{1}$$

If ${}^{H}S$ is the symmetry factor of the space group for the equipoint occupied by the heavy atom (Chapter 10), then the heavy-atom contribution is

$${}^{H}F_{hkl} = {}^{H}f_{hkl}\,{}^{H}S_{hkl}. \tag{2}$$

This can be readily computed. Thus (1) can be rewritten

$$F_{hkl} = {}^{H}f_{hkl}\,{}^{H}S_{hkl} + {}^{R}F_{hkl}. \tag{3}$$

This is a general relation. If the crystal is centrosymmetrical and if a symmetry center is chosen as origin, then all values in (3) are real, and consequently are either positive or negative quantities. The maximum value of ${}^{R}F_{hkl}$ is the sum of f_{hkl}'s pertaining to the residue of the structure. This is also readily computed. Whenever the heavy-atom contribution is greater than this $\Sigma\,{}^{R}f_{hkl}$ term, then the sign of F_{hkl} can be taken as that for the contribution of the heavy atom. This result is useful in preparing a preliminary Fourier synthesis of the structure. Such a preliminary Fourier synthesis has as coefficients only those F's whose signs are controlled by the contributions of the heavy atom (other F's being omitted).

When the heavy atoms are not in special positions in a centrosymmetrical structure, their contributions to the general reflections are ordinarily neither a maximum (in-phase contribution) nor zero, but rather have variable values. In these instances, (2) should be computed for each reflection to see if the contribution of the heavy atoms dominates the right side of (3). When the first term in the right of (3) is greater than the maximum value of the second term, the sign of F_{hkl} can be said to be certainly determined, and can be taken as that of the contribution of the heavy atom. In such instances a preliminary Fourier synthesis, such as mentioned above, can be prepared, using only those F's whose

† See M. J. Buerger, *Vector space and its application in crystal-structure investigation*. (John Wiley and Sons, New York, 1959) Chapters 6 and 7.

signs are controlled by the metal atoms. Actually, except in the rare circumstance that all the atoms of the residue scatter in phase, the contributions of all the individual atoms of the residue never attain their maximum values. In fact, in the more complicated structures, the atoms of the residue tend to be uniformly, but randomly, distributed throughout the cell. The phasal relations of the waves scattered by this kind of distribution is such that the waves contributed by the individual atoms tend to annul one another. Thus, even a moderate contribution by the set of heavy atoms tends to dominate F_{hkl} and hence usually determines its sign.

Example: platinum phthalocyanine. A simple example of the determination of a crystal structure by utilizing phases based upon a heavy atom is afforded by Robertson and Woodward's[4] analysis of platinum phthalocyanine, $PtC_{32}H_{16}N_8$. This compound crystallizes in the monoclinic system, space group $P\,2_1/a$. The cell chosen by Robertson and Woodward has the following dimensions:

$$a = 23.9\text{ Å},$$
$$b = 3.81,$$
$$c = 16.9,$$
$$\beta = 129.6°.$$

The cell contains two molecules of $PtC_{32}H_{16}N_8$. The only 2-fold equipoints in $P\,2_1/a$ are sets of inversion centers, so that the molecule must occupy one of these, which may be taken as the origin and $\frac{1}{2}\,0\,0$. The space group $P\,2_1/a$ projects on (010) as plane group $p2$. The projected cell has half the a axis of the space cell, so that there is only one molecule per cell in projection, with its platinum atom at the origin. Accordingly, all permissible reflections have a maximum positive contribution from the heavy atom.

Table 1
Approximate maximum scattering powers of heavy atom and residue in platinum phthalocyanine, $PtC_{32}H_{16}N_8$

Atom	Atomic number, Z	Number of atoms in asymmetrical unit	Total number of scattering electrons	
Pt	78	1	78	$= {}^{H}Z$
N	7	8	56	
C	6	32	192	264 $= {}^{R}Z$ (N + C + H)
H	1	16	16	

Table 1 shows the analysis of the maximum possible contributions of the heavy atom and the residue. These are in the ratio 0.29. In spite of this small ratio, the residue atoms are distributed through the cell in a non-regular fashion and therefore never attain more than a fraction of their maximum possible contribution. On the other hand, the heavy atom is always making a maximum positive contribution. On this basis it was assumed that the signs of all the 302 $h0l$ spectra were positive. These were used to compute the Fourier synthesis of $\rho_{(xz)}$, which is shown in Fig. 1. On the basis of the atom locations revealed by this electron-density map, all the F_{h0l}'s were recomputed. The signs of all of these were found to have been correctly given by the assumption that the heavy atom controlled the phases.

Example: KH_2PO_4. If the heavy atoms occupy one or more sets of special equipoints, two possible cases arise. The heavy atoms may contribute to all reflections as in the case of the example just given, or they may contribute only to a certain class of reflections. Which of these two actually occurs can usually be determined in advance by examining the form of the contribution by the heavy atoms. The analysis of the structure of KH_2PO_4 by West[1] furnishes an illustration of the second case.

KH_2PO_4 is tetragonal, symmetry $I\bar{4}2d$. Its cell has the following dimensions:

$$a = 7.43 \text{ Å},$$

$$c = 6.97.$$

The cell contains 4 KH_2PO_4.

Space group $I\bar{4}2d$ has two 4-fold, four 8-fold, and a general 16-fold equipoint. The 4-fold equipoints are

$$\begin{array}{llllll} 4a: & 0\,0\,0, & 0\,\tfrac{1}{2}\,\tfrac{1}{4}, & \tfrac{1}{2}\,\tfrac{1}{2}\,\tfrac{1}{2}, & \tfrac{1}{2}\,0\,\tfrac{3}{4}; \\ 4b: & 0\,0\,\tfrac{1}{2}, & 0\,\tfrac{1}{2}\,\tfrac{3}{4}, & \tfrac{1}{2}\,\tfrac{1}{2}\,0, & \tfrac{1}{2}\,0\,\tfrac{1}{4}. \end{array}$$

The 4 K and 4 P atoms must necessarily fill the two sets of equipoints $4a$ and $4b$.

Space group $I\bar{4}2d$ projects on (001) as plane group $c4mg$. This projection has superposed K and P atoms at $0\,0$, $\tfrac{1}{2}\,0$, $0\,\tfrac{1}{2}$, and $\tfrac{1}{2}\,\tfrac{1}{2}$. The contribution of these atoms to reflections $hk0$ is

$$\begin{aligned} {}^{H}F_{hk0} &= (f_K+f_P)\{e^{i2\pi(h\cdot 0+k\cdot 0)} + e^{i2\pi(h\cdot \frac{1}{2}+k\cdot 0)} \\ &\quad + e^{i2\pi(h\cdot 0+k\cdot \frac{1}{2})} + e^{i2\pi(h\cdot \frac{1}{2}+k\cdot \frac{1}{2})}\} \qquad (4) \\ &= (f_K+f_P)\{1 + e^{i\pi h} + e^{i\pi k} + e^{i\pi(h+k)}\}. \qquad (5) \end{aligned}$$

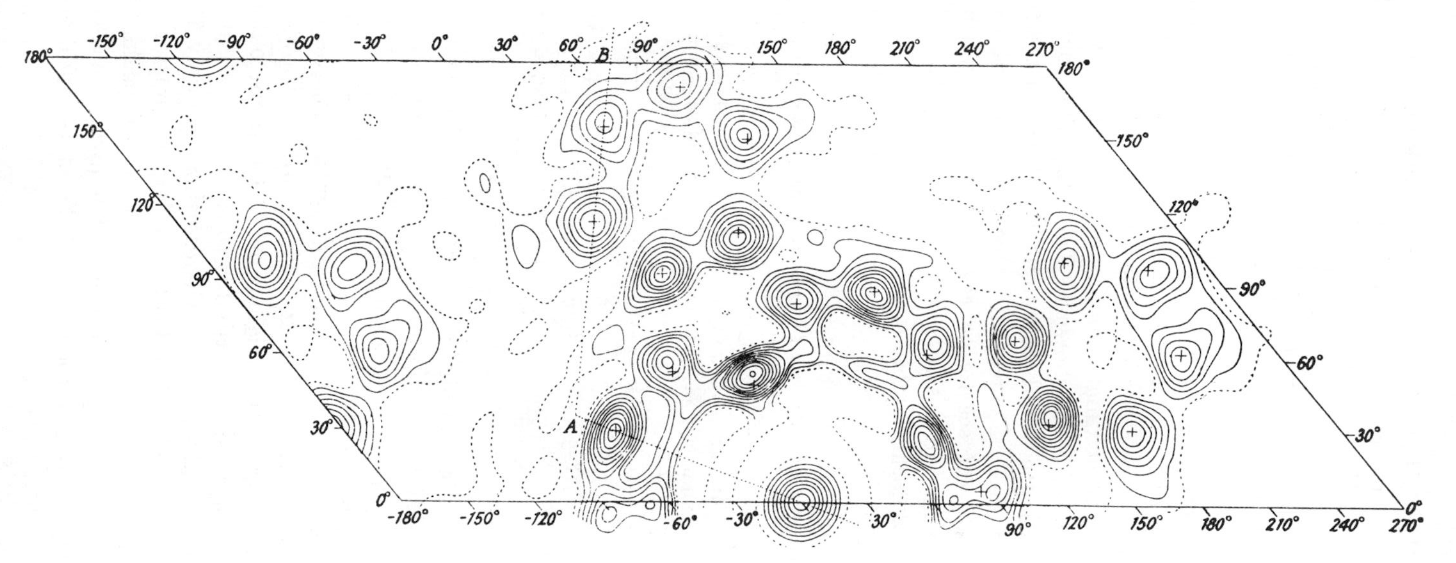

Fig. 1.

Platinum phthalocyanine, $PtC_{32}H_{16}N_8$; electron-density projection, $\rho(xz)$. The contour interval is 1 electron per Å² except for the platinum atom (lower center) where it is 20 electrons per Å². The locations accepted for the atom centers are indicated by crosses. (After Robertson and Woodward.[4])

Recalling that $e^{i(2n)\pi} = +1$ and $e^{i(2n+1)\pi} = -1$, one finds that expression (5) has different values according as h and k are even or odd, as follows:

	$1 + e^{i\pi h}$	$+ e^{i\pi k}$	$+ e^{i\pi(h+k)}$	
h even, k even:	$1 + 1$	$+ 1$	$+ 1$	$= 4$
h odd, k odd:	$1 - 1$	$- 1$	$+ 1$	$= 0$

The combinations h even, k odd and h odd, k even also give zero, but this is a space-group extinction for $h+k$ odd. Thus the heavy atoms give no contribution for permissible reflections which have h odd.

An analysis of the approximate maximum scattering powers of the heavy atoms and the residue is shown in Table 2. The analysis is for neutral atoms; the balance would shift slightly in favor of the residue if the atoms were regarded as ionized. Table 2 shows that the heavy atoms safely dominate the scattering regardless of the distribution of oxygen atoms. The margin is stronger than shown by the table since the f curve for oxygen declines more rapidly with $\sin\theta$ than the f curves for the metals.

Table 2
Approximate maximum scattering powers of heavy atoms and residue in KH_2PO_4

Atom	Atomic number Z	Number of atoms in cell	Total number of scattering electrons	
K	19	4	76	$144 = {}^HZ$
P	17	4	68	
H	1	8	8	$136 = {}^RZ$
O	8	16	128	

This means that the signs of all $hk0$ reflections for h even can be fixed with certainty. The $hk0$ reflections for h odd have no contributions from the metal atoms and depend only on the distribution of oxygen atoms, which is unknown. A Fourier synthesis can be prepared, using only $hk0$ terms for which h is even. An area of this synthesis corresponding to $\frac{1}{16}$ of the entire cell is given in Fig. 2. This synthesis shows not only high peaks corresponding to the original metal locations, but also two very low peaks. These must be attributed to the oxygen atoms. There are, however, 32 such peaks per cell, whereas there are only 16 oxygen atoms per cell, and the duplication displays a diagonal mirror plane not present in the space group. Only one of the oxygen peaks shown in Fig. 2 is actually a correct oxygen location. The mirror image is caused by using only half the Fourier coefficients. The false additional symmetry arises because the Fourier coefficients used have this high symmetry, which is caused, in turn, by the more symmetrical positions of the heavy atoms.

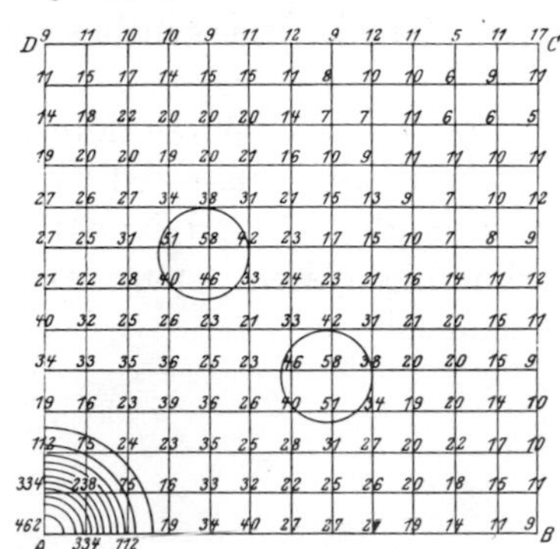

Fig. 2.
KH_2PO_4; Fourier synthesis made using only F_{hk0}'s with h even. The two circles represent alternative possible locations of oxygen atom. (Projection of $\frac{1}{16}$ full cell.)

(After West.[1])

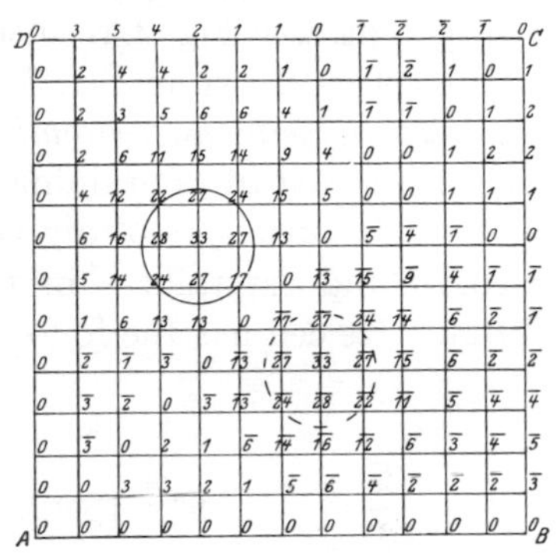

Fig. 3.
KH_2PO_4; Fourier synthesis made using only F_{hk0}'s with h odd, with signs based upon the upper-left oxygen location shown in Fig. 2. (Projection of $\frac{1}{16}$ full cell.)

(After West.[1])

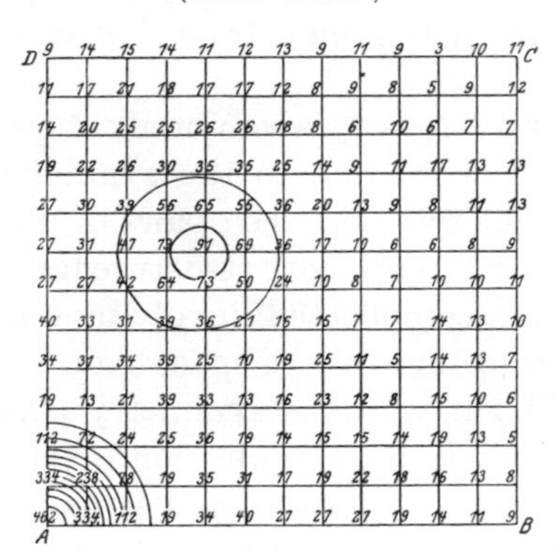

Fig. 4.
Sum of Figs. 2 and 3, i.e., Fourier synthesis made using all F_{hk0}'s. (Projection of $\frac{1}{16}$ full cell.)

(After West.[1])

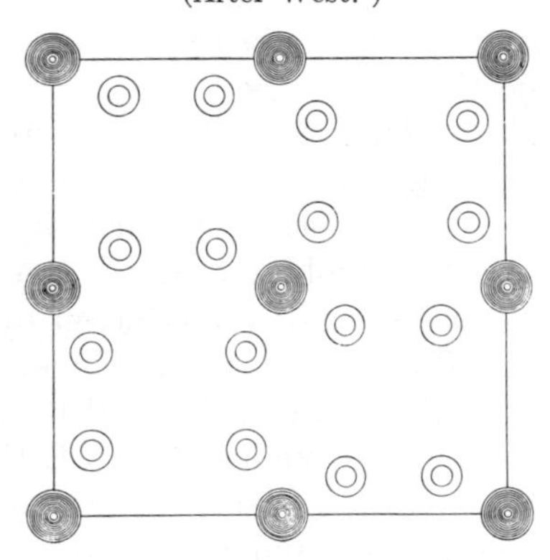

Fig. 5.
KH_2PO_4; $\rho_{(xy)}$ on $\frac{1}{16}$ the scale of Figs. 2, 3, and 4. (Projection of full cell.)

(After West.[1])

Either of the two possible oxygen locations can be arbitrarily accepted. It is then possible to compute the phases for reflection $hk0$ having h odd for this choice. A separate synthesis using only these terms is shown in Fig. 3. Note that this synthesis shows no metal atoms (since these do not contribute to $hk0$ with h odd). It does show a peak at the oxygen location chosen to compute phases, plus a corresponding depression at the false oxygen location of Fig. 2. When Figs. 2 and 3 are added, this corresponds to the full Fourier synthesis using all F_{hk0}'s. This full synthesis is given in Fig. 4. Note that the depression in Fig. 3 cancels the false oxygen peak in Fig. 2, resulting in Fig. 4. The Fourier synthesis for the entire cell is shown in Fig. 5.

Example: diglycine hydrobromide. The general case of phase determination for a structure containing a heavy atom is well illustrated by the analysis of the structure of diglycine hydrobromide,[19] $2(C_2H_5NO_2)HBr$. This compound crystallizes in the orthorhombic system, space group $P2_1\,2_1\,2_1$. Its cell has the following dimensions:

$$\begin{aligned} a &= \ 8.21\text{Å}, \\ b &= 18.42, \\ c &= \ 5.40, \\ Z &= \ 4\ C_4H_{11}N_2O_4Br \text{ per cell.} \end{aligned}$$

The only equipoint in space group $P2_1\,2_1\,2_1$ is the general position, which is 4-fold. Thus the 4 Br atoms must occupy one set of general positions. Since the Br atoms are relatively heavy, and since there is only one equivalent set, their locations can be readily determined from Patterson projections.†

The origin of coordinates for space group $P2_1\,2_1\,2_1$ is commonly chosen in such a way that it bears a symmetrical relation to the symmetry elements; specifically it is usually chosen halfway between pairs of nonintersecting 2-fold screws. When this origin is used, both structure factor and expression for the electron density are complicated, involving both A and B components. However, the projections of the space group on each of the planes (100), (010), and (001) have plane symmetry $p2gg$, which is centrosymmetrical. If the 2-fold axis of the projection is taken as the origin, the symmetry factor and electron-density expressions lack B components and are comparatively simple. The relation between coordinates of an atom in projection, referred to a 2-fold axis as origin, and also referred to the usual space-group origin, is given in Table 3 and illustrated in Fig. 6. The coordinates of the Br atoms referred to these several coordinate systems are shown in Table 4.

The maximum scattering powers of the various atoms in the cell are proportional to f_j and this is approximately proportional to the atomic number Z. Thus the quantities in (1) are approximately as shown in Table 5. This shows that

† See M. J. Buerger, *Vector space and its application in crystal-structure investigation.* (John Wiley and Sons, New York, 1959) Chapter 6.

$$\frac{{}^{H}F_{\max}}{{}^{R}F_{\max}} \simeq \frac{{}^{H}Z}{{}^{R}Z} = \tfrac{35}{81} = 0.43.$$

Thus, even when the 4 Br atoms are scattering in phase, they attain less than half the maximum scattering power of the residue. In spite of this,

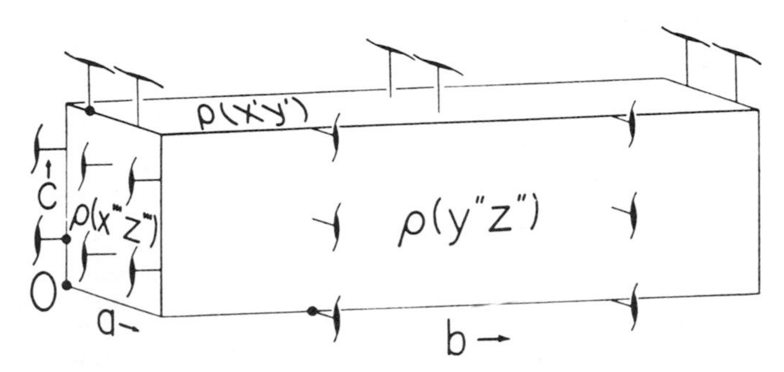

Fig. 6.
Relation of the origin of space group $P2_1\,2_1\,2_1$ (at O), to the origins of its projection on (001), (010), and (100), indicated by dots.
(After Buerger, Barney, and Hahn.[19])

Table 3
Relations between coordinates of $P2_1\,2_1\,2_1$, with origin chosen halfway between pairs of screw axes, and the coordinates of the projections of $P2_1\,2_1\,2_1$ on (100), (010), and (001), using projections of 2-fold screw axes as origins

Projection coordinates from space-group coordinates	Space-group coordinates from projection coordinates
$\rho(x'y')$ from $\rho(xyz)$:	$\rho(xyz)$ from $\rho(x'y')$:
$x' = x-\frac{1}{4}$	$x = x'+\frac{1}{4}$
$y' = y$	$y = y'$
$\rho(y''z'')$ from $\rho(xyz)$:	$\rho(xyz)$ from $\rho(y''z'')$:
$y'' = y-\frac{1}{4}$	$y = y''+\frac{1}{4}$
$z'' = z$	$z = z''$
$\rho(x'''z''')$ from $\rho(xyz)$:	$\rho(xyz)$ from $\rho(x'''z''')$:
$z''' = z-\frac{1}{4}$	$z = z'''+\frac{1}{4}$
$x''' = x$	$x = x'''$

Table 4
The coordinates of the Br atoms referred to the several coordinate schemes

	Referred to projected 2-fold axis as origin in $\rho(x'y')$	$\rho(y''z'')$	$\rho(x'''z''')$	Referred to standard space-group origin
x	0.178		0.072	0.072
y	0.035	0.215		0.035
z		0.167	0.083	0.167

Table 5
Approximate maximum scattering powers of heavy atom and residue in diglycine hydrobromide, $C_4H_{11}N_2O_4Br$

Atom	Atomic number, Z	Number of atoms in asymmetrical unit	Total number of scattering electrons	
Br	35	1	35	$= {}^HZ$
C	6	4	24	
H	1	11	11	81 $= {}^RZ$
N	7	2	14	
O	8	4	32	

the signs due to the bromine atoms alone were computed and assigned as signs to the observed $|F_{hkl}|$'s. A first set of electron-density maps $\rho_{(xy)}$ and $\rho_{(yz)}$ was computed with these signs. These are shown in Fig. 7. A set of coordinates was derived from these maps, and from them a new set of F's was computed. In all, 137 terms were used to compute $\rho(xy)$. When the signs were recomputed on the basis of atomic positions

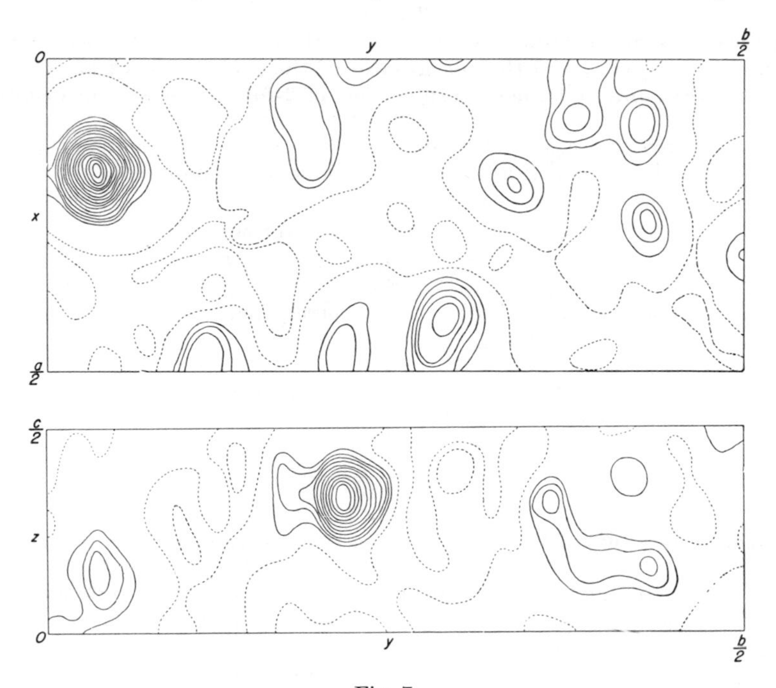

Fig. 7.
Diglycine hydrobromide. First electron-density projections using signs determined by bromine atoms alone; $\rho_{(xy)}$ above, $\rho_{(yz)}$ below.

Table 6
Unrefined coordinates of atoms of diglycine hydrobromide, as determined from projections

	Projection $\rho(x'y')$		Projection $\rho(y''z'')$		Projection $\rho(x'''z''')$		For $\rho(xyz)$, referred to space-group origin		
	x'	y'	y''	z''	x'''	z'''	x	y	z
Br	0.178	0.035	0.785	0.833	0.428	0.583	0.428	0.035	0.833
C_1	.095	.380	.130	.833	.345	.583	.345	.380	.833
C_2	.104	.428	.178	.070	.354	.820	.354	.428	.070
C_3	.450	.267	.017	.517	.700	.267	.700	.267	.517
C_4	.072	.170	.920	.397	.322	.147	.322	.170	.397
N_1	.278	.430	.180	.158	.528	.908	.528	.430	.158
N_2	.149	.194	.944	.158	.399	.908	.399	.194	.158
O_1	.462	.110	.860	.315	.712	.065	.712	.110	.315
O_2	.200	.338	.088	.842	.450	.592	.450	.338	.842
O_3	.472	.213	.963	.638	.722	.388	.722	.213	.638
O_4	.418	.283	.033	.266	.668	.016	.668	.283	.266

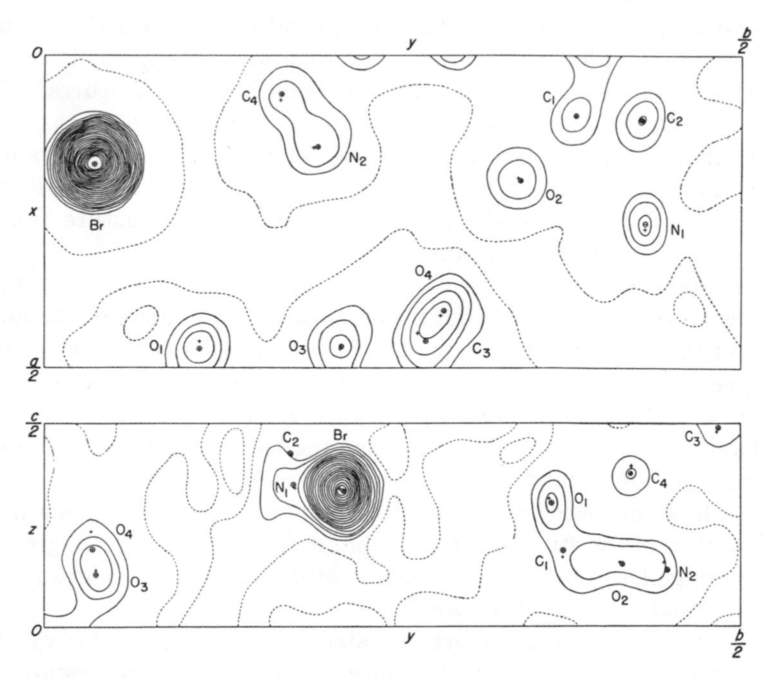

Fig. 8.
Diglycine hydrobromide. Final electron-density projections; $\rho_{(xy)}$ above, $\rho_{(yz)}$ below. The small crosses show the atom locations derived from Fig. 7, and the small circles show the final locations.

revealed by this first electron-density map, only 5 of the 137 signs computed on the basis of the position of the Br atom alone were reversed, specifically the signs of 0·22·0, 320, 370, 610, and 940. The resulting electron-density maps are shown in Fig. 8, and the coordinates derived from them are also shown in Fig. 8 and Table 6. When these new coordinates were used to compute new F's, the new set had signs identical with the set used to compute Fig. 8, so that if another set of Fourier syntheses had been made with the new F's, they would have been identical with Fig. 8. The process had thus converged to Fig. 8.

Remarks on the heavy-atom method. The heavy-atom method can also be followed if the initial information concerns the positions of several located atoms. The extent to which the sum of the atomic numbers of the located atoms overshadows the sum of the atomic numbers of the unlocated atoms is a measure of the probable reliability of the first electron-density projection.

This discussion makes it plain that, except in simple symmetry-fixed structures, the contributions of the heavy atom tend to dominate the phases of the F's except when their contribution is small or actually zero. As a consequence it is usually a worthwhile procedure to carry out a first Fourier synthesis in which are included only those F_{hkl}'s whose signs can be regarded as certainly or probably correct. This usually gives a map of the structures which reveals its chief features. This should include the location of some or all of the residue atoms. The F's can then be recomputed and the signs of many of the uncertain F's can be transformed into probable, or even safe signs. A second Fourier synthesis then usually provides a close approach to the structure unless it is quite complicated and contains unresolved peaks.

Lipson and Cochran† suggest that the heavy-atom method is most successful when the sums of the squares of the atomic numbers of the heavy atom and the light atoms are equal. Under these circumstances, about three-quarters of the signs are correct: Half are correct because, on the average, in half the cases the contributions of the residue and the heavy atom have the same sign. Of the remaining half, the residue and heavy atoms have opposite signs, but of this group, in only half the cases, on the average, does the residue have a greater net amplitude. Thus about three-quarters of the structure factors have their signs correctly given by the scattering phase of the heavy atom. More exact relationships are discussed by Luzzati[14] and Woolfson.[20]

In many investigations of organic structures, using the heavy-atom method, the shape of the organic molecule has already been established by chemists, and the probable location of attachment of a heavy atom,

† H. Lipson and W. Cochran. *The determination of crystal structures* (G. Bell and Sons, London, 1953) 207.

such as iodine or a metal, is also predictable on chemical grounds. When the first or both of these features are known, the crystal-structure analysis degenerates into fitting the known molecular configuration onto the peaks of a crude electron-density map. Such a procedure can be successful even if quite a number of signs of the Fourier synthesis are wrong. These investigations comprise a degenerate form of crystal-structure analysis. For, when analysis is being made of a structure in which the shape of the molecules is unknown (as in organic crystals with complicated molecules), or in which the crystal does not contain discrete molecules (as in the case of most inorganic crystals), then many incorrect signs of the F's cannot be tolerated, except for F's of small absolute magnitude.

Phase determination for crystals having a set of replaceable atoms

J. M. Cork[22] was the first to make use of a replaceable atom to determine phases of F's for Fourier synthesis. There are a number of other examples of the use of this phase-determining strategy in the literature for both centrosymmetrical and non-centrosymmetrical cases.[22–54]

General relations. Consider two crystals of compositions MABCD · · · and NABCD · · · which have the same structure except that the set of atoms of species M in one crystal is replaced by a set of atoms of species N in the other crystal. The atom in the M, N site may be said to be replaceable, and the rest of the structure, ABCD · · · , may be described as the residue, R. The composition of the two crystals can thus be described as MR and NR. The diffraction amplitudes of these two crystals are related.

Let superscripts refer to compositions. Then relations like (1) can be written as follows:

$$^{\mathrm{MR}}F_{hkl} = {}^{\mathrm{M}}F_{hkl} + {}^{\mathrm{R}}F_{hkl}, \tag{6}$$

and

$$^{\mathrm{NR}}F_{hkl} = {}^{\mathrm{N}}F_{hkl} + {}^{\mathrm{R}}F_{hkl}, \tag{7}$$

so that

$$^{\mathrm{MR}}F_{hkl} - {}^{\mathrm{NR}}F_{hkl} = {}^{\mathrm{M}}F_{hkl} - {}^{\mathrm{N}}F_{hkl}. \tag{8}$$

The terms on the right are simple expressions given by relations similar to (2):

$$^{\mathrm{M}}F_{hkl} = {}^{\mathrm{M}}f_{hkl}\ {}^{\mathrm{M}}S_{hkl}, \tag{9}$$

$$^{\mathrm{N}}F_{hkl} = {}^{\mathrm{N}}f_{hkl}\ {}^{\mathrm{N}}S_{hkl}. \tag{10}$$

Since M and N occupy identical coordinates,

$$^{\mathrm{M}}S_{hkl} = {}^{\mathrm{N}}S_{hkl}, \tag{11}$$

so

$$ {}^{\mathrm{M}}F_{hkl} - {}^{\mathrm{N}}F_{hkl} = ({}^{\mathrm{M}}f_{hkl} - {}^{\mathrm{N}}f_{hkl})S_{hkl} \tag{12} $$

and the detailed form of (8) is

$$ {}^{\mathrm{MR}}F_{hkl} - {}^{\mathrm{NR}}F_{hkl} = ({}^{\mathrm{M}}f_{hkl} - {}^{\mathrm{N}}f_{hkl})S_{hkl}. \tag{13} $$

If (13) is arranged so that M is an atom of higher atomic number than N, then the term in parentheses is a positive number, i.e., a scale constant. Relation (13) then shows that the phase of the left side is controlled by the symmetry factor of the replaceable atom.

This relation is perfectly general, and, as will be seen in succeeding sections, useful. To use it requires a knowledge of the F's on an absolute scale and a knowledge of the location of the replaceable atom. A note on the determination of scale is given in a subsequent section.

In order that (13) should hold rigorously, it is necessary that the coordinates of all atoms of both crystals MR and NR be the same. Under ordinary circumstances this is reasonably well obeyed in isomorphous series. Robertson[25] has pointed out that whether this is obeyed or not can sometimes be tested by comparing reflections from both crystals for which the M and N atoms make no contribution. If the circumstances provide such reflections, they are due to the residue alone, and only if the coordinates of the residue atoms are the same in both crystals will the reflections be the same. When this requirement is not obeyed, the reflections of smallest intensity are most likely to be seriously affected.

Centrosymmetrical case. For centrosymmetrical crystals the F's and S's of (8) through (13) are all real, and therefore are merely positive or negative quantities. If the set of replaceable atoms can be located, for example because they occupy special positions, or by interpretation of a Patterson synthesis,† then S is readily computed for all spectra. In this event both the magnitudes and signs of the quantities on the right side of (12) are known for all reflections. The two quantities on the left are known in magnitude only, and may have either a positive or negative sign. To determine these signs it is necessary to assume the four possible combinations of signs and see which combination causes the F's to span the quantity on the right.

Geometrical interpretations of (8) are shown in Figs. 9 and 10. In these illustrations, the symmetry factor of the replaced atom is taken as positive. Figure 9 shows the relation when the contribution of the residue is negative, Fig. 10 when it is positive.

† See M. J. Buerger. *Vector space and its application in crystal-structure investigation.* (John Wiley and Sons, New York, 1959) Chapter 6.

Example: phthalocyanine. One of the first crystal structures to be solved by finding the phases for a two-dimensional Fourier synthesis by the replaceable-atom method was phthalocyanine, which was solved by Robertson[25] in 1936. This classical solution is actually a degenerate case so that it forms a simple example for first consideration.

The general crystallographic features of the metal phthalocyanine have been described in an earlier section of this chapter. The crystals are monoclinic, space group $P\,2_1/a$, and the $\rho_{(xy)}$ projection can be referred to a cell of plane symmetry $p2$. In the projection the metal atom occupies the origin, and therefore contributes a wave of amplitude ${}^{\mathrm{M}}f$ to all $h0l$ spectra.

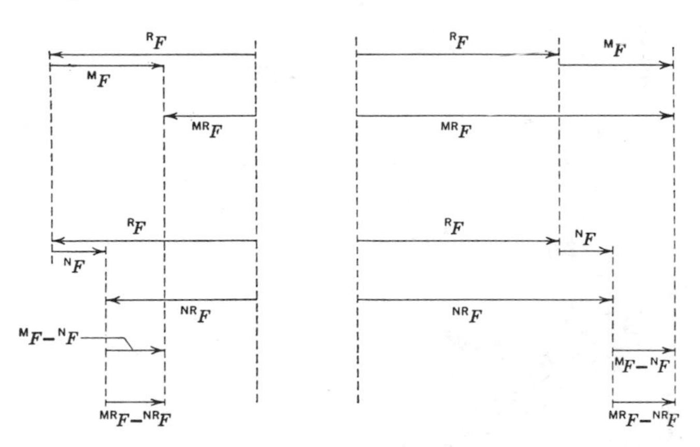

Fig. 9. Fig. 10.
Geometrical interpretations of equation (8) under the condition that ${}^{\mathrm{M}}F - {}^{\mathrm{N}}F$ is positive. Figure 9 shows ${}^{\mathrm{R}}F$ negative, while Fig. 10 shows ${}^{\mathrm{R}}F$ positive.

Both metal phthalocyanine and the metal-free phthalocyanine are known. Thus M in relation (13) may be taken as a metal atom (nickel in Robertson's investigation) and N may be regarded as an empty space. Thus (13) assumes the special form

$$ {}^{\mathrm{NiR}}F - {}^{\mathrm{R}}F = ({}^{\mathrm{Ni}}f - 0)S_{hkl}. \tag{14}$$

Since the metal atom is at the origin, S is always $+1$ and (14) becomes

$$ {}^{\mathrm{NiR}}F - {}^{\mathrm{R}}F = {}^{\mathrm{Ni}}f. \tag{15}$$

Robertson assumed the four possible sign combinations for the two F's of (15), and plotted these against $\sin\theta$, Fig. 11. The combination falling closest to the f curve of Ni was accepted as the correct one.

Example: Br- and Cl-camphor. An excellent example of the more general use of a replaceable atom in phase determination is provided by Wiebenga and Krom's analysis[30] of the structures of d-α-Br-cam-

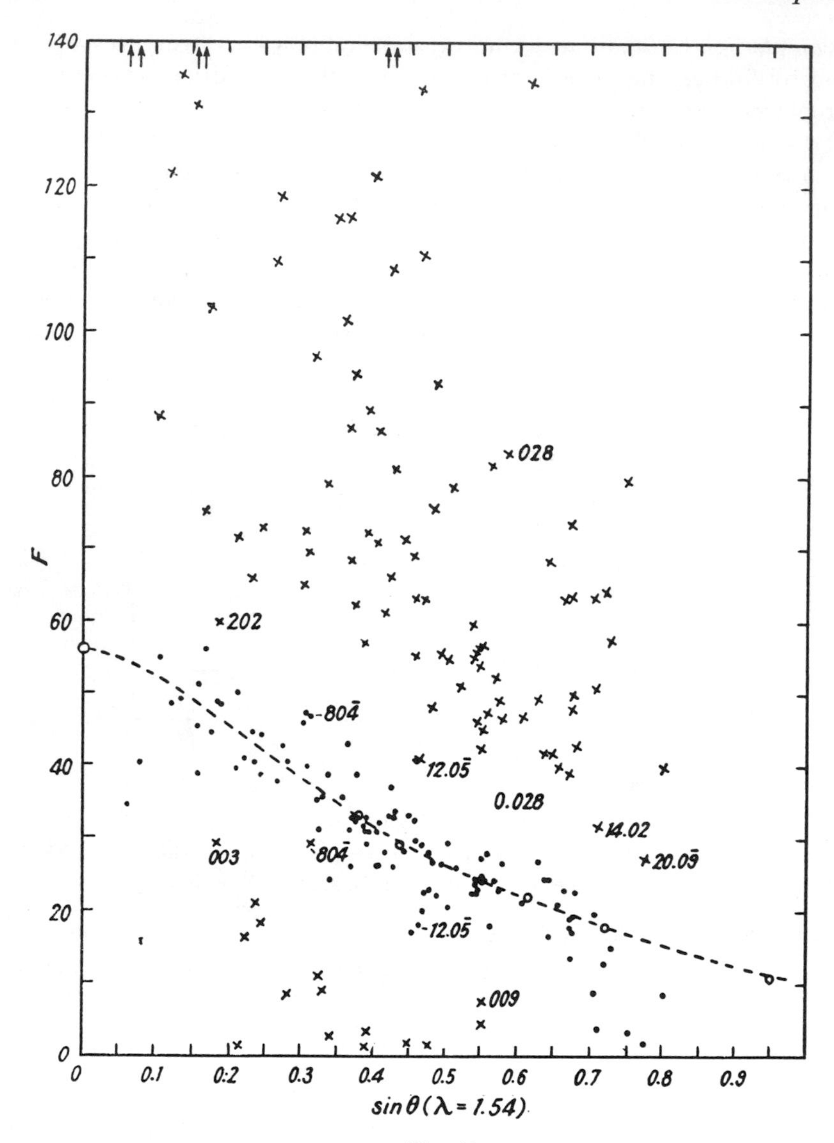

Fig. 11.
Plot of the possible values of the left of (15) against sin θ for nickel phthalocyanine and metal-free phthalocyanine. The correct sign combination is the one best fitting the f curve of nickel, according to (15). This is shown as the dashed line in this figure.

(After Robertson.[25])

phor and *d*-α-Cl-camphor. These belong to space group $P2_1$ and have cell dimensions

	Br derivative	Cl derivative
$a =$	7.38 Å	7.25 Å
$b =$	7.57	7.51
$c =$	9.12	9.04
$\beta =$	94°0′	93°15′.

This cell contains 2 molecules. Although the space group is noncentrosymmetrical, it projects on (010) as the centrosymmetrical plane group $p2$. Thus the $\rho_{(xz)}$ projection can be handled as a centrosymmetrical case. The cell dimensions of the two crystals are sufficiently close so that the positions of the atoms can be assumed to be approximately the same in both structures.

The intensities were obtained by comparing the reflections with those given by a standard sodium chloride crystal, as recorded on Weissenberg photographs. The location of the replaceable atom was found from Patterson syntheses $P_{(xz)}$. These showed the Br position to be $x = 0.284$, $z = 0.164$. The Cl positions were substantially the same. With these coordinates, the Br and Cl contributions, ${}^{Br}F$ and ${}^{Cl}F$, can be computed for each reflection according to (9) and (10).

The determination of the signs of the F_{h0l}'s using relation (8) are illustrated in Table 7. The second column shows the measured values of the $|F|$'s. The third column shows the differences between these quantities assuming the four possible sign combinations. These give four possible values for the left side of (8). One of these must be equal to the right side of (8), which is easily computed by (12). The results of this computation are listed in column 4. By comparing the value in this column with the four possibilities in column 3, one should find one comparison to be acceptable. On this basis the signs in the last column were selected. Note that when the observed F's are small an ambiguity may exist. This is true for $60\bar{7}$ and 606, for example.

By this method the signs of all but 44 of the 160 F_{h0l}'s could be determined; the 44 undetermined signs were associated with small $|F|$'s. The preliminary Fourier synthesis $\rho_{(xz)}$ prepared by using the F's with determined signs are shown in Figs. 12 and 13. These syntheses revealed positions of atoms with sufficient exactness so that the undetermined signs could be computed. A final Fourier synthesis based upon all F's could then be made and the coordinates of the atoms correspondingly refined.

Other examples of phase determination for centrosymmetrical crystals are cited in the literature list.[22–53]

Table 7
Determination of phases for chlorine-camphor and bromine-camphor
(After Wiebenga and Krom[30])

1	2		3				4	5	
			${}^{\mathrm{BrR}}F - {}^{\mathrm{ClR}}F$						
$h0l$	$\lvert {}^{\mathrm{BrR}}F \rvert$	$\lvert {}^{\mathrm{ClR}}F \rvert$	$(++)$	$(+-)$	$(-+)$	$(--)$	Computed ${}^{\mathrm{Br}}F - {}^{\mathrm{Cl}}F$	Deduced signs ${}^{\mathrm{BrR}}F$	${}^{\mathrm{ClR}}F$
001	45	36	$+9$	$+81$	-81	-9	$+18$	$+$	$+$
100	13	10	$+3$	$+23$	-23	-3	-3	$-$	$-$
$10\bar{1}$	15	35	-20	$+40$	-40	$+20$	$+27$	$-$	$-$
103	32	34	-2	$+66$	-66	$+2$	$+1$		
300	7	<4	$>+3$	$<+11$	>-11	<-3	$+7$	$+$	
$20\bar{3}$	19	11	$+8$	$+30$	-30	-8	$+25$	$+$	$-$
$30\bar{1}$	27	10	$+17$	$+37$	-37	-17	-18	$-$	$-$
301	40	19	$+21$	$+59$	-59	-21	$+24$	$+$	$+$
004	22	12	$+10$	$+34$	-34	-10	-14	$-$	$-$
$60\bar{7}$	11	7	$+4$	$+18$	-18	-4	-6	$-$	
606	9	7	$+2$	$+16$	-16	-2	-6	$-$	
$80\bar{1}$	9	7	$+2$	$+16$	-16	-2	$+5$	$+$	$+$
800	8	4	$+4$	$+12$	-12	-4	$+4$	$+$	$+$

Establishing an absolute scale. To use the replaceable-atom method one must have the measured $|F|$'s on an absolute scale. Beevers and Cochran[32] have suggested an ingenious method of doing this: The measured value of F is on an arbitrary scale, and also may contain some factors not allowed for, such as absorption factor, temperature factor, etc. But these are usually functions of $\sin\theta$. Let G be the measured value. Then

$$F = gG, \tag{16}$$

and instead of (13) one can write

$${}^{\mathrm{M}}g\,{}^{\mathrm{M}}G - {}^{\mathrm{N}}g\,{}^{\mathrm{N}}G = ({}^{\mathrm{M}}f - {}^{\mathrm{N}}f)S. \tag{17}$$

When one of the left-hand terms accidentally equals zero, then the approximate value of g for that $\sin\theta$ region can be determined. If the values of g are plotted against $\sin\theta$, a graph like that shown in Fig. 14 results.

Extension to general substitution. The replacement method can be used with pairs of crystals having several atoms simultaneously replaceable. Suppose one crystal contains atoms M and P which are replaced in a second crystal by N and Q. Relation (13) can be generalized to include this double replacement as follows:

$${}^{\mathrm{MPR}}F - {}^{\mathrm{NQR}}F = ({}^{\mathrm{M}}f - {}^{\mathrm{N}}f)\,{}^{\mathrm{M,N}}S + ({}^{\mathrm{P}}f - {}^{\mathrm{Q}}f)\,{}^{\mathrm{P,Q}}S. \tag{18}$$

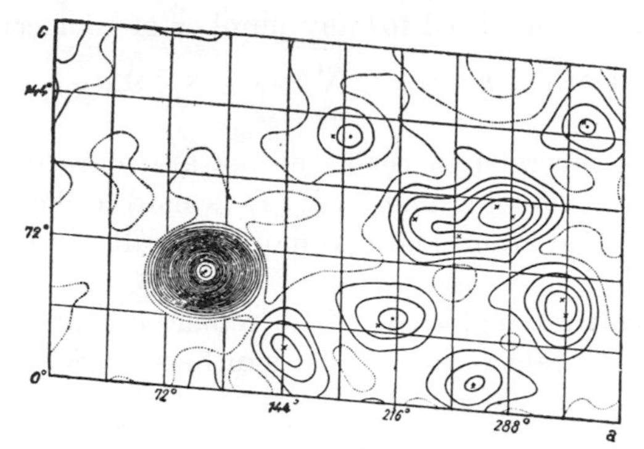

Fig. 12.
Preliminary Fourier synthesis $\rho_{(xz)}$ for Br camphor, using signs predicted by (13). (After Wiebenga and Krom.[30])

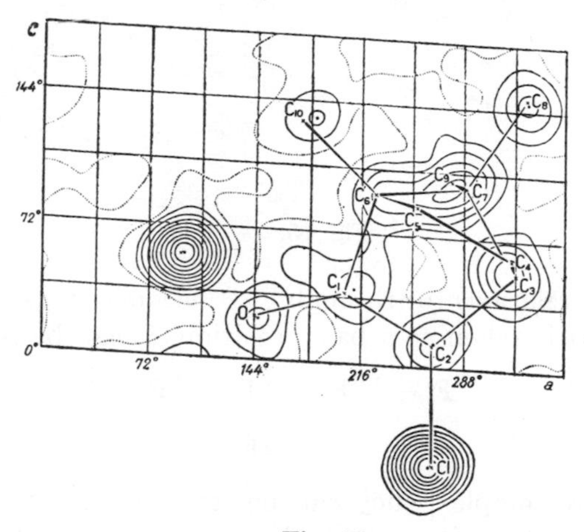

Fig. 13.
Preliminary Fourier synthesis $\rho_{(xz)}$ for Cl camphor, using signs predicted by (13). (After Wiebenga and Krom.[30])

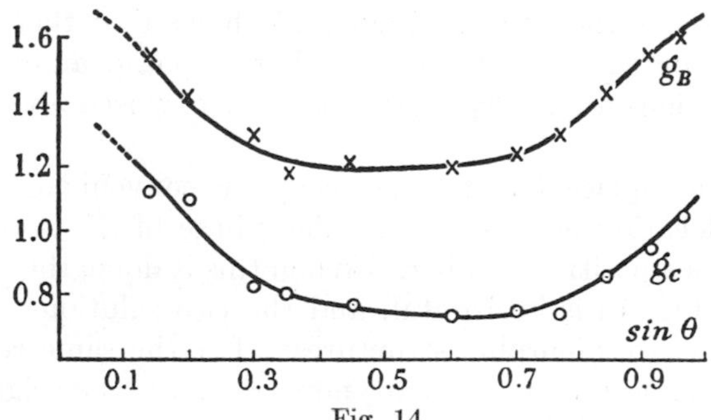

Fig. 14.
Variation of ^{M}g and ^{N}g of (17) with sin θ, for sucrose sodium bromide dihydrate. (After Beevers and Cochran.[32])

This can be further generalized for any number of replacements as

$$^{R+\Sigma M}F - {}^{R+\Sigma M}F = \sum ({}^{M}f - {}^{N}f)\ {}^{M,N}S. \tag{19}$$

Cases involving several replaceable atoms are more difficult in two respects than those involving one atom: It is more difficult to find the locations of the several atoms and it is more difficult to identify them as to species.

Non-centrosymmetrical case. The replaceable-atom method can also be used for non-centrosymmetrical syntheses. When this is done

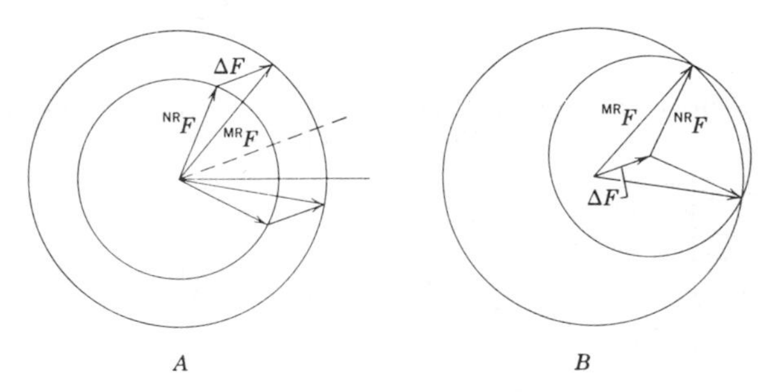

Fig. 15.

the F's and S's of relations (6) through (13) become, in general, complex quantities. If (8) is taken as a basic relation, it can be represented as

$$\begin{aligned} {}^{MR}\mathbf{F} - {}^{NR}\mathbf{F} &= {}^{M}\mathbf{F} - {}^{N}\mathbf{F} \\ &= \Delta\mathbf{F}. \end{aligned} \tag{20}$$

Since these **F**'s are complex, they can be represented by vectors in the complex plane, Fig. 15. In general, the magnitudes of the two vectors on the left of (20) are known, the magnitude and phase of $\Delta\mathbf{F}$ are known, and it is required to find positions for each of the vectors ${}^{MR}\mathbf{F}$ and ${}^{NR}\mathbf{F}$ such that $\Delta\mathbf{F}$ spans their ends. Figure 15 shows that there are always two solutions for such a relation, so that a phase ambiguity exists. These two solutions are always symmetrically disposed about the phase direction of $\Delta\mathbf{F}$.

If there is one replaceable atom per cell, it is convenient to choose the origin at the location of this atom. The phase of $\Delta\mathbf{F}$ is then zero, so that it is always a positive number. When this is done, the vector representing $\Delta\mathbf{F}$ in Fig. 15 is horizontal, and the two solutions for ${}^{MR}\mathbf{F}$ and ${}^{NR}\mathbf{F}$ represent enantiomorphic structures. For the same reason, if the replaceable atom occurs in a centrosymmetrical set, the origin should be chosen at the center of the set.

Example: strychnine sulfate pentahydrate. Bokhoven, Schoone, and Bijvoet[35–37] used this method in determining the structure of strychnine sulfate pentahydrate from the sulfate and selenate. The space group of these crystals is $C2$. The S or Se atom is known to occupy the origin. The projection on (010) is centrosymmetrical, but the other projections are not. The projection along [001], for example, has only a projected mirror line parallel to b. The Fourier synthesis for this projection has the form

$$\rho_{(xy)} = \frac{1}{A} \sum_h \sum_k |F_{hk0}| \cos 2\pi(hx+ky-\phi). \tag{21}$$

The Fourier synthesis of the inverse projection with opposite phases in Fig. 15 has the form

$$\bar{\rho}_{(xy)} = \frac{1}{A} \sum_h \sum_k |F_{hk0}| \cos 2\pi(hx+ky+\phi). \tag{22}$$

Since $+\phi$ and $-\phi$ cannot be distinguished in Fig. 15, Bokhoven et al. synthesized (21) and (22) at the same time in the form

$$\begin{aligned} \rho_{(xy)} + \bar{\rho}_{(xy)} &= \frac{1}{A} \sum_h \sum_k |F_{hk0}| \{\cos 2\pi(hx+ky-\phi) \\ &\qquad + \cos 2\pi(hx+ky+\phi)\} \\ &= \frac{1}{A} \sum_h \sum_k |F_{hk0}| 2 \cos 2\pi(hx+ky) \cos \phi. \end{aligned} \tag{23}$$

This synthesis contains a false mirror plane which relates (21) and (22). To rid the solution of the unwanted inverse structure it was necessary to consider a number of possible structures suggested by the peaks of the synthesis and compute intensities for them. The one best fitting the observed data was retained.

This example shows the nature of the ambiguity for the non-centrosymmetrical case, and how it is treated. The ambiguity can be eliminated if there are two heavy atoms. The use of two replaceable atoms, the difficulties attending their use, and the precautions which should be observed when applying the method, are discussed by Harker.[44]

Literature

Phase determination using heavy atoms

[1] J. West. *A quantitative x-ray analysis of the structure of potassium dihydrogen phosphate* (KH_2PO_4). Z. Krist. **74** (1930) 306–332.

[2] W. A. Wooster. *On the crystal structure of gypsum*, $CaSO_4 \cdot 2H_2O$. Z. Krist. **94** (1936) 375–396.

[3] E. G. Cox. *Molecular crystals.* Ann. Repts. Prog. Chem. **34** (1938) 176–195.

[4] J. Monteath Robertson and Ida Woodward. *An x-ray study of the phthalocyanines. Part IV. Direct quantitative analysis of the platinum compound.* J. Chem. Soc. (1940) 36–48.

[5] G. Huse and H. M. Powell. *The crystal structure of picryl iodide.* J. Chem. Soc. (1940) 1398–1402.

[6] G. A. Jeffrey. *The structure of the polyisoprenes. I. The crystal structure of geranylamine hydrochloride.* Proc. Roy. Soc. (London) (A) **183** (1945) 388–404.

[7] C. H. Carlisle and D. Crowfoot. *The crystal structure of cholesteryl iodide.* Proc. Roy. Soc. (London) (A) **184** (1945) 64–83.

[8] M. F. Perutz and O. Weisz. *The crystal structure of tribromo(trimethylphosphine)gold.* J. Chem. Soc. (1946) 438–442.

[9] Dorothy Crowfoot and J. D. Dunitz. *Structure of calciferol.* Nature **162** (1948) 608–609.

[10] J. H. Robertson and C. A. Beevers. *Crystal structure of strychnine hydrobromide.* Nature **165** (1950) 690–691.

[11] J. D. Dunitz. *The crystal structure of chloramphenicol and bromamphenicol.* J. Am. Chem. Soc. **74** (1952) 995–999.

[12] R. G. Curtis, J. Fridrichsons, and A. McL. Mathieson. *Structure of lanostenol.* Nature **170** (1952) 321–322.

[13] A. McL. Mathieson, D. P. Mellor, and N. C. Stephenson. *The crystal structure of potassium hydroxy-chlororuthenate,* $K_4Ru_2Cl_{10}O \cdot H_2O$. Acta Cryst. **5** (1952) 185–186.

[14] V. Luzzati. *Resolution d'une structure cristalline lorsque les positions d'une partie des atomes sont connues: Traitment statistique.* Acta Cryst. **6** (1953) 142–152.

[15] J. Fridrichsons and A. McL. Mathieson. *Triterpenoids. The crystal structure of lanostenyl iodoacetate.* J. Chem. Soc. (1953) 2159–2167.

[16] Cecily Darwin Littleton. *A structure determination of the gluconate ion.* Acta Cryst. **6** (1953) 775–781.

[17] R. Brill and A. P. de Bretteville. *On the chemical bond type in* $AlPO_4$. Acta Cryst. **8** (1955) 567–570

[18] A. McL. Mathieson. *The direct determination of molecular structure and configuration of moderately complex organic compounds.* Rev. Pure and Appl. Chem. **5** (1955) 113–142.

[19] M. J. Buerger, Elsa Barney, and Theo Hahn. *The crystal structure of diglycine hydrobromide.* Z. Krist. **108** (1956) 130–144.

[20] M. M. Woolfson. *An improvement of the "heavy atom" method of solving crystal structures.* Acta Cryst. **9** (1956) 804–810.

[21] J. A. Bland and D. Clark. *Studies in aluminum-rich alloys with the transition metals manganese and tungsten. I. The crystal structure of* ϵ $(W{-}Al){-}WAl_4$. Acta Cryst. **11** (1958) 231–236.

Phase determination using replaceable atoms

[22] J. M. Cork. *The crystal structure of some of the alums.* Phil. Mag. (7) **4** (1927) 688–698.

[23] H. Lipson and C. A. Beevers. *The crystal structure of the atoms.* Proc. Roy. Soc. (London) (A) **148** (1935) 664–680.

[24] J. Monteath Robertson. *An x-ray study of the structure of the phthalocyanines. Part. I. The metal-free, nickel, copper, and platinum compounds.* J. Chem. Soc. (1935) 615–621.

[25] J. Monteath Robertson. *An x-ray study of the phthalocyanines. Part. II. Quantitative structure determination of the metal-free compound.* J. Chem. Soc. (1936) 1195–1209.

[26] J. Monteath Robertson and Ida Woodward. *An x-ray study of the phthalocyanines. Part III. Quantitative structure determination of nickel phthalocyanine.* J. Chem. Soc. (1937) 219–230.

[27] E. G. Cox and G. A. Jeffrey. *Crystal structure of glucosamine hydrobromide.* Nature **143** (1939) 894–895.

[28] C. A. Beevers and W. Hughes. *The crystal structure of Rochelle salt (sodium potassium tartrate tetrahydrate* $NaKC_4H_4O_6{\cdot}4H_2O$*).* Proc. Roy. Soc. (London) (A) **177** (1941) 251–259.

[29] J. M. Bijvoet and E. H. Wiebenga. *Eine direkte röntgenographische Molekülstrukturbestimmung durch Vergleich isomorpher Kristallstrukturen.* Naturwiss. **32** (1944) 45–46.

[30] E. H. Wiebenga and C. J. Krom. *X-ray investigation of* d-α-*Br, Cl and CN-camphor. A direct determination of a molecular structure by comparison of isomorphous crystal structures.* Rec. trav. chim. **65** (1946) 663–681.

[31] A. Hargreaves. *Crystal structure of zinc* p-*toluenesulphonate.* Nature **158** (1946) 620–621.

[32] C. A. Beevers and W. Cochran. *The crystal structure of sucrose sodium bromide dihydrate.* Proc. Roy. Soc. (London) (A) **190** (1947) 257–272.

[33] Joy Boyes-Watson and M. F. Perutz. *An x-ray study of horse methaemoglobin. I.* Proc. Roy. Soc. (London) (A) **191** (1947) 83–132.

[34] J. M. Bijvoet. *De Fourier-methode in de Röntgenanalyse van kristallen.* Proc. Koninkl. Ned. Akad. Wetenschap. **50** (1947) 823–825.

[35] C. Bokhoven, J. C. Schoone, and J. M. Bijvoet. *On the crystal structure of strychnine sulfate and selenate. I. Cell dimensions and space group.* Proc. Koninkl. Ned. Akad. Wetenschap. **50** (1947) 825.

[36] C. Bokhoven, J. C. Schoone, and J. M. Bijvoet. *On the crystal structure of strychnine sulfate and selenate. II.* [010] *projection and structure formula.* Proc. Koninkl. Ned. Akad. Wetenschap. **51** (1948) 990.

[37] C. Bokhoven, J. C. Schoone, and J. M. Bijvoet. *On the crystal structure of strychnine sulfate and selenate. III.* [001] *projection.* Proc. Koninkl. Ned. Akad. Wetenschap. **52** (1949) 120–121.

[38] J. M. Bijvoet. *Phase determination in direct Fourier synthesis of crystal structures.* Proc. Koninkl. Ned. Akad. Wetenschap. **52** (1949) 313–314.

[39] C. Bokhoven, J. C. Schoone, and J. M. Bijvoet. *The Fourier synthesis of the crystal structure of strychnine sulphate pentahydrate.* Acta Cryst. **4** (1951) 275–280.

[40] P. M. deWolff and L. Walter-Lévy. *The crystal structure of* $Mg_2(OH)_3(Cl, Br){\cdot}4H_2O$*.* Acta Cryst. **6** (1953) 40–44.

[41] J. Monteath Robertson and G. Todd. *The structure of β-caryophyllene alcohol chloride and bromide.* Chem. and Ind. (1953) 437.

[42] R. H. Moffett and D. Rogers. *The molecular configuration of longifolene hydrochloride.* Chem. and Ind. (1953) 916–917.

[43] J. Monteath Robertson and G. Todd. *X-ray studies in the caryophyllene series. The chloride and bromide from β-caryophyllene alcohol.* J. Chem. Soc. (1955) 1254–1263.

[44] David Harker. *The determination of the phases of the structure factors of non-centrosymmetric crystals by the method of double isomorphous replacement.* Acta Cryst. **9** (1956) 1–9.

[45] M. F. Perutz. *Isomorphous replacement and phase determination in non-centrosymmetric space groups.* Acta Cryst. **9** (1956) 867–873.

[46] F. H. C. Crick and Beatrice S. Magdoff. *The theory of the method of isomorphous replacement for protein crystals. I.* Acta Cryst. **9** (1956) 901–908.

[47] D. June Sutor. *The isomorphous-replacement method applied to molecules containing like atoms.* Acta Cryst. **9** (1956) 969–970.

[48] A. Hargreaves. *The crystal structure of zinc* p-*toluene sulphonate hexahydrate.* Acta Cryst. **10** (1957) 191–195.

[49] A. Hargreaves. *The application of the isomorphous-replacement method in the determination of centrosymmetric structures.* Acta Cryst. **10** (1957) 196–199.

[50] Rosalind E. Franklin and K. C. Holmes. *Tobacco mosaic virus: Application of the method of isomorphous replacement to the determination of the helical parameters and radial density distribution.* Acta Cryst. **11** (1958) 213–220.

[51] James Trotter. *The crystal and molecular structure of diindenyl iron.* Acta Cryst. **11** (1958) 355–360.

[52] J. C. Kendrew, G. Bodo, H. M. Dintzis, R. G. Parrish, and H. Wyckoff. *A three-dimensional model of the myoglobin molecule obtained by x-ray analysis.* Nature **181** (1958) 662–666.

[53] M. M. Bluhm, G. Bodo, H. M. Dintzis, and J. C. Kendrew. *The crystal structure of myoglobin. IV. A Fourier projection of sperm-whale myoglobin by the method of isomorphous replacement.* Proc. Roy. Soc. (London) (A) **246** (1958) 369–389.

[54] H. A. Weakliem and J. L. Hoard. *The structures of ammonium and rubidium ethylenediaminetetraacetatocobaltate (III).* J. Am. Chem. Soc. **81** (1959) 549–555.

20

Phase determination for certain special cases

In the last chapter it was seen that the phases of the F's required for Fourier synthesis can often be determined provided that the crystal is specialized in that it contains a heavy or replaceable atom. A large number of crystal structures have been solved by taking advantage of such specialization. There also exist other types of specializations which lend themselves to phase determination. The best known of these are noted in this chapter.

Crystals with regions of uniform, low density

In crystals composed of molecules, the spaces between molecules are often nearly empty. If the distribution of such regions can be guessed from the geometrical relations between the chemical molecule and the unit cell, the conditions of low densities of these regions can be used to place restrictions on the phases of the F's, as pointed out by Tesche.[1,2] Tesche considered specifically the case of a long-chain molecule. Between chains, and at chain ends, the density should be low and uniform. The cell can usually be chosen so that these regions occur along certain cell borders.

To see how this restricts the phases, consider the case of a centrosymmetrical projection $\rho_{(xy)}$. The Fourier synthesis of the electron-density projection is given by

$$\rho_{(xy)} = \frac{1}{S} \sum_h \sum_k F_{hk0} \, e^{-i2\pi(hx+ky)} \tag{1}$$

According to (35) of Chapter 17, for centrosymmetry this can be recast into the following form:

$$\rho_{(xy)} = \frac{2}{S} \sum_{\substack{0 \\ h}}^{\infty}{}' \sum_{\substack{0 \\ k}}^{\infty}{}' \{(F_{hk0}+F_{\bar{h}k0}) \cos 2\pi hx \cos 2\pi ky$$

$$- (F_{hk0}-F_{\bar{h}k0}) \sin 2\pi hx \sin 2\pi ky\}. \quad (2)$$

It will be recalled that the primes following the summation signs call attention to the fact that the proper multiplicities must be observed for the borders of the representative field in reciprocal space. These multiplicities can be specifically recognized by writing the summation for the borders separately, as follows:

$$\rho_{(xy)} = \frac{2}{S} \Big\{ \tfrac{1}{2}F_{000}$$

$$+ \sum_{\substack{1 \\ h}}^{H} F_{h00} \cos 2\pi hx$$

$$+ \sum_{\substack{1 \\ k}}^{K} F_{0k0} \cos 2\pi ky$$

$$+ \sum_{\substack{1 \\ h}}^{H} \sum_{\substack{1 \\ k}}^{K} (F_{hk0}+F_{\bar{h}k0}) \cos 2\pi hx \cos 2\pi ky$$

$$- \sum_{\substack{1 \\ h}}^{H} \sum_{\substack{1 \\ k}}^{K} (F_{hk0}-F_{\bar{h}k0}) \sin 2\pi hx \sin 2\pi ky \Big\}. \quad (3)$$

Now suppose it is known that along the cell border $0y$ the electron density is constant. In the first place the last line of (3) vanishes when $x = 0$. In the second place, to meet the conditions, everything in the other four lines must yield a constant value for $x = 0$. The first line is always constant, and the second line becomes constant when $x = 0$. This furnishes the relation

$$\tfrac{1}{2}F_{000} + \sum_{\substack{1 \\ h}}^{H} F_{h00} = \text{const.} \quad (4)$$

The conditions require the third and fourth lines to give no fluctuation as

y varies. This can occur if these two lines vanish, that is, for the condition $x = 0$:

$$\sum_{k=1}^{K} F_{0k0} \cos 2\pi ky + \sum_{h=1}^{H} \sum_{k=1}^{K} (F_{hk0}+F_{\bar{h}k0}) \cos 2\pi ky = 0, \tag{5}$$

or

$$\sum_{k=1}^{K} \left\{F_{0k0} + \sum_{1}^{H} (F_{hk0}+F_{\bar{h}k0})\right\} \cos 2\pi ky = 0. \tag{6}$$

This can occur when the term in braces is zero, which provides the relation

$$F_{0k0} + \sum_{h=1}^{H} (F_{hk0}+F_{\bar{h}k0}) = 0, \tag{7}$$

for any value of k.

The only non-vanishing terms in (3) are those of (4). If the density along the line $0y$ is not only uniform, but also very low, then the value of the constant on the right of (4) is substantially zero, giving

$$\tfrac{1}{2}F_{000} + \sum_{h=1}^{H} F_{h00} \approx 0. \tag{8}$$

Corresponding results can be found for the condition that uniform density occurs along $x0$. It can be extended to planes of uniform density in the three-dimensional electron-density function. Some of the simple variations are shown in Table 1.

An example of the use of these relations is provided by Tesche[1]: A set of $0kl$ reflections had the following amplitudes:

$\lvert F_{005}\rvert$	$\lvert F_{015}\rvert$	$\lvert F_{0\bar{1}5}\rvert$	$\lvert F_{025}\rvert$	$\lvert F_{0\bar{2}5}\rvert$	$\lvert F_{035}\rvert$	$\lvert F_{0\bar{3}5}\rvert$	$\lvert F_{045}\rvert$	$\lvert F_{0\bar{4}5}\rvert$	(9)
7.9	2.3	2.7	0	0	0	0	0	0	

The density is presumed to be uniform and nearly zero along $y = 0$. According to Table 1, this calls for $F_{00l} + \sum_{l} (F_{0kl}+F_{0\bar{k}l}) \approx 0$, for any value of l. It is evident that this can occur for (9) only if F_{005} is opposite in sign to both F_{015} and $F_{0\bar{1}5}$.

Crystals with subperiods within the cell

The atoms of a crystal are periodic on the lattice of the crystal. The number of periods in this pattern is indefinitely large. Some crystals

Table 1
Phase conditions resulting from certain loci of zero density

Condition	Synthesis		
	$\rho_{(xy)}$	$\rho_{(yz)}$	$\rho_{(xz)}$
$\rho_{(x00)} = 0$	$\frac{1}{2}F_{000} + \sum_{k=1}^{K} F_{0k0} = 0$ $F_{h00} + \sum_{k=1}^{K} (F_{hk0} + F_{\bar{h}k0}) = 0$		$\frac{1}{2}F_{000} + \sum_{l=1}^{L} F_{00l} = 0$ $F_{h00} + \sum_{l=1}^{L} (F_{h0l}+F_{\bar{h}0l}) = 0$
$\rho_{(0y0)} = 0$	$\frac{1}{2}F_{000} + \sum_{h=1}^{H} F_{h00} = 0$ $F_{0k0} + \sum_{h=1}^{H} (F_{hk0}+F_{\bar{h}k0}) = 0$	$\frac{1}{2}F_{000} + \sum_{l=1}^{L} F_{00l} = 0$ $F_{0k0} + \sum_{l=1}^{L} (F_{0kl}+F_{0k\bar{l}}) = 0$	
$\rho_{(00z)} = 0$		$\frac{1}{2}F_{000} + \sum_{k=1}^{K} F_{0k0} = 0$ $F_{00l} + \sum_{k=1}^{K} (F_{0kl}+F_{0k\bar{l}}) = 0$	$\frac{1}{2}F_{000} + \sum_{h=1}^{H} F_{h00} = 0$ $F_{00l} + \sum_{h=1}^{H} (F_{h0l}+F_{\bar{h}0l}) = 0$

have, in addition, a pattern containing repetitions within each cell. Subperiods of this kind are found, for example, in the zigzag chains of long-chain carbon compounds. An example is shown in Fig. 1*A*. The periodicity is emphasized by dividing the cell into subcells, as in Fig. 3.

The relation between cells, subcells, and their reciprocals is discussed in tensor notation by Vand.[3] The geometry involved follows simple reciprocal-lattice theory, and has been discussed from this viewpoint by the author.[†]

That a crystal may have a substructure can be suspected from its reciprocal structure. The reason for this is as follows: The corresponding atoms in the various cells of the substructure scatter in phase in the

[†] M. J. Buerger. *Vector space and its application in crystal-structure investigation.* (John Wiley and Sons, New York, 1959) Chapter 14.

various orders of reflection of the subcells in much the same way as the corresponding atoms in different unit cells scatter in phase. Therefore, in the reciprocal structure, those points reciprocal to the subcell tend to be strong, and thus to mark out reciprocal supercells. The reciprocal of this is the subcell of the crystal. An actual example, observed in β trilaurin by Vand and Bell,[4] is shown in Figs. 2 and 3.

In instances like these, the approximate structure can be crudely guessed in advance, since the shape of the molecule is known from

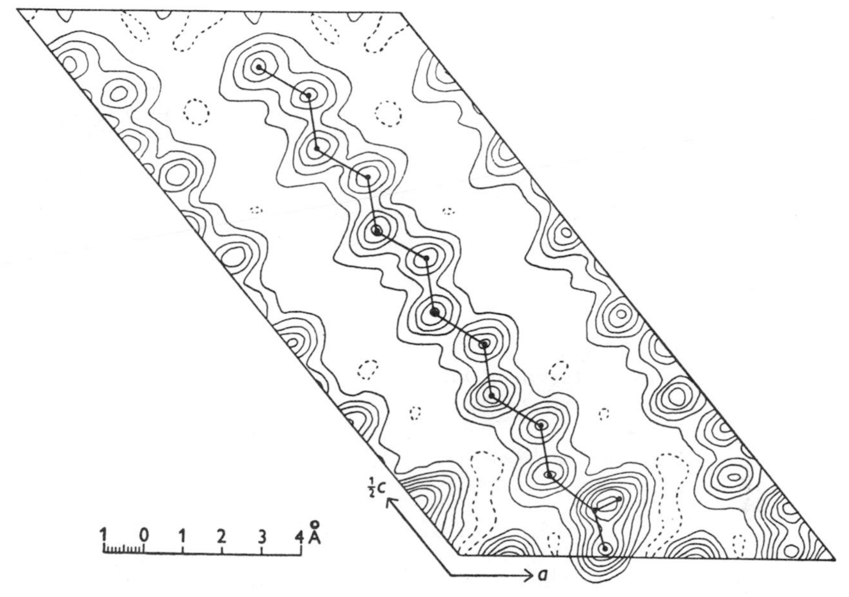

Fig. 1.
Lauric acid, $\rho_{(xz)}$.
(After Vand, Morley, and Lomer, Acta Cryst, **4** 324.)

chemical considerations. The known molecule can be plotted along the cell so that with possible tilting it spans the length of the cell. The repetition in the zigzag of the carbon chain then comprises the substructure.

More exact locations of the carbons within a subcell can be found by trial-and-error calculation of intensities. When the trial parameters duplicate the intensities of the reciprocal supercell, a fit is obtained. In making this computation it is assumed that the contribution to the F's of atoms not in the substructure can be neglected. The resulting distribution found by Vand and Bell[4] for the subcell of β trilaurin is shown in Fig. 3.

So far the few reflections from the substructure alone have been used. The other reflections are, however, different samples of the same Fourier transform of the molecule. Since the atoms of the molecule have been approximately located by using the reflections from the substructure, the Fourier transform of the entire molecule can be computed. If this is now laid on the reciprocal structure, the phases of all the remaining reflections can be read off.

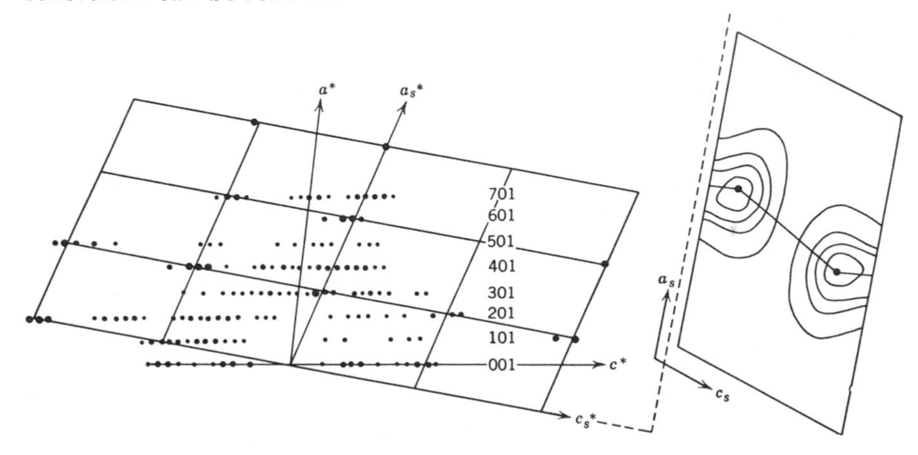

Fig. 2.
Example of the reciprocal structure of a crystal having a substructure; β trilaurin. The grid marks out a superlattice in reciprocal space.
(After Vand and Bell.[4])

Fig. 3.
Distribution of electron density in the subcell of β trilauin.
(After Vand and Bell.[4])

The nodal method

In Chapter 15 it was shown that the Fourier transform of a crystal can be regarded as a sampling at the points of the reciprocal lattice of the Fourier transform of the unit cell. If the crystal is centrosymmetrical, and the origin taken at the center, then all these Fourier transforms are real, and can therefore attain only positive and negative values.

The Fourier transform of the unit cell is a continuously variable function, and consists of positive and negative topography separated by nodal lines having zero value. If one could examine a map of the absolute values of this transform, it would be an easy matter to assign signs to its regions simply because the sign changes on opposite sides of each zero contour. Since x-ray diffraction supplies merely the sampling of the square of the transform at reciprocal lattice points, the lines of zero value are not obvious, and so signs cannot, offhand, be assigned to the sampling regions.

The author became interested in some possibilities suggested by this relation about 1943. An obvious possibility is to guide the contouring of the reciprocal by the contribution of a somewhat heavy atom. The

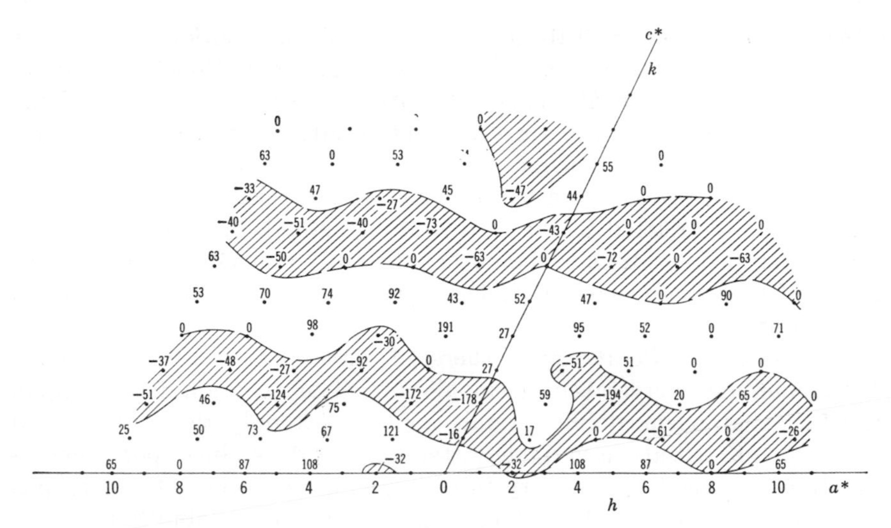

Fig. 4*A*.
The contributions by the metal atoms to the F_{h0l}'s of orthoclase, $KAlSi_3O_8$, separated into negative and positive regions.

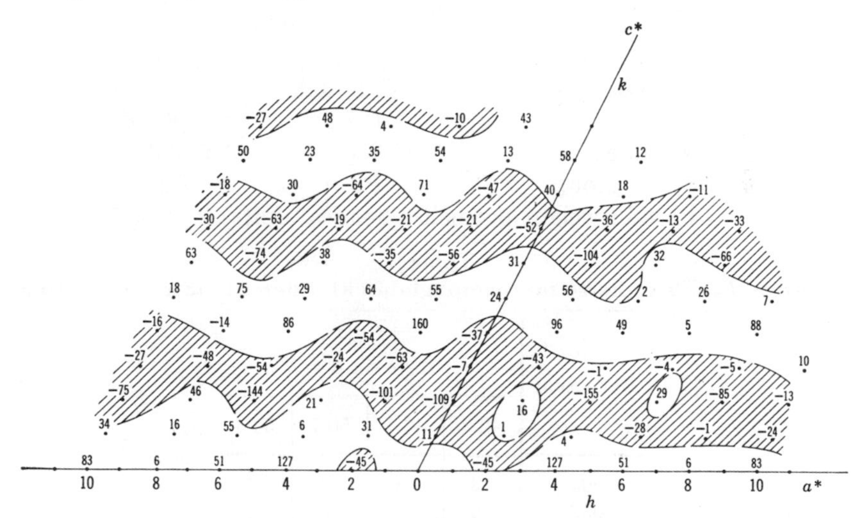

Fig. 4*B*.
The F_{h0l}'s of orthoclase, $KAlSi_3O_8$, separated into negative and positive regions.

relation is illustrated in Figs. 4*A* and 4*B*. Fig. 4*B* shows a contouring of the F_{h0l}'s of orthoclase, $KAlSi_3O_8$. The signs of these F's are known from the structure analysis.† The orthoclase structure contains moderately heavy atoms K, as well as the lighter metals Si and Al, located in

† S. H. Chao, A. Hargreaves, and W. H. Taylor. *The structure of orthoclase.* Mineral Mag. **25** (1940) 498–512.

general positions of the [010] projection. The contribution to the F_{h0l}'s of these metal atoms are shown in Fig. 4*A*. It is seen that the phases of these metal-atom contributions dominate the pattern of phases of the entire reciprocal structure, Fig. 4*B*. The situation can be used in reverse to determine many of the signs of the F's on the basis of the pattern of distribution of signs of the contribution of a moderately heavy atom. This scheme is related to the heavy-atom method, but requires only a moderately heavy atom. The author used it to check the signs of the F_{hk0}'s of nepheline,† $KNa_3Al_4Si_4O_{16}$, after most of the metal atoms had been located.

Boyes-Watson, Perutz, and others[12–20] have applied a neat variation of this method to crystals of haemoglobin. The variation depends upon the fact that haemoglobin crystals contain not only the molecules of haemoglobin, but also a variable water content, depending upon preparation and treatment. These crystals have enormous cells. For example, the cell characteristics of one kind of horse methaemoglobin[13] are:

	"Dry"	"Wet"
a	←—109 Å—→	
b	←— 63.2 Å—→	
c	53.5 Å	54.6 Å
$d_{001} = c \sin\beta$	42.3 Å	54.4 Å
β	127.5°	84.5°
Volume	292,000 Å^2	375,000 Å^2
Space group	$P2$	$P2$
Protein weight per cell (chemical units) $2 \times 66{,}700$		

Table 2
Relative $|F_{00l}|^2$'s of horse methaemoglobin at different states of swelling

Reflection	$\lvert F_{00l}\rvert^2$ for $c \sin\beta =$			
	42.3	46.1	50.7	54.4
001	3	4	10	5
002	0	6	10	7
003	17	21	2	0
004	17	11	5	20
005	45	11	21	24
006	0	32	46	28
007	4	0	17	16

† M. J. Buerger and Gilbert Klein. *The crystal structure of nepheline.* Am. Soc. for X-ray and Electron Diffraction, Lake George Meeting, June 10–14, 1946. Also, M. J. Buerger, Gilbert E. Klein, and Gabrielle Donnay. *Determination of the crystal structure of nepheline.* Am. Mineralogist **39** (1954) 805–818.

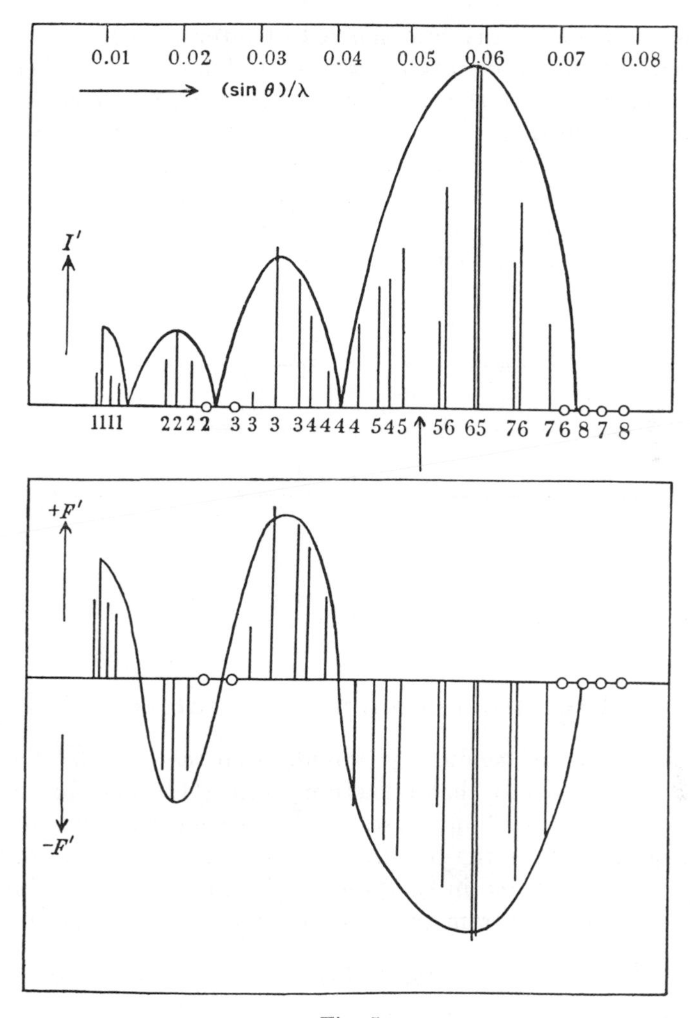

Fig. 5.

A. Vertical lines are samples of the intensities of various orders of $00l$ with several stages of shrinkage, for horse methaemoglobin. The envelopes suggest the zero points.

B. The regions of positive and negative F_{00l}, deduced from *A*.

(After Boyes-Watson, Davidson, and Perutz.[13])

It is evident that as the crystal takes on water its spacing d_{001} increases. Accordingly the translation, $t_{001}{}^*$, of the reciprocal cell shrinks. As this occurs, the sampling locations of the Fourier transform of the cell shifts. This permits a study of the transform at nearby points. The values of $|F_{00l}|^2$ for four values of the shrinkage are listed in Table 2. These values are plotted as vertical lines against $(\sin\theta)/\lambda$ in Fig. 5*A*. It can be seen that the transform is now sampled at intervals sufficiently closely spaced so that the locations of the zeros are well indicated. There appear to be four such nodes in the transform. Assuming that F_{001} is positive, signs can be assigned to all F_{00l}'s for any particular state of hydration.

Unfortunately there are few crystals for which this variety of the nodal method can be used. For most crystals, any change in cell geometry is accompanied by a corresponding change in the contents of the cell, so that the sampling of the Fourier transform does not change.

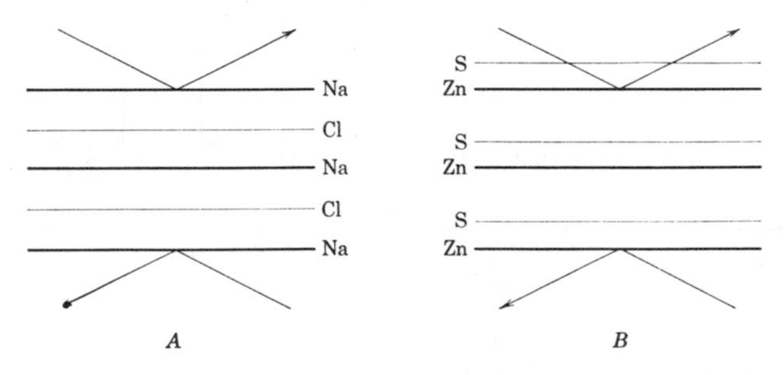

Fig. 6.

Utilization of anomalous scattering

For centric crystals, symmetry provides that (hkl) and $(\bar{h}\bar{k}\bar{l})$ are indistinguishable. It follows that reflections from (hkl) and $(\bar{h}\bar{k}\bar{l})$ are indistinguishable, Fig. 6*A*. For crystals lacking centers of symmetry, the sequence of atoms from the side (hkl) is the reverse of that from the side $(\bar{h}\bar{k}\bar{l})$, Fig. 6*B*. As noted in Chapter 16, the consequence of this is that the phases of reflection from the two different sides are opposite:

$$\begin{aligned} F_{hkl} &= |F_{hkl}|e^{i\phi}, \\ F_{\bar{h}\bar{k}\bar{l}} &= |F_{hkl}|e^{-i\phi}. \end{aligned} \tag{10}$$

Since phases cannot be measured in x-ray diffraction experimentation, it follows that reflections from (hkl) and $(\bar{h}\bar{k}\bar{l})$ have the same intensities. This is *Friedel's law*, and for most crystal-structure work it can be accepted as valid.

The situation is different if any atom in the crystal has an absorption edge just on the long-wavelength side of the radiation used for diffraction. Under these circumstances the wave scattered by an atom has an anomalous phase shift. The resultant wave, f, can be expressed in terms of the normal wave, f_0, with the aid of a real and imaginary correction f' and if'', as follows:

$$f = f_0 + f' + if''. \tag{11}$$

Since i has the function of advancing the phase by $\pi/2$, a positive phase shift occurs in the scattering of such atoms. Of course this has an effect on the intensity scattered by the crystal and, as will be seen, is generally different for hkl and $\bar{h}\bar{k}\bar{l}$ for non-centric crystals.

The breakdown of Friedel's law was tested in 1930 by Coster, Knol, and Prins[21] for ZnS, Fig. 6*B*. They used Au$L\alpha$, whose wavelength is just shorter than the K absorption edge of zinc. If the structure factors of reflections from (111) and ($\overline{111}$) of Fig. 6*B* are computed, taking account of the phase shift caused by the anomalous scattering of zinc, it turns out that the upper face, (111), reflects more intensely than the lower face ($\overline{111}$). Thus the asymmetric sequence can be detected.

Although this effect has been known experimentally since 1930, it has only recently been applied as a routine method for determining *absolute configuration* by Bijvoet[22–25, 27] and his school. The method gained in popularity when it became evident that the ordinary experimental technique of crystal-structure analysis is adequate to observe the inequality between $|F_{hkl}|$ and $|F_{\bar{h}\bar{k}\bar{l}}|$.

The description of Fig. 6*B* can be changed a little to bring out its relation to absolute configuration. In connection with Fig. 6*B* it proved possible to distinguish between an asymmetrical sequence of planes Zn·S–Zn·S–Zn·S, and its inverse S·Zn–S·Zn–S·Zn. More generally, it is possible to compute a set of diffraction intensities to be expected from an acentric molecule and from its enantiomorph. When the wavelength of the radiation is not near an absorption edge, these two sets are the same. But if the molecule contains an atom with an absorption edge just to the long-wavelength side of the radiation used, the two sets are different. They can be computed, compared with the observed intensities, and an identification of absolute configuration made.

The principle involved is simple, and is illustrated in Fig. 7. For simplicity, suppose that there is one atom in the crystal which scatters anomalously. The crystal can be regarded as made up of this anomalous scatterer, A, plus a residue, R. The amplitude of the diffracted wave is the complex sum of these:

$$F_{hkl} = {}^{R}F_{hkl} + {}^{A}F_{hkl}. \qquad (12)$$

The residue scatters with the same magnitude but opposite phases for hkl and $\bar{h}\bar{k}\bar{l}$, since it is a normal scatterer. But the anomalous scatterer contains two components, a "normal" component ${}^{A}F'$, and an "abnormal" component ${}^{A}F''$. The phase of the latter is $\pi/2$ ahead of the "normal" component. As a result, the part ${}^{A}F'$ has equal magnitudes and opposite phases for hkl and $\bar{h}\bar{k}\bar{l}$, but the part ${}^{A}F''$ does not have this relation. If it were not for the component ${}^{A}F''$, the resultant (12) would have the same magnitude for hkl and $\bar{h}\bar{k}\bar{l}$, and Friedel's law would be obeyed. But the part $^{A}F''$ always has a direction related to that of $^{A}F'$ by $+\pi/2$. If this direction is not at right angles to that of the resultant of $^{R}F + {}^{A}F'$, then the resultants

$$F_{hkl} = {}^{R}F_{hkl} + {}^{A}F'_{hkl} + {}^{A}F''_{hkl}$$

and (13)

$$F_{\bar{h}\bar{k}\bar{l}} = {}^{R}F_{\bar{h}\bar{k}\bar{l}} + {}^{A}F'_{\bar{h}\bar{k}\bar{l}} + {}^{A}F''_{\bar{h}\bar{k}\bar{l}}$$

have different magnitudes, as illustrated in Fig. 7.

In general, every reflection of a non-centric crystal must give different magnitudes for $|F_{hkl}|$ and $|F_{\bar{h}\bar{k}\bar{l}}|$ when the diffraction experiment is undertaken with a radiation whose wavelength is just shorter than the absorption edge of an atom in the structure. The effect becomes greater with increasing scattering power of the anomalous scatterer A. The effect may be inappreciable for certain reflections which have $^{A}F''$ normal to the resultant F.

This inequality for a whole set of reflections was first observed by Peerdeman, van Bommel, and Bijvoet.[23] They had determined the structure of NaRb tartrate, except that the molecule and its enantiomorph had not been distinguished. To make use of anomalous scattering they used monochromated Zr$K\alpha$ radiation ($\lambda_{\alpha_1} = 0.79010$ Å, $\lambda_{\alpha_2} = 0.78588$ Å) in connection with the absorption edge of Rb ($\lambda = 0.81549$ Å). With this radiation they took a c-axis, first-layer Weissenberg photograph. They observed the inequalities in intensities noted in the second column of Table 3. Now the intensities to be expected for each of the two enantiomorphs can be determined either graphically or analytically along the lines indicated in Fig. 7. The results for enantiomorphs A and B are listed in the right column of Table 3. It is evident that the computed inequalities for enantiomorph A are the same as those observed. This establishes the absolute configuration of the tartrate molecule in this crystal.

It was later discovered by Peterson[26] that the effects of anomalous scattering could be observed even when the wavelengths involved were not very near the absorption edge of an element in the structure. Accordingly Dauben and Templeton[30] computed tables of the effects to be expected using Mo, Cu, and Cr$K\alpha$ radiations. Their results are given in Table 4. Peerdeman[34] determined the absolute configuration of strychnine hydrobromide dihydrate using ordinary Cu$K\alpha$ radiation, Ramachandran and Chandrasekaran[37] determined the absolute configuration of $NaClO_3$ with the same radiation, while de Vries[43] determined the absolute configuration of quartz using Cr$K\alpha$ radiation.

Ramachandran and Raman[35] have pointed out that the phase angle of a

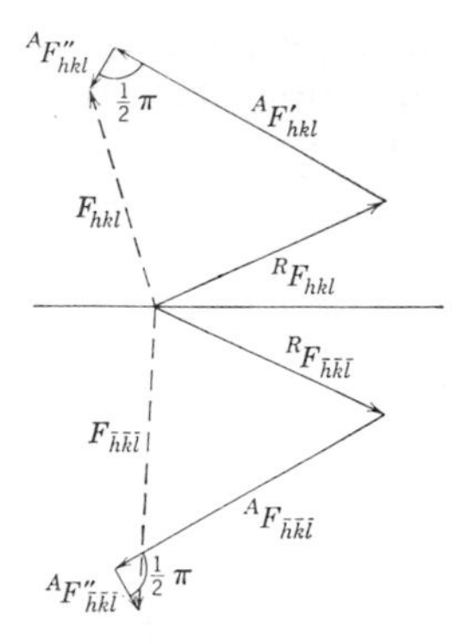

Fig. 7.

Table 3
Results of anomalous scattering experiment for NaRb tartrate[23]

Indices	Observed relation		Computed intensities			
			Enantiomorph *A*		Enantiomorph *B*	
	hkl	$\bar{h}\bar{k}\bar{l}$	I_{hkl}	$I_{\bar{h}\bar{k}\bar{l}}$ ($\simeq I_{\bar{h}kl}$)	I_{hkl}	$I_{\bar{h}\bar{k}\bar{l}}$ ($\simeq I_{\bar{h}kl}$)
141	(?)		361	377	377	361
151	(?)		337	313	313	337
161	>		313	241	241	313
171	<		65	78	78	65
181	>		185	148	148	185
191	>		65	46	46	65
1·10·1	>		248	208	208	248
1·11·1	<		27	41	41	27
261	>		828	817	817	828
271	>		18	8	8	18
281	>		763	716	716	763
291	(?)		170	166	166	170
2·10·1	<		200	239	239	200
2·11·1	(?)		159	149	149	159
2·12·1	<		324	353	353	324

reflection can be determined by making use of the computed difference in anomalous scattering for F_{hkl} and $F_{\bar{h}\bar{k}\bar{l}}$. Raman[41] verified that the method worked in the case of L-ephedrine hydrochloride using Cu$K\alpha$ radiation, but the results are not as clear cut as one might wish.

Table 4
Values of $\Delta f'$ and $\Delta f''$ for correcting f for effects of anomalous scattering according to $f = f_0 + \Delta f' + i\Delta f''$
(After Dauben and Templeton[30])

Atomic number	Element	Radiation					
		Mo$K\alpha$		Cu$K\alpha$		Cr$K\alpha$	
		$-\Delta f'$	$\Delta f''$	$-\Delta f'$	$\Delta f''$	$-\Delta f'$	$\Delta f''$
20	Ca	−0.2	0.2	−0.2	1.4	0.2	2.7
21	Sc	−0.2	0.3	−0.2	1.6	0.7	3.2
22	Ti	−0.3	0.4	−0.2	1.9	1.7	3.8
23	V	−0.3	0.5	−0.2	2.3	4.4†	0.6
24	Cr	−0.4	0.6	0.1	2.6	2.2	0.7
25	Mn	−0.4	0.8	0.5	3.0	1.8	0.8
26	Fe	−0.4	1.0	1.1	3.4	1.6	0.9
27	Co	−0.4	1.1	2.2	3.9	1.4	1.0
28	Ni	−0.4	1.2	3.1†	0.6	1.2	1.2
29	Cu	−0.4	1.4	2.1	0.7	1.1	1.3
30	Zn	−0.3	1.6	1.7	0.8	1.0	1.5
31	Ga	−0.2	1.7	1.5	0.9	0.9	1.7
32	Ge	−0.2	1.9	1.3	1.1	0.8	1.9
33	As	−0.1	2.2	1.2	1.2	0.7	2.2
34	Se	0.1	2.4	1.0	1.3	0.7	2.4
35	Br	0.3	2.6	0.9	1.5	0.6	2.7
36	Kr	0.6	2.9	0.9	1.7	0.6	3.0
37	Rb	0.9	3.2	0.8	1.9	0.6	3.4
38	Sr	1.4	3.6	0.7	2.1	0.6	3.8
39	Y	2.3	3.9	0.7	2.3	0.6	4.2
40	Zr	2.8	0.8	0.6	2.5	0.7	4.6
41	Nb	2.1	0.9	0.6	2.8	0.8	5.1
42	Mo	1.7	0.9	0.5	3.0	0.9	5.6
43	Tc	1.4	1.0	0.5	3.3	1.0	6.2
44	Ru	1.2	1.1	0.5	3.6	1.2	6.7
45	Rh	1.1	1.2	0.5	4.0	1.3	7.3
46	Pd	1.0	1.3	0.5	4.3	1.6	7.9
47	Ag	0.9	1.4	0.5	4.7	1.9	8.6
48	Cd	0.8	1.6	0.6	5.0	2.2	9.2
49	In	0.7	1.7	0.6	5.4	2.7	10
50	Sn	0.6	1.9	0.7	5.8	3.2	11
51	Sb	0.6	2.0	0.8	6.3	4.0	12
52	Te	0.5	2.2	0.9	6.7	5.0	12
53	I	0.5	2.4	1.1	7.2	7.2†	14
54	Xe	0.4	2.5	1.4	7.8	§	11
55	Cs	0.4	2.7	1.7	8.3	12†	12
56	Ba	0.4	2.9	2.1	8.9	11†	8
57	La	0.3	3.1	2.5	9.6	14†	3
58	Ce	0.3	3.3	3.0	10	10	3
59	Pr	0.3	3.6	3.5	11	9	4

Table 4 (*Continued*)

Atomic number	Element	Radiation					
		MoKα		CuKα		CrKα	
		$-\Delta f'$	$\Delta f''$	$-\Delta f'$	$\Delta f''$	$-\Delta f'$	$\Delta f''$
60	Nd	0.3	3.8	4.3	12	8	4
61	Pm	0.3	4.0	5.2	12	7	4
62	Sm	0.3	4.2	6.7	13	7	5
63	Eu	0.3	4.5	§	11	6	5
64	Gd	0.3	4.7	12†	12	6	5
65	Tb	0.4	5.0	11†	8	6	6
66	Dy	0.4	5.2	10	8	6	6
67	Ho	0.4	5.5	13†	4	6	7
68	Er	0.5	5.8	9†	4	5	7
69	Tm	0.5	6.2	8	4	5	8
70	Yb	0.6	6.4	8	4	5	9
71	Lu	0.7	6.8	7	5	5	8
72	Hf	0.8	7.1	7	5	5	9
73	Ta	0.9	7.4	6	5	5	10
74	W	1.1	7.8	6	6	5	11
75	Re	1.3	8.1	6	6	5	11
76	Os	1.5	8.5	6	6	5	12
77	Ir	1.8	8.9	5	7	5	13
78	Pt	2.0	9.3	5	7	5	13
79	Au	2.4	9.8	5	8	6	14
80	Hg	2.8	10	5	8	6	15
81	Tl	3.4	11	5	8	6	16
82	Pb	4.0	11	5	9	7	16
83	Bi	4.8	12	5	9	7	17
84	Po	5.6	12	5	10	8	18
85	At	§	10	5	10	9	20
86	Rn	8†	10	5	11	9	20
87	Fr	8†	7	5	11	10	21
88	Ra	7	7	5	12	11	23
89	Ac	7	8	5	12	13	24
90	Th	7	8	5	13	15	25
91	Pa	7	8	5	14	17†	26
92	U	8†	8	5	15	19†	19
93	Np	§	5	5	15	§	§
94	Pu	§	5	5	16	§	§
95	Am	§	5	6	17	§	§
96	Cm	§	5	6	18	§	§

† These values are especially uncertain because of proximity to absorption edges.
§ Not calculated.

Literature

Crystals with regional low uniform density

[1] O. A. Tesche. *A contribution to the determination of signs in the Fourier analysis of crystals.* Acta Cryst. **6** (1953) 564.

[2] O. A. Tesche. *The crystal structure of ximenynic acid.* Acta Cryst. **7** (1954) 737–739.

Crystals containing long-chain molecules

[3] Vladimir Vand. *Method for determining the signs of the structure factors of long-chain compounds.* Acta Cryst. **4** (1951) 104–105.

[4] V. Vand and I. P. Bell. *A direct determination of the crystal structure of the β form of trilaurin.* Acta Cryst. **4** (1951) 465–569.

[5] Erik von Sydow. *On the structure of the crystal form* B′ *of* n-*pentadecanoic acid.* Acta Cryst. **7** (1954) 823–826.

[6] Sixten Abrahamsson. *On the crystal structure of* 16DL-*methyloctadecanoic acid.* Acta Cryst. **11** (1958) 270–273.

[7] Erik von Sydow. *The crystal structure of* (−)-*2-methyl-2-ethyleicosanoic acid.* Acta Chem. Scand. **12** (1958) 777–779.

[8] T. Brotherton, B. Craven, and G. A. Jeffrey. *The crystal structure of trans* D,L 9-10 *methylene octadecanoic acid.* Acta Cryst. **11** (1958) 546–551.

[9] Sixten Abrahamsson. *On the crystal structure of* 14 DL-*methyloctadecanoic acid.* Acta Cryst. **12** (1959) 206–209.

[10] Sixten Abrahamsson. *On the crystal structure of* 2 DL-*methyloctadecanoic acid.* Acta Cryst. **12** (1959) 301–304.

[11] Sixten Abrahamsson. *On the crystal structure of* 2 D-*methyloctadecanoic acid.* Acta Cryst. **12** (1959) 304–309.

The nodal method

[12] Joy Boyes-Watson and M. F. Perutz. *X-ray analysis of haemoglobin.* Nature **151** (1943) 714–716.

[13] Joy Boyes-Watson, Edna Davidson, and M. F. Perutz. *An x-ray study of horse methaemoglobin. I.* Proc. Roy. Soc. (London) (A) **191** (1947) 83–132, especially 117–120.

[14] Kate Dornberger-Schiff. *Patterson and Fourier projections of single molecules of haemoglobin.* Acta Cryst. **3** (1950) 143–146, especially 144–145.

[15] Lawrence Bragg and M. F. Perutz. *The structure of haemoglobin.* Proc. Roy. Soc. (London) (A) **213** (1952) 425–435.

[16] W. L. Bragg, E. R. Howells, and M. F. Perutz. *Arrangement of polypeptide chains in horse methaemoglobin.* Acta Cryst. **5** (1952) 136–141.

[17] W. L. Bragg and M. F. Perutz. *The external form of the haemoglobin molecule. I.* Acta Cryst. **5** (1952) 277–283. *II.* Acta Cryst. **5** (1952) 323–328.

[18] Lawrence Bragg, E. R. Howells, and M. F. Perutz. *The structure of haemoglobin. II.* Proc. Roy. Soc. (London) (A) **222** (1954) 33–44.

[19] M. F. Perutz. *The structure of haemoglobin. III. Direct determination of the molecular transform.* Proc. Roy. Soc. (London) (A) **225** (1954) 264–286.

[20] D. W. Green, V. M. Ingram, and M. F. Perutz. *The structure of haemoglobin. IV. Sign determination by the isomorphous replacement method.* Proc. Roy. Soc. (London) (A) **225** (1954) 287–307.

Utilization of anomalous scattering

[21] D. Coster, K. S. Knol, and J. A. Prins. *Unterschiede in der Intensität der Röntgenstrahlenreflexion an den beiden* 111-*Flächen der Zinkblende.* Z. Phys. **63** (1930) 345–369.

[22] J. M. Bijvoet. *Phase determination in direct Fourier-synthesis of crystal structures.* Proc. Koninkl. Ned. Akad. Wetenschap. (B) **52** (1949) 313–314.

[23] A. F. Peerdeman, A. J. van Bommel, and J. M. Bijvoet. *Determination of the absolute configuration of optically active compounds by means of x-rays.* Proc. Koninkl. Ned. Akad. Wetenschap. (B) **54** (1951) 16–19.

[24] J. M. Bijvoet. *Structure of optically active compounds in the solid state.* Nature **173** (1954) 888–891.

[25] J. Trommel and J. M. Bijvoet. *Crystal structure and absolute configuration of the hydrochloride and hydrobromide of* D(−)-*isoleucine.* Acta Cryst. **7** (1954) 703–709.

[26] S. W. Peterson. *Anomalous x-ray scattering at wave-lengths far from an absorption edge.* Nature **176** (1955) 395.

[27] J. M. Bijvoet. *Determination of the absolute configuration of optical antipodes.* Endeavor **14** (1955) 71–77.

[28] Yoshihiko Saito, Kazumi Nakatsu, Motoo Shiro, and Hisao Kuroyo. *Determination of the absolute configuration of optically active complex ion* $[Coen_3]^{3+}$ *by means of x-rays.* Acta Cryst. **8** (1955) 729–730.

[29] Y. Okaya, Y. Saito, and R. Pepinsky. *New method in x-ray crystal structure determination involving the use of anomalous dispersion.* Phys. Rev. **98** (1955) 1857–1858.

[30] Carol H. Dauben and David H. Templeton. *A table of dispersion corrections for x-ray scattering of atoms.* Acta Cryst. **8** (1955) 841–842.

[31] David H. Templeton. *X-ray dispersion effects in crystal-structure determination.* Acta Cryst. **8** (1955) 842.

[32] A. McL. Mathieson. *The determination of absolute configuration by the use of an internal reference asymmetric centre.* Acta Cryst. **9** (1956) 317.

[33] Yoshiharu Okaya and Ray Pepinsky. *New formulation and solution of the phase problem in x-ray analysis for noncentric crystals containing anomalous scatterers.* Phys. Rev. **103** (1956) 1645–1657.

[34] A. F. Peerdeman. *The absolute configuration of natural strychnine.* Acta Cryst. **9** (1956) 824.

[35] G. N. Ramachandran and S. Raman. *A new method for the structure analysis of non-centro-symmetric crystals.* Current Sci. **25** (1956) 348–351.

[36] A. F. Peerdeman and J. M. Bijvoet. *The indexing of reflections in investigations involving the use of the anomalous scattering effect.* Acta Cryst. **9** (1956) 1012–1015.

[37] G. N. Ramachandran and K. S. Chandrasekaran. *The absolute configuration of sodium chlorate.* Acta Cryst. **10** (1957) 671–675.

[38] R. Pepinsky and Y. Okaya. *Comparison of two procedures for solution of non-*

centric crystal structures utilizing anomalous dispersion. Phys. Rev. **108** (1957) 1231–1232.

[39] S. Ramaseshan, K. Venkatesan, and N. V. Mani. *The use of anomalous scattering for the determination of crystal structures—$KMnO_4$.* Proc. Indian Acad. Sci. **46** (1957) 95–111.

[40] S. Ramaseshan and K. Venkatesan. *The use of anomalous scattering without phase change in crystal structure analysis.* Current Sci. **26** (1957) 352–353.

[41] S. Raman. *Anomalous dispersion method of determining structure and absolute configuration of crystals.* Proc. Indian Acad. Sci. **47** (1958) 1–11.

[42] A. J. van Bommel and J. M. Bijvoet. *The crystal structure of ammonium hydrogen* D*-tartrate.* Acta Cryst. **11** (1958) 61–70.

[43] A. de Vries. *Determination of the absolute configuration of α-quartz.* Nature **181** (1958) 1193.

[44] D. M. Blow. *The structure of haemoglobin. VII. Determination of phase angles in the non-centrosymmetric* [100] *zone.* Proc. Roy. Soc. (London) (A) **247** (1958) 302–336, especially 323–328.

[45] J. M. Bijvoet and A. F. Peerdeman. *De bepaling van de absolute configuratie van optische antipoden in de Röntgenanalyse en enige verdere toepassingen van anomale Röntgenverstrooiing.* Ned. Tijdschr. Natuurk. **24** (1958) 140–155.

[46] P. Harrison, G. A. Jeffrey, and J. R. Townsend. *An experimental study of the anomalous scattering of MoKα radiation by single crystals of ZnO.* Acta Cryst. **11** (1958) 552–556.

[47] A. W. Hanson and F. R. Ahmed. *The crystal structure and absolute configuration of the monoclinic form of* d*-methadone hydrobromide.* Acta Cryst. **11** (1958) 724–728.

[48] S. Raman. *Determination of the structure and absolute configuration of* L(+)*-lysine hydrochloride dihydrate by the anomalous-dispersion method.* Z. Krist. **111** (1959) 301–317.

21

Direct determinations

The phase problem

In Chapter 13 it was seen that the electron density of a crystal can be supplied by a Fourier synthesis, and that the coefficients of the Fourier summations are the F_{hkl}'s of the various diffraction spectra. In general, each F_{hkl} is a complex quantity. This can be emphasized by writing F_{hkl} as $|F_{hkl}|e^{i\phi_{hkl}}$; that is, each F is characterized by both a magnitude and a phase. Now the magnitude of the F's can be readily derived from the measured intensities, but no experimental means has been found for observing the phases, ϕ_{hkl}. Thus there are not sufficient data at hand from the diffraction experiment for performing the Fourier synthesis. In fact, only half the required data are provided by the experiment. At first sight, therefore, it appears that crystal structures are indeterminate due to the lack of phase data. This constitutes the *phase problem* of x-ray crystallography. If there were some way of learning these phases, then it would be a routine matter to find the arrangement of atoms in any crystal whatever, no matter how complicated.

In the absence of direct phase information, crystal structures are solved by finding the phases in an indirect way. In the early days of crystal-structure analysis the phases were found by what amounted to guessing the structure. This could often be done for the simpler structures, especially when one or more of the atoms were symmetry fixed.

More complicated structures can also be solved without direct phase information, provided special conditions make the phases available. Some of these conditions, and how they can be used to furnish the phases indirectly, were discussed in Chapters 19 and 20. Most of the structures which have been solved, including some very complicated structures, have been solved with the aid of the heavy-atom method or the replaceable-atom method. Indeed, the arrangements of atoms in many organic molecules have been revealed by deliberately preparing crystals containing the molecules in question, plus heavy atoms, and then solving the crystal structures by the heavy-atom method.

Of course, such methods can only be used when special conditions obtain, or when they can be caused to occur. But if one desires to determine the structure of some arbitrarily chosen crystal he is faced with the phase problem.

There now exist solutions, within limits to be noted, to the phase problem. These come as the culmination of a series of developments which started in 1934. In that year Patterson published a paper in which he presented what is now known as the *Patterson function.*[23] This is a Fourier series whose coefficients are $|F_{hkl}|^2$'s. This development did not solve the phase problem, but did permit an insight into what limitations are imposed by the lack of phase information. Two years later Harker showed that certain sections (corresponding to electron-density sections) of the three-dimensional Patterson function have useful properties in helping to locate atoms for crystals having rotational symmetry. These became known as *Harker sections.*[23] The Second World War intervened shortly thereafter, but immediately after the war the author presented implication theory,[23] which is concerned with the systematic interpretation of Harker sections. One of the immediate consequences of this theory was that for crystals having certain symmetries, the Harker section could be transformed to a map of the crystal structure by a simple geometrical transformation. Since this was all based upon the Patterson function, which involves no knowledge of the phases, it became evident that at least some crystal structures could be determined without a direct knowledge of the phases, or, what amounted to the same thing, that phase information was contained in the set of phaseless amplitudes. While none of this development was concerned specifically with phases, it nevertheless cast new light on the phase problem, which then lost its hopeless aspect.

Within a year after the presentation of implication theory, Harker and Kasper[24] discovered some explicit relations between amplitudes and phases. This began a new era in the field of the theory of crystal-structure analysis. Many have contributed to the problem, and the literature is voluminous.

It should not be supposed that the phase problem is solved. At the present time a sufficient number of the phases can be found for structure analyses, at least in not-unfavorable cases, for crystals having between a half-dozen to a score of atoms per projected asymmetric unit. Crystals of much more complexity cannot be solved unless some favorable circumstance obtains, like the presence of a set of heavy atoms. So again it may be said that every crystal structure cannot be solved by methods currently known.

The variety of the solutions

As soon as it was demonstrated that there existed relations between the phases and the intensities of the x-ray reflections, the matter of finding solutions to the phase problem became a popular pursuit. Many have made contributions to this field, and its literature is already so voluminous that it occupies many times the number of pages in this book. In this literature a number of different kinds of solution have been presented, each subject to certain limitations. To give a reasonably complete account of the several methods would require several books, and indeed one book has already been devoted to one class of solutions.

It is plain that it is out of the question to give an adequate discussion of these solutions in a small compass. This chapter, therefore, deals only with the highlights of the various methods. The solutions of the phase problem which have been offered fall into several categories, as follows:

Algebraic methods,
Vector-space methods,
Inequality methods,
Sign relations,
Statistical methods,
Permutation methods.

These are outlined in the following sections.

Algebraic methods

The earliest attempts at direct solutions of crystal structures were by algebraic methods. These methods have the following basis. Each of the structure factors has a similar form:

$$|F_{hkl}|e^{i\phi_{hkl}} = \sum_{j=1}^{N} f_j \, e^{i2\pi(hx_j+ky_j+lz_j)}. \tag{1}$$

This equation contains as unknowns the $3N$ values of the coordinates, and the phase ϕ_{hkl}. The magnitude $|F_{hkl}|$ is known from measurement. Ordinarily there are many more equations available than unknowns, so that it should be possible by algebraic means to solve for either the unknown phases ϕ, or the coordinates xyz.

Many attempts have been made to solve sets of equations based upon (1) for either the phases or coordinates, but none of the schemes has come into common use. There appear to be several reasons for this. In the first place, equations (1) are not linear, so that simple elimination procedures cannot be used. Furthermore, whereas it would appear, offhand, that, for N atoms per cell, $3(N-1)$ equations should be sufficient for a solution, actually several solutions are allowed by this number of equations. For example, for three atoms in a one-dimensional cell, there are four solutions.[14] More equations must be included to make the solution unique. In any event, the use of a high-speed digital computer would probably be required if N is reasonably large.

The method of Banerjee[2] appears to be most promising, and reasonably simple. It was revived and generalized by Hughes.[11]

Vector-space methods[23]

The Patterson synthesis. The science of x-ray crystallography is indebted to A. L. Patterson for initiating a major branch of development in the study of the phase problem. The phase problem can be stated in two ways: (*a*) What are the phases of the F's? (*b*) What is the maximum information which can be obtained from the phaseless $|F|$'s or $|F|^2$'s? Patterson sought to answer the second question along the following lines: If two Fourier series, each representing the electron density of a crystal, are multiplied together, terms like $F_{hkl}F_{\bar{h}\bar{k}\bar{l}}$ occur. This is the product of complex conjugates, and equal to $|F_{hkl}|^2$, which is the desired phaseless intensity. On this basis, Patterson was able to show that if a Fourier synthesis whose coefficients are $|F_{hkl}|^2$ is prepared, this synthesis has a meaning in terms of the electron-density synthesis. More exactly, if the two Fourier syntheses are compared:

$$\rho_{(xyz)} = \frac{1}{V} \sum_h \sum_k \sum_l F_{hkl}\, e^{i2\pi(hx+ky+lz)}, \tag{2}$$

$$A_{(uvw)} = \frac{1}{V^2} \sum_h \sum_k \sum_l |F_{hkl}|^2\, e^{i2\pi(hu+kv+lw)}, \tag{3}$$

then, (2) has peaks at atom locations, while (3) has peaks at the ends of vectors between atom locations in (2). An equivalent statement is that the coordinates of the rth peak in (3) are related to the coordinates of the ith and jth peaks in (2), by

$$\begin{aligned} u_r &= x_j - x_i, \\ v_r &= y_j - y_i, \\ w_r &= z_j - z_i. \end{aligned} \tag{4}$$

Thus to every pair of peaks in (2) there is a specific peak in (3). Synthesis (3) is called the *Patterson synthesis*, or the *Patterson function*.

In crystal-structure analysis, one can always prepare a Patterson synthesis. This is a synthesis of the original data, in its least degraded form, obtainable from the diffraction experiment. The analyst would like to transform the information into the electron density of the crystal. To do this requires a knowledge of the relationship between *vector space* and ordinary space (or crystal space). This background and how to make practical use of it in the "solution" of a Patterson synthesis cannot be adequately contained in a single chapter of a book. Accordingly the author has separated out this whole branch of direct crystal-structure analysis into a companion volume.[23]

In addition to the solution discussed in the book just mentioned, there are two classes of solutions for the phases of the F's which are based upon relations derived from vector space. These solutions are merely transformations into phase information of the geometry which results from vector-space considerations.

Product space. The earliest such results were presented by the author[15,16] to reveal the general similarity of the results obtained from the older implication theory[23] and the newer inequalities[24] (see next section). The results were obtained by mapping the geometry of Patterson space in *product space*. This space makes use of the fact that the phase diffracted from a point at xyz in orders hkl is $2\pi(hx+ky+lz)$, so that the products hx, ky, and lz control the phase. The space whose coordinates are the products hx, ky, and lz is called product space, and a map of a set of atoms diffracting in orders hkl defines the phase of the diffracted beam. On such a map it is possible to discover relations between phases of various diffracted beams. In particular, if a map in product space is prepared for both F_{hkl} and $|F_{hkl}|^2$ for a set of atoms which are symmetrical with respect to some axial symmetry element, it is evident that geometrical relations exist between these two figures. Thus, in Fig. 1, F_{hkl} and $|F_{hkl}|^2$ are plotted for a set of points related by a 4_1 screw. It can be seen that the projection of the map of F_{hkl} is a set of points on the vertices of a square. The section of $|F_{hkl}|^2$ at level $\frac{1}{2}l$ is also a set of points on the vertices of a square having the same orientation but twice the scale. Therefore the section of $|F_{hkl}|^2$ at level $\frac{1}{2}l$ has the same map as the projection of $F_{2h\,2k\,2l}$. The section at level $\frac{1}{4}l$ is another square, rotated 45°, and the section at level zero is a single point. Taking account of the weightings of these levels, and considering the points in F_{hkl} as unit scatterers, the relation between F and $|F|^2$ can be written

$$|F_{hkl}|^2 = F_{000} + F_{(h+k)(\bar{h}+k)0} + 2F_{2h\,2k\,0}. \tag{5}$$

Furthermore, if the Fourier representation of projections and sections of Fig. 1 are prepared, the following further results are obtained:

$$\text{Level } \tfrac{1}{2}: \qquad \sum_{l=-\infty}^{\infty} |F_{hkl}|^2 \cos 2\pi \frac{l}{2} = 2F_{2h\ 2k\ 0}. \tag{6}$$

$$\text{Level } \tfrac{1}{4}: \qquad \sum_{l=-\infty}^{\infty} |F_{hkl}|^2 \cos 2\pi \frac{l}{4} = F_{(h+k)(\bar{h}+k)0}. \tag{7}$$

In a similar way relations can be written for the various F's in terms of the set of $|F|^2$'s for a symmetrical set of unit scatterers for each symmetry element, and for each combination of symmetry elements or space group.

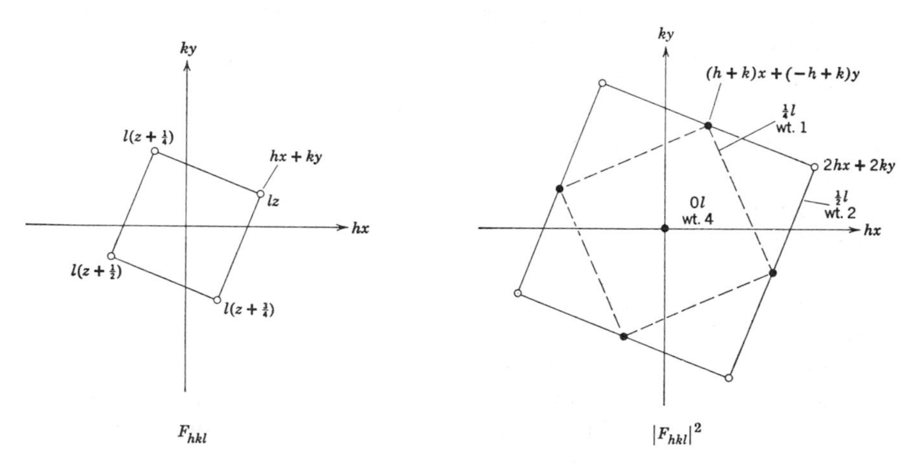

Fig. 1.
Relation between F_{hkl} and $|F_{hkl}|^2$ as brought out by plotting in product space.

Results of types (6) and (7) can be written for several symmetrical sets (such as occur in actual crystals), and also extended to non-unit scatterers by adding terms to the simpler expressions[15,16]. These results have not been put to use in crystal-structure determination, but they do indicate the relations between vector space and crystal space in terms of the F's involved.

Image-seeking theory. A second class of solutions descending from vector-space considerations is derived from converting image-seeking theory,[23] in one of its forms, into phase information. This has been done by McLachlan[18] for the sum function, and by McLachlan and Harker,[19] and by Vaughan[22] for the product function.

Harker-Kasper inequalities

Harker and Kasper[24, 25] effected a major change of direction in the investigation of the phase problem by showing that there exist relations between certain F's and $|F|^2$'s for symmetrical crystals which could be expressed in the form of inequalities. To do this they made use of the well-known Cauchy's inequality

$$\left|\sum_{j=1}^{N} a_j b_j\right|^2 \leqslant \left(\sum_{j=1}^{N} |a_j|^2\right)\left(\sum_{j=1}^{N} |b_j|^2\right), \tag{8}$$

and the related Schwartz's inequality

$$\left|\int fg\, d\tau\right|^2 \leqslant \left(\int |f|^2\, d\tau\right)\left(\int |g|^2\, d\tau\right). \tag{9}$$

The fundamental scattering relations can be written

$$\begin{aligned} F_{hkl} &= \frac{V}{abc} \int_0^a \int_0^b \int_0^c \rho_{(XYZ)}\, e^{i2\pi(hX/a+kY/b)+lZ/c)}\, dX\, dY\, dZ \\ &= V \int_0^1 \int_0^1 \int_0^1 \rho_{(xyz)}\, e^{i2\pi(hx+ky+lz)}\, dx\, dy\, dz. \end{aligned} \tag{10}$$

For Schwartz's inequality, let

$$\begin{aligned} f &= (V\rho_{(xyz)})^{1/2}, \\ g &= (V\rho_{(xyz)})^{1/2}\, e^{i2\pi(hx+ky+lz)}, \\ d\tau &= dx\, dy\, dz. \end{aligned} \tag{11}$$

Then, substitution in (9) gives

$$\begin{aligned} |F_{hkl}|^2 \leqslant V^2 &\left[\int_0^1 \int_0^1 \int_0^1 \rho_{(xyz)}\, dx\, dy\, dz\right] \\ &\times \left[\int_0^1 \int_0^1 \int_0^1 \rho_{(xyz)} |e^{i2\pi(hx+ky+lz)}|^2\, dx\, dy\, dz\right]. \end{aligned} \tag{12}$$

Since $|e^x| = 1$, it follows that

$$|F_{hkl}|^2 \leqslant V^2 \left[\int_0^1 \int_0^1 \int_0^1 \rho_{(xyz)}\, dx\, dy\, dz\right]^2. \tag{13}$$

The quantity $V\, dx\, dy\, dz$ is the volume element of the coordinate system xyz. Thus

$$V \int_0^1 \int_0^1 \int_0^1 \rho_{(xyz)}\, dx\, dy\, dz = Z, \tag{14}$$

so that (13) means

$$|F_{hkl}|^2 \leqslant Z^2. \tag{15}$$

This initial result corresponds to the well-known fact that no scattering amplitude can exceed Z.

It is convenient to utilize a scale of F's for a particular crystal such that the maximum value of F is 1. This maximum occurs when all atoms scatter in phase. This can be done by defining a normalized F, to be called the *unitary structure factor*, and designated U, such that

$$U_{hkl} = \frac{F_{hkl}}{Z} \tag{16}$$

for a particular crystal. When this is done, (15) may be written as

$$|U_{hkl}|^2 \leqslant 1. \tag{17}$$

If the value of F given by (10) is substituted into (16), the general value of the unitary structure factor is seen to be

$$U_{hkl} = \frac{V}{Z} \int_0^1 \int_0^1 \int_0^1 \rho_{(xyz)}\, e^{i2\pi(hx+ky+lz)}\, dx\, dy\, dz. \tag{18}$$

Limitations due to symmetry. Although (15) and (17) express a result which is known from other considerations, the application of Schwartz's inequality to amplitudes arising from symmetrical sets of atoms provides new information.

Inversion center. If a crystal has inversion centers, and if one of them is taken as an origin, then

$$\rho_{(xyz)} = \rho_{(\bar{x}\bar{y}\bar{z})} \tag{19}$$

and (18) becomes

$$U_{hkl} = \frac{V}{Z} \int_0^1 \int_0^1 \int_0^1 \rho_{(xyz)} \cos 2\pi(hx+ky+lz)\, dx\, dy\, dz. \tag{20}$$

Now, in Schwartz's inequality (9), let

$$f = \left(\frac{V}{Z} \rho_{(xyz)}\right)^{1/2},$$

$$g = \left(\frac{V}{Z} \rho_{(xyz)}\right)^{1/2} \cos 2\pi(hx+ky+lz), \tag{21}$$

$$d\tau = dx\, dy\, dz.$$

If these are inserted in Schwartz's inequality, there results

$$|U_{hkl}|^2 \leqslant \left[\frac{V}{Z} \int_0^1 \int_0^1 \int_0^1 \rho_{(xyz)}\, dx\, dy\, dz\right] \times \left[\frac{V}{Z} \int_0^1 \int_0^1 \int_0^1 \rho_{(xyz)} \cos^2 2\pi(hx+ky+lz) dx\, dy\, dz\right]. \tag{22}$$

If the relation

$$\cos^2 a = \tfrac{1}{2}(1 + \cos 2a) \tag{23}$$

is used, (22) can be rewritten

$$|U_{hkl}|^2 \leqslant \frac{V}{Z}\left[\int_0^1\int_0^1\int_0^1 \rho_{(xyz)}\,dx\,dy\,dz\right]\left[\frac{V}{2Z}\int_0^1\int_0^1\int_0^1 \rho_{(xyz)}\,dx\,dy\,dz + \frac{V}{2Z}\int_0^1\int_0^1\int_0^1 \rho_{(xyz)}\cos 2\pi(2hx+2ky+2lz)\,dx\,dy\,dz\right]. \tag{24}$$

Let (14) be substituted for the first two integrals in (24). The last integral, according to (20) is $U_{2h\,2k\,2l}$. Therefore (24) can be reduced to

$$|U_{hkl}|^2 \leqslant [1][\tfrac{1}{2} + \tfrac{1}{2}U_{2h\,2k\,2l}]. \tag{25}$$

2-fold axis. If the crystal has 2-fold axes parallel to c, and if and origin is taken as one of them,

$$\rho_{(xyz)} = \rho_{(\bar{x}\bar{y}z)}. \tag{26}$$

Under these circumstances the unitary structure factor (18) reduces to

$$U_{hkl} = \frac{V}{Z}\int_0^1\int_0^1\int_0^1 \rho_{(xyz)}\,e^{i2\pi lz}\cos 2\pi(hx+ky)\,dx\,dy\,dz. \tag{27}$$

For substitution in Schwartz's inequality, let

$$f = \left(\frac{V}{Z}\right)\rho_{(xyz)}{}^{1/2},$$

$$g = \left(\frac{V}{Z}\right)\rho_{(xyz)}{}^{1/2}\,e^{i2\pi lz}\cos 2\pi(hx+ky), \tag{28}$$

$$d\tau = dx\,dy\,dz.$$

When inserted in Schwartz's inequality, there results

$$|U_{hkl}|^2 \leqslant \left[\frac{V}{Z}\int_0^1\int_0^1\int_0^1 \rho_{(xyz)}\,dx\,dy\,dz\right] \times\left[\frac{V}{Z}\int_0^1\int_0^1\int_0^1 \rho_{(xyz)}|e^{i2\pi(2l)z}|\cos^2 2\pi(hx+ky)\,dx\,dy\,dz\right]. \tag{29}$$

Using (23), this can be rewritten

$$|U_{hkl}|^2 \leqslant \left[\frac{V}{Z}\int_0^1\int_0^1\int_0^1 \rho_{(xyz)}\,dx\,dy\,dz\right]\left[\frac{V}{2Z}\int_0^1\int_0^1\int_0^1 \rho_{(xyz)}\,dx\,dy\,dx + \frac{V}{2Z}\int_0^1\int_0^1\int_0^1 \rho_{(xyz)}\cos 2\pi(2hx+2ky+20z)\,dx\,dy\,dz\right]. \tag{30}$$

With the aid of (14) and (20), it is evident that this is equivalent to

$$|U_{hkl}|^2 \leqslant [1][\tfrac{1}{2} + \tfrac{1}{2}U_{2h\,2k\,0}]. \tag{31}$$

Other symmetry elements. Proceeding in a similar way the limitations of the amplitude due to each of the possible symmetry elements can be investigated. The results are listed in Table 1.

Table 1
Harker-Kasper inequalities: limitations on amplitudes due to symmetry elements

Symmetry element $\parallel c$	Limitation
1	$\|U_{hkl}\|^2 \leqslant 1$
$\bar{1}$	$U_{hkl}{}^2 \leqslant \frac{1}{2} + \frac{1}{2}U_{2h\,2k\,2l}$
2	$\|U_{hkl}\|^2 \leqslant \frac{1}{2} + \frac{1}{2}U_{2h\,2k\,0}$
2_1	$\|U_{hkl}\|^2 \leqslant \frac{1}{2} + \frac{1}{2}(-1)^l\,U_{2h\,2k\,0}$
$\bar{2} = m$	$\|U_{hkl}\|^2 \leqslant \frac{1}{2} + \frac{1}{2}U_{0\,2k\,0}$
a	$\|U_{hkl}\|^2 \leqslant \frac{1}{2} + \frac{1}{2}(-1)^h\,U_{0\,0\,2l}$
3	$\|U_{hkl}\|^2 \leqslant \frac{1}{3} + \frac{2}{3}\|U_{(h-k)\,(h+2k)\,0}\|\cos 2\pi\phi_{(h-k)\,(h+2k)\,0}$
3_1, 3_2	$\|U_{hkl}\|^2 \leqslant \frac{1}{3} + \frac{2}{3}\|U_{(h-k)\,(h+2k)\,0}\|\cos 2\pi(\phi_{(h-k)\,(h+2k)\,0} + \frac{1}{3}l)$
$\bar{3} = 3 + \bar{1}$	$U_{hkl}{}^2 \leqslant \frac{1}{6} + \frac{1}{6}U_{2h\,2k\,2l} + \frac{1}{3}U_{h\,k\,2l} + \frac{1}{3}U_{(h-k)\,(h+2k)\,0}$
4	$\|U_{hkl}\|^2 \leqslant \frac{1}{4} + \frac{1}{4}U_{2h\,2k\,0} + \frac{1}{2}U_{(h-k)\,(h+k)\,0}$
4_1, 4_3	$\|U_{hkl}\|^2 \leqslant \frac{1}{4} + \frac{1}{4}(-1)^l\,U_{2h\,2k\,0} + \frac{1}{2}(\cos 2\pi\frac{1}{4}l)U_{(h-k)\,(h+k)\,0}$
4_2	$\|U_{hkl}\|^2 \leqslant \frac{1}{4} + \frac{1}{4}U_{2h\,2k\,0} + \frac{1}{2}(-1)^l\,U_{(h-k)\,(h+k)\,0}$
$\bar{4}$	$\|U_{hkl}\|^2 \leqslant \frac{1}{4} + \frac{1}{4}U_{2h\,2k\,0} + \frac{1}{2}\|U_{(h-k)\,(h+k)\,2l}\|\cos 2\pi\phi_{(h-k)\,(h+k)\,2l}$
6	$\|U_{hkl}\|^2 \leqslant \frac{1}{6} + \frac{1}{6}U_{2h\,2k\,0} + \frac{1}{3}U_{(h-k)\,(h+2k)\,0} + \frac{1}{3}U_{hk0}$
6_1, 6_5	$\|U_{hkl}\|^2 \leqslant \frac{1}{6} + \frac{1}{6}(-1)^l\,U_{2h\,2k\,0} + \frac{1}{3}(\cos 2\pi\frac{1}{3}l)U_{(h-k)\,(h+2k)\,0} + \frac{1}{3}(\cos 2\pi\frac{1}{6}l)U_{hk0}$
6_2, 6_4	$\|U_{hkl}\|^2 \leqslant \frac{1}{6} + \frac{1}{6}U_{2h\,2k\,0} + \frac{1}{3}(\cos 2\pi\frac{1}{3}l)U_{(h-k)\,(h+2k)\,0} + \frac{1}{3}(\cos 2\pi\frac{1}{3}l)U_{hk0}$
6_3	$\|U_{hkl}\|^2 \leqslant \frac{1}{6} + \frac{1}{6}(-1)^l\,U_{2h\,2k\,0} + \frac{1}{3}U_{(h-k)\,(h+2k)\,0} + \frac{1}{3}(-1)^l\,U_{hk0}$
$\bar{6} = 3/m$	$\|U_{hkl}\|^2 \leqslant \frac{1}{6} + \frac{1}{6}U_{0\,0\,2l} + \frac{1}{3}\|U_{(h-k)\,(h+2k)\,0}\|\cos 2\pi\phi_{(h-k)\,(h+2k)\,0} + \frac{1}{3}\|U_{(h-k)\,(h+2k)\,2l}\|\cos 2\pi\phi_{(h-k)\,(h+2k)\,2l}$

Application to phase determination. These relations can be used to derive phase information. For example, in (25), suppose that a value for a particular reflection $|U_{hkl}|^2$ is measured. If $|U_{hkl}|^2$ turns out to be greater than $\frac{1}{2}$, then the inequality is only satisfied if $U_{2h\,2k\,2l}$ is positive.

If $|U_{hkl}|^2$ turns out to be less than $\frac{1}{2}$, then no conclusions can be drawn, because

$$\tfrac{1}{2} \leqslant \tfrac{1}{2} + \tfrac{1}{2}U_{2h\,2k\,2l}$$

can be satisfied by either a positive or negative $U_{2h\,2k\,2l}$.

There is an interesting physical interpretation of this. It has been noted that when $|U_{hkl}|^2$ in (25) exceeds $\frac{1}{2}$, $U_{2h\,2k\,2l}$ must be positive. Now, if $|U_{hkl}|$ is a strong reflection, many of the atoms of the structure must lie near the crests of the Fourier waves of spacing $d_{hkl} \equiv d_H$ in Fig. 2. When this is so, then they must also be near the crests of those Fourier

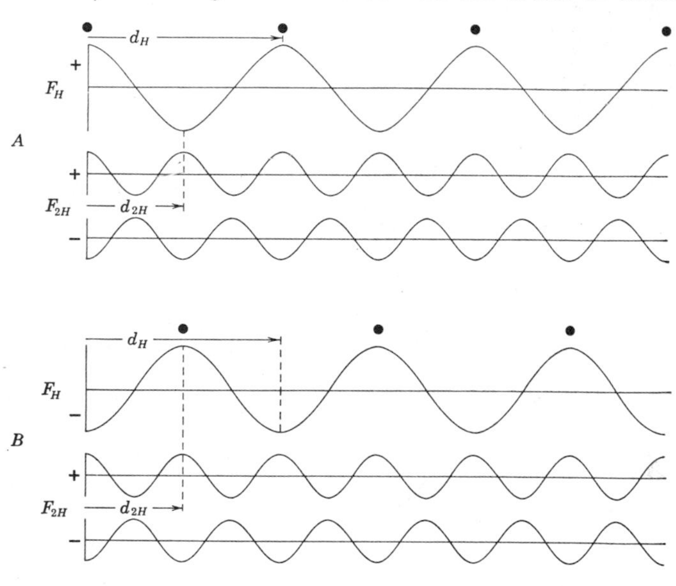

Fig. 2.
Geometrical interpretation of a simple Harker-Kasper inequality.

waves having half the spacing, namely those having spacing d_{2H}, Fig. 2. This can only occur if F_{2H} is +, regardless if F_H is +, Fig. 2*A*, or −, Fig. 2*B*. There is no way in which information can be obtained about F_H from F_{2H}, however.

Inequalities for specific space groups. The inequalities shown in Table 1 display the limitation on phase due to one symmetry element. In general a space group contains more than one symmetry element. By separating the structure factor for the space group in various ways, many inequalities can be found for each space group. Up to the present time there has been no systematic attempt to tabulate the possible inequalities for each space group.

Approximate compensation for volumes of atoms. One of the practical difficulties in making use of the Harker-Kasper inequalities is that, because of the decline of the f curves with $(\sin\theta)/\lambda$, the amplitudes of the reflections also decline with $(\sin\theta)/\lambda$. When $(\sin\theta)/\lambda$ reaches 0.6, even an amplitude representing all atoms scattering in phase is already less than $\frac{1}{2}$. The relations given in Table 1, however, are good even for point scatterers. Now it is possible, as pointed out in more detail elsewhere,† to transform measured amplitudes to approximately what they would be if they had been scattered by point atoms. To do this, each F

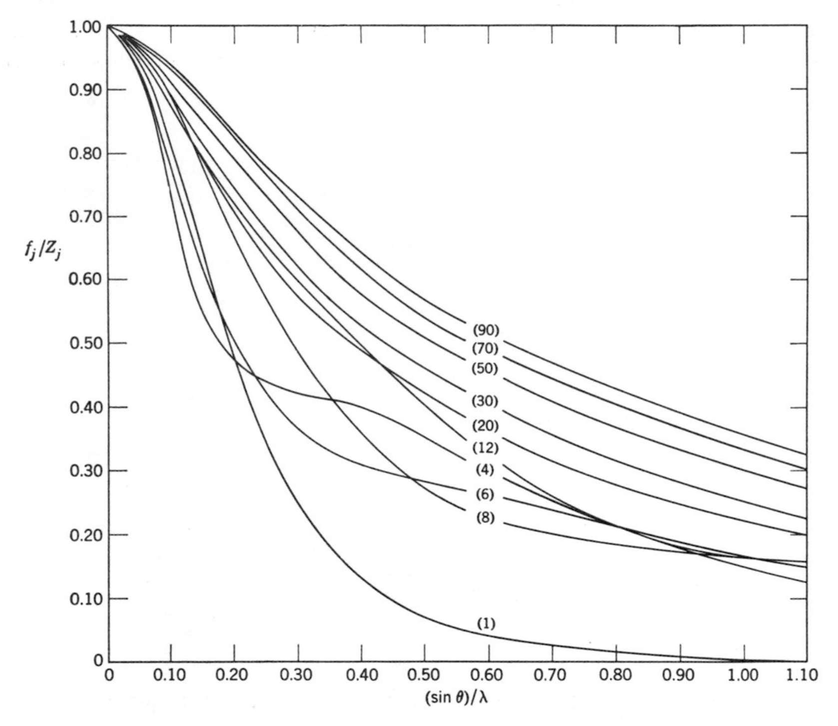

Fig. 3.
The shapes of the f curves for atoms of various atomic numbers.
(After Harker and Kasper.[25])

is normalized to the average f of the sin-θ region at room temperature, instead of to Z as in (16). This is justified as an approximation because f curves have approximately the same shape, as shown by Harker and Kasper,[25] Fig. 3.

Another approach to this normalizing process is to establish an average f curve normalized to unity at $(\sin\theta)/\lambda = 0$. Let the normalized f be designated $\hat{f}$ ("f-hat"). By using this, the f curve for any atom j is approximately

$$f_j = Z_j\hat{f}. \tag{32}$$

† See pages 59–64 of reference 23.

If the crystal structure is assumed to be composed of spherically symmetrical atoms, (10) can be written

$$F_{hkl} = \sum_{j=1}^{N} f_j \, e^{i2\pi(hx+ky+lz)}. \tag{33}$$

Then, utilizing (32) this becomes

$$F_{hkl} = \hat{f} \sum_{j=1}^{N} Z_j \, e^{i2\pi(hx+ky+lz)}. \tag{34}$$

If the normalized unitary structure factor is now defined as

$$U_{hkl} = \frac{F_{hkl}}{Z\hat{f}}, \tag{35}$$

then, using (34),

$$U_{hkl} = \sum_{j=1}^{N} \frac{Z_j}{Z} \, e^{i2\pi(hx+ky+lz)}. \tag{36}$$

The fraction in (36),

$$\frac{Z_j}{Z} = q_j, \tag{37}$$

is the fraction of the electrons of the unit cell which are in the jth atom.

When Cauchy's inequality, (8), is applied to (36), the same inequalities result as when Schwartz's inequality is applied to (10). To the degree that approximation (32) is valid, the inequalities of Table 1 are valid when applied to amplitudes normalized to an average f curve. Normalizing F to the approximate f of the $(\sin\theta)/\lambda$ region strengthens the inequalities and permits their practical use in phase determination, provided that $|U_{hkl}|^2$ exceeds $\frac{1}{2}$. Note, again, that in this process, the signs of certain U's can be proven to be positive. None can be proven to be negative.

Inequalities involving three reflections. Although the inequalities of Table 1 give information about a fraction of the reflections, other inequalities can be derived which give information about other reflections. For example, (25) only gives information about even orders. To get information about odd orders, Cauchy's inequality can be applied to the sum of two reflections. Letting $H \equiv hkl$, the normalized unitary structure factors for these reflections are

$$U_H = \sum_j q_j \cos 2\pi Hx, \tag{38}$$

$$U_{H'} = \sum_j q_j \cos 2\pi H'x. \tag{39}$$

Then,

$$
\begin{aligned}
U_H + U_{H'} &= \sum_j q_j(\cos 2\pi Hx + \cos 2\pi H'x) \\
&= \sum_j q_j \, 2 \cos \frac{2\pi(H+H')x}{2} \cos \frac{2\pi(H-H')x}{2} \\
&= \sum_j q_j \, 2 \left[\frac{1 + \cos 2\pi(H+H')x}{2}\right]^{1/2} \left[\frac{1 + \cos 2\pi(H-H')x}{2}\right]^{1/2} \\
&= \sum_j [q_j + q_j \cos 2\pi(H+H')x]^{1/2} \, [q_j + q_j \cos 2\pi(H-H')x]^{1/2}.
\end{aligned} \tag{40}
$$

When Cauchy's inequality is applied to this, there results

$$
|U_H + U_{H'}|^2 \leqslant \left\{\sum_j [q_j + q_j \cos 2\pi(H+H')x]\right\}\left\{\sum_j [q_j + q_j \cos 2\pi(H-H')]\right\}. \tag{41}
$$

According to (37), $\Sigma\, q_j = 1$. Utilizing this and also (38), (41) can be reduced to

$$
|U_H + U_{H'}|^2 \leqslant (1 + U_{H+H'})(1 + U_{H-H'}). \tag{42}
$$

Some other inequalities involving three reflections are shown in Table 2.

Table 2
Some inequalities involving three reflections

Centrosymmetrical structures	Non-centrosymmetrical structures
$\lvert U_H + U_{H'}\rvert^2 \leqslant (1+U_{H+H'})(1+U_{H-H'})$	$\lvert U_H + U_{H'}\rvert^2 \leqslant 2 + 2 \operatorname{Re} U_{H-H'}$
$\lvert U_H - U_{H'}\rvert^2 \leqslant (1-U_{H+H'})(1-U_{H-H'})$	$\lvert U_H - U_{H'}\rvert^2 \leqslant 2 - 2 \operatorname{Re} U_{H-H'}$
$\lvert U_{H+H'} - U_{H-H'}\rvert^2 \leqslant (1+U_{2H})(1-U_{2H'})$	

The origin may be chosen in several locations for each space group.[36, 69] If a centrosymmetrical structure factor is used as the basis of an inequality relation, then the origin can be taken at 00, $\frac{1}{2}0$, $0\frac{1}{2}$, or $\frac{1}{2}\,\frac{1}{2}$ in projection. If the origin is shifted from 00 to $\frac{1}{2}0$, the reflections F_{hk0} change sign for h odd, etc. This means that some signs (usually two for a projection, or three for three dimensions) can be fixed arbitrarily. The arbitrary choice selects a particular symmetry center in the structure as origin.

Linear inequalities. Okaya and Nitta[38] have devised some linear structure-factor inequalities. Some of these are shown in Table 3. Each bears a relation to, but is less powerful than, the corresponding

Table 3
Some linear structure-factor inequalities[28]

$$4U_H \leqslant 3 + 2U_{2H}$$
$$2|U_H + U_{H'}| \leqslant 2 + U_{H+H'} + U_{H-H'}$$
$$2|U_H - U_{H'}| \leqslant 2 - U_{H+H'} - U_{H-H'}$$
$$4|U_H + U_{H'}| \leqslant 5 + U_{H+H'} + 4U_{H-H'}$$
$$4|U_H - U_{H'}| \leqslant 5 - U_{H+H'} - 4U_{H-H'}$$
$$4|U_H + U_{H'}| \leqslant 5 + 4U_{H+H'} + U_{H-H'}$$
$$4|U_H + U_{H'}| \leqslant 5 - 4U_{H+H'} - U_{H-H'}$$

Harker-Kasper inequality.[40] Since they are linear relations they are somewhat easier to apply.

Limitations on the success of sign determination by inequalities. The limitations attending the use of inequalities has been discussed by Hughes.[28] If all the atoms are assumed to be the same, then (36) becomes

$$U_H = \sum_{j=1}^{N} \frac{1}{N} e^{i2\pi H x_j}, \tag{43}$$

so that the magnitude of U_{hkl} is provided by

$$\begin{aligned} |U_H|^2 &= U_H \tilde{U}_H \\ &= \sum_{i=1}^{N} \sum_{j=1}^{N} \frac{1}{N^2} e^{i2\pi H(x_i - x_j)}. \end{aligned} \tag{44}$$

If terms for which $i = j$ are separated, this may be written

$$|U_H|^2 = \frac{1}{N} + \sum_{i}^{N} \sum_{\substack{j \\ i \neq j}}^{N} \frac{1}{N^2} e^{i2\pi H(x_i - x_j)}. \tag{45}$$

If this is averaged over all reflections, H, the double summation tends to zero, since positive and negative values of all magnitudes are equally probable, so that

$$\overline{|U_H|^2} = \frac{1}{N}. \tag{46}$$

The root-mean-square value is

$$\hat{\sigma}_H = (\overline{|U_H|^2})^{1/2} = \frac{1}{N^{1/2}}. \tag{47}$$

If the symmetry requires p symmetrical companions, then $N = pn$, and

$$|U_H|^2 = \frac{1}{N} = \frac{1}{pn}, \tag{48}$$

$$\hat{\sigma}_H = \frac{1}{N^{1/2}} = \frac{1}{(pn)^{1/2}}. \tag{49}$$

For reflections on symmetrical locations of the reciprocal structure, $\overline{|U_H|^2}$ may be larger by an integer corresponding to the symmetry of the location.

Now if N is sufficiently large, assuming a Gaussian distribution of amplitudes, only about 10% of the $|U|$'s may be expected to exceed $1.7\hat{\sigma}$, and 0.1% may be expected to exceed $3.3\hat{\sigma}$. Harker-Kasper inequalities have been used successfully to solve oxalic acid dihydrate, with $n = 4$, and decaborane, with $n = 5$. Hughes solved by other methods melamine, $n=9$, but when reinvestigated by inequalities only two signs of second order U's could be fixed for some 150 U's for $h0l$ and $0kl$ reflections, and two more signs could be fixed by using inequalities involving three reflections. For β carotine, $n = 20$, $\hat{\sigma}_{hkl} = 0.112$, $\hat{\sigma}_{h0l} = 0.158$. It is unlikely that any $|U_{h0l}|$ will exceed 0.520. Actually none was found to exceed 0.24, and no signs could be determined. Thus Harker-Kasper inequalities have been used successfully with $n = 4$, $n = 5$, but fail with $n = 9$ and $n = 20$. It appears, therefore, that solution of structures with the aid of sign determination by inequalities is limited to crystals with comparatively few atoms per asymmetrical unit.

Karle-Hauptman inequalities. Karle and Hauptman[48, 49] investigated the relations between phases and magnitudes of a set of F's (or U's) on a more general basis. They pointed out that the electron density is positive, and this fact places certain restrictions on a Fourier series. If the first n Fourier coefficients are given, then the $n+1$ coefficient is restricted to narrow bounds if a positive function is to be represented.

When Friedel's law holds, $\tilde{F}_{hkl} = F_{\bar{h}\bar{k}\bar{l}}$. Therefore the F's can be arranged in a Hermitian matrix

$$\begin{bmatrix} F_0 & F_{\bar{H}_1} & F_{\bar{H}_2} & \cdot \\ F_{H_1} & F_0 & F_{\bar{H}_3} & \cdot \\ F_{H_2} & F_{H_3} & F_0 & \cdot \\ \cdot & \cdot & \cdot & \cdot \end{bmatrix}. \tag{50}$$

Now, let the electron density be expressed by

$$\rho_{(x)} = \frac{1}{V} \sum F_H \, e^{-i2\pi Hx}. \tag{51}$$

The conditions on the F_H's that $\rho_{(x)}$ be positive had already been investigated by mathematicians, and is stated in Herglotz's theorem. This requires that the determinant of any matrix like (50) be greater than zero. For example

$$\begin{vmatrix} F_{000} & F_{00\bar{1}} & F_{\bar{1}\bar{1}\bar{1}} \\ F_{001} & F_{000} & F_{\bar{1}\bar{1}0} \\ F_{111} & F_{110} & F_{000} \end{vmatrix} \geqslant 0. \tag{52}$$

If the determinant of (50) is manipulated it can be shown that a particular coefficient is restricted to lie within a certain limited region of the complex plane, that is, the real and imaginary parts of F are limited. This result can be expressed as an inequality. This is obtained without any resort to symmetry considerations, and is based purely on the fact that the electron density is positive.

The introduction of symmetry strengthens the inequalities derived by means of this method. In particular, the Harker-Kasper inequalities can be so derived, and, in addition, infinitely more inequalities.

Sign relations

Sayre's squaring method. Sayre[57] examined the relations between the F's of a structure and the F's of the same structure type in which the atoms were replaced by "squared atoms." This led to a fruitful new development in the determination of phases from a set of intensities.

The electron density at location x is given by

$$\rho_{(x)} = \frac{1}{V} \sum_H F_H \, e^{-i2\pi Hx}. \tag{53}$$

Let the density at each point be squared. This squared density can also be represented by a Fourier series

$$\rho_{(x)}{}^2 = \frac{1}{V} \sum_H {}^{sq}F_H \, e^{-i2\pi Hx}. \tag{54}$$

The new Fourier series has different F's but, provided that all atoms are equal and resolved, the F's of (53) and (59) must be related by some factor g which takes account of the change in shape; that is

$${}^{sq}F_H = g_H \, F_H. \tag{55}$$

For centrosymmetrical crystals,† $F_H = F_{-H}$, so the square of the electron density can be written

$$
\begin{aligned}
\rho_{(x)}{}^2 = \rho_{(x)}\rho_{(x)} &= \left[\frac{1}{V}\sum_H e^{-i2\pi Hx}\right]\left[\frac{1}{V}\sum_H e^{-i2\pi(-H)x}\right] \\
&= \frac{1}{V^2}\sum_{H_1}\sum_{H_2} F_{H_1}F_{H_2}e^{-i2\pi(H_1-H_2)x} \\
&= \frac{1}{V}\sum_{H_1}\left[\frac{1}{V}\sum_{H_2} F_{H_1}F_{H_2}\right]e^{-i2\pi(H_1-H_2)x}. \qquad (56)
\end{aligned}
$$

Now, let $H_1 - H_2 = H$, and let $H_2 = H'$. Then, since the summations run over all integers, (56) can be rewritten

$$
\rho_{(x)}{}^2 = \frac{1}{V}\sum_H\left[\frac{1}{V}\sum_{H'} F_{H'}F_{H+H'}\right]e^{-i2\pi Hx}. \qquad (57)
$$

---

† This follows Sayre's original development. The relation $F_H = F_{-H}$ is only valid for centrosymmetrical crystals. But for both centrosymmetrical and non-centrosymmetrical crystals the square of the electron density can be written in an alternative form

$$
\begin{aligned}
\rho_{(x)}{}^2 = \rho_{(x)}\rho_{(x)} &= \left[\frac{1}{V}\sum_H F_H\, e^{-i2\pi Hx}\right]\left[\frac{1}{V}\sum_H F_H\, e^{-i2\pi Hx}\right] \\
&= \frac{1}{V}\sum_{H_1}\left[\frac{1}{V}\sum_{H_2} F_{H_1}F_{H_2}\right]e^{-i2\pi(H_1+H_2)x}. \qquad (56A)
\end{aligned}
$$

Now let $H_1 + H_2 = H$, and let $H_1 = H'$. Then, since the summations run over all integers, (56A) can be rewritten

$$
\rho_{(x)}{}^2 = \frac{1}{V}\sum_H\left[\frac{1}{V}\sum_{H'} F_H F_{H-H'}\right]e^{-i2\pi Hx}. \qquad (57A)
$$

Following the argument of (58), it is found that

$$
F_H = \frac{1}{g_H V}\sum_{H'} F_{H'}F_{H-H'}. \qquad (59A)
$$

For centrosymmetrical crystals, this leads to the sign relation

$$
S_H = S_{H'}S_{H-H'}, \qquad (60A)
$$

$$
\text{or} \qquad S_{H-H'} = S_H S_{H'}. \qquad (61A)
$$

For non-centrosymmetrical crystals this leads to a relation between the phases, ϕ, of the three reflections:

$$
\phi_H = \phi_{H'} + \phi_{H-H'}, \qquad (60B)
$$

$$
\text{or} \qquad \phi_{H-H'} = \phi_H - \phi_{H'}. \qquad (61B)
$$

If (54), (55), and (57) are compared, it is found that

$$ {}^{sq}F_H = g_H F_H = \frac{1}{V} \sum_{H'} F_{H'} F_{H+H'}, \tag{58} $$

so

$$ F_H = \frac{1}{g_H V} \sum_{H'} F_{H'} F_{H+H'}. \tag{59} $$

This is a general relation which must be obeyed by crystals composed of equal, resolved atoms. For such a crystal, the signs can only be correct if (59) holds.

The right of (59) involves the sum of products like $F_1 F_{10+1}$, $F_2 F_{10+2}$, $F_3 F_{10+3}$, etc. Sayre set up an artificial one-dimensional structure, and by an arithmetic study of a matrix of such products, was able to show that only one set of signs of the F's was consistent with such sums. He next applied the method to the projection of hydroxyproline, whose structure had already been worked out. This structure has symmetry $P2_1\,2_1\,2_1$, and contains 4 molecules of $C_5H_9NO_3$ per cell. It took about a week to investigate the interproducts of the 153 F_{0kl}'s. Signs for 19 F's, including the 10 strongest, could be explained by eight different combinations of signs, and one of these combinations was selected as most likely. The sign determination was extended in two stages to 31 terms, and then to 53 terms. All the signs at the 19-term stage later proved correct; one sign was wrong at the 31-term stage, and 9 were incorrect at the 53-term stage. The Fourier synthesis made with the 19 correct terms was sufficient to reveal the general shape of the molecule.

In using the method it became obvious that in a sum of products, such as occur on the right of (59), if one term is so large that it dominates the sum, the sign of the sum is controlled by that product. If S_H is used to indicate the sign of F_H, then this conclusion can be expressed by

$$ S_H = S_{H'} S_{H+H'}, \tag{60} $$

That is, the sign of F_H tends to be the same as the product of the signs of two other F's, related as in (60). This can be rewritten

$$ S_{H+H'} = S_H S_{H'}. \tag{61} $$

This rule certainly holds when F_H, $F_{H'}$, and $F_{H+H'}$ are all large. But it appears to express a deeper principle, perhaps that, if F_H and $F_{H'}$ are both large, they impose a characteristic pattern upon the entire array of signs.

Sayre's relation has an interesting physical interpretation which is illustrated in Fig. 4. In reciprocal space, the relation between points

H, H' and $H+H'$ is shown in Fig. 4A. In crystal space, Fig. 4B, the geometrical arrangement is similar but rotated 90° (if the arrangement is regarded as two-dimensional). In Fig. 4B, if the wave is positive at the origin, the crests are shown as full lines, while if the wave is negative at the origin, the crests are shown as broken lines. Now if the magnitudes of both F_H and $F_{H'}$ are large, it is because atoms are located near to the crests of both. For clearness, suppose this condition is represented by a single atom exactly on the crests of both. This can occur in only four ways with respect to H and H', shown in the four atom locations in Fig.

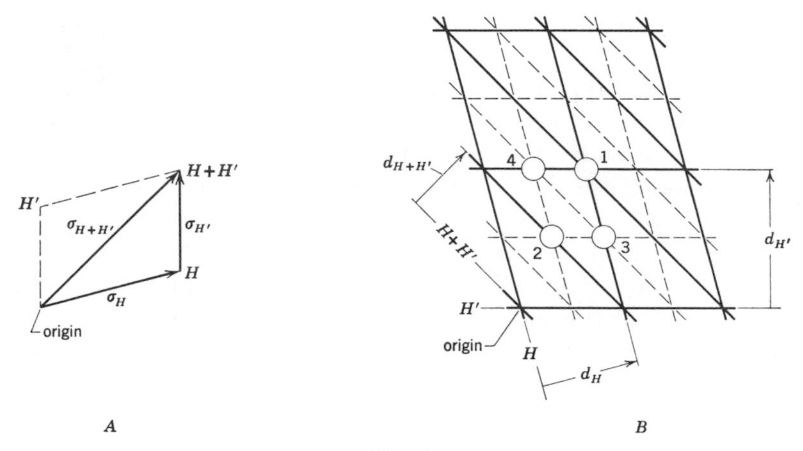

Fig. 4.
Geometrical interpretation of Sayre's sign relationship.

4B. But whichever way occurs, it is also on a crest of $H+H'$. The signs involved (+ when the circle is on a full line, − when it is on a broken line) are

location	S_H	$S_{H'}$	$S_{H+H'}$
1	+	+	+
2	−	−	+
3	+	−	−
4	−	+	−

It will be seen that, in every case, (61) is satisfied.

Sayre's relation holds when the U's involved are large. It is also probably valid[60] when $\sqrt{3n}\ \bar{U} > 1$, where n is the symmetry number and $\bar{U}$ is the r.m.s. value of U_H, $U_{H'}$, and $U_{H+H'}$. Cochran and Douglas[62] have noted that the extent to which

$$\chi = \sum_H \sum_{H'} U_H\, U_{H'}\, U_{H+H'} \tag{62}$$

is positive may be taken as a measure of the plausibility of a particular set of signs.

Cochran[58] and others investigated Sayre's squaring method, especially in respect to a limited set of large F's, and justified Sayre's sign relation on more rigorous grounds. The sign relation has been used to solve several structures,[60, 64] and programs adapting it to a digital computer have been arranged.[62, 75, 78, 84, 91]

Zachariasen's method. Zachariasen started with an equation from which Schwartz's inequality can be derived, and devised a new method of determining signs.[59] The initial relation is

$$\left(\sum |a_i|^2\right)\left(\sum |b_i|^2\right) = \left|\sum a_i\, b_i\right|^2 + \tfrac{1}{2}\sum_i \sum_j \begin{vmatrix} a_i & a_j \\ b_i & b_j \end{vmatrix}^2. \tag{63}$$

From this Schwartz inequality (8) can be derived. If the additional term is included, a relation similar to (42) can be derived, which can be written

$$(|U_H| + |U_{H'}|)^2 = 1 - S_H\, S_{H'}(U_{H+H'} + U_{H-H'}) + U_{H+H'}U_{H-H} - D_{HH'}. \tag{64}$$

The final term is given by

$$D_{HH'} = 8 \sum_1^{\frac{1}{2}N} \sum_1^{\frac{1}{2}N} n_i n_j \begin{vmatrix} \cos\,(\mathbf{H}+\mathbf{H}')\cdot\mathbf{r}_i & \cos\,(\mathbf{H}+\mathbf{H}')\cdot\mathbf{r}_j \\ \cos\,(\mathbf{H}-\mathbf{H}')\cdot\mathbf{r}_i & \cos\,(\mathbf{H}-\mathbf{H}')\cdot\mathbf{r}_j \end{vmatrix}^2. \tag{65}$$

If this term is omitted, the result is

$$(|U_H| + |U_{H'}|)^2 \leqslant 1 + S_H\, S_{H'}(U_{H+H'} + U_{H-H'}) + U_{H+H'}\, U_{H-H'}, \tag{66}$$

which is an alternative form of (42), a relation which had been derived by using inequality (8) rather than equality (63).

Returning to (65), this tends approximately[59] to 1 if an average is taken over all H and H'. Applying this to (64), a good approximation under these circumstances is

$$(|U_H| + |U_{H'}|)^2 \approx S_H\, S_{H'}(U_{H+H'} + U_{H-H'}) + U_{H+H'}\, U_{H-H'}. \tag{67}$$

From this it would follow that

$$S_{H+H'} = S_H\, S_{H'}, \tag{68}$$

which is Sayre's relation. This can be expected to hold rather well for sets of three strong reflections.

Now in (64), let H' be replaced by H_i' and H by $H+H_i'$, giving

$$(|U_{H+H_i'}| + |U_{H_i'}|)^2 \approx 1 - S_{H+H_i'}\, S_{H_i'}(U_{H+2H_i'} + U_H) + U_{H+2H_i'}\, U_H - D_{H_i'\,H+H_i'}. \tag{69}$$

Suppose that $U_{H_i'}\, U_{H+H_i'}$ be pairs of large values of U, and that for each H, (69) is averaged over all H_i'. Then, since $\overline{D_{H_i'H+H_i'}} = 1$, (69) becomes

$$\overline{(|U_{H_i'}| + |U_{H+H_i'}|)^2} = \overline{S_{H_i'}\, S_{H+H_i'}}\; U_H + \overline{S_{H_i'}\, S_{H+H_i'}\, U_{H+2H_i'}} + \overline{U_{H+2H_i'}}\; U_H. \quad (70)$$

The known quantities on the left are large and positive. The last two quantities on the right can be shown to be less than the left member down to $\sigma = 0.07$. Thus,

$$S_H = S(\overline{S_{H_i}\, S_{H+H_i'}}), \quad (71)$$

which is the statistical equivalent of (68).

Hughes has derived Zachariasen's general relation in a simpler fashion.[65] For a centrosymmetrical crystal

$$U_H = 2 \sum_j^{N/2} q_j \cos 2\pi H\cdot r_j, \quad (72)$$

$$U_H\, U_{H'} = 4 \sum_i^{N/2} \sum_j^{N/2} q_i\, q_j \cos 2\pi H\cdot r \cos 2\pi H'\cdot r \quad (73)$$

$$= 2 \sum_i^{N/2} \sum_j^{N/2} q_i\, q_j[\cos 2\pi(H\cdot r_i + H'\cdot r_j) + \cos 2\pi(H\cdot r_i - H'\cdot r_j)] \quad (74)$$

$$= 2 \sum_i^{N/2} q_i^2[\cos 2\pi(H+H')\cdot r_i + \cos 2\pi(H-H')\cdot r_j]$$

$$+ \; 2 \sum_{i}^{N/2} \sum_{\neq j}^{N/2} q_i\, q_j[\cos 2\pi(H\cdot r_i + H'\cdot r_j) + \cos 2\pi(H\cdot r_i - H'\cdot r_j)]. \quad (75)$$

If an average is taken, keeping $H+H'$ constant, the double sum, and terms with $H-H'$, average zero, as in Wilson's method, and there results

$$\overline{U_H\, U_{H'}}^{H+H'} = 2 \sum_i^{N/2} q_i^2 \cos 2\pi(H+H')\cdot r_i. \quad (76)$$

When all the atoms are the same, $q_i = 1/N$, and may be factored out, giving

$$\overline{U_H\, U_{H'}}^{H+H'} = 2\frac{1}{N} \sum_i^{N/2} q_i \cos 2\pi(H+H')\cdot r_i. \quad (77)$$

If this is compared with (72), it is seen to be

$$\overline{U_H\, U_{H'}}^{H+H'} = \frac{1}{N}\, U_{H+H'}. \quad (78)$$

Thus the average over $H+H'$ of $U_H\, U_{H'}$ has the same sign as $U_{H+H'}$ or

$$S_{H+H'} = \overline{S(S_H S_{H'})}^{H+H'} \quad (79)$$

To follow Zachariasen's method, inequality (66) is first used to establish as many signs as possible. Next, one expresses the signs of the largest structure factors in terms of symbols a, b, c, d $\cdots$. Then the inequality relation is used to deduce signs of further structure factors in terms of these. This provides a list of large structure factors in terms of signs a, b, c, d. Next (71) is used with this list to expand it, and to determine which letters are $+1$ or -1.

Zachariasen used his method to solve the structure of metaboric acid directly.[59] Precautions to be taken in respect to the assignment of signs have been discussed by Lonsdale and Grenville-Wells.[69]

Statistical methods

In a series of papers, Hauptman and Karle have devised a statistical approach to the phase problem. Their discussion[95] of their method is given in mathematical form which, though not inherently difficult, is cumbersome and difficult to follow. Before considering the statistical approach itself, some remarks on relations between phases, which are emphasized by Hauptman and Karle[95] are first given.

Not all phases are controlled entirely by the structure. Some are based upon the choice of origin.[63, 69] For a centrosymmetrical crystal

$$F_{hkl} = 2 \sum_{j=1}^{N/2} f_j \cos 2\pi(hx_j+ky_j+lz_j). \tag{80}$$

Suppose that the origin is changed to $\frac{1}{2}\frac{1}{2}\frac{1}{2}$. Then, in (80) $xyz \rightarrow x-\frac{1}{2}$, $y-\frac{1}{2}$, $z-\frac{1}{2}$, and (80) becomes

$$\begin{aligned} F_{hkl} &= 2 \sum f_j \cos 2\pi[h(x-\tfrac{1}{2}) + k(y-\tfrac{1}{2}) + l(z-\tfrac{1}{2})] &\quad (81)\\ &= 2 \sum f_j \cos 2\pi[hx+ky+lz - \tfrac{1}{2}(h+k+l)] \\ &= 2 \sum f_j \cos 2\pi(hx+ky+lz) \cos \pi\,(h+k+l) \\ &\qquad + \sin 2\pi(hx+ky+lz) \sin \pi\,(h+k+l) \\ &= 2 \sum f_j \cos 2\pi(hx+ky+lz) \cos \pi\,(h+k+l) \\ &= F_{hkl}(-1)^{h+k+l}. &\quad (82) \end{aligned}$$

The change of origin, therefore, does not alter the sign of F_{hkl} if $h+k+l = 2n$. The phase of this type of F is *structure invariant*, which means that the phase does not depend on the choice of origin. When $h+k+l = 2n + 1$, the sign of F_{hkl} depends on the choice of origin, and the phase is said to be *structure semivariant*. The relations between signs of some structure factors can be expressed very compactly as follows:

(1) If $H = h+k+l$ is even, $H \cong 0$ mod 2 [i.e. H is congruent to zero, mod 2].

(2) If $H_a - H_b \cong H_c$, then $H_a \cong H_b$ mod 2.

(3) If $\Sigma\, a_j\, H_j \cong 0$ mod 2, the set is linearly dependent, mod 2.

(4) If $H = \Sigma\, a_j\, H_j$ mod 2, H is linearly dependent, mod 2, on the set $\Sigma\, a_j H_j$.

Any set of four H's in three-dimensional space is linearly dependent, mod 2. The maximum number of linearly independent vectors in three-dimensional is three. Any three H's which are not linearly dependent can be given arbitrary signs. When this has been done, a particular origin has been chosen.

Hauptman and Karle[95] investigated the probability that a magnitude of an F lies between certain limits, as a function of the whole set of magnitudes. This enabled them to set down relations which provide the sign of the F. They worked with a normalized structure factor, E, which is related to F in the following way:†

$$E = \frac{1}{\epsilon \sum_{j=1}^{N} f_{j,hkl}^2} F. \tag{83}$$

The factor ϵ simply takes care of the degeneracy of an F if it lies on a symmetry location in the reciprocal structure. Naturally the sign of E is the same as that of the corresponding F. Some average values of E for centrosymmetrical crystals are[124]

$$\begin{aligned} \langle E^2 \rangle &= 1, \\ \langle |E^2 - 1| \rangle &= \frac{4}{\sqrt{2\pi e}} = 0.968, \\ \langle |E| \rangle &\approx \sqrt{\frac{2}{\pi}} = 0.798. \end{aligned} \tag{84}$$

To use Hauptman and Karle's method the E's are arranged in order of decreasing magnitude and the signs are determined in this order. The method cannot be discussed in detail, nor can directions for its use be given in this brief description. The general idea of procedure can be grasped for the instructions (for $P\bar{1}$) for determining the signs of the F's whose signs are linearly dependent mod 2. The sign of E_H, where H is even, is the sign of

$$\sum = \sum\nolimits_1 + \sum\nolimits_2 + \sum\nolimits_3 + \sum\nolimits_4, \tag{85}$$

† This is equivalent to

$$E = \frac{1}{(\overline{F^2})^{1/2}} F. \tag{83A}$$

where the various $\sum$'s are as follows:†

$$\sum_1 = \sum_{H=2H_\mu} \frac{\sum_j f_{jH} f_{jH_\mu}{}^2}{4\left(\sum_j f_{jH}\right)^{1/2}\left(\sum_j f_{jH_\mu}{}^2\right)} (E_{H_\mu}{}^2 - 1). \tag{86}$$

(This bears a resemblance to the Harker-Kasper inequality $U_H{}^2 \leqslant \frac{1}{2} - \frac{1}{2}U_{2H}$ if it is rewritten $U_{2H} \geqslant 2U_H{}^2 - 1$.)

$$\sum_2 = \sum_{H=H_\mu \pm H_\nu} \frac{\sum_j f_{jH} f_{jH_\mu} f_{jH_\nu}}{2\left(\sum_j f_{jH}{}^2\right)^{1/2}\left(\sum_j f_{jH_\mu}{}^2\right)\left(\sum_j f_{jH_\nu}{}^2\right)^{1/2}} E_{H_\mu} E_{H_\nu}. \tag{87}$$

(This bears a resemblance to the Sayre sign relation $S_{H+H} = S_H S_{H'}$.)

$$\sum_3 = \sum_{H=H_\mu \pm 2H_\nu} \frac{\sum_j f_{jH} f_{jH_\mu} f_{jH_\nu}{}^2}{4\left(\sum_j f_{jH}{}^2\right)^{1/2}\left(\sum_j f_{jH_\mu}{}^2\right)^{1/2}\left(\sum_j f_{jH_\nu}{}^2\right)} E_{H_\mu} (E_{H_\nu}{}^2 - 1), \tag{88}$$

$$\sum_4 = \sum_{H=2H_\mu \pm 2H_\nu} \frac{\sum_j f_{jH} f_{jH_\mu}{}^2 f_{iH_\nu}{}^2}{8\left(\sum_j f_{jH}{}^2\right)^{1/2}\left(\sum_j f_{jH_\mu}{}^2\right)\left(\sum_j f_{jH_\nu}\right)} \times (E_{H_\mu}{}^2 - 1)(E_{H_\nu}{}^2 - 1). \tag{89}$$

† Alternative forms of (86) through (89) are:[101]

$$\sum_1 = \frac{\sum_j z_j{}^3}{4\left(\sum_j z_j{}^2\right)^{3/2}} \sum_{H=2H_\mu} (E_{H\mu}{}^2 - 1), \tag{86A}$$

$$\sum_2 = \frac{\sum_j z_j{}^3}{2\left(\sum_j z_j{}^2\right)^{3/2}} \sum_{H=H_\mu+H_\nu} E_{H\mu} E_{H\nu}, \tag{87A}$$

$$\sum_3 = \frac{\sum_j z_2{}^4}{4\left(\sum z_j{}^2\right)^2} \sum_{H=H_\mu+2H_\nu} E_{H\mu}(E_{H\mu}{}^2 - 1), \tag{88A}$$

$$\sum_4 = \frac{\sum_j z_j{}^5}{8\left(\sum z_j{}^2\right)^{5/2}} \sum_{H=2H_\mu+2H_\nu} (E_{H\mu}{}^2 - 1)(E_{H\nu}{}^2 - 1). \tag{89A}$$

Initially only the magnitudes of the E's are known, so that, at first only Σ_1 (which contains only one summand) and Σ_4 can contribute to Σ. As some signs become known, Σ_3 and Σ_4 play increasing roles. There are corresponding formulae for other categories of H's.

The Hauptman-Karle method of sign determination has been too little used to permit a valid appraisal of its power. Nevertheless a number of published objections to the method have appeared. A recurring objection is the following.[96, 102, 105]

Only Σ_1 and Σ_4 can be used in starting the sign determination, since they alone depend on E^2's. If the signs of the E's were controlled entirely by Σ_1, this would be the same as making a Fourier synthesis using $(E_H{}^2 - 1)$ as coefficients. This should give essentially a sharpened Patterson map with the origin removed. The solution, using this as a start, converges to the high peaks on the Patterson map. Similarly, for Σ_4, the solution converges to the high peaks on the map of the square of the Patterson density.

The Hauptman-Karle method has been used to solve the structure of colemanite, $CaB_3O_4(OH)_3 \cdot H_2O$, which is monoclinic, $P\,2_1/a$, $Z = 4$. This was not a very severe test since the Ca acts as a heavy atom and, once located (for example, by Σ_1 or by Patterson maps) the domination of the F's by Ca assures convergence to the correct structure. The method has also been used,[118] however, on p,p'-dimethoxybenzophenone, which imposed a much more severe test. This crystal is monoclinic, $P\,2_1/a$, $Z = 8$. The asymmetric unit is two molecules each of formula $C_{15}O_3H_{28}$ so that 36 distinct non-hydrogen atoms were located by the method. This appears to be the most complicated structure solved by direct methods to date. The signs of the E_{hkl}'s were used to compute a three-dimensional E map from which the shape of the molecule was recognized.

Permutation syntheses

An unsophisticated, yet direct, method of solving crystal structures is to select a limited number of the most intense reflections, and then using all permutations of their signs, make as many Fourier syntheses. If practical, such a method would be useful if the shape of the molecule could be recognized when roughly synthesized. The method might be useful if little effort were used in the Fourier synthesis. It is a practical procedure with X-RAC,† for which a sign is changed by throwing a switch and the new Fourier synthesis is instantly visible.

† Ray Pepinsky. *X-RAC and S-FAC: Electric analogue computors for x-ray analysis.* Computing methods and the phase problem in x-ray analysis. (The Pennsylvania State College, 1952) 167–277.

For a centrosymmetrical crystal, each F can be either + or −. If there are p measured $|F|$ values, the number of possible combinations is 2^p. Of these, two can be chosen arbitrarily, for a projection, leaving 2^{p-2} possible sign combinations for the projection of the structure. If there were only 20 F's, this would come to 262,144 combinations. It is manifestly out of the question to perform all the Fourier syntheses corresponding to these, or to examine them for plausibility such as the revelation of the shape of a chemical molecule.

Woolfson[128] suggested a way of essentially trying many combinations with a minimum of effort. In his scheme, the seven greatest F's are selected. This would formally call for $2^7 = 128$ combinations. But if one wrong sign can be tolerated, these 128 can be explored using the 16 combinations shown in Table 4. If a second group of seven F's is treated

Table 4
Sign combinations
(After Woolfson[128])

Term	1	2	3	4	5	6	7	8	9	10	11	12	13	14	15	16
1	+	+	+	+	+	+	+	+	−	−	−	−	−	−	−	−
2	−	−	−	−	+	+	+	+	−	−	−	−	+	+	+	+
3	+	+	−	−	+	+	−	−	+	+	−	−	+	+	−	−
4	−	+	−	+	−	+	−	+	−	+	−	+	−	+	−	+
5	+	−	−	+	+	−	−	+	−	+	+	−	−	+	+	−
6	+	+	−	−	−	−	+	+	−	−	+	+	+	+	−	−
7	−	+	−	+	+	−	+	−	+	−	+	−	−	+	−	+

in the same way, combining the results is equivalent to using 14F's and tolerating two possibly incorrect signs. This method has been further developed by Good.[129]

Literature

Algebraic methods

[1] H. Ott. *Zur Methodik der Strukturanalyse.* Z. Krist. **66** (1927) 136–153.

[2] K. Banerjee. *Determination of the signs of the Fourier terms in complete crystal structure analysis.* Proc. Roy. Soc. (London) (A) **141** (1933) 188–193.

[3] Melvin Avrami. *Direct determination of crystal structure from x-ray data.* Phys. Rev. **54** (1938) 300–303.

[4] Melvin Avrami. *A method for the direct determination of crystal structure from x-ray data.* Z. Krist (A) **100** (1939) 381–393.

[5] S. H. Yu. *An improved algebraic method for the determination of crystal structure from x-ray data.* Sci. Records, Acad. Sinica **1** (1942) 109–110.

[6] S. H. Yu and C. P. Ho. *A new method of analysis of x-ray data for the determination of crystal structure—its application to iron pyrites.* Sci. Records, Acad. Sinica **1** (1942) 111–115.

[7] S. H. Yü. *X-ray analysis of the structure of KH_2PO_4 by the improved algebraic method for crystal analysis by x-rays.* Sci. Records, Acad. Sinica **1** (1942) 361–363.

[8] S. H. Yü. *A new synthesis of x-ray data for crystal analysis.* Nature **149** (1942) 638–639, 729.

[9] S. H. Yü. *Determination of absolute from relative x-ray intensity data.* Nature **150** (1942) 151–152.

[10] S. H. Yü. *Determination of absolute intensities of x-ray reflexions from relative intensity data.* Nature **163** (1949) 375–376.

[11] E. W. Hughes. *An extension of Banerjee's method for determining signs of Fourier coefficients.* Acta Cryst. **2** (1949) 37–38.

[12] H. Hauptman and J. Karle. *A geometric approach to the crystal structure problem.* Acta Cryst. **3** (1950) 478.

[13] J. Karle and H. Hauptman. *A note on the solution of the structure-factor equations.* Acta Cryst. **4** (1951) 188–189.

[14] H. Hauptman and J. Karle. *Multiplicity of solutions of crystal-structure equations.* Acta Cryst. **4** (1951) 383.

Vector-space methods

[15] M. J. Buerger. *Phase determination with the aid of implication theory.* Phys. Rev. **73** (1948) 927–928.

[16] M. J. Buerger. *Some relations between the* F's and F^2's *of x-ray diffraction.* Proc. Nat. Acad. Sci. U. S. **34** (1948) 277–285.

[17] Félix Bertaut. *Relations entre la structure et la fonction de Patterson symétrisée.* Compt. rend. **231** (1950) 1320–1322.

[18] Dan McLachlan, Jr. *The determination of crystal structures from x-ray data without a knowledge of the phases of the Fourier coefficient.* Proc. Nat. Acad. Sci. U. S. **37** (1951) 115–124.

[19] Dan McLachlan, Jr. and David Harker. *Finding the signs of the* F's *from the shifted Patterson product.* Proc. Nat. Acad. Sci. U. S. **37** (1951) 846–849.

[20] Caroline H. MacGillavry. *On the relations between Harker-Kasper inequalities and Buerger equalities.* Computing methods and the phase problem in x-ray crystal analysis (X-Ray Analysis Laboratory, The Pennsylvania State College, 1952) 57–60.

[21] E. F. Bertaut. *Relations entre la fonction de Patterson symétrisée et la structure.* Bull. soc. franç. minéral. cristal. **75** (1952) 401–418.

[22] Philip A. Vaughan. *A phase-determining procedure related to the vector-coincidence method.* Acta Cryst. **11** (1958) 111–115.

[23] M. J. Buerger. *Vector space and its application in crystal-structure investigation.* (John Wiley and Sons, New York, 1959)

Inequalities

[24] David Harker and J. S. Kasper. *Phases of Fourier coefficients directly from crystal structure data.* J. Chem. Phys. **15** (1947) 882–884.

[25] D. Harker and J. S. Kasper. *Phases of Fourier coefficients directly from crystal diffraction data.* Acta Cryst. **1** (1948) 70–75.

[26] J. Gillis. *Structure-factor relations and phase determination.* Acta Cryst. **1** (1948) 76–80.

[27] J. Gillis. *The application of the Harker-Kasper method of phase determination.* Acta Cryst. **1** (1948) 174–179.

[28] E. W. Hughes. *Limitations on the determination of phases by means of inequalities.* Acta Cryst. **2** (1949) 34–37.

[29] Caroline H. MacGillavry. *On the derivation of Harker-Kasper inequalities.* Acta Cryst. **3** (1950) 214–217.

[30] J. S. Kasper, C. M. Lucht, and D. Harker. *The crystal structure of decaborane,* $B_{10}H_{14}$. Acta Cryst. **3** (1950) 436–455.

[31] Robinson D. Burbank. *The crystal structure of α-monoclinic selenium.* Acta Cryst. **4** (1951) 140–148.

[32] Beatrice S. Magdoff. *Forbidden reflections in the Harker-Kasper inequalities.* Acta Cryst. **4** (1951) 268–269.

[33] R. Pepinsky and Caroline H. MacGillavry. *Phase-limiting relations following from a known maximum value of the electron density.* Acta Cryst. **4** (1951) 284.

[34] Emmanuel Grison. *De l'usage des inégalités de Harker-Kasper.* Acta Cryst. **4** (1951) 489–490.

[35] Robinson D. Burbank. *The crystal structure of β-monoclinic selenium.* Acta Cryst. **5** (1952) 236–246.

[36] Y. Okaya and I. Nitta. *Some remarks on J. Gillis's paper on phase determination by the Harker-Kasper method.* Acta Cryst. **5** (1952) 291.

[37] Kiichi Sakurai. *A graphical method for applying Harker-Kasper inequalities to structure determination.* Acta Cryst. **5** (1952) 546–547.

[38] Yoshiharu Okaya and Isamu Nitta. *Linear structure-factor inequalities and their application to the structure determination of tetragonal ethylenediamine sulphate.* Acta Cryst. **5** (1952) 564–570.

[39] Yoshiharu Okaya and Isamu Nitta. *Application of our linear inequalities and some remarks on B. S. Magdoff's paper on 'Forbidden reflections in the Harker-Kasper inequalities.'* Acta Cryst. **5** (1952) 687–688.

[40] Kiichi Sakurai. *Some remarks on the relation between the Harker-Kasper inequality and the Okaya-Nitta linear inequalities.* Acta Cryst. **5** (1952) 697.

[41] Louis R. Lavine. *Corrections to Grison's paper on the Harker-Kasper inequalities and to Zachariasen's paper on the 'statistical method.'* Acta Cryst. **5** (1952) 846–847.

[42] M. M. Qurashi. *An extension of the inequalities method of sign determination by means of negative-density transforms.* Acta Cryst. **6** (1953) 103.

[43] P. M. de Wolff and J. Bouman. *A fundamental set of structure-factor inequalities.* Acta Cryst. **7** (1954) 328–333.

[44] Edward W. Hughes. *A new type of inequality relationship between unitary structure factors.* Acta Cryst. **10** (1957) 376–377.

[45] M. M. Woolfson. *The critical examination of a weak sign relationship between structure factors.* Acta Cryst. **10** (1957) 635–638.

[46] J. A. Bland. *Studies of aluminum-rich alloys with the transition metals manganese and tungsten. II. The crystal structure of* $\delta(Mn\text{-}Al)\text{-}Mn_4Al_{11}$. Acta Cryst. **11** (1958) 236–244.

[47] J. Gillis. *An application of electronic computing to x-ray crystallography.* Acta Cryst. **11** (1958) 833–834.

Karle-Hauptman inequalities

[48] J. Karle and H. Hauptman. *The phases and magnitudes of the structure factors.* Acta Cryst. **3** (1950) 181–187.
[49] H. Hauptman and J. Karle. *Relations among the crystal structure factors.* Phys. Rev. **80** (1950) 244–248.
[50] J. A. Goedkoop. *Remarks on the theory of phase limiting inequalities and equalities.* Acta Cryst. **3** (1950) 374–378.
[51] R. Pepinsky and Caroline H. MacGillavry. *Phase-limiting relations following from a known maximum value of the electron density.* Acta Cryst. **4** (1951) 284.
[52] Gérard von Eller. *Inégalités de Karle-Hauptman et géométrie Euclidienne.* Acta Cryst. **8** (1955) 641–645.
[53] J. Bouman. *A general theory of inequalities.* Acta Cryst. **9** (1956) 777–780.
[54] Masao Atoji. *Some new relations following from the Herglotz theorem.* Acta Cryst. **10** (1957) 464.
[55] Inao Taguchi and Shigeo Naya. *Matrix-theoretical derivation of inequalities.* Acta Cryst. **11** (1958) 543–545.

Sign relations

[56] Ray Pepinsky and William Cochran. *New methods of phase determination in x-ray crystal analyses* (Abstract). Phys. Rev. **83** (1951) 226.
[57] D. Sayre. *The squaring method: a new method for phase determination.* Acta Cryst. **5** (1952) 60–65.
[58] W. Cochran. *A relation between the signs of structure factors.* Acta Cryst. **5** (1952) 65–67.
[59] W. H. Zachariasen. *A new analytical method for solving complex crystal structures.* Acta Cryst. **5** (1952) 68–73.
[60] W. Cochran and Bruce R. Penfold. *The crystal structure of L-glutamine.* Acta Cryst. **5** (1952) 644–653.
[61] Louis R. Lavine. *Corrections to Grison's paper on the Harker-Kasper inequalities and to Zachariasen's paper on the 'statistical method.'* Acta Cryst. **5** (1952) 846–847.
[62] W. Cochran and A. S. Douglas. *A new application of EDSAC to crystal structure analysis.* Nature **171** (1953) 1112–1113.
[63] J. Krogh-Moe. *A remark on some new methods of phase determination.* Acta Cryst. **6** (1953) 568–569.
[64] Bruce R. Penfold. *The crystal structure of α-thiopyridone.* Acta Cryst. **6** (1953) 707–713.
[65] Edward W. Hughes. *The signs of products of structure factors.* Acta Cryst. **6** (1953) 871.
[66] Edward W. Hughes. *A generalization of the Patterson function.* Acta Cryst. **6** (1953) 872.
[67] M. M. Woolfson. *The statistical theory of sign relationships.* Acta Cryst. **7** (1954) 61–64.
[68] E. F. Bertaut and R. Pepinsky. *On the basic assumptions and the validity of Zachariasen's sign relations in x-ray crystal analysis.* Acta Cryst. **7** (1954) 214–215.

[69] K. Lonsdale and H. J. Grenville-Wells. *Sign determination in crystal structure analysis.* Acta Cryst. **7** (1954) 490–491.

[70] W. Cochran. *The determination of signs of structure factors from intensities.* Acta Cryst. **7** (1954) 581–583.

[71] M. M. Woolfson. *Sign determination for* pgg, p4g *and certain other space groups.* Acta Cryst. **7** (1954) 721–725.

[72] M. F. Perutz and V. Scatturin. *A test of the usefulness of direct mathematical methods in the structure analysis of a protein.* Acta Cryst. **7** (1954) 799–800.

[73] A. I. Kitaigorodskii. *The theory of the distribution of the signs of structure amplitudes* (In Russian). Doklady Akad. Nauk. S. S. S. R. **101** (1955) 73–76.

[74] A. I. Kitaigorodskii. *The theory of the relation between structure amplitudes* (In Russian). Doklady Akad. Nauk. S. S. S. R. **105** (1955) 482–484.

[75] W. Cochran and A. S. Douglas. *The use of a high-speed digital computer for the direct determination of crystal structures. I.* Proc. Roy. Soc. (London) (A) **227** (1955) 486–500.

[76] W. Nowacki and H. Bürki. *Die Kristallstruktur des Monohydrates der purinanalogen Verbindung Xanthazol,* $C_4H_3N_5O_2 \cdot H_2O$. Z. Krist. **106** (1955) 339–387.

[77] W. Cochran. *Relations between the phases of structure factors.* Acta Cryst. **8** (1955) 473–478.

[78] V. Vand and R. Pepinsky. *The constraint matrix method for application of Sayre relationships to the solution of crystal structures.* Z. Krist. **107** (1956) 202–224.

[79] J. Gillis. *Note on the statistical approach to the phase problem.* Acta Cryst. **9** (1956) 616.

[80] M. M. Woolfson. *An efficient process for solving crystal structures by sign relationships.* Acta Cryst. **10** (1957) 116–120.

[81] D. F. Grant, R. G. Howells, and D. Rogers. *A method for the systematic application of sign relations.* Acta Cryst. **10** (1957) 489–497.

[82] A. I. Kitaigorodskii. *The theory of the relation between structure amplitudes, and methods of direct analysis of crystal structures* (In Russian). Cristallographia, **2** (1957) 352–357.

[83] A. I. Kitaigorodskii. *The theory of structure analysis.* (Academy of Sciences, U. S. S. R., Moscow, 1957.)

[84] W. Cochran and A. S. Douglas. *The use of a high-speed digital computer for the direct determination of crystal structure. II.* Proc. Roy. Soc. (London) (A) **243** (1957) 281–288.

[85] M. M. Woolfson. *The utilization of relationships between sign relationships.* Acta Cryst. **11** (1958) 4–6.

[86] M. M. Woolfson. *An equation between structure factors for structures containing unequal or overlapped atoms. I. The equation and its properties.* Acta Cryst. **11** (1958) 277–283.

[87] K. H. Jost. *Zur Anwendung der Tripelproduktmethoden.* Acta Cryst. **11** (1958) 392–393.

[88] M. M. Woolfson. *An equation between structure factors for structures containing unequal or overlapped atoms. II. An application to structure determination.* Acta Cryst. **11** (1958) 393–397.

[89] W. B. Wright. *The crystal structure of glutathione.* Acta Cryst. **11** (1958) 632–642.

[90] W. B. Wright. *A comparison of the methods used in the attempt to determine the crystal structure of glutathione.* Acta Cryst. **11** (1958) 642–653.

[91] W. Cochran and E. J. McIver. *The use of Edsac II for the direct determination of crystal structures.* Acta Cryst. **11** (1958) 892.

Statistical methods

[92] H. Hauptman and J. Karle. *Crystal-structure determination by means of a statistical distribution of interatomic vectors.* Acta Cryst. **5** (1952) 48–59.

[93] J. Karle and H. Hauptman. *The probability distribution of the magnitude of a structure factor. I. The centrosymmetric crystal.* Acta Cryst. **6** (1953) 131–135.

[94] H. Hauptman and J. Karle. *The probability distribution of the magnitude of a structure factor. II. The non-centrosymmetric crystal.* Acta Cryst. **6** (1953) 136–141.

[95] Herbert Hauptman and Jerome Karle. *The solution of the phase problem. I. The centrosymmetric crystal.* Am. Cryst. Assoc. Monograph No. 3 (Edwards Brothers, Ann Arbor, Michigan, 1953).

[96] Vladimir Vand and Ray Pepinsky. *The statistical approach to x-ray structure analysis.* (X-ray and Crystal Analysis Laboratory, Department of Physics, Pennsylvania State University, 1953) 1–98.

[97] V. Vand. *Tables for direct determination of crystal structures.* (Chemistry Department, The University of Glasgow, Glasgow, 1953) 1–110.

[98] V. Vand. *A direct approach to the determination of crystal structures.* Acta Cryst. **7** (1954) 343–346.

[99] H. Hauptman and J. Karle. *Solution of the phase problem for space group* $P\bar{1}$. Acta Cryst. **7** (1954) 369–374.

[100] J. Karle and H. Hauptman. *Probability distribution for atomic coordinates.* Acta Cryst. **7** (1954) 375–376.

[101] W. Cochran and M. M. Woolfson. *Have Hauptman & Karle solved the phase problem?* Acta Cryst. **7** (1954) 450–451.

[102] V. Vand and R. Pepinsky. *The statistical approach of Hauptman & Karle to the phase problem.* Acta Cryst. **7** (1954) 451–452.

[103] H. Hauptman and J. Karle. *A note on the solution of the phase problem.* Acta Cryst. **7** (1954) 452–453.

[104] C. L. Christ, Joan R. Clark, and H. T. Evans, Jr. *The structure of colemanite,* $CaB_3O_4\ (OH)_3{\cdot}H_2O$, *determined by the direct method of Hauptman & Karle.* Acta Cryst. **7** (1954) 453–454.

[105] R. K. Bullough and D. W. J. Cruickshank. *Comments on probability distributions for interatomic vectors and atomic coordinates.* Acta Cryst. **7** (1954) 598–599.

[106] W. Cochran and M. M. Woolfson. *The theory of sign relations between structure factors.* Acta Cryst. **8** (1955) 1–12.

[107] R. K. Bullough and D. W. J. Cruickshank. *Some relations between structure factors.* Acta Cryst. **8** (1955) 29–31.

[108] H. Hauptman and J. Karle. *A relationship among the structure-factor magnitudes for* P1. Acta Cryst. **8** (1955) 355.

[109] E. F. Bertaut. *La méthode statistique en cristallographie. I.* Acta Cryst. **8** (1955) 537–543.

[110] E. F. Bertaut. *La méthode statistique en cristallographie. II. Quelques applications.* Acta Cryst. **8** (1955) 544–548.

[111] Félix Bertaut. *Sur la probabilité des valeurs de fonctions. Application à la crystallographie.* Compt. rend. **240** (1955) 152–154.

[112] Félix Bertaut. *Fonctions de répartition. Paramètres les plus probables. Application à la détermination directe des structures atomiques en cristallographie.* Compt. rend. **240** (1955) 272–274.

[113] E. F. Bertaut. *Fonctions de répartition: application à l'approche directe des structures.* Acta Cryst. **8** (1955) 823–832.

[114] J. Karle and H. Hauptman. *A theory of phase determination for the four types of non-centrosymmetric space groups* 1P222, 2P22, $3P_12$, $3P_22$. Acta Cryst. **9** (1956) 635–651.

[115] H. Hauptman and J. Karle. *A unified algebraic approach to the phase problem. I. Space group* $P\bar{1}$. Acta Cryst. **10** (1957) 267–270.

[116] J. Karle and H. Hauptman. *A unified algebraic approach to the phase problem. II. Space group* P1. Acta Cryst. **10** (1957) 515–524.

[117] H. Hauptman and J. Karle. *Phase determination from new joint probability distributions: Space group* $P\bar{1}$. Acta Cryst. **11** (1958) 149–157.

[118] I. L. Karle, H. Hauptman, J. Karle, and A. B. Wing. *Crystal and molecular structure of* p,p′*-dimethyloxybenzophenone by the direct probability method.* Acta Cryst. **11** (1958) 257–263.

[119] J. Karle and H. Hauptman. *Phase determination from new joint probability distributions: Space group* P1. Acta Cryst. **11** (1958) 264–269.

[120] E. F. Bertaut. *La probabilité élémentaire des positions atomiques.* Acta Cryst. **11** (1958) 405–412.

[121] A Klug. *Joint probability distribution of structure factors and the phase problem.* Acta Cryst. **11** (1958) 515–543.

[122] W. Cochran. *Structure factor relations and the phase problem.* Acta Cryst. **11** (1958) 579–585.

[123] E. F. Bertaut. *Das phasenproblem.* Fortschr. Min. **36** (1958) 119–148.

[124] J. Karle, H. Hauptman, and C. L. Christ. *Phase determination for colemanite,* $CaB_3O_4(OH)_3 \cdot H_2O$. Acta Cryst. **11** (1958) 757–761, 761–770.

[125] H. Hauptman and J. Karle. *Seminvariants for centrosymmetric space groups with conventional centered cells.* Acta Cryst. **12** (1959) 93–97.

[126] E. F. Bertaut, P. Blum, and A. Sagnières. *Structure du ferrite bicalcique et de la brownmillerite.* Acta Cryst. **12** (1959) 149–159.

[127] J. Karle and H. Hauptman. *A unified program for phase determination, type* 1P. Acta Cryst. **12** (1959) 404–410.

Permutation syntheses

[128] M. M. Woolfson. *Structure determination by the method of permutation syntheses.* Acta Cryst. **7** (1954) 65–67.

[129] I. J. Good. *On the substantialization of sign sequences.* Acta Cryst. **7** (1954) 603–604.

Phase relations for crystals in which some atom locations are known

[130] J. A. Goedkoop, Caroline H. MacGillavry, and Ray Pepinsky. *Phase-determining relations based on a knowledge of the electron-density function in parts of the unit cell.* Acta Cryst. **4** (1951) 491–492.

[131] J. D. Bernal. *Phase determination in the x-ray diffraction patterns of complex crystals and its application to protein structure.* Nature **169** (1952) 1007–1008.

[132] Yoshiharu Okaya and Isamu Nitta. *On an application of inequality methods to centrosymmetrical crystals with partly known structures.* Acta Cryst. **5** (1952) 687.

[133] V. Luzzati. *Resolution d'une structure cristalline lorsque les positions d'une partie des atomes sont connues: Traitement statistique.* Acta Cryst. **6** (1953) 142–152.

[134] Harry L. Yakel, Jr. and Edward W. Hughes. *The crystal structure of* N,N′-*diglycyl-*L*-cystine dihydrate.* Acta Cryst. **7** (1954) 291–297.

[135] G. A. Sim. *The fraction of structure factors determined in sign by a selected atom or group of atoms in a molecule.* Acta Cryst. **10** (1957) 177–179.

[136] G. A. Sim. *The distribution of phase angles for structures containing heavy atoms. I. Space group* P1 *with one heavy atom in the asymmetric unit.* Acta Cryst. **10** (1957) 536–537.

[137] E. F. Bertaut. *La méthode statistique dans le cas d'une structure partiellement connue.* Acta Cryst. **10** (1957) 670–671.

[138] G. A. Sim. *The probability distribution of x-ray intensities: The effect of one heavy atom in a triclinic cell containing a number of light atoms.* Acta Cryst. **11** (1958) 123–124.

22

Refinement

Once a model of a structure has been proposed, or when one has been found by one of the direct methods of Chapters 19, 20, or 21, it is necessary to improve the preliminary coordinates by some process of refinement. The several methods of refinement are outlined in this chapter.

It is sometimes difficult to say exactly when the process of structure determination leaves off and the process of refinement begins. Sometimes it is possible to find the approximate locations of some of the atoms, and yet the locations of the remaining atoms are in doubt, or are completely unknown. In such cases, the locating of the remaining atoms is not only a part of the structure determination, but it grades into the preliminary stages of refinement.

Measures of the correctness of a structure

Two crystal structures which are in every way identical produce identical sets of diffraction maxima. Let one of these crystal structures be the actual structure being investigated, and its set of diffraction maxima the set actually observed; let the other crystal structure be the model being proposed, and its set of diffraction maxima that set which is calculated for the model. Then a necessary condition for the proposed model to be correct is that the calculated maxima duplicate those observed (due account being taken in the calculation of the appropriate physical conditions involved, such as corrections for absorption, "extinction," temperature, etc.). With due care in the experimental conditions, the identity of these sets of data can be rather closely approached. Discrepancies between the two sets of data can be ascribed to errors in measurement of the intensities, and to failure to allow, in the calculation, for all features present in the experiment.

It is evident that the general departure of the computed from the observed intensities must increase with the departure of the model from the actual structure. In other words, a discrepancy in intensities follows a discrepancy in the proposed structure. In attempting to relate one discrepancy to the other, it is difficult to know how discrepancies in different F_{hkl}'s are to be weighted, or combined. No really satisfactory weighting has been proposed. Nevertheless it has become common practice to utilize a *residual* of the following form for this purpose

$$R = \frac{\sum \left| |F_o| - |F_c| \right|}{\sum |F_o|}. \tag{1}$$

The numerator is the sum of all discrepancies in the $|F_{hkl}|$'s. The denominator is the sum of all observed $|F_{hkl}|$'s. Thus R is a measure of the relative discrepancies of the $|F_{hkl}|$'s for the model. The value of R is a comparatively small fraction when the structure is correct, and a large fraction when the structure is incorrect.

Wilson[4] has shown that an entirely wrong arrangement of atoms of the same symmetry has $R = 0.828$ for a centric crystal and $R = 0.586$ for an acentric crystal. Centric models for which $R < 0.5$ are therefore usually regarded as worth attempting to refine. It is common experience to begin with a model having $R = 0.4$, which later refinement shows was the correct model except that the coordinates required comparatively small corrections. Correct structures usually have $R < 0.25$, and very well refined structures may have R in the neighborhood of 0.05.

One of the difficulties with accepting R at its face value is that it comes from a mixture of reflections. Those who have made structure-factor computations in trial-and-error structure determination realize that a finer adjustment is required to obtain agreement for high-order reflections than for low-order reflections. Luzzati[5,6] has formulated this relation for R. If $|\sigma_{hkl}| = (2 \sin \theta)/\lambda$ is the distance of a point hkl from the origin of the reciprocal lattice, and if $\overline{|\Delta r|}$ is the mean of the absolute errors in atomic position, then the values of R are related to the values of $\overline{|\Delta r|} \cdot |\sigma_{hkl}|$ as shown in Table 1. If the R's are determined for different σ ranges of the reciprocal lattice they can be plotted, as in Fig. 1, and from their alignment, the mean absolute deviation of the atoms in the model can be estimated. In actual practice, experimental errors increase the R values beyond those noted in Table 1.

It should be pointed out that the value of R deduced for any model depends on how the very small values of $|F_o|$ are treated. There is a lower

Table 1
Relation of R to mean absolute displacement $\overline{|\Delta r|}$, and to σ_{hkl}
(From Luzzati[5])

$\overline{\lvert\Delta r\rvert}\cdot\lvert\sigma_{hkl}\rvert$	R			
	Centric		Acentric	
	2 dimensions	3 dimensions	2 dimensions	3 dimensions
0.00	0.000	0.000	0.000	0.000
.01	.050	.039	.031	.025
.02	.098	.078	.062	.050
.03	.145	.115	.093	.074
.04	.191	.152	.124	.098
.05	.234	.188	.155	.122
.06	.276	.223	.185	.145
.07	.317	.256	.214	.168
.08	.356	.288	.243	.191
.09	.394	.320	.270	.214
.10	.430	.350	.296	.237
.12	.494	.410	.342	.281
.14	.552	.462	.384	.319
.16	.603	.510	.420	.353
.18	.647	.554	.452	.385
.20	.686	.595	.480	.414
.22	.715	.631	.504	.440
.24	.742	.662	.524	.463
.26	.763	.689	.540	.483
.28	.781	.713	.552	.502
.30	.794	.735	.560	.518
.35	.816	.776	.574	.548
.40	.824	.802	.580	.564
.45	—	.817	—	.574
.50	—	.823	—	.580
∞	.828	.828	.586	.586

limit below which $|F_o|$ cannot be observed, especially when diffraction is recorded photographically. If an unobserved reflection is recorded as $|F_o| = 0$, then $\big||F_o| - |F_c|\big|$ has a non-zero value, and so makes a small contribution to the numerator, although $|F_o|$ makes none to the denominator. The result is a fictitiously high value of R. On the other hand, if both $\big||F_o| - |F_c|\big|$ and $|F_o|$ are omitted in computing R in such cases, any error in F_c is neglected, so the value of R is too low. A third possible treatment is to regard $|F_o|$ in such cases as half the least observable value. This is the most probable value if the unobserved $|F|$ of the crystal is centrosymmetrical.

In comparing observed and computed $|F|$'s, it should be remembered that "extinction" tends to severely affect $|F|$'s of high intensity and low $\sin\theta$. For this reason, before computing R, it is advisable to search the reciprocal structure in the neighborhood of the origin to see if the $|F_o|$'s of high value tend to be systematically smaller than the $|F_c|$'s. If so, it is likely that they are too low due to "extinction." Since these reflections make a high contribution to R it is advisable, when computing R, to omit those $|F|$'s within a small sphere of the reciprocal structure near the origin within which the effect is noticeable.

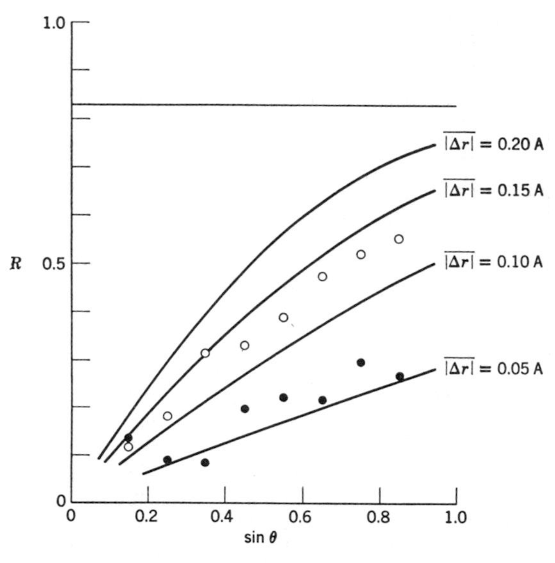

Fig. 1.
The variation of $R = \Sigma||F_o - F_c||/|F_o|$ with $\sin\theta$ for various mean values of the absolute deviation of coordinates, $\overline{|\Delta r|}$. The two sets of points are sets of experimental values obtained in the refinement of $HNO_3 \cdot 2H_2O$.
(After Luzzati.[5])

After the refinement of coordinates, by one of the methods to be described, the value of R normally converges to some lower limit. Assuming that this value of R is low enough to assure that the structure is correct, then the level of the R can be taken as a rough quantitative accuracy of the observed $|F|$'s. The final value of R is about 2% higher than the percentage inaccuracy of the data.[109]

Booth[1] has recommended the use of a different residual factor, namely

$$R_2 = \frac{\sum (|F_o| - |F_c|)^2}{\sum |F_o|^2}. \tag{2}$$

This is mathematically more tractable than (1). By using various reasonable assumptions, Booth was able to reduce this to a function of the mean departure, $\bar{\delta}$, of the coordinates from their true values. Some values of $\bar{\delta}$ for various values of R_2 are listed in Table 2. Booth's R_2 residual is closely related to the quantity which is minimized by least-squares refinement.

Table 2

Root-mean-square error $\bar{\delta}$ as a function of $R_2 = \dfrac{\Sigma(|F_0| - |F_c|)^2}{\Sigma|F_0|^2}$

(After Booth[1])

Two dimensions		Three dimensions	
R_2	$\bar{\delta}$	R_2	$\bar{\delta}$
0.029	0.00 Å	0.05	0.000 Å
.079	.07	.10	.062
.129	.10	.15	.088
.179	.13	.20	.108
.229	.15	.25	.124
.279	.17	.30	.139

Residuals of several other related forms have been suggested.[2,3] None of these has come into common use.

Preliminary scale and temperature-factor adjustments

Before the first computations of structure factors are made, the $|F_o|$'s are known on an absolute scale, but without allowance for the temperature motion of the atoms. A preliminary notion of both of these factors can be found by use of Wilson's method, as outlined in Chapter 8.

As refinement proceeds it becomes possible to determine the scale factor and temperature factor more and more closely. The F_o's are actually known in their scaled form KF_o. If ${}^0F_c$ is the structure factor at absolute zero, then the value of F_c at room temperature (which is required) is ${}^0F_c\, e^{-(B \sin^2 \theta)/\lambda^2}$. When observed and computed structure factors agree, then

$$KF_o = {}^0F_c\, e^{-(B \sin^2 \theta)/\lambda^2}, \tag{3}$$

so

$$\frac{F_o}{{}^0F_c} = \frac{1}{K} e^{-(B \sin^2 \theta)/\lambda^2} \tag{4}$$

and

$$\ln\left(\frac{F_o}{{}^0F_c}\right) = \ln\left(\frac{1}{K}\right) - \left(\frac{B}{\lambda^2}\right)\sin^2\theta. \tag{5}$$

This has the form

$$y = c + ax, \tag{6}$$

where

$$\begin{aligned} y &= \ln\left(\frac{F_o}{{}^0F_c}\right), \\ c &= \ln\left(\frac{1}{K}\right), \\ a &= -\frac{B}{\lambda^2}, \\ x &= \sin^2\theta. \end{aligned} \tag{7}$$

Thus, if $\ln(F_o/{}^0F_c)$ is plotted against $\sin^2\theta$ for the various reflections *hkl*, the points should fall on a straight line. The intercept of the line provides the value of K and the slope of the line provides B.

After each preliminary computation of F_c, this procedure should be undertaken, thus providing a new scale for F_o and a new temperature correction for ${}^0F_c$, which yields the F_c necessary for the computation of R. It should be noted that this method of obtaining a temperature factor is approximate in that (*a*) all atoms are assumed to have the same temperature factor, and (*b*) the thermal motion is assumed to be isotropic. In general, neither of these assumptions is correct, and a detailed refinement should allow for anisotropic thermal motions which differ for each atom. But for preliminary work, the method noted above is useful. After preliminary refinement the relation $K\Sigma F_o = \Sigma F_c$ can be used to improve the scale factor.

Successive Fourier syntheses

Except for trial-and-error adjustment of coordinates, the oldest method of refinement of a structure is that of successive Fourier syntheses. This method often grades imperceptibly from structure determination into refinement of the structure.

Procedure for centric syntheses. The process of refinement starts with some model of the crystal structure which is usually known only partially or imperfectly. For example, it may start with a knowledge of the locations of a set of heavy atoms, as discussed in Chapter 19. Alternatively it may start with the known arrangement of atoms in a chemical molecule, and a proposal as to the way the molecule fits in the cell. In any case, what is known about the structure can be used to compute the expected F_c's for the model.

First consider the nature of the refinement when the required Fourier synthesis (either a projection or full three-dimensional case) is centrosymmetrical. Then the calculated F's are either positive or negative. If the R for the model is 0.5 or less, the model may be retained as promising. With this value of R, there should be a fair tendency for the $|F_c|$'s to follow the $|F_o|$'s, that is, if $|F_{hkl}|$ is observed to be large, then the same $|F_{hkl}|$ computed for the model should be large. Presumably the model is similar to the actual structure except for small differences in the coordinates of the atoms. Therefore it can be assumed that reflections for which F_o and F_c are both large have the same sign. For such reflections, then, the actual structure provides the magnitudes of the F's while the model (though it supplies but crude magnitudes) provides the unobservable signs. If R is as high as 0.5, the number of reflections for which the sign can be reasonably inferred is only a fraction of the total, but may include most of the strong reflections.

Data are at hand, then, for a Fourier synthesis in which many, if not most, of the largest Fourier coefficients F_{hkl} are available, at least for small values of h, k, and l. In making such a preliminary synthesis, the F's for weak reflections should not be included. For such reflections the contributions from the several atoms to the calculated F's are in near balance, and a small change in coordinates of even one atom may cause the F_c to change sign. Furthermore, if an F with incorrect sign is included in the Fourier synthesis, it gives rise to twice the electron-density error that would have occurred in the result had the doubtful term been omitted entirely.

The resulting Fourier synthesis, though incomplete, should not only return the atoms of the model, but it should improve the coordinates of their location. The reason for this is that the coordinates of the atoms of the model determined the signs of certain F_{hkl}'s, but these signs would have remained the same within a certain range of change of the coordinates of the atoms (although a change in coordinates would vary the relative magnitudes of the F_{hkl}'s). But with the correct magnitudes of the F_{hkl}'s available from the experimental measurements, the Fourier synthesis automatically reveals the locations of the coordinates most consistent with the observed magnitudes and computed signs. Thus the Fourier synthesis leads to refined coordinates.

Not only does the Fourier synthesis lead to refined coordinates of the atoms used for computing the structure factors, but it also tends to reveal the positions of any atoms not included in computing the structure factors. For example, in the heavy-atom method discussed in Chapter

19, only the heavy atoms are needed to determine the signs of most of the structure factors, yet the Fourier synthesis not only shows the heavy atoms used to compute the phases, but the lighter atoms as well. This additional information is not a part of the phases, but is contained in the magnitudes[†] of the $|F|$'s.

The improved coordinates revealed in the preliminary Fourier synthesis define a somewhat refined structure. The new structure almost certainly has smaller discrepancies between observed and computed $|F|$'s and, if so, has a lower R value. Because of the improved coordinates, many of the F's whose signs were doubtful in the first computation of F_c's can now be regarded as certain; others, especially those with small values of $|F|_c$, may still be uncertain. But with the availability of additional signs for the F's, a new Fourier synthesis can be computed which will reveal the atom positions with additional improvement. In this way, each structure-factor computation followed by a Fourier synthesis comprises a refinement cycle. Each cycle brings in new signs which can be regarded as reasonably certain. These cycles converge on a final set of atom locations which are the correct ones, except for small corrections noted in a subsequent section. This kind of refinement process can be regarded as complete when (*a*) all signs have been determined and included in the last synthesis, (*b*) the structure-factor computation *after* the last Fourier synthesis produces the same set of signs which were computed by the structure-factor computation *before* the last Fourier synthesis, and which were used in that synthesis. This implies that the R for the structure-factor computation before and after the last Fourier synthesis is essentially the same, so that R has converged to a low value. (Minor improvements in R can still be made, as discussed later.)

The map of the final Fourier synthesis should show certain features which tend to authenticate the structure as correct. In the first place, the peaks should represent the expected chemistry in number of atoms, in the electron count of each, and in their interatomic distances. Each peak should be round, and without distortion. Sharp distortions usually imply one or more F's with incorrect phases, but may also be due to poor intensity data, for example because of uncorrected absorption. Furthermore, the spaces between peaks should be reasonably uniform and without intense negative regions.

Procedure for acentric syntheses. Refinement for acentric syntheses is more difficult than for centric syntheses for two reasons. In

[†] This implies that two structures can have the same phases, but somewhat different intensities; for example a structure containing a set of heavy atoms alone, and this structure plus the residue, have the same phases for all but the weaker spectra. But the second structure has different intensities caused by the superposition of the Fourier density waves due to the residue structure.

the acentric case, both the structure factors and Fourier syntheses have real and imaginary parts and so are more tedious to compute. But, in addition, the refinement converges more slowly. The reason for this is that phases of the F_{hkl}'s are not merely 0 or π, but are continuously variable. Each phase does not, therefore, refine quickly to the correct sign, but gradually drifts toward the correct value of ϕ, which it approaches but does not actually attain.

In studying this situation, Donohue[15] noted that, in acentric syntheses, peaks appeared at positions halfway between the initial structure and that later found to be final. In other words, the correction suggested by the new peak locations of a Fourier synthesis are only halfway to the correct location. This suggests that the coordinate corrections found from the Fourier synthesis should *not* be used for the phases of the F's of the next synthesis, but that *double* the corrections be used instead. The second Fourier synthesis should then show little or no change. This is the *double-shift rule* used for entirely acentric syntheses.

A complication occurs in the actual application of this rule. Only for plane group $p1$ and space group $P1$ do all F's have general phase angles. In other space groups, some F's lie on symmetry elements in reciprocal space and consequently must have phases which can only be 0 or π. For such syntheses, the double-shift rule must be replaced by an *n-shift rule*, where n lies between 1 and 2. The value of n depends on the fraction, p, of the F's which are restricted to 0 or π, and the remainder which may have any value. The value of n is then $1p+2(1-p)$.

The n-shift rule can only be safely applied in cases where a particular atom does not dominate the phases.[21]

Models which cannot be refined. It is often discovered that a model cannot be refined to below some fairly high value of R. This is usually an indication that the initial model was incorrect, and cannot be modified by small atom shifts to fit the observed $|F|$'s. In such instances the model must be discarded and a new one tried. Two common reasons for incorrect models are noted below:

(*a*) If the incorrect chemical model is assumed, the structure usually cannot be refined. An example was supplied by Furberg and Hassel[17] in studying the structure of bi-1,3-dioxacyclopentyl (2). This is monoclinic with

$$a = 4.46 \text{ Å}, \quad b = 7.76 \quad \beta = 121.5°, \quad c = 11.64, \quad Z = 2\ C_6H_{10}O_8 \text{ per cell},$$

space group: $P\,2_1/c$.

Since there are two molecules per cell, the molecule must occupy one of the special positions, all of which are symmetry centers. There are two chemical possibilities, represented in the upper parts of Figs. 2*A* and 2*B*. Since the *a* axis is relatively short, the structure was studied in projection $\rho_{(yz)}$. A reasonable fit for the molecule was found, shown in the lower part of Fig. 2*A*. This started with $R = 0.55$, but could not be refined to less than $R = 0.38$, at which point no further sign changes occurred. The final map is shown in the lower part of Fig. 2*A*. The failure to

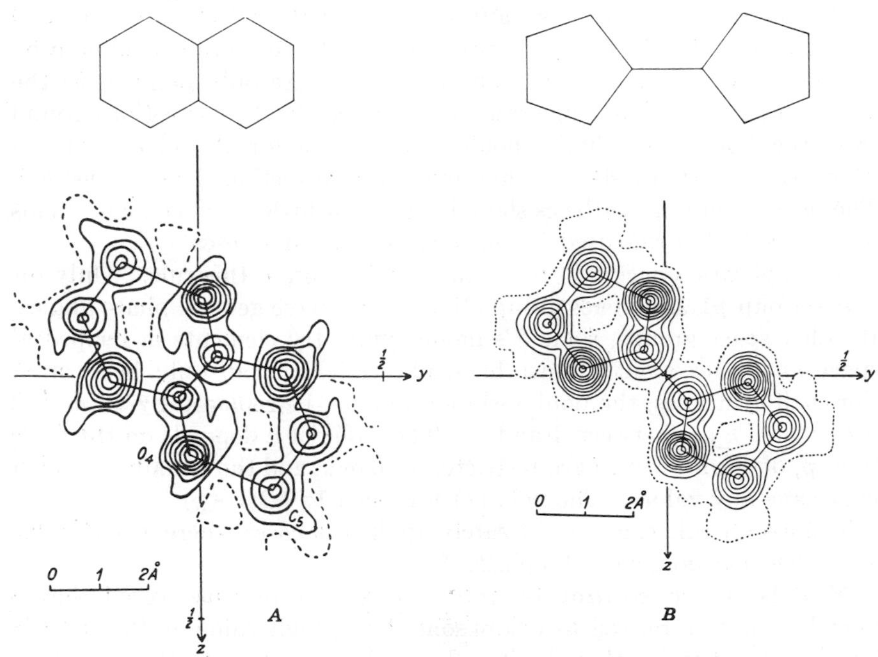

Fig. 2.
Incorrect model, *A*, and correct model, *B*, of bi-1, 3-dioxacyclopentyl(2). Refinement of *A* stopped at $R = 0.38$, but that of *B* continued to $R = 0.13$. The initial models differ chiefly by the arrangement of the pair of atoms nearest the center.
(After Furberg and Hassel.[17])

permit further refining indicates that this is an incorrect model. Furthermore the distribution of the peaks in the final Fourier synthesis makes the results suspect. The distance O_4–C_5 is 1.80 Å, which is chemically unacceptable.

The first model was accordingly discarded and the chemical molecule in the upper part of Fig. 2*B* was tried. This model also started with $R = 0.55$, but was readily refined to $R = 0.13$. The final electron-density map is shown in the lower part of Fig. 2*B*. It lacks the distortions of Fig. 2*A*, has round peaks, and has acceptable interatomic distances. Note that the major difference in the models of Fig. 2 is the initial locations of one pair of carbon atoms in the neighborhood of the symmetry center.

(*b*) In the interpretation of symmetrical Patterson projections and of Harker sections, an n-fold ambiguity is involved.† If the ambiguity is not resolved, the model initially used may be incorrect, and not capable of refinement.

Bunn's error synthesis. When a model cannot be refined, the error in the proposed model can sometimes be found with the aid of Bunn's error synthesis.[12] This was first used in solving the structure of the penicillin, and was the forerunner of the difference synthesis, discussed later. To understand Bunn's error synthesis, suppose that two Fourier syntheses are made, one with F_o and the other with F_c. The former has peaks where the atoms really are; the latter has peaks where the postulated model has atoms. If the second is subtracted from the first, the composite synthesis has interesting properties. If an atom has been correctly placed, its peak vanishes in the difference of the two component syntheses, and the atom location appears on topography of low relief in the difference map. But if an atom has been incorrectly placed, its incorrect location is given by the F_c component and its correct location by the F_o component. On the composite $F_o - F_c$ map the postulated location is in a hole, and the correct location is on a peak. In this case the postulated location can be corrected by shifting it from the hole to the peak. If the postulated location is only somewhat in error, the postulated location is on a steep gradient between hole and peak. In this event the postulated position can be improved by shifting it somewhat up the slope.

Difference syntheses are commonly made for projections. They can, of course, be prepared by subtracting, point by point, the results of the syntheses using F_o and F_c for coefficients of the Fourier syntheses. But this is equivalent to a single synthesis using $(F_o - F_c)$ as coefficients, because

$$\rho_o = \frac{1}{S} \sum_h \sum_k F_o \, e^{-i2\pi(hx+ky)}, \tag{8}$$

$$\rho_c = \frac{1}{S} \sum_h \sum_k F_c \, e^{-i2\pi(hx+ky)}; \tag{9}$$

therefore

$$\rho_o - \rho_c = \frac{1}{S} \sum_h \sum_k (F_o - F_c) e^{-i2\pi(hx+ky)}. \tag{10}$$

† M. J. Buerger. *Vector space and its application in crystal-structure investigation.* (John Wiley and Sons, New York, 1959) Chapter 7.

Unfortunately, in the initial stages of a structure determination, (10) cannot be carried out because, although the phases of the F_c's are known, the phases of the F_o's are not. But there are *some* terms $(F_o - F_c)$ which can be used, namely those for which F_o is zero (or very small) while F_c is reasonably large. Bunn's error synthesis consists in making up a Fourier synthesis of such terms only. This comprises only part of the full Fourier synthesis of (10) but it may be enough to give a clue to what is wrong with the proposed structure. An example of the original error synthesis used by Bunn[12] is shown in Fig. 3*C*, which is an error map of sodium benzylpenicillin. The postulated structure is shown in Fig. 3*A*. One of the important errors of Fig. 3*A* revealed by the error map was that the position attributed to sulfur in the postulated structure was found to be occupied by incompletely resolved carbon atoms. The sodium atoms were also found on steep gradients. After these corrections were made the Fourier synthesis of the corrected structure was found to be shown in Fig. 3*B*.

Series-termination effects. In Chapter 13 it was seen that a Fourier series with an infinite number of terms can give a perfect representation of a point. On the other hand, if the same Fourier series is used, but the terms beyond a given number are discarded, the representation of the point becomes a peak of finite width, surrounded by a diffraction ripple. The radius of the peak for two-dimensional syntheses is $r = 0.61 d_{\text{min}}$ and $r = 0.715 d_{\text{min}}$ for three dimensional syntheses,[10] where d_{min} is the wavelength of the shortest Fourier wave in the Fourier series. Since $d_{\text{min}} = \lambda/(2 \sin \theta_{\text{max}})$, the radius of the peak can also be expressed as $r = 0.61\lambda/(2 \sin \theta_{\text{max}})$ for two-dimensional syntheses, and $r = 0.715\lambda/(2 \sin \theta_{\text{max}})$ for three-dimensional syntheses, where θ_{max} is maximum glancing angle of the reflections used for the Fourier series. Points which are closer than r appear as a single peak. Accordingly peaks in a Fourier synthesis which are closer than $r = 0.61\lambda/(2 \sin \theta)$ are unresolved in projection, and appear as a single peak. For copper radiation $\lambda = 1.54$ Å, so using all reflections within the possible range $\theta_{\text{max}} = 90°$, r is $0.61 \times 1.54/2 = 0.47$ Å for two dimensions,† and 0.55 Å for three dimensions. Since atoms are always farther apart than this, there is no trouble with resolution in three dimensions. In two dimensions, however, atoms can be closer than 0.47 Å in projection, and, in fact, they can overlap completely. In such instances the two (or more) atoms appear as a single peak, although the shape and volume of the peak may well suggest more than one atom. The increase in resolving power with the increase in the number of Fourier terms used is well illustrated

† Practical experience sets a higher limit. With Cu$K\alpha$ the practical limit[48] is 0.90 Å, and with Mo$K\alpha$ it is about 0.65 Å.

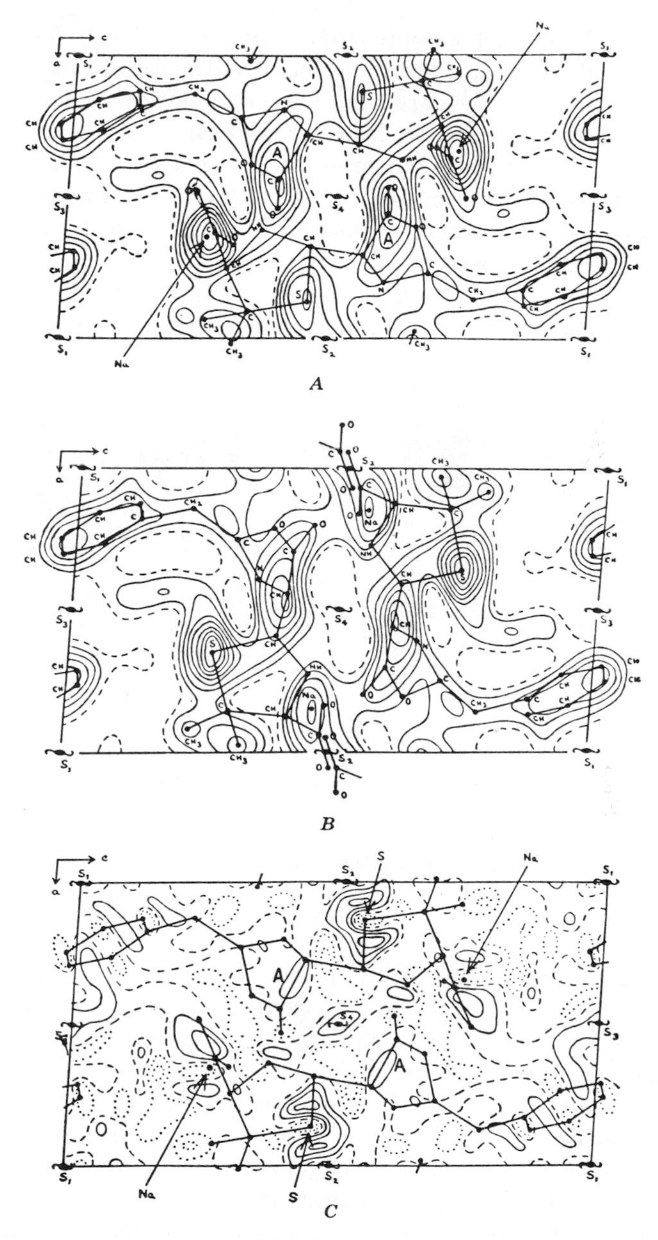

Fig. 3.

Bunn's error synthesis, as used in sodium benzylpenicillin.

A. Postulated model.

B. Model later found to be correct.

C. Error synthesis.

(After Crowfoot, Bunn, Rogers-Low, and Turner-Jones.[12])

in Fig. 4. Figure 4*A* was prepared with the relatively few Fourier coefficients available with the longer-wavelength Cu$K\alpha$ radiation. Fig. 4*B* was prepared with the larger number of Fourier coefficients available with the shorter-wavelength Mo$K\alpha$ radiation.

Even when peaks do not overlap, it is sometimes difficult to judge the center of the peak in order to fix accurately the coordinates of the atom it represents. A number of authors[13, 22, 23, 26, 31] have suggested methods for locating the centers.

The artificial termination of the series has another effect. The diffraction ripple from one atom adds itself to the peaks of the other atoms;

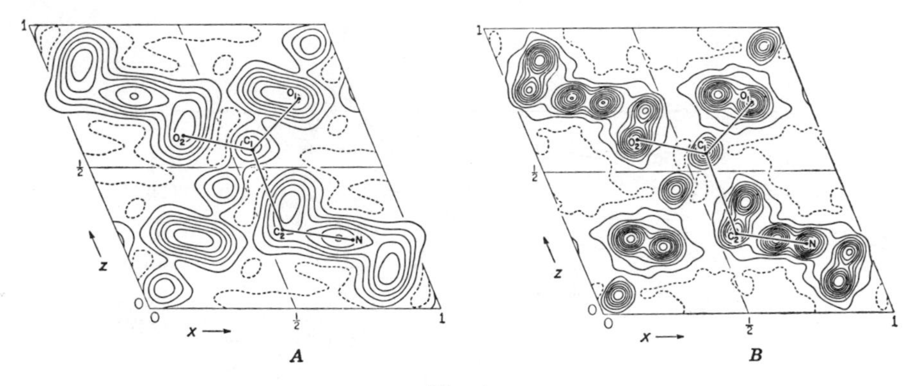

Fig. 4.

Resolving power as a function of the number of terms used in the Fourier series; $\rho_{(xz)}$ for glycine.

A. Fourier synthesis made using the F_{h0l}'s available with Cu$K\alpha$ radiation.

B. Fourier synthesis made using the larger number of F_{h0l}'s available with Mo$K\alpha$ radiation.

(After Marsh.[30])

the resultant disturbance in the peaks is particularly marked on nearest neighbors. As a result, the apparent coordinates of every atom are shifted due to the diffraction ripples of the others. The effect is not easy to allow for directly, but may be estimated and corrected for by a method due to Booth[77,78] and developed by Donohue[15] and others.[16, 27] When the process of refinement is complete, the last Fourier synthesis (made using F_o) is known to have its peak locations displaced by unknown amounts due to these series-termination effects. But another Fourier synthesis can be made with F_c, and it will be subject to substantially the same series-termination errors. Specific coordinates are inserted into this F_c synthesis, and the same coordinates *should be* deduced from it. Due to series-termination errors, however, these two sets are found to differ by Δx_j, Δy_j, Δz_j for atom j. This error may then be subtracted from the coordinates of atom j as derived from the original F_o synthesis.

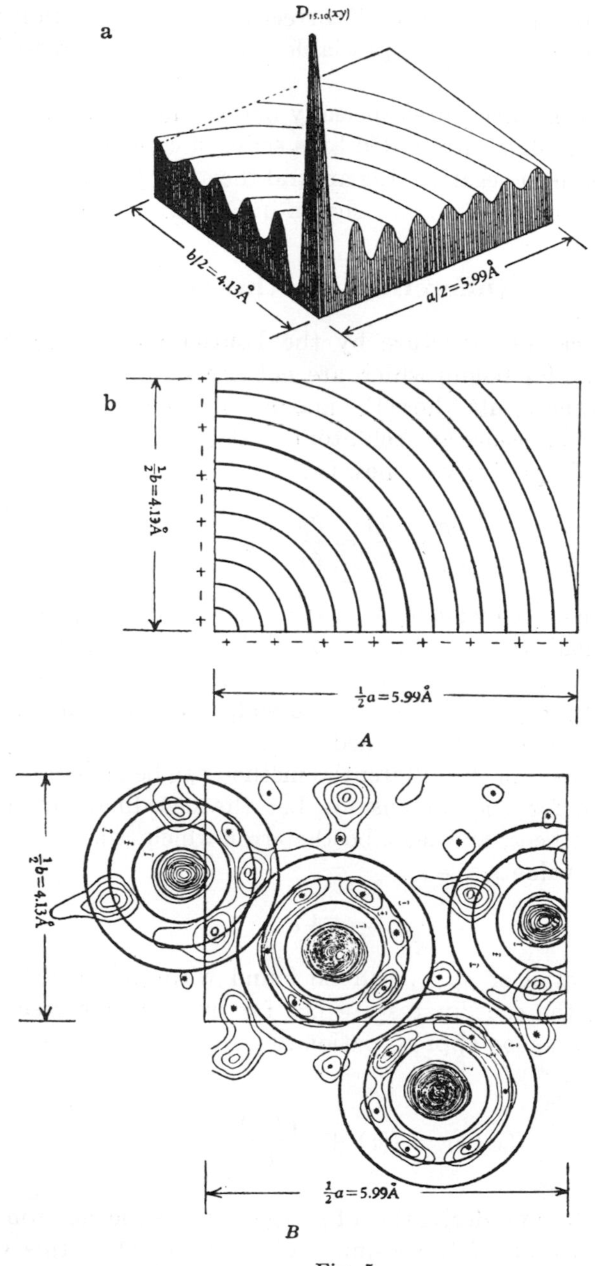

Fig. 5.

Series-termination effect.

A. Fluctuations to be expected about a point represented by a terminated Fourier series.

B. Fourier synthesis, $\rho_{(xy)}$, of $CeOSO_4 \cdot H_2O$. About each cerium atom appears a sequence of circular ripples, as in *A*, producing some false peaks within the first circular crest.

This correction is called a *back-shift* correction. The corrections due to series-termination effects are appreciable and may amount to 0.01 to 0.04 Å.

Series-termination effects are especially noticeable about heavy atoms. The heavy-atom peak is surrounded by a series of circular ripples. This provides a rather deep negative ring around the atom peak, followed by a somewhat smaller positive ring, as shown in Fig. 5.

Differential syntheses

In refining a crystal structure by the Fourier method, many summations are made for points which are not needed; all that is actually desired is to find the points where the maxima of the function $\rho_{(xyz)}$ occur. One such maximum occurs at each atom. At these points of maximum density the first derivative vanishes, i.e.

$$\frac{\partial \rho}{\partial x} = \frac{\partial \rho}{\partial y} = \frac{\partial \rho}{\partial z} = 0. \tag{11}$$

Booth[32–35] devised a refinement method to find the departures in coordinates from the final structure by the use of these derivatives. It will be seen that refinement by this *differential method* involves much less computation than by the standard Fourier method, especially since the three coordinates xyz can be readily refined.

The general scheme of the differential method can be understood by the following outline for one dimension: Let the coordinate of an atom before refinement be x, and let ϵ be the error which, when added to x, gives the correct value x_o, i.e.,

$$x_o = x + \epsilon. \tag{12}$$

Now, the differential of $\rho_{(x+\epsilon)}$ can be found with the aid of Taylor's theorem. Since the coordinates are not far from their correct values, ϵ is small, and it is a good approximation to use only the first two terms of Taylor's theorem:

$$f(x+\epsilon) = f(x) + \epsilon \frac{df(x)}{dx} \cdots . \tag{13}$$

Here $f(x+\epsilon)$ is the first derivative of $\rho_{(x+\epsilon)}$, which is the electron density at the correct location of the atom. According to (11), this vanishes so that

$$f(x) + \epsilon \frac{df(x)}{dx} = 0. \tag{14}$$

In this application,

$$f(x) = \frac{d\rho_{(x)}}{dx}, \tag{15}$$

so that (14) becomes

$$\frac{d\rho_{(x)}}{dx} + \epsilon \frac{d^2\rho_{(x)}}{dx^2} = 0. \tag{16}$$

Assuming that the first and second derivations of $\rho_{(x)}$ can be computed, (16) can be solved for the error ϵ.

In three dimensions, partial differentials are used. The first two terms of Taylor's series become

$$f(x+\epsilon_x, y+\epsilon_y, z+\epsilon_z) = f(xyz) + \epsilon_x \frac{\partial f(xyz)}{\partial x} + \epsilon_y \frac{\partial f(xyz)}{\partial y} + \epsilon_z \frac{\partial f(xyz)}{\partial z}. \tag{17}$$

Here $f(xyz)$ is the partial derivative of $\rho_{(xyz)}$ with respect to x or y or z, and $f(x+\epsilon_x\ y+\epsilon_y\ z+\epsilon_z)$ is the corresponding derivative at the peak location. According to (11) this must be zero. Therefore three equations corresponding to (16), and of the following type, can be written:

$$\frac{\partial \rho}{\partial x} + \epsilon_x \frac{\partial^2 \rho}{\partial x^2} + \epsilon_y \frac{\partial^2 \rho}{\partial x\, \partial y} + \epsilon_z \frac{\partial^2 \rho}{\partial x\, \partial z} = 0. \tag{18}$$

Assuming that the partial derivatives can be found, this gives three equations with three unknowns ϵ_x, ϵ_y, and ϵ_z. These unknowns can accordingly be determined.

The known parts in (18) are the derivatives of the electron density. For a centrosymmetrical structure the electron density is

$$\rho_{(xyz)} = \frac{1}{V} \sum_h \sum_k \sum_l F_{hkl} \cos 2\pi(hx+ky+lz). \tag{19}$$

The required types of derivatives are

$$\frac{\partial \rho}{\partial x} = -\frac{2\pi}{V} \sum_h \sum_k \sum_l hF_{hkl} \sin 2\pi(hx+ky+lz), \tag{20}$$

$$\frac{\partial^2 \rho}{\partial x^2} = -\frac{4\pi^2}{V} \sum_h \sum_k \sum_l h^2 F_{hkl} \cos 2\pi(hx+ky+lz), \tag{21}$$

$$\frac{\partial^2 \rho}{\partial x\, \partial y} = -\frac{4\pi^2}{V} \sum_h \sum_k \sum_l hkF_{hkl} \cos 2\pi(hx+ky+lz). \tag{22}$$

As an abbreviation, let

$$\frac{\partial \rho}{\partial x} = A_h,$$
$$\frac{\partial^2 \rho}{\partial x\, \partial y} = A_{hk}, \text{ etc.} \tag{23}$$

Then the three simultaneous equations of type (18) can be written

$$\begin{aligned} A_{hh}\,\epsilon_x + A_{hk}\,\epsilon_y + A_{hl}\,\epsilon_z + A_h &= 0,\\ A_{hk}\,\epsilon_x + A_{kk}\,\epsilon_y + A_{kl}\,\epsilon_z + A_k &= 0,\\ A_{hl}\,\epsilon_x + A_{kl}\,\epsilon_y + A_{ll}\,\epsilon_z + A_l &= 0. \end{aligned} \tag{24}$$

The A's are constants; each is a series of type (20), (21), or (22). Each requires a summation at one point xyz only, so the work of summation is limited. One set of equations like (24) is required for each atom whose location is to be refined.

The procedure can be further simplified by assuming that, near the maxima, the density distributions in the atoms have spherical symmetry. In this event, the electron density is a function only of the distances from the center of the sphere, i.e.,

$$\rho(xyz) = f(s). \tag{25}$$

For triclinic coordinates, the distance† of point xyz from the center of the atom, located at $x_0y_0z_0$, is given by

$$\begin{aligned} s = (x-x_0)^2\,a^2 + (y-y_0)^2\,b^2 + (z-z_0)^2\,c^2 \\ + 2(y-y_0)(z-z_0)bc\cos\alpha \\ + 2(z-z_0)(x-x_0)ca\cos\beta \\ + 2(x-x_0)(y-y_0)ab\cos\gamma. \end{aligned} \tag{26}$$

Thus near the center of the atom, $x_0\,y_0\,z_0$, neglecting second-order terms,

$$\frac{\partial^2 \rho}{\partial x^2} = 2a^2\left(\frac{\partial f}{\partial s}\right)_0; \quad \frac{\partial^2 \rho}{\partial y^2} = 2b^2\left(\frac{\partial f}{\partial s}\right)_0; \quad \frac{\partial^2 \rho}{\partial z^2} = 2c^2\left(\frac{\partial f}{\partial s}\right)_0, \tag{27}$$

$$\frac{\partial^2 \rho}{\partial x\,\partial y} = 2\left(\frac{\partial f}{\partial s}\right)_0 ab\cos\gamma, \tag{28}$$

$$\frac{\partial^2 \rho}{\partial y\,\partial z} = 2\left(\frac{\partial f}{\partial s}\right)_0 bc\cos\alpha, \tag{29}$$

$$\frac{\partial^2 \rho}{\partial z\,\partial x} = 2\left(\frac{\partial f}{\partial s}\right)_0 ca\cos\beta. \tag{30}$$

† See Chapter 23.

Using the abbreviation

$$a_h = \frac{A_h}{2\left(\frac{\partial f}{\partial s}\right)_0}, \text{ etc.,} \tag{31}$$

then, if (27) through (30) are substituted into (24), the result is

$$\begin{aligned} a^2 \quad \epsilon_x + ab \cos \gamma \, \epsilon_y + ac \cos \beta \, \epsilon_z + a_h &= 0, \\ ab \cos \gamma \, \epsilon_x + b^2 \quad \epsilon_y + bc \cos \gamma \, \epsilon_z + a_k &= 0, \\ ac \cos \beta \, \epsilon_x + bc \cos \alpha \, \epsilon_y + c^2 \quad \epsilon_z + a_l &= 0. \end{aligned} \tag{32}$$

This complexity is required only for triclinic axes. For orthogonal systems it reduces to

$$\begin{aligned} x &= -\frac{a_h}{a^2}, \\ y &= -\frac{a_k}{b^2}, \\ z &= -\frac{a_l}{c^2}, \end{aligned} \tag{33}$$

and even for the monoclinic system the solution of (32) reduces to

$$\begin{aligned} x &= \frac{a_l \, a \cos \beta - a_h \, c}{a^2 \, c \sin^2 \beta}, \\ y &= -\frac{a_k}{b^2}, \\ z &= \frac{a_h \, c \cos \beta - a_l \, a}{a^2 \, c \sin^2 \beta}. \end{aligned} \tag{34}$$

Booth's method of differential synthesis can also be extended to noncentric structures, for which the correct phase-angle errors must be sought as well as the correct *xyz*'s.

Difference syntheses

In a foregoing section, Bunn's error synthesis was discussed. This is related to the full difference synthesis outlined in (8), (9), and (10). In the earlier stages of a structure analysis, the phases of the F_o's are not known, so that only those terms $(F_o - F_c)$ can be used in the error synthesis for which F_o is zero or very small. But in the later stages of the refinement of a structure the phases of most or all of the F_o's becomes known and the full synthesis can be carried out. The full *difference synthesis*, also known as the $(F_o - F_c)$ *synthesis* or as the $\rho_o - \rho_c$ *synthesis*, was first suggested as a device for refinement by Booth[38] and its properties were exploited by Cochran.[39, 40]

The difference synthesis has a fundamental property which makes it useful in refinement: When the proposed model exactly matches the actual crystal structure, the difference map is characterized by a flat topography whose only features are minor and random undulations caused by errors of observation. If the proposed structure deviates in any way from the actual structure, the difference map reveals the nature of the deviation by a topography characteristic of the deviation. It follows that the proposed structure ought to be modified in such a way as to produce a nearly featureless difference map.

In addition to this characteristic property, difference maps have two other advantages. When the proposed and actual structures are nearly the same, their series-termination errors are substantially identical. This error, therefore, vanishes on subtraction, so that difference syntheses are substantially free from series-termination errors. For this reason, accurate values of the atomic coordinates can be derived from them. From a technical point of view, a difference synthesis is one of the easiest to compute because the Fourier coefficients are all differences, and therefore small numbers. The synthesis can therefore be computed rapidly, and simple analogue machines are sufficiently accurate for the small Fourier coefficients involved.

Correction of errors in location. If an atom of the proposed structure is in almost the correct position, but its location still requires a small correction, ϵ, then the map $\rho_o - \rho_c$ contains a characteristic feature in the neighborhood of the proposed location. The nature and origin of this feature should be clear from Fig. 6. The point representing the proposed location lies on a sharp gradient of the difference map. The direction of ϵ is normal to the contours and *up* the gradient.

The magnitude of ϵ can be determined from the geometry involved. The simplest way to determine ϵ is to prepare duplicate profiles of the atom as in Fig. 6*A* (for example, from a map of the projected electron density, if available), and then displace the two until the observed gradient is achieved.

The magnitude of ϵ can also be estimated, with certain assumptions, by computation.[39] Near the center of an atom the electron density at a distance r from the center is given very closely[39] by

$$\rho_{(r)} = \rho_{(0)} \, e^{-pr^2} \tag{35}$$

where $\rho_{(0)}$ is the maximum density. For very small values of x, e^x

can be approximated by $1+x$, so that, for small values of r,

$$e^{-pr^2} \approx 1 - pr^2. \tag{36}$$

A good approximation for (35), provided r is small, is therefore

$$\rho_{(r)} \approx \rho_{(0)}(1 - pr^2). \tag{37}$$

Now, the gradient of $\rho_o - \rho_c$ is the gradient of ρ_o less the gradient of ρ_c. At the peak of ρ_c, its gradient is zero, so that

$$\frac{d(\rho_o - \rho_c)}{dr} = \frac{d\rho_o}{dr} \tag{38}$$

$$= \frac{d(\rho_{o(0)} - \rho_{o(0)}\, pr^2)}{dr} \tag{39}$$

$$= 2\rho_{o(0)}\, pr. \tag{40}$$

Therefore

$$\epsilon = r = \frac{\dfrac{d(\rho_o - \rho_c)}{dr}}{2\rho_{o(0)}\, p}. \tag{41}$$

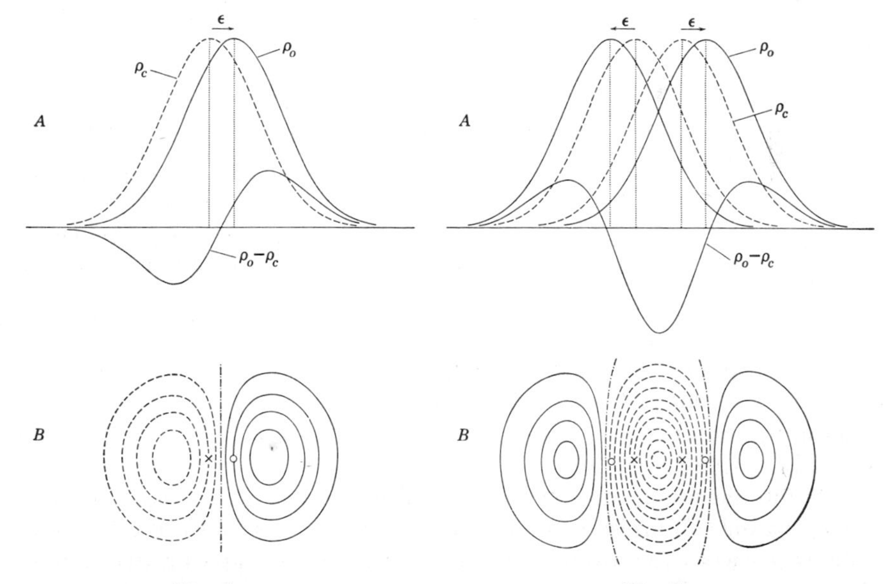

Fig 6.
The result of a small error in atom location.
A. Section of ρ_o, ρ_c and $\rho_o - \rho_c$, parallel to the direction of error.
B. Appearance of contours on $\rho_o - \rho_c$.

Fig. 7.
The result of placing two equal atoms somewhat too close together in a proposed structure.
A. Section of ρ_o, ρ_c and $\rho_o - \rho_c$, parallel to the direction of error.
B. Appearance of contours on $\rho_o - \rho_c$.

The value of ρ_o is given approximately by

$$\rho_o = Z\left(\frac{p}{\pi}\right)^{3/2}, \tag{42}$$

where p is about 5.0. It is evident from (41) that the required correction in coordinates is proportional to the gradient, and inversely proportional to the density of the atom, or, according to (42), inversely proportional to the atomic number of the atom.

It should be observed that this derivation is independent of the scale of ρ_o, since ρ_c enters the derivation only in that the gradient at its peak is

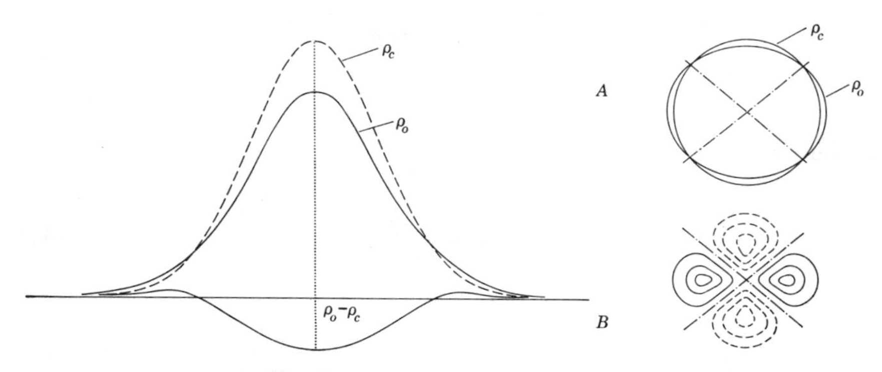

Fig. 8.
The effect on $\rho_o - \rho_c$ if the temperature correction in ρ_c has been underestimated.

Fig. 9.
The effect of assuming isotropic thermal vibration when it is actually anisotropic.
A. The same contour on ρ_c (isotropic) and ρ_o (anisotropic).
B. Contours on $\rho_o - \rho_c$.

zero. For this reason, the final location of each atom should be at or near zero gradient of the difference map. If this happens to occur on a small hill or depression, this has a further significance which will be discussed in the next section.

The difference map is especially useful in indicating errors in the projected distance of unresolved atoms. Figure 7 shows that if the assumed distance between the atoms is too close, the pair occupies a depression. If the assumed distance between atoms is too large, the reverse is true, and the pair occupies a hill in the difference map.

Temperature correction. Temperature motion has the effect of spreading the electrons of an atom over a larger volume. Accordingly it reduces the peak intensity and broadens the base of a peak on the electron-density map. After the position of an atom has been corrected, then if the temperature correction has been underestimated, $\rho_o < \rho_c$ at the peak, Fig. 8, and the atom location appears in a difference synthesis in a depression surrounded by a raised ring. On the other hand, if the temperature correction has been overestimated, then $\rho_o > \rho_c$ at the peak, and the atom location appears on a small hill surrounded by a ring-shaped depression.

Fig. 10.

Example of the use of the difference map; diglycine hydrochloride.

A. Electron-density projection $\rho_{(xy)}$.

B. Corresponding difference map $\rho_o - \rho_c$. The Cl atom shows the anomaly characteristic of anisotropic thermal motion, and the anomaly at the C_4 and N_2 positions shows that they had been placed too close together. Some atoms are still on sharp gradients.

The thermal motion has been assumed to be isotropic in the discussion just given. If it is anisotropic, then the observed electron distribution is drawn out in the direction of maximum vibration, and narrowed in a direction at right angles to it, as indicated in Fig. 9*A*. As a result, a characteristic pattern, shown in Fig. 9*B*, occurs in the difference map. The line drawn between the positive anomalies in the difference map marks out the direction of greatest thermal motion.

Examples of some of the effects just discussed can be seen in Fig. 10. The Cl atom at the left border of Fig. 10*A* shows the anomaly characteristic of anisotropic thermal motion, with the maximum amplitude directed somewhat west of north. Just to the left of the center are shown two atoms, C_4 and N_2, which, according to Fig. 7, are too close together in the proposed model. In several other instances atoms are on sharp gradients and should be moved up gradient.

From the quantitative aspects of the temperature anomaly in the difference map it is possible to compute an improved temperature correction.

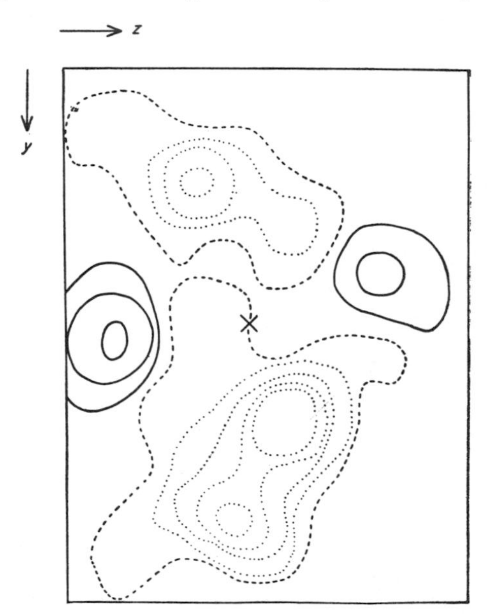

Fig. 11.
Anomaly on difference map due to improper anisotropic absorption allowance; muscarine iodide. The specimen was approximately elliptical.
(After Jellinek.[47])

If the temperature anomaly is isotropic, as in Fig. 8, this amounts to finding a better value of the temperature-corrected scattering factor

$$f_c = f_c' \, e^{-(B \sin^2 \theta)/\lambda^2}, \tag{43}$$

by adjusting B so that it will increase or reduce the ρ_c hill in Fig. 8. If the temperature anomaly is anisotropic, as in Fig. 9, then B varies with the angle ϕ between the direction of greatest thermal motion and the direction of the reciprocal-lattice point. Hughes[48] suggests that, to take care of this angular variation, (43) be modified to

$$f_c = f_c' \, e^{-(A+B \cos^2 \phi)(\sin^2 \theta)/\lambda^2}. \tag{44}$$

The details of improving the temperature correction from difference maps is discussed by Cochran[39, 40] and others.[46]

Observational errors. Jellinek[47] has called attention to the fact that certain observational errors have characteristic appearances on the difference map. Absorption correction follows a law rather close to that of temperature correction. If an absorption error remains uncorrected in the observational data, its appearance on a difference map resembles that of a residual temperature error. If a crystal specimen is elliptical, but its absorption is approximated by assuming that it is cylindrical, the heavy atoms show an absorption anomaly such as is illustrated in Fig. 11. This resembles a temperature anomaly, but has a greater extension.

"Extinction" has the effect of reducing the intensity of strong reflections of small sin θ value. Roughly, this is what too low a temperature correction does. "Extinction" therefore gives an effect crudely approximating Fig. 8, that is, the atom location appears to be in a hole surrounded by a small positive ring.

Least squares

The refinement of a structure calls for making small variations in the parameters of the several atoms of the cell, so that the collection of computed F's fit the collection of observed F's. Unfortunately the observed F's are subject to errors of observation, so that refinement consists in finding the most acceptable fit of the set of F_c's with the set of F_o's. This process thus comes into the general realm of finding the values of a set of variables which best satisfy a set of somewhat inconsistent observations. This matter was first treated by Legendre,† who proposed the principle that the most acceptable values of the variables were such as to make the sum of the squares of the errors a minimum. The principle was later shown to be consistent with several different sets of acceptable axioms.§

The general nature of least-squares determination of appropriate variables is as follows: Suppose an observable quantity, q, is a linear function of a set of variables $x, y, z \cdots$,

$$q = ax + by + cz + \cdots . \tag{45}$$

Suppose, further, that other somewhat similar experiments can be performed in which the constants, $a, b, c \cdots$ are different but known. If there were no error in observation of q, then if there are n variables $x, y, z \cdots$, it would be possible to determine these from n equations like (45), that is, from n experiments. But because of different errors of observation, E, associated with each q, the situation is more complicated. When the number of observations, m, is greater than the number of variables, n, then the equations taken n at a time do not yield quite the same solution. According to Legendre's principle, the most acceptable

† A. M. Legendre. *Nouvelles méthodes pour la détermination des orbites des comètes.* (Courcier, Paris, 1806), 72.

§ E. T. Whittaker and G. Robinson. *The calculus of observations.* 4th Ed. (Blackie and Son, London and Glasgow, 1949) Chapter 9.

solution is the one which makes the sum of the squares of the errors, E, a minimum. This has the following consequences:

If E is the error associated with q of (45), then

$$q + E = ax+by+cz + \cdots, \tag{46}$$

so that the error in each observation is

$$E = ax+by+cz + \cdots - q. \tag{47}$$

The set of m such equations is

$$\begin{aligned} E_1 &= a_1 x + b_1 y + c_1 z + \cdots - q_1, \\ E_2 &= a_2 x + b_2 y + c_2 z + \cdots - q_2, \\ E_3 &= a_3 x + b_3 y + c_3 z + \cdots - q_3, \\ &\cdot \\ &\cdot \\ &\cdot \\ E_m &= a_m x + b_m y + c_m z + \cdots - q_m. \end{aligned} \tag{48}$$

According to Legendre's principle

$$E_1{}^2 + E_2{}^2 + E_3{}^2 + \cdots E_m{}^2 = \sum_j E_j{}^2 \tag{49}$$

must be made a minimum. If the equations in (48) are added, this sum is seen to be

$$\sum_j E_j{}^2 = \sum_j (a_j x + b_j y + c_j z + \cdots - q_j)^2. \tag{50}$$

This is a minimum when its partial derivatives with respect to x, y, $z \cdots$ vanish, that is, when

$$\begin{aligned} \frac{\partial \sum_j E_j{}^2}{\partial x} &= 2 \sum_j (a_j x + b_j y + c_j z + \cdots - q_j) a_j = 0, \\ \frac{\partial \sum_j E_j{}^2}{\partial y} &= 2 \sum_j (a_j x + b_j y + c_j z + \cdots - q_j) b_j = 0, \\ \frac{\partial \sum_j E_j{}^2}{\partial x} &= 2 \sum_j (a_j x + b_j y + c_j z + \cdots - q_j) c_j = 0 \\ &\cdot \qquad\qquad \cdot \qquad\qquad \cdot \\ &\cdot \qquad\qquad \cdot \qquad\qquad \cdot \\ &\cdot \qquad\qquad \cdot \qquad\qquad \cdot \end{aligned} \tag{51}$$

The middle and right sides of (51) can be rewritten in the form (omitting the j under Σ)

$$\begin{aligned}
(\Sigma a_j^2\ )x + (\Sigma a_j\, b_j)y + (\Sigma a_j\, c_j)z + \cdots &= \Sigma a_j\, q_j, \\
(\Sigma b_j\, a_j)x + (\Sigma\, b_j^2\ )y + (\Sigma b_j\, c_j)z + \cdots &= \Sigma b_j\, q_j, \\
(\Sigma c_j\, a_j)x + (\Sigma c_j\, b_j)y + (\Sigma c_j^2\ )z + \cdots &= \Sigma c_j\, q_j \qquad (52) \\
\vdots \qquad \vdots \qquad \vdots \qquad &\quad \vdots
\end{aligned}$$

These are known as the *normal equations*. They comprise a system of n equations in n unknowns, and from them the desired values of x, y, $z \cdots$, which best satisfy Legendre's principle, can be determined.

It may be that in making observations like (45), some observations may be considered more trustworthy than others. The various observations q_j may be assigned weights w_j, which indicate relative estimates of their reliabilities. When this is done, both sides of (46) are multiplied by w_j. When followed through, this replaces each quantity in (52) by its product with w_j.

This scheme can be applied in the following way to finding the values of the coordinates of the atoms in a structure which best fit the set of observed F's: Each structure factor is computed by a relation which is, in general,

$$F_c = \sum_r f_r\, e^{i2\pi(hx_r+ky_r+lz_r)}. \qquad (53)$$

Here the variables are exponentials in x, y, and z, and these do not supply the desired linear equations similar to (48). Linear relations can, however, be devised by using the first two terms in Taylor's series, as noted in (17). In this application the function "f" is treated as follows: Each atom in the proposed structure is in a slightly incorrect location xyz. The correct location can be found by adding small corrections, ϵ, to xyz, giving $x+\epsilon_x\ y+\epsilon_y\ z+\epsilon_z$ as the correct locations. In (17),

$$f(x+\epsilon_x\ y+\epsilon_y\ z+\epsilon_z) \rightarrow F_o$$

while

$$f(xyz) \rightarrow F_c. \qquad (54)$$

Then, applying (17), one obtains

$$\begin{aligned}
\Delta F &= F_o - F_c \\
&= F_c + \sum_r \left(\epsilon_x \frac{\partial F_c}{\partial x_r} + \epsilon_y \frac{\partial F_c}{\partial y_r} + \epsilon_z \frac{\partial F_c}{\partial z_r} \right) - F_c. \qquad (55)
\end{aligned}$$

Therefore

$$\Delta F = \sum_r \left(\epsilon_x \frac{\partial F_c}{\partial x_r} + \epsilon_y \frac{\partial F_c}{\partial y_r} + \epsilon_z \frac{\partial F_c}{\partial z_r} \right). \tag{56}$$

This summation is taken over the R atoms of the structure. For each observed reflection there exists an observational equation like (56). When the observational error, E, is added to each equation like (56), the set of equations can be recast into the form of (48), from which the normal equations like (52) can be derived.

If there are R atoms in the crystal, there are $3R$ unknowns. Therefore the normalizing of the observational equations is tedious, and the solution of the normal equation like (52) is a tedious and complicated process. It has been found[16, 53] in practice, however, that if a large number of equations are used, the off-diagonal terms of the normal equations (52) are small, and, to first approximation, can be neglected. If so, the solutions (including a weighting factor which estimates the reliability of each observation) reduce to

$$\epsilon_{x_r} = \frac{\displaystyle\sum_{i=1}^{m} w \left(\frac{\partial F_i}{\partial x} \right)_r \Delta F}{\displaystyle\sum_{i=1}^{m} w \left(\frac{\partial F_i}{\partial x} \right)_r^2}. \tag{57}$$

This approximation is not valid if the atoms overlap. It is therefore valid for three-dimensional refinement, but not usually valid for refinement of projections.

One of the most tedious parts of the least-squares procedure is the derivation of the exact form of the partial differentials in, for example, (57). These must be derived from the particular form of (53) which pertains to the space group of the crystal. This involves looking up the trigonometric form of the structure factor in suitable tables, and differentiating it.

Refinement by least squares has a number of advantages. It is free from the series-termination errors which characterize Fourier methods. It is also possible to use less than all the F's in the refining process, which is impossible with any Fourier method. Thus any F's of doubtful nature (for example, those suspected of being affected by "extinction") can be omitted. This also permits a modified procedure of refinement. McDonald and Beevers[53] used only the F's with greatest discrepancies in the initial stages of refinement. These were found to give corrections which were too large, but the refinement process converged rapidly.

It is also possible to include a temperature factor and scale factor in the least-squares refinement process. In its simplest form, each atom is assumed to have the same isotropic temperature correction of the form $e^{-(B \sin^2 \theta)/\lambda^2}$. A somewhat more elaborate device is to include in (56) an isotropic temperature factor which is different for each atom. Finally, the thermal motion of each atom can be assumed to be not only different but anisotropic. In this case a temperature correction of the form[100]

$$t = e^{-(b_{11}h^2+b_{22}k^2+b_{33}l^2+b_{12}hk+b_{23}kl+b_{31}lh)} \tag{58}$$

is applied to each atom's contribution to F_c of (56).

When a weighting function, w, is used, the least-squares refinement minimizes $\Sigma w(|F_o| - |F_c|)^2$. Cochran[8] showed that Fourier synthesis minimizes the same residual provided the peaks are resolved and $w = 1/f$. Cruickshank[52] investigated the relation when overlap is allowed, and showed that, for general weighting function w, the differential synthesis minimizes the same function and therefore gives the same results.

Since high-speed digital computers have come into common use, the refinement of a structure by the method of least squares is no longer a formidable undertaking. A number of programs for high-speed digital computers have been written for the refinement of crystal structures by least squares,[59, 60, 63] and at least one is available commercially.[59] In other words, when a structure has been determined, the tedious, time-consuming details of its refinement can be ignored, since the service can be purchased.

Steepest descents

Booth[64] called attention to the possibility of refining structures by the method of *steepest descents*. The theory and procedure has been developed by Booth,[65, 71] Vand,[66, 70, 73] Cochran,[69] and Qurashi.[72] The method has been actually used for refinement by Qurashi,[74] Lomer,[75] Steeple,[76] and others.

The general idea of the method of steepest descents is as follows. In refinement an attempt is made to minimize a residual such as

$$\begin{aligned} R &= \big|\, |F_o| - |F_c| \,\big|, \\ R' &= (F_o - F_c)^2, \\ R'' &= (|F_o| - |F_c|)^2, \\ R''' &= (|F_o|^2 - |F_c|^2)^2. \end{aligned} \tag{59}$$

The process of minimization can be illustrated by the simplified problem in which F is determined by two variables x and y. This is equivalent to saying that the $\rho_{(xy)}$ projection of the crystal structure contains only one atom with undetermined coordinates xy. The situation is graphically illustrated in Fig. 12. The variation of one of the R's of (59) with x and

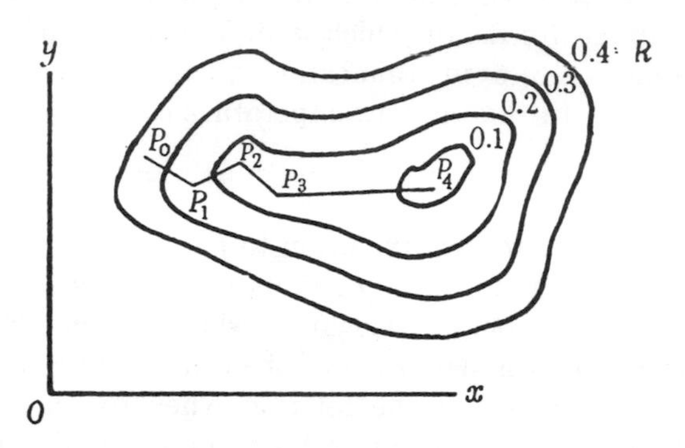

Fig. 12.
The course of refinement by the method of steepest descents. In this simplified example, the value of a residual, R, is contoured as a function of the coordinates x and y of the one variable atom of the structure.
(After Booth.[71])

y can be represented by the contours. The details of the map are unknown when refinement is started. The atom is initially placed at P_0, and the R is observed to be high, about 0.35. The method of steepest descents requires that an improvement in R can be obtained by moving the atom normal to the contours, in a direction down gradient. The movement cannot be indefinite, for this is only a local gradient, and does not lead directly toward the lowest contour of R. A short shift of parameter, therefore, is made down gradient to P_1; then another shift is made directly down gradient at P_2. This process is repeated until, eventually, the lowest level of R is attained at P_4.

This method would not appear to be practical when a large number of atoms, N, and their $3N$ coordinates $x_1\, y_1\, z_1,\ x_2\, y_2\, z_2,\ x_2\, y_2\, z_2 \cdots x_N\, y_N\, z_N$ are involved. Actually there are mathematical methods of handling the situation which are satisfactory. To see this, note that R is a function of these $3N$ coordinates:

$$R = \phi(x_{1,1}\, x_{1,2}\, x_{1,3} \cdots x_{r,1}\, x_{r,2}\, x_{r,3} \cdots x_{N,1}\, x_{N,2}\, x_{N,3}) \tag{60}$$

$$= \phi(x_{r,j}). \tag{61}$$

The locus of $R =$ constant can be regarded as a surface in $3N$-dimensional space. The method of steepest descents requires a shift of coordinates at right angles to this surface. The direction cosines of the normal n, to this surface are

$$\cos\,(n,x_{r,j}) = \frac{\dfrac{\partial R}{\partial x_{rj}}}{\dfrac{\partial R}{\partial n}}, \qquad \begin{pmatrix} r = 1 \cdot \cdot \cdot N \\ j = 1, 2, 3 \end{pmatrix} \tag{62}$$

where
$$\frac{\partial R}{\partial n} = |\text{grad } \phi| = \left\{ \sum_{r,j} \left(\frac{\partial \phi}{\partial x_{rj}} \right)^2 \right\}^{1/2}. \tag{63}$$

A desired shift $\epsilon_{x,r}$, in the x direction of the rth atom should be proportional to the gradient $\partial R/\partial x_r$ at the original location:

$$\epsilon_{x_r} = K \frac{\partial R}{\partial x_{rj}} \tag{64}$$

$$= -2K \sum (|F_o| - |F_c|) \frac{\partial F_c}{\partial x_{rj}}. \tag{65}$$

More generally, let ${}_0R$ be the initial value of R for coordinate x_{rj}; let R' be the value of R for corrected coordinates $x_{rj}+\epsilon_n$. Then, using Taylor's expansion,

$$R' = {}_0R + \epsilon_n \frac{\partial \phi}{\partial n} + \frac{1}{2!} \epsilon_n{}^2 \frac{\partial^2 \phi}{\partial n^2} \cdot \cdot \cdot . \tag{66}$$

If higher differentials in (66) are neglected, then to first approximation the value of ϵ_n which makes R' zero is

$$\epsilon_n = - \frac{{}_0R}{\dfrac{\partial \phi}{\partial n}} \tag{67}$$

$$= - \frac{{}_0R}{\left\{ \sum_{r,j} \left(\dfrac{\partial \phi}{\partial x_{rj}} \right)^2 \right\}^2}. \tag{68}$$

The use of (67) or (68) suffers from the disadvantage that the gradient continues at the same value as at the starting point. This point has been studied in detail by Qurashi,[72] who recommends an improved correction:

$$\epsilon_{x_{rj}} = \frac{\sum_{hkl} w^2 (F_o - F_c) \dfrac{\partial F_c}{\partial x_{rj}}}{\sum_{hkl} w^2 \left(\dfrac{\partial F_c}{\partial x_{rj}} \right)^2}, \tag{69}$$

where w is a weight assigned to an observation. Qurashi has pointed out that the solutions are equivalent to those of the least-squares method when the non-diagonal terms are ignored.

Assessment of accuracy

After the structure has been refined it is desirable to know how large the errors are likely to be which still remain. These errors have their origins in other errors, the nature of which will be considered presently. To define them, the usual methods of statistics are used.

Parameters of a set of measurements. When a measurement is repeated a number of times, the same value is not obtained, but rather a

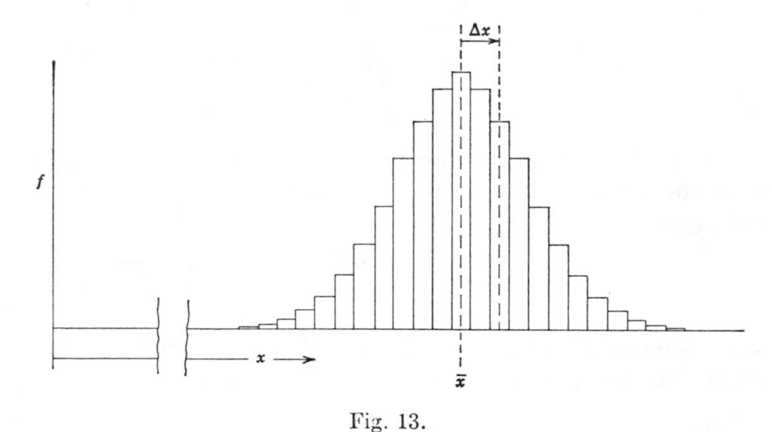

Fig. 13.

set of values which diverge somewhat from one another. There are standard ways of deriving a "best" value from this set, and of characterizing the amount of variation or spread of the measurement from this "best" value.

Suppose n measurements of a quantity x are made. Ordinarily certain measurements of x recur, as illustrated in the *histogram*, Fig. 13. If the number (or frequency) of recurrence of measurement x_i is f_i, then the *mean* value, $\bar{x}$, of the measurements is defined by

$$\bar{x} = \frac{1}{n} \sum f_i x_i \qquad (\Sigma f_i = n). \qquad (70)$$

An alternative form of (70) is obtained by defining the probability, p_i, of making a measurement x_i, as

$$p_i = \frac{f_i}{n} \qquad (\Sigma p_i = 1). \qquad (71)$$

Substituting this in (70) gives

$$\bar{x} = \sum_i p_i\, x_i. \tag{72}$$

The spread of the measurements from the mean value is characterized by the *variance*, σ^2, of the set of values. This is defined as

$$\sigma^2 = \frac{1}{n}\sum_i f_i(x_i - \bar{x})^2 = \frac{1}{n}\sum_i f_i(\Delta x)^2 \tag{73}$$

$$= \sum_i p_i(x_i - \bar{x})^2 = \sum_i p_i(\Delta x)^2. \tag{74}$$

The *standard deviation*, σ, is the square root of the variance. This is a kind of weighted root-mean-square deviation from the mean value.

Sources of errors. The errors which remain in the refined structure depend upon other errors which descend from two different sources. These are the errors in the original measurements and the errors in the theory used to interpret them.

The errors in measurement are those inherent in measuring the original x-ray diffraction maxima. These include errors in measurement of the positions of the maxima, and errors in the measurement of the intensities.

Errors in the measurements of the positions of the maxima lead to errors in the cell geometry. Cell-edge errors can be reduced to the order of 0.001 Å, which is less than the other errors to be corrected, by using simple precautions.†

Greater errors are ordinarily made in the measurement of intensities. While these errors must be estimated in each instance, photographic intensity determination may be expected to have standard deviations in the neighborhood of 12% to 25%, whereas counter measurements tend to be much lower, 1% to 2% in favorable cases. These lead to standard deviations of about 4% for $|F|$'s obtained from photographic measurements, but only about 1% for $|F|$'s obtained from counter measurement.

The estimation of errors in the $|F_o|$'s has been considered by Ibers.[90] In three-dimensional structure determinations the reflections measured can be classed in four categories:

† See M. J. Buerger. *X-ray crystallography.* (John Wiley and Sons, New York, 1942) Chapters 20 and 21.

1. Reflections measured more than once,[†] say from Weissenberg photographs for both *a*-axis and *b*-axis rotations, or from different precession photographs corresponding to two different dial settings, as noted in Chapter 6.
2. Reflections measured only once.
3. Reflections too weak to be observed.
4. Reflections missing because of space-group extinctions.

Even in category 1, the number of measurements is small indeed, but the theory of very small samples can be used. For such cases the standard deviation can be estimated from the range $y_{max} - y_{min}$ of measured values y:

$$\sigma(y) = c|y_{max} - y_{min}|. \tag{75}$$

Here c is a function of the number of samples n:

n	c
2	0.89
3	0.59
4	0.49

Thus, for the usual two observations,

$$\bar{F} = \tfrac{1}{2}(F_1 + F_2), \tag{76}$$

and

$$\sigma(F_o) = 0.89|F_1 - F_2|. \tag{77}$$

This assumes that F_1 and F_2 are equally reliable. If one value is regarded as considerably less reliable than the other, it should be discarded, thus placing the reflection in category 2.

To handle category 3, a plot of $\sigma(F_o)$ against $|F_o|$ is first made for the reflections of category 1. Then it is assumed that category 1 follows the same scheme, and $\sigma(F_o)$ is estimated as a function of $|F_o|$ from the plot.

In category 3, use is made of Wilson's statistical results.[§] Assuming that the minimum observable value of $|F|$, called $|F_{min}|$, is much smaller than the root mean square value of the $|F|$'s, then

[†] Ibers' remarks assume photographic methods, by which certain reflections are measured twice with different settings of the crystal. Using modern counter methods (Chapter 6), all reflections are measured with one setting. Ibers' theory can be applied, however, to the standard deviations of measurements of symmetrically equivalent reflections.

[§] A. J. C. Wilson. *The probability distribution of x-ray intensities.* Acta Cryst. **2** (1949) 318–321.

For centric crystals:

$$\bar{F}_{\text{unobs}} \approx \tfrac{1}{2}F_{\text{min}},$$

$$\sigma(F_{\text{unobs}}) \approx \frac{1}{\sqrt{12}} F_{\text{min}}; \qquad \text{(for the same region of } \theta \text{ on the film)} \tag{78}$$

For acentric crystals:

$$\bar{F}_{\text{unobs}} \approx \tfrac{2}{3}F_{\text{min}},$$

$$\sigma(F_{\text{unobs}}) \approx \frac{1}{\sqrt{18}} F_{\text{min}}.$$

In addition to these random errors of measurement, there are systematic errors in attempting to reduce the intensities to $|F|$'s. These are the errors in allowing for physical factors affecting intensities, especially absorption and "extinction," discussed in Chapter 8. The allowance for absorption is difficult, but can often be closely approximated by the use of a rod-shaped specimen. There appears to be no really satisfactory method of allowing for "extinction."

Estimation of errors in electron density. When the absolute value of the electron density is important, its standard deviation can be estimated with the aid of the following argument[83]: The standard deviation of a sum y of weighted deviations is

$$\sigma(y) = \left\{\sum w^2 \sigma^2\right\}^{1/2}. \tag{79}$$

Now the electron density is a sum:

$$\rho_{(xyz)} = \frac{1}{V} \sum_h \sum_k \sum_l F_{hkl}\, e^{-i2\pi(hx+ky+lz)}. \tag{80}$$

This can also be expressed in such a way that the phases of the F_{hkl}'s are incorporated with the phases corresponding to the exponentials, as follows:

$$\rho_{(xyz)} = \frac{1}{V} \sum_h \sum_k \sum_l |F_{hkl}| \cos 2\pi(hx+ky+lz-\phi). \tag{81}$$

According to (79) the standard deviation, due to deviations in the F's, is

$$\sigma(\rho_{(xyz)}) = \frac{1}{V} \left\{\sum_h \sum_k \sum_l \sigma^2(F_{hkl}) \cos^2 2\pi(hx+ky+lz-\phi)\right\}^{1/2}. \tag{82}$$

This varies with the coordinate of the location, xyz. But if the number of terms in the summation is large, the sum over the trigonometric terms in the last part of (82) is constant. The specific value of this constant can be evaluated by noting that, if $hkl = 000$ is excluded, the average value of the cosine is zero. For the particular reflection 000 it is unity, which is the only residue. Thus, for a reasonably large set of F's, (82) reduces to

$$\sigma(\rho) = \frac{1}{V}\left\{\sum_h \sum_k \sum_l \sigma^2(F_{hkl})\right\}^{1/2}. \tag{83}$$

When applied to acentric structures this is too small, and the right of (83) should be multiplied by the n of the n shift, noted earlier in this chapter under *Procedure for acentric crystals.*

If Ibers' values in (78) are used in (83) for $\sigma(F_{hkl})$, the result may be somewhat low since only random errors in the measurement of F are included. Cruickshank[83] takes the view that the only estimate of the deviation of F available is

$$\sigma(F_{hkl}) = \sigma(\Delta F_{hkl}) = \Delta F_{hkl} = F_o - F_c, \tag{84}$$

and that this estimate contains all sources of deviations in both F_o and F_c. If this is used, the standard deviation of the electron density can be derived with the aid of

$$\sigma(\rho) = \frac{1}{V}\left\{\sum_h \sum_k \sum_l (\Delta F_{hkl})^2\right\}^{1/2}. \tag{85}$$

A further alternative is offered by Lipson and Cochran,[84] who point out that it can be shown that

$$\frac{1}{V}\left(\sum_h \sum_k \sum_l (\Delta F_{hkl})^2\right)^{1/2} = \{\overline{(\rho_o - \rho_c)^2}\}^{1/2}, \tag{86}$$

that is, that the average value of the difference synthesis is the same as the standard deviation, (83). This can be taken as the average over the whole cell, in which case it may lead to a slight overestimate of the error. Alternatively, it may be taken over those points (between atoms) where the density is expected to be zero. In this case the error may be slightly underestimated.

Estimation of errors in coordinates. The errors in coordinates can be estimated by making use of Booth's equation (24). Consider first, crystals with orthogonal axes. In these cases the cross terms like

A_{hk} vanish, leaving

$$\epsilon_x = -\frac{A_h}{A_{hh}},$$
$$\epsilon_y = -\frac{A_k}{A_{kk}}, \quad (87)$$
$$\epsilon_z = -\frac{A_l}{A_{ll}}.$$

Now, since $\sigma(\epsilon_x) = \sigma(x)$, the standard deviations in (87) are

$$\sigma(x) = \frac{\sigma(A_h)}{A_{hh}},$$
$$\sigma(y) = \frac{\sigma(A_k)}{A_{kk}}, \quad (88)$$
$$\sigma(z) = \frac{\sigma(A_l)}{A_{ll}}.$$

The terms A_{hh}, etc., are second derivatives, and therefore curvatures of the electron-density function. Relations (88) require an evaluation of the deviation in first derivatives, like $\sigma(A_h)$. This is the first derivative of the electron-density function, and its value was given in (20). If the phases of the F's are treated as in (81), then (20) can be written as

$$A_h = \frac{\partial\rho}{\partial x} = -\frac{2\pi}{V}\sum_h\sum_k\sum_l h|F_{hkl}| \sin 2\pi(hx+ky+lz-\phi). \quad (89)$$

Following the reasoning between (81) and (82),

$$\sigma(A_h) = \frac{2\pi}{V}\left\{\sum_h\sum_k\sum_l h^2\,\sigma^2(F_{hkl}) \sin^2 2\pi(hx+ky+lz-\phi)\right\}^{1/2}, \quad (90)$$

which reduces in a manner similar to (82), to

$$\sigma(A_h) = \frac{2\pi}{V}\left\{\sum_h\sum_k\sum_l h^2\,\sigma^2(F_{hkl})\right\}^{1/2}. \quad (91)$$

If (84) is utilized, this becomes

$$\sigma(A_h) = \frac{2\pi}{V}\left\{\sum_h\sum_k\sum_l h^2\,(\Delta F_{hkl})^2\right\}^{1/2}. \quad (92)$$

If the complete and more exact form of Booth's equation (24) is required, then the deviations of second derivations like A_{hk} are needed. These can be derived from (22) following the arguments used above, and have the form

$$\sigma(A_{hk}) = \frac{4\pi^2}{abV}\left\{\sum_h \sum_k \sum_l hk\sigma^2(F)\right\}^{1/2}. \qquad (93)$$

The more general forms of (88) for monoclinic crystals is[83]

$$\begin{aligned}
\sigma(x) &= \frac{\{\sigma(A_h)^2 - \cos^2\beta\,\sigma(A_l)^2\}^{1/2}}{A_{hh}\sin^2\beta}, \\
\sigma(y) &= \frac{\sigma(A_k)}{A_{hh}}, \qquad (94) \\
\sigma(z) &= \frac{\{\sigma(A_l)^2 - \cos^2\beta\,\sigma(A_h)^2\}^{1/2}}{A_{hh}\sin^2\beta}.
\end{aligned}$$

Except for very exact work, and for certain special problems, the values provided by (88), which holds for orthogonal axes and well-resolved peaks in a centrosymmetrical structure, are usually sufficiently close for all practical purposes. Ordinarily, also, $\sigma(x) \simeq \sigma(y) \simeq \sigma(z)$. In this case the root-mean-square radial error of position is

$$\sigma(r) = \sqrt{3}\,\sigma(x). \qquad (95)$$

Literature

Measures of the correctness of a structure

[1] A. D. Booth. *An expression for following the process of refinement in x-ray structure analysis using Fourier series.* Phil. Mag. **36** (1945) 609–615.

[2] V. Vand. *Method of steepest descents: improved formula for x-ray analysis.* Nature **161** (1948) 600–601.

[3] V. Vand. *Method of steepest descents in x-ray analysis.* Nature **163** (1949) 129–130.

[4] A. J. C. Wilson. *Largest likely values for the reliability index.* Acta Cryst. **3** (1950) 397–398.

[5] V. Luzzati. *Traitement statistique des erreurs dans la determination des structures cristallines.* Acta Cryst. **5** (1952) 802–810.

[6] V. Luzzati. *Structure cristalline de $HNO_3 \cdot 3H_2O$. I. Resolution de la structure; utilization de la fonction de Patterson.* Acta Cryst. **6** (1953)152–157.

[7] Ephraim Segerman. *Scale-factor adjustment for optimum comparison of observed and calculated structure factors.* Acta Cryst. **11** (1958) 374–375.

Successive Fourier synthesis

[8] W. Cochran. *The Fourier method of crystal structure analysis.* Nature **161** (1948) 765.

[9] W. Cochran. *A critical examination of the Beevers-Lipson method of Fourier series summation.* Acta Cryst. **1** (1948) 54–56.

[10] R. W. James. *False detail in three-dimensional Fourier representations of crystal structures.* Acta Cryst. **1** (1948) 132–134.

[11] W. Cochran. *The Fourier method of crystal-structure analysis.* Acta Cryst. **1** (1948) 138–142.

[12] D. Crowfoot, C. W. Bunn, B. W. Rogers-Low, and A. Turner-Jones. *The x-ray crystallographic investigation of the structure of penicillin.* The chemistry of penicillin. (Princeton University Press, 1949) 310–367.

[13] G. S. Parry and G. J. Pitt. *The derivation of atomic co-ordinates from planar and linear Fourier syntheses.* Acta Cryst. **2** (1949) 145–147.

[14] D. W. J. Cruickshank. *The convergence of the least-squares and Fourier refinement methods.* Acta Cryst. **3** (1950) 10–13.

[15] Jerry Donohue. *The crystal structure of* DL-*Alanine. II. Revision of parameters by three-dimensional Fourier analysis.* J. Am. Chem. Soc. **72** (1950) 949–953.

[16] David P. Shoemaker, Jerry Donohue, Verner Schomaker, and Robert B. Corey. *The crystal structure of* L_s-*threonine.* J. Am. Chem. Soc. **72** (1950) 2328–2349.

[17] S. Furberg and O. Hassel. *Crystal structure of* "cis-*naphthodioxane*"–*bi*-1,3-*dioxacyclopentyl* (2). Acta Chem. Scand. **4** (1950) 1584–1596.

[18] V. Luzzati. *Sur la convergence et l'erreur dans les structures non-centrosymétriques.* Acta Cryst. **4** (1951) 367–369.

[19] Andrew D. Booth. *A "zero" synthesis for Fourier refinements.* Nature **171** (1953) 168.

[20] V. Luzzati. *Resolution d'une structure cristalline lorsque les positions d'une partie des atoms sont connues: Traitement statistique.* Acta Cryst. **6** (1953) 142–152.

[21] Mary R. Truter. *Refinement of a non-centrosymmetrical structure: Sodium nitrate.* Acta Cryst. **7** (1954) 73–77.

[22] Joshua Ladell and J. Lawrence Katz. *An analytic method for the determination of shape and location of Fourier peaks.* Acta Cryst. **7** (1954) 460–461.

[23] Helen D. Megaw. *Location of atomic centers in an electron-density synthesis.* Acta Cryst. **7** (1954) 771.

[24] Kenji Doi. *On the termination effects in the Fourier analysis of a crystal.* Proc. Japan Acad. **30** (1954) 180–182.

[25] M. M. Qurashi. *An analysis of the efficiency of convergence of different methods of structure determination. II. The method of Fourier synthesis: Centrosymmetric case.* Acta Cryst. **8** (1955) 445–456.

[26] D. M. Burns. *An analytic method of dealing with unresolved peaks in Fourier projections.* Acta Cryst. **8** (1955) 517–518.

[27] Kenzi Doi. *A new method for estimating the termination effect in Fourier analysis.* Mineral. Jour. (Japan) **1** (1955) 329–336.

[28] V. Vand and R. Pepinsky. *Weighting of Fourier series for improvement of*

efficiency of convergence in crystal analysis: Space group P1. Acta Cryst. **10** (1957) 563–567.

[29] D. M. Burns. *Interpolation errors in Fourier projections.* Acta Cryst. **11** (1958) 88–92.

[30] Richard E. Marsh. *A refinement of the crystal structure of glycine.* Acta Cryst. **11** (1958) 654–663.

[31] J. Lawrence Katz. *The validity of the elliptic paraboloid approximation for peaks on electron density maps.* Acta Cryst. **11** (1958) 749–750.

Differential synthesis

[32] A. D. Booth. *A differential Fourier method for refining atomic parameters in crystal structure analysis.* Trans. Faraday Soc. **42** (1946) 444–448.

[33] A. D. Booth. *The simultaneous differential refinement of co-ordinates and phase angles in x-ray Fourier synthesis.* Trans. Faraday Soc. **42** (1946) 617–619.

[34] A. D. Booth and F. J. Llewellyn. *The crystal structure of pentaerythritol tetranitrate.* J. Chem. Soc. (London) (1947) 837–846.

[35] A. D. Booth. *Fourier technique in x-ray organic structure analysis.* (Cambridge University Press, Cambridge, England, 1948) 46–52.

[36] F. R. Ahmed and D. W. J. Cruickshank. *A refinement of the crystal structure analyses of oxalic acid dihydrate.* Acta Cryst. **6** (1953) 385–392.

[37] Edgar L. Eichhorn. *On the structure of 4-nitropyridine-*N*-oxide.* Acta Cryst. **9** (1956) 787–793.

Difference syntheses

[38] Andrew D. Booth. *A new Fourier refinement technique.* Nature **161** (1948) 765–766.

[39] W. Cochran. *The structures of the pyrimidines and purines. V. The electron distribution in adenine hydrochloride.* Acta Cryst. **4** (1951) 81–92.

[40] W. Cochran. *Some properties of the* $(F_o - F_c)$*-synthesis.* Acta Cryst. **4** (1951) 408–411.

[41] D. M. Donaldson and J. Monteath Robertson. *The crystal and molecular structure of octamethylnaphthalene: A non-planar naphthalene derivative.* J. Chem. Soc. (London) (1953) 17–24.

[42] V. Luzzati. *Structure cristalline de* $HNO_3 \cdot 3H_2O$. *II. Localisation des atomes d'hydrogéne; discussion des erreurs; description de la structure.* Acta Cryst. **6** (1953) 157–164.

[43] Edgar L. Eichhorn. *On the use of slope and curvature maps in refinement of crystal structures.* Acta Cryst. **8** (1955) 63–64.

[44] E. Harnik. *A simplified computation technique for structure refinement by means of two-dimensional* $F_o - F_c$ *synthesis.* Acta Cryst. **8** (1955) 362–363.

[45] Theodor Hahn and M. J. Buerger. *The crystal structure of diglycine hydrochloride,* $2(C_2H_5O_2N) \cdot HCl$. Z. Krist. **108** (1957) 419–453.

[46] Yuen C. Leung, Richard E. Marsh, and Verner Schomaker. *The interpretation of difference maps.* Acta Cryst. **10** (1957) 650–652.

[47] F. Jellinek. *On some errors in x-ray analysis. I. Systematic errors in observed structure factors.* Acta Cryst. **11** (1958) 677–679.

Least squares

[48] E. W. Hughes. *The crystal structure of melamine.* J. Am. Chem. Soc. **63** (1941) 1737–1752.

[49] D. W. J. Cruickshank. *The convergence of the least-squares and Fourier refinement methods.* Acta Cryst. **3** (1950) 10–13.

[50] David P. Shoemaker, Jerry Donohue, Verner Schomaker, and Robert B. Corey. *The crystal structure of* L_s*-threonine.* J. Am. Chem. Soc. **72** (1950) 2328–2349.

[51] V. Luzzati. *Sur la convergence et l'erreur dans les structures non-centrosymétriques.* Acta Cryst. **4** (1951) 367–369.

[52] D. W. J. Cruickshank. *On the relation between Fourier and least-squares methods of structure determination.* Acta Cryst. **5** (1952) 511–518.

[53] T. R. R. McDonald and C. A. Beevers. *The crystal and molecular structure of* α*-glucose.* Acta Cryst. **5** (1952) 654–659.

[54] M. M Qurashi and V. Vand. *Weighting of the least-squares and steepest-descents methods in the initial stages of the crystal-structure determination.* Acta Cryst. **6** (1953) 341–349.

[55] M. M. Qurashi. *An analysis of the efficiency of convergence of different methods of structure determination. I. The methods of least squares and steepest descents: Centrosymmetric case.* Acta Cryst. **6** (1953) 577–588.

[56] D. M. Burns. *The symmetry of the normal equations for least squares.* Acta Cryst. **8** (1955) 63.

[57] Walter C. Hamilton. *On the treatment of unobserved reflexions in the least-squares adjustment of crystal structures.* Acta Cryst. **8** (1955) 185–186.

[58] S. C. Abrahams. *The crystal and molecular structure of orthorhombic sulfur.* Acta Cryst. **8** (1955) 661–671.

[59] P. H. Friedlander, W. Love, and D. Sayre. *Least-squares refinement at high speed.* Acta Cryst. **8** (1955) 732.

[60] L. R. Lavine and J. S. Rollett. *Crystal structure refinement by least-squares with the ElectroData computer.* Acta Cryst. **9** (1956) 269–273.

[61] James A. Ibers and Don T. Cromer. *The least-squares refinement of the crystal structure of* $Ce(IO_3)_4{\cdot}H_2O$*, ceric iodate monohydrate.* Acta Cryst. **11** (1958) 794–798.

[62] C. Aravindakshan. *An accurate redetermination of the structure of potassium chlorate,* $KClO_4$. Z. Krist. **111** (1958) 35–45.

[63] V. Vand and R. Pepinsky. *Modification of Sayres IBM 704 machine least-squares program NY XR1 for refinement of crystal structures, and some additional computational procedures.* Z. Krist. **111** (1958) 46–52.

Steepest descents

[64] Andrew D. Booth. *A new refinement technique for x-ray structure analysis.* J. Chem. Phys. **15** (1947) 415–416.

[65] A. D. Booth. *Application of the method of steepest descents to x-ray structure analysis.* Nature **160** (1947) 196.

[66] V. Vand. *Method of steepest descents: Improved formula for x-ray analysis.* Nature **161** (1948) 600–601.

[67] A. D. Booth. *A new Fourier refinement technique.* Nature **161** (1948) 765–766.

[68] G. J. Pitt. *The crystal structure of 4,6-dimethyl-2-hydroxypyramidine. I.* Acta Cryst. **1** (1948) 168–174.

[69] W. Cochran. *X-ray analysis and the method of steepest descents.* Acta Cryst. **1** (1948) 273.

[70] V. Vand. *Method of steepest descents in x-ray analysis.* Nature **163** (1949) 129.

[71] A. D. Booth. *The refinement of atomic parameters by the technique known in x-ray crystallography as 'the method of steepest descents.'* Proc. Roy. Soc. (London) (A) **197** (1949) 336–355.

[72] M. M. Qurashi. *Optimum conditions for convergence of steepest descents as applied to structure determination.* Acta Cryst. **2** (1949) 404–409.

[73] Vladimir Vand. *A simplified method of steepest descents.* Acta Cryst. **4**(1951) 285–286.

[74] I. G. Edmunds and M. M. Qurashi. *The structure of the ζ phase in the silver-zinc system.* Acta Cryst. **4** (1951) 417–425.

[75] T. R. Lomer. *The application of Booth's method of steepest descents to the determination of the structure of potassium caproate.* Acta Cryst. **5** (1952) 14–17.

[76] H. Steeple. *The crystal structure of the cadmium-magnesium alloy CdMg.* Acta Cryst. **5** (1952) 247–249.

Assessment of accuracy

[77] A. D. Booth. *Accuracy of atomic co-ordinates derived from Fourier synthesis.* Nature **156** (1945) 51.

[78] A. D. Booth. *The accuracy of atomic coordinates derived from Fourier series in x-ray structure analysis.* Proc. Roy. Soc. (London) (A) **188** (1946) 77–92.

[79] A. D. Booth. *Accuracy of atomic co-ordinates derived from x-ray data.* Nature **157** (1946) 517.

[80] Edward W. Hughes and William N. Lipscomb. *The crystal structure of methylammonium chloride.* J. Am. Chem. Soc. **68** (1946) 1970–1975.

[81] E. G. Cox and D. W. J. Cruickshank. *The accuracy of electron-density maps in x-ray structure analysis.* Acta Cryst. **1** (1948) 92–93.

[82] D. W. J. Cruickshank. *The accuracy of electron-density maps in x-ray analysis with special reference to dibenzyl.* Acta Cryst. **2** (1949) 65–82; corrections: **3** (1950) 72–73, **7** (1954) 519.

[83] D. W. J. Cruickshank. *The accuracy of atomic co-ordinates derived by least-squares or Fourier methods.* Acta Cryst. **2** (1949) 154–157.

[84] H. Lipson and W. Cochran. *The determination of crystal structures.* (G. Bell and Sons, London, 1953) 308.

[85] F. R. Ahmed and D. W. J. Cruickshank. *A refinement of the crystal structure analyses of oxalic acid dihydrate.* Acta Cryst. **6** (1953) 385–392.

[86] D. W. J. Cruickshank and A. P. Robertson. *The comparison of theoretical and experimental determinations of molecular structures, with applications to naphthalene and anthracene.* Acta Cryst. **6** (1953) 698–705.

[87] D. W. J. Cruickshank and J. S. Rollett. *Electron-density errors at special positions.* Acta Cryst. **6** (1953) 705–707.

[88] J. H. Bryden and J. D. McCullough. *The crystal structure of benzeneseleninic acid.* Acta Cryst. **7** (1954) 833–838.
[89] Howard T. Evans, Jr. and Mary E. Mrose. *A crystal chemical study of montroseite and paramontroseite.* Am. Mineralogist **40** (1955) 861–875.
[90] James A. Ibers. *Estimates of the standard deviations of the observed structure factors and of the electron density from intensity data.* Acta Cryst. **9** (1956) 652–654.
[91] Raymond B. Roof, Jr. *The crystal structure of ferric acetylacetonate.* Acta Cryst. **9** (1956) 781–786.

Anisotropic temperature factor

[92] D. R. Holmes. *Anisotropic temperature factors in* hkl *structure-factor calculations.* Acta Cryst. **6** (1953) 301–302.
[93] W. Cochran. *The effect of anisotropic thermal vibration on the atomic scattering factor.* Acta Cryst. **7** (1954) 503–504.
[94] J. S. Rollett and David R. Davis. *The calculation of structure factors for centrosymmetrical monoclinic systems with anisotropic atom vibration.* Acta Cryst. **8** (1955) 125–128.
[95] David R. Davis and J. J. Blum. *The crystal structure of parabanic acid.* Acta Cryst. **8** (1955) 129–136.
[96] J. S. Rollett. *The crystal structure of phenyl-propiolic acid.* Acta Cryst. **8** (1955) 487–494.
[97] Jürg Waser. *The anisotropic temperature factor in triclinic coordinates.* Acta Cryst. **8** (1955) 731.
[98] Henri A. Levy. *Symmetry relations among coefficients of the anisotropic temperature factor.* Acta Cryst. **9** (1956) 679.
[99] H. J. Grenville-Wells. *Anisotropic thermal vibrations in crystals. II. The effect of changes in atomic scattering factors and temperature parameters on the accuracy of the determination of the structure of urea.* Acta Cryst. **9** (1956) 709–721.
[100] D. W. J. Cruickshank. *The determination of the anisotropic thermal motion of atoms in crystals.* Acta Cryst. **9** (1956) 747–753.
[101] D. W. J. Cruickshank. *The analysis of the anisotropic thermal motion of molecules in crystals.* Acta Cryst. **9** (1956) 754–756.
[102] D. W. J. Cruickshank. *A detailed refinement of the crystal and molecular structure of anthracene.* Acta Cryst. **9** (1956) 915–923.
[103] D. W. J. Cruickshank. *A detailed refinement of the crystal and molecular structures of naphthalene.* Acta Cryst. **10** (1957) 504–508.
[104] William R. Busing and Henri A. Levy. *Determination of the principal axes of the anistropic temperature factor.* Acta Cryst. **11** (1958) 450.
[105] Mary R. Truter. *A detailed refinement of potassium ethyl sulfate.* Acta Cryst. **11** (1958) 680–685.

Miscellaneous

[106] A. D. Booth. *A new "averaging" method for the location of atomic positions from x-ray data.* Trans. Faraday Soc. **44** (1948) 282–285.
[107] V. Luzzati. *Traitement statistique des erreurs dans le cas d'une structure pseudo-centrosymétrique.* Acta Cryst. **6** (1953) 550–552.

[108] A. I. Kitajgorodskij. *Is 'super-refinement' legitimate in x-ray crystal analysis?* Nature **179** (1957) 410–411.

[109] Kathleen Lonsdale, Ronald Mason, and Judith Grenville-Wells. *Is 'super-refinement' legitimate in x-ray crystal analysis?* Nature **179** (1957) 856–857.

[110] E. G. Cox and D. W. J. Cruickshank. *Is 'super-refinement' legitimate in x-ray crystal analysis?* Nature **179** (1957) 857–859.

[111] S. Caticha-Ellis and A. Rimsky. *Critere de fin de l'affinement d'une structure atomique.* Acta Cryst. **11** (1958) 481–484.

23

Calculation of interatomic distances and angles

When the problem of finding the locations of the atoms in the cell has been solved, and when the coordinates defining these locations have been refined, there remains the task of interpreting the meaning of this arrangement of atoms. Often the general plan of the arrangement of atoms becomes clear as the structure study approaches completion. For example, the investigator of the structure of an organic crystal may expect to find it based upon the packing of molecules of a certain shape; or, in the case of an inorganic crystal, certain coordination numbers may be expected for the several atoms. But sometimes the nature of the structure is not obvious and a three-dimensional model must be made in order that the nature of the arrangement of atoms can be interpreted.

But even when the general plan of the structure is understood, present-day crystal chemistry requires that the detailed environment of each atom must be studied. This calls for calculation of the interatomic distances, and of the angles between these distance vectors. The calculation of interatomic distances should be carried out for the distances from each atom to all its nearest neighbors, and often it is advisable to include distances to neighbors of the next-nearest set. Bond angles are ordinarily computed only between vectors to the nearest neighbors.

Such calculations are more complicated than ordinary calculations of distances and angles in engineering problems because the coordinates of the atoms are not, in general, the ordinary Cartesian coordinates, but rather are coordinates referred to the natural axes of the crystal, namely the edges of the unit cell, which may be unequal and oblique to one another. The nature of these calculations, their precision and their significance, are considered in this chapter.

Calculation of interatomic distances

Let the coordinates of two atoms be $x_1\, y_1\, z_1$ and $x_2\, y_2\, z_2$. Then these atoms are at the ends of vectors

$$\begin{aligned} \mathbf{s}_1 &= x_1\,\mathbf{a} + y_1\,\mathbf{b} + z_1\,\mathbf{c}, \\ \mathbf{s}_2 &= x_2\,\mathbf{a} + y_2\,\mathbf{b} + z_2\,\mathbf{c}. \end{aligned} \tag{1}$$

The vector $\mathbf{s}_{12}$ between these two atoms is the difference between the two vectors of (1), namely

$$\begin{aligned} \mathbf{s}_{12} &= (x_2\,\mathbf{a} + y_2\,\mathbf{b} + z_2\,\mathbf{c}) - (x_1\,\mathbf{a} + y_1\,\mathbf{b} + z_1\,\mathbf{c}) \\ &= (x_2 - x_1)\mathbf{a} + (y_2 - y_1)\mathbf{b} + (z_2 - z_1)\mathbf{c}. \end{aligned} \tag{2}$$

The square of the magnitude of a vector can be found by forming the scalar product of the vector with itself:

$$\begin{aligned} \mathbf{s}_{12}\cdot\mathbf{s}_{12} = \\ &(x_2 - x_1)(x_2 - x_1)\mathbf{a}\cdot\mathbf{a} + (x_2 - x_1)(y_2 - y_1)\mathbf{a}\cdot\mathbf{b} + (x_2 - x_1)(z_2 - z_1)\mathbf{a}\cdot\mathbf{c} \\ &+ (y_2 - y_1)(x_2 - x_1)\mathbf{b}\cdot\mathbf{a} + (y_2 - y_1)(y_2 - y_1)\mathbf{b}\cdot\mathbf{b} + (y_2 - y_1)(z_2 - z_1)\mathbf{b}\cdot\mathbf{c} \\ &+ (z_2 - z_1)(x_2 - x_1)\mathbf{c}\cdot\mathbf{a} + (z_2 - z_1)(y_2 - y_1)\mathbf{c}\cdot\mathbf{b} + (z_2 - z_1)(z_2 - z_1)\mathbf{c}\cdot\mathbf{c}. \end{aligned} \tag{3}$$

The terms symmetrical in the main diagonal are equal. If the terms of the main diagonal are written first, the pairs of symmetrical terms next, and if the values of the scalar product are expanded, (3) can be rewritten in the form

$$\begin{aligned} s_{12}^2 = (x_2 - x_1)^2\, a^2 + (y_2 - y_1)^2\, b^2 + (z_2 - z_1)^2\, c^2 \\ + 2(x_2 - x_1)(y_2 - y_1)ab \cos\gamma \\ + 2(z_2 - z_1)(x_2 - x_1)\, ca \cos\beta \\ + 2(y_2 - y_1)(z_2 - z_1)\, bc \cos\alpha. \end{aligned} \tag{4}$$

This is a general form for computing an interatomic distance in a triclinic crystal. Since many such computations are normally necessary for the same crystal structure, it should be noted that the six terms of (4) concerned with the cell geometry are constant for the whole set of calculations. It is worthwhile computing these terms in advance:

$$
\begin{aligned}
k_1 &= a^2, \\
k_2 &= b^2, \\
k_3 &= c^2, \\
k_4 &= ab \cos \gamma, \\
k_5 &= ca \cos \beta, \\
k_6 &= bc \cos \alpha.
\end{aligned}
\tag{5}
$$

If the differences in coordinates are abbreviated, Δx, etc., then, utilizing (5), the interatomic distance calculations can be written

$$
\begin{aligned}
s_{12} = \{&(\Delta x)^2 k_1 + (\Delta y)^2 k_2 + (\Delta z)^2 k_3 \\
&+ 2\,\Delta x\,\Delta y\,k_4 \\
&+ 2\,\Delta z\,\Delta x\,k_5 \\
&+ 2\,\Delta y\,\Delta z\,k_6\}^{1/2}.
\end{aligned}
\tag{6}
$$

Table 1
Reduced forms of the interatomic-distance computations for the several crystal systems

Crystal system	Distance, s_{12}, from atom 1 to atom 2
Triclinic	$\{(\Delta x)^2 a^2+(\Delta y)^2 b^2+(\Delta z)^2 c^2$ $+ 2\Delta x\,\Delta y\,ab \cos \gamma$ $+ 2\Delta z\,\Delta x\,ac \cos \beta$ $+ 2\Delta y\,\Delta z\,bc \cos \alpha\}^{1/2}$
Monoclinic	$\{(\Delta x)^2 a^2+(\Delta y)^2 b^2+\Delta z)^2 c^2$ $+ 2\Delta x\,\Delta y\,ab \cos \gamma\}^{1/2}$ (first setting) or $\{(\Delta x)^2 a^2+(\Delta y)^2 b^2+(\Delta z)^2 c^2$ $+ 2\Delta x\,\Delta z\,ac \cos \beta\}^{1/2}$ (second setting)
Hexagonal	$\{[(\Delta x)^2+(\Delta y)^2-\Delta x\,\Delta y]a^2+(\Delta z)^2 c^2\}^{1/2}$ (hexagonal axes) $\{[(\Delta x)^2+(\Delta y)^2+(\Delta z)^2+(\Delta x\,\Delta y+\Delta x\,\Delta z+\Delta y\,\Delta z)\,2 \cos \alpha]a^2\}^{1/2}$ (rhombohedral axes)
Orthorhombic	$\{(\Delta x)^2 a^2+(\Delta y)^2 b^2+(\Delta z)^2 c^2\}^{1/2}$
Tetragonal	$\{[(\Delta x)^2+(\Delta y)^2]a^2+(\Delta z)^2 c^2\}^{1/2}$
Isometric	$\{[(\Delta x)^2+(\Delta y)^2+(\Delta z)^2]a^2\}^{1/2}$

Again it should be noted that this form is for the most general case, triclinic crystals. For monoclinic and hexagonal crystals, two of the last three lines of (5) vanish, and for the three orthogonal systems, all three of the last lines vanish. A reduced form of (6) for each of the several crystal systems is given in detail in Table 1. The reduced forms are suited to hand computation. For computation by digital computer it is, perhaps, more convenient to have a single program using the general form in (6).

Calculation of interatomic angles

Suppose an atom, 1, has two neighbors, 2 and 3. The lengths of the vectors $\mathbf{s}_{12}$ and $\mathbf{s}_{13}$ can be computed with the aid of (6). The angle between these vectors may also be required.

The angle between two vectors can be determined by using the expansion of the scalar product of the two vectors. In this case, the two vectors are $\mathbf{s}_{12}$ and $\mathbf{s}_{13}$. Let ψ be the angle between these. Then

$$\mathbf{s}_{12}\cdot\mathbf{s}_{13} = s_{12}\, s_{13} \cos \psi, \tag{7}$$

so that

$$\cos \psi = \frac{\mathbf{s}_{12}\cdot\mathbf{s}_{13}}{s_{12}\, s_{13}}. \tag{8}$$

The denominator of (8) is the arithmetic product of the lengths of the two vectors, both of which are available from computation like (6). The numerator can be formed by writing the scalar product of two vectors like (2). This is as follows:

$$\begin{aligned} \mathbf{s}_{12} &= (x_2-x_1)\mathbf{a} + (y_2-y_1)\mathbf{b} + (z_2-z_1)\mathbf{c}, \\ \mathbf{s}_{13} &= (x_3-x_1)\mathbf{a} + (y_3-y_1)\mathbf{b} + (z_3-z_1)\mathbf{c}; \end{aligned} \tag{9}$$

$$\begin{aligned} \mathbf{s}_{12}\cdot\mathbf{s}_{13} = {} & \\ & (x_2-x_1)(x_3-x_1)\mathbf{a}\cdot\mathbf{a} + (x_2-x_1)(y_3-y_1)\mathbf{a}\cdot\mathbf{b} + (x_2-x_1)(z_3-z_1)\mathbf{a}\cdot\mathbf{c}, \\ & (y_2-y_1)(x_3-x_1)\mathbf{b}\cdot\mathbf{a} + (y_2-y_1)(y_3-y_1)\mathbf{b}\cdot\mathbf{b} + (y_2-y_1)(z_3-z_1)\mathbf{b}\cdot\mathbf{c}, \\ & (z_2-z_1)(x_3-x_1)\mathbf{c}\cdot\mathbf{a} + (z_2-z_1)(y_3-y_1)\mathbf{c}\cdot\mathbf{b} + (z_2-z_1)(z_3-z_1)\mathbf{c}\cdot\mathbf{c}. \end{aligned} \tag{10}$$

If this is expanded in the same way (4) was derived from (3), it can be rewritten

$$\begin{aligned} \mathbf{s}_{12}\cdot\mathbf{s}_{13} = {} & (x_2-x_1)(x_3-x_1)a^2 + (y_2-y_1)(y_3-y_1)b^2 + (z_2-z_1)(z_3-z_1)c^2 \\ & + \{(x_2-x_1)(y_3-y_1) + (y_2-y_1)(x_3-x_1)\}ab \cos \gamma \\ & + \{(z_2-z_1)(x_3-x_1) + (x_2-x_1)(z_3-z_1)\}ca \cos \beta \\ & + \{(y_2-y_1)(z_3-z_1) + (z_2-z_1)(y_3-y_1)\}bc \cos \alpha. \end{aligned} \tag{11}$$

As in (4), the terms a, b, c, α, β, and γ are constant for a particular crystal structure. Thus the quantities in (5) need be computed only once for each structure. If $\Delta_2 x$ is used to replace $(x_2 - x_1)$, etc., and if substitutions (5) are made, then (11) can be rewritten

$$\begin{aligned}\mathbf{s}_{12}\cdot\mathbf{s}_{13} &= \Delta_2 x\,\Delta_3 x\,k_1 + \Delta_2 y\,\Delta_3 y\,k_2 + \Delta_2 z\,\Delta_3 z\,k_3 \\ &\quad + \{\Delta_2 x\,\Delta_3 y + \Delta_2 y\,\Delta_3 x\}k_4 \\ &\quad + \{\Delta_2 z\,\Delta_3 x + \Delta_2 x\,\Delta_3 z\}k_5 \\ &\quad + \{\Delta_2 y\,\Delta_3 z + \Delta_2 z\,\Delta_3 y\}k_6. \end{aligned} \tag{12}$$

Again it should be pointed out that this full form is required for triclinic crystals only. For monoclinic and hexagonal crystals, two of the last three lines vanish, and for the three orthogonal crystal systems, all lines but the first vanish. A reduced form of (12) for each of the several crystal systems is given in Table 2. As in the case of (6), the reduced forms of (12) are suited to hand computation. For computation by digital computer it is, perhaps, more convenient to have a single program using the general form in (12).

Table 2
Reduced forms of the scalar product of two vectors for the several crystal systems

Crystal system	Scalar product $\mathbf{s}_{12}\cdot\mathbf{s}_{13}$	
Triclinic	$\Delta_2 x\,\Delta_3 x\,a^2 + \Delta_2 y\,\Delta_3 y\,b^2 + \Delta_2 z\,\Delta_3 z\,c^2$ $+ (\Delta_2 x\,\Delta_3 y + \Delta_2 y\,\Delta_3 x)ab\cos\gamma$ $+ (\Delta_2 z\,\Delta_3 x + \Delta_2 x\,\Delta_3 z)ac\cos\beta$ $+ (\Delta_2 y\,\Delta_3 z + \Delta_2 z\,\Delta_3 y)bc\cos\alpha$	
Monoclinic	$\Delta_2 x\,\Delta_3 x\,a^2 + \Delta_2 y\,\Delta_3 y\,b^2 + \Delta_2 z\,\Delta_3 z\,c^2$ $+ (\Delta_2 x\,\Delta_3 y + \Delta_2 y\,\Delta_3 x)ab\cos\gamma$	(first setting)
	$\Delta_2 x\,\Delta_3 x\,a^2 + \Delta_2 y\,\Delta_3 y\,b^2 + \Delta_2 z\,\Delta_3 z\,c^2$ $+ (\Delta_2 x\,\Delta_3 z + \Delta_2 z\,\Delta_3 x)ac\cos\beta$	(second setting)
Hexagonal	$(\Delta_2 x\,\Delta_3 x + \Delta_2 y\,\Delta_3 y - \frac{1}{2}\Delta_2 x\,\Delta_3 y - \frac{1}{2}\Delta_2 y\,\Delta_3 x)a^2 + \Delta_2 z\,\Delta_3 z\,c^2$	(hexagonal axes)
	$[(\Delta_2 x\,\Delta_3 x + \Delta_2 y\,\Delta_3 y + \Delta_2 z\,\Delta_3 z)$ $+ (\Delta_2 x\,\Delta_3 y + \Delta_2 y\,\Delta_3 x$ $+ \Delta_2 z\,\Delta_3 x + \Delta_2 x\,\Delta_3 z$ $+ \Delta_2 y\,\Delta_3 z + \Delta_2 z\,\Delta_3 y)\cos\alpha]a^2$	(rhombohedral axes)
Orthorhombic	$\Delta_2 x\,\Delta_3 x\,a^2 + \Delta_2 y\,\Delta_3 y\,b^2 + \Delta_2 z\,\Delta_3 z\,c^2$	
Tetragonal	$(\Delta_2 x\,\Delta_3 x + \Delta_2 y\,\Delta_3 y)a^2 + \Delta_2 z\,\Delta_3 z\,c^2$	
Isometric	$(\Delta_2 x\,\Delta_3 x + \Delta_2 y\,\Delta_3 y + \Delta_2 z\,\Delta_3 z)a^2$	

Relations (12) and (6), or their reduced forms from Tables 2 and 1, provide appropriate numerical values for the numerator and denominator of (8). Data are therefore available for the computation of the angle ψ between the desired vectors.

The accuracy of interatomic distances and angles

In the latter part of Chapter 22 some elementary considerations were given to the accuracy of the determinations of the coordinates of the atoms of a structure. The accuracy of the interatomic distances and the angles, calculated as outlined in the foregoing sections of this chapter, are dependent upon the accuracy of the original coordinates, in the following manner.[2]

The bond length s is computed from the coordinates of two atoms, 1 and 2. If the atoms are not related by symmetry, and if the variances of their positions are σ_1^2 and σ_2^2 in the direction of the vector **s** joining them, then the variance in the length of this vector is

$$\sigma_s^2 = \sigma_1^2 + \sigma_2^2. \tag{13}$$

On the other hand, if the atoms 1 and 2 are equivalent by an inversion, a 2-fold rotation axis, or reflection, then their variances are dependent, and

$$\sigma_s^2 = 4\sigma_1^2. \tag{14}$$

The variance σ_ψ^2 of an angle ψ between vectors $\mathbf{s}_{12}$ and $\mathbf{s}_{13}$ is given by[2]

$$\sigma_\psi^2 = \frac{\sigma_2^2}{s_{12}^2} + \frac{\sigma_3^2}{s_{13}^2} + \sigma_1^2\left(\frac{1}{s_{12}^2} + \frac{1}{s_{13}^2} - \frac{2}{s_{12}\,s_{13}}\cos\beta\right). \tag{15}$$

Here

σ_1^2 = the variance of atom 1 in the direction of the center of the circle containing the three atoms,

σ_2^2 = the variance $\perp s_{12}$ of atom 2,

σ_3^2 = the variance $\perp s_{13}$ of atom 3.

The accuracy of interatomic distances and angles has been discussed in considerable detail by Cruickshank and Robertson,[2] and the reader especially interested in this subject is advised to consult their paper on the subject.

Literature

Accuracy of interatomic distances and bond angles

[1] D. W. J. Cruickshank. *The accuracy of electron-density maps in x-ray analysis with special reference to dibenzyl.* Acta Cryst. **2** (1949) 65–82.

[2] D. W. J. Cruickshank and A. P. Robertson. *The comparison of theoretical and experimental determinations of molecular structure, with applications to nephthalene and anthracene.* Acta Cryst. **6** (1953) 698–705.

[3] David H. Templeton. *Accuracy of bond distances in oblique coordinate systems.* Acta Cryst. **12** (1959) 771–773.

Index